Dedication

This work is dedicated to the memory of my parents, Harvey Carl Gorham (1896–1983) and Martha May Foster Gorham (1896–1986), and of my granddaughter, Suzanne Michelle Mason (1979–1989).

ECOLOGY AND MANAGEMENT OF FOOD-INDUSTRY PESTS

J. Richard Gorham, Editor

Food and Drug Administration
Public Health Service
U.S. Department of Health and Human Services
Washington, DC   20204

FDA Technical Bulletin 4

ASSOCIATION OF OFFICIAL ANALYTICAL CHEMISTS
Arlington, Virginia   1991

Designer: **Joe Phillips**

Production Editor and Indexer: **Julie Plumstead Phillips**

Foreword

We live in a paradoxical world, a world in which a large proportion of agricultural production is lost between the farm and the consumer, while millions of people suffer from hunger and malnutrition, and some—far too many—simply starve to death. These dramatic losses can be largely attributed to depredations by pests, spoilage by microorganisms, and spillage and wastage of food during handling, storage, transportation, and processing. Given these circumstances, the need for expertise in the ecology and management of stored-food pests and microorganisms becomes obvious, especially in the developing world where countries are continuously faced with food security problems.

The Food and Agriculture Organization, in addition to the regular work of improving crop production, has contributed to the post-harvest sector in developing countries through the following major activities: Food Security Assistance Scheme, whose goal is to ensure an adequate, timely, and steady supply of food through an efficient distribution system; the Global Informa-tion and Early Warning System, providing informa-tion on impending food shortages or supplies in order to be able to assess food supply prospects through a continuous monitoring operation; and Action Pro-gramme for Preventing Food Losses, whose main ob-jectives are to increase food availability and safeguard food quality, thereby improving the living standards of rural populations and assisting developing countries in the implementation of sustainable post-harvest pro-grammes.

This manual, *Ecology and Management of Food-Industry Pests*, produced under the direction of the U.S. Food and Drug Administration, is designed to advance our knowledge in this vital field. It will cer-tainly help to adapt and utilize the wealth of infor-mation available, mainly from the developed world, and transform it into action to preserve and provide adequate food supplies for the undernourished mil-lions of the Third World. I heartily recommend it to your attention.

Edouard Saouma, Director General
Food and Agriculture Organization
of the United Nations
Rome

Preface

This volume and its companion publication, FDA Technical Bulletin 3 (also known as Agriculture Handbook 655), *Insect and Mite Pests in Food: An Illustrated Key,* were developed simultaneously and are intended to be used together. Both publications cover insects and mites, but this volume, FDA Technical Bulletin 4, also includes major sections on the vertebrate pests of the food industry.

In most cases I have omitted the authors of scientific names. Most author names can be found in Agriculture Handbook 655. In those instances where a taxon is likely to be especially unfamiliar, as in the parasitic Hymenoptera, or where the taxon does not appear in Agriculture Handbook 655, I have retained the authors' names for the convenience of the reader.

Within this volume there is great diversity in the quantity of space devoted to the various pests. This diversity is largely a reflection of the relative importance of the several pest groups. The big three—rats, roaches, and beetles—warrant, and get, a lot of attention. Other pest groups get less attention, but the aim in every case is to give the reader sufficient information to make responsible and informed decisions regarding the management of pest populations.

Unless otherwise noted, all temperatures are in degrees Celsius.

Proprietary names are used only for the convenience of the reader and their mention should in no way be construed as endorsement by any governmental entity.

The opinions or assertions contained in this publication represent the private views of the authors and are not to be construed as official or as reflecting the view of the U.S. government, or of any state or local government, or of any governmental department or agency.

The support given by Dr. Douglas L. Archer, Director, Division of Microbiology, and Mr. Paris M. Brickey, Jr., Chief, Microanalytical Branch, Food and Drug Administration, was essential to the completion of this project and is gratefully acknowledged.

J. Richard Gorham
Food and Drug Administration
Washington, DC 20204

Contents

V. Health Considerations

VI. Regulation and Inspection

VII. Management

VIII. Glossary and Indexes

I
Introduction

1

Introduction to Food-Pest Ecology and Management

Robert Davis

One need only glance at the Table of Contents to obtain an idea of the wealth of subject material to be found in this Technical Bulletin. Those responsible for food-pest management in the food industry will have a strong interest in this material. This concern with food begins with the selection of appropriate plant varieties and continues through growing, harvesting, storing, processing, and transporting to the consumer. Since food is a value-added product following harvest, the food industry places its greatest efforts on pest management during storing, processing, and distribution. Therefore, the primary objectives of those in the industry involve sanitation, pest control, and quality assurance to ensure that insects and other pests are kept out of the food supply that is to be presented to the consumer.

This assurance of a clean food supply is the "law" which is enforced by federal as well as state agencies. Clean food is also the desire of the industries and their associations. They support this position by adhering to Good Manufacturing Practices statements and by maintaining well-trained staffs of sanitarians and pest control managers. However, in the food industry the prevailing philosophy of pest management differs from that of some agrarian systems where the coexistence of a pest population is considered necessary to secure some degree of natural regulation of pest populations. But in most cultures, the consumer's interest is not whether the insect in his or her food is a primary consumer or a "beneficial" parasite or predator. It is recognized only as a contaminant and an indicator of

Davis: Stored-Product Insects Research and Development Laboratory, Agricultural Research Service, U.S. Department of Agriculture, P.O. Box 22909, Savannah, GA 31403. Retired.

further unseen contamination. Therefore, the food industry strives for complete elimination of all food-industry pests.

The value of the food lost to pests and the costs of pest management or control are so difficult to calculate that the sum arrived at usually represents someone's "best professional judgment." In 1966, A. H. Moseman reported the following: "The Agricultural Research Service of the U.S. Department of Agriculture (USDA) has estimated the losses to agriculture from various hazards, the cost of controlling these hazards, and the cost of various inspections and quarantine programs to contain or restrict them. On the basis of information from 1951 through 1960, it is estimated that the average annual loss of crops, forest trees, and forest nurseries, and of livestock, poultry, and their products during production is about $14.3 billion. Additional losses during storage, marketing, and processing of agricultural commodities and forest products are estimated at $2.3 billion." These figures can easily be doubled today, and the loss to postharvest agricultural commodities is estimated to exceed $5 billion. This value does not include food losses or rejections by the consumer during and following preparation for consumption.

In developed nations, the costs of losses and control are proportioned about 60% and 40%, respectively. However, in emerging and developing nations, the costs of losses are much higher because of the lack of funds for management or control interventions. For postharvest losses in developing nations, the National Research Council (NRC) (1) reports that, for planning purposes, 10% should be used as an average minimum overall loss figure for cereal grains and legumes. A figure of 20% is set as a minimum for perishables and fish. In developing nations, food losses are important

3

not only because of the quantities lost, but also because of losses in quality and in nutritional and economic value. These factors are also important to those living in developed countries, but here the damage is usually removed from the food supply, and because it is a smaller portion of the total production, it does not represent such an immense economic impact. In the final analysis, everyone will agree that worldwide there is a very significant loss of much needed food, and a large expenditure of monies will be required to stem these losses.

In this technical bulletin, the Editor has selected and brings together a group of scientists who have reviewed the current data on techniques to manage or control insects and other pest populations attacking our food as it moves from the farm to the consumer. One is struck with how much information is available and at the same time one is awed as to the knowledge yet needed to prevent these losses.

In 1966, A. G. Norman (2), introducing an NRC report of the Symposium on Scientific Aspects of Pest Control, wrote: "the word 'pesticide' has acquired connotations and inflections ranging from the menacing to the magical, or from baleful threats to the welfare of man to the beneficent agent insuring his health and food supply." Similar statements can be made about most of mankind's inventions from the bow and arrow to ionizing radiation. All have served us well until we have begun to rely too heavily on chemical inventions. This bulletin provides a wealth of acceptable alternatives for use in any food-pest management program. It is here, through diversity of pest control approaches, that we will achieve a quality food supply.

Cited References

1 National Research Council. 1978. *Postharvest Food Losses in Developing Countries*. National Academy of Sciences, Washington, DC.

2 Norman, A. G. 1966. Introduction. In *Scientific Aspects of Pest Control*. Publ. No. 1402. National Academy of Sciences, Washington, DC.

2

World Resources and Food Losses to Pests

David Pimentel

Pest insects, microorganisms, weeds, birds, and mammals annually destroy nearly one-half of the world's potential food supply (29). This waste of food is taking place despite the annual use of about 2.3 billion kilograms (kg) of pesticides along with numerous nonchemical control measures (21). Before food losses to pests can be assessed in some detail, a perspective must be provided on the dimensions of the world food problem and the interrelationships of human population growth, food production, and land, water, and energy resources.

World Population Growth

At no time in history have humans so dominated the earth by their sheer numbers as they do now. This phenomenon, however, is a fairly recent event. For 99.9% of the more than one million years that humans have evolved and inhabited the earth, the maximum world population numbered less than 6 million, less than the current population of New York City. For most of that period, population growth was only about 0.001% per year (Figure 2.1). That comparatively slow growth rate explains why the world's population took nearly one million years to reach less than 6 million.

During that long period, people were hunter-gatherers and depended exclusively on natural ecosystems for their food. Then, about 10,000 years ago, humans began to cultivate food crops and establish settlements. With the beginning of agriculture came an increased and more stable food supply. This contributed to the increased growth rate of world population (Figure 2.1).

The next major increase in the world's population growth rate coincided with the discovery and use of stored fossil energy resources such as coal, oil, and gas. Since that time, a mere 300 to 400 years ago, rapid population growth has closely paralleled the increased use of fossil energy (Figures 2.1, 2.2) because humans have been able to use energy to manipulate and manage their environment. In particular, energy has been used to increase yields from crops and to improve public health, the prime contributory factors to rapid population growth in the world.

At present, world population stands at more

Pimentel: Department of Entomology, Cornell University, Ithaca, NY 14853.

Figure 2.1
Human population trends over the past million years.

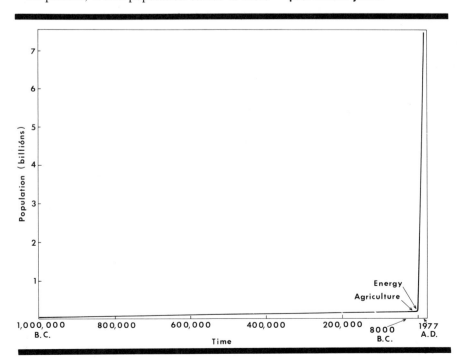

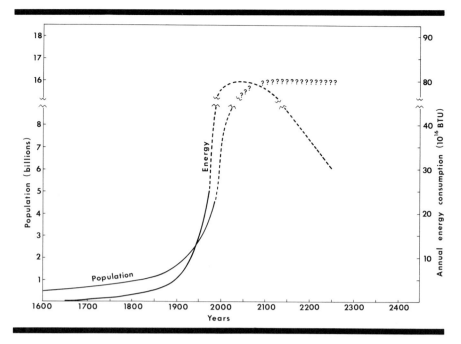

Figure 2.2
World population estimated from 1600 to 1975 (—) and projected (----???) to the year 2250. Estimated fossil fuel consumption from 1650 to 1975 (—) and projected to the year 2250 (-------).
SOURCES: Environmental Fund. 1979. *World Population Estimates.* Washington, DC; Linden, H. R. 1980. Importance of natural gas in the world energy picture. Paper presented at the International Institute of Applied Systems Analysis Conference on Conventional and Unconventional World Natural Gas Resources (Laxenburg, Austria, June 30 to July 4).

than 5 billion, but what is more alarming is the high growth rate of the population, nearly 1.7% (*37*). This is 1,700 times greater than the growth rate for the entire first 990,000 years of human existence on earth. This rate of increase adds about 230,000 people per day to the world's population. Demographers project that the world's population will reach more than 6 billion by the year 2000 and 10 billion to 16 billion by 2100 (*23*). There appears to be no generally accepted way to limit this rate of growth (*35*).

Contributing to the high growth rate is the young age structure of many populations in many nations. For example, 52% of the population of Honduras is 15 years old or younger (*22*). This means that a large proportion of the population is within the age range for childbearing. This age structure has long-term implications for human population growth. For example, in India, where population density is already high, if all Indian families were to adopt a zero-population-growth policy and begin to limit their family size to an average of 2.3 children, the population would continue to increase for the next 70 years (*14*), from the present 600 million to more than 900 million. The same would be true of all countries with a high proportion of young in their population, even if they restricted rates to just two children per family.

But what happens in countries such as China which now has a population of 1.07 billion and manages to support 22% of the world's population on only 7% of the arable land? Even if 70% of the married couples were to have only one child, the population could still be expected

to reach 1.2 billion by the year 2000 (*47*). With more than half the Chinese population below the age of 21, one must be skeptical that the aims of the State Family Planning Commission can be met. Similar problems of population growth exist in India and other countries of Asia, South America, and Africa.

Increased economic development is often cited as a means to slow birth rates. But, economic development alone cannot be depended on for population control because the environment has already reached the limits of its biological carrying capacity in many parts of the world (*23*). There are insufficient resources available for economic development similar to that in Europe and North America (*18*). This is evidenced by the fact that the populations in many parts of the world are already so large that they can no longer be fed without importing substantial amounts of food from other nations. Of the total of 164 nations in the world today, only about 10 are net food exporters. Of these 10, the United States (USA) is the largest food exporter (*23*).

There is an immediate need for all of us to take serious action to control our numbers before the vital resources of our environment are further reduced and become even less adequate to support the world's population. Indeed, if humans do not control their numbers, nature will, and with serious consequences.

Food Supplies

Food is a vital resource. Therefore, it is relevant to analyze the current status of food production and also to project how the future world population can be fed. For 99% of the time humans have existed, they have been hunter-gatherers, living off the land (*8*). Existence and survival were precarious, being influenced by human ability to forage for food over large areas of land and by the many fluctuations in climate and moisture that normally occurred. Many factors that influenced the abundance of a food supply were beyond human control. Hunter-gatherer adults had to supply their own needs as well as those of young children and elderly adults. Thus, obtaining food, an essential resource for survival of the family group, was a vital activity and one that required a major investment of the human energy resources of primitive societies.

Today, cereal grains are the primary source of food for humans. About 80% of the plant foods consumed by people comes from grains (wheat, rice, corn, millet, sorghum, rye, barley) and legumes (beans, soybeans, and peanuts).

About 90% of all food for human consumption comes from these grains and legumes plus cassava, sweet potatoes, coconuts, and bananas. In some developing nations, the annual per-capita consumption is about 182 kg (400 pounds) of grains or about 0.5 kg per day (23). This is just barely enough to support human life. In fact, an estimated one-half billion humans in the world today are protein and calorie malnourished (12, 42). The incidence of debilitating diseases is high in these food-deficient parts of the world.

On a worldwide basis, scarcity of calories and essential nutrients is a major nutritional problem. Some reports indicate that there is "famine in various developing nations, and death rates are reported rising in at least 12 and perhaps 20 nations, largely in Central Africa and Southern Asia" (23). Many other estimates support this view and suggest that up to one-quarter of the human population, or one billion people, are malnourished (12, 42). In some areas of the world there is simply not enough food to sustain normal human activity. In these areas the food intake, consisting primarily of grain, may be as meager as 0.5 kg per person per day (23).

The World Food Council at its 1982 meeting in Mexico concluded that imbalances in food supplies are causing hunger and severe malnutrition (24). Further, the Council predicted that the number of malnourished would increase from the present 400 million to 600 million by the year 2000 *unless* food supplies were shared. This emphasizes that poor distribution of food is one of the major world problems. Many of the countries needing food the most are the very ones that lack the economic resources required to purchase it. The average increase in life expectancy that has occurred in the past 30 years (36) suggests that the world is better fed now than in previous times, but better fed does not necessarily mean well nourished. The increase in lifespan may be due as much to improvements in environmental sanitation and the development of drugs and other chemicals (e.g., pesticides) as to a more adequate food supply.

The full dimensions of the current food problem are difficult to document, but the disparities in caloric intake are well known. In a few countries, such as the USA, daily per-capita intake is about 3,500 kilocalories (kcal); about 70% of the 103 grams of protein consumed per day comes from animal sources. In contrast, most of the world's population lives on about 2,100 kcal per person per day and obtains most of its protein from plants, especially grains (31). Thus, it would seem that even at current production

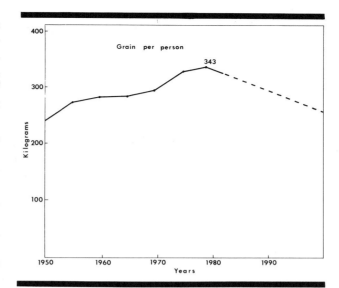

Figure 2.3
Declining world grain production per capita.
  SOURCE: Environmental Fund. 1979. *World Population Estimates.* Washington, DC.

levels, not all humans are receiving sufficient food. At the very best, it is obvious that the demand for food in many nations is increasing faster than food production (Figure 2.3).

Unfortunately, crop yields per hectare in certain regions such as Africa are declining rather than increasing (3). Even in the USA, with its high-yield agriculture, setbacks have recently taken place. Heavy rains, floods, droughts, crop pests, and other problems have reduced crop productivity. Other nations such as India, China, and some countries in Western Europe have experienced similar difficulties. Even without floods, droughts, and pest problems, increasing crop yields has become a more difficult problem than in the past because the prime resources for production—arable land, water, and fossil energy—are becoming scarcer (30). Further, the long-range effect of changes in climatic conditions is surely a factor to be considered in the future (4).

Land Resources

First, consider that at present most of the land suitable for cultivation is now in use. Of the total 13.4 billion hectares of land area in the world, only about 11% (1.5 billion hectares) is considered suitable for cultivation (23). The remaining land is either too dry, too cold, or too steep, or it lacks suitable soil structure for planting.

When the world's population numbered about 500 million (about twice the current population of the USA) during the year 1650, the available arable land per person was about 3 hectares. Today, with a world population of more than 5 billion, this has shrunk to less than 0.3 hectare or about one-tenth of what it had been. Obviously, attempting to produce the same amount of food on 0.3 hectare as on 3 hectares requires more intensive management of the land. This means larger inputs of fertilizers, pesticides, and water for irrigation, all of which require fossil energy inputs.

The estimate of 0.3 hectare of arable land per capita assumes increased population growth but does not include a net reduction in agricultural land or soil degradation due to soil erosion. But, in fact, with such a population expansion we logically can expect a reduction in available land. This has already happened in the USA where, from 1945 to 1975, over 29 million hectares were lost to highways and urbanization, about half of which was valuable cropland (25, 41, 42).

Soil erosion also seriously reduces the productivity of land. As a result, an estimated 80 million hectares in the USA have been either totally ruined for crop production or have been so badly eroded that the land is only marginally suitable for production (1, 32, 45). The rate of soil erosion per hectare of cropland in the USA is estimated to be about 18 metric tons annually (15, 43, 44, 46), resulting in the removal of one-half the topsoil of cropland in use today in Iowa (38) and in other regions of the USA. The rate at which land is lost due to soil erosion in developing countries is estimated to be two to three times greater than it is in the USA (13). The reduced productivity of cropland due to soil erosion in the USA has been offset by the increased use of fossil energy in the form of fertilizers, pesticides, and other inputs (15).

About 22% (3.0 billion hectares) of the world's land area is pasture and range used to graze a livestock population about equal to the weight of the present human population on earth (11). Some livestock are also pastured in the world's forests which occupy 30% of the land area (11). Livestock, particularly the polygastric ruminants, are important converters of forage grasses and shrubs into food suitable for humans.

Water Resources

The availability of water is critical not only for agriculture but for industrial and home use. In the USA during the twentieth century, the total "withdrawal" of water for all uses has been doubling at about 20-year intervals until, at present, about 7,200 liters (1,900 gallons) per person per day are withdrawn from surface and underground sources. Some of this water is consumed directly by individuals and used in industry, but most (about 83%) is used for

agriculture (27, 34). A major share of that used for agriculture is for irrigation, especially in the western agricultural region. For example, the amounts of irrigation water required to produce one kilogram of the following food and fiber products are (in liters) 1,400 for corn, 1,910 for sugar beets, 4,650 for rice, and 17,100 for cotton (39).

With regard to land resources, suggestions have been made that the world's potential arable land might be doubled with irrigation and other significant alterations of the ecosystem (17). Any answer to this suggestion is complicated by the future availability of both water and energy. Only about 12% of the world's cultivated land is now irrigated (10). Unfortunately, irrigation and other major environmental manipulations require the expenditure of enormous amounts of energy. For example, about 6 million kcal of fossil energy are required to produce a hectare of corn under conditions of normal rainfall in the USA, whereas about 20 million kcal are necessary to produce one hectare of irrigated corn (28). Thus, irrigated corn production is three times as energy intensive as corn watered by rain.

The problems of water supply and allocation are by no means limited to the USA. As many as 80 other countries, which account for nearly 40% of the world's population, are now seriously affected by droughts (19). Although the current amount of water withdrawn per capita on a global basis is less than a third of the amount withdrawn per capita in the USA, the growth in world population can be expected to double water needs by the year 2000 (6). At that time, world agricultural production will need an estimated 64% of all the water withdrawn from aquatic systems.

The conflicting needs for available water among agriculture, growing urban populations, industry, and fossil energy mining portend major changes in water use throughout the world. Among the four competing groups, evidence suggests that the proportion of water allocated to agriculture will decline because the economic yields from agriculture per unit of water used are far less than yields from industry and mining. If so, the potential shift in water use could seriously curtail agricultural productivity.

Energy Resources

Energy and food are both in short supply today primarily for the same reason—the growing number of humans. In fact, energy use has been increasing faster than world population. For example, world population doubled in the past 30 years, but world energy consumption doubled in the past 10 years. To date, we have met our energy needs by using our nonrenewable resources (fossil energy) with the result that known supplies are rapidly dwindling. Known world reserves of petroleum and natural gas are expected to be more than half depleted by the year 2000, and more than half of the coal reserves are estimated to be depleted about the year 2100 (16). Concurrently, the world's population is expected to increase nearly four times.

Agriculture, in its most productive form, depends upon large expenditures of many kinds of energy. For example, one-third of the 6 million kcal of fossil energy (about 600 liters of fuel equivalents) required to produce a hectare of corn in the USA can be attributed directly to fuel and the production of nitrogen fertilizer (28). Substitution of fossil-fuel-consuming machinery for human labor has reduced the labor input to only 10 person-hours per hectare. This is less than 1/100th of the 1,144 person-hours per hectare needed to produce corn primarily by hand labor (30). Although the fossil fuel input (6 million kcal) to produce a hectare of corn is only about 10% of the solar energy input (54 million kcal), our concern focuses on fossil energy because it is a finite resource.

Energy input into agriculture in the USA accounts for nearly 6% of the national expenditure of fossil energy (30). An additional 6% is used for food processing and another 5% for distribution and preparation. Therefore, about 17% of fossil energy expenditures is attributed to the food system. Converting this to oil equivalents, an estimated 1,500 liters (400 gallons) of oil equivalents of energy are expended in the food system to feed one person per year in the USA (35). When projected to a world population of 5 billion consuming a USA-style diet, the equivalent of 7,500 billion liters of fuel would be expended per year. Then the question is: How long would known petroleum reserves last balanced against this kind of use? Assuming petroleum were the only source of energy for food production, this known reserve would last a mere 11 years. Obviously, the real outlook is not this grim, but critical problems do face us in our use of fossil fuel for agriculture and in the many other ways in which we depend on this finite resource. Further, this assessment shows that we cannot rely solely on techniques of the past to increase food supplies.

Losses of Crops to Pests

On a worldwide basis, food losses to pests are high. At present, world crop losses to pests are estimated to be about 35% (7, 26). These losses include destruction by insects (about 12%), pathogens (12%), weeds (10%), and mammals and birds (about 1%). Losses to mammals and birds appear to be more severe in the tropics and subtropics than in the temperate regions, but these losses are still low compared to those to the three major pest groups—insects, pathogens, and weeds. At certain times on a given crop, one particular pest group may cause extremely heavy losses. For example, rats have been observed in the Philippines destroying nearly 40% of a rice crop.

Preharvest pest losses in the USA are estimated to be about 37% (5, 33), with pest insects causing a 13% loss, plant pathogens 12%, and weeds 12%. The larger preharvest losses in the USA compared to the world average (35%), even with the USA's use of sophisticated pest control technologies, may be attributed to differences in cosmetic standards and other factors. Many fruits, vegetables, and other foodstuffs sold commercially in other countries would be classified as ''unsatisfactory'' or a total loss in the USA.

Postharvest losses are estimated to range from 9% in the USA (39) to 20% in some of the developing nations, especially in the tropics. The prime pests of harvested foods are microorganisms, insects, and rodents. When postharvest losses are added to preharvest losses, total food losses due to pests are estimated to be as high as 48%. Thus, the pest populations are consuming and/or destroying nearly one-half the world's food supply. Surely this is a loss that society cannot afford as humankind faces world food shortages and a rapidly growing world population.

Conclusions

The world's population now exceeds the biological carrying capacity of the earth's land, water, and energy resources. This is especially true if the goal is to feed the world's population a healthy diet and provide everyone with a standard of living comparable to that of the United States and Europe. The improvement of human nutrition around the world will require controlling human population growth and improving the control of pests. Insects, birds, mammals, diseases, and weeds now destroy nearly one-half of all food produced in the world in spite of the use of pesticides and other technologies to control them. If pest losses could be eliminated, we could produce the same amount of food with half the land, half the water, and half the energy resources. More important, the elimination of current food wastage would mean more food to help reduce the malnutrition that is a daily reality for more than one billion human beings.

Cited References

1 Bennett, H. H. 1939. *Soil Conservation*. McGraw-Hill, New York, NY.

2 Brown, L. R. 1983. *Population Policies for a New Economic Era*. Worldwatch Paper 53. Worldwatch Institute, Washington, DC.

3 Brown, L. R., et al. 1985. *State of the World 1985*. Norton, New York, NY.

4 Bryson, R. A. 1977. The how and why of climatic change. Paper presented at Food Editors Symposium, Pineapple Growers' Association of Hawaii (Honolulu, April 3).

5 Chandler, J. M. 1981. Estimated losses of crops to weeds. In *Handbook of Pest Management in Agriculture*, D. Pimentel, ed. CRC, Boca Raton, FL.

6 Council on Environmental Quality. 1980. *The Global 2000 Report to the President*, vol. 2. Government Printing Office, Washington, DC.

7 Cramer, H. H. 1967. Plant protection and world crop production. *Pflanzenschutz-Nachrichten "Bayer"* 20(1)1–524.

8 Deevey, E. S., Jr. 1960. The human population. *Scientific American* 203(3)195–204.

9 Environmental Fund. 1979. *World Population Estimates*. Washington, DC.

10 Food and Agriculture Organization. 1970. *Production Yearbook 1969*. United Nations, Rome.

11 Food and Agriculture Organization. 1973. *Production Yearbook 1972*. United Nations, Rome.

12 Food and Agriculture Organization. 1974. *Assessment of the World Food Situation*. United Nations World Food Conference, Rome.

13 Food and Agriculture Organization. 1977. *Soil Conservation and Management in Developing Countries*. FAO Soils Bulletin 33. United Nations, Rome.

14 Gulhati, K. 1977. Compulsory sterilization: The change in India's population policy. *Science* 195(4284)1300–1305.

15 Hargrove, T. R. 1972. *Agriculture Research: Impact on Environment*. Special Report 69. Agriculture and Home Economics Experiment Station, Iowa State University of Science and Technology, Ames.

16 Hubbert, M. K. 1972. Man's conquest of energy: Its ecological and human consequences. In *The Environmental and Ecological Forum 1970–1971*. Office of Information Services, U.S. Atomic Energy Commission, Oak Ridge, TN.

17 Kellogg, C. E. 1967. World food prospects and potentials: A long-run look. In *Alternatives for Balancing World Food Production Needs*, E. O. Heady, ed. Iowa State University, Ames.

18 Keyfitz, N. 1976. World resources and the world middle class. *Scientific American* 235(1)28–35.

19 Kovda, V. A., B. G. Rozanov, and S. K. Onishenko. 1978. On probability of droughts and secondary salinisation of world soils. In *Arid Land Irrigation in Developing Countries*, E. B. Worthington, ed. Pergamon, London.

20 Linden, H. R. 1980. Importance of natural gas in the world energy picture. Paper presented at the International Institute of Applied Systems Analysis Conference on Conventional and Unconventional World Natural Gas Resources (Laxenburg, Austria, June 30 to July 4).

21 Munnecke, D. M. 1979. Chemical, physical, and biological methods for the disposal and detoxification of pesticides. *Residue Reviews* 7:1–26.

22 National Academy of Sciences. 1971. *Rapid Population Growth*. Johns Hopkins, Baltimore, MD.

23 National Academy of Sciences. 1975. *Population and Food: Crucial Issues*. Washington, DC.

24 *New York Times*. 1982. Wider hunger foreseen by World Food Council. June 22, p. A5.

25 Office of Technology Assessment. 1982. *Impacts of Technology on U.S. Cropland and Rangeland Productivity*. Government Printing Office, Washington, DC.

26 Pimentel, D. (ed.). 1981. *Handbook of Pest Management in Agriculture* (3 vols.). CRC, Boca Raton, FL.

27 Pimentel, D. 1986. Water resources for food, fiber and forest production. *Ambio* 15(6)335–340.

28 Pimentel, D., and M. Burgess. 1980. Energy inputs in corn production. In *Handbook of Energy Utilization in Agriculture*, D. Pimentel, ed. CRC, Boca Raton, FL.

29 Pimentel, D., and M. Pimentel. 1978. Dimensions of the world food problem and losses to pests. In *World Food, Pest Losses, and the Environment*, D. Pimentel, ed. Westview, Boulder, CO.

30 Pimentel, D., and M. Pimentel. 1979. *Food, Energy and Society*. Arnold, London.

31 Pimentel, D., et al. 1975. Energy and land constraints in food protein production. *Science* 190(4216)754–761.

32 Pimentel, D., et al. 1976. Land degradation: Effects on food and energy resources. *Science* 194(4261)149–155.

33 Pimentel, D., et al. 1980. Environmental and social costs of pesticides: A preliminary assessment. *Oikos* 34(2)126–140.

34 Pimentel, D., et al. 1982. Water resources in food and energy production. *BioScience* 32(11)861–867.

35 Pimentel, M. 1983. Food for people. In *Food and Energy Resources*, D. Pimentel and C. W. Hall, eds. Academic Press, New York, NY.

36 Poleman, T. T. 1982. *World Hunger: Extent, Causes and Cures*. A. E. Research 82–17. College of Agriculture and Life Sciences, Cornell University, Ithaca, NY.

37 Population Reference Bureau. 1986. *1986 World Population Data Sheet*. Washington, DC.

38 Risser, J. 1981. A renewed threat of soil erosion: It's worse than the Dust Bowl. *Smithsonian* 11(12)120–131.

39 Ritschard, R. L., and K. Tsao. 1978. Energy and water use in irrigated agriculture during drought conditions. Lawrence Berkeley Laboratory, University of California, Berkeley.

40 U.S. Department of Agriculture. 1965. *Losses in Agriculture*. Agriculture Handbook 291. Agricultural Research Service, Washington, DC.

41 U.S. Department of Agriculture. 1971. *Agriculture and the Environment*. Report No. 481. Economic Research Service, Washington, DC.

42 U.S. Department of Agriculture. 1974. *Our Land and Water Resources, Current and Prospective Supplies and Uses*. Misc. Publ. 1290. Economic Research Service, Washington, DC.

43 U.S. Department of Agriculture. 1981. Soil, Water, and Related Resources in the United States. Status, Condition and Trends. 1980 Appraisal, part II, Soil and Water Resources Conservation Act. Washington, DC.

44 U.S. Department of Agriculture. 1982. *An Analysis of the Timber Situation in the United States 1952–2030*. Forest Resources Report 23. Forest Service, Washington, DC.

45 U.S. National Resources Board. 1935. *Soil Erosion, a Critical Problem in American Agriculture*. Supplemental Report. Land Planning Committee, Washington, DC.

46 Wadleigh, C. H. 1968. *Wastes in Relation to Agriculture and Forestry*. Misc. Publ. 1065. U.S. Department of Agriculture, Washington, DC.

47 Wren, C. S. 1982. China plans a new drive to limit birth rate. *New York Times*, Nov. 7, p. 8.

3
Origins of Insects as Storage Pests

R. W. Howe

The presence of storage insects in a building can be examined from two viewpoints—the practical one of how the pest invaded that particular building and the more general one of why the species thrives in an artificial, man-made environment. Although the practical question may be difficult to answer for a particular pest population, the principles that govern the spread of storage species are now well understood, and most storage entomologists have seen how readily these insects disperse. Answers to the more general historic question can only be inferred; although new species are still being added to our list of minor pests, their origins are seldom known with any certainty.

Speculation About Prehistoric Origins

The earliest identifiable references to storage species in Britain date from about 1825, but in the USA there is some information for the previous century (6). There are also some archeological records from Egypt (5) and Roman Britain (17), for example, but the history of human beings, their agricultural pursuits, and the insects associated therewith, extend much further into the past.

Thus, speculation on the origins of storage pests must depend principally on our knowledge of their biology and on that of their closest wild relatives. True storage pests (9) fall into two simple categories: (a) those that breed in seeds, fruits, and other fleshy parts of plants; and (b) scavengers (16). The former include the weevils, bruchids, and a few moths; the

Howe: Slough Laboratory, Ministry of Agriculture, Fisheries, and Food, Slough SL3 7HJ, England. Current address: 28 Crossbush Road, Bognor Regis PO22 7LT, West Sussex, England.

latter, all other species except the obligate predators and parasitoids that attack the true pests. These latter can be ignored because they can invade any ecological niche.

Pest Biology

Storage pests have the ability to conserve water, enabling them to feed and breed in dry foodstuffs, often at a low relative humidity. Often they cannot cope with wetter foods even though adults as well as larvae drink readily. The egg retains water, and the rate of development of the embryo inside it is not influenced by humidity. However, the shell may get very hard when the humidity is low, and hatching may be delayed by the time taken for the larva to pierce the shell. The growth of larvae is slowed and eventually prevented by a sufficiently low relative humidity, the minimum varying by species from zero to about 70%.

Generally, scavengers are omnivorous. They eat vegetable matter (fruits, seeds, pollen, fungi, yeast, algae, living and dead leaves, roots, stems, tubers) and animal matter (honey, corpses, feces, feathers, skin, and living immobile forms including the eggs and pupae of insects and juvenile stages of vertebrates).

Diapause

Some storage insects have annual life cycles, especially in undisturbed places, irrespective of climate. Other species go into a resting stage (diapause), usually in the autumn, as a reaction to decreasing day length. Such species probably originated in a Mediterranean climate. Many phycitine moths have such a diapause but many pass two or three generations per year before it supervenes (19). Spider beetles may

13

diapause as adults or larvae; these species produce only one generation per year (*12*). Dermestid beetles have a range of complex life cycles and benefit if the adults are able to visit flowers. *Anthrenus verbasci* enters diapause as a larva every year during its life cycle, which may occupy one, two, or even three years (*3*). *Attagenus* species may diapause as adults or larvae. *Trogoderma* species undergo peculiar delays in development which may be retrogression (*2*) or diapause (*4*).

Past and Present Environments
Probable Original Habitats

The original habitats of storage pests were probably dry, sheltered ones where plant and animal debris collected. Before humankind evolved, some of these niches were provided by other animals that built nests or collected food and kept it in sheltered places. Likely habitats were the nests of termites, ants, wasps, and bees; bird nests built in sheltered places; bat roosts; and the dens of rats, mice, squirrels, badgers, foxes, and many others. All of these animals would collect nest materials and food for their young, produce feces and other wastes, and often raise the temperature and humidity of the nest above that of the surroundings. Caves provided shelter for early humans and are still used by various birds and mammals; consequently, storage insects are often found in them.

Other sheltered sites are the forest floor and the ground under dense tussocks of grass where the humidity is often raised by dew. Such places are rich in decaying vegetation. More specialized sites in trees are the restricted spaces under the bark, holes caused by decay, and the tunnels of wood-boring organisms. Some of these can be invaded only by the flatter species that can move in the narrow spaces and by species related to the wood-boring families that feed on decaying wood and lichens.

Hatch (*9*) was probably wrong to suppose that the origins of seed eaters were obvious. Neither should we assume that species that now attack the seeds of modern strains of wheat, maize, and other cereals originally fed on the seeds of their ancestral grasses; only very tiny adults can develop in such small seeds. It is highly probable that the two grain borers from the wood-boring family Bostrichidae, *Prostephanus truncatus* and *Rhyzopertha dominica*, initially fed in fleshy stems, roots, or tubers, as they can still do today. It is possible that *Sitophilus* species did so likewise but also that they bred in larger wild seeds such as acorns (*13, 15*).

It is, perhaps, dangerous to speculate about the bruchids that have evolved along with the legumes and their protective alkaloids. Many breed in large wild seeds. The need for an open mind, however, is illustrated by the moth, *Sitotroga cerealella*, because Joubert (*15*) reported this species developing in the tiny seeds of *Setaria nigrirostris* and producing adults resembling gnats—adults so small that they would not be expected to lay many eggs. It is possible but unlikely that the size of individuals of storage species has increased so considerably.

Foragers

These remarks so far have dealt with species that live in the food materials whether in storage or under the plants from which fruits, seeds, or dead leaves have fallen. There remains a group that builds nests in refuges away from the food and either forages for food or browses on fungi and lichens. Included here are the ants, the geographic range of which has been increased by human action through the provision of warmer, protected nest sites inside buildings. Buildings also provide niches for cockroaches, silverfish, and psocids, but there is no reason to suppose that artificial refuges differ in any important way from natural ones.

Agricultural Development

The most obvious feature of the nest and underbark environments is that they are small and most could support only a few hundred insects (*21*). The vegetative carpet environment was large but poor in nutrients and probably patchy. When human beings started to practice agriculture and then food storage, the amount of food available to pests eventually increased. But up to 1850, and much later over most of the world, most storage environments were still relatively small and were comprised of sacks or small bins. In recent years, however, the quantities in store have increased enormously, often to thousands of tons. Such large quantities so alter the storage environment that much of the food there is not available to pests.

Avoiding Overpopulation

We do not know if storage insects evolved in relation to slow changes in the environments available to them over the millenia, but it is probable that they did. Nevertheless, the main hazard faced by opportunistic storage species in a tiny niche was overpopulation—the food supply could be so easily exhausted. The insects had to be able to find small caches yet avoid overexploitation. The regular daily activity typical of storage species would ensure that adults finding food one evening would leave it the next without laying too many eggs. It is probable that they would leave behind some olfactory signal that would inhibit others of their own species from utilizing the same food.

A few species, such as *Callosobruchus maculatus*, appear to have a mechanism triggered by high temperature which ensures that a form capable of flight develops when the host plant is ripening in the field or the infested legumes are overcrowded (20). This adaptation appears to be restricted to the Bruchidae, but in warmer, drier climates, conditions at harvest time should allow any storage insect that can fly to reach ripening seeds and fruits.

Dispersal of Pests

Pest species spread either by their own powers or by carriage on some other object. A few storage species—e.g., certain cockroaches, *Dermestes* spp., and *Necrobia* spp.—are strong fliers; they may move downwind many miles or even upwind toward attractive odors. Other species fly weakly but regularly (usually at dusk) in warm climates. In any case, flight, whether strong or weak, does no more than facilitate local dispersion. Most of the important species, however, had a worldwide distribution by 1800, before precise identification was possible, so we do not know their original homes with certainty. It is obvious that they have been dispersed by trade.

This dispersal of storage species around the world continues today (7, 8). Even now it is a process that attracts little attention unless a feared species (such as *Trogoderma granarium*) subject to quarantine is found. In the past it caused concern only when the transported species found a favorable niche in which it could undergo a damaging population explosion. The most spectacular example of this happened a century ago when roller flour mills were introduced and *Anagasta kuehniella* invaded them. The species was not described until it became abundant in a mill in Germany (22), but then it spread extremely rapidly to mills in Europe and North America, as can be seen by accounts in the literature (10, 18). Yet this is a species that is seldom found nowadays in shipboard cargoes.

Howe (12) reviewed the less spectacular import into Britain of a series of spider beetles. *Gibbium aequinoctiale* and *Mezium affine* arrived before 1830, *Niptus hololeucus* came in by 1837, *Ptinus clavipes* and its triploid form *mobilis* arrived around 1850, *Ptinus tectus* (=*P. ocellus*) reached Britain in 1892, *Trigonogenius globulum* in 1900, *Ptinus pusillus* in 1906, and *Pseudeurostus hilleri* in 1939. Oddly enough, like *Anagasta kuehniella*, spider beetles are seldom found in ship cargoes nowadays and their successful colonization must be attributed to their finding an unexploited niche. In contrast, species that have frequently been imported into Britain have not become established, either because of an unsuitable climate or because, as with *Cryptolestes pusilloides* (14), they could not compete with related species that were already resident.

One of the most recent examples of a species invading a new niche is that of *Alphitobius diaperinus*. These beetles are relatively common in cargoes (especially those from India and Southeast Asia), being found in nearly 2% of those inspected in Britain in the 1960s (1). The beetles need warm, damp conditions that were not available in storage situations until the poultry industry introduced deep-pit houses where the accumulated manure provided both this environment and an ideal food. The beetles presumably gained entry on the ingredients used for poultry feed.

Bulk Storage

The sources of infestation for a newly built warehouse or factory handling food are (a) local stores from which the pests can fly; (b) produce brought into the premises; (c) vehicles, sacks, and other containers and the clothing of personnel bringing produce into the store; and (d) birds, bats, and rodents nesting in the buildings, especially those that bring in food and nesting materials.

By increasing quantities in storage, human beings have allowed larger insect populations to develop without removing the risk of overpopulation and have created a new hazard for the pests. Populations living on the surface of large bulks can still exhaust their food supply and those that penetrate into the bulks can raise the temperature therein by their metabolism and render the core of the bulk untenable or lethal.

Conclusions

The presence of insects in animal nests has attracted much attention in relation to their origins as storage pests; in particular, Hicks (11) and Linsley (16) have provided numerous examples of known storage pests utilizing animal nests. There can be no doubt that animal nests have been the main stepping stones for storage insects to warehouses.

Cited References

1 Aitken, A. D. 1975. *Insect Travellers,* vol. 1. *Coleoptera.* Pest Infestation Control Laboratory Technical Bulletin 31. Her Majesty's Stationery Office, London.

2 Beck, S. D. 1971. Growth and retrogression in larvae of *Trogoderma glabrum* (Coleoptera: Dermestidae). 1. Characteristics under feeding and starvation conditions. *Annals of the Entomological Society of America* 64(1)149–155.

3 Blake, G. M. 1958. Diapause and the regulation of development in *Anthrenus verbasci* (Col., Dermestidae). *Bulletin of Entomological Research* 49(4)751–755.

4 Burges, H. D. 1959. Studies on the dermestid beetle, *Trogoderma granarium* Everts. II. The occurrence of diapause larvae at a constant temperature, and their behaviour. *Bulletin of Entomological Research* 50(2)407–422.

5 Chadwick, P. R., and F. F. Leek. 1972. Further specimens of stored products insects found in ancient Egyptian tombs. *Journal of Stored Products Research* 8(1)83–86.

6 Cotton, R. T. 1941. *Insect Pests of Stored Grain and Grain Products*. Burgess, Minneapolis, MN.

7 Freeman, J. A. 1976. Problems of stored products entomology in Britain arising out of the import of tropical products. *Annals of Applied Biology* 84(1)120–124.

8 Freeman, J. A., and H. Piltz. 1975. Storage pests: Lists of dangerous pests in commodities particularly liable to infestation. *Plant Health Newsletter*. EPPO Publ. B80:9–18.

9 Hatch, M. H. 1942. The biology of stored grain insects. *Bulletin of the Association of Operative Millers* 42(July)1207–1211.

10 Heinrich, C. 1956. *American Moths of the Subfamily Phycitinae*. Bulletin 207. U.S. National Museum, Washington, DC.

11 Hicks, E. A. 1959. *Checklist and Bibliography on the Occurrence of Insects in Birds' Nests*. Iowa State College, Ames.

12 Howe, R. W. 1959. Studies on beetles of the family Ptinidae. XVII. Conclusions and additional remarks. *Bulletin of Entomological Research* 50(2)287–326.

13 Howe, R. W. 1965. *Sitophilus granarius* (L.) (Coleoptera, Curculionidae) breeding in acorns. *Journal of Stored Products Research* 1(1)99–100.

14 Howe, R. W., and L. P. Lefkovitch. 1957. The distribution of the storage species of *Cryptolestes* (Col., Cucujidae). *Bulletin of Entomological Research* 48(4)795–809.

15 Joubert, P. C. 1966. Field infestations of stored-product insects in South Africa. *Journal of Stored Products Research* 2(2)159–161.

16 Linsley, E. G. 1944. Natural sources, habitats, and reservoirs of insects associated with stored food products. *Hilgardia* 16(4)187–224.

17 Osborne, P. J. 1977. Stored product beetles from a Roman site at Droitwich, England. *Journal of Stored Products Research* 13(2)203–204.

18 Richards, O. W., and W. S. Thomson. 1932. A contribution to the study of the genera *Ephestia*, Gn. (including *Strymax*, Dyar), and *Plodia*, Gn. (Lepidoptera, Phycitidae), with notes on parasites of the larvae. *Transactions of the Entomological Society of London* 80(2)169–250, 8 pl.

19 Tenhet, J. N., and C. O. Bare. 1951. *Control of Insects in Stored and Manufactured Tobacco*. Circular 869. U.S. Department of Agriculture, Washington, DC.

20 Utida, S. 1972. Density dependent polymorphism in the adult of *Callosobruchus maculatus* (Coleoptera, Bruchidae). *Journal of Stored Products Research* 8(2)111–125.

21 Woodroffe, G. E. 1953. An ecological study of the insects and mites in the nests of certain birds in Britain. *Bulletin of Entomological Research* 44(4)739–772, 3 pl.

22 Zeller, P. C. 1879. Lepidopterologische Bemerkungen. *Stettiner Entomologische Zeitung* 11:462–473.

4
Storage Ecosystems

R. N. Sinha

In modern society, storage of food has become part of an elaborate system involving production, transportation, storage, marketing, and consumption (1). Because food grains provide nutrition to build and sustain the bodies of human beings and their livestock, the quality of stored food remains a prime concern throughout the system.

Storage of grain and other foods generally reflects a dynamic state, even though the term "storage" implies a static state. Cereal grains are living materials with all the properties of life (56, 57); therefore, they cannot be treated as inanimate objects. Although alive, stored seeds may not manifest some of their metabolic activities, mainly because they remain in a dormant state. The organisms that attack or become associated with stored grains and foods are also alive and commonly interact with these stored materials. Understanding both the nature of and the interactions among these agents is crucial to understanding the progressive deterioration of stored food.

Goal of Proper Food Storage

Although often alive and sound at the time of initial storage, all stored grains and foods gradually deteriorate and lose quality. The aim of proper management of stored food is to slow down or temporarily arrest these processes. Deterioration, decomposition, and decay are func-

Sinha: Research Station, Agriculture Canada, 195 Dafoe Road, Winnipeg R3T 2M9, Canada.

Contribution No. 905 from the Research Station, Agriculture Canada, Winnipeg.

tional characteristics of the stored-grain ecosystem (or storage ecosystem), not isolated events involving only one or two biotic and abiotic agents.

In the past, most food-storage management strategies have been directed toward one or two of the decay-causing organisms. For example, one group of food managers has often worked on the bacteriological aspect of food spoilage while ignoring mycological, entomological, or acarological aspects. It is not that flour beetles are not as important and relevant as coliform bacteria or *Salmonella,* but as a food passes from one niche with one moisture-temperature-time combination to another, one group of harmful organisms becomes more important than others that had dominated the scene earlier.

The challenge to the food manager is not to lose the holistic view of the decay process of a particular food as it moves from one phase to another. To have such a view, one must be familiar with some of the concepts and definitions of ecology and use an ecological approach in devising and applying food protection measures. Without this kind of approach, the results of many carefully applied measures could be short-lived, even when they are effective.

What Is Ecology?

Ecology is the systematic study of interactions among plants, animals, and their abiotic environments that determine the distribution and abundance of organisms. An ecologist is involved with one or more of the following levels of organization: individual, population, community, ecosystem, biome, and biosphere. Energy to sustain life processes flows through all

17

these levels from individual to biosphere.

The most important and only free energy source is radiant energy from the sun. This energy is used in photosynthesis by green plants (autotrophs) whereby carbon dioxide (CO_2) is assimilated into energy-rich carbon compounds. The plants also use nutrients from the mineral reservoir of the soil which are cycled and recycled as part of the biogeochemical cycle. Cattle eat green plants and insects eat stored seeds of green plants to make their own tissues. Thus, energy from different sources flows from one level to another.

The lowest of these trophic levels is that of green plants, the true "producers" of food. The second level is that of the heterotrophs or "consumers" such as cows or insects. At each level of transfer, considerable energy is lost through respiration and other metabolic activities. Numbers of trophic levels can be two through five, depending on the stability of the interrelations in an ecosystem. Orderly energy transfer is crucial for survival, propagation, and decay of all living organisms.

Four important terms relative to the following discussion of an ecological approach to food system management are defined as follows: A *population* is a group of organisms of the same species occupying a particular space at a particular time. A *community* consists of groups of coexisting, interdependent populations of plants and animals in a given place; the groups may be of various sizes and degrees of integration. The *ecosystem* is the biotic community and abiotic environment (e.g., communities of microflora, insects, and mites) as they live together and respond to changing temperature and relative humidity of the air; an assembly of interlocking life systems. A *life system* is a component of an ecosystem that determines the existence, abundance, and evolution of a population (*11*); it has a subject population and its effective environment is composed of all external agencies affecting the population. Interlocking life systems make up the framework of an ecosystem.

Simple and Complex Ecosystems

This chapter deals mainly with the community and ecosystem levels of organization, using Margalef's hypothesis of ecosystem development as the theoretical basis for community ecology studies of human food systems. Margalef (*39, 40*) judges an ecosystem on the basis of its level of maturity (complexity). His dynamic concept is related to structure, function, and energy flow within an ecosystem.

In an undisturbed ecosystem, maturity increases with time. Examples of relatively stable, more mature ecosystems with considerable species diversity and involved food webs with many successions are grasslands or deciduous forests. Stability implies persistence when disturbed and minimal fluctuations in populations.

An example of a less mature type of ecosystem with a simple food web, a low species diversity, and only a few successional changes is large-acreage cropland. An individual species lacks a high degree of specialization and many of its populations succumb to abiotic population control forces. Although populations can rebuild their numbers quickly, climatic stress can prevent communities within such ecosystems from progressing toward a higher level of maturity. A similar hypothesis suggests that the succession of species populations allows undisturbed ecosystems to develop high species diversity and greater stability (homeostasis), thereby providing protection from environmental disturbance (*46*). Although neither hypothesis has been validated with data from natural and manmade ecosystems, Margalef's maturity concept provides a convenient working hypothesis to explain the structure and function of a man-made ecosystem such as the postharvest stored-grain ecosystem.

The Stored-Grain Ecosystem

To understand the ecology of storage, one should consider a grain bulk as an ecosystem in which the abiotic (physicochemical) environment and the biotic assemblage (flora and fauna) interact with each other. A grain bulk is a manmade, immature ecosystem with a relatively simple structure and a nonregenerating energy supply. Energy captured by the cereal plants (producers) and concentrated in the seeds is used to drive the storage ecosystem. Because of frequent interference from the physical environment and from humans, such an ecosystem is unstable.

Only species with high growth and reproductive rates and low specialization seem to do well in this type of ecosystem. The concentration of energy per unit of biomass is relatively high. Unlike mature and stable natural ecosystems, patterns of relationships among biotic elements and between biotic and abiotic elements are vague and often too transitory to quantify and define. Only by understanding the functional dynamics of these valuable ecosystems can pest management specialists hope to develop practical man-

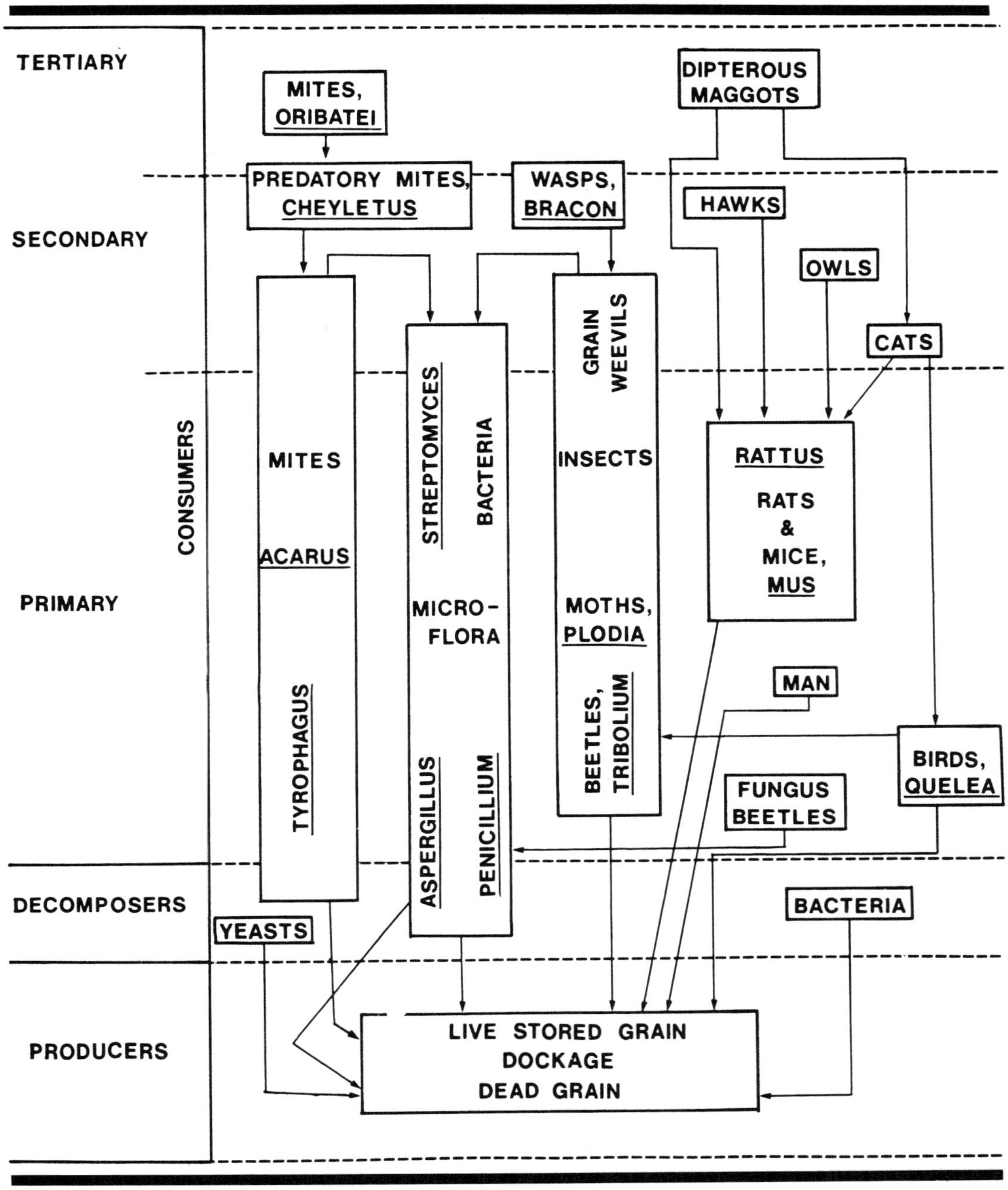

Figure 4.1

A simplified food web in a stored-grain ecosystem. Arrows indicate broad relationship between consumers and kinds of food and prey. Trophic levels are on the left side.

agement strategies and at the same time exploit fully the resources of these ecosystems.

Energy Flow

The pattern of energy flow in grain bulk ecosystems, i.e., transfer of energy in a food chain from the plant seeds through a succession of organisms, conforms to fundamental thermodynamic principles: (a) energy is neither created nor destroyed, but it can be transformed; (b) in real processes, energy cannot be quantitatively converted into another state without some loss of energy in a randomized and dissipated form. At each of two to five transfers—or trophic levels—in an ecosystem, a large proportion of the potential energy is lost as heat. Heat is generated whenever an organism respires due to oxidation processes involving liberation of CO_2. For example, a 2-day culture of mold liberates 1,750 to 1,870 mg CO_2 per gram of dry matter in 24 hours, whereas the respiration intensity of dry wheat in the same period is less than 0.1 mg (73). Grain weevils liberate 130,000 times more CO_2 than the equivalent weight of grain (47).

Food webs created by interlocking food chains in a hypothetical, aging grain bulk are diagrammed in Figure 4.1.

Consumers

Microbial consumers (decomposers) flourish and decompose organic matter successfully only when the food has a carbon/nitrogen (C/N) ratio ranging from 15/1 to 30/1. Most cereals have a C/N ratio within the optimum range of microbial activity (wheat, 15/1 to 26/1; barley, 18/1; oats, 20/1; and rice, 25/1). However suitable the substrate may be, most microorganisms (except some *Aspergillus* species) are inactive unless the relative humidity (rh), and consequently the moisture content, of the grain is high (>70% rh for most fungi and actinomycetes, and 90% for most bacteria).

Fungal infection is faciliated by seed injury (75). First-level consumer insects damage the grain and thereby provide routes of entry for various microflora and other secondary invaders such as insects and mites. In addition, the nitrogenous waste products and metabolic moisture from insects provide more favorable C/N ratios and moisture levels and thereby facilitate further microbial infection. Rodents and birds, as first-level consumers, not only consume a sizeable quantity of the grain but also foul a larger quantity with their excrement, hairs, and feathers.

Most mites and psocids feed on grain debris and on the fungi that grow on the grain (52, 55). But *Acarus siro*, which feeds directly on wheat germ and often multiplies to large numbers (66), is more a first- than a second-level consumer. The first-level consumers produce food for the second-level consumers. The second-level consumers include insectivorous birds and rodents, mycophagous insects (e.g., *Ahasverus, Latheticus, Typhaea*), and mycophagous mites (e.g., *Tarsonemus, Caloglyphus, Tyrophagus, Aleuroglyphus*). Also at the second level are the mite and insect predators and parasites: *Cheyletus* and bdellids, that feed on acarid mites; parasitic wasps such as *Bracon* and *Cephalonomia* that feed on larvae of stored-grain moths and beetles; and *Blattisocius* mites that feed on the eggs and larvae of various pyralid moths.

Third-level consumers include pseudoscorpions, tydeid and mesostigmatid mites, and maggots parasitic on rodents and birds. Because the feeding specificities of many insects and mites are still unknown, the third-level consumers are difficult to separate from those at the second level. Excrement from the primary consumers serves as a substrate for luxuriant microbial growth and is also eaten by certain secondary and tertiary consumers (scavengers). Other nutrients for decomposer microflora are provided by the dead bodies of the various animals.

Recycling of nitrogen and other essential food components from one organism to another continues. Prolonged successions of organisms and recycling of nutrients gradually make the grain unsuitable for human use. Useful information on the biological productivity and the energy flow has been obtained for natural ecosystems (32, 45), but even though a grain bulk is both relatively less complex and closed, no energy-flow model or energy budget has been made for its bioenergetic relationships. Such studies are needed to develop long-term solutions to grain storage problems.

Climate as a Main Regulator

Climate and human beings are the two most important regulators of storage ecosystems. In this section the characteristics of storage organisms are broadly described as they relate to major climatic subdivisions of the world. Several hundred species of microorganisms, insects, and mites are capable of causing spoilage, but only a small proportion of these species causes most of the damage. For example, although 100 species of microorganisms (38, 75), over 80 insect species (34, 53, 79), and about 75 species of mites (unpublished data) have been found in stored grain in the Prairie Provinces of Canada, only 20 species have been implicated in heating bulk oats in Manitoba (see section on Mites for further discussion of hot spots).

In a survey of insects in farm-stored grain during July–September 1945–50, only *O. surinamensis, Cryptolestes* spp., *Tenebroides mauritanicus, Tribolium castaneum, Sitophilus granarius, S. oryzae,* and *Rhyzopertha dominica* were found (74). Similar conclusions can be drawn from various other studies (7, 20, 26, 29, 54, 83).

Climates of the World

Climates of the world have been divided into four major classes, each having two further subdivisions (modified version of Köppen's scheme) (72) as shown in the margin.

The relationships of the microflora, insects, and mites to stored cereal grains in each climatic region can be generalized as follows (56). Deteriorating stored grain, especially wheat, in regions with temperate climates has a relatively more dominant mite component and fewer species of insects and microorganisms. In contrast, on similar substrates in humid tropical climates, one can expect to find insects primarily, microbes to a lesser extent, and mites least of all. In dry climates, insects are dominant. In subtropical climates the three groups of organisms

A. Tropical humid climate
 a. Tropical wet
 b. Tropical dry

B. Subtropical climate
 a. Subtropical dry summer
 b. Subtropical humid

C. Dry climate
 a. Desert or arid
 b. Steppe or semiarid

D. Temperate climate
 a. Temperate oceanic
 b. Temperate continental

are often equally abundant, but their relative abundance may be influenced by the type of grain and by the harvesting, drying, and storage practices in a given country within the subtropical region. For example, bacteria could cause most of the damage to stored grain in humid tropical and subtropical climates, but it is a common practice for farmers or grain handlers to use either sunlight or artificial heat to dry grain down to a moisture content that limits damage to that caused by insects and fungi.

Whether an insect reproduces in a given climatic zone at a rate that causes economic damage is discussed below in the section on Climatic Plasticity Index.

Animal and Plant Communities

As stated above, depending upon its geographical (climatic) location, an ecosystem may contain an assemblage of several communities of plant and animal forms. For convenience of discussion, the following are considered as distinct communities: (a) fungi (including actinomycetes), (b) bacteria (including yeasts), (c) mites, (d) insects, (e) rodents, and (f) birds. Member-species populations of some of these communities interact conspicuously. For example, most storage mites and fungus beetles depend on certain types of fungi that commonly grow on stored grain (52, 55). Some other members of the same two communities, however, simply share the environment. For example, the lesser grain borer (*Rhyzopertha dominica*) feeds only sparingly on the fungi that occur on the grain it infests (55).

Ideally, storage ecosystems should be investigated using basic principles of all three branches of modern ecology. Population ecology involves biotic potential (intrinsic rate of increase in the face of environmental resistance), size, growth, population regulation mechanisms, and evolution amd extinction of species. Storage pests have been the subject of numerous studies from the view point of population ecology (2, 5, 10, 17, 21, 23, 24, 28, 60).

Community ecology involves the study of the structure (species composition, diversity), the function (physiological processes, interactions), the succession of species or species groups, and competition (intra- or interspecific). Relatively few studies from the community ecology viewpoint have been undertaken (14, 43, 44, 61, 64).

Ecosystem ecology includes the study of structure, function, energy flow, and maturity.

Studies in this area are scarce (6, 9).

The investigator must be particularly aware of three main characteristics of species invading food storage systems: (a) the ability to adapt with changing human societies, cultures, and civilizations; (b) the ability to multiply quickly; and (c) the capacity to use a wide range of man-made and natural foods. When these basic concepts and characteristics shared by all biotic agents are considered against the background of living bulk grain or nonliving processed food, one can acquire a truly holistic view of the real food-loss problem. Such an ambitious attempt to integrate concepts and facts related to all known components from all levels of organization is possible in our computer era. Enormous amounts of data on many variables can be processed, analyzed, and summarized with the aid of multivariate statistical techniques (59).

The food system manager must remember that the ecosystem is a real entity in which the abiotic environment and the biotic assemblage of plant and animal forms interact in a basically energy-dependent fashion and are governed by the fundamental laws of thermodynamics. Since a thorough review of this complex subject would far exceed the scope of this chapter, only a few examples are given here to elucidate the community roles of microflora, mites, insects, rodents, and birds.

Microflora

The concept of a biotic community is crucial to our understanding of the ecology of the microflora in storage ecosystems. This concept emphasizes that diverse organisms with unity of taxonomic composition and somewhat uniform appearance not only live together harmoniously within a physical habitat but also function as a group in a dynamic way. Such an assemblage of populations has its characteristic trophic organization and metabolic pattern (46). The best way to control a particular organism, such as mycotoxin-producing strains of *Aspergillus flavus* in stored grain, would be to modify the microfloral community instead of attacking the organism directly (see Chapter 5, this volume). Because people arbitrarily superimpose their cultural and management practices on plant and animal communities and ecosystems, their ultimate well-being depends largely on how well they understand the workings of such systems and how well they can manipulate them to their advantage.

Microfloral Succession Most studies on the ecology of the microflora of stored grain have described the community structure (species diversity or kinds of microbial species and their succession) at one time, and have occasionally included seasonal and annual variations. For example, there is a succession among preharvest (field) fungi and postharvest (storage) fungi similar to that observed in the development of various seres (stages that follow one another in an ecological succession) progressing toward a climax stage (final stage of succession that permits the component organisms to live on in a balanced environment) containing plants and animals (31, 37). Regular sampling of maize from harvest through postharvest storage has shown that the field fungi died and made room for the postharvest storage fungi. At the end of storage, only *Penicillium cyclopium*, *P. chrysogenum*, and *Aspergillus versicolor* remained (37).

A similar succession of microflora in two 13.6-tonne farm-stored bulks of wheat was demonstrated in Manitoba during an 8-year study. Using descriptive (76) as well as multivariate statistical analyses (64), my colleagues and I found that *Alternaria* and its associates *Gonatobotrys*, *Cochliobolus*, and *Nigrospora*, which are carried in with seeds from the field, gradually disappear and are replaced by *Aspergillus versicolor*, *A. glaucus*, *A. candidus*, *Penicillium* spp., *Absidia*, *Chaetomium*, *Rhizopus*, and *Streptomyces*. Interrelationships among fungi and between some fungi and some insects and mites in grain-bulk ecosystems were confirmed statistically.

The common granary species of *Aspergillus*, *Penicillium*, *Fusarium*, and *Alternaria* alternate their dominance pattern according to the substrate, the length of storage, and the treatment of the grain. At a moisture content (mc) below 15%, *Aspergillus niger* was dominant in wheat, maize, and sorghum. At higher moisture levels, *A. niger* was replaced in time by *P. citrinum* and *A. sydowi* on wheat, by *P. citrinum* and *A. terreus* on maize, and by *A. terreus* on sorghum. At a higher moisture content at 8° (46.4°F), *Penicillium* species seemed to dominate. *A. ochraceus* preferred wheat over other seeds (42).

There is considerable preharvest contamination of barley and wheat grains by *Penicillium*, *Aspergillus*, and other fungi (19). When combine-harvested barley and wheat were plated directly on malt-salt agar, up to 250 times more kernels were contaminated by *Penicillium* species than when the same crop was hand threshed. Both crops, however, contained similar levels of *A. glaucus*. Possible sources for these contaminants during combine harvesting were soil and residues in the combine harvesters, sacks used in bagging, and the vegetative parts of the plants (19).

The typical pattern of relationships between seeds and fungi, from the preharvest period through postharvest storage over several years, involves contamination or infection, or both, by *Alternaria*, *Cladosporium*, *Aspergillus glaucus*, and *Penicillium*. Probably, fungal spores or mycelia accompany grain from the preharvest period through the early phase of postharvest storage (19). In storage, preharvest fungi gradually disappear, and the contamination by certain *Penicillium* species gradually turns into infection (76). This type of succession of internal seed microflora has also been observed in stored barley in a partially sealed silo (43). *Alternaria*, *Cladosporium*, and *Mucor* predominated in the incoming grain; *Penicillium cyclopium* and *Absidia* became dominant after 2 months; and *P. cyclopium*, *P. expansum*, *P. roqueforti*, and yeasts were dominant after 4 months.

In England, a typical succession of fungi and actinomycetes developed in barley (25–28% mc) stored in six top-unloading, unsealed concrete silos (30). The grain at the top became increasingly heated and turned moldy as the rate of unloading decreased. At an unloading rate of about 7.5 m³ per day, only yeasts such as *Pichia burtonii* (*Endomycopsis chodatii*) and *Hansenula anomala* were abundant, but *Penicillium* species became common at a slightly slower rate of unloading. With further slowing of unloading, both groups of microflora became scarce and were replaced by *Absidia* species and *Mucor pusillus*, then by *Aspergillus fumigatus*, *Thermomyces lanuginosus* (=*Humicola lanuginosa*), and *Micropolyspora faeni*; *Thermoactinomyces vulgaris* appeared as the grain temperature rose.

A health hazard exists in handling, transporting, and consuming stored grain that has undergone extensive qualitative and quantitative changes. Undisturbed, the silos contained 10^6 to 10^7 airborne spores/m³ of air, but when moldy grain was unloaded, concentrations rose to a maximum of $2,860 \times 10^6$ spores/m³ of air; over half of these were bacteria and actinomycetes and one-fourth were *A. flavus* (30).

Differences from Other Communities Members of a microbial community that includes various species of fungi, actinomycetes, yeasts, and bacteria are just as opportunistic and as adaptable to changing environments as the members of other biotic communities sharing the storage ecosystems. They differ from the communities of insects, mites, and birds in that (a) they are ubiquitous; (b) most reproduce asexually; (c) they proliferate faster than other organisms under optimal conditions; (d) they can withstand most adverse changes in the ecosystem caused by climate and humans; and (e) they produce substances either highly beneficial (e.g., antibiotics) or highly toxic (e.g., aflatoxin, ochratoxin) to humans.

Mites

Mites constitute an important part of stored-food ecosystems, especially where food is stored cool at intermediate moisture levels (14–17% mc) and where farming and storage practices favor rapid multiplication of the mites. Over 50 species of mites have been found in storage areas for cereals, oilseeds, flour, and livestock feed (24). Various foods in many agricultural and food storage areas of countries within 40- to 60-degree latitude in the Northern Hemisphere are most susceptible to infestation by mites, some of which also occur in bird or rodent nests or in soil.

Because most species of mites feed and reproduce readily on the fungi with which they are often associated in stored foods, it is popularly believed that the mites are dependent on microflora alone. But detailed, long-term quantitative studies of decaying bulk grain have shown that mite populations are regulated mainly by the moisture content, temperature, and changes in the nutritional quality of the grain or related food substrates (62, 64).

Of course, the microfloral component is foremost among the variables influencing the proliferation of mites. Other variables are the size and type of the stored-food mass, seed cracks, amount of dockage, predators, and parasites. Various kinds of mites act as energy transformers, granivores, herbivores, fungivores, predators, parasites, and scavengers. As energy transformers along the pathways of deterioration of stored food, mites such as *Tyrophagus putrescentiae* and *Rhizoglyphus echinopus* are more efficient than most harmful stored-product insects (9, 68, 69).

Hot Spots in Grain Bulks   The role of mites in the development of insect-induced, moldy hot spots in a stored-grain ecosystem was studied in Manitoba (61). As grain temperature and moisture rose in 262 tonnes (17,000 bushels) of oats as a result of multiplication of the beetles *Oryzaephilus surinamensis* and *Cryptolestes ferrugineus*, the grain was invaded successively by postharvest microflora, e.g., *Penicillium*, *Aspergillus*, *Absidia*, and *Streptomyces*; my-cophagous insects, e.g., *Ahasverus* and *Cryptophagus*; and mites, e.g., *Acarus siro*, *Androlaelaps casalis*, *Cheyletus eruditus*, and *Cheletomorpha lepidopterorum*. Moderate populations of *Lepidoglyphus destructor* were maintained before and after development of the hot spot. When hot spots develop, variability in density of insect and mite populations is usually caused by seasonal changes in atmospheric temperature (61).

A Mite Community in Japan   An acarine community was studied from samples of rice, wheat, barley, grain debris, and rice straw from fields, areas adjacent to granaries, granary floors, straw bags, and metal drums in two farmhouses and three rural granaries in Shiga Prefecture during June, September, and October 1966, and February 1967 (54). Thirty-four genera with 49 kinds of mites, several undescribed, were identified. *L. destructor, Cheyletus eruditus, Tyrophagus putrescentiae, Cheletomorpha lepidopterorum, Tarsonemus granarius,* and *Tydeus* species were the commonest mites on granary floors and were the most abundant during summer. During autumn and winter, only *L. destructor* maintained large populations in all granaries; in addition, large numbers of *Tyrophagus putrescentiae* and *Tarsonemus* species were observed in one granary. Nineteen kinds of mites occurred in the stored grain. The three commonest mites, i.e., those present usually in small numbers in 10 to 100% of the cereals in bags and drums, were, in descending frequency, on rice, *L. destructor, Tyrophagus putrescentiae,* and *Cheyletus eruditus*; on wheat, *C. eruditus, L. destructor,* and *T. putrescentiae*; on barley, *L. destructor, Tarsonemus granarius,* and *Tyrophagus putrescentiae*. All these mites appeared in grain samples throughout the year, although the numbers per sample generally decreased from summer to winter. The six commonest mites on granary floors were represented in rice straw in the field. The infestation of stored grain and grain products appeared to have originated primarily in the field and only secondarily from dirty granary floors in rural areas.

Insects

Insects that can attack relatively dry stored grains and foods (above 10% mc) at 15–42° (50–107°F) in all parts of the world comprise one of the most successful groups of organisms in the storage ecosystem (23, 57). Having originated in the tropical or subtropical parts of the world, most storage insects do not hibernate. They persist in the cooler parts of the world only by finding or creating a favorable environment, either within bulk food or in a warehouse (*65*).

Analysis of Canadian Prairie Provinces Ecosystem The impact of insect infestation on certain biotic and abiotic variables of bulk wheat ecosystems exposed to the extreme climate of the Canadian Prairie was studied during 1969 and 1970. An insect-free ecosystem (control), one set of bulks artificially infested with *Cryptolestes ferrugineus* and *Oryzaephilus surinamensis,* and another infested with *Sitophilus granarius* and *Tribolium castaneum* were used (*6*). Changes in temperature, moisture, viability of the grain, grain weight, dust weight, species of microorganisms and insects, fat acidity value, and uric acid content were measured monthly and analyzed separately by conventional descriptive and multivariate statistical methods (principal-component analyses). The analyses gave a clear picture of the pathways of biological and chemical deterioration of the grain in the systems.

In the control system, *Alternaria alternata* decreased while *Aspergillus versicolor* and *Penicillium* species increased, probably causing a moderate rise in the fat acidity of the grain and a slight drop in the grain weight. In the second system, *C. ferrugineus* and *O. surinamensis* attacked few sound kernels, but polluted the environment with excreta and exuviae and appeared to accelerate the kinds of deteriorative processes observed to be proceeding more slowly in the control system. In the third ecosystem, *S. granarius* and *T. castaneum* thrived, creating new pathways that involved the production of large amounts of frass, invasion by *Streptomyces* and bacteria, and further acceleration of the deteriorative process.

That study showed that time and the physical environment (especially temperature) dominate both the insect-free and the insect-infested systems, particularly the one infested with *Cryptolestes* and *Oryzaephilus*; the system was only slightly influenced by changes in microflora and free fatty acid levels. These insects, however, polluted the wheat by excreting uric acid. The system infested with *Sitophilus* and *Tribolium* was affected by a metabolic moisture variable created by the insects. Addition of this moisture to the system accelerated development of *Aspergillus, Penicillium, Streptomyces,* and bacterial populations. The main conclusion from this ecosystem analysis was that within a favorable range of moisture conditions, temperature sets the limits for survival and multiplication of the stored-product insects in the Prairie Provinces.

Climatic Plasticity Index Whether or not a stored-product insect or mite causes economic loss when there is adequate suitable food depends largely upon the physical limits and optimal ranges of temperature and relative humidity within which the species can thrive and also on its intrinsic rate of increase. Although several major insect species occur in most food-producing and populated parts of the world, some species are particularly destructive in certain climatic zones but unimportant in others.

To cause outbreaks in any climatic zone a species must be able to reproduce quickly during favorable periods of the year and survive well during unfavorable periods. Laboratory life-history studies of 51 species of stored-product insects have shown that their maximum rate of increase (r_m) occurs within a narrow (3–4°) temperature range, usually at 60% rh or higher (*23*). The maximum temperature at which these insects can multiply is about 4° above the optimum; most species do not reproduce fast enough to cause problems until the temperature is about 3–5° above the laboratory minimum and the relative humidity is at least 10% above the minimum (*23*).

The climatic, agricultural, and socioeconomic conditions of various parts of the world are the reasons for the present distribution of the various insect species; clues for their potential ranges of distribution in relatively unprotected environments can be found mostly in the inherent ability of each species to adapt and multiply. By integrating the laboratory data on the biology of each of 51 species, a plasticity index statistic (I_p) was developed to describe the potential ability of a species to adapt in various climatic zones (*58, 63*). The higher the I_p value, the greater the climatic adaptability of the species.

The formula for calculating the climatic plasticity index is

$$I_p = (\lambda/2)(t_3 - t_0 + t_2 - t_1)(h_1 - h_0 - 5)$$

where

t_0 = minimum temperature for reproduction of a storage species,

t_1 = lowest limit of optimal range,

t_2 = highest limit of optimal range,

t_3 = highest limit of reproduction is assumed as $t_2 + 4°$,

λ = rate of increase for a lunar month,

h_0 = minimum relative humidity required by the species to reproduce, and

h_1 = maximum relative humidity below which the species can multiply.

In the expression $(h_1 - h_0 - 5)$, 5 is subtracted to account for reduced reproduction as the relative humidity decreases to its minimum level.

The data for computation were taken from the published information of various workers as summarized by Howe (23). When a group of species is considered tolerant of low relative humidity, 80 or 90% rh was used in the equation; similarly, for other species groups that need moderate and high relative humidity, 90% and 100% rh, respectively, were used for h_1.

A ranking of 51 species on the basis of their I_p values is given in Table 4.1. Species with some of the highest I_p values—*Tribolium castaneum* ($I_p = 700$) and *T. confusum* ($I_p = 570$)—are in fact the most cosmopolitan pests of stored cereals and other foods. By contrast, the rice moth (*Corcyra cephalonica*), a specialized but dominant pest of rice only in low-latitude, wet-and-dry lands, has an I_p value of only 110. *Trogoderma granarium*, which is restricted to a hot and dry climate, also has a low value ($I_p = 131$). The most destructive stored-product species, such as *Sitophilus oryzae* ($I_p = 275$) and *S. granarius* ($I_p = 172.5$), are not necessarily the most flexible species with regard to their ability to flourish in various climatic zones of the world.

Rodents

Of all the rodents, only three highly successful, cosmopolitan, commensal species are commonly associated with storage ecosystems, usually as periodic invaders from agricultural field-crop ecosystems: the house mouse, *Mus musculus*; the Norway rat, *Rattus norvegicus*; and the roof rat, *R. rattus*. During the Neolithic Age, these species probably became associated with humans, especially if Stone Age man introduced them to his home as part of his own diet (50). From then on, these rodents became intruders and spoilers of food in storage ecosystems (25). Their success, in spite of extensive campaigns to destroy them, lies in their extreme adaptability, high reproductive rate, and omnivorous habits. Rodents cause enormous damage to stored grains; for example, 2 million tonnes of grain are eaten by rodents in India every year (50).

Rat Communities   Several ethological and ecological studies (mainly autecological) on mice and rats have been conducted (17, 22) but, to my knowledge, no ecosystem analysis using rodents with other animals, plants, and microflora

Table 4.1

Climatic plasticity index (I_p) values for species of stored-product insects

I_p	Species	I_p	Species
700	*Tribolium castaneum*	172.5	*Sitophilus granarius*
600	*Anagasta kuehniella*	165	*Ephestia elutella*
575	*Cadra cautella*	150	*Cryptolestes ugandae*
570	*Cryptolestes ferrugineus*	131	*Trogoderma granarium*
570	*Tribolium confusum*	110	*Corcyra cephalonica*
550	*Callosobruchus maculatus*	105	*Cryptolestes capensis*
550	*Sitotroga cerealella*	105	*Cryptolestes pusilloides*
550	*Carpophilus hemipterus*	100	*Cryptolestes pusillus*
500	*Oryzaephilus surinamensis*	95	*Latheticus oryzae*
475	*Cryptolestes turcicus*	67.5	*Stegobium paniceum*
360	*Dermestes maculatus*	42	*Gibbium aequinoctiale*
360	*Araecerus fasciculatus*	42	*Caryedon serratus*
330	*Plodia interpunctella*	42	*Ptinus ocellus*
330	*Endrosis sarcitrella*	24	*Mezium affine*
315	*Callosobruchus chinensis*	23	*Ptinus fur*
285	*Dermestes frischii*	21	*Stethomezium squamosum*
275	*Acanthoscelides obtectus*	21	*Niptus hololeucus*
275	*Sitophilus oryzae*	19	*Hofmannophila*
250	*Necrobia rufipes*		*pseudospretella*
230	*Callosobruchus rhodesianus*	15	*Trigonogenius globulum*
230	*Cadra calidella*	15	*Pseudeurostus hilleri*
210	*Oryzaephilus mercator*	10	*Ptinus clavipes*
200	*Lasioderma serricorne*	9.5	*Ptinus sexpunctatus*
190	*Zabrotes subfasciatus*	9	*Tipnus unicolor*
180	*Gnatocerus cornutus*	5	*Ptinus pusillus*

in stored-grain ecosystems has been attempted. Davis (14–16) performed extensive autecological studies of *R. norvegicus* populations on farms in Maryland. His main findings were that the growth pattern of the rat population follows a sigmoid curve, and the upper level beyond which no major changes can occur is regulated by food supply and available habitat. When conditions do not change, rat populations can remain at the carrying-capacity level for as long as 7 years; or the population level can fluctuate because of changes in the carrying capacity.

The maximum natality of the black rat is 37 young per female per year; the average is 10. About 95% of the population dies every year. A balance between natality and mortality, not the dispersal pattern, determines the population density. Reduction in rat populations through intraspecific competition, predation, and killing by humans increases proportionally with increasing population size.

Mouse Communities   Studies of population regulation mechanisms of *M. musculus* have shown that the establishment of a social hierarchy is crucial in determining the optimum density of the house mouse population (8, 13, 35). When the food supply is limited, the territoriality of the mouse is enhanced. Most researchers conclude that the establishment of the social organization is differentially affected by the regulation of the house mouse population density which in turn depends on the interaction of variables such as food supply, water resources, the location of these resources, and weather (3, 44, 67). In a favorable habitat, such as a well-sheltered stored-grain ecosystem, house mice can breed throughout the year even though they have a restricted breeding period (49). Common physiological changes in house mice under conditions of a rising population density are an increase in fetal resorption, delay in sexual maturity, inhibition of estrous cycles, and a reduction in the number of weaned young per lactating female (18, 36, 41, 70).

In a study of the population dynamics and energy utilization strategies of *M. musculus* populations feeding on corn and insects in experimental field conditions in Ohio, it was found that the assimilation efficiency for a male mouse is about 87.45 ± 2.04%. In kilojoules per gram of body weight per day, the ingestion rate was 3.72 ± 1.09; fecal output was 0.33 ± 0.04; and urine production was 0.08 ± 0.04 (4.184 kilojoules = 1.0 kcal) (71). Granivorous mice consumed minimal amounts of corn during periods of insect abundance. It was concluded that the resource architecture of a habitat significantly affects the functional and demographic characteristics of house mouse populations, suggesting that both the quantitive (e.g., caloric amount) and qualitative (e.g., availability) aspects of habitat resource are important in the population regulation of the house mouse.

Rodents as Consumers Rodents are both primary and secondary consumers of stored grain, i.e., they eat the grain and the insects associated with it. To understand reproduction and population growth patterns of rodents with a view to managing their populations, one must know (a) the average litter size, (b) the gestation period, (c) the growth rate, (d) the age at sexual maturity of each species, and (e) the duration of the reproductive life of an individual (80; see also Chapter 18, this volume). Rodent population fluctuations are dependent on the sex ratio, the proportion of the population in different age classes, and the proportion reproducing during the year. Of course, these variables are influenced by other variables (members of other communities and the abiotic environment) within the stored-grain ecosystem. To analyze and manage storage ecosystems, particularly in the semitropical and tropical zones of the world where the rodent community is one of the most important destructive components, long-term multivariate studies must be conducted. By knowing the role of various abiotic and biotic variables in causing rodent population fluctuations in a storage ecosystem, better ways can be found to reduce rodent numbers and rodent-borne diseases (50).

Birds

Granivorous components of the avian community such as blackbirds, queleas, house and tree sparrows, common grackles, and starlings are active in natural as well as man-made ecosystems. The man-made ecosystem includes preharvest agricultural crops and to a lesser extent the postharvest storages. In the USA, the redwing blackbird and a few other species of birds cause crop damage of about $100 million annually. In Ohio, direct bird damage to corn, together with associated losses such as molding, was estimated at $15 million in 1966; feedlot losses in Georgia were about $1,000 per day (4). Although the blackbirds that feed on corn, oats, and other ripening cereal crops as well as weed seeds and insects, are not directly involved in postharvest storage ecosystems, their extensive feeding on seeds makes the subsequent stored crop highly vulnerable to infection by microflora and infestation by insects and mites. Partly eaten, bird-damaged seeds are readily infested by fungal and arthropod pest species in storage ecosystems.

Keil (27), upon analyzing the stomach contents of house sparrows in a large cereal-growing district in Germany, found that the main foods were wheat, oats, and weed seeds. The stomach contents consisted of 66% gravel and 34% food. The proportions among five types of house-sparrow foods were cereals, 48%; weed seeds, 36%; glumes and plant fibers, 12%; other vegetable parts, 3%; and animal foods, 1%.

Crop Losses to Queleas in Africa One of the most devastating problems faced by many producers in the drier parts of Africa is destruction of their ripening cereal crops by flocks of the red-billed quelea. From ancient times, these birds have sporadically invaded fields of millet and guinea corn. They may also cause up to 100% loss of rice and wheat in at least 16 countries of western, eastern, and southern Africa (78).

The attack pattern and dietary habits of these birds follow simple ecological principles (77). The preference shown by queleas for their natural food of wild grass seeds is at the root of their population dynamics. The great majority of these birds, in their 2-year lifespan, normally do not eat any cultivated grain, but attack grain only when they cannot obtain wild grass seeds. As the supply of small grass seeds disappears in the early dry season, the birds change gradually to larger seeds.

In years when this shifting of food habits begins early, it is likely to coincide with the ripening of cereal crops; the result is severe loss caused by birds. The dry-season food supply disappears as soon as the rainy season begins, and seeds germinate simultaneously over large areas. The birds prepare for the lean period by producing fat reserves. Then they migrate south to areas where rain has fallen for several weeks. During the brief period spent in the south, the birds feed on ripening grass seeds including those of cultivated millet. Because there is considerable bird mortality, their population is probably limited by the food supply.

These birds form enormous nesting colonies at the end of the rainy season, and dense roosts numbering millions of individuals. As the feeding flocks raid ripening fields of millet, sorghum, rice, or wheat, they cause devastating losses (48). Other birds such as the golden sparrow and the village weaver selectively attack crop seeds growing in desert, semidesert, and tropical zones of Africa.

Stored-Grain Losses to Pest Birds In stored-grain ecosystems, birds such as English sparrows, pigeons, and barnyard fowl are the only casual invaders (12). The impact of their consumption and contamination of stored grain (by excreta and transfer of stored-product mites from birds) becomes appreciable only when proper storage practices are consistently ignored. This happens when doors, windows, and ventilators are left open, thus providing easy access to the interiors of the granaries. Because there is a similarity in the species composition of insects and mites between the bird nests and the stored-grain ecosystem (33, 81, 82), it has been suggested that birds are the carriers of these harmful arthropod pests to storage ecosystems. However, after monitoring the seasonal abundance of mites in sparrow nests and grain bins in Poland, Sandner and Wasylik (51) concluded that mites were not regularly brought to stored products by tree sparrows even though acarine species in the two habitats were very similar.

Conclusions

The unstable, often short-lived, man-made, stored-food ecosystem contains an unusual assortment of highly prolific and opportunistic animal and plant species. Odum (46) calculated

the following rates of natural increase of some of the most successful of these species, as well as the human species, under optimum conditions. The number of times the population would multiply in a year under different optimun conditions and without environmental interference was for *Sitophilus oryzae* at 29° (84.2°F), 1.58×10^{16}; for *Tribolium castaneum* at 28.5° (83.3°F) and 65% rh, 1.06×10^{15}; for *Rattus norvegicus*, 221; and for humans, 1.0055.

When compared with field crops, stored-food ecosystems, even though they are closed systems, usually provide an abundant food supply and structural protection (both through the granary building material and the bulk grain per se) from the direct effects of the climate. The indefinite length of the storage period during which the system will remain intact and undisturbed provides the greatest obstacle to the success of pest species in destroying the stored food. This indefinite aspect may keep the stored-grain ecosystem from becoming mature. If storage managers possess sufficient information about the interrelationships and checks and balances within the storage ecosystem, then they can protect stored food under various conditions of storage with a minimal loss in quality and quantity.

Cited References

1 Anderson, J. A. 1973. Problems of controlling quality in grain. In *Grain-Storage—Part of a System*. R. N. Sinha and W. E. Muir, eds. AVI, Westport, CT.

2 Anderson, J. A., and A. W. Alcock. 1954. *Storage of Cereal Grains and Their Products*. American Association of Cereal Chemists, St. Paul, MN.

3 Anderson, P. K., and J. L. Hill. 1965. *Mus musculus*, experimental induction of territory formation. *Science* 148(3678)1753–1755.

4 Anonymous. 1967. Blackbird depredation in agriculture. A report on the 1967 North American Conference. *Agricultural Science Review* 5(2)15–22.

5 Beuchat, L. R. 1978. *Food and Beverage Mycology*. AVI, Westport, CT.

6 Bronswijk, J. E. M. H. van, and R. N. Sinha. 1971. Interrelations among physical, biological, and chemical variates in stored-grain ecosystems—a descriptive and multivariate study. *Annals of the Entomological Society of America* 64(4)789–803.

7 Burges, H. D., and N. J. Burrell. 1964. Cooling bulk grain in the British climate to control storage insects and to improve keeping quality. *Journal of the Science of Food and Agriculture* 15(1)32–50.

8 Calhoun, J. B. 1963. The social use of space. In *Physiological Mammalogy*, M. V. Mayer and R. G. Van Gelder, eds. Academic Press, New York, NY.

9 Campbell, A., and R. N. Sinha. 1978. Bioenergetics of granivorous beetles, *Cryptolestes ferrugineus* and *Rhyzopertha dominica*. *Canadian Journal of Zoology* 56(4)624–633.

10 Christensen, C. M., and H. H. Kaufmann. 1969. *Grain-storage—The Role of Fungi in Quality Loss*. University of Minnesota, Minneapolis, MN.

11 Clark, L. R., P. W. Geier, R. D. Hughes, and R. E. Morris. 1967. *The Ecology of Insect Populations in Theory and Practice*. Methuen, London.

12 Cotton, R. T. 1963. *Pests of Stored Grain and Grain Products*. Burgess, Minneapolis, MN.

13 Crowcroft, P., and F. P. Rowe. 1963. Social organization and territorial behavior in the wild house mouse. *Proceedings of the Zoological Society of London* 140(3)517–531, 2 pl.

14 Davis, D. E. 1948. The survival of wild brown rats on a Maryland farm. *Ecology* 29(4)437–448.

15 Davis, D. E. 1951. A comparison of the reproductive potential of two rat populations. *Ecology* 32(3)469–475.

16 Davis, D. E. 1953. The characteristics of rat populations. *Quarterly Review of Biology* 28(4)373–401.

17 Delany, M. J. 1974. *The Ecology of Small Mammals*. Arnold, London.

18 DeLong, K. 1967. Ecology of feral house mice. *Ecology* 48(4)611–634.

19 Flannigan, B. 1978. Primary contamination of barley and wheat grain by storage fungi. *Transactions of the British Mycological Society* 71(1)37–42.

20 Graham, W. M. 1970. Warehouse ecology studies of bagged maize in Kenya. II. Ecological observation of an infestation by *Ephestia (Cadra) cautella* (Walker) (Lepidoptera, Phycitidae). *Journal of Stored Products Research* 6(2)157–167.

21 Hall, D. W. 1970. *Handling and Storage of Food Grains in Tropical and Subtropical Areas*. Agricultural Development Paper 90. Food and Agriculture Organization, Rome.

22 Hanney, P. W. 1975. *Rodents: Their Lives and Habits*. David and Charles, North Vancouver, Canada.

23 Howe, R. W. 1965. A summary of estimates of optimal and minimal conditions for population increase of some stored products insects. *Journal of Stored Products Research* 1(2)177–184.

24 Hughes, A. M. 1976. *The Mites of Stored Food and Houses*. Her Majesty's Stationery Office, London.

25 Jackson, W. B. 1977. Evaluation of rodent depredations to crops and stored products. *EPPO Bulletin* 7(2)439–458.

26 Joffe, A. 1958. Moisture migration in horizontally-stored bulk maize: The influence of grain-infesting insects under South African conditions. *South African Journal of Agricultural Science* 1(2)175–193.

27 Keil, W. 1973. Investigations on food of house- and tree-sparrows in a cereal-growing area during winter. In *Productivity, Population Dynamics and Systematics of Granivorous Birds*, S. C. Kendeigh and J. Pinowski, eds. PWN-Polish Scientific, Warsaw.

28 Kendeigh, S. C., and J. Pinowski (eds.). 1973. *Productivity, Population Dynamics and Systematics of Granivorous Birds*. PWN-Polish Scientific, Warsaw.

29 Kiritani, K. 1958. On the distribution and seasonal prevalence of stored grain insects in a farm premises. *Botyu-Kagaku* 23(4)164–172 (in Japanese, English summary).

30 Lacey, J. 1971. The microbiology of moist barley storage in unsealed silos. *Annals of Applied Biology* 69(3)187–212.

31 Lagrandeur, G., and J. Poisson. 1968. La microflore du maïs, son évolution en fonction des conditions hydriques et thermiques de stockage en atmosphère renouvelée. *Industries Alimentaires Agricoles* 85:775–788.

32 Lindeman, R. L. 1942. The trophic-dynamic aspect of ecology. *Ecology* 23(4)399–418.

33 Linsley, E. G. 1944. Natural sources, habitats, and reservoirs of insects associated with stored food products. *Hilgardia* 16(4)187–224.

34 Liscombe, E. A. R., and F. L. Watters. 1962. Insect and mite infestations in empty granaries in the Prairie Provinces. *Canadian Entomologist* 94(4)433–441.

35 Lloyd, J. A. 1975. Social structure and reproduction in two freely growing populations of house mice (*Mus musculus* L.). *Animal Behaviour* 23(2)413–424.

36 Lloyd, J. A., and J. J. Christian. 1967. Relationship of activity and aggression to density in two confined populations of house mice (*Mus musculus*). *Journal of Mammalogy* 48(2)262–269.

37 Lutey, R. W., and C. M. Christensen. 1963. Influence of moisture content, temperature and length of storage period upon survival of fungi in barley kernels. *Phytopathology* 53(6)713–717.

38 Machacek, J. E., W. J. Cherewick, H. W. Mead, and W. C. Broadfoot. 1951. A study of some seed borne diseases of cereals in Canada. II. Kinds of fungi and prevalence of disease in cereal seed. *Scientific Agriculture* 31(May)193–206.

39 Margalef, R. 1963. On certain unifying principles in ecology. *American Naturalist* 97(897)357–374.

40 Margalef, R. 1968. *Perspectives in Ecological*

Theory. University of Chicago.

41 McClure, T. J. 1958. Temporary nutritional stress and infertility in mice. *Nature* 181(4616)1132.

42 Moubasher, A. H., M. A. Elnaghy, and S. I. Abdel-Hafez. 1972. Studies on the fungus flora of three grains in Egypt. *Mycopathologia et Mycologia Applicata* 47(3)261–274.

43 Mulinge, S. K., and C. G. C. Chesters. 1970. Ecology of fungi associated with moist stored barley grain. *Annals of Applied Biology* 65(2)277–284.

44 Newsome, A. 1969. A population study of house mice temporarily inhabiting a South Australian wheat field. *Journal of Animal Ecology* 38(2)341–377.

45 Odum, E. P. 1957. The ecosystem approach in the teaching of ecology illustrated with sample class data. *Ecology* 38(3)531–535.

46 Odum, E. P. 1969. The strategy of ecosystem development. *Science* 164(3877)262–270.

47 Oxley, T. Z. 1948. *The Scientific Principles of Grain Storage*. Northern, Liverpool, England.

48 Park, P. O. 1974. Granivorous bird pests in Africa. Towards integrated control. *Span* 17(3)126–128.

49 Pearson, O. P. 1963. History of two local outbreaks of feral house mice. *Ecology* 44(3)540–549.

50 Rowe, F. P., and K. D. Taylor. 1970. Rodent biology. In *Food Storage Manual*, part I. World Food Programme, Rome.

51 Sandner, H., and A. Wasylik. 1973. The mites of sparrow nests and the danger of infestation of granaries by them. *Ekologia Polska* 21(22)323–338.

52 Sinha, R. N. 1964. Ecological relationships of stored-products mites and seed-borne fungi. Proceedings of the First International Congress of Acarology (Fort Collins, 1963). *Acarologia* 6(fasc. h. s.)372–389.

53 Sinha, R. N. 1965. Mites of stored grain in Western Canada—ecology and methods of survey. *Proceedings of the Entomological Society of Manitoba* 20(1964)19–33.

54 Sinha, R. N. 1968. Seasonal changes in mite populations in rural granaries in Japan. *Annals of the Entomological Society of America* 61(4)938–949.

55 Sinha, R. N. 1971. Fungus as food for some stored product insects. *Journal of Economic Entomology* 64(1)3–6.

56 Sinha, R. N. 1973. Ecology of storage. *Annales de Technologie Agricole* 22(3)351–369.

57 Sinha, R. N. 1973. Interrelations of physical, chemical, and biological variables in the deterioration of stored grain. In *Grain-Storage–Part of a System*. R. N. Sinha and W. E. Muir, eds. AVI, Westport, CT.

58 Sinha, R. N. [1975.] Climate and the infestation of stored cereals by insects. *Proceedings of the First International Working Conference on Stored-product Entomology* (Savannah, GA, 1974), pp. 117–141.

59 Sinha, R. N. 1977. Use of multivariate analyses in the study of stored-grain ecosystems. *Environmental Entomology* 6(2)185–192.

60 Sinha, R. N., and W. E. Muir (eds.). 1973. *Grain-Storage—Part of a System*. AVI, Westport, CT.

61 Sinha, R. N., and H. A. H. Wallace. 1966. Ecology of insect-induced hot spots in stored grain in Western Canada. *Researches on Population Ecology* 8(2)107–132.

62 Sinha, R. N., and H. A. H. Wallace. 1973. Population dynamics of stored-products mites. *Oecologia* 12(4)315–327.

63 Sinha, R. N., and F. L. Watters. 1985. *Insect Pests of Flour Mills, Grain Elevators, and Feed Mills and Their Control*. Publ. 1776E. Research Branch, Agriculture Canada, Ottawa.

64 Sinha, R. N., H. A. H. Wallace, and F. S. Chebib. 1969. Principal-component analysis of interrelations among fungi, mites, and insects in grain bulk ecosystems. *Ecology* 50(4)536–547.

65 Smith, C. V. 1969. *Meteorology and Grain Storage*. Technical Note 101. World Meterological Organization, Geneva, Switzerland.

66 Solomon, M. E. 1969. Establishment, growth, and decline of populations of the grain mite *Acarus siro* L. on a handful of wheat. *Proceedings of the Second International Congress of Acarology* (Sutton Bonington, England, 1967), pp. 255–260.

67 Southwick, C. H. 1955. The population dynamics of confined house mice supplied with unlimited food. *Ecology* 36(2)212–225.

68 Stępień, Z. A. 1970. The energy budget of *Rhizoglyphus echinopus* (Acarina, Acaridae) during its development. Dissertation, Agricultural University of Warsaw (in Polish).

69 Stępień, Z. A., W. Goszcynski, and J. Boczek. 1973. The energy budget of *Tyrophagus putrescentiae* (Schr.) (Acaridae). *Proceedings of the Third International Congress of Acarology* (Prague, 1971), pp. 373–378.

70 Strecker, R. L., and J. T. E. Emlen. 1953. Regulatory mechanisms in house-mouse populations: The effect of limited food supply on a confined population. *Ecology* 34(2)375–385.

71 Stueck, K. L., and G. W. Barrett. 1978. Effects of resource partitioning on the population dynamics and energy utilization strategies of feral house mice (*Mus musculus*) populations under experimental field conditions. *Ecology* 59(3)539–551.

72 Trewartha, G. T. 1968. *An Introduction to Climate*. McGraw-Hill, New York, NY.

73 Trisvyatskii, L. A. 1969. *Storage of Grain*, vol.

3. National Lending Library for Science and Technology, Boston Spa (translated from Russian by D. M. Keane).

74 Walkden, H. H. 1951. Farm storage of cereal grains. *Milling Production* 16(10)1, 7, 23–25.

75 Wallace, H. A. H. 1973. Fungi and other organisms associated with stored grain. In *Grain-Storage–Part of a System*. R. N. Sinha and W. E. Muir, eds. AVI. Westport, CT.

76 Wallace, H. A. H., R. N. Sinha, and J. T. Mills. 1976. Fungi associated with small wheat bulks during prolonged storage in Manitoba. *Canadian Journal of Botany* 54(12)1332–1343.

77 Ward, P. 1965. Feeding ecology of the black-faced dioch *Quelea quelea* in Nigeria. *Ikis* 107(2)173–214.

78 Ward, P. 1973. A new strategy for the control of damage by queleas. *Pans* 19(1)97–106.

79 Watters, F. L. 1955. Entomological aspects of bulk grain storage in the Prairie Provinces. *Proceedings of the Entomological Society of Manitoba* 11:28–37.

80 World Health Organization Scientific Group. 1974. *Ecology and Control of Rodents of Public Health Importance*. WHO Technical Report Series No. 553. Geneva, Switzerland.

81 Woodroffe, G. E. 1953. An ecological study of insects and mites in the nests of certain birds in Britain. *Bulletin of Entomological Research* 44(4)739–772.

82 Woodroffe, G. E., and B. J. Southgate. 1951. Birds' nests as source of domestic pests. *Proceedings of the Zoological Society of London* 121(1)55–62, 2 pl.

83 Yoshida, T., and K. Kawano. 1958. Seasonal fluctuation in the number of insects in grain stored in a farm house. 1. Ecological studies of the pests infesting stored grain. Part 2. *Memoirs of the Faculty of Liberal Arts and Education, Miyazaki University, Natural Science* 5:11–23 (in Japanese, English summary).

II
Ecology

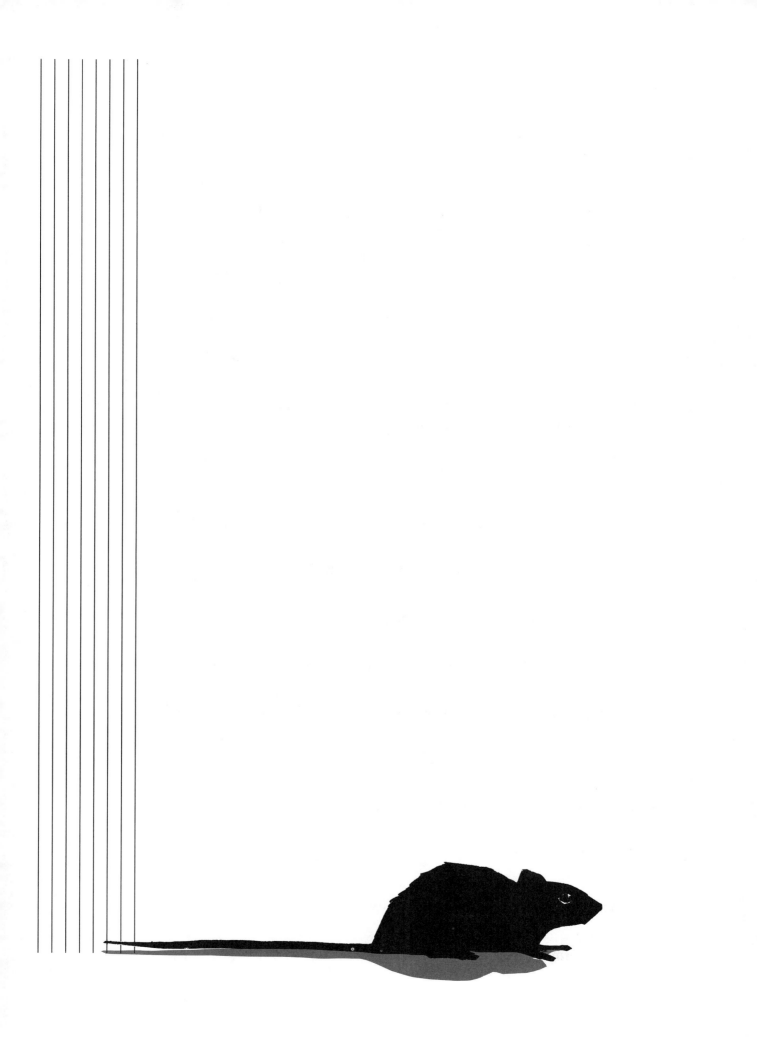

5

Ecology and Control of Microorganisms Decomposing Stored Foods

John T. Mills

The aim of this chapter is to describe the ecological principles that govern development of microorganisms in stored foods and to show, using practical examples, how a knowledge of these principles can be used to predict, prevent, and control microbial decomposition. It is directed toward food managers as an aid in the recognition and control of microbial decomposition of foods in a wide variety of storage situations with the aim of ensuring safe storage with minimum quality loss.

The scope of the chapter includes storage situations on the farm and plantation; at the elevator, warehouse, and processing plant; in trucks, trains, and ships; and also in the supermarket, restaurant, and family kitchen. Most emphasis is placed on storage situations on the farm, particularly in regard to the decomposition of unprocessed cereals and oilseeds by fungi. In other areas, because of the diversity of the topic, much reliance has been placed on published reviews. Examples cited in the text are selected from the world literature to ensure wide application.

Terms used to describe aspects of microbial decomposition of stored foods are defined in the Glossary at the end of this volume. Further information may be found in a technical glossary (8).

Microbial Decomposers

Molds, Yeasts, and Bacteria

Molds, yeasts, and bacteria are all associated with spoilage of stored foods (Tables 5.1 through 5.3). Bacteria and yeasts usually grow much more quickly than molds and, therefore, under broadly favorable conditions, they make food unfit to eat before mold development is evident (134). Species of bacteria (148) are of particular importance in spoilage of vegetables, fruits, milk, meat, poultry, and fish (Table 5.2); yeasts (100) often spoil foods such as cured meats, catsup, mayonnaise, and pickles in brine (Table 5.3). Conditions that may favor mold growth over other microorganisms are high acidity, limited available water, and low incubation temperatures. There are several useful guides to the classification of microbial decomposers: molds and yeasts (95, 128); bacteria (52); spoilage molds (134); spoilage bacteria (148); and spoilage yeasts (100).

Foods Subject to Mold Spoilage

Molds are the main cause of market diseases (141) of fresh fruits (132) and vegetables (91) and are largely responsible for spoilage from the time of harvest until the product is consumed (129, 134). Molds are the main cause of deterioration and spoilage of bulk-stored cereal grains (19) and oilseeds, and are the principal spoilage agents of bread and other bakery products (106). Molds attack fresh and cured meats (61), poultry, and seafood (60). In particular, molds develop on fresh beef aging in cold storage and on bacon that has not received sufficient heat to destroy spores.

Molds may cause spoilage of cheese and but-

Mills: Research Station, Agriculture Canada, 195 Dafoe Road, Winnipeg R3T 2M9, Canada.

Contribution No. 912 from Research Station, Agriculture Canada, Winnipeg.

Table 5.1

Some important mold genera causing deterioration of various classes of foods

Bakery products	Cereal grains	Dairy products	Fruits and vegetables	Meats
Aspergillus	Aspergillus	Alternaria	Alternaria	Alternaria
Monilia	Fusarium	Cladosporium	Aspergillus	Aspergillus
Mucor	Mucor	Geotrichum	Botrytis	Botrytis
Penicillium	Penicillium	Monilia	Cladosporium	Cladosporium
Rhizopus	Rhizopus	Oospora	Colletotrichum	Fusarium
Sporotrichum		Penicillium	Diplodia	Monilia
			Fusarium	Mucor
			Monilia	Oospora
			Mucor	Penicillium
			Oospora	Rhizopus
			Penicillium	Sporotrichum
			Phomopsis	Thamnidium
			Phytophthora	
			Rhizopus	
			Sclerotinia	
			Trichoderma	

SOURCES: Splittstoesser, D. F., and D. B. Prest. 1976. Molds. In *Food Microbiology: Public Health and Spoilage Aspects*, M. P. Defiguerido and D. F. Splittstoesser, eds. AVI, Westport, CT; Weiser, H. H., G. J. Mountney, and W. A. Gould. 1971. *Practical Food Microbiology and Technology*, 2nd ed. AVI, Westport, CT.

Table 5.2

Spoilage bacteria and yeasts often associated with various classes of foods

Class of food products	Genera dominating when spoilage occurs during standard conditions of storage	
	Bacteria[a]	Yeasts
Milk and milk products	Streptococcus, Lactobacillus, Microbacterium, Gram-negative rods, Bacillus	
Fresh meat, poultry	Gram-negative rods, Micrococcus	
Sausage, bacon, ham	Micrococcus, Lactobacillus, Streptococcus	Debaryomyces
Fish, shrimp, shellfish	Gram-negative rods, Micrococcus	
Eggs	Pseudomonas	
Vegetables	Gram-negative rods, Lactobacillus, Bacillus	
Fruits and fruit juices	Acetobacter, Lactobacillus	Saccharomyces, Torulopsis (=Candida)
Bread	Bacillus	Endomyces

[a]Gram-negative rods include particularly strains of *Achromobacter*, *Pseudomonas*, and *Flavobacterium*, with exclusion of the coliform-aerogenes group.

SOURCE: Weiser, H. H., G. J. Mountney, and W. A. Gould. 1971. *Practical Food Microbiology and Technology*, 2nd ed. AVI, Westport, CT.

Effects of Mold Spoilage

Spoilage of foods by molds is often clearly visible, e.g., totally rotted fruit or compacted, molded cereal grains in a grain bin, both of which involve total economic loss. Other visible spoilage may be less severe, but nevertheless represent economic loss in terms of salability or through costs of labor to trim off affected parts. Examples are the occurrence of partially rotted fruits in a box contaminated from an original infected fruit; discoloration of butter, eggs, or aging meat; or unsightly mold growth on bread, jam, or cheese. Food losses due to molds occur in developed as well as developing countries (9, 151).

Molds are also responsible for biochemical changes in foods, resulting in loss of food quality. The biochemical behavior of microorganisms in food, and thus the type of metabolic end product, is largely determined by (a) the chemical composition of the food, (b) the type of organism involved, (c) the environmental conditions of both food and microorganisms, and (d) the changes occurring in the food and the environmental conditions during the course of the spoilage process (37, 148). Objectionable odors and sliminess (148), rotting and softening (91), and rancidity, soapiness, and bitterness (108) can all result from microbiological action in various foods. Increased free fatty acid levels, decreased germination, heating and caking (19, 79), and loss of food energy and nutrients (Table 5.4) occur during spoilage of stored cereal grains by molds.

Certain molds associated with stored foods are known to cause diseases, namely fungal allergies and mycotoxicoses. Bronchial asthma and farmer's lung (hypersensitivity pneumonitis) have been associated with the handling of moldy grain (30); other types of allergies affect animals (3). Aflatoxin, an important, naturally occurring toxin in certain human and animal foods, was discovered in 1960 when 100,000 turkey poults died after eating peanut meal contaminated with *Aspergillus flavus* Link (114). The discovery that aflatoxin B_1 is extremely carcinogenic and is produced by a very common mold occurring on many foods has since promoted many investigations on the occurrence of other toxigenic fungi in stored foods, on conditions leading to toxigenesis, and on prevention and control measures (1, 11, 98, 116, 150).

Isolation and Identification of Food Spoilage Molds

The isolation of molds from foods by specific

ter, resulting in discolorations and off-flavors including rancidity (40). Mold development is favored by the acidity and relatively low moisture content of many finished cheeses. Many other foods are occasionally affected by certain mold species—meat pies stored for long periods in freezers with significant temperature fluctuations (44), canned fruit products (53), low-moisture foods such as prunes and jellies (28), and the surface of fruit juices (27) and of vegetables in brine (134).

techniques (*19, 56, 59, 138*) permits identification of the contaminant or spoilage organism by comparison with known species so that the probable source can be determined, human health hazards identified, and preventive measures taken to stop recurrence. By knowing the type of mold involved, its probable source, and the history of the food from farm or processing plant, blame for financial loss can be more easily apportioned among producer, shipper, wholesaler, and others. The use of standardized procedures to determine mold levels in human foods (*149*) permits comparison with published action levels (*14*). Action levels for mold are usually based on a percentage, either by weight or by count, of the food containing the mold or a combination of mold and other defects. The routine determination of mold levels in food and beverage products after manufacture permits regulation of quality before human consumption.

Techniques for isolation of molds from stored cereal grains are given by Tuite (*138*) and Christensen and Kaufmann (*19*). Wallace (*145*) lists *Aspergillus* spp., *Penicillium* spp., and various phycomycete species isolated from stored products in various countries. *Aspergillus* and *Penicillium* species are the most important organisms associated with spoilage of stored grains and many other foods. Pertinent references on various mold groups and their identification include the following: *Aspergillus* (*109*); *Fusarium* (*10, 12*); *Penicillium* (*63, 104, 110*); dematiaceous hyphomycetes (*36*); genera sporulating in pure culture (*142*); xerophilic molds and food spoilage (*103*); and identification manuals (*95, 128*). It is essential that the isolated fungus be correctly identified; therefore, if in doubt, consult a specialist.

Factors Influencing Microbial Activity

Occurrence and Distribution of Microorganisms

All substances in which carbon exists alongside an adequate amount of moisture provide a home for microbial life (*140*). Thus, microorganisms are found in abundance in soil, on vegetation, and in or on other organic matter. The principal means of dispersal for molds to new locations is the transportation of asexual spores by air currents, water, or animals (*55, 64*). The nonsporing types of bacteria are susceptible to desiccation. Therefore, water and insects or other animal vectors constitute more important means of dispersal of these microbes than air currents (*140*).

Table 5.3
Yeasts and food spoilage

Food	Activity	Yeast
Catsup	gaseous fermentation	*Saccharomyces bailii* (Lindner)
Mayonnaise, blue cheese dressing	gaseous fermentation	*S. bailii*
Cured meats	acidification	*Debaryomyces, Torulopsis* (=*Candida*)
Sausages	slime	*Debaryomyces*
Pickles in brine	acid loss, softening, discoloration	*Debaryomyces*
Olives in brine	gas pocket formation, softening	*Hansenula, Saccharomyces*
Tomatoes	fermentation	*Hanseniaspora uvarum* (Niehaus) Shehata et al., *Kloeckera apiculata* (Rees emend. Klöcker) Janke, *Pichia kluyveri* Bedford
Margarine	rancidity	*Saccharomycopsis lipolytica* (Wickerham et al.) Yarrow
Lard	increased alkanals and methyl ketones	*S. lipolytica*
Beer	turbidity, off-flavors	strains of *Saccharomyces cerevisiae* Hansen (=*S. bayanus* Saccardo); *Dekkera intermedia* van der Walt
Wine	off-flavor, turbidity	*Brettanomyces, Candida, Pichia, Hansenula, Kloeckera*
Carbonated soft drinks	gaseous fermentation	osmophilic yeasts
Yogurt	fruity flavor, fermentation	*Kluyveromyces marxianus* (Hansen) van der Walt [=*K. fragilis* (Jorgensen) van der Walt]
Fruit juice concentrates, honey	ester formation, fermentation	*Hansenula, S. rouxii* [=*S. bisporus* (Naganishi) Lodder & Van Rij var. *mellis* (Fabian & Quinet) van der Walt]
Syrups	maple sap fermentation	*Trichosporon* sp., *Cryptococcus albidus* (Saito) Skinner
Figs, strawberries	fermentation, softening	*Kloeckera apiculata, Hanseniaspora* sp.
Dates	souring, fermentation	*Hanseniaspora* sp., *Candida guilliermondii* (Castellani) Langeron & Guerra, *C. krusei* (Castellani) Berkhout, *C. tropicalis* (Castellani) Berkhout

SOURCE: Peppler, H. J. 1976. Yeasts. In *Food Microbiology: Public Health and Spoilage Aspects*, M. P. Defiguerido and D. F. Splittstoesser, eds. AVI, Westport, CT.

Table 5.4
Deterioration in food value of wheat by *Aspergillus flavus*[a]

Wheat	Carbohydrate (%)	Protein (%)	N₂ (%)	Fat (%)	Niacin (mg/ 100 g)	Approximate spore count/g of wheat
Sterilized wheat	64.2	13.7	2.2	1.3	5.2	
2 days infection	50.0	12.5	2.0	1.0	4.8	19,000
10 days infection	20.0	1.1	0.2	0.4	1.9	4,000,000

[a] Initial inoculum as suspension of spores, 0.5 mL of which contained about 50,000 to 60,000 spores. Incubation was on sterile wheat in flasks at 25°.
SOURCE: Mehrotra, B. S. 1976. *Investigations of Selected Microorganisms Associated with Cereal Grains and Flours in India, to Provide Basic Information Related to the Utilization of Cereal Grains in Foods and Feedstuffs*. Report of Research Project FG-In-410. University of Allahabad, India.

Ecology and Control of Microorganisms Decomposing Stored Foods **35**

Mold spores in the air and on raw products are sources of contamination in food-processing plants. Good-quality raw materials harbor fewer molds than those of poor quality (71, 134). Sanitation procedures in the food plant, such as the removal of waste organic matter from the working area and from drains (48), the prevention of water condensation on ceilings or machinery, or the purchase of better-quality raw products (128), tend to reduce the numbers of mold colonies and airborne spores that might contaminate the finished product. Ultraviolet light is often used to kill airborne mold spores and also spores that have settled on walls or equipment (134).

Factors Affecting Microbial Growth

Spoilage of foods by particular microorganisms is determined initially by the presence of spores or mycelia in or on the food. Germination and growth of the microorganisms are subsequently influenced by six major environmental factors: moisture, temperature, oxygen, nutrients, pH, and growth inhibitors (148). Most methods used to preserve and protect foods from microbial spoilage (105) are based on a knowledge of the occurrence and distribution of microorganisms in the food environment and on the known effects of these six environmental factors on microbial development.

Moisture

Bacteria generally require higher moisture levels for development than molds and yeasts. Some species of molds and yeasts can grow in atmospheres of little more than 70% relative humidity (rh) (130).

Microorganisms are dependent on water for growth. Nutrients are carried into the cells by water. Water also serves as a means of eliminating waste products that have accumulated as the result of metabolic activity. Water is present in all foods in varying amounts: fresh fruits contain 75–90%, tomatoes 95%, milk 87%, fresh eggs 70–75%, and dry beans and cereals 5–15%. Pure water has a water activity (A_w) of 1.00 and is in equilibrium with a relative humidity of 100%. Because some water exists in bound form in the food, the water activity of a system is not the same as its water content (148). The A_w ranges of various foods and the minimum growth limits for the main groups of microorganisms are shown in Table 5.5. Yeasts, molds, and halophilic bacteria in salt-preserved foods are the only organisms likely to grow at levels below 0.85 A_w; their ability to tolerate reduced A_w exceeds that of any other group. The A_w minima for different strains of the yeast, *Saccharomyces rouxii* Boutroux, vary widely (96); most molds and yeasts have characteristic A_w minima for growth (25, 103, 137).

Freezing, freeze-drying, and drying remove much of the moisture or change it to a solid state, making it unavailable to microorganisms for their normal metabolic activity (148). Apart from physical removal of the water, drying can also be achieved by adding high concentrations of sodium chloride or sucrose (25). Addition of salt or sugar to food, such as in curing and preserving, also lowers microbiological activity on the food. The resulting solution is denser than the solution within the bacterial cell; since water passes from the less dense area to the more dense, water moves out of the bacterial cell and it shrivels (148). Packaging of foods—e.g., wrapping in paper or polythene film—influences the relative humidity of the air within the package, and the relative humidity in turn influences microbial development.

Temperature

Microorganisms can be classified into psychrophilic, mesophilic, and thermophilic groups on the basis of their thermal requirements (117) (Table 5.6). Microorganisms grow faster and sporulate sooner as the temperature approaches their optimum, and they usually sporulate within narrower ranges of temperature than those that permit their growth. Moisture, availability of nutrients, oxygen concentration, and other factors interact with temperature as additional growth parameters. Usually, molds have lower maximum growth temperatures than bacteria. Microorganisms perish at temperatures outside the range conducive to growth. Some die rapidly when temperatures are above the maximum. Most molds and bacteria die within 10 minutes at 55°; a few, usually thermophiles, survive 65° for the same length of time (117).

Since psychrophiles are able to grow at very low temperatures, they are important spoilage organisms of refrigerated and, occasionally, of frozen products. Some species of *Cladosporium* and *Sporotrichum* can grow at −6.7° (75). Development at a temperature below the freezing point of water requires an ability to grow at a low A_w (134). Growth is, of course, very slow at low temperatures; for example, 9 months were required for turkey pies to develop visible mold colonies when stored at −6.7° (46). The slow growth of psychrophilic molds and bacteria on stored food in cold storage often results in flavor defects, particularly in milk products and eggs (140). Refrigeration and freezing of food decreases the rates of growth and multiplication of microorganisms. Heating and smoking of foods kills microorganisms if the temperature attained exceeds the maximum temperature for their growth; heating also removes moisture, thus lowering the A_w of the food to safer levels.

Available Oxygen

Most forms of life require oxygen to carry on metabolic activity. Free atmospheric oxygen (O_2) is readily available to certain groups of microorganisms. Certain organisms have a mechanism that enables them to utilize combined oxygen in compounds such as carbohydrates (148). The biochemical behavior of microorganisms in food is largely determined by the amount of oxygen present. If atmospheric oxygen is present, most food materials are completely, though slowly, oxidized. Carbon dioxide and water are the end products when sugar is oxidized. However, if the oxygen tension is lowered, more intermediate products (e.g., alcohol and lactic, acetic, or formic acids) and fewer end products are formed by microbial action.

As aerobic organisms, molds usually require free oxygen for growth (24). They rarely cause food spoilage under anaerobic conditions (148), but sometimes the presence of only a small amount of oxygen does support sufficient growth for spoilage (76, 134). Thus, the degree of anaerobiosis attained under commercial processing conditions often is not adequate to prevent growth of molds in some foods. However, storage of vegetables and fruits under controlled atmospheres, and of cereal grains anaerobically in silos, permits maintenance of quality with minimum spoilage. And, canning of food ideally provides unfavorable conditions for mold growth in airtight containers (148).

Nutrients

Nutrients provide both a source of energy and essential growth elements for microorganisms. Energy sources vary from simple carbon compounds available to many species to complex compounds (e.g., cellulose, pectin) utilizable by a few. Because fungi and bacteria rely upon absorption through the cell membrane for their intake of food, nutrients must be soluble and of fairly low molecular weight.

Before protein or cellulose substrates can be used as energy sources, they must be broken down into easily assimilable forms. For this to

Table 5.5

Minimum water activities (A_w) for growth of microorganisms and range of A_w of various foods

Microorganisms	Minimum A_w permitting growth	Food
	1.0 —	Highly perishable foods (fresh fruits and vegetables, meat, fish, milk)
Salmonellas and most Gram-negative bacteria; *Clostridium botulinum*	—	Cured meats (ham), cooked sausage (liver and blood sausage), some cheeses, bread
Most Gram-positive bacteria	0.9 —	Fermented sausage (salami), dry cheeses, sponge cakes, maple and fruit syrups
Most yeasts and molds; *Staphylococcus aureus*	—	Fruit cakes, sweetened condensed milk / Jams and marmalade, marshmallows, marzipan, salt-preserved foods, intermediate moisture foods
Most penicillia; many less xerotolerant yeasts; most mycotoxigenic aspergilli	0.8 —	
Halophilic bacteria; *Wallemia sebi*	—	
		Honey
Most members of the *Aspergillus glaucus* group; xerotolerant yeasts	0.7 —	Rolled oats, fudge, Turkish delight, some cereals
Exceptionally xerotolerant yeasts and molds (*Saccharomyces rouxii, Aspergillus echinulatus, Monascus bisporus*)	—	Dried fruits, some toffees, and caramels
Supports no microbial proliferation	0.6 —	Cookies, dehydrated foods (soups, vegetables, milk powder, egg powder), spices, corn flakes, butterscotch, chocolate, some toffee, refined sugar

SOURCE: Correy, J. 1978. Relationships of water activity to fungal growth. In *Food and Beverage Mycology*, L. R. Beuchat, ed. AVI, Westport, CT.

Table 5.6

Minimum, optimum, and maximum temperatures required for growth of microorganisms

Microorganism	°C Minimum	Optimum	Maximum
Psychrophiles	−8 to 0	10 to 20	25 to 30
Mesophiles	5 to 25	20 to 40	40 to 45
Thermophiles	25 to 40	50 to 60	70 to 80

SOURCE: Semeniuk, G. 1954. Microflora. In *Storage of Cereal Grains and Their Products*, J. A. Anderson and A. W. Alcock, eds. American Association of Cereal Chemists, St. Paul, MN.

occur, conditions have to be favorable for enzymatic degradation, namely, the ability to produce the required exoenzymes and a temperature and pH range suitable for the reaction to proceed. Thus, microorganisms cannot grow in the absence of moisture, and only those species that

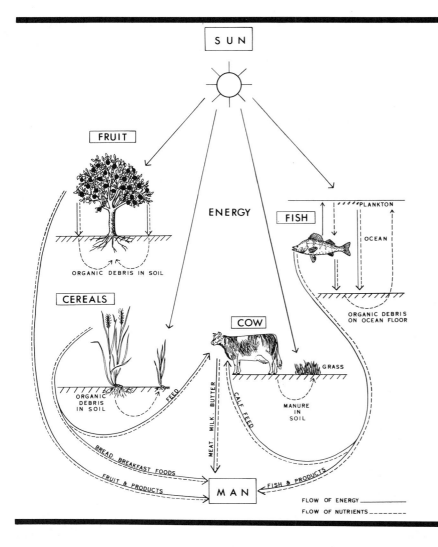

SUN

FRUIT

ENERGY

FISH

PLANKTON

OCEAN

ORGANIC DEBRIS IN SOIL

ORGANIC DEBRIS
ON OCEAN FLOOR

CEREALS

COW

GRASS

ORGANIC
DEBRIS
IN SOIL

FEED

MEAT, MILK, BUTTER

CALF FEED

MANURE
IN
SOIL

BREAD, BREAKFAST FOODS

FRUIT & PRODUCTS

MAN

FISH & PRODUCTS

FLOW OF ENERGY _____
FLOW OF NUTRIENTS _ _ _ _ _ _ _

Figure 5.1
Flow of energy and nutrients in the production of foods used by humans.

from germinating (*39*). These inhibitors are added directly to foods; others are used as dips for treating exterior food surfaces or are impregnated into packaging materials such as paper wrappings. Spices added to meats also prevent mold growth, probably through action of oils.

Ecology of Stored Foods

Food Ecosystems

Foods such as fruits, cereals, and fish are energy- and nutrient-rich products of particular ecosystems (Figure 5.1). Fruits, for example, contain energy-rich substances such as carbohydrates and chemical nutrients (e.g., potassium and phosphorus). The source of the energy is the sun; solar energy, transformed by a chemical process known as photosynthesis, is stored in the food as carbohydrates and other substances. The chemical nutrients are derived from plant and animal tissues, from artificial fertilizers in the soil, or through microbial decomposition of organic matter. There is a flow of energy and nutrients within and among ecosystems. The energy flows only once through an ecosystem and is not recycled but is transformed to heat and ultimately lost to the system. Only the continual input of new solar energy keeps the ecosystem working.

Chemical nutrients, on the other hand, can be recycled many times through an ecosystem. Thus, a molecule of phosphorus may be taken up by a grass root, used in a grass leaf, eaten by a cow, and reenter the soil when the cow dies and the molecule is released by bacterial decomposition. Microbial decomposition makes available these chemical nutrients that are otherwise locked up in foods such as fruits; it should be regarded as the normal ending of a sequence of events rather than as an abnormal situation (*58*).

The food ecosystems depicted in Figure 5.1 can be divided into two main types: (a) those in which the food remains alive in storage, e.g., unprocessed cereal grains and vegetables; (b) those in which the food is not alive, i.e., processed to a form suitable for human consumption (e.g., meat, milk, butter, or fish). The foods of the first type are subject to decay mostly by fungi and yeasts; those of the second type, mainly by bacteria.

have enzyme systems adapted to dealing with a particular substrate can grow on that substrate (*140*). Drying, freezing, curing, preserving, and smoking all indirectly affect the level of nutrients for microbial development through reduced solubilization.

pH

The acidity of the medium in which the organisms live may vary considerably. The molds and yeasts thrive best under acidic conditions and have a pH range of 2 to 8 with an optimum of 5. Bacteria generally have a more restricted pH range for growth, and develop best at pH 7.5. However, many acid-loving types such as the acetic and lactic acid producers may flourish down to pH 2 (*140*). The decay of fruits, which are mainly highly acidic, is caused by acid-tolerant yeasts and molds, not by bacteria.

Inhibitors

Propionic, sorbic, and benzoic acids and their salts either kill mold spores or prevent them

Unprocessed Foods

Foods move from the producer to the consumer by many routes, and as they progress they may be stored in a few or many locations. This movement of food is also a movement of energy (originally derived from the sun) and of nutrients. The main food storage situations encountered from producer to consumer, linked by a flow of energy and of nutrients, are depicted in Figure 5.2. At each storage location, depending on the storage conditions and length of time, microbial decomposition of the food may occur, resulting in a loss of energy and nutrients to the ultimate consumer. It is the duty of responsible food managers at each storage location to protect the foods in their care from microbial decomposition. The ecological principles that govern development of microorganisms in stored, unprocessed foods, and the methods used to predict, prevent, and control microbial decomposition in such foods, are described below.

Processed Foods

The routes by which food moves from the processing plant to the customer are shown in Figure 5.2; the arrows indicate both the routes of movement and the flow of nutrients and energy. Mold spoilage of processed food occurs much less frequently than spoilage of unprocessed foods, because the food is treated and packaged in ways that do not favor mold growth on the product, the packaged portions are generally small, and rigid sanitary procedures are employed in the processing plant to reduce the level of mold inoculum. Nevertheless, some mold spoilage of stored, processed food occurs and the spoilage is frequently initiated in the processing plant. The main sources of contamination are molds naturally present on incoming raw materials and spores that develop on waste organic matter or on improperly cleaned equipment. The ecological principles that govern development of microorganisms in stored processed foods are described below.

Ecology and Control of Microorganisms Decomposing Unprocessed Cereals, Pulses, and Oilseeds

Ecology and Energy Relationships of Stored-Grain Ecosystems

Grain bulks are man-made ecosystems in which living organisms, the most important being the grain itself, and their nonliving environment interact with each other. Deterioration of stored grains results from interaction among physical, chemical, and biological variables. Important

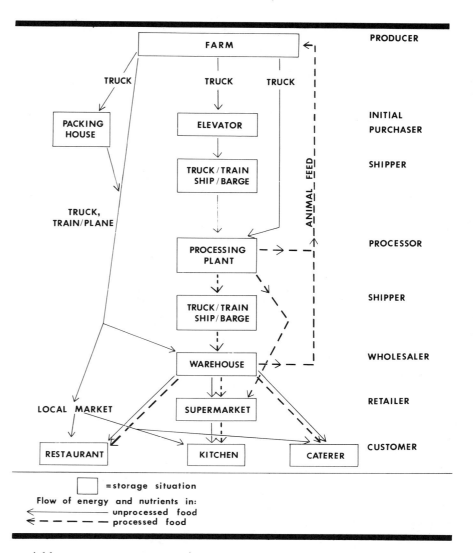

variables are temperature; moisture; oxygen; geographical location; storage structure; physical, chemical, and biological properties of the grain bulks; microorganisms; insects; mites; rodents; and birds. Although many variables affect grain and its quality, they seldom act alone or all at once. They interact with the grain and, in groups, among themselves (78). Although this deterioration is usually slow and less dramatic in the beginning, complete loss of the grain can occur if the correct combinations of variables are maintained in an undisturbed bulk (122).

A grain bulk has a nonregenerating energy supply. Energy captured by the cereal plant "producers" is concentrated in the seeds (123). The pattern of energy flow in grain-bulk ecosystems—i.e., transfer of food energy in a food chain from the plant seeds through a succession of organisms—involves the transformation and loss of energy. There are usually several trophic levels in an ecosystem and at each there is a loss of a portion of the potential energy as heat during respiration (123).

Figure 5.2
Main agricultural food-storage situations and flow of energy and nutrients from producer to consumer.

Climate and Incidence of Microflora and Arthropods in Stored Grains

Climate sets the basic limits within which food plants can be grown, harvested, stored, and distributed. Climate changes the quality of stored grains indirectly by influencing harvest and storage conditions and the rate of increase of various biotic agents that invade stored grains. The degree of climatic effect depends primarily on the extent of protection provided in terms of the design of granary structures, trucks, railcars, and ship holds, and bulk or bag size (124).

Although insects, mites, and microorganisms are found almost anywhere grain is stored, the frequency of their occurrence and their virulence depend largely on the regional climate (122). A generalized pattern emerges from a study of the world literature on grain storage problems. This pattern indicates that deteriorating stored grain in a temperate climate—mainly in wheat-producing countries—has relatively more dominant species of mites and fewer dominant species of insects and microorganisms (20, 145). A tropical humid climate (121) features primarily insects and to a lesser extent microbial species; least dominant are the mites. In a dry climate, insects are dominant and mites and molds relatively less important. In a subtropical climate, all three types of organisms are equally abundant (123, 136).

Fungi Associated with Stored Grain

More than 150 species of fungi have been found on cereal grains (19, 73). These fungi can be placed into two main groups: preharvest (field) and postharvest (storage). Preharvest fungi gradually die out with time because they require a continuously high moisture content for development, a condition not normally present in stored seeds. This decline is usually accompanied by a gradual increase in postharvest fungi which require only relatively low moisture levels for initial growth, although their later development is accelerated by higher moisture levels.

Preharvest Fungi

The major preharvest fungi on most stored grains are species of *Alternaria, Cladosporium,* and *Helminthosporium,* although *Curvularia* and *Stemphylium* also occur. *Alternaria* species are abundant in cereal grains; in temperate zones usually 60–70% of the grains are infected, but in drier areas such as western Australia infection

is lower, ranging from 13 to 23% (118). Also present on seeds are *Fusarium* species, some of which—e.g., *F. culmorum* (W. G. Smith) Saccardo—are preharvest pathogens.

Postharvest Fungi

Aspergillus Aspergilli are ubiquitous. Of the 80 known species (109), at least 26 occur on stored grain (145) (Table 5.7). Aspergilli vary considerably in color, including shades of white, blue, green, yellow, brown, and black. Grains stored under low-moisture conditions (rh < 78–80%) are first invaded by the osmophilic *A. restrictus* Smith and *A. glaucus* group species (20). At higher seed moisture content, these species invariably appear first and may be followed by *A. candidus* Link *ex* Fries, *A. ochraceus* (Wilhelm), *A. versicolor* (Vuillemin) Tiraboschi, *A. flavus,* and *Penicillium* species. An ecological succession often occurs: a slow or moderately rapid increase in *A. restrictus* and *A. glaucus* in the first stages of deterioration, followed by *A. candidus* and *A. flavus.* By the time more than 5–10% of the kernels have been invaded by *A. candidus* or *A. flavus,* spoilage is well under way and heating of the grain follows. Each of the species listed above has its own sharply defined lower limit of moisture content for growth on particular seeds (19). The occurrence and relative abundance of particular *Aspergillus* species in a representative seed sample give a most useful indication of the past history, present condition, and future storability of a crop.

Penicillium Samples of stored grain have yielded 66 of the 137 known species (110, 145). Unlike the aspergilli, *Penicillium* species vary little in color, ranging through shades of blue to green and grey and are frequently referred to as the "blue" or "green" molds. *Penicillium* species are difficult to identify. They require a higher minimum relative humidity for germination of conidia than *Aspergillus* species. They occur in cooler climates than the aspergilli (145). Injury predisposes seeds to colonization by *Penicillium* species (144). Some *Penicillium* species are preharvest fungi (87). The lower limits of moisture for growth of *Penicillium* species on corn, wheat, sorghum, and soybeans have been determined (20).

Other Postharvest Fungi Generally associated with moist, rotting conditions, *Absidia, Mucor,* and *Rhizopus* require a minimum rh of 88%; they are, therefore, not the initiators of cereal deterioration in storage (145). However, if there are many broken kernels and sufficient

Table 5.7

Aspergillus species recorded on stored grains in various countries

Species	Cereals France	Romania	Italy	Australia	Canada	Corn, USA	Rice, Pakistan
A. glaucus group							
A. amstelodami (Mangin) Thom & Church	+			+	+	+	
A. chevalieri (Mangin) Thom & Church	+		+	+		+	
A. echinulatus (Delacroix) Thom & Church	+		+		+		
A. repens de Bary	+			+	+	+	
A. ruber (König, Spieckermann & Bremer) Thom & Church	+			+	+	+	
A. candidus Link	+		+	+	+	+	+
A. clavatus group							
A. clavatus Desmazières		+			+		
A. giganteus Wehmer					+		
A. flavus group							
A. flavus Link	+	+	+	+	+	+	+
A. oryzae (Ahlburg) Cohn			+				
A. tamarii Kita	+				+		+
A. flavipes (Bainier & Sartory) Thom & Church	+		+			+	
A. fumigatus Fresenius	+	+	+		+	+	+
A. nidulans (Eidam) Winter	+	+			+		+
A. niger van Tieghem	+	+	+		+	+	+
A. ochraceus group							
A. alliaceus Thom & Church						+	
A. ochraceus Wilhelm	+	+	+	+	+	+	
A. sulphureus (Fresenius) Thom & Church						+	
A. restrictus group							
A. conicus Blochwitz	+						
A. restrictus Smith	+					+	
A. terreus Thom	+	+	+		+	+	+
A. ustus (Bainier) Thom & Church	+			+		+	
A. versicolor group							
A. sydowi (Bainier & Sartory) Thom & Church			+		+	+	+
A. versicolor (Vuillemin) Tiraboschi	+		+	+	+	+	+
A. wentii group							
A. terricola Marchal					+		
A. wentii Wehmer	+					+	

NOTE: + indicates presence of fungus on or in seeds.
SOURCE: Wallace, H. A. H. 1973. Fungi and other organisms associated with stored grain. In *Grain Storage: Part of a System*, R. N. Sinha and W. E. Muir, eds. AVI, Westport, CT.

moisture, the theromophilic species, *Absidia lichtheimi* (Lucet & Costantin) Lendner and *Mucor pusillus* Lindt, may occur. Species of *Absidia, Mucor, Rhizopus*, and other phycomycetes have been found on grain (*145*). *Chaetomium* species have been isolated from grain in France (*99*) and Canada. *Wallemia sebi* (Fries) von Arx is a xerophilic species commonly found on stored seeds of low-moisture content but apparently causes little damage (*18*).

Recognition of Mold Problems in Stored Grains

One of the first indications of mold spoilage in a bulk is a rise in grain temperature when compared to that of the grain at binning. Mold-induced hot spots normally occur in grain of high-moisture content; those due to insect activity occur in dry grain binned at high temperature. In freshly harvested grain, the rise in temperature as a result of mold activity is often accompanied by a rise in CO_2 production. Other later signs of mold activity include visible mold growth resulting in aggregation or caking of grains, a moldy odor, sprouting of surface grains, decreased germination, and high free fatty acid levels (80).

Prediction, Prevention, and Control of Postharvest Fungi in Stored Grains

Prediction The probable development of postharvest fungi in a seed lot during storage frequently can be predicted on the basis of the kinds of seed-borne fungi present and on a knowledge of the condition, source, and type of seeds involved. Decisions can then be made on how long to keep the particular lot, whether it or others should be disposed of first, or which preventive or control measures should be taken to ensure safe storage with minimal deterioration.

Fungi The abundant occurrence of particular species of postharvest fungi—e.g., *Aspergillus candidus* or *A. flavus*—on surface-sterilized seeds is evidence of poor storage conditions. The subsequent increase in fungal abundance signals an emergency situation (20).

Condition Samples representative of the bulk should be properly obtained (97), and the variation in moisture content and temperature and levels of dockage, broken kernels, free fatty acids, and germination determined. The age of the seeds should also be determined. Freshly harvested lots may be actively respiring and contain immature seeds or dockage, whereas lots several years old often have minimal respiratory activity and probably more postharvest infection than newer lots.

Source The types of postharvest fungi present in a seed lot frequently vary with the country of origin (146). The climate of the export and import countries and means of transportation may influence fungal development. The movements of grains from Illinois to Georgia or from Canada to southern Japan are from temperate to subtropical climates but passage from Argentina to the United Kingdom is the reverse and is likely more favorable to development of postharvest fungi.

Type of Crop Rapeseed/canola, for example, requires special care in storage. This is because postharvest fungi can develop quickly on the very small seeds, especially when the seeds vary in maturity and moisture content and are accompanied by green dockage. Seeds of various crops differ in physical and chemical properties which, in turn, influence storage characteristics. The manager should be aware of these differences and know the maximum periods for safe storage of particular crops when binned at different moistures and temperatures. Also, any specific procedures required for aeration and drying of the crops should be known.

Prevention and Control Development of postharvest fungi on grains in storage is influenced primarily by seed moisture, seed temperature, and the level of available oxygen. Most prevention and control measures against postharvest fungi are based on a knowledge of these factors (139).

Seed Moisture Certain preventive measures may be taken on the farm at or just before binning to reduce likelihood of later spoilage problems. These include sufficient air drying in the field before combine operations; separation of the often wetter first-cut crops from the later drier, binned crops; ensuring that the bin is weathertight; using a spreader to minimize pockets of fines, green weed seeds, and immature crops in the bin; removing excess dockage before binning; mixing bin contents by reaugering after 2–3 days to ensure even moisture distribution; and taking numerous samples for accurate moisture content.

Preventive measures that may be taken during storage on the farm, in elevators, or on ships include frequent monitoring of grain moisture in many locations, particularly near top center and any doors or hatches; a knowledge of patterns of moisture migration in bins, silos, elevators, or ships (89, 127); removal of wind-driven snow on top of farm and elevator bins (danger of hot spots after melting), and checking for leaks in bins and in hatch covers on ships. Control measures include aeration, artificial drying (6, 41, 139), and admixture with grain of lower moisture content.

Seed Temperature Preventive measures that may be taken at or just before binning include combining and binning on cooler days or early evenings, and frequent measurement of temperature by electrical probes, particularly in the center of the bulk. Preventive measures that may be taken during farm and elevator storage include frequent monitoring to detect microbial activity and the use of special storages (e.g., underground). Control measures include aeration (42) and turning or moving stocks from bin to bin on cool days with low relative humidity.

Available Oxygen Airtight storages (*54*) are used to prevent development of postharvest fungi on high-moisture feed grains on farms in Europe and the United States and on normal-moisture grains in warm countries such as Kenya and Cyprus.

Examples of Mold Problems and Prevention Strategies

Corn, wheat, soybeans, and, to a lesser extent, sorghum, barley, and oats were the principal crops produced in the United States during 1983; wheat and barley were the main crops produced in Canada during that year (*38*; Table 5.8). The principal oilseed crop in Canada—rapeseed/canola—was worth about $655 million (Canadian) in 1982–83 (*31*). The seed moisture contents of various major crops, in equilibrium with relative humidities at 25–30°, are presented in Table 5.9 (*18*). The following practical examples reflect the importance of the above crops and illustrate the application of ecological principles (previously described) to the management of stored grains.

Shipment of Grain to Japan from Various Countries

Maize, wheat, sorghum, and barley shipments arriving in Japan from Argentina, Australia, Canada, China, and the United States were studied during 1966 and 1967 (*146*). The grain, usually graded dry, yielded a moisture content below 13.7% (wet-weight basis). Al-

Table 5.8

Production (1,000 metric tons) of principal cereal, oilseed, and pulse crops in the United States and Canada during 1983

Crop	USA	Canada
Corn	106,781	5,875
Wheat	66,010	26,914
Soybeans	43,421	721
Sorghum	12,270	NA
Oats	6,928	2,773
Barley	11,300	10,616
Rye	715	831
Beans (dry)	692	39
Flaxseed	296	734
Rapeseed/canola	NA	2,681
Sunflower seed	2,661	94
Mustard seed	NA	90
Peanuts	1,485	NA

NOTE: NA = Data not available or crop not grown.
SOURCE: Food and Agriculture Organization. 1984. *FAO Production Yearbook 1983*. United Nations, Rome.

Table 5.9

Moisture content of grains and seeds in equilibrium with various relative humidities at 25–30°C

Relative humidity (%)	Moisture content (%) Wheat, corn, sorghum	Soybeans	Sunflower seeds	Rapeseed
65	12.5–13.5	11.5	8.5	—
70[a]	13.5–14.5	12.5	9.5	8.3
75	14.5–15.5	13.5	10.5	—
80	15.5–16.5	16.0	11.5	10.6
85	18.0–18.5	18.0	13.5	—

[a] Lower limit at which spoilage molds generally develop.
SOURCES: Christensen, C. M. 1978. Storage fungi. In *Food and Beverage Mycology*, L. R. Beuchat, ed. AVI, Westport, CT; Pixton, S. W., and S. Warburton. 1977. The moisture content/equilibrium relative humidity relationship and oil composition of rapeseed. *Journal of Stored Products Research* 13(2)77–81.

though the temperature of grain in the holds of ships on arrival in Japan varied with the grain, country of origin, and season, it was probably determined by the temperature of the grain when it was loaded into the ships. The temperature of the grain in 4 ships out of 62 exceeded 30°. The cargoes were sorghum from Argentina (42°), maize from Thailand (38°), soybeans from the United States (38°), and millet from Australia (35°). Preharvest fungi seemed to have died in transit; postharvest fungi were more prevalent on the large-seeded grains than on small-seeded grains.

More *Aspergillus* species were found on grains originating in the United States and Australia than on those originating in Canada (Table 5.10). The levels of dockage, including dust, were generally higher in shipments from the United States than from other countries, particularly with grains such as maize, soybeans, and sorghum. The large amount of dockage in some stored-grain bulks could be a source of the contaminant fungi on imported cereal grains. Handling procedures and higher pressures in large-bulk storage may increase breakage, especially if the grain is unusually dry and brittle.

Harvest of High-Moisture Corn

The practice of using combines and picker-shellers, rather than picking corn by the ear, results in storage of shelled corn with a high moisture content and frequently with mechanically damaged kernels, thus permitting access of postharvest fungi. Guidelines (*2*) are available to assist producers in minimizing the amount of fungal damage to wet corn. These include reducing the moisture and temperature of shelled corn whenever possible (fungal growth is negligible below 13% mc; temperatures below 4°

slow the growth of most fungi). Also, damage is minimized by never exceeding the maximum storage time for a given combination of corn moisture and air temperatures (Table 5.11), and by drying wet corn to 12–14% mc immediately after harvest.

Shipment of Maize

From 1962 to 1966, the yearly export of maize from Argentina to Europe often approached 2 million tons. Heating damage frequently occurred in bulk cargoes arriving in the United Kingdom, particularly during protracted voyages of 60–70 days (83–85). Examination of ship cargoes on arrival often revealed evidence of microbial changes. On the surface, where there was abundant air, mold growth was apparent; deeper down, at higher temperatures of up to 60–70°, yeasts and bacteria had been active as indicated by the presence of fruity or cheesy odors. Once heating occurs in a grain bulk, little can be done to remove heat by surface ventilation (85). In this respect, a cargo in a vessel differs from a bulk in a silo where heating often can be controlled by transfer or ventilation. To ensure a safe ocean voyage, the grain must be dispatched in a condition that will preclude heating in transit (83). Maize shipped from hot climates must be of a lower moisture content than that shipped from cooler climates if safe transit is to be ensured (83). Shipments should be dried to below the legally allowed export limits; if the maize is too warm or too moist, spoilage cannot be prevented whether ventilation is used or not (85).

Investigation of a shipment of bagged No. 2 white corn showed that spoilage occurred after a voyage of 17 days from New Orleans to El Salvador (19). By comparing the kinds of molds present on an uninvaded control sample of seed with those on the spoiled seed, it was determined that the corn had to have been exposed to 85–90% rh for at least 2 weeks at a moderately high temperature. By examining the ship's log, it was shown that the corn was in fact exposed to warm moist air through the scoop ventilators during the voyage, thus facilitating the development of mold. Settlement was made in favor of the shipper.

Spoilage of Corn

Several aspects of corn spoilage have been studied and reported: mold flora of shelled corn (66, 73, 87, 115) and effect of moisture content, temperature, and duration of storage upon the subsequent storability of shelled corn (101).

Development of a Fungus-Induced Hot Spot in Wheat

The ecology of an artificially induced hot spot was studied from samples collected from two 500-bushel wheat bulks at Winnipeg during 1959–61 (125). Heating was initiated in winter primarily by the activity of the low-temperature fungi, *Penicillium cyclopium* Thom and *P. funiculosum* Thom, growing in a 4-month-old grain

Table 5.10
Mean percentage of seed yielding *Aspergillus* species

Grain	Source	Number of bulks sampled	*Aspergillus* amstelodami	candidus	flavus	niger	ochraceus	terreus	versicolor
Wheat	Canada	10	0.8	—	0.4	—	—	—	1.2
Wheat	Australia	4	9.0	1.0	2.0	—	—	—	5.0
Wheat[a]	USA	10	8.0	1.2	7.2	—	0.4	—	2.4
Wheat[b]	USA	6	1.3	—	4.6	—	—	—	1.3
Wheat[c]	USA	2	—	—	—	—	—	—	6.0
Barley	Australia	4	—	—	1.0	—	—	—	—
Barley	USA	2	—	—	2.0	—	—	—	—
Sorghum	Australia	1	4.0	—	4.0	24.0	—	4.0	—
Sorghum	Argentina	1	—	4.0	12.0	40.0	—	4.0	4.0
Sorghum	USA	6	6.7	—	9.3	0.7	—	1.3	6.7
Soybeans	USA	6	—	2.7	2.7	—	0.7	—	9.4
Maize	USA	8	1.3	—	5.0	1.3	1.3	—	7.5
Maize	Thailand	3	—	—	10.4	5.3	1.3	1.3	—
Maize	China	2	—	—	—	6.0	—	—	6.0

NOTE: Seeds were imported from various countries and collected from ships in Kobe, Japan, in 1966 and 1967.
[a] Hard red winter.
[b] White winter.
[c] Red northern spring.
SOURCE: Wallace, H. A. H., and R. N. Sinha. 1975. Microflora of stored grain in international trade. *Mycopathologia* 57(3)171–176.

pocket at −5° to 8° and 18.5 to 21.8% mc. The hot spot reached a maximum of 64°, and cooled in 2 weeks.

Storage of Wheat

Wheat, previously harvested at 27–32° and stored in a large bin at Cairo, Illinois, had an average moisture content, according to the records, of 13.2%; however, some grain was binned at or near 14% and even at 16% mc because of a moisture meter of uncertain reliability. During the subsequent cool autumn, rapid moisture transfer probably occurred in the bulk. Slow, then rapid, heating occurred, resulting in 40% germ damage, reduction to sample grade, and considerable monetary loss. The spoilage was due to development of postharvest fungi, and the warehouseman was judged responsible (19).

Other aspects of wheat spoilage have been studied, including the biological, physical, and chemical changes occurring in stored wheat and the establishment of safe storage guidelines (147); the comparative storage characteristics of three types of farm granaries containing wheat at Winnipeg, Canada (90); and the fungi associated with stored wheat in India (74) and France (99).

Spoilage and Drying of Oilseeds

If oilseeds are to be kept for any length of time without deterioration, then several environmental conditions must be established (45, 88), the most important being seed moisture content. Moisture levels generally considered safe for storage are: soybeans, 13%; flax, 10.5%; sunflower, 8.5%; rapeseed/canola, 7%, and peanuts, 11%. Good warehousing practices include (a) controlled segregation or placement of various moisture-level materials; (b) temperature measurement at least once a week; (c) aeration of higher moisture seed; and (d) good recordkeeping of the condition of the material going to storage and of successive temperature checks. If permitted to continue heating unchecked, almost all oilseeds will char, smolder, and then break into open flame when a hot spot is uncovered and exposed to air.

Storage of Soybeans

Existing reports cover most aspects of this topic: storage and quality control of soybeans on the farm (7); commercial storage and handling (62); maintenance of quality in storage (17); incidence of toxic and nontoxic fungal species (86, 120); and microbiological changes occurring on board ship (50).

Table 5.11

Maximum time[a] (in days) for safe storage of shelled corn at various corn moistures and air temperatures

Storage air temperature (°C)	Moisture content (%)			
	15	20	25	30
24	116	12.1	4.3	2.6
21	155	16.1	5.8	3.5
18	207	21.5	7.8	4.6
16	259	27	9.6	5.8
13	337	35	12.5	7.5
10	466	48	17	10
7	725	75	27	16
4	906	94	34	20
2	1,140	118	42	25

[a] Times given are those during which mold growth will cause enough loss in corn quality to bring about a lower grade.
SOURCE: Agricultural Research Service. 1969. *Guidelines for Mold Control in High-Moisture Corn.* Farmers' Bulletin 2238. Agricultural Engineering Research Division, U.S. Department of Agriculture, Washington, DC.

Combustion in Stored Soybeans

There is a genuine danger of spontaneous combustion in soybeans and other oilseeds because, unlike heating cereals which usually do not exceed 58°, temperatures during heating of soybeans can exceed 150°. To prevent such dangerous situations from reaching the flash point, the stored crop should be aerated for 72 hours. However, if after 72 hours there has not been a noticeable cooling in the hot spot, the aeration should be discontinued, the affected seeds removed, or the bulk turned (62). If the aeration system is well designed and the heating area is small, cool air may lower the temperature of the hot spot and eventually solve the problem. However, if the air cannot reduce the heat fast enough, the atmospheric oxygen will feed and spread the fire that could affect the structure or lead to a dust explosion.

Peanuts

This crop differs from cereals, pulses, and oilseeds in that the flower is fertilized above ground and the developing fruit bends down and develops in the soil. Peanut kernels can be invaded by aerial flora, terrestrial flora, and an intermediate flora, including *Aspergillus flavus*, above as well as below the ground. Careful harvesting and storage procedures are required to reduce fungal infestation by *A. flavus* and the development of aflatoxins (73). The degree of toxin production has been reduced by use of mechanical drying equipment; fast drying rather than slow has been recommended (*57, 102*).

Rapeseed/Canola

In Canada, extreme care is needed to safely store rapeseed/canola. The legal upper limit of safe storage of so-called dry rapeseed/canola is presently 10% mc (in equilibrium with about 77% rh). This level of moisture, however, permits growth of *Penicillium*. Growth of *Aspergillus glaucus* group species occurs at 70% rh (equivalent to 8.3% mc at 25°) (*107*). However, if the rapeseed/canola is binned at temperatures higher than 25–30°, if pockets of immature seeds are present, or if many green seeds are present, moisture levels above 8.3% may be too high for long-term safe storage (*81*).

Preventive measures, therefore, should include binning at no less than two percentage points below the 10% mc maximum, using a deflector under the auger or in the bin, using an aeration unit, and monitoring the temperature with an electrical probe every few days during the fall and every 2 weeks during the winter (*70*). Turning the rapeseed/canola after it has been in storage 3–5 days breaks up any pockets of green weed seeds and dockage that could facilitate heating. Withdrawing a load by auger from below reshapes the upper bin contents from a cone to an inverted cone which has less depth and improved air access.

Safe storage guidelines for rapeseed/canola, based on seed moisture and temperature at binning, have been developed (*13, 77, 81, 126*). The appearance and smell of crushed seeds—indicators used by elevator managers in western Canada—are practical and reliable methods for determining rapeseed/canola quality (*82*). In the laboratory, the most reliable criteria for judging low-quality rapeseed/canola are higher-than-average fat acidity value, weathered seed surfaces, brown seeds having a sour spoilage smell when crushed, presence of dead seeds, and a high frequency of *Aspergillus* species in seed samples.

Sunflower Seeds

Two types are commonly produced in North America (especially Minnesota, North Dakota, and Manitoba): oilseeds and seeds for consumption by humans and birds. Information is available on the following aspects of sunflower seed storage: safe storage (*23*); invasion of seeds by postharvest fungi (*15*); seed condition and storability (*16*); and the effects of heat damage and fungi on seed quality (*112*).

Other Grains

Information has been compiled on the incidence and risk of spoilage and heating in 35 major grains and grain products, together with suggested prevention, detection, and control methods (*80*).

Ecology and Control of Microorganisms Decomposing Unprocessed Vegetables and Fruits

Ecology of Stored Vegetables and Fruits

Reports in the literature on the ecology of stored vegetables and fruits are scarce. Many reports exist, however, on the storage behavior of particular crops, crop losses, and the development, prevention, and control of certain diseases or disorders. Optimal storage quality of vegetables and fruits should be considered from a holistic, or total, viewpoint rather than on a piecemeal basis. The development of specific spoilage organisms is closely involved with many other factors and thus necessitates an ecological approach to the prediction, prevention, and control of storage diseases of fruits and vegetables.

Vegetables and fruits are the energy- and nutrient-rich end products of particular ecosystems, man-made or natural (Figure 5.1). These crops deteriorate in situ, fall to the ground, or are transported by animals or birds to other sites nearby. Humans, however, have interfered with these processes by gathering and keeping the food in "safe" locations. Most vegetables and fruits, however, especially in the tropics, are improperly stored, resulting in serious losses (*26*).

Deterioration is a normal process, and the role of fungi as deteriorative agents is to liberate the nutrients and energy locked up in the fruits or vegetables, making such nutrients available to the next trophic level or generation of life. Humans, through manipulation of the storage environment, have sought to minimize the effects of deteriorative agents such as molds. And, at least in the developed countries, the supermarket shopper is usually supplied with an unblemished fruits and vegetables. This achievement is not without a price—the addition of pesticides to the food. Because they are omnipresent and adaptable to changing conditions, molds still cause losses in stored fruits and vegetables despite the use of modern handling and storage techniques. Examples of this are the development of molds after temperature malfunction in a railroad car and of new mold strains resistant to a particular fungicide.

Fresh vegetables and fruits in storage differ from stored cereals and oilseeds in that they are not normally stored in bulk, they have a higher moisture content, and the temperature of storage and the physiological crop condition are relatively more important than for bulk grains. Pathogenic fungi and bacteria attacking nonsenescent tissues are more important in the storage of fruits and vegetables than in bulk grains. The high value of these crops permits implementation of expensive and sophisticated prevention and control measures. The end product that the consumer finally sees is determined by a succession of steps going back to the field and orchard. The length of the storage period is usually much shorter, and the rate of fungal and bacterial spoilage is much faster, in stored vegetables and fruits than in bulk cereals and oilseeds.

Physiology of Stored Vegetables and Fruits

Each vegetable and fruit goes through several stages (Figure 5.3): development, prematuration, maturation, ripening, and senescence (113). The terminal phase of maturation is the most crucial period for developing the edible form and may last a few hours or several months. Senescence, characterized by the dominance of degradative changes, ends when the fruit or vegetable is no longer usable as human food. The term senescence applies only to normal physiological changes caused by the passage of time, e.g., changes in composition, flavor, texture, color, or mode of growth. Deterioration encompasses all aspects of quality loss, e.g., senescence, physiological disorders, diseases caused by fungi and bacteria, or the aftermath of mechanical injury. Deterioration can start at any time during senescence and continue beyond the end of usefulness of the food for human consumption (113). Fruits differ from vegetables in one important aspect—they are not destined by nature to grow; they are meant to deteriorate and supply an appropriate seed bed for germination (32).

Kinds of Postharvest Losses

The three main categories of postharvest loss are (a) mechanical injuries including cuts and abrasions; (b) nonparasitic disorders including physiological responses to the postharvest environment; and (c) parasitic diseases or decay caused by fungi, bacteria, and viruses (47, 106). Certain vegetables and fruits are more subject to particular types of loss. Table 5.12 summarizes the types of loss associated with selected vegetables and fruits sampled at retail and consumer levels in New York over many months (47). Parasitic diseases predominated in oranges, Bartlett pears, White Rose potatoes, strawberries, sweet potatoes, and tomatoes.

A study of deterioration in fresh fruits and vegetables arriving in the United Kingdom by ship from 1974 to 1977 showed that 22% of problems were caused by preharvest factors such as field disease; 43% by postharvest handling conditions (e.g., wounding); 8% through ex-

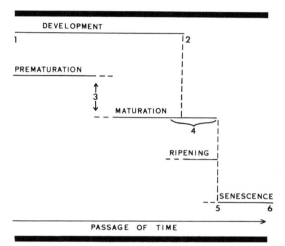

Figure 5.3
Stages during the life-span of fresh vegetables and fruits: (1) Initiation of edible part; (2) Termination of natural or desirable growth in size or type; (3) Start of period of usefulness but too immature for most users; (4) Period of maximum usefulness; (5) Degradative changes become predominant; (6) End of usefulness for human consumption.
SOURCE: Ryall, A. L., and W. J. Lipton. 1972. *Handling, Transportation and Storage of Fruits and Vegetables*, vol. 1, 2nd ed. *Vegetables and Melons*. AVI. Westport, CT.

Table 5.12
Kinds of postharvest loss in selected fruits and vegetables sampled at retail and consumer level in New York

Commodity	Origin in USA	Mechanical injury (%)	Nonparasitic disorder (%)	Parasitic disease (%)
Apples, Red Delicious	West	1.8	1.3	0.5
Apples, Red Delicious	East	1.1	0.4	0.2
Cucumbers	South	1.2	3.4	3.3
Grapes, Emperor	West	4.2[a]	0.9[a]	0.4[a]
Grapes, Thompson	West	8.3[a]	1.6[a]	0.6[a]
Lettuce, Iceberg	West	5.8	3.2	2.7
Oranges, Navel	West	0.8	0.3	3.1
Oranges, Valencia	South	0.2	0.4	2.6
Peaches		6.4	—	6.2
Pears, Bartlett	West	2.1	0.7	3.1
Pears, Bose	West	4.1	2.2	3.8
Pears, d'Anjou	West	1.6	0.8	1.7
Peppers, Bell	South	2.2	4.4	4.0
Potatoes, Katahdin	East	2.5[b]	1.0[b]	1.4[b]
Potatoes, White Rose	West	1.5[b]	0.4[b]	2.4[b]
Strawberries	West	7.7	—	15.2
Sweet potatoes		1.7	4.2	9.2
Tomatoes, packaged		2.5	0.9	10.7
Tomatoes, bulk		2.7	1.1	7.6

[a] Retail losses only.
[b] Wholesale and consumer losses only.
SOURCE: Harvey, J. M. 1978. Reduction of losses in fresh market fruits and vegetables. *Annual Review of Phytopathology* 16:321–341.

porters issuing wrong instructions about the carriage temperature; 22% through mismanagement by the carrier (e.g., incorrect storage in the holds); and 5% for delay for any reason (*131*). Yams were the subject of a similar study (*92*).

Pathogens Causing Postharvest Diseases

The major postharvest diseases of fresh fruits and vegetables and the pathogens involved are shown in Table 5.13 (*34, 91, 132*). The pathogens listed are of two types: (a) the specialized, virulent plant pathogens, e.g., *Colletotrichum musae* (Berkeley & Curtis) von Arx, causing anthracnose of banana (a field infection); and (b) the less specialized, ubiquitous, opportunistic pathogens that can invade only weakened or damaged tissues, e.g., *Rhizopus stolonifer* (Ehrenberg *ex* Fries) Vuillemin, causing *Rhizopus* rots of peaches, cherries, strawberries, and sweet potatoes.

Mechanisms of Infection

Several distinct mechanisms for crop infection are known; they vary with the pathogen involved (*33, 34*). Fruits attached to the plant may be infected by direct penetration of certain fungi through the intact cuticle or through wounds and natural openings in the surface of the fruit (preharvest infection). Alternatively, many postharvest diseases are initiated through injuries created during and subsequent to harvest such as cut stems and mechanical damage to the surface of the product incurred during handling and transportation (*33*).

Preharvest infection occurs when spores of some true pathogenic fungi, transported by wind and rain to flowers and developing fruits, germinate in the presence of free water and develop to a limited extent before growth is halted by the resistance of the cells of the immature fruit. Spores of some fungi (e.g., *Colletotrichum*) can germinate anywhere on the surface of developing banana, citrus, mango, and papaya fruits and produce a structure called an appressorium and a few hyphae, but further development is halted. This type of preharvest infection is called latent infection.

Spores of other pathogenic fungi (e.g., *Phlyctaena vagabunda* Desmazières) penetrate the lenticular cavities of European apples during periods of relatively high temperature and high humidity late in summer. The fungus develops to a very limited extent in the lenticular cavity, remains quiescent, and then produces a lesion around the lenticel as the fruit begins to ripen

in storage. This type of preharvest infection is called lenticular infection.

Harvest and postharvest infections occur at natural openings or at the site of injuries incurred during and after harvest when *Penicillium*, *Rhizopus*, and *Geotrichum* gain entry and often cause devastating rots. Such fungi are not capable of directly penetrating the cuticle/epidermis of the host (*33, 34*). *Rhizopus* species have recently been implicated as a probable cause of slip-skin maceration disorder of dried French prunes (*119*).

Prevention and Control of Spoilage Organisms

Prevention and control of spoilage by molds are achieved through careful selection of the initial product for storage; careful handling and customized packing; rigid controls of levels of potential mold inoculum; use of mold inhibitors; and manipulation of temperature, moisture content or relative humidity, and levels of O_2 and CO_2 during storage. Most of the measures described below are also used to ensure that the product remains in good physical and physiological condition.

Initial Quality

Postharvest diseases can be prevented by (a) controlling frugivorous flies and fruit-sucking moths capable of inoculating fruit with pathogenic fungi (*35*); (b) applying protective treatments in the field or orchard and postharvest treatments in the packing houses; (c) carefully harvesting and storing the sound product at the correct stage of maturity; (d) removing soil and fungal propagules through washing and other processes; and (e) selling first those fruits and vegetables that have a high decay potential.

Handling and Packaging

Spoilage by organisms is lessened by adopting careful handling and packaging procedures that minimize bruising. Customized packing of particular products prevents recontamination of the product with spoilage organisms and also retards moisture loss. Considerable information is available on the methods used for protecting foods transported by air (*22*), truck (*5*), train (*111*), and ship (*51*).

Inoculum Levels

Spore loads are reduced in the field by preharvest fungicide applications and in the packing house mainly by good sanitation procedures, e.g., removal of organic waste debris, washing the product with water, and washing the work areas and equipment with broad-spectrum antimicrobial chemicals.

Inhibitors

Preharvest infections are controlled by application of protectant or eradicant fungicides at and after flowering. Postharvest infections are prevented by spraying fruits and vegetables just before or after harvest (33, 34, 94).

Temperature

Prevention of postharvest diseases of fruits and vegetables in storage is primarily by refrigeration at temperatures that restrict development of particular pathogens but also maintain optimal product quality. Field heat is often removed from the product by precooling soon after harvest. Control of some organisms already within the outer layers of fruit is possible by heat treatment; for example, anthracnose on papaya is controlled by hot water treatment for 20 minutes at 49° (4).

Moisture Content and Relative Humidity

Prevention of growth of certain microorganisms can be achieved by a period of curing, e.g., holding potatoes and onions in warm air. During curing, the potato skin becomes suberized, preventing *Fusarium* dry rot; the outer onion scales are thoroughly dried, preventing development of *Botrytis* neck rot (143). On the other hand, prevention of moisture loss in fruits and vegetables is achieved by accurate control of relative humidity in storage and by wrapping the products in plastic film (47).

Levels of O_2 or CO_2

Certain mixtures of O_2 and CO_2 inhibit the activity of decay microorganisms and slow the respiration rates of fruits and vegetables (67). Thus, when apples are stored in an atmosphere containing 5% CO_2, the extent of decay caused by *Gloeosporium* species is decreased to half that occurring during storage in an uncontrolled atmosphere (68).

Mold and Bacterial Problems Encountered During Storage

Food Warehouse Management

During 1978, average spoilage and shrinkage losses incurred at a large fruit and vegetable warehouse in Winnipeg, Canada, were approximately 0.1%. This low figure reflects good management that included the use of modern storage and handling techniques, anticipation of

Table 5.13

Major postharvest diseases of fresh fruits and vegetables

Crop	Disease	Pathogen
Apples and pears	Lenticel rot	*Cryptosporiopsis curvispora* (Peck) Gremmen [= *C. malicorticis* (Cordley & Nannfeldt) = *Gloeosporium perennans* Zeller & Childs] *Phlyctaena vagabunda*
	Eye rot	*Nectria galligena* Bresadola
	Blue mold rot	*Penicillium expansum* Thom
	Gray mold rot	*Botrytis cinerea* Persoon *ex* Fries
Bananas	Crown rot	*Colletotrichum musae* (= *Gloeosporium musarum* Massee) *Fusarium* spp. *Verticillium theobromae* (Turconi) Mason & Hughes *Ceratocystis paradoxa* (Dade) Moreau (= *Thielaviopsis paradoxa* de Seynes)
	Anthracnose	*Colletotrichum musae*
Carrots	Gray mold rot	*Botrytis cinerea*
	Centrospora rot	*Centrospora acerina* (Hartig) Newhall
	Watery soft rot	*Sclerotinia sclerotiorum* (Libert) de Bary
Citrus fruits	Stem-end rot	*Phomopsis citri* Fawcett *Diplodia natalensis* Pole-Evans *Alternaria citri* Ellis & Pierce
	Green mold rot	*Penicillium digitatum* Saccardo
	Blue mold rot	*Penicillium italicum* Wehmer
Grapes	Gray mold rot	*Botrytis cinerea*
Papayas	Anthracnose	*Colletotrichum gloeosporioides* Penzig
Peaches and cherries	Brown rot	*Monilinia fructicola* (Winter) Honey (= *Sclerotinia fructicola* (Winter) Rehm
	Rhizopus rot	*Rhizopus stolonifer*
Pineapples	Black rot	*Ceratocystis paradoxa*
Potatoes	Bacterial soft rot	*Erwinia carotovora* and other species
	Dry rot	*Fusarium* spp.
	Gangrene	*Phoma exigua* var. *foveata* (Foister) Boerema
	Skin spot	*Oospora pustulans* Owen & Wakefield
	Silver scurf	*Helminthosporium solani*
Strawberries	Gray mold rot	*Botrytis cinerea*
	Rhizopus rot	*Rhizopus stolonifer*
Sweet potatoes	Black rot	*Ceratocystis fimbriata* Ellis & Halsted [= *Endoconidiophora fimbriata* (Ellis & Halsted) Davidson]
	Rhizopus soft rot	*Rhizopus stolonifer*
Leafy vegetables	Bacterial soft rot	*Erwinia carotovora* and other species
	Gray mold rot	*Botrytis cinerea*
	Watery soft rot	*Sclerotinia sclerotiorum*

SOURCE: Eckert, J. W. 1977. Control of post harvest diseases. In *Antifungal Compounds*, vol. 1, M. R. Siegel and H. D. Sisler, eds. Marcel Dekker, New York, NY.

future markets, and knowledge of alternative sources for particular vegetables and fruits. If spoilage was found in a consignment on arrival, government food inspectors were brought in to assess the loss, and the consignment was then salvaged. Generally, more losses occur in the field, processing plant, supermarket, and home than in efficiently run warehouses.

Rail Shipments

Fruits and vegetables are normally transported from California to Winnipeg in refrigerated trucks, the journey taking 5–6 days. Twenty years ago, shipment was mainly in top-iced, refrigerated railcars, the time involved was about 18 days, and decay was frequently observed on arrival. Responsibility for loss was determined by examining the temperature records for various parts of the car and noting any equipment malfunction, the presence and amount of bunker ice on car arrival, the type of spoilage organisms present, the extent of disease development, and the condition of properly obtained product samples.

The presence of *Aspergillus niger* van Tieghem in a consignment of onions arriving by ship/rail from the West Coast indicated that the damage had occurred sometime before train shipment since temperatures in the railcars on arrival were sufficiently low to have prevented its development. This disease develops in onions when they are shipped somewhat damp, as might occur with immature or improperly cured bulbs or if the humidity and temperature during shipment are too high.

Green and Blue Molds of Citrus

These diseases are of considerable importance and their prevention and control require good management in the field as well as in the packing house. Spores are produced on rotting fruit lying under the trees or in packing houses and are transported by air currents to sound fruit. Germination occurs only on injured fruits, and infection takes 24–48 hours at 20–25°; fungicidal treatments must be applied before completion of the infection process. Many fungicidal and cosmetic operations are presently used to prevent and control these diseases (*35*).

Potatoes

Chipping potatoes must be stored at a higher temperature than table stock or processing potatoes to prevent a buildup of sugars which leads to darkening of the chips. The higher temperature combined with high relative humidity causes them to be affected by silver scurf disease (caused by *Helminthosporium solani* Durieu & Montagne). Silver scurf and dry rot can be prevented by a postharvest spray of thiabendazole. In Canada, soft rot caused by *Erwinia carotovora* (Jones) Holland is also a problem in new potatoes packaged in polythene bags. Nonsuberized skins, excess humidity, and a buildup of CO_2 in the bags favor development of the disease. Addition of chlorine to wash water, use of porous packaging materials, and rapid cooling after pack-

aging are measures used to reduce incidence of soft rot.

Molds in Processed Foods

Prevention and Control

Mold growth in and on stored processed foods is influenced primarily by temperature, relative humidity, available O_2, mold inhibitors, and the amount of available inoculum.

Temperature

Most mold spores are inactivated during high-temperature procedures, e.g., baking of cereal products, heat treatment of dairy products, and canning. However, the spores of some molds—e.g., *Byssochlamys nivea* Westling and *B. fulva* (Olliver & Smith)—are heat resistant and may cause spoilage of canned fruit. Mold growth on many products is prevented by storage at particular temperatures, e.g., by rapidly cooling meats to 0° immediately after butchering.

Relative Humidity

Mold growth is prevented by careful manipulation of relative humidity in storage rooms. In the absence of surface moisture, minimal mold growth occurs on freshly killed meat animals stored below 90% rh and on cheese stored below 40% rh.

Available O_2

Packaging of dairy products (*72*) as well as canning (*148*) essentially eliminate O_2 from the atmosphere, thus creating unfavorable conditions for mold development.

Inhibitors

Mold growth on dairy and bakery products is prevented by addition of sorbates, propionates, or other substances (*72, 108*). Sometimes microbial inhibitors are added to packing materials (*32*).

Available Inoculum

Mold inoculum entering the processing plant is reduced by discarding diseased material (e.g., rotten fruits) before entry and by washing (e.g., sound fruits and vegetables). Bakery products are generally free of contamination when taken out of the oven but are contaminated (during cooling and before packaging) by airborne spores of the black [*Rhizopus stolonifer* (Ehrenberg *ex* Fries) Vuillemin] or red (*Neurospora sitophila* Shear & B.O. Dodge) bread molds. A further source of contamination of bakery products is through icing, glazing, and addition of sugars and spices. Levels of airborne spores in processing plants can be reduced by air filtration, use of ultraviolet light (e.g., in meat storage areas), and removal of organic debris. Washing of processing equipment and machinery with antimicrobial agents also reduces mold levels. *Geotrichum candidum* Link (machinery mold) is the principal mold contaminant of processing equipment (*132*), and its presence in canned fruit may be an indicator of unsanitary conditions within a processing plant (*21*).

Further Information

Details of the occurrence, prevention, and control of molds and/or bacteria causing spoilage in particular processed foods are available in the following published sources: spoilage of jams, jellies, and prunes by xerophilic fungi (*103, 106*); spoilage of bakery products (*106, 108*); spoilage of dairy products (*65, 93*); spoilage and control of microorganisms in vegetables (*106, 133*); spoilage of meat and meat products (*60, 69, 106, 135*); spoilage of canned foods (*29, 43, 49, 106*).

Summary

In this chapter, the ecological principles that govern development of microorganisms in stored foods are reviewed and practical examples are used to show how a knowledge of the principles can be used to predict, prevent, and control microbial decomposition. Foods in storage should be considered man-made ecosystems. At present, no single comprehensive study has been made of the various components in a food ecosystem, and the ancillary technical operations involved have scarcely been tested. Until such a study is undertaken from a multidisciplinary viewpoint, our reliance on preservatives and unidisciplinary management practices will continue. It is to be hoped that eventually most stocks of stored foods will be managed from a total ecological viewpoint because such man-made ecosystems are important for our economic and physical survival.

Cited References

1 Abramson, D., and J. T. Mills. 1985. Mycotoxin production during storage. In *Mycotoxins: A Canadian Perspective*. Publ. No. 22848. National Research Council of Canada, Ottawa.

2 Agricultural Research Service. 1969. *Guidelines for Mold Control in High-Moisture Corn*. Farmers' Bulletin 2238. Agricultural Engineering Research Division, U.S. Department of Agriculture, Washington, DC.

3 Ainsworth, G. C., and P. K. C. Austwick. 1973. *Fungal Diseases of Animals,* 2nd ed. Commonwealth Agricultural Bureaux, London.

4 Akamine, E. K., and T. Arisumi. 1953. Control of postharvest storage decay of fruits and papaya (*Carica papaya* L.) with special reference to the effect of hot water. *Proceedings of the American Society for Horticultural Science* 61:270–274.

5 Ashby, B. H. 1970. *Protecting Perishable Foods During Transport by Motortruck*. Agriculture Handbook 105. Agricultural Research Service, U.S. Department of Agriculture, Washington, DC.

6 Bailey, J. E. 1982. Whole grain storage. In *Storage of Cereal Grains and Their Products*, 3rd ed., C. M. Christensen, ed. American Association of Cereal Chemists, St. Paul, MN.

7 Barre, H. J. 1976. Storage and quality of soybeans on the farm. In *World Soybean Research Conference Proceedings*, L. D. Hill, ed. Interstate, Danville, IL.

8 Bender, A. E. 1975. *Dictionary of Nutrition and Food Technology*, 3rd ed. Butterworths, London.

9 Board on Science and Technology for International Development. 1978. *Postharvest Food Losses in Developing Countries*. National Academy of Sciences, Washington, DC.

10 Booth, C. 1977. *Fusarium—Laboratory Guide to the Identification of the Major Species*. Commonwealth Mycological Institute, Kew, England.

11 Bullerman, L. B. 1978. Methods of detecting mycotoxins in foods and beverages. In *Food and Beverage Mycology*, L. R. Beuchat, ed. AVI, Westport, CT.

12 Burgess, L. W., and C. M. Liddell. 1983. *Laboratory Manual for* Fusarium *Research Incorporating a Key and Description of Common Species Found in Australia*. University of Sydney.

13 Burrell, N. J., et al. 1980. Determination of the time available for drying rapeseed before the appearance of surface molds. *Journal of Stored Products Research* 16(3–4)115–118.

14 Center for Food Safety and Applied Nutrition. 1985. *The Food Defect Action Levels*. HHS Publ. No. (FDA)85-2199. Food and Drug Administration, Washington, DC.

15 Christensen, C. M. 1969. Factors affecting invasion of sunflower seeds by storage fungi. *Phytopathology* 59(11)1699–1702.

16 Christensen, C. M. 1971. Evaluating condition and storability of sunflower seeds. *Journal of Stored Products Research* 7(3)163–169.

17 Christensen, C. M. 1976. Maintaining quality of soybeans during storage. In *World Soybean*

Research Conference Proceedings, L. D. Hill, ed. Interstate, Danville, IL.

18 Christensen, C. M. 1978. Storage fungi. In *Food and Beverage Mycology*, L. R. Beuchat, ed. AVI, Westport, CT.

19 Christensen, C. M., and H. H. Kaufmann. 1969. *Grain-Storage—The Role of Fungi in Quality Loss*. University of Minnesota, Minneapolis.

20 Christensen, C. M., and D. B. Sauer. 1982. Microflora. In *Storage of Cereal Grains and Their Products*, 3rd ed., C. M. Christensen, ed. American Association of Cereal Chemists, St. Paul, MN.

21 Cichowicz, S. M., and W. V. Eisenberg. 1974. Collaborative study of the determination of *Geotrichum* mold in selected canned fruits and vegetables. *Journal of the Association of Official Analytical Chemists* 57(4)957–960.

22 Claypool, L. L., et al. 1958. *Air Transportation of Fruits and Vegetables and Cut Flowers: Temperature and Humidity Requirements and Perishable Nature*. AMS-280. Biological Sciences Branch, Agricultural Marketing Service, U.S. Department of Agriculture, Washington, DC.

23 Cobia, D. W., and D. E. Zimmer. 1978. *Sunflower. Production and Marketing*. Extension Bulletin 25. North Dakota State University of Agriculture and Applied Science, Fargo.

24 Cochrane, V. W. 1958. *Physiology of Fungi*. Wiley, New York, NY.

25 Correy, J. 1978. Relationships of water activity to fungal growth. In *Food and Beverage Mycology*, L. R. Beuchat, ed. AVI, Westport, CT.

26 Coursey, D. G., and R. H. Booth. 1972. The postharvest phytopathology of perishable tropical produce. *Review of Plant Pathology* 51(12)751–765.

27 Dakin, J. C., and A. C. Stolk. 1968. *Moniliella acetoabutans*: Some further characteristics and industrial significance. *Journal of Food Technology* 3(1)49–53.

28 Dallyn, H., and J. R. Everton. 1969. The xerophilic mold, *Xeromyces bisporus*, as a spoilage organism. *Journal of Food Technology* 4(4)399–403.

29 Davidson, P. M., L. J. Pflug, and G. M. Smith. 1981. Microbiological analysis of food product in swelled cans of low-acid foods collected from supermarkets. *Journal of Food Protection* 44(9)686–691.

30 Dennis, C. A. R. 1973. Health hazards of grain storage. In *Grain Storage: Part of a System*, R. N. Sinha and W. E. Muir, eds. AVI, Westport, CT.

31 Department of External Affairs. 1983. *Fats and Oils in Canada. Annual Review 1983*. Grain Marketing Bureau, Ottawa, Canada.

32 Desrosier, N. W. 1970. *The Technology of Food Preservation*, 3rd ed. AVI, Westport, CT.

33 Eckert, J. W. 1975. Postharvest diseases of fresh fruits and vegetables—etiology and control. In *Symposium: Postharvest Biology and Handling of Fruits and Vegetables*, N. F. Haard and D. K. Salunkhe, eds. AVI, Westport, CT.

34 Eckert, J. W. 1977. Control of post harvest diseases. In *Antifungal Compounds*, vol. 1, M. R. Siegel and H. D. Sisler, eds. Marcel Dekker, New York, NY.

35 Eckert, J. W. 1978. Post-harvest diseases of citrus fruits. *Outlook on Agriculture* 9(5)225–232.

36 Ellis, M. B. 1971. *Dematiaceous Hyphomycetes*. Commonwealth Mycological Institute, Kew, England.

37 Eskin, N. A. M., H. M. Henderson, and R. J. Townsend. 1971. *Biochemistry of Foods*. Academic Press, London.

38 Food and Agriculture Organization. 1984. *FAO Production Yearbook 1983*. United Nations, Rome.

39 Food Protection Committee. 1965. *Chemicals Used in Food Processing*. Publ. 1274. National Academy of Sciences, Washington, DC.

40 Foster, E. M., et al. 1983. *Dairy Microbiology*. Ridgeview, Atascadera, CA.

41 Foster, G. H. 1982. Drying cereal grains. In *Storage of Cereal Grains and Their Products*, 3rd ed., C. M. Christensen, ed. American Association of Cereal Chemists, St. Paul, MN.

42 Foster, G. H., and J. Tuite. 1982. Aeration and stored grain management. In *Storage of Cereal Grains and Their Products*, 3rd ed., C. M. Christensen, ed. American Association of Cereal Chemists, St. Paul, MN.

43 Gilbert, R. J., K. L. Kolvin, and D. Roberts. 1982. Canned foods—the problems of food poisoning and spoilage. *Health and Hygiene* 4:41–47.

44 Gunderson, M. F. 1961. Mold problem in frozen foods. In *Proceedings of Low Temperature Microbiology Symposium*. Campbell Soup Company, Camden, NJ.

45 Gustafson, E. H. 1976. Loading, unloading, storage, drying, and cleaning of vegetable oil–bearing materials. *Journal of the American Oil Chemists' Society* 53(6)248–250.

46 Hanson, H. L., and L. R. Fletcher. 1958. Time-temperature tolerance of frozen foods. XII. Turkey dinners and turkey pies. *Food Technology* 12(1)40–43.

47 Harvey, J. M. 1978. Reduction of losses in fresh market fruits and vegetables. *Annual Review of Phytopathology* 16:321–341.

48 Heldman, D. R., T. I. Hedrick, and C. W. Hall. 1965. Sources of airborne microorganisms in food processing areas—drains. *Journal of Milk*

52 John T. Mills

and Food Technology 28(1)41–45.

49 Hersom, A. C., and E. D. Hulland. 1969. *Canned Foods—An Introduction to Their Microbiology*, 6th ed. Churchill, London.

50 Hesseltine, C. W., R. F. Rogers, and R. J. Bothast. 1978. Microbiological study of exported soybeans. *Cereal Chemistry* 55(3)332–340.

51 Hinds, R. H., Jr. 1970. *Transporting Fresh Fruits and Vegetables Overseas*. ARS 52-39. U.S. Department of Agriculture, Washington, DC.

52 Holt, J. D. (ed.). 1977. *The Shorter Bergey's Manual of Determinative Bacteriology*, 8th ed. Williams & Wilkins, Baltimore, MD.

53 Hull, R. 1939. Study of *Byssochlamys fulva* and control measures in processed fruits. *Annals of Applied Biology* 26(4)800–822.

54 Hyde, M. B., and N. J. Burrell. 1982. Controlled atmosphere storage. In *Storage of Cereal Grains and Their Products*, 3rd ed., C. M. Christensen, ed. American Association of Cereal Chemists, St. Paul, MN.

55 Ingold, C. T. 1953. *Dispersal in Fungi*. Oxford University Press, London, England.

56 International Commission on Microbiological Specifications for Foods. 1978. *Microorganisms in Foods, Vol. 1. Their Significance and Methods of Enumeration*, 2nd ed. University of Toronto, Canada.

57 Jackson, C. R. 1967. Influence of drying and harvesting procedures on fungus populations and aflatoxin production in peanut in Georgia. *Phytopathology* 57(5)458–462.

58 Janzen, D. H. 1977. Why fruits rot, seeds mold, and meat spoils. *American Naturalist* 111(980)691–713.

59 Jarvis, B. 1978. Methods for detecting fungi in foods and beverages. In *Food and Beverage Mycology*, L. R. Beuchat, ed. AVI, Westport, CT.

60 Jay, J. M. 1978. Meats, poultry and seafoods. In *Food and Beverage Mycology*, L. R. Beuchat, ed. AVI, Westport, CT.

61 Jensen, L. B. 1954. *Microbiology of Meats*, 3rd ed. Garrard, Urbana, IL.

62 Kaufmann, H. H. 1976. Commercial storage and handling of soybeans. In *World Soybean Research Conference Proceedings*, L. D. Hill, ed. Interstate, Danville, IL.

63 Kulik, M. M. 1968. *A Compilation of Descriptions of New* Penicillium *Species*. Agriculture Handbook 351. U.S. Department of Agriculture, Washington, DC.

64 Lacey, J., and P. H. Gregory. 1980. The ecological significance of microbial dispersal systems. In *Contemporary Microbial Ecology*, D. C. Ellwood, et al., eds. Academic Press, London.

65 Ledford, R. A. 1979. Dairy products. In *Controlling Microorganisms in Food Processing*. 13th Annual Symposium. Special Report 31. New York State Agricultural Experimental Station, Geneva.

66 Lichtwardt, R. W., G. L. Barron, and L. H. Tiffany. 1958. Mold flora associated with shelled corn in Iowa. *Iowa State College Journal of Science* 33(1)1–11.

67 Lipton, W. J. 1975. Controlled atmospheres for fresh vegetables and fruits—why and when. In *Symposium: Postharvest Biology and Handling of Fruits and Vegetables*, N. F. Haard and D. K. Salunkhe, eds. AVI, Westport, CT.

68 Lockhart, C. L. 1969. Effect of CA storage on storage rot pathogens. *Michigan State University Horticulture Report* 9:113–121.

69 Lowry, P. D., and C. O. Gill. 1984. Temperature and water activity minima for growth of spoilage moulds for meat. *Journal of Applied Bacteriology* 56(2)193–199.

70 Lyster, B. 1978. How to store rapeseed. *Country Guide* (West ed.) 97(7)34–35.

71 Marshall, C. R., and V. T. Walkley. 1951. Some aspects of microbiology applied to commercial apple juice production. I. Distribution of microorganisms on the fruit. *Food Research* 16:448–456.

72 Marth, E. H. 1978. Dairy products. In *Food and Beverage Mycology*, L. R. Beuchat, ed. AVI, Westport, CT.

73 Martin, P. M. D. 1976. *A Consideration of the Mycotoxin Hypothesis with Special Reference to the Mycoflora of Maize, Sorghum, Wheat, and Groundnuts*. Publ. G105. Tropical Products Institute, London.

74 Mehrotra, B. S. 1976. *Investigations of Selected Microorganisms Associated with Cereal Grains and Flours in India, to Provide Basic Information Related to the Utilization of Cereal Grains in Foods and Feedstuffs*. Report of Research Project FG-In-410. University of Allahabad, India.

75 Michener, H. D., and R. P. Elliott. 1964. Minimum growth temperatures for food-poisoning, fecal-indicator, and psychrophilic microorganisms. *Advances in Food Research* 13:349–396.

76 Miller, D. D., and N. S. Golding. 1949. The gas requirements of molds. V. The minimum oxygen requirements for normal growth and for germination of six mold cultures. *Journal of Dairy Science* 32(2)101–110.

77 Mills, J. T. 1976. Spoilage of rapeseed in elevator and farm storage in western Canada. *Canadian Plant Disease Survey* 56(3)95–103.

78 Mills, J. T. 1983. Insect-fungus associations influencing seed deterioration. *Phytopathology* 73(2)330–335.

79 Mills, J. T. 1985. Characteristics of mold-dam-

aged stored grains. *Official Proceedings, Thirty-Third Annual Meeting, American Association of Feed Microscopists* (Ottawa), pp. 38–42.

80 Mills, J. T. 1989. *Spoilage and Heating of Stored Agricultural Products. Prevention, Detection and Control.* Publ. 1823E. Agriculture Canada, Ottawa.

81 Mills, J. T., and R. N. Sinha. 1980. Safe storage periods for farm-stored rapeseed based on mycological and biochemical assessment. *Phytopathology* 70(6)541–547.

82 Mills, J. T., R. N. Sinha, and H. A. H. Wallace. 1978. Assessment of quality criteria of stored rapeseed—a multivariate study. *Journal of Stored Products Research* 14(2–3)121–133.

83 Milton, R. F., and K. J. Jarrett. 1969. Storage and transport of maize—1. Temperature, humidity and microbiological spoilage. *World Crops* 21(6)356–357.

84 Milton, R. F., and K. J. Jarrett. 1970. Storage and transport of maize—2. *World Crops* 22(1)48–49.

85 Milton, R. F., and K. J. Jarrett. 1970. The storage and transport of maize—3. *World Crops* 22(2)96, 98, 99.

86 Mislivec, P. B., and V. R. Bruce. 1977. Incidence of toxic and other mold species and genera in soybeans. *Journal of Food Protection* 40(5)309–312.

87 Mislivec, P. B., and J. Tuite. 1970. Species of *Penicillium* occurring in freshly-harvested and in stored dent corn kernels. *Mycologia* 62(1)67–74.

88 Moysey, E. B. 1973. Storing and drying of oilseeds. In *Grain Storage: Part of a System*, R. N. Sinha and W. E. Muir, eds. AVI, Westport, CT.

89 Muir, W. E. 1973. Temperature and moisture in grain storages. In *Grain Storage: Part of a System*, R. N. Sinha and W. E. Muir, eds. AVI, Westport, CT.

90 Muir, W. E., R. N. Sinha, and H. A. H. Wallace. 1977. Comparison of the storage characteristics of three types of farm granaries. *Canadian Agricultural Engineering* 19(1)20–24.

91 Mundt, J. O. 1978. Fungi in the spoilage of vegetables. In *Food and Beverage Mycology*, L. R. Beuchat, ed. AVI, Westport, CT.

92 Noon, R. A., and J. Colhoun. 1979. Market and storage diseases of yams imported into the United Kingdom. *Phytopathologische Zeitschrift* 94(4)289–302.

93 Northolt, M. D. 1984. Growth and inactivation of pathogenic micro-organisms during manufacture and storage of fermented dairy products. A review. *Netherlands Milk and Dairy Journal* 38(3)135–150.

94 Ogawa, J. M., B. T. Manji, and A. H. El Behadli. 1976. Chemical control of postharvest diseases. In *Proceedings of the Third International Biodegradation Symposium* (Kingston, RI, 1975), J. M. Sharpley and A. M. Kaplan, eds. Applied Science, London, pp. 561–575.

95 Onions, A. H. S., D. Allsopp, and H. O. W. Eggins. 1981. *Smith's Introduction to Industrial Mycology*, 7th ed. Arnold, London.

96 Onishi, H. 1963. Osmophilic yeasts. *Advances in Food Research* 12:53–94.

97 Parker, P. E., G. R. Bauwin, and H. L. Ryan. 1982. Sampling, inspection, and grading of grain. In *Storage of Cereal Grains and Their Products*, 3rd ed., C. M. Christensen, ed. American Association of Cereal Chemists, St. Paul, MN.

98 Patterson, D. S. P. 1982. Mycotoxins. In *Environmental Chemistry*, vol. 2. Royal Society of Chemistry, London, pp. 205–233.

99 Pelhâte, J. 1968. Inventaire de la mycoflore des blés de conservation. *Bulletin Trimestriel de la Société Mycologique de France* 84(1)127–143.

100 Peppler, H. J. 1976. Yeasts. In *Food Microbiology: Public Health and Spoilage Aspects*, M. P. Defiguerido and D. F. Splittstoesser, eds. AVI, Westport, CT.

101 Perez, R. A., J. Tuite, and K. Baker. 1982. Effect of moisture, temperature, and storage time on subsequent storability of shelled corn. *Cereal Chemistry* 59(3)205–209.

102 Pettit, R. E., and R. A. Taber. 1968. Factors influencing aflatoxin accumulation in peanut kernels and the associated mycoflora. *Applied Microbiology* 16(8)1230–1234.

103 Pitt, J. I. 1975. Xerophilic fungi and the spoilage of foods of plant origin. In *Water Relations of Foods*, R. B. Duckworth, ed. Academic Press, London.

104 Pitt, J. I. 1979. *The Genus Penicillium and its Teleomorphic States* Eupenicillium *and* Talaromyces. Academic Press, London.

105 Pitt, J. I. 1981. Food spoilage and biodeterioration. In *Biology of Conidial Fungi*, vol. 2, G. T. Cole and B. Kendrick, eds. Academic Press, London.

106 Pitt, J. I., and A. D. Hocking. 1985. *Fungi and Food Spoilage*. Academic Press, New York, NY.

107 Pixton, S. W., and S. Warburton. 1977. The moisture content/equilibrium relative humidity relationship and oil composition of rapeseed. *Journal of Stored Products Research* 13(2)77–81.

108 Ponte, J. G., Jr., and G. C. Tsen. 1978. Bakery products. In *Food and Beverage Mycology*, L. R. Beuchat, ed. AVI, Westport, CT.

109 Raper, K. B., and D. I. Fennell. 1965. *The Genus* Aspergillus. Williams & Wilkins, Baltimore, MD.

110 Raper, K. B., and C. Thom. 1949. *A Manual*

of the Penicillia. Williams & Wilkins, Baltimore, MD.

111 Redit, W. H. 1969. *Protection of Rail Shipments of Fruits and Vegetables.* Agriculture Handbook 195. Agricultural Research Service, U.S. Department of Agriculture, Washington, DC.

112 Robertson, J. A., R. G. Roberts, and G. W. Chapman, Jr. 1985. An evaluation of "heat damage" and fungi in relation to sunflower seed quality. *Phytopathology* 75(2)142–145.

113 Ryall, A. L., and W. J. Lipton. 1979. *Handling, Transportation and Storage of Fruits and Vegetables, Vol. 1. Vegetables and Melons.* AVI, Westport, CT.

114 Sargeant, K., et al. 1961. Toxicity associated with certain samples of groundnuts. *Nature* 192(4807)1096–1097.

115 Sauer, D. B., and C. M. Christensen. 1968. Germination percentage, storage fungi isolated from, and fat acidity values of export corn. *Phytopathology* 58(10)1356–1359.

116 Scott, P. M. 1978. Mycotoxins in feeds and ingredients and their origin. *Journal of Food Protection* 41(5)385–398.

117 Semeniuk, G. 1954. Microflora. In *Storage of Cereal Grains and Their Products,* J. A. Anderson and A. W. Alcock, eds. American Association of Cereal Chemists, St. Paul, MN.

118 Shipton, W. A., and S. C. Chambers. 1966. The internal microflora of wheat grains in western Australia. *Australian Journal of Experimental Agriculture and Animal Husbandry* 6(23)432–436.

119 Sholberg, P. L., and J. M. Ogawa. 1983. Relation of postharvest decay fungi to the slip-skin maceration disorder of dried French prunes. *Phytopathology* 73(5)708–713.

120 Shotwell, O. L., et al. 1969. Survey of cereal grains and soybeans for the presence of aflatoxin. II. Corn and soybeans. *Cereal Chemistry* 46(5)454–463.

121 Singh, T., and R. P. S. Tyagi. 1984. Fungal diseases and spoilage of cereals and millets in storage. In *Review of Tropical Plant Pathology,* vol. 1, S. P. Raychaudhuri and J. P. Verma, eds. Today and Tomorrow's, New Delhi, India.

122 Sinha, R. N. 1973. Ecology of storage. *Annales de Technologie Agricole* 22(3)351–369.

123 Sinha, R. N. 1973. Interrelations of physical, chemical, and biological variables in the deterioration of stored grains. In *Grain Storage: Part of a System,* R. N. Sinha and W. E. Muir, eds. AVI, Westport, CT.

124 Sinha, R. N. [1975.] Climate and the infestation of stored cereals by insects. *Proceedings of the First International Working Conference on Stored-Product Entomology* (Savannah, GA, 1974), pp. 117–141.

125 Sinha, R. N., and H. A. H. Wallace. 1965. Ecology of a fungus-induced hot spot in stored grain. *Canadian Journal of Plant Science* 45(1)48–59.

126 Sinha, R. N., and H. A. H. Wallace. 1977. Storage stability of farm-stored rapeseed and barley. *Canadian Journal of Plant Science* 57(2)351–365.

127 Smith, C. V. 1969. *Meteorology and Grain Storage.* Technical Note 101. World Meteorological Organization, Geneva, Switzerland.

128 Smith, G. 1969. *An Introduction to Industrial Mycology,* 6th ed. Arnold, London.

129 Smoot, J. J., L. G. Houck, and H. B. Johnson. 1971. *Market Diseases of Citrus and Other Subtropical Fruits.* Agriculture Handbook 398. U.S. Department of Agriculture, Washington, DC.

130 Snow, D. 1949. The germination of mould spores at controlled humidities. *Annals of Applied Biology* 36(1)1–13.

131 Snowdon, A. L. 1979. Diseases and disorders of imported fruits and vegetables. *Annals of Applied Biology* 91(3)404–409.

132 Splittstoesser, D. F. 1978. Fruits and fruit products. In *Food and Beverage Mycology,* L. R. Beuchat, ed. AVI, Westport, CT.

133 Splittstoesser, D. F. 1979. Vegetables. In *Controlling Microorganisms in Food Processing.* 13th Annual Symposium. Special Report 31. New York State Agricultural Experimental Station, Geneva.

134 Splittstoesser, D. F., and D. B. Prest. 1976. Molds. In *Food Microbiology: Public Health and Spoilage Aspects,* M. P. Defiguerido and D. F. Splittstoesser, eds. AVI, Westport, CT.

135 Tompkin, R. B. 1979. Meat and poultry products. In *Controlling Microorganisms in Food Processing.* 13th Annual Symposium. Special Report 31. New York State Agricultural Experimental Station, Geneva.

136 Trewartha, G. T., and L. H. Horn. 1980. *An Introduction to Climate,* 5th ed. McGraw-Hill, New York, NY.

137 Troller, J. A. 1980. Influence of water activity on microorganisms in foods. *Food Technology* 34(5)76–80, 82.

138 Tuite, J. 1969. *Plant Pathological Methods. Fungi and Bacteria.* Burgess, Minneapolis, MN.

139 Tuite, J., and G. H. Foster. 1979. Control of storage diseases of grain. *Annual Review of Phytopathology* 17:343–366.

140 Turner, J. N. 1967. *The Microbiology of Fabricated Materials.* Churchill, London.

141 Vaughn, R. H. 1963. Microbial spoilage problems of fresh and refrigerated foods. In *Microbiological Quality of Foods,* L. W. Slanet, et al., eds. Academic Press, New York, NY.

142 von Arx, J. A. 1974. *The Genera of Fungi Spor-*

ulating in Pure Culture, 2nd ed. J. Cramer, Vaduz, Liechtenstein.

143 Walker, J. C. 1969. *Plant Pathology,* 3rd ed. McGraw-Hill, New York, NY.

144 Wallace, H. A. H. 1960. Factors affecting subsequent germination of cereal seeds sown in soils of subgermination moisture content. *Canadian Journal of Botany* 38(3)287–306.

145 Wallace, H. A. H. 1973. Fungi and other organisms associated with stored grain. In *Grain Storage: Part of a System,* R. N. Sinha and W. E. Muir, eds. AVI, Westport, CT.

146 Wallace, H. A. H., and R. N. Sinha. 1975. Microflora of stored grain in international trade. *Mycopathologia* 57(3)171–176.

147 Wallace, H. A. H., et al. 1983. Biological, physical and chemical changes in stored wheat. *Mycopathologia* 82(2)65–76.

148 Weiser, H. H., G. J. Mountney, and W. A. Gould. 1971. *Practical Food Microbiology and Technology,* 2nd ed. AVI, Westport, CT.

149 Williams, S. (ed.). 1984. *Official Methods of Analysis,* 14th ed. Association of Official Analytical Chemists, Arlington, VA (*Note:* See the 15th edition for the latest information.)

150 Wyllie, T. D., and L. G. Morehouse. 1977–78. *Mycotoxin Fungi, Mycotoxins, Mycotoxicoses,* 3 vols. Marcel Dekker, New York, NY.

151 Zaehringer, M. V., and J. O. Early. 1976. *Proceedings of National Food Loss Conference.* College of Agriculture, University of Idaho, Moscow.

6

Mite Pests in Stored Food

Jan Boczek

All kinds of stored-food products, as well as other organic materials, can be infested by mites when such items are stored for a long time. Pest mites are most abundant in temperate and humid climates. They are very serious pests in Canada, Czechoslovakia, Great Britain, Japan, Poland, Portugal, and in some regions of the Soviet Union. Products with a moisture content (mc) above 13% are seldom completely free from fungi, mites, and insects in the geographic areas just mentioned. Some products such as cheese, pet food, and oilseeds may become infested in regions with warmer and drier climates. Modern management techniques, such as drying, special construction of storage facilities, protective packaging, and use of pesticides, all aimed at reducing losses caused by pests, are seldom completely effective.

Stored-food products are often simultaneously infested by various species of mites, insects, and microflora. Each pest species has specific environmental requirements and specific biological characteristics. Mite reproductive rates vary greatly and mite populations fluctuate irregularly. The factors affecting mite populations are very complex. Although our knowledge of these factors is still incomplete, we are able to prevent explosive outbreaks in some situations.

Ecological Notes

In this section, the subject families (Acaridae, Ascidae, Carpoglyphidae, Cheyletidae, Glycy-

Boczek: Department of Applied Entomology, Warsaw Agricultural University, 02766 Warsaw, Poland.

phagidae, Lardoglyphidae, Pyroglyphidae, and Tarsonemidae) are presented in alphabetical order. The species within a given family are also arranged in alphabetical order. Note that mite pests as well as mites that prey on mite pests are included in this review. Definitions of technical terms, especially those applied to life stages and morphological features, can be found in the Glossary as well as in a number of reference volumes (2, 37, 44, 94). [Hughes (37) also discusses many other species of stored-food mites that are not mentioned here.]

Acaridae

Acarus farris

A. farris is commonly found in the field in moist plant remnants, in haystacks, on the foliage of cereal crops, and in nests of ants, birds, and rodents. This species is present during all seasons, but the hypopus (deutonymph) is more common during autumn. A. farris comes into grain stores on plant remnants, but populations tend to decline in storage situations. Its development on grain requires a moisture content no lower than 16% [corresponds to 80% relative humidity (rh)]; 18% mc (89% rh) and temperatures from 15° to 25° are optimum (41). The physical limits for development are temperatures between 3° and 31° and relative humidity greater than 74%.

Developmental time for one generation at 89% rh requires 90 days at 5°, 24 days at 15°, and 10.5 days at 30°. Longevity is highest (135 days) at 8°; at 15°, it is 35 days; at 30°, 19 days (Table 6.1). Egg production from a single female on

57

Table 6.1

Effect of temperature and relative humidity on *Acarus farris*

Temperature (°C)	8	15	25	25	25	25	30
Relative humidity (%)	89	89	75	81	85	89	89
Fecundity (eggs/female)							
Yeast	158	106	8	8	28	112	38
Wheat germ	—	—	—	—	—	28	—
Developmental time for one generation (days)							
Yeast	—	—	—	—	—	12	—
Wheat germ	61	24	—	—	—	13	10.5
Powdered milk	—	—	—	—	—	18	—
Cheese	—	—	—	—	—	20	—
Mortality during development (%)							
Yeast	—	—	—	—	—	27	—
Wheat germ	—	—	—	—	—	57	—
Flour	—	—	—	—	—	93	—
Powdered milk	—	—	—	—	—	30	—
Longevity (days)							
Wheat germ	110	36	10	7	14	22	17

SOURCE: Jakubowska, J. 1971. Biology and ecology of *Acarus farris* (Oud.) (Acarina, Acaridae). *Roczniki Nauk Rolniczych* E1(2)75-94 (in Polish, English summary).

yeast reaches 158 eggs at 8°, 112 at 25°, and 38 at 30°. One to six eggs are laid per day. Females lay only a few eggs at 15° and 80% rh, but at 25° and 85% rh they lay an average of 28 eggs. In one series of experiments at 15°, females lived 8 days at 80% rh and 13 days at 85% rh (*42*).

A. farris thrives on yeast, powdered milk, or wheat germ. Exposure to several antibiotics and antimicrobial agents does not affect developmental time but does increase the rate of mortality during development. Several fungal species are suitable foods for *A. farris*: *Alternaria tenuis* (Wiltshire) [=*A. alternata* (Fries) Keissler], *Hormodendrum cladosporioides* (Fresenius) Saccardo, *Penicillium chrysogenum* Thom, *Aspergillus repens* de Bary, and *Aspergillus echinulatus* (Delacrois) Thom & Church (*26, 74*). Medicinal herbs are only moderately attractive media for this species: *A. farris* reached adulthood on 39 of 52 kinds of medicinal herbs and laid eggs on 26 (*27*).

A motile hypopus is often formed in this species, much more often than in *A. siro*. The hypopal stage occurs throughout the year in field and storage situations. The hypopi (=hypopodes) are slightly more tolerant of lower humidities than are the other stages. Hypopi are commonly found in nests of ants, birds, and rodents; they cling to the bodies of these and other animals including insects and mites (*41*). Hypopi become more numerous when the temperatures reach 20–25° and the relative humidity is 85–90%. Low temperatures and low humid-

ities apparently cause the hypopi to disappear. Numerous hypopi may be observed on wheat germ and yeast when the general population density is high. Transformation of hypopi to tritonymphs (homeomorphic deutonymphs) occurs only at high relative humidities in the presence of suitable food (*21*).

Acarus immobilis

This field species is largely confined to the temperate regions where it occurs in bird nests, humus, and ground-level vegetation. It has been found on occasion in raw cereal grains on farms and in granaries, in spilled grain, and on cheese (*37*). Moist stored grain sometimes supports small populations of *A. immobilis*, but it is much less common than *A. farris*, especially in field situations.

Inert hypopi often occur in populations of this species. Hypopal formation is favored by malnutrition (notably, deficient quantities of B vitamins and ergosterol) resulting from exhaustion of the food supply. In a culture, hypopus formation is suppressed if the blend of nutrients resembles that found in yeast. The preadult stages are apparently incapable of digesting complex, water-insoluble proteins; the assimilation of amino acids proceeds faster in the presence of a good supply of RNA (*33*). The probability of the production of the hypopal form increases as population density increases and as temperature and humidity approach optimal levels. The hypopus is very resistant to desiccation and very tolerant of low humidities. Hypopi of *A. immobilis*, *A. farris*, and *A. siro* are often carried about by insects and other animals (*21*).

Acarus siro, grain mite

A. siro, one of the most important pests of storage premises, is found in cereals and processed cereal products, medicinal herbs, hay, cheese, deep litter of poultry houses, and in abandoned beehives (*37*). A common inhabitant of house dust, it is strongly allergic. If the humidity and temperature are favorable, *A. siro* can infest almost any food used by man or beast. Since it can live on fungi, materials otherwise unsuitable as food may be infested if they are even slightly moldy. *A. siro* eats out the germ of grains. The endosperm may be eaten, too, but only if it is moldy. None of the acarid mites can penetrate undamaged grains or seeds (*79*).

A. siro is cosmopolitan but is more common in temperate regions with cool, moist climates. Grain mites infest foods in storage and in transport, but not grain in the field. They are serious pests of stored grain and cereal products in Great

Britain, the Soviet Union, Poland, and Czechoslovakia. They have been found on cheese in Australia and on pet food in the United States. Grain mites are generally common in Spain, Portugal, Canada, and Japan.

All females of the genus *Acarus* retain the spermatophores acquired from the males. Since each mating results in the acquisition of one spermatophore, the number of matings can be determined by microscopic examination. Females of *A. siro* mate 16–40 times (*34*). Egg laying begins the day after eclosion and continues until death. Average egg production is about 230, but reaches 670 when the female feeds on powdered milk or wheat germ under favorable conditions (20° and 80% rh). Daily egg production ranges from 1 to 24 (*11*).

Populations of *A. siro* die out when grain moisture drops below 13.4%. The mites thrive at 14% mc. Grain moisture content in the range of 15–18% (or higher) provides conditions suitable for serious infestation (*22*) (Figure 6.1). Feeding and other activities cease at 0°. In tropical regions there is at least one season of the year when conditions are too hot or too dry to permit survival or development of the grain mite. *A. siro* is, therefore, virtually unknown as a pest of stored grain in those regions. In most northern European countries and in large areas of the USSR, infestation by *A. siro* is a major problem. In those countries, wheat may be harvested with a moisture content of about 16–20%. Even when dried, its moisture content seldom drops below 14%, and the atmospheric relative humidity may be 80% or more. *A. siro* may become a serious problem under those conditions.

Environmental conditions may be considered optimum for *A. siro* when the temperature stands between 20° and 25° with a relative humidity in the range of 75–80%. Grain mites tend to avoid atmospheric humidities below 75% and above 85%. Evidence of their sensitivity to variations in relative humidity in the optimal range is minimal, but their reaction to differences of 0.1% near the lower limit of tolerance and to differences of 0.25% near the upper limit of tolerance has been documented (*43*). Uptake of water vapor cannot occur below 71% rh. Longevity is indirectly proportional to the saturation deficit of the air (*43*). In common with *Tyrophagus putrescentiae* and *Dermatophagoides farinae*, *A. siro* secretes hygroscopic salts (potassium chloride) from its supracoxal glands. Water from the ambient air may be added to the water pool of the mite through the medium of these hygroscopic salt secretions (*91*).

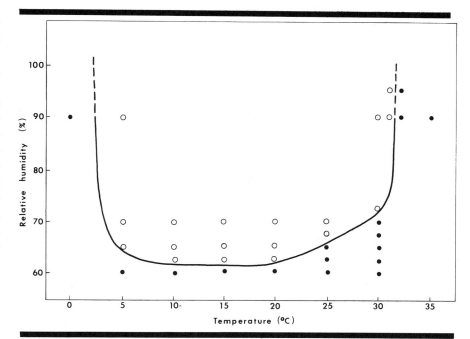

When exposed to −18° for 24–168 hours, at least 1% of the adult, larval, and nymphal mites survived. Developmental time for one generation ranges from 78 days at 4° to 9.2 days at 28° and 80% rh (*11*) (Figure 6.2).

Longevity depends upon temperature, relative humidity, food, and reproductive activity. Under optimal conditions, females reared on wheat germ live 42–51 days (average) (maximum, 63 days); at 28°, they live 30 days. Male longevity is usually a few days less. The content of lipids and proteins in the diet does not affect

Figure 6.1
Physical limits for complete development of the grain mite, *Acarus siro*.
SOURCE: Cunnington, A. M. 1965. Physical limits for complete development of the grain mite, *Acarus siro* L. (Acarina, Acaridae) in relation to its world distribution. *Journal of Applied Ecology* 2(2) 295–306.

Figure 6.2
Effect of relative humidity on the duration of development of one generation of *Acarus siro*.

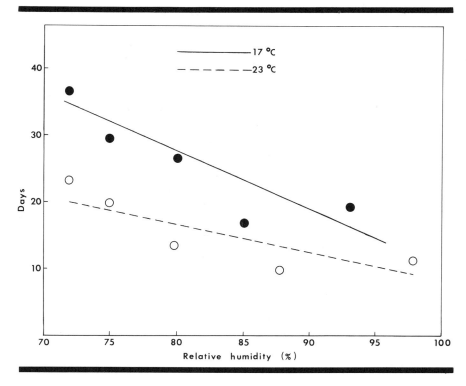

longevity. Rye germ with added minerals and vitamins increases longevity of both sexes. Females live significantly longer when fed on the best foods (yeast, casein, hazel nuts); males live longest on rye germ. The respiration rate of females is generally higher than that of males, but aging of mites may alter these rates (15). Virgin females live longest: 83 days, average; 103 days, maximum. Virgin males usually live about 50 days (11).

Egg production and reproductive success reach their highest levels when A. siro feeds on wheat germ, yeast, cheese, powdered milk, flour, or seeds. In finely comminuted products such as flour or powdered milk, infestations are usually confined to the exposed surface layer. Coarser foods (e.g., grains or nuts) may be infested throughout the bulk. Population explosions are associated with a congruence of favorable environmental factors (especially temperature, humidity, and fungal growth); simple aging of the food product is largely unrelated to population expansions.

A. siro feeds on fungi associated with stored foods. The results of choice tests indicate that the mites favor *Penicillium camemberti* Thom, *Aspergillus flavus* Link, *A. repens* de Bary, *A. ruber* (König et al.) Thom & Church, and *A. amstelodami* (Mangin) Thom & Church. *Trichoderma lignorum* (Tode) Harz and *Fusarium moniliforme* Sheldon seem to be minimally attractive to grain mites (35, 69, 87). Grain mites can also live on *Alternaria* species and *Aspergillus candidus* Link *ex* Fries, but egg production decreases on these hosts (26). Populations of A. siro die out when restricted to a diet of *Sporendonema sebi* Fresenius and *Aspergillus restrictus* Smith (81).

Grain mites are sensitive to antibiotics and antimicrobial agents. Minute dietary doses of the preservatives potassium sorbate, methyl paraben, and calcium propionate inhibit their development. Higher doses increase mortality and decrease fecundity. High doses of antibiotics diminish egg production (14).

Some kinds of medicinal herbs support populations of A. siro; others do not (27). Seeds and fruits are most suitable; leaves and flowers less so; roots, rhizomes, and bark are least suitable. When volatile oils, alkaloids, and glycosides derived from medicinal herbs are mixed with the rye germ diet of the mites, prolonged developmental time and increased mortality result (27).

Grain mites may be anesthetized by a 30 + % concentration of carbon dioxide; a 3-day exposure at 100% CO_2 kills them (38).

Hypopi often fail to appear in laboratory colonies of A. siro. Motile hypopi sometimes appear in natural populations when ambient conditions are favorable (20–25°, 85–95% rh), the population is dense, and the food supply is deficient either in quantity or quality, especially with regard to B vitamins (21, 32). The chain of events (possibly cumulative in effect) that eventually results in the formation of a hypopus begins in the active larval and protonymphal stages. The process is apparently reversible through the larval stage but becomes irreversible during the protonymphal stage. Developmental time for individuals that become hypopi is much longer than that required by individuals that become tritonymphs. This slower development is associated with the physiological and morphological peculiarities of hypopus formation, not with starvation.

Aleuroglyphus ovatus, brownlegged grain mite

A. *ovatus* is widely distributed (Europe, Japan, Turkey, United States, Soviet Union; probably cosmopolitan) in grains, bran, animal feeds, dried fish, and flour. It may produce very large populations. It has been found in chicken houses and nests of rodents and moles. On a diet of wheat germ, the life cycle lasts 2–3 weeks at 23° and 87% rh or 16.5 days at 25° and 75% rh. The fertilized females lay about 60 eggs. Optimal conditions are 30° and 85% rh. The lower limits for development are reported to be 22° and 80% rh (96), but development has been observed to occur at 64% rh (48).

Several species of fungi, *Neurospora crassa* Shear & Dodge, *Nigrospora sphaerica* (Saccardo) Mason, and *Trichothecium roseum* (Persoon) Link *ex* S. F. Gray, support rapid rates of population growth of A. *ovatus* (37). A 1:1 mixture of starch and yeast produces the highest reproductive rate. Glucose and lactose cannot be assimilated. Amylase- and protease-mediated reactions are common but saccharase activity is minimal (50, 51).

Caloglyphus anomalus

These mites are found on onions, fruits, and vegetables where they feed on semifluid materials associated with decomposition. Mating takes place the day after adult emergence. The female starts to lay eggs after a preoviposition period of about 2 days when the temperature is 20°. Oviposition continues for 2–4 weeks. Each female lays from 457 to 1,179 eggs (average,

737). The number of eggs laid per day varies from 1 to 133 (average, 34.5). Optimal conditions for growth occur at temperatures between 20° and 25° with the relative humidity close to 100%.

The life cycle requires 9–10 days at 25°, 14.8 days at 15°, and 6 days at 30°. No hatching occurs at 35°. Mated males live for 45–49 days at 20°; mated females, 31–45 days (58). Pleomorphic as well as homeomorphic males occur, the ratio of one kind to the other apparently being a function of dietary quality (92).

Caloglyphus berlesei

This cosmopolitan species is found occasionally on damp, often moldy commodities: wheat, copra, flaxseed, and peanuts. *C. berlesei* also often invades and damages insect, bacterial, and fungal cultures; infests chicken litter and mushroom beds (37, 39); and is a major pest of stored foods in Japan (72).

The mites begin mating as soon as they become adults; they mate repeatedly thereafter. Females begin to oviposit about 36 hours after copulation. The oviposition period varies from 3 to 26 days (average, 12.5 days); the postoviposition period varies from 9 to 32 days (average, 9.1 days). Egg production ranges from 60 to 1,174 per female (average, 588). The mean number of eggs laid per female per day is 26.7. Occasional males in populous cultures have massive legs III.

Average longevity of reproductive females is significantly shorter than that of virgin females as well as mated males. The time from oviposition to adult emergence averages 8 days at 27° and 80% rh for mites reared on wheat germ or on *Neurospora crassa*. The optimal temperature range for *C. berlesei* is 22–30° (minimum, 16.5°). The sex ratio is close to 1:1. The survival rate for the immature stages stands at about 70.5%, but these mites cannot survive even one-hour exposure to a temperature of 18° (70).

C. berlesei and *Tyrophagus putrescentiae* have very similar nutritional requirements; the essential amino acid requirements for the two species are identical (63). Malonic acid, cholesteryl chloride, sodium propionate, potassium sorbate, and 3,3'-thiodipropionic acid inhibit various metabolic or developmental processes (65).

The net reproductive rate (R_o) of *C. berlesei* reared on a special diet is relatively high—192.45 (62). Theoretically, then, a population can increase by a factor of 192.45 during the generation time of 14 days. The intrinsic rate of natural increase is also high ($r_m = 0.3755$) because of the high survival and fecundity rates of young females. On the basis of a finite rate of natural increase ($\lambda = 1.456$), a population of *C. berlesei* can multiply itself by 1.456 times every day (66). Mites of this species consume food at higher rates than *Rhizoglyphus echinopus* or *Tyrophagus putrescentiae*. Adult *C. berlesei* assimilate about 43% of ingested food. Cumulative egg production is about 2.7 times greater than cumulative body mass production (84).

Nigrospora sphaerica and *Scopulariopsis brevicaulis* (Saccardo) Bainier readily support populations of *C. berlesei* (69). Mean weight per mated female and fecundity are reported to be lower on fungi (*Aspergillus, Cladosporium, Fusarium, Penicillium*) than on malt extract agar (28). Even so, populations of *C. berlesei* increased 30 times in 21 days on *Aspergillus wentii* Wehmer and *Penicillium viridicatum* Thom (=*P. verrucosum* Dierckx var. *verrucosum* Samson). This suggests that fungivory serves a more significant function than merely as a survival mechanism in the absence of other food.

Rhizoglyphus echinopus, bulb mite

Mites of the genus *Rhizoglyphus* occur widely throughout the world in cultivated soils, decaying plant and animal matter, on diseased roots, and in mushroom beds. Bulbs and onions are attacked under field as well as storage conditions. The mites feed between the scales, causing damage and introducing pathogens. *R. echinopus* infests mainly onion, gladiolus, and hyacinth bulbs, often appearing in large populations. *R. robini* also infests these same kinds of bulbs, as well as those of *Freesia* and *Narcissus*, but only in relatively smaller numbers.

The development of *R. echinopus* can proceed at temperatures between 3° and 31° if the relative humidity is greater than 86%. The life cycle, excluding the hypopal stage, lasts 8–14 days at 30°, 38–58 days at 10°, and 123 days at 5°. A 10° increase in temperature shortens development by at least half (9, 10; Figure 6.3). Temperatures between 20° and 30° and humidities greater than 85% appear to be optimum for growth and development. Mortality increases at temperatures above and below the optimum (Table 6.2). Increase in temperature and humidity shortens longevity; at temperatures of 20–30°, the mites live an average of 35 days; at 5°, 90 days or longer. Temperatures above and humidities below the optimum also decrease longevity.

When reared on yeast or wheat germ under optimal conditions, females of *R. echinopus* lay

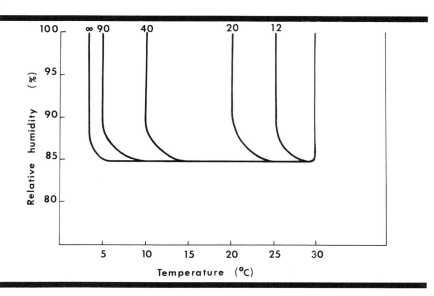

Figure 6.3
Duration of time (days) needed to produce one generation of *Rhizoglyphus echinopus* on rye germ at various temperatures and relative humidities.
SOURCES: Bielska, I. 1975. Life cycle and ecology of *Rhizoglyphus echinopus* (F. & R.) and *R. robini* Clap. (Acarina: Acaridae). Dissertation, Agricultural University of Warsaw (in Polish); Bielska, I. 1975. Demographic parameters of *Rhizoglyphus echinopus* (F. & R.) and *Rhizoglyphus robini* C. depending on temperature, humidity and food. *Zeszyty Problemowe Postępów Nauk Rolniczych* 171:179–188 (in Polish, English summary).

400–700 eggs. Egg production drops off markedly at lower temperatures (*9, 10*; Figures 6.4 through 6.6). Females produce about 100 eggs when reared on onions, bulbs, and corms (Figure 6.7). Most eggs are laid by females between 8 and 20 days old. Each female lays 2–26 eggs/day throughout her life.

Bulb mites reared on wheat germ at 94% rh and on yeast at 87% rh generated the highest net reproductive rate (Table 6.2). This rate is much lower on media consisting of onions and bulbs of ornamental plants (16.45 on *Crocus* bulbs; 66.71 on hyacinth bulbs) (*10*).

Compared to *R. echinopus*, *R. robini* has a

slower developmental time for one generation (11–21 days at 30°, 40–50 days at 10°, and 83–96 days at 5°). R_o ranges from 0.027 to 76.66. Each female lays about 252 eggs. The mites live about 25 days at 20–30° and about 4 months at 5° (*9*).

Under natural conditions, *R. echinopus* and *R. robini* feed on fungi growing on stored products. Numerous kinds of fungi can serve as food for *R. echinopus* (*26*). Fecundity is highest (358 eggs per female) on *Aspergillus repens*, *Botrytis cinerea* Persoon *ex* Fries (342 eggs), and *Trichothecium roseum* (350 eggs). Only single eggs were laid on *Aspergillus chevalieri* (Mangin) Thom & Church and *Cladosporium herbarum* Link *ex* Fries. Fecundity, developmental time, and reproductive rate are comparable on fungi, yeast, or wheat germ.

Bulb mites can to develop and reproduce on many kinds of medicinal herbs. Development proceeds fairly well on wheat germ mixed with several volatile oils, alkaloids, and glycosides. Only a few active extracts from medicinal herbs inhibit development and egg laying. *R. echinopus* also feeds on plant parasitic nematodes (*85*).

Motile hypopi appear mainly in the spring (*46*). They are often found in high-density populations; temperature and humidity apparently do not influence their appearance. The hypopi are carried by flies (*Scatopsis* and *Eumerus* spp.) and by other small insects that occupy similar habitats.

Figure 6.4
Fecundity (average number of eggs/female) of *Rhizoglyphus echinopus* and *R. robini* at 10° and 100% relative humidity when reared on yeast or rye germ.
SOURCES: Bielska, I. 1975. Life cycle and ecology of *Rhizoglyphus echinopus* (F. & R.) and *R. robini* Clap. (Acarina: Acaridae). Dissertation, Agricultural University of Warsaw (in Polish); Bielska, I. 1975. Demographic parameters of *Rhizoglyphus echinopus* (F. & R.) and *Rhizoglyphus robini* C. depending on temperature, humidity and food. *Zeszyty Problemowe Postępów Nauk Rolniczych* 171:179–188 (in Polish, English summary).

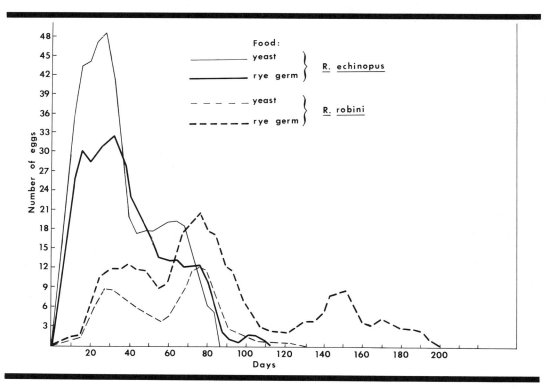

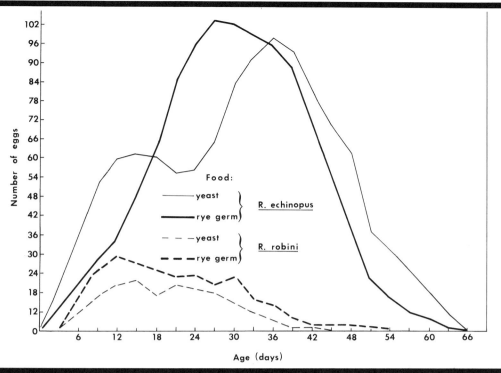

Figure 6.5
Fecundity (average number of eggs/female) of *Rhizoglyphus echinopus* and *R. robini* at 20° and 90% relative humidity when reared on yeast or rye germ.
SOURCES: Bielska, I. 1975. Life cycle and ecology of *Rhizoglyphus echinopus* (F. & R.) and *R. robini* Clap. (Acarina: Acaridae). Dissertation, Agricultural University of Warsaw (in Polish); Bielska, I. 1975. Demographic parameters of *Rhizoglyphus echinopus* (F. & R.) and *Rhizoglyphus robini* C. depending on temperature, humidity and food. *Zeszyty Problemowe Postępów Nauk Rolniczych* 171:179–188 (in Polish, English summary).

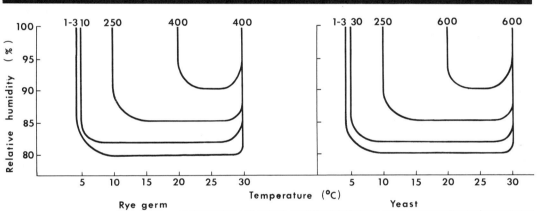

Figure 6.6
Fecundity (average number of eggs/female) of *Rhizoglyphus echinopus* at various temperatures and relative humidities.
SOURCES: Bielska, I. 1975. Life cycle and ecology of *Rhizoglyphus echinopus* (F. & R.) and *R. robini* Clap. (Acarina: Acaridae). Dissertation, Agricultural University of Warsaw (in Polish); Bielska, I. 1975. Demographic parameters of *Rhizoglyphus echinopus* (F. & R.) and *Rhizoglyphus robini* C. depending on temperature, humidity and food. *Zeszyty Problemowe Postępów Nauk Rolniczych* 171:179–188 (in Polish, English summary).

Table 6.2
Effects of temperature and relative humidity (rh) on *Rhizoglyphus echinopus*

Temp. (°C)	5	5	5	5	10	10	10	10	25	25	25	25	30	30	30
rh (%)	81	86	90	100	81	86	90	100	80	86	90	100	80	90	100
Fecundity (eggs/female)															
Yeast	—	1	9	15	1	105	232	250	2	151	311	247	1	360	365
Wheat germ	—	1	28	27	1	144	229	276	3	187	634	372	2	604	661
Developmental time (days)															
Yeast	—	114	87	80	—	55	40	39	—	26	15	14	—	13	8
Wheat germ	—	124	88	88	—	59	42	39	—	21	17	14	—	10	8
Mortality (%)															
Yeast	—	77	75	55	—	71	69	56	—	47	26	27	—	31	29
Wheat germ	—	84	80	50	—	78	72	58	—	43	47	50	—	38	30
Longevity (days)															
Yeast	34	28	108	108	40	48	48	55	22	39	38	24	19	39	34
Wheat germ	21	32	111	76	30	48	58	68	21	27	39	23	19	34	30
R_o															
Yeast	—	0.001	0.6	1.7	—	6	13	25	—	74	318	209	—	—	320
Wheat germ	—	0.001	0.2	0.5	—	4	10	18	—	52	122	110	—	—	149
r_m															
Yeast	—	−0.051	—	0.003	—	0.05	—	0.05	—	—	0.16	0.19	—	—	0.22
Wheat germ	—	−0.04	—	0.004	—	0.03	—	0.05	—	—	0.12	0.18	—	—	0.22

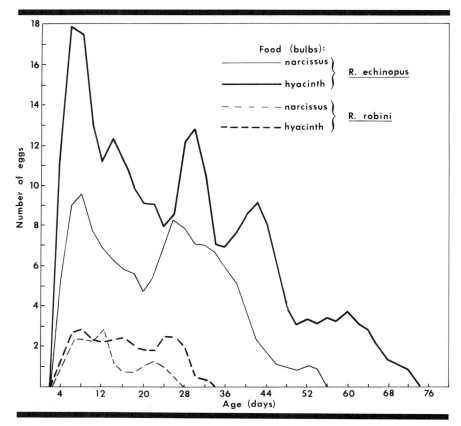

Figure 6.7
Fecundity (average number of eggs/female) of *Rhizoglyphus echinopus* and *R. robini* at 20° and 95% relative humidity when reared on narcissus and hyacinth bulbs.

SOURCES: Bielska, I. 1975. Life cycle and ecology of *Rhizoglyphus echinopus* (F. & R.) and *R. robini* Clap. (Acarina: Acaridae). Dissertation, Agricultural University of Warsaw (in Polish); Bielska, I. 1975. Demographic parameters of *Rhizoglyphus echinopus* (F. & R.) and *Rhizoglyphus robini* C. depending on temperature, humidity and food. *Zeszyty Problemowe Postępów Nauk Rolniczych* 171:179–188 (in Polish, English summary).

Females usually outnumber males in most populations. The sex ratio, 1.5:1 to 1.9:1 on wheat germ and 1.3:1 to 2.1:1 on yeast, depends neither on temperature nor humidity (9). Bulb mites are generally efficient in the assimilation of food (82, 84) (Figure 6.8). In small colonies, occasionally single heteromorphic (andropolymorphic) males appear; they do not live as long as homeomorphic males nor do they compete with them (93).

Thyreophagus entomophagus

Flour, cereal products, medicinal herbs, animal feeds, spices, and insect collections are commonly infested by *T. entomophagus*. The mites mate a few hours after becoming adults. When the temperature is 25° and the humidity is 87%, the first eggs are laid on the fourth day after mating. An average of 76 eggs (maximum, 170) per female are laid when wheat germ is the substrate. The oviposition period lasts 32 days. Female longevity is 46 days; male, 41 days (20). Under optimal conditions the life cycle takes 15.5 days; at 75% rh, about 30 days. The mortality rate for the immature stages runs about 30%. The male to female ratio is 48.8 to 51.2.

Growth and development can take place at temperatures between 3° and 32° if the relative humidity is between 75 and 100%. Tempera-

tures between 15° and 25°, with relative humidity of 85%, are optimum. If the relative humidity is adequate, the mites can live 2 years at 3–5° or 9 months at 6–10°. Egg production is high on rye germ (156 eggs per female) and bakers' yeast (107 eggs per female) at 25° and 95% rh. Starving mites survive 69 days at 25° and 85% rh; 168 days at 15°; 235 days at 7–10°; and 318 days at 0–3°. Most eggs kept for 100 hours at −20° are likely to hatch. Many of the mites in food eaten by birds and mice survive passage through the digestive tract (18). There are no hypopi in this species.

Tyrolichus casei, cheese mite

T. casei is a cosmopolitan but relatively uncommon fungivore. Numerous kinds of fungi associated with stored grains can support populations of this mite (71). In Iceland, this species makes up 2.8% of the mite fauna in farm-stored barley and 35% of the mite populations in flour mills (25). Other sources of cheese mites include stored foods, cheese, grain, damp flour, old honeycombs, dog meal, dead tree bark, rodent nests, and insect collections. The life cycle is 15–18 days at 23° and 87% rh (37).

Tyrophagus longior

T. longior is a cosmopolitan species, probably mycophagous, associated with a variety of habitats. As a rule, it prefers substrates with a high-moisture content but not patently decomposed or moldy. In the field it is associated with haystacks and strawstacks and is usually the most common of the mites found in those habitats. *T. longior* is an important pest of stockfish in Iceland. Also in Iceland, it made up 55.4% of the mites collected from residues in bakery trucks, 41.1% from farm-stored oats, 14.2% from farm-stored milo (*Sorghum bicolor*), 12.3% from farm-stored barley, 8.8% from whole grains in commercial granaries, 5.2% from debris in farm buildings, and 3.3% from debris in railcars (25). This species may also infest poultry litter, mushroom compost, cheese, beets, tomatoes, spinach, seeds of florist's cyclamen, sugar beet seeds, and bird nests (37). Its life cycle lasts 2–3 weeks at 23° and 87% rh.

Tyrophagus putrescentiae, mold mite

The mold mite is a cosmopolitan, synanthropic species infesting many kinds of stored products—grain, peanuts, medicinal herbs, cheese, nuts, copra, dried eggs, cottonseed, rapeseed, sunflower seed, dried bananas, wheat spillage, tobacco, and flour. This is the species most frequently found in the flour of bagging-off spouts in mills. *T. putrescentiae* is often present in mushroom houses (*40*). Mold mites are only rarely recovered from field habitats such as rodent and bird nests. They are often found in stored foods with a high content of fat and protein. Mold mites often infest laboratory cultures of insects and microorganisms. These mites are common stored-product and household pests in Japan, sometimes reaching prodigious numbers in the rice straw mats on the floors of houses (*36, 68*). The mites also are commonly found in house dust, and they are known to cause allergic reactions.

Adults mate soon after eclosion; starved females can mate but cannot lay eggs. A steady, adequate diet leads to vigorous sexual activity on the part of the female mites. Mating usually occurs after a chance meeting of a male and a female. Females are attracted by tactile and olfactory stimuli; males respond only to olfactory cues and recognize a female only when she is very close. Copulation lasts an average of 1.5 hours. If copulation is interrupted any time during the first 45 minutes, the female fails to lay eggs. Copulatory activity alone does not stimulate oviposition (*13*).

Males and females of *T. putrescentiae* often mate repeatedly. One male can fertilize as many as 450 females. Females kept on yeast or wheat germ at 20° and 85% rh lay their first eggs within 24 hours after mating. Such females are able to lay eggs throughout their life but must mate no less than monthly to maintain fertility. A female that mates at least twice a month often produces about 500 eggs. A female mold mite is capable of laying 60 eggs/day, but the average is 4 per day. About 70% of the eggs are produced during the first 3 weeks of life.

Egg laying can occur at temperatures above 8°; optimal temperatures are between 22° and 26° (*31*) (Figure 6.9). Oviposition stops at relative humidities below 62% or about 13.5 to 14% mc in the food medium. Mold mites can develop normally at 9° and 75% rh or at 25° and 85% rh (*3*).

Rapeseed with a moisture content of 8% supports the mites but without population growth; at 10% mc or higher, however, the population explodes (*30*). Females reared on wheat germ or yeast produce the most eggs, about 500 per mite; on dried milk the production is about 85 eggs per mite; on dried plums, 12; on rolled oats, 8. Females of an Asian strain produce about 184 eggs (13 eggs/day) when reared on cottonseed (*1*).

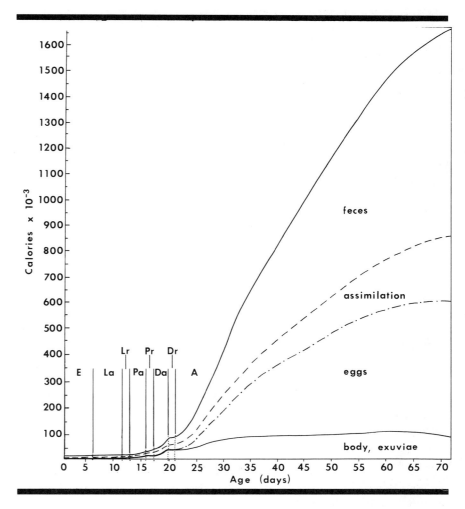

Figure 6.8
Elements of the cumulative energy budget for *Rhizoglyphus echinopus*. E, egg; La, active larva; Lr, resting larva; Pa, active protonymph; Pr, resting protonymph; Da, active deutonymph; Dr, resting deutonymph; A, adult.

Development cannot proceed at temperatures below 8.5° or above 36° (*23*) (Figure 6.10). The life cycle can be completed at 65% rh but only if the temperature is kept between 15° and 25°. Higher relative humidities are required if the temperature drops below 15° or rises above 25°; for example, 80% rh is required at 10° and 35°. The temperature range that allows complete development in an Asian strain is 5.1° to 35°; 60% is the minimum level of relative humidity (*1*). Mold mites are very sensitive to changes in relative humidity in the range of 22–78% with optimal development at the high end of that range (*97*).

Eggs of *T. putrescentiae* are killed in less than one day when exposed to a temperature of −15°, but eggs survive 21 days at −10°, and 24 days at −5° (*88*). Developmental time for one generation ranges from 9.5 days under optimal conditions (25°, 85% rh, with yeast as food) to 130 days at 8.5°. At 20° and 67% rh, one generation can be produced in 75 days (*31*). The duration of developmental time also varies in response to the kind of food available. On wheat germ at 20° and 85% rh, developmental time is 10

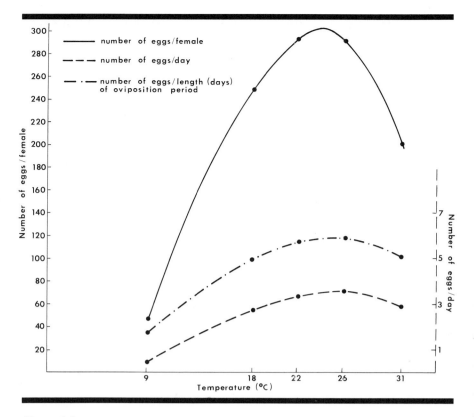

Figure 6.9

Effect of temperature on fecundity of *Tyrophagus putrescentiae*.
SOURCE: Gołębiowska, Z. 1963. *Tyrophagus putrescentiae* (Schrank, 1781) (= *T. noxius* Zachwatkin 1935), morphology, biology and ecology. *Prace Naukowe Instytutu Ochrony Roślin* 5(2)29-88 (in Polish, English summary).

days, with a mortality rate of 12%; on cheese, 21 days (32% mortality); on fish meal, 24 days (24% mortality); and on powdered milk, 18 days (67% mortality). About 35% of the time between hatching and the emergence of the adults is spent in an immobile stage.

Longevity of adult mites depends on temperature, relative humidity, and food. At 25° and 85% rh, females reared on wheat germ live about 66 days; males, about 59 days. Longevity decreases with increasing temperature: 115 days at 9.3° and 42.8 days at 31°. Longevity increases as the relative humidity increases from 75 to 85%: 53–80 days at 26°. At relative humidities higher than 85%, longevity decreases. Longevity is greatest on wheat germ (99 days, females; 120 days, males); on pumpkin seeds it is 80 days; on powdered milk, 75 days; and on rolled oats, 25 days (*31*).

Medicinal herbs are often heavily infested by *T. putrescentiae*; most of those tested have proved to be suitable foods for these mites. Mold mites respond favorably to many volatile oils and alkaloids and to robinin and rutin when these substances are mixed with their food (rye germ). Colchicine, caffeine, quinine, digitoxin, and some volatile oils inhibit development (*27*). Some drugs are acaricidal (*47*). Ten amino acids are essential (*64*). Two kinds of chemicals [methyl ketones and 3-methyl-1-butanol (isoamyl alcohol)] in dairy products exert a synergistic attraction

for mold mites; either compound alone is much less attractive (*95*). Under optimal conditions, an individual mold mite can consume the equivalent of its body weight (6–8 μg) in food each day (*98*). This food is used efficiently, most of the energy going into egg production in the case of the females (*83*).

T. putrescentiae can feed on insect eggs (*7, 86*), but it is essentially a fungivore. Numerous kinds of fungi can support normal growth and development (*77*): *Polyporus* spp., *Poria* spp., *Stereum* spp., *Flammula* spp., *Alternaria tenuis, Aspergillus versicolor* (Vuillemin) Tiraboschi, *A. candidus, A. amstelodami, Fusarium moniliforme, Helminthosporium sativum* Pammel et al., *Mucor racemosus* f. *sphaerosporus* (Hagern) Schipper, *Nigrospora sphaerica, Stemphylium botryosum* Wallroth, *Trichothecium roseum, Botrytis cinerea, Penicillium purpureum* Stolk & Samson, and *P. chrysogenum.* With *Fusarium culmorum* (W. G. Smith) Saccardo as the food medium, egg production is minimal. Developmental rates are retarded when the mites are restricted to diets of *Aspergillus fumigatus* Fresenius, *Penicillium verrucosum* Dierckx var. *cyclopium* Westling, and *Mucor racemosus*. Mold mites cannot feed or reproduce on *Chaetomium* species or *Streptomyces griseus* (Krainsky) Waksman & Henrici (*26, 74*).

T. putrescentiae appears generally unaffected by various antibiotics and antioxidants added to its diet (*14*). However, sodium propionate, 3,3′-thiodipropionic acid, malonic acid, and cholesteryl chloride all appear to retard development (*14, 65*). Larvae and nymphs of mold mites succumb relatively quickly to starvation, but the adults are able to tolerate prolonged starvation. At 24° and 100% rh, they can live without food for at least 2 months; at lower temperatures (10–12°), up to 5 months. Hypopi do not occur in the genus *Tyrophagus*.

Females predominate in populations of *T. putrescentiae*. The sex ratio is not dependent on food, parental age, or frequency of copulation. Neither single nor polygamous matings influence the sex ratio (*13*). Fecundity is dependent on the number of males per female. Females lay fewer eggs if they are courted by numerous males. Fecundity is reduced by 50% when the male:female ratio is 4:1 (*16*).

The age structure of mold mite populations is influenced by environmental parameters. With optimal food, temperature, and relative humidity, adults make up about 10% of the population, eggs about 50%, immobile larvae and nymphs about 10%, and mobile immatures about 20%.

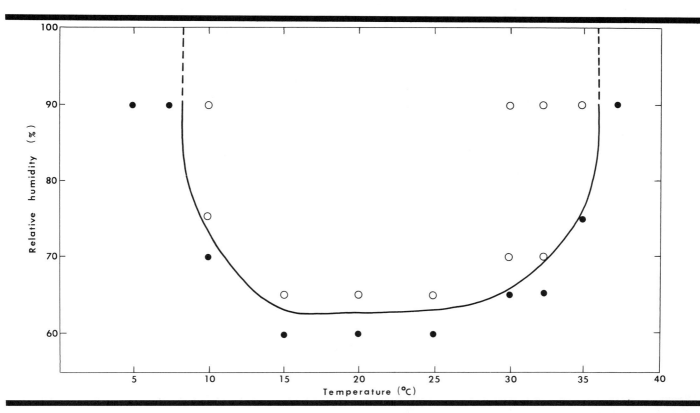

The percentages of eggs and mobile immatures decrease under less-than-optimal conditions. Demographic parameters for *T. putrescentiae* are rather high. R_o for mites reared on rye germ at 25° and 85% rh is 99.03 (*17*); on yeast at 30° and 98–100% rh, 204 (Table 6.3). The population may multiply 204 times during a single generation time of 20 days (*3*). Population growth is much slower on dried fruits ($R_o = 2–32$) and fungi ($R_o \leq 71$). Neryl formate, an alarm pheromone produced by mold mites (*45*), is somewhat fungistatic and is moderately repellent to four other species of mites (*Dermatophagoides farinae*, *Carpoglyphus lactis*, *Lardoglyphus konoi*, and *Aleuroglyphus ovatus*).

Ascidae

Blattisocius dentriticus

This predatory species is often associated with acarid mites. It has been recorded from wheat,

Figure 6.10
Relative humidity and temperature limits for complete development of *Tyrophagus putrescentiae*.
SOURCE: Cunnington, A. M. 1969. Physical limits for complete development of the copra mite, *Tyrophagus putrescentiae* (Schrank) (Acarina, Acaridae). In *Proceedings of the Second International Congress of Acarology* (Sutton Bonington, England, 1967), pp. 241–248.

Table 6.3

Effects of temperature and relative humidity on *Tyrophagus putrescentiae*

Temperature (°C)	10			15–18			20–22			25			30		
Relative humidity (%)	70	80	95	70	80	95	70	80	95	70	80	95	70	80	95
Fecundity (eggs/female)															
Wheat germ	2	88	33	43	234	260	76	310	327	36	317	306	42	102	383
Developmental time for one generation (days)															
Wheat germ	—	130	76	—	75	—	50	32	26	—	—	16	—	18	47
Cottonseed	55	53	76	—	37	38	23	21	17	—	20	14	14	12	11
Mortality (%)															
Wheat germ	—	—	98	—	—	90	—	—	48	—	—	38	—	—	57
Longevity (days)															
Wheat germ	62	125	159	80	85	81	75	64	52	53	80	53	53	39	36
Powdered milk	—	—	—	—	—	—	—	75	—	—	—	—	—	—	—
Rolled oats	—	—	—	—	—	—	—	25	—	—	—	—	—	—	—
R_o															
Yeast	—	—	—	—	—	—	—	—	119	—	—	113	—	—	204
r_m															
Yeast	—	—	—	—	—	—	—	—	0.15	—	—	0.2	—	—	0.26
λ															
Yeast	—	—	—	—	—	—	—	—	0.17	—	—	1.22	—	—	1.30
Dried plums	—	—	—	—	—	—	—	—	—	—	1.1	—	—	—	—

flour, moldy coconuts, sprouting potatoes, and dry, medicinal herbs. Females of *B. dentriticus* fed on larval *Tyrophagus putrescentiae* usually start to lay eggs on the third day after mating. Survival and rate of development of the immature stages increase slightly with increasing humidity. Average time of development from egg to adult at 20° takes 11.8 days at 90–100% rh and 12.6 days at 70–80% rh. Adult females survive well at various humidities. The oviposition period is much shorter at 70% rh than at higher levels. Approximately the same rate of egg laying (about 36 eggs per female) occurs at 70, 80, and 90% rh. More eggs (e.g., 44) are laid at 100% rh.

On the average, each of these predators attacks about six larval *T. putrescentiae* per day (about 55 prey per predator). Consumption of prey increases with increasing density of prey. More eggs are laid when prey consumption increases, and mites develop faster when prey is available in large numbers. It is probable that prey populations may be more effectively suppressed at humidities lower than 70% (*60, 61*).

Blattisocius keegani

B. keegani, a cosmopolitan, predatory species, is found in stored foods, nests of birds and rodents, insect cultures, grains and grain products, medicinal herbs, seeds of various plants, and on citrus trees in Florida (*37*). This species often appears in Canadian grain bulks at locations where the grain is warm and dry.

Eggs are laid singly or in groups of two or three. At 27° and 70–75% rh, each female lays a maximum of 5 eggs/day. The egg-to-adult time interval is 6 days at 27° and 70–75% rh, 14 days at 22° and 70–75% rh, and 8.7 days at 27° and 95–100% rh. Immature stages survive better at 70–75% rh than at 95–100%. Mating takes place as soon as the adults appear. The first eggs are usually laid the next day. Total egg production varies from 11 to 61 eggs during 14 days of life at 27° and 70–75% rh. The sex ratio is close to 1:1 (*5*).

Adults, protonymphs, deutonymphs, and larvae prey on eggs of stored-product insects of the genera *Cryptolestes, Tribolium, Trogoderma,* and *Oryzaephilus,* and on eggs or mobile stages of some stored-product mites, especially members of the genera *Glycyphagus* and *Acarus*. The predatory stages can live 2–7 days without food, but starving mites become cannibalistic. At 27° and 70–75% rh, each mite requires a daily average of three eggs of *Cryptolestes turcicus* to reach adulthood. Females that consume three such eggs per day produce a maximum of three eggs/day. Daily consumption of a single prey egg ensures growth, maturation, and oviposition.

Populations of *B. keegani* can increase 12.4 times per generation time of 15 days. The innate capacity for increase (r_m) is 1.172 per day; the finite rate of increase (λ) is 1.19 female offspring per female per day. Under very special circumstances, these mites can be used to control other pests of stored foods (*5*). After observing populations of *B. keegani* for 8 years, Sinha et al. (*76*) concluded that although this predator is dependent for its own survival upon prey populations (*Glycyphagus* sp. and *Lepinotus* sp.), it does not effectively control the prey populations.

Blattisocius tarsalis

This predatory species is frequently found in stored grain, medicinal herbs, various seeds, and animal feeds infested with insect pests. The larvae and nymphs feed on the eggs and larvae of insects, notably food-infesting pyralid moths. The female mites are often carried about by insects, and lay their eggs on the silk webbing covering caterpillars. *B. tarsalis* is capable of controlling some insect pests in the surface layer of stored products (*29*). The life cycle (egg to adult) lasts about 10 days at 27°. Adults live about 10 days. Normal bisexual reproduction follows feeding by these mites on the eggs of *Ephestia (Anagasta) kuehniella.*

Carpoglyphidae

Carpoglyphus lactis, driedfruit mite

Driedfruit mites are found in a wide variety of foods—milk products, dried fruits, honey, beer, wine, and animal feeds. They also infest medicinal herbs and beehives. The mites are generally synanthropic but hypopi have been found in nests of birds, rodents, and ants. *C. lactis* is a serious pest of miso in Japan.

Mating begins a few hours after the adults emerge and lasts 20–25 minutes; it is repeated several times during the lifespan. Under optimal conditions (15–25°, 80–85% rh), the first eggs are laid on the third day after mating. Most of the eggs are laid on the seventh to tenth days of life, with the oviposition period usually lasting 14 days. Fecundity varies greatly: 278 eggs per female on the average (maximum, 444). Daily production runs from 20 to 56 eggs per female. Since the eggs are in various stages of embryonic development at the moment of oviposition, the incubation period varies from 2–4 days. Developmental time for one generation under optimal conditions is about 9–11 days. The lower threshold temperature for development is 3°; the upper, 35°. These mites cannot survive in humidity lower than 60%.

Yeast appears to be the best food for *C. lactis*. On this medium, fecundity is high (278 eggs per female) and mortality during development is low (24%) (Table 6.4). On dried plums the fecundity is 118 eggs and the mortality rate is 38%. The mites live 20 days and oviposit for 14 days on yeast; on dried plums they live 65 days and lay eggs for 33 days (*19*). Cultures also do well on a mixture of two parts sugar (glucose or sucrose) and three parts yeast (*49, 50*).

In mixed cultures of *C. lactis* and *T. putrescentiae*, the population density of *C. lactis* is always greater when the food medium is composed of 0–60% yeast. *T. putrescentiae* becomes dominant only on a medium of pure dried yeast. Ratios of the two species approach 1:1 when the substrate consists of 20% sugar and 80% yeast.

Ten out of 52 medicinal herbs have proved suitable as growth media for *C. lactis*, but mortality rates on such media are high. Glycosides, volatile oils, and alkaloids added to the diet of rye germ inhibited development to varying degrees (*27*).

Fungi play a minor role as food for the dried-fruit mite. Development is possible on *Penicillium divaricata* Thom (= *Paecilomyces variotii* Bainier), *Penicillium chrysogenum*, *Mucor silvaticus* Hagern, *Nigrospora sphaerica*, and *Trichoderma viride* Persoon *ex* Fresenius (*26*).

Adult longevity is greatly influenced by temperature, humidity, and food. When the relative humidity is 85%, the mites live an average of 192 days (maximum, 630 days) at 0–3°; 25 days at 15°; 20 days at 25°; and 11 days at 35°. When the temperature is 25°, the mites live for 15 days at 60% rh, 19 days at 75% rh (Table 6.4), and 29 days at 95% rh. Females usually live longer than males. At 0° and 85% rh, adult driedfruit mites can live without food for 97–350 days; at 20°, 10–57 days; and at 35°, 3–7 days. At 25° and 50% rh, the mites live 3–12 days; at 80% rh, 3–92 days; and at 100% rh, 8–30 days. The immature stages are much less tolerant of starvation than the adults.

The adult mites are the most active stage in the process of colonizing new substrates. The larvae and nymphs rarely move away from a good source of food. The adults adapt more readily to a wider variety of foods than do the immature stages (*99*). Motile hypopi appear in some populations. This stage is transported and distributed by insects and other animals. Hypopi sometimes remain attached to animals for as long as 3 months. Under suitable conditions, they become tritonymphs.

Hypopi are most numerous when the mites are reared on wheat germ at 25° and 70–85% rh. At humidities below the developmental threshold, no hypopi appear before the whole population dies out. Hypopi appear often when the colony feeds on yeast that has accumulated a layer of mite fecal matter. When hypopi are transferred to subtropical conditions, they transform into tritonymphs.

Hypopal forms die quickly when exposed to low humidity (<50%) or low temperature (<5°); they are only slightly more resistant to these adverse conditions than are the other developmental stages. Food is not essential for the transformation of the hypopus into the next stage, but the tritonymphs must feed before they can become adults. More than 70% of mites having a hypopal stage in their life cycles originate from parents that also passed through the hypopal stage.

Table 6.4

Effects of temperature and relative humidity on *Carpoglyphus lactis*

Temperature (°C)	5	15	25	25	25	30
Relative humidity (%)	85	85	60	75	85	85
Fecundity (eggs/female)						
Yeast	105	254	57	133	278	76
Dried plums	—	—	—	—	118	—
Developmental time for one generation (days)						
Yeast	77	14	11	9	9	9
Dried plums	—	—	—	—	11	—
Mortality (%)						
Yeast	64	18	72	65	24	30
Dried plums	—	—	—	—	38	—
Longevity (days)						
Yeast	66	25	15	19	20	11
Dried plums	—	—	—	—	65	—
R_0						
Yeast	20.4	114.19	6.16	26.24	118.06	27.72
Dried plums	—	—	—	—	35.35	—
r_m						
Yeast	0.029	0.187	0.089	0.199	0.29	0.231
Dried plums	—	—	—	—	0.117	—
Duration of starvation tolerance (days)	86	36	15	30	29	12
Hypopus occurrence (%)						
Yeast	0	11	11	27	25.8	13
Plum jam	—	—	—	—	6.3	—
Yeast + plum jam	—	—	—	—	50	—
Dried plums + figs	—	—	—	—	0.8	—

SOURCE: Chmielewski, W. 1971. Morphology, biology and ecology of *Carpoglyphus lactis* (L., 1758) (Glycyphagidae, Acarina). *Prace Naukowe Instytutu Ochrony Roślin* 13(2)63–166 (in Polish, English summary).

The demographic parameters (Table 6.4) of *C. lactis* are strongly dependent on environmental factors. The highest net reproductive rate is seen when the mites are reared on yeast at 15–25° and 85% rh. Under these conditions, the population can multiply 118 times over a period of 16 days. At 5° and 85% rh, $R_o = 20.4$; at 35° and 85% rh, $R_o = 1.92$; at 25° and 60% rh, $R_o = 6.16$; at 25° and 100% rh, $R_o = 48.03$. Under optimal conditions, the net reproductive rate is much lower on dried plums ($R_o = 35.35$) than on yeast ($R_o = 118.06$) (*19*).

Cheyletidae

Cheyletus eruditus

C. eruditus is a cosmopolitan, predatory species found commonly in grain stores, farm detritus, and in house and mattress dust. It is also a regular inhabitant of bird and mammal nests where it feeds on the scavenging mites that live there. This species is regularly associated with stored products infested with mites.

Growth and development occur at temperatures between 8° and 31° and at humidities of 52% and higher. Optimal conditions are about 20° and 80% rh. Feeding activity stops at 6°. Usually, only the parthenogenetic females are found (males are rare); 35 consecutive generations have been produced parthenogenetically without negative effect on fecundity or body size (*12*). The eggs, laid in heaps, are guarded by the female. In spite of this maternal solicitude, eggs, larvae, and nymphs may be eaten by a hungry female if she is confined without food.

Egg laying may begin one day after the female emerges. Five to 13 eggs are laid per day, with most eggs being laid during the first 4 weeks of life. Average fecundity of this species is 71 eggs laid by one female (maximum, 137). Usually, two or three clutches of eggs are formed (65 eggs have been found in one clutch) (*12*).

When the relative humidity is held at 85%, one generation takes 84 days at 9.4°, 17 days at 20.2°, and 10 days at 31.5°; mortality rates are 70, 4.8, and 47.3%, respectively, at these temperatures. When the humidity is held at 70%, mortality is 42% at 20.2° and 100% at 31.5° (*12*) (Table 6.5). All stages soon die out at temperatures less than −2° or greater than 40°. This explains why winter survival is poor in temperate climates. The pest mites *Acarus siro* and *Lepidoglyphus destructor* are more tolerant of winter temperatures than is *C. eruditus*, but if the moisture content of the food drops to about 11 or 12%, the pest species are more likely to be attacked by *C. eruditus*.

The developmental time of one generation is shortened by 0.5 day when the relative humidity is increased by 1% at 13.7°; by 3.25 days when the temperature is increased by 1° at 70% rh. Longevity ranges from one to 2 months. The mites feed only on living animal food, preferably stored-product mites, but they also accept early-instar insect larvae. Cannibalism also occurs. Fecundity of the predators fed on psocids or young caterpillars is very low, and the rate of mortality during development is very high.

Larvae and nymphs of *C. eruditus* feed mostly on eggs, larvae, and nymphs of acarid mites;

Table 6.5

Effects of temperature and relative humidity on *Cheyletus eruditus*

Temperature (°C)	10.6			17.1		20.2			26.8			31.5		
Relative humidity (%)	70	85	100	70	85	70	85	100	70	85	100	70	85	100
Fecundity (eggs/female) on prey of														
Acarus siro	—	—	—	—	—	—	107	—	—	—	—	—	—	—
Lepidoglyphus destructor	—	—	—	—	—	—	89	—	—	—	—	—	—	—
Psocids	—	—	—	—	—	—	28	—	—	—	—	—	—	—
Developmental time for one generation (days) on prey of *A. siro*	82	79	—	27	23	24	17	15	16	11	10	—	10	—
Mortality (%) on prey of *A. siro*	19	40	—	35	7	42	5	8	61	14	3	100	47	—
Longevity (days) on prey of *A. siro*	—	62	60	—	—	41	38	48	—	—	31	30	29	27
R_0 on prey of *A. siro*	—	—	—	—	—	—	24	—	—	—	—	—	—	—
Duration of starvation tolerance (days)	111	—	89	98	—	69	70	83	48	—	—	51	—	31

SOURCE: Boczek, J. 1961. Biology and ecology of *Cheyletus eruditus* (Schrank) (Acarina: Cheyletidae). *Prace Naukowe Instytutu Ochrony Roślin* 1:175–230 (in Polish, English summary).

Table 6.6
Effects of temperature and relative humidity on *Glycyphagus domesticus*, with yeast + wheat germ as food

Temperature (°C)	11.7	15.6	15.6	18.9	18.9	20.6	20.6	23.3
Relative humidity (%)	95–100	73	98	73	98	73	98	98
Fecundity (eggs/female)	—	—	92	—	86	—	135.4	—
Developmental time for one generation (days)	56	49	36	35	28	32	25	23
Longevity (days)[a]	—	—	26	—	23	—	31	—
R_0	—	—	27.26	—	24.38	—	32.51	—
r_m	—	—	0.066	—	0.081	—	0.097	—

[a] Preoviposition period plus oviposition period.

SOURCE: Barker, P. S. 1968. Bionomics of *Glycyphagus domesticus* (de Geer) (Acarina: Glycyphagidae), a pest of stored grain. *Canadian Journal of Zoology* 46(1)89–92.

adults consume all stages. A female may consume one to three adult acarid mites daily, or in a lifetime, an average of 21 adult mites (maximum, 58). Each predatory larva consumes about 35 eggs and about 40 first- and second-stage acarid nymphs. Neither inert nor motile hypopi are attacked by *C. eruditus*. These predators in the adult stage can survive without food for 7 months at 6.5° and 100% rh, but only for 82 days at 20–27°. Nymphs may live up to 3 months without food.

Populations of prey as well as predators increase with increasing humidity. Population increase is much faster for *A. siro* and *L. destructor* than for *C. eruditus*. Predator populations increase fastest in the presence of a moderate population level of acarid mites (very high or very low prey population densities inhibit population growth of the predator). Generally, populations of *C. eruditus* reach only moderate levels, up to a few thousands of mites per kilogram of product. Reproductive success is strongly dependent on temperature, humidity, and prey population density. The highest numbers of *C. eruditus* are obtained at the predator:prey ratio of 1:25 to 1:100 (*12*).

Estimates of the effect of these predatory mites on acarid populations must take into account the changing age distributions of the two populations. Juvenile predators, which may generate the major pressure that depresses prey population growth, tend to aggregate in regions of high prey activity and are generally much more active than the nest-guarding adults (*8*). According to successful biological control experiments in Czechoslovakia, the predatory population should be placed in the grain bulk in spring or autumn when temperatures are 5–10°; the predator:prey ratio should be 1:100 to 1:1,000 (*59*).

Glycyphagidae

Aeroglyphus robustus, warty grain mite

The geographic distribution of the warty grain mite is, so far, restricted to North America and the Soviet Union. One generation lasts a little more than 3 months, with a mortality rate of 33%, under optimal conditions (30° and 70–75% rh). The lethal temperature for this species is just under 34°. Females lay 8–15 eggs during the ovipositional period of 6–8 days. The mites live 7–11 days. The net reproductive rate under optimal conditions, is 1.74 to 2.5. This suggests that populations of *A. robustus* seldom explode (*4*). Grain held in storage too long seems most suitable as a food medium for *A. robustus* (*75*). Several species of fungi are cropped as food by the warty grain mite (*69*), at least under laboratory conditions. *Penicillium verrucosum* var. *cyclopium* can support small populations of *A. robustus*.

A. robustus has been placed in the family Aeroglyphidae (*56*).

Glycyphagus domesticus, house mite

A cosmopolitan species, *G. domesticus* infests a wide variety of dried plant and animal materials including flour, wheat, cheese, ham, and fish. On occasion it is found in medicinal herbs, animal nests, and beehives (*37*). House mites are often carried about by honeybees, other insects, and birds. They are commonly found in house dust and have been associated with house dust allergy (*55*).

The optimal temperature for postembryonic development in *G. domesticus* is 22° (*6*). The life cycle lasts for 56.5 days at 11.7° and 95–100% rh; 25 days at 20.6° and 98% rh; or 22.5 days at 23.3° and 98% rh (Table 6.6). Oviposition goes on for 20–29 days. At 98% rh, egg production averages 91.5 per female at 15.6° and 135.4 per female at 20.6°.

House mites get along very well in culture on media containing yeast with either wheat germ or rolled oats. Some of the fungi that can support feeding and reproduction of *G. domesticus* are *Nigrospora sphaerica*, *Hormodendrum cladosporioides*, and *Scopulariopsis brevicaulis* (*74*).

Inert hypopi enclosed in the protonymphal cuticle are often found in populations of house mites. As many as 50% of the protonymphs may pass through this stage. The hypopus is very resistant to drying out and may remain in this inert stage for several years. Numerous hypopi are found in dense populations at temperatures of 25–30° and humidities of 65–85%. They become tritonymphs when the temperature drops to the 15–20° range and the relative humidity climbs to 75–95%. Development from hypopus to adult takes about 4–5 weeks (*21*). R_o at 95–100% rh equals 27.26 at 15.6°, 24.38 at 18.9°, and 32.51 at 20.6°. The population may multiply nearly 30 times during one generational period (about 3 weeks) (*6*).

Lepidoglyphus destructor

L. destructor, a cosmopolitan species, is one of the most common stored-product pests. It occurs in agricultural soils of Canada, Japan, and the Soviet Union; in grain stacks, hay, and straw in open fields; in nests of rodents, bumblebees, and ants; in stored grain, cereals, cereal products, flaxseed, sunflower seed, dried fruits, and medicinal herbs; and in mattress stuffing, house dust, chicken litter, and animal feeds.

L. destructor cannot develop at or below 60% rh or in stored products in equilibrium with atmospheric humidities in this range. The temperature range within which *L. destructor* can complete its life cycle is comparable to that of *Acarus siro*. Threshold temperatures are about 3° and 34°. Under optimal conditions (25° and 90% rh or 15–17% mc in food), development of one generation takes 12 days (Table 6.7). Rates of population increase reach maximum levels—about fourfold per week—at 25° and 90% rh (*24*). When a culture of *L. destructor* was exposed to −18° for 7 days, 4% of the population survived (*70*).

Population fluctuations are usually relatively small and gradual. Periodic outbreaks, when they occur, are probably related to the growth of some fungus. *L. destructor* crops the abundant spores and mycelia of *Aspergillus alternata* (Stennis) (*75*). Many kinds of fungi can support

populations of these mites (*74*). Conditions in bulk grain that favor an increase in populations of scavenger and fungivorous arthropods are at the same time unfavorable to *L. destructor* (*76*).

Hypopal formation is most common in dense populations held at 20–25° and 85% rh. Inert hypopi enclosed in the protonymphal cuticle occur commonly. If food supplies are adequate, transformation of hypopi into tritonymphs and adults takes place under conditions of 25–30° and 65–95% rh. Lower temperatures and higher humidities are even more favorable. Development from hypopus to adult takes about 6 weeks. Hypopi are carried about by birds and insects (*21*).

Lepidoglyphus zachari

L. zachari is a pest of dried meat products in temperate regions (*37*).

Gohieria fusca, brown flour mite

The brown flour mite is widely distributed around the world in flour, grain, rice, corn, bran, poppy seeds, animal feeds, medicinal herbs, and spices. It is often abundant in mixed cereal dust and spillage in mills. From there it spreads to warehouses. Females typically lay 11–24 eggs during their lifetime. At 24° or 25°, the life cycle takes 11–23 days (*37*).

Lardoglyphidae

Lardoglyphus konoi

L. konoi is a serious pest of dried or salted fish in India, Japan, and several European countries. Development and reproduction occur at relative humidities above 75%. Optimal conditions are 30° and 83% rh, with fish meal as food (*52*). The shortest developmental time for one generation is 8.5 days at 35°. At 23° and 87% rh, the life cycle takes 9–11 days (*37*).

Motile hypopi often appear at 20–30° and 76–94% rh. In colonies where the diet (fish meal) is restricted and the population density is high, hypopi are more common (*53*). Most of the hypopi transform into tritonymphs under conditions of favorable temperature and humidity in the presence of adequate food (*21*). The hypopi are carried about by insects, especially *Dermestes* and *Necrobia* species.

Pyroglyphidae

Dermatophagoides farinae, American house dust mite

A cosmopolitan species, *D. farinae* occurs in animal feeds, flour, insect cultures, and house

Table 6.7

Ecological parameters for some important stored-product mites

Species	Physical limits Temperature (°C) Range	Optimum	Relative humidity (%) Minimum	Optimum	Results under optimum conditions Fecundity (number of eggs/ female)	Develop- mental time (days)	Longevity (days)	R_0	r_m	Sex ratio (males: females)
Acarus										
farris	3–31	20	74	89	137	10	19	32.9	0.176	—
siro	2.5–31	20	63	80	230	9	30	—	—	47:53
Aeroglyphus										
robustus	—	30	—	75	12	90	—	2.5	0.03	—
Aleuroglyphus										
ovatus	—	25	—	75	60	16	—	—	—	—
Caloglyphus										
anomalus	?–34	23	—	98	738	9.5	30	—	—	—
berlesei	16–35	27	—	95	588	8	24	192	0.375	50:50
Carpoglyphus										
lactis[a]	3–35	20	60	80	278	10	20	118	—	45:55
Cheyletus										
eruditus[a]	8–31	20	56	80	71	17	35	—	—	0:100
Glycyphagus										
domesticus	—	22	—	—	135	22	31	32	0.097	50:50
Lardoglyphus										
konoi	—	30	75	82	—	8	—	—	—	—
Lepidoglyphus										
destructor	3–34	25	62	90	150	27	—	—	—	—
Rhizoglyphus										
echinopus	3–31	25	86	90	600	8	35	336	0.162	40:60
Thyreophagus										
entomophagus[a]	3–32	20	75	85	156	16	47	—	—	49:51
Tyrophagus										
putrescentiae[a]	8.5–36	25	69	85	500	10	62	30	0.210	45:55

[a]Duration of starvation (days) is 29 for *Carpoglyphus lactis*, 69 for *Cheyletus eruditis* and *Thyreophagus entomophagus*, and 60 for *Tyrophagus putrescentiae*.

dust. In North America and Czechoslovakia, this species is more common in house dust and mattress dust than *D. pteronyssinus* (*67, 73, 90*).

American house dust mites require a relative humidity of about 70% for reproduction and survival. Development from egg to egg takes 30 days under optimal conditions (25°, 75% rh), the egg stage lasting 6–8 days. At 26.6° and 32.2°, developmental time is 23.2 and 18.8 days, respectively, but the mortality rate is higher in both cases than at 25°. At 15.5°, development from egg to adult takes 389 days, with 50% of that time being spent in the tritonymph stage and 31% as actively feeding larvae or nymphs. At optimal temperatures, 65% of developmental time is spent in an active stage. Although no hypopi have been observed, crowded cultures contain immobile, quiescent protonymphs that are tolerant of adverse conditions and are functionally comparable to nonphoretic hypopi of free-living Acaridae (*90*).

There were no survivors when a population of American house dust mites was exposed to −18° for 48 hours; less than 5% of the adult mites survived a 24-hour exposure. Fifteen percent of the adults survived 168 hours at 2°. The mites can better withstand exposure to low temperatures when they are reared at 21° instead of at 27°. Male and female mites do not differ significantly in their tolerance to low temperatures (*57*).

D. farinae seems to thrive on high-protein diets that contain animal fats. Yeast is an important dietary supplement. Culture media composed of house dust or whisker dust in dog food have proved to be most suitable. Gelatin with yeast or dog food is another very good medium. Fish, albumin, and bean flour are satisfactory sources of protein (*89*). Various fats in the food medium favor reproductive success in cultured populations of *D. farinae* (*54*). Cannibalism is common (*90*).

Multiple matings occur. Only fertilized females lay eggs. These they lay singly at the rate of one to three per day for at least 30 days, producing during their lifetime some 200–300 eggs (*90, 91*). Adults live about 2 months at optimal temperatures; unmated mites live about 10 weeks at 21°. All instars are represented in cultured populations. It is postulated that several behavioral patterns are pheromone moderated, including aggregation behavior, density-dependent migration of larvae from cultures, and cessation of egg laying in crowded cultures. Potent allergenic extracts have been obtained from *D. farinae* and *D. pteronyssinus*.

Tarsonemidae

Tarsonemus granarius

This scavenger-fungivore is found predominantly in moldy, damp, or old grain bulks. It is largely restricted to granaries. Populations of this mite have persisted for years in Canadian granaries (*71*). These mites prefer wheat over oats; they are not found in barley. In Japan, the mites are more often found on granary floors than in the bulks of wheat or rice. Cereal dust or spillage appears to be a preferred niche. The occurrence of *Tarsonemus* species is correlated with moisture, temperature, and the presence of certain fungi. The age of the spilled grain and of the decaying organic debris and the composition and succession of the fungal flora play larger roles than temperature in the population dynamics of these mites (*71*).

These mites thrive on fungi, especially those species of the genera *Penicillium*, *Aspergillus*, *Chaetomium*, and *Hormodendrum* that appear during the succession of microflora in stored grain (*69*). Since they are tolerant of low temperatures, the mites are present throughout the year (*71*). Reproduction is probably by parthenogenesis.

Concluding Comments

Fungi grow and mites and insects reproduce in pockets in bulk grain where the moisture content is greater than 12.8%. The infestations gradually spread from these foci to most parts of the grain bulk, but the surface layer of the bulk is typically the region most heavily infested. To be effective, techniques of sampling and control must take into account these basic facts.

Many mites can tolerate starvation for long periods. Under conditions of low temperatures and high atmospheric humidity, the mites can survive for several months in cracks in floors and walls and in processing and packaging machinery. Except in parthenogenetic species, multiple mating is the rule, and the sex ratio is usually close to 1:1 and does not depend upon conditions of temperature, relative humidity, and availability and moisture content of food.

There is a close relationship between mites and the microorganisms that live in deteriorating stored products. The majority of the mite species infesting stored products are fungivores. Their main or supplementary food consists of fungal mycelia. Those mite species, such as *Tyrophagus putrescentiae*, *Tyrolichus casei*, and *Lepidoglyphus destructor*, capable of feeding and multiplying on many different kinds of seed-borne fungi, are less specialized than the ones, such as *Carpoglyphus lactis*, that can make use of only a few species of fungi.

Some species of mites are synanthropic in temperate zones (e.g., *Acarus siro*, *Carpoglyphus lactis*, *Lardoglyphus konoi*, *Tyrophagus putrescentiae*, *Gohieria fusca*, and *Glycyphagus domesticus*). Those mites live and reproduce in storage premises where they infest food products and associated waste materials. Many other species (e.g., *Thyreophagus entomophagus*, *Lepidoglyphus destructor*, *Rhizoglyphus echinopus*, and *Cheyletus eruditus*) can infest food crops in the field. They can reproduce in storage premises with varying degrees of success.

Several species are fairly tolerant of low temperatures. *Acarus siro*, *A. farris*, *Rhizoglyphus echinopus*, *Lepidoglyphus destructor*, and *Carpoglyphus lactis* can survive a temperature of 3°. These are the dominant pests in cool regions. Some other species are less tolerant of low temperatures. *Cheyletus eruditus* and *Tyrophagus putrescentiae* succumb to temperatures lower than 8°. The minimum threshold for *Calogyphus berlesei* is 16.5°; for *Aleuroglyphus ovatus* it is 22°. Some mites can survive very low temperatures (−18°). For most species, the upper temperature threshold for survival is between 30° and 36° (Table 6.7).

Humidity requirements for development vary from species to species but generally lie in the range of 52–100% rh. Moisture content of food affects habitat choice and reproductive potential.

Hypopi are unknown in the genera *Tyrophagus* and *Gohieria*. Motile hypopi appear in *Acarus siro*, *A. farris*, *Carpoglyphus lactis*, *Lardoglyphus konoi*, and *Rhizoglyphus echinopus*. *Acarus immobilis*, *Lepidoglyphus destructor*, and *Glycyphagus domesticus* have immobile hypopi. Hypopi generally appear when conditions of temperature and moisture content are optimum but also when the population is dense and food is inadequate. Field populations are more likely to produce hypopi than are laboratory cultures. Only certain populations of *Acarus siro* and *Carpoglyphus lactis* produce

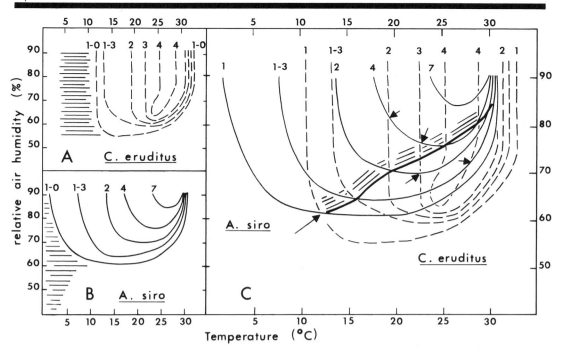

Figure 6.11
Population growth of *Acarus siro* and *Cheyletus eruditus*. (A), net increase per week of *C. eruditus* at various temperatures and relative humidities; (B), net increase per week of *A. siro* at various temperatures and relative humidities; (C), (B) superimposed on (A), with a line drawn through the conditions in which the calculated net rates are equal; the shading suggests some upward movement of the line to allow for a predator-mediated reduction of the rate of increase of the prey.
SOURCE: Solomon, M. E. 1961. Interaction of a predator and physical factors in the control of a grain mite. *Proceedings of the Eleventh International Congress of Entomology* (Vienna, 1960) 1:768–772.

hypopi. Habitat variability is not associated with hypopus formation. Intraspecific variations among strains of the same species commonly occur.

Cheyletus eruditus and other predatory mites often appear in populations of pest acarids (*80*). The predators are more active when the substrate is somewhat warm and dry. They can effectively suppress pest mite populations only when favorable environmental conditions prevail and when the predators outnumber the prey (*78*) (Figure 6.11).

Four main factors regulate mite numbers in stored products: First and most important is temperature (3–35°); second is moisture content of the medium, which must be above 12.8%; third is food; and fourth is the intrinsic rate of increase or net reproductive rate. Population explosions are common in species with very high reproductive rates. These major population expansions are, however, dependent upon the convergence of several favorable events involving temperature, moisture content of the substrate, and adequacy of food supplies. Other factors that influence population levels include predatory mites, microflora, dockage, cracked seeds or grains, cannibalism, and migration behavior (*75*).

Prevention of all infestations can be achieved by drying products to 12–13% mc and then protecting them from taking on additional moisture during storage.

Cited References

1 Alimuchamedov, S. N. 1973. *Acaroid Mites—Pests of Stored Cotton and other Agricultural Products in Middle Asia.* Federal Academy of Science of the Uzbeckistan SSR, Tashkent (in Russian).

2 Baker, E. W., and G. W. Wharton. 1952. *An Introduction to Acarology.* Macmillan, New York, NY.

3 Barker, P. S. 1967. The effects of high humidity and different temperatures on the biology of *Tyrophagus putrescentiae* (Schrank) (Acarina: Tyroglyphidae). *Canadian Journal of Zoology* 45(1)91–96.

4 Barker, P. S. 1967. Effect of humidity and temperature on the biology of *Aëroglyphus robustus* (Banks) (Acarina: Glycyphagidae). *Canadian Journal of Zoology* 45(4)479–483.

5 Barker, P. S. 1967. Bionomics of *Blattisocius keegani* (Fox) (Acarina: Ascidae), a predator on eggs of pests of stored grains. *Canadian Journal of Zoology* 45(6, pt. 1)1093–1099.

6 Barker, P. S. 1968. Bionomics of *Glycyphagus domesticus* (de Geer) (Acarina: Glycyphagidae), a pest of stored grain. *Canadian Journal of Zoology* 46(1)89–92.

7 Bass, J. A., and S. B. Hays. 1976. Predation by the mite *Tyrophagus putrescentiae* on eggs of the imported fire ant. *Journal of the Georgia Entomological Society* 11(1)16.

8 Berreen, J. 1976. An analysis of feeding in *Cheyletus eruditus*, a predator of storage mites. *Annals of Applied Biology* 82(1)190–192.

Mite Pests in Stored Food   **75**

9 Bielska, I. 1975. Life cycle and ecology of *Rhizoglyphus echinopus* (F. & R.) and *R. robini* Clap. (Acarina: Acaridae). Dissertation, Agricultural University of Warsaw (in Polish).

10 Bielska, I. 1975. Demographic parameters of *Rhizoglyphus echinopus* (F. & R.) and *Rhizoglyphus robini* C. depending on temperature, humidity and food. *Zeszyty Problemowe Postępów Nauk Rolniczych* 171:179–188 (in Polish, English summary).

11 Boczek, J. 1957. The grain mite (*Tyrophagus farinae* L.): Morphology, biology and ecology, estimation of damages and tests on control measures. *Roczniki Nauk Rolniczych* 75A(4)559–644 (in Polish, English summary).

12 Boczek, J. 1961. Biology and ecology of *Cheyletus eruditus* (Schrank) (Acarina: Cheyletidae). *Prace Naukowe Instytutu Ochrony Roślin* 1:175–230 (in Polish, English summary).

13 Boczek, J. [1975.] Reproduction biology of *Tyrophagus putrescentiae* (Schr.) (Acarina: Acaridae). In *Proceedings of the First International Working Conference on Stored-Product Entomology* (Savannah, GA, 1974), pp. 154–159.

14 Boczek, J., and B. Czajkowska. 1968. The effect of antimicrobial agents and antibiotics on some stored products mites (Acaroidea). *Roczniki Nauk Rolniczych* 93A(4)597–612 (in Polish, English summary).

15 Boczek, J., and B. Czajkowska. 1973. A study of some aspects of ageing in *Acarus siro* L. (Acarina: Acaridae). *Ekologia Polska* 21(11)173–184.

16 Boczek, J., and B. Czajkowska. 1976. Studies on the fecundity of acarid mites (Acarina: Acaridae). *EPPO Bulletin* 6(4)323–330.

17 Boczek, J., and T. Legat. 1973. Demographic parameters of *Tyrophagus putrescentiae* (Schr.) reared on various foods. In *Proceedings of the Third International Acarological Congress* (Prague, 1971), pp. 353–356.

18 Chmielewski, W. 1970. Experimental passage of mites through the alimentary canal of vertebrates. *Zeszyty Problemowe Postępów Nauk Rolniczych* 109:233–238 (in Polish, English summary).

19 Chmielewski, W. 1971. Morphology, biology and ecology of *Carpoglyphus lactis* (L., 1758) (Glycyphagidae, Acarina). *Prace Naukowe Instytutu Ochrony Roślin* 13(2)63–166 (in Polish, English summary).

20 Chmielewski, W. 1975. Observations on *Thyreophagus entomophagus* (Laboulbéne). *Zeszyty Problemowe Postępów Nauk Rolniczych* 171:253–259 (in Polish, English summary).

21 Chmielewski, W. 1977. Formation and importance of hypopus stage in the life of mites belonging to the superfamily Acaroidea. *Prace Naukowe Instytutu Ochrony Roślin* 19(1)5–94 (in Polish, English summary).

22 Cunnington, A. M. 1965. Physical limits for complete development of the grain mite, *Acarus siro* L. (Acarina, Acaridae) in relation to its world distribution. *Journal of Applied Ecology* 2(2)295–306.

23 Cunnington, A. M. 1969. Physical limits for complete development of the copra mite, *Tyrophagus putrescentiae* (Schrank) (Acarina, Acaridae). In *Proceedings of the Second International Congress of Acarology* (Sutton, Bonington, 1967), pp. 241–248.

24 Cunnington, A. M. 1976. The effect of physical conditions on the development and increase of some important storage mites. *Annals of Applied Biology* 82(1)175–178.

25 Cusack, P. D., G. O. Evans, and P. A. Brennan. 1975. A survey of the mites of stored grain and grain products in the Republic of Ireland. *Scientific Proceedings of the Royal Dublin Society* B3(20)273–329.

26 Czajkowska, B. 1970. The development of stored products mites on fungi as food. *Zeszyty Problemowe Postępów Nauk Rolniczych* 109:219–227 (in Polish, English summary).

27 Czajkowska, B. 1972. Influence of active substances of medicinal herbs on stored products mites. *Zeszyty Problemowe Postępów Nauk Rolniczych* 129:197–232.

28 Daneshvar, H., J. G. Rodriguez, and L. D. Rodriguez. 1977. Monoxenic culture of *Caloglyphus berlesei* (Acarina: Acaridae) on fungi. *Journal of Invertebrate Pathology* 30(2)175–180.

29 Flanders, S. E., and M. E. Badgley. 1963. Prey-predator interactions in self-balanced laboratory populations. *Hilgardia* 35(8)145–183.

30 Fleurat-Lessard, F., and P. Anglade. 1973. Influence de la teneur en eau des grains stockés sur le développement des populations d'acariens. *Annales de Technologie Agricole* 22(3)531–540 (English summary).

31 Golębiowska, Z. 1963. *Tyrophagus putrescentiae* (Schrank, 1781) (= *T. noxius* Zachwatkin 1935), morphology, biology and ecology. *Prace Naukowe Instytutu Ochrony Roślin* 5(2)29–88 (in Polish, English summary).

32 Griffiths, D. A. 1966. Nutrition as a factor influencing hypopus formation in the *Acarus siro* species complex (Acarina, Acaridae). *Journal of Stored Products Research* 1(4)325–340.

33 Griffiths, D. A. 1969. The influence of dietary factors on hypopus formation in *Acarus immobilis* Griffiths (Acari, Acaridae). In *Proceedings of the Second International Congress of Acarology* (Sutton, Bonington, 1967), pp. 419–432.

34 Griffiths, D. A., and J. Boczek. 1977. Spermatophores of some acaroid mites (Astigmata: Acarina). *International Journal of Insect Morphology and Embryology* 6(56)231–238.

35 Griffiths, D. A., A. C. Hodson, and C. M. Christensen. 1959. Grain storage fungi associated with mites. *Journal of Economic Entomology* 52(3)514–518.

36 Hirakoso, S., et al. 1971. Experimental studies on the control of *Tyrophagus putrescentiae* (Acarina: Acaridae) in straw mats with various pesticides. *Japanese Journal of Sanitary Zoology* 22(1)62–65 (in Japanese, English summary).

37 Hughes, A. M. 1976. *The Mites of Stored Food and Houses,* 2nd ed. Ministry of Agriculture, Fisheries and Food Technical Bulletin 9. Her Majesty's Stationery Office, London.

38 Hughes, T. E. 1943. The respiration of *Tyrophagus farinae. Journal of Experimental Biology* 2C(1)1–5.

39 Hussey, N. W. 1964. Mites as pests of cultivated mushrooms. *Reports of the Glasshouse Crops Research Institute* 1963:114–117.

40 Hussey, N. W., W. H. Read, and J. J. Hesling. 1969. *The Pests of Protected Cultivation.* American Elsevier, New York, NY.

41 Jakubowska, J. 1971. Distribution and ecology of mites of genus *Acarus* (Acarina, Acaridae) in Poland. *Roczniki Nauk Rolniczych* E1(2)57–73 (in Polish, English summary).

42 Jakubowska, J. 1971. Biology and ecology of *Acarus farris* (Oud.) (Acarina, Acaridae). *Roczniki Nauk Rolniczych* E1(2)75–94 (in Polish, English summary).

43 Knülle, W. 1965. Die Sorption und Transpiration des Wasserdampfes bei der Mehlmilbe (*Acarus siro* L.). *Zeitschrift für Vergleichende Physiologie* 49(6)586–604.

44 Krantz, G.W. 1978. *A Manual of Acarology,* 2nd ed. O.S.U. Book Stores, Corvalis, OR.

45 Kuwahara, Y. 1976. Alarm pheromones produced by several grain mites. Pest Research Institute, College of Agriculture, Kyoto University, pp. 65–76.

46 Kuznecov, N. N. 1970. Root mite and its control. Part 1, *Abstracts of the Second Conference on Mites,* Naukova Dumka Kiev, pp. 267–269 (in Russian).

47 Matsumoto, K. 1962. Studies on the environmental factors for the breeding of grain mites. Part III. On the breeding of *Tyrophagus dimidiatus* in various drug samples. *Japanese Journal of Sanitary Zoology* 13(2)105–111 (in Japanese, English summary).

48 Matsumoto, K. 1963. Studies on the environmental factors for the breeding of grain mites. Part IV. Comparison of the effects of humidity and temperature on the breeding of the grain mites, *Tyrophagus dimidiatus, Aleuroglyphus ovatus* and *Glycyphagus destructor. Japanese Journal of Sanitary Zoology* 14(2)82–88 (in Japanese, English summary).

49 Matsumoto, K. 1964. Studies on the environmental factors for the breeding of grain mites. Part V. Comparison of the breeding rate of *Carpoglyphus lactis* and *Tyrophagus dimidiatus. Japanese Journal of Sanitary Zoology* 15(1)17–24 (in Japanese, English summary).

50 Matsumoto, K. 1965. Studies on the environmental factors for the breeding of grain mites. Part VI. Digestive enzymes of the grain mites, *Carpoglyphus lactis, Aleuroglyphus ovatus* and *Tyrophagus dimidiatus. Japanese Journal of Sanitary Zoology* 16(1)86–89 (in Japanese, English summary).

51 Matsumoto, K. 1965. Studies on the environmental factors for the breeding of grain mites. Part VII. Relationship between reproduction of mites and kind of carbohydrates in the diet. *Japanese Journal of Sanitary Zoology* 16(2)118–122 (in Japanese, English summary).

52 Matsumoto, K. 1968. Studies on the environmental factors for the breeding of grain mites. Part IX. The effect of relative humidity on the age composition of the population of *Lardoglyphus konoi. Japanese Journal of Sanitary Zoology* 19(3)196–203 (in Japanese, English summary).

53 Matsumoto, K. 1973. Studies on the environmental factors for the breeding of grain mites. XI. The effect of nutrients on hypopus formation in *Lardoglyphus konoi. Japanese Journal of Sanitary Zoology* 24(1)1–7 (in Japanese, English summary).

54 Matsumoto, K. 1975. Studies on the environmental requirements for breeding the dust mites, *Dermatophagoides farinae* Hughes, 1961. Part 3. Effect of the lipids in the diet on the population growth of the mites. *Japanese Journal of Sanitary Zoology* 26(2–3)121–127 (in Japanese, English summary).

55 Maunsell, K., D. G. Wraith, and A. M. Cunnington. 1968. Mites and house-dust allergy in bronchial asthma. *Lancet* i(7555)1267–1270.

56 O'Connor, B. M. 1982. Acari: Astigmata. In *Synopsis and Classification of Living Organisms,* vol. 2, S. B. Parker, ed. McGraw-Hill, New York, NY.

57 Paul, T. C., and R. N. Sinha. 1972. Low-temperature survival of *Dermatophagoides farinae. Environmental Entomology* 1(5)547–549.

58 Pillai, P. R. P., and P. W. Winston. 1969. Life history and biology of *Caloglyphus anomalus* Nesbitt (Acarina: Acaridae). *Acarologia* 11(2)295–303.

59 Pulpán, J., and P. H. Verner. 1965. Control of tyroglyphoid mites in stored grain by the predatory mite *Cheyletus eruditus* (Schrank). *Canadian Journal of Zoology* 43(3)417–432.

60 Rivard, I. 1962. Influence of humidity on the predaceous mite *Melichares dentriticus* (Berlese)

(Acarina: Aceosejidae). *Canadian Journal of Zoology* 40(5)761–766.

61 Rivard, I. 1962. Some effects of prey density on survival, speed of development, and fecundity of the predaceous mite *Melichares dentriticus* (Berl.) (Acarina: Aceosejidae). *Canadian Journal of Zoology* 40(7)1233–1236.

62 Rodriguez, J. G. 1972. Inhib.:ion of acarid mite development by fatty acids. In *Insect and Mite Nutrition*, J. G. Rodriguez, ed. North-Holland, Amsterdam.

63 Rodriguez, J. G. 1973. Feeding behavior and nutritional requirements of some Acari. *Proceedings of the Third International Congress of Acarology* (Prague, 1971), pp. 739–743.

64 Rodriguez, J. G., and L. M. Lasheen. 1971. Axenic culture of *Tyrophagus putrescentiae* on a chemically defined diet and determination of essential amino acids. *Journal of Insect Physiology* 17(6)979–985.

65 Rodriguez, J. G., and M. F. Potts. [1975.] Nutritional requirements and metabolic inhibition of some acarid mites. In *Proceedings of the First International Working Conference on Stored-Product Entomology* (Savannah, GA, 1974), pp. 160–161.

66 Rodriguez, J. G., and Z. A. Stępień. 1973. Biology and population dynamics of *Caloglyphus berlesei* (Michael) (Acarina: Acaridae) in xenic diet. *Journal of the Kansas Entomological Society* 46(2)176–183.

67 Sămsiňák, K., F. Dusbabek, and E. Vobrázkova. 1972. Note on the house dust mites in Czechoslovakia. *Folia Parasitologica (Praha)* 19:383–384.

68 Sasa, M., et al. 1967. On the identity of *Tyrophagus putrescentiae* (Schrank) and *T. dimidiatus* (Hermann) (Acarina: Acaridae) in Japan. *Japanese Journal of Sanitary Zoology* 18(4)216–217.

69 Sinha, R. N. 1964. Ecological relationships of stored-product mites and seed-borne fungi. *Proceedings of the First International Congress of Acarology* (Ft. Collins, 1963). *Acarologia* 6(fasc. h. s.)372–389.

70 Sinha, R. N. 1964. Effect of low temperature on the survival of some stored products mites. *Acarologia* 6(2)336–341.

71 Sinha, R. N. 1968. Adaptive significance of mycophagy in stored-product Arthropoda. *Evolution* 22(4)785–798.

72 Sinha, R. N. 1968. Climate and potential range of distribution of stored-product mites in Japan. *Journal of Economic Entomology* 61(1)70–75.

73 Sinha, R. N., and T. C. Paul. 1972. Survival and multiplication of two stored-product mites on cereal and processed foods. *Journal of Economic Entomology* 65(5)1301–1303.

74 Sinha, R. N., and H. A. H. Wallace. 1966. Association of granary mites and seed-borne fungi in stored grain and in outdoor and indoor habitats. *Annals of the Entomological Society of America* 59(6)1170–1181.

75 Sinha, R. N., and H. A. H. Wallace. 1973. Population dynamics of stored-product mites. *Oecologia* 12(4)315–327.

76 Sinha, R. N., H. A. H. Wallace, and F. S. Chebib. 1969. Principal-component analysis on interrelationships among fungi, mites, and insects in grain bulk ecosystems. *Ecology* 50(4)536–547.

77 Sinha, R. N., and R. D. Whitney. 1969. Feeding and reproduction of the grain and the mushroom mite on wood-inhabiting Hymenomycetes. *Journal of Economic Entomology* 62(4)837–840.

78 Solomon, M. E. 1961. Interaction of a predator and physical factors in the control of a grain mite. *Proceedings of the Eleventh International Congress of Entomology* (Vienna, 1960) 1:768–772.

79 Solomon, M. E. 1962. Ecology of the flour mite, *Acarus siro* L. (= *Tyroglyphus farinae* DeG.). *Annals of Applied Biology* 50(1)178–184.

80 Solomon, M. E. 1969. Experiments on predator-prey interactions of storage mites. *Acarologia* 11(3)484–503.

81 Solomon, M. E., et al. 1964. Storage fungi antagonistic to the flour mite (*Acarus siro* L.). *Journal of Applied Ecology* 1(1)119–125.

82 Stępień, Z. A. 1970. Energy budget of *Rhizoglyphus echinopus* (F. & R.) (Acarina: Acaridae) during its development. Dissertation, Agricultural University of Warsaw.

83 Stępień, Z. A., W. Goszczynski, and J. Boczek. 1973. The energy budget of *Tyrophagus putrescentiae* (Schr.) (Acaridae). *Proceedings of the Third International Congress of Acarology* (Prague, 1971), pp. 373–378.

84 Stępień, Z. A., and J. G. Rodriguez. 1972. Food utilization by acarid mites. In *Insect and Mite Nutrition*, J. G. Rodriguez, ed. North-Holland, Amsterdam.

85 Sturhan, D., and G. Hampel. 1977. Pflanzenparasitische Nematoden als Beute der Wurzelmilbe *Rhizoglyphus echinopus* (Acarina, Tyroglyphidae). *Anzeiger für Schädlingskunde, Pflanzenschutz, Umweltschutz* 50(8)115–118 (English abstract).

86 Ter-Grigorian, M. A. 1976. On the life-history of the Ararat cochineal insect *Porphyrophora hamelii* Brandt (Homoptera, Coccoidea, Margarodidae). *Entomologicheskie Obozrenie* 55(2)300–307 (in Russian).

87 Thomas, C. M., and R. J. Dicke. 1971. Response of the grain mite, *Acarus siro* (Acarina: Acaridae), to fungi associated with stored-food commodities. *Annals of the Entomological Society of America* 64(1)63–68.

88 Ushatinskaya, R. S. 1954. Biological basis for

the use of low temperature for the control of stored-grain pests (insects and mites). Academy of Sciences USSR, Moscow (in Russian).

89 Waki, S., and K. Matsumoto. 1973. Studies on the environmental requirements for the breeding of the dust mite, *Dermatophagoides farinae* Hughes, 1961. Part 2. Observations on the mode of breeding in various kinds of food. *Japanese Journal of Sanitary Zoology* 24(1)117–121 (in Japanese, English summary).

90 Wharton, G. W. 1976. House dust mites. *Journal of Medical Entomology* 12(6)577–621.

91 Wharton, G. W., and R. T. Furumizo. 1977. Supracoxal gland secretions as a source of fresh water for Acaridei. *Acarologia* 19(1)112–116.

92 Woodring, J. P. 1969. Environmental regulation of andropolymorphism in tyroglyphids (Acari). In *Proceedings of the Second International Congress of Acarology* (Sutton, Bonington, 1967), pp. 433–440.

93 Woodring, J. P. 1969. Observations on the biology of six species of acarid mites. *Annals of the Entomological Society of America* 62(1)102–108.

94 Woolley, T. A. 1988. *Acarology: Mites and Human Welfare*. Wiley, New York, NY.

95 Yoshizawa, T., I. Yamamoto, and R. Yamamoto. 1971. Synergistic attractancy of cheese components for cheese mites, *Tyrophagus putrescentiae*. *Scientific Pest Control (Kyoto)* 36:1–7.

96 Zakhvatkin, A. A. 1941. Tyroglyphoidea [Acari]. In *Fauna of USSR*, Vol. 6, No. 1 (English translation by A. Ratcliffe and A. M. Hughes published in 1959 by American Institute of Biological Sciences, Washington, DC).

97 Žďárková, E. 1973. Orientation of *Tyrophagus putrescentiae* (Schrank) towards olfactory stimuli. In *Proceedings of the Third International Congress of Acarology* (Prague, 1971), pp. 385–389, 1 fig.

98 Žďárková, E., and M. Reska. 1976. Weight losses of groundnuts (*Arachis hypogaea* L.) from infestation by the mites *Acarus siro* L. and *Tyrophagus putrescentiae* (Schrank). *Journal of Stored Products Research* 12(2)101–104.

99 Żyromska-Rudzka, H. 1959. Preliminary observations on the occupation of a new substrate by *Carpoglyphus lactis* (L.) (Acarina, Glycyphagidae). *Ekologia Polska* A7(13)339–356 (in Polish, English summary).

7
Insect Pests of Minor Importance

William H Robinson

When compared to such important pests as the warehouse beetle (*Trogoderma variabile*) or the Indianmeal moth (*Plodia interpunctella*), it is reasonable to classify silverfish and the other insects considered in this chapter as "minor." But perhaps it is also reasonable to say that in the food industry, there is no such thing as a "minor" pest. The appearance of any kind of pest, even a single specimen, should stimulate immediate concern and investigation.

Silverfish (Thysanura)

These small, fast-moving insects have long, thin bodies that taper toward the rear, giving a fishlike appearance. They are usually covered with shiny scales. They are about one-half inch (1.3 cm) long and are most active at night or in the dark. Silverfish adults and nymphs have chewing mouthparts, long antennae, and three similar caudal appendages (bristles) at the posterior end of the body. They often dart for cover when disturbed by movement or sudden light.

Pest Status

Silverfish feed on a variety of human foods; in addition, they may feed on starch, glue, wallpaper paste, starched cotton, linen, and paper products with a large amount of sizing. They are often serious pests in granaries, mills, bakeries, and similar food-processing locations. Of course, they are common household pests (6). Silverfish nymphs, adults, and their eggs can be transported from place to place in cardboard cartons, books, paper, and various food materials. Silverfish are not known to transmit any disease to humans.

Silverfish are "domestic" insects, often associated with houses and various foods. They have adapted to damp and warm locations in homes and other buildings around the world. There are three cosmopolitan species of economic importance: *Lepisma saccharina* (common silverfish), *Thermobia domestica* (firebrat), and *Ctenolepisma lineata pilifera* (fourlined silverfish).

Biology and Habits

Silverfish are unusual in preferring locations where humidity and temperature are high, and are unique in continuing to molt their skin periodically after they become adults (*1*). At each molt the lining of the spermatheca is lost, together with its contents, so that copulation must occur in each adult instar if fertile eggs are to be produced. Females mate and lay one batch of eggs in each instar. Each female usually lays fewer than 100 eggs, singly or in small groups, in cracks and crevices and under objects. The eggs hatch in 10–16 days. Development of the nymphs is slow, ranging from 90 days to 3 years. Nymphs closely resemble adults, except for size. Molting continues throughout the long adult life, and the reproductive potential is great.

Common Silverfish, *Lepisma saccharina*

At temperatures ranging from 22° to 32°, nearly all reproduction occurs at relative humidities (rh) above 75% (*5, 13*). The highest rate of oviposition occurs at 84–100% rh. The egg stage lasts from 19 days at 32° to 43 days at 22°. The nymphal stage lasts 90–120 days at 27°, but under less favorable conditions

Robinson: Urban Pest Control Research Center, Department of Entomology, Virginia Polytechnic Institute and State University, Blacksburg, VA 24061.

it could last as long as 3 years. Adults can live as long as 3.5 years at 27°, 2 years at 29°, and 1.5 years at 32° (6).

Firebrat, *Thermobia domestica*

This species prefers temperatures above 32° with an optimum of 37–41° (12). Firebrats are found more often in bakeries, kitchens, and around stoves than are silverfish. Incubation takes from 77 days at 25° to 9 days at 44°. The nymphal stage lasts from 330 days at 27° to 47 days at 42°. The adult lifespan is from 2 to 2.5 years at 32° and from 1 to 1.5 years at 37°.

Fourlined Silverfish, *Ctenolepisma lineata pilifera*

This species occurs in most parts of the USA and it is especially common in the Southwest. The distribution of fourlined silverfish in a building is not so strictly limited by temperature and humidity as is the case with the other two species mentioned here (4).

Predators, Parasites, and Pathogens

Spiders are the principal predators of silverfish.

Springtails (Collembola)

Springtails are soft-bodied, wingless insects, usually 2–3 mm long, but a few species reach 10 mm. Their color varies from white to black. Their bodies are small, compact (6-segmented), cylindrical, and frequently covered with setae and scales. Most species can jump by means of a forked structure (furcula) extending downward from the posterior end (1). When it springs backward, the furcula propels the insect into the air. Springtails have rasping-sucking mouthparts. They require dark, moist conditions; respiration is entirely through the skin in most species. They are sensitive to light; when disturbed, they often propel themselves into a new position.

Pest Status

Springtails are found most commonly on or near the soil surface, especially in decaying plant material. They feed on algae, fungi, and decaying plant parts. They are not specifically associated with humans, but if their natural environment becomes dry, then in the course of their search for moisture, springtails may invade buildings. They are pests by virtue of their presence, since they do not cause physical damage and apparently do not transmit diseases to humans. Species in the genus *Entomobrya* have been found associated with man and stored food (10). *E. atrocinta* is a pest of dried milk (10). Decaying fruit and fungal growth attract springtails. Mushrooms are highly susceptible to attack by springtails.

Biology and Habits

Springtails prefer locations with high humidity and abundant organic debris. Populations tend to increase during hot, humid weather, and decrease during cold weather. The female usually lays eggs singly on the substrate. The young look like the adults; there are several preadult instars. Adults may molt up to 50 times, usually without much additional growth.

Predators, Parasites, and Pathogens

Various spiders and beetles prey on collembolans.

Psocids (Psocoptera)

Psocids range from less than 1 to almost 10 mm in length and have a characteristic appearance—round head, long antennae, and large abdomen. Their color varies from creamy white to light brown. Wings are usually absent from species found indoors, but many other species possess two pairs. Psocids have biting-chewing mouthparts. They prefer damp, warm, undisturbed places, and can run quickly when disturbed.

Pest Status

Psocids are found most commonly outdoors on the foilage or branches of trees, under bark, and in leaf litter. A few species are regularly associated with buildings and stored food (3). Psocids can be serious pests of stored foods when conditions are humid. The species associated with stored food sometimes develop large populations. They occasionally occur in large numbers in houses, where they are a nuisance rather than destructive.

Species commonly associated with houses, stored foods, and the food industry (2) include *Liposcelis bostrychophilus* (14, 16); *Lepinotus patruelis,* and *Trogium pulsatorium* (7); and *Lachesilla pedicularia* (cosmopolitan grain psocid) (7).

Biology and Habits

Psocids prefer dark, humid conditions when they infest buildings or stored food (7). Most species cannot survive more than 2 or 3 weeks at less than 58% rh. Mating is usually preceded by a short courtship dance. Females lay eggs singly and produce fewer than 100 in their lifetime. The sticky eggs adhere to the substrate. The eggs hatch in several days, and the nymphs generally resemble the adults. There are normally six nymphal instars, but the number may vary. There may be one to many generations per year. Some species are parthenogenetic (males unknown).

Some kinds of psocids commonly infest cereal products, especially when poor storage has lead to high-moisture conditions. The eggs hatch in about 11 days at 25° and 75% rh, and the nymphal stage lasts about 15 days. Adults live about 6 months. Females produce about 200 eggs. Many kinds of psocids cannot survive at relative humidities of 55% at 25° or 65% rh at 35° (6). In a laboratory study of the biotic potential of two species of psocids at 20°, one female *Liposcelis bostrychophilus* produced 288 offspring over a 12-week period whereas one female *Lepinotus patruelis* produced 27 progeny (9). Laboratory experiments have demonstrated that *Liposcelis bostrychophilus* can, under certain conditions, survive starvation for 2 months and can resume normal fecundity when food becomes available (15).

Predators, Parasites, and Pathogens

Psocids are preyed upon by spiders, pseudoscorpions, ants, wasps, and thrips. They are attacked by parasitic nematodes and entomophagous fungi (11).

Thrips (Thysanoptera)

Thrips are very small (0.5–5 mm) insects, with slender bodies, short antennae, and narrow wings (when present). Their body color varies but usually is dark brown, red, or black. Thrips have rasping-piercing mouthparts, and most species are plant feeders (1). Some thrips are predators on small insects and mites, and a few feed on fungal spores. These insects sometimes occur in large numbers, especially on flowers and shrubs. The small size, light weight, and weak flight of thrips make them susceptible to strong winds, and they are frequently carried into buildings and other unlikely places.

Pest Status

Several species of thrips are serious plant pests. Many fruit and vegetable crops are damaged by the feeding of all active stages on the floral parts, young buds, and ripe fruits. With the exception of *Limothrips cerealium*, no thrips species are known to infest buildings regularly or to be attracted to stored food or food-processing operations. Because of their microscopic size, they are easily carried with raw materials into food-processing plants and incorporated in the finished products. With the onset of warm weather in Great Britain, populations of *L. cerealium* take to the air. Upon landing they seek confined spaces such as buildings and packaged foods (8). Of course, these are dead-end destinations, since they find nothing in these places to support life.

Biology and Habits

Thrips are active insects. During the day they can be found moving over the surface of plants, in leaf litter, or debris. Most thrips lay fertilized eggs, but parthenogenesis may occur and result in production of only males or females (1). In some species the eggs are laid singly in a slit cut into the plant; in others the eggs are scattered on the surface of the plant. During warm weather, hatching occurs in about 7 days. The pale-colored nymphs feed for another 10 days or so before the onset of two resting stages. Transformation into the adult form follows the completion of several preadult instars. There are usually several generations per year.

Predators, Parasites, and Pathogens

Several species of wasps provision their nests with thrips. Other natural enemies include fungi, predatory insects, and spiders.

Aphids (Homoptera)

Aphids are small, soft-bodied insects with a characteristic pearlike shape to the body, a pair of projections (cornicles) at the posterior end of the abdomen, and fairly long antennae. There are winged as well as wingless forms. Nymphs and adults have piercing-sucking mouthparts that they use to suck the sap from plant stems and leaves (1). Most species are parthenogenetic during part of their life cycle; females bear live young. These unusual features ensure large populations of aphids wherever they occur. Their small size and weak flight make them susceptible to strong winds; consequently, they can be carried into buildings and other unlikely places.

Pest Status

This group of insects contains a number of serious pests of cultivated plants. Aphids cause leaf curling, wilting, and galls on the host plants. No aphid species are known to infest buildings regularly or to be attracted to stored food or food-processing plants. Because of their small size, they are easily introduced into food-processing operations and incorporated in the finished products.

Biology and Habits

The life cycle of most aphids is rather unusual and complex (1). Most species overwinter in the egg stage on the bark of trees and shrubs. In the spring the eggs hatch into females that reproduce parthenogenetically and give birth to live young. Several generations are produced during the summer, with only females being produced and the young born alive. The early generations (spring and summer) are usually wingless; in late summer and fall, winged individuals appear. These winged aphids migrate to the overwintering host plants, and a generation of males and females appears. The individuals of this bisexual generation mate, and the females lay eggs that overwinter.

Predators, Parasites, and Pathogens

Aphids have numerous parasites and predators. The principal parasites are small wasps. The dominant predators are ladybird beetles, wasps, lacewings, and the larvae of certain syrphid flies.

Scale Insects (Homoptera)

This group contains some of the most unusual forms of insects. Scale insects are generally microscopic in size and specialized for a sessile or nearly sessile life attached to the roots, leaves, or branches of a host plant. These insects are most commonly known by their scalelike covering which protects them while they are attached to the host (*1*). The scale covering is formed of wax secreted by the insect and varies according to species: they may be circular or oval, smooth or rough, and variously colored. The nymphs and female adults have piercing-sucking mouthparts which they use to suck sap from plants. Females are wingless, usually legless, and most remain attached to the host. Males have one pair of wings (when present); they lack mouthparts and do not feed.

Pest Status

These insects injure plants by sucking sap. When numerous, scale insects may kill the host plant. The group contains a number of serious pests of cultivated plants. No scale insects regularly infest stored foods or are attracted to food-processing operations. They are often associated with leaves, fruits, and berries and may become incorporated in finished products made from these foods.

Family and species classification and identification of most scale insects are extremely difficult, requiring removal of the scale, then clearing, staining, and mounting the insect on slides. However, species identification is rarely necessary in connection with food problems.

Biology and Habits

Development of scales varies according to species, but in most cases it is rather complex (*1*). Females remain under the scale covering when they become adults; they produce eggs or bear live young. The first-instar nymphs have legs and antennae and are fairly active; they are often called crawlers. After the first molt, the legs and antennae are discarded and the insect secretes a waxy or scalelike covering over the body. The males develop much like the females; they are winged, have well-developed legs and antennae, and do not produce a scale covering. Usually one or two generations appear each year. Scale insects overwinter in the egg stage.

Predators, Parasites, and Pathogens

Scale insects are parasitized by several kinds of minute wasps and flies.

Cited References

1 Borror, D. J., C. A. Triplehorn, and N. F. Johnson. 1989. *An Introduction to the Study of Insects.* Saunders College Publishing, Philadelphia, PA.

2 Dodd, G. D. 1984. Introduction to booklice . . .: Survey of the problem. In *Aspects of Pest Control in the Food Industry* (Long Eaton, September 27), G. D. Dodd, ed. Society of Food Hygiene Technology, Potters Bar, Hertsfordshire, England, pp. 37–42.

3 Downing, F. S. 1984. Introduction to booklice . . .: The SOFHT survey of the booklouse problem. In *Aspects of Pest Control in the Food Industry* (Long Eaton, September 27), G. D. Dodd, ed. Society of Food Hygiene Technology, Potters Bar, Hertsfordshire, England, pp. 43–56.

4 Ebeling, W. 1975. *Urban Entomology.* Agriculture and Natural Resources, University of California, Oakland.

5 Hedges, S. 1989. Silverfish and bristletails: By any name they're a nuisance. *Pest Control Technology* 17(11)43, 46, 48.

6 Hickin, N. E. 1974. *Household Insect Pests.* St. Martin's, New York.

7 Obr, S. 1978. Psocoptera of food-processing plants and storages, dwellings and collections of natural objects in Czechoslovakia. *Acta Entomologica Bohemoslovaca* 75(4)226–242.

8 Palmer, J. M., L. A. Mound, and G. J. du Heaume. 1989. *Thysanoptera*, Vol. 2 of *CIE Guides to Insects of Importance to Man.* CAB International, Oxon, United Kingdom.

9 Pinninger, D. B. 1984. Introduction to booklice . . .: Recent research on psocids. In *Aspects of Pest Control in the Food Industry* (Long Eaton, September 27), G. D. Dodd, ed. Society of Food Hygiene Technology, Potters Bar, Hertsfordshire, England, pp. 57–65.

10 Pratt, H. D., K. S. Littig, and H. G. Scott. 1975. *Household and Stored-Food Insects of Public Health Importance and Their Control.* DHEW Publ. No. (CDC)75-8122. Center for Disease Control, Atlanta, GA.

11 Smithers, C. N. 1964. The Myopsocidae (Psocoptera) of Australia. *Proceedings of the Royal Entomological Society of London* 1333(7–8)133–138.

12 Sweetman, H. L. 1938. Physical ecology of the firebrat, *Thermobia domestica* (Packard). *Ecological Monographs* 8(2)285–311.

13 Sweetman, H. L. 1939. Responses of the silverfish, *Lepisma saccharina* L., to its physical environment. *Journal of Economic Entomology* 32(5)698–700.

14 Turner, B. D. 1987. Forming a clearer view of *L. bostrychophilus. Environmental Health* 95(5)9, 11, 13.

15 Turner, B. D., and H. Maude-Roxby. 1988. Starvation survival of the stored product pest *Liposcelis bostrychophilus* Badonnel (Psocoptera, Liposcelidae). *Journal of Stored Products Research* 24(1)23–28.

16 Turner, B., and H. Maude-Roxby. 1989. The prevalance of the booklouse *Liposcelis bostrychophilus* Badonnel (Liposcelidae, Psocoptera) in British domestic kitchens. *International Pest Control* 31(4)93–97.

8
Ecological and Behavioral Aspects of Cockroach Management

Walter Ebeling

Cockroaches are among the most primitive of winged insects. They were already very abundant during the Carboniferous Period, 250 million years ago, a time of high temperature and humidity. Cockroaches generally continue to seek warm and humid environments, although some species can survive temperatures between $-7°$ and $47°$ (146). Most of the principal domiciliary species probably originated in northern and tropical Africa (148). Species names such as *americana*, *australasiae*, *germanica*, and *orientalis*, and the corresponding common names, are therefore misleading.

Distribution

The principal domestic species in the USA are the German cockroach (*Blattella germanica*), the oriental cockroach (*Blatta orientalis*), the brownbanded cockroach (*Supella longipalpa*), the American cockroach (*Periplaneta americana*), the smokybrown cockroach (*P. fuliginosa*), and the brown cockroach (*P. brunnea*). The normal geographic range of the domiciliary species is, of course, modified in areas where they occupy buildings kept at temperatures comfortable for human occupants throughout the year. For example, severe infestations of the German cockroach have been reported in Alaska (38). In occupied buildings, cockroaches find microclimates similar to those of the outdoor subtropics or tropics.

Survival Strategies

The success of the domiciliary cockroach species in inhabiting human dwellings and other

Ebeling: Department of Biology, University of California, Los Angeles, CA 90024.

buildings can be attributed to certain biological characteristics: omnivory, high reproductive potential, secretive habits that protect them from detection and destruction, and great speed in escaping to their out-of-sight and often very narrow harborages. Among the domiciliary cockroaches, the German cockroach has the shortest life cycle, giving it the greatest reproductive potential and also resulting in the most rapid development of resistance to insecticides. It is the most important insect pest in the urban environment throughout most of the USA (116), and it is the animal "least liked" by the American public (111).

Reproductive Potential

It is commonly assumed that stored-food insects are the most important pest problem of the food industry, but some leading consultants consider the German cockroach to be more important. Food plant sanitarians are said to fear it most because of its great reproductive capacity (129). Under optimal conditions, one fertilized female German cockroach could theoretically produce over 10 million females within one year (about 3.5 generations) and over 10 billion females in just 1.5 years (five generations) (90). The life histories, bionomics, and biologies, including details of structure, physiology, and behavior, of most of the important domiciliary cockroach species have been well studied (44, 81, 82, 96) and summarized (115).

Food and Harborage

Human habitations and workplaces not only provide ideal microclimates for pest cockroaches, but also provide food and moisture for these omnivorous insects. Cockroaches consume any human food or beverage as well as dead animal material, including individuals of their own species or their cast skins, live or dead plant material, leather, glue, hair, wallpaper, fabrics, and the starch in bookbindings. Most buildings also provide the dark and secluded daytime harborage and breeding areas that cockroaches require. The habitats and habits of cockroaches make them prime suspects as carriers of disease organisms (2, 32, 77, 83, 164, 165).

Ships also provide good harborages and breeding places, and this accounts for the worldwide distribution of domiciliary species. Transport over long distances on ocean liners may in some instances be facilitated by the ability of cockroaches to survive for long periods without food and water (202), but certainly the vast majority of such vessels could readily satisfy the needs of any cockroaches that happened to be on board. Ships probably remain the principal means of intercontinental distribution of cockroaches, although their transport in the baggage compartments and galleys of aircraft has also been well documented (165).

This brief introduction emphasizes the role of the most obvious ecological factors in the survival and distribution of cockroaches. These and more subtle factors, and the way in which they can be utilized in a cockroach management program, are discussed in the following pages.

Cockroach management is to a large extent an exercise in applied ecology.

Temperature, Humidity, and Habitat

From an ecological standpoint, temperature is a more meaningful concept when discussed in relation to humidity and air movement, for these three factors are interrelated (44). When a sample of air is warmed, its relative humidity (rh) decreases and its drying power (saturation deficit) increases; it is able to hold more water. Air movement may displace moist air with drier air, causing a loss of moisture from, and a cooling of, the insect's body surface. As with all insects, the activity of a cockroach decreases with decreasing temperature.

Temperature and Moisture Preferences

In a choice test, American and oriental cockroaches were placed in a chamber, 6 feet long, 4 inches wide, and 4 inches deep, in which temperatures ranged from 4.4° at one end to 32.2° at the other end (81). In each test, 20 insects were allowed 30 minutes to congregate in a specific temperature zone. Both species preferred the range of 26.1 to 27.8°.

In other experiments (91, 93), oriental cockroaches tended to settle in areas of lower temperatures in dry air rather than in moist air (Figure 8.1), thus providing themselves with an atmosphere of lower saturation deficit. The difference in the preferred temperature in dry and moist air increases with hours of exposure. At 30 hours of exposure, the preferred temperature was 24.5° in moist air and 20.7° in dry air. As the amount of water in the insects decreased, they moved to cooler air that provided a lower saturation deficit, thus reducing the rate of water loss. As the period of exposure increased, change in preferred temperature was greater in dry than in moist air. The insects returned to their normal temperature preference zone when given access to water. This zone for German and American cockroaches ranged from 24° to 33°; for oriental cockroaches, from 20° to 29°.

Temperature Effects

Low Temperatures

When nymphal and adult American cockroaches were reared at 27° and held at low temperatures for one hour, mortality increased from 2% at freezing point to 71% at −5°, 98% at −10°, and 100% at −15° (114). The minimum temperature at which a cockroach can survive

Figure 8.1
Change in temperature preferences of adult male oriental cockroaches on the basis of moisture gradient.
SOURCE: Gunn, D. L. 1931. Temperature and humidity relations of the cockroach. *Nature* 128(3222)186–187.

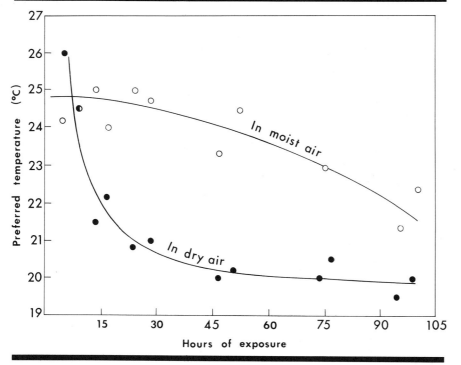

is affected by its previous temperature experiences. Oriental cockroaches that had been living at 15° survived for 3 weeks after the temperature was lowered to the range of 2–5°; those moved from 30° to these same low temperatures began to die within 5 days (119).

It follows that in centrally heated dwellings in which heating is discontinued, cockroaches will not survive as well in winter as in buildings that have not been heated. Likewise, cockroaches quickly transported from a warm to a cold geographic region in the heated parts of jet aircraft would not be expected to survive as well as those transported from a cold region if the environment were equally unfavorable upon arrival in either case (44).

High Temperatures

Critical Thermal Maximum The critical thermal maximum (CTM) has been determined for 10 cockroach species (7). This is the temperature producing knockdown, after which the insects are capable of recovery within an hour when returned to ambient conditions. CTMs ranged from 51.4° for *Supella longipalpa* to 47.6° for *Leucophaea maderae* (Madeira cockroach). The three species with the highest CTMs (*S. longipalpa*, *Blaberus craniifer*, and *Periplaneta brunnea*) are found in habitat types referred to as ''hot,'' ''hot-warm,'' and ''warm,'' respectively, whereas the two species with the lowest CTMs (*L. maderae* and *Blatta orientalis*) are in habitat types referred to as ''cool.''

The brownbanded cockroach is about the same size as *Blattella germanica* but is more successful in very warm habitats because of its high CTM. However, a reduction in temperature of only 5° (from 30° to 25°) nearly doubles the developmental time of *S. longipalpa*, from 37.6 to 73.3 days (98). Also, this is the temperature at which its parasitoid, *Comperia merceti*, is at its maximum rate of development (Lynn M. LeBeck and Arthur G. Appel, unpublished data). Under these conditions, the German cockroach is generally the more abundant of the two species.

Population Pressures Population pressures may result in the presence of large numbers of the insects throughout a building, even in areas with temperature and humidity far from ideal for the species. A PCO in Houston, Texas, described a midsummer infestation in a house in which the number of insects was so high that the attic, at 37.8°, was heavily infested, the voids of outside as well as partition walls were ''caked'' with cockroaches; after treatment with

diazinon plus dichlorvos (highly repellent), the insects ''boiled out'' from under the sidings (159).

Considerations in Management and Prevention

An environmental variable the pest control operator (PCO) may see fit to consider as a facet of integrated pest management (IPM) programs is the choice of weather for treatment. Busvine (34) and Hayes (99) emphasized the complexity of the interactions between temperature and the required dose of a pesticide. Involved are such matters as penetration, distribution, excretion, metabolism, and action of the pesticide or other temperature-dependent toxins within the insect body. Reichenbach and Collins (149) made multiple logit analyses of the effect of temperature and humidity on the toxicity of topically applied propoxur to German cockroaches. At 23° and 100% rh, as much as fivefold more propoxur was needed to kill 50% of the insects than at 10° and low humidity (20%). The researchers suggest that weather conditions might be chosen to maximize the effect of treatment. They point out, however, that at low temperatures the cockroaches might be less active and their contact with the treated surfaces might be decreased.

The pyrethroids, some of which show great promise for the control of the German cockroach, are generally also adversely affected by high temperature. They have distinct temperature-activity profiles, some retaining insecticidal activity at higher temperatures than others (D. A. Reierson, personal communication).

Moisture Effects and Sources

Water Loss Studies

When percentages of water loss from three species of cockroaches were plotted against temperature (and saturation deficit of the air), the German cockroach lost water more rapidly than the oriental cockroach, and the latter lost water more rapidly than the American cockroach (93). This may account for the preference of the German cockroach for relatively moist, warm, steamy kitchen areas (in Great Britain, this species is often called the steamfly). However, German cockroaches also thrive in kitchens with relatively low humidity when water is available for them to drink. Cockroaches quickly restore their optimal levels of body moisture when they have access to water (92).

Cuticle permeability

Additional studies (4) of the comparative water

relations and temperature sensitivity of 10 species of cockroaches have confirmed earlier findings (93) regarding the relative rates of water loss of German, oriental, and American cockroaches, but also showed that cuticle permeabilities were in the opposite ratio, that of the American cockroach being the highest. However, the more important consideration is body water reserves, which would increase with body size. The German cockroach is the most ubiquitous and numerous of the domiciliary cockroach species. Its small body size, low cuticle permeability, and nocturnal habits give it a competitive advantage. Retention and water regulation of the ootheca by the female, practically until time of egg hatch, is another advantage (44).

The smokybrown cockroach has a higher rate of cuticular water loss than any of the major cockroach pests—over 6.5 times that of the German cockroach—and this severely limits its tolerance to dry conditions. Adult males of this species die in 6–8 hours at 32° and 33% rh. This species requires large amounts of drinking water to maintain water balance (5, 6).

Water Sources

Cockroaches can obtain sufficient water from moist foods. When reared on dry food (e.g., dry dog chow), water must be supplied to any cockroach species. In buildings, cockroaches obtain moisture from such sources as water used to wash equipment, from "hosing down" such areas as food preparation rooms in restaurants, from damp mops, from wooden surfaces and crevices that hold water for long periods (and also provide humid microclimates), from leaky faucets and toilet bowls, and from condensation on utility pipes or under refrigerators.

Habitats of Cockroaches

In southern California, German cockroaches can sometimes be found outdoors in summer, usually close to foundation walls or garbage dumpsters. Migration of this species from vacant buildings to adjacent occupied buildings has been observed. Insecticide treatment may stimulate migration, but migration can also occur as a response to overcrowding.

The German cockroach is usually the most common species in dwellings and other buildings, particularly where food is processed, packaged, stored, and served. It is generally found in greatest numbers in areas that provide both warmth and seclusion, such as beneath and behind gas ranges and refrigerators (and in the insulation in these appliances), beneath and behind sinks which may be warm and also provide water of condensation or moisture from drainboards or loosely fitted splashboards, under hot water tanks, behind steam heaters, in television and radio sets, and in electric clocks.

Although the German cockroach is much more abundant in dwellings than the American cockroach, in southern California the latter may be more abundant in warehouses and other buildings where there is less moisture. Productivity of field populations of German cockroaches may be even more severely restricted by water and/or humidity than by food (161).

The oriental cockroach, with a lower range of temperature preference than the other two species, appears to be the one best able to adapt to outdoor conditions in temperate regions. In Kansas City, Missouri, oriental cockroaches survived outdoors under a light snow cover during 13 weeks of almost continuous freezing temperatures (179). About 2 weeks after the arrival of the first warm weather of spring, some of these insects invaded a nearby house. The oriental cockroach prefers situations below ground level, as in crawl spaces, and on ground floors of buildings, but in Kansas City this species has been found with increasing frequency in the upper floors and in hot, dry locations, provided the insects had occasional access to water (179).

Oriental cockroaches are relatively more abundant in north temperate regions. In Great Britain, they were found in 87% of 4,000 infested premises and German cockroaches were found in 21%. Densely populated areas such as London and Glasgow favored German cockroaches regardless of the type of property, and this species was increasing in numbers relative to oriental cockroaches. Buildings with food readily available, especially industrial kitchens, favored German cockroaches; cooler, less humid environments tended to favor oriental cockroaches (45).

Reports on 2,321 cockroach infestations in buildings in West Germany revealed that in larger cities the incidence of German cockroaches was 5 times greater than the incidence of oriental cockroaches; in towns, 2.1 times greater; and in rural areas, 1.6 times greater (45). German cockroaches occurred throughout infested premises, often on the upper floors, whereas oriental cockroaches were seen more frequently in cellars and boiler rooms.

In two 4-year surveys on a military installation in North Carolina, German, American, brownbanded, and oriental cockroaches were found most abundantly in housing facilities; food preparation buildings, "miscellaneous" buildings, and stores were infested in descending frequency (208). Cockroaches are generally less abundant in individual houses than in apartment-style, low-rent housing projects; they are usually less abundant in new than in old buildings (153).

Even in such a relatively mild climate as that of coastal southern California, American cockroaches are almost exclusively found in sewerage systems, especially in manholes (197). In desert areas, such as Palm Springs, California, and Phoenix, Arizona, they are fairly common outdoors. In the southern states, they can be seen outdoors flying into palm trees or around electric lights on warm summer evenings. Even as far north as the District of Columbia, American cockroaches have been seen in large numbers living under bark and feeding on sap running down tree trunks (3).

Considerations in Management and Prevention

Smokybrown Cockroaches Outdoors In detailed studies of the population ecology of the smokybrown cockroach in an outdoor urban environment of Texas, it was found that the dispersal area of tagged adult males and females was not extensive, the mean distance moved being less than 10 meters (74). Females were found outdoors up to 363 days, and males up to 267 days, after the initial capture in traps designed for outdoor use. However, ootheca-bearing females, presumably seeking protected winter harborages, regularly invaded urban structures during October and November (143). In the Texas study cited above (74), it was concluded that insecticide treatments directed against the smokybrown cockroach should be restricted to aggregation sites, foundations, and potential entry areas of structures and should be applied before the usual June surge of activity and movement and again in October and November. Applied during these periods, the insecticides should greatly reduce the size of summer and winter cockroach populations.

Oviposition Habits Unlike the related *Periplaneta americana,* which tends to deposit its oothecae in concealed sites, *P. fuliginosa* deposits its oothecae mostly on the outside surfaces of urban structures. Therefore, interior insecticide treatment will not eliminate the source of infestation. Since movements of marked *P. fuliginosa* have been monitored for distances up to 36.6 meters, it is assumed that the species can reinfest a treated area from at least that

distance. This species also readily flies. For effective insecticidal control, it is essential to treat the outside of buildings and the substrates on which cockroaches might oviposit or find harborage for as far as 37 meters from the building. *P. fuliginosa* has a greater tendency than *P. americana* to attach its oothecae to substrates; it attaches them to the vertical surfaces of substrates whereas *P. americana* does not (72).

It has been suggested that prevention of American, oriental, and smokybrown cockroach infestations in buildings depends primarily on suppression or elimination of outdoor populations and that any long-term program to keep these species out of urban structures will be difficult, if not impossible, unless outdoor populations are greatly suppressed or eliminated (143). Closing avenues of entry of the insects into buildings is also important.

Field Cockroach The species best able to inhabit outdoor areas that are both hot and dry is the field cockroach, *Blattella vaga.* This species can be found, sometimes abundantly, in irrigated fields and in yards in southern California, Arizona, and Texas, but it also sometimes appears in the desert, far from inhabited or cultivated areas (75). In 1967, a severe infestation was found in northern California (35). Although the field cockroach is usually found outdoors, where it feeds on decaying vegetation during the drier seasons, it may invade houses in large numbers.

The field cockroach and the newly introduced Asian roach (*B. asahinai*) (124; see also Chapter 9, this volume) closely resemble the German cockroach (*B. germanica*). Adult field cockroaches are slightly smaller, grayish, and can be most readily distinguished by a wide, blackish-brown, vertical stripe that extends from the mouthparts to between the eyes; these stripes are absent or less distinct in the German cockroach (75, 194). The two parallel longitudinal stripes on the pronotum of the field cockroach are similar to those of the German cockroach, but are very dark and more sharply defined.

A PCO in Texas who believed that he was treating a house for the German cockroach failed to control the infestation. When the insects were correctly identified as *B. vaga,* he sprayed the yard, which was the source of infestation, and obtained good control (198). Infestations also have been eliminated by removing decomposing plant material from around the foundations of houses (75, 139).

Brownbanded Cockroach The brownbanded cockroach is usually abundant only in buildings in which temperatures are higher than most people would consider to be comfortable. They are often widely scattered about as isolated individuals. At lower room temperatures they move about very little, some remaining in one position for days. For this reason, residual insecticides sometimes perform poorly against brownbanded cockroaches because of the minimal contact of the insects with the residues.

Mobility in Apartment Houses A trap survey of cockroaches in a four-story building with 48 apartments in southern California yielded a few brownbanded cockroaches from most of the apartments, but no German cockroaches (*60*). This was a most unusual situation for southern California, where about 95% of the domiciliary cockroaches are the latter species. The cockroaches decreased in numbers with increase in distance from a focus on the second floor, a cluttered apartment left virtually undisturbed and with an uncomfortably high temperature. This extremely heavily infested apartment had served as a source of infestation for the entire building, demonstrating the ability of cockroaches to spread for great distances in an apartment building, vertically and horizontally, apparently via the wall voids and spaces around utility pipes, electrical conduits, and loosely fitting baseboards. Probably some also crawl under exit doors of apartments and down hallways.

In experiments in which German cockroaches were marked and recaptured, it was demonstrated that these insects are more mobile than previously recognized (*25, 106*). Plumbing connections to adjacent apartments were an important avenue of movement. In adjacent apartments with common plumbing, the movement of German cockroaches between them was 7% per week, compared to less than 2% between adjacent apartments not sharing common plumbing. It would be interesting to set up a similar experiment to measure the rate of movement between two adjacent apartments before and after one of these is treated with an insecticide, particularly one containing pyrethrins or synthetic pyrethroids, well known for their flushing action.

An even higher rate of movement has been found *within* an apartment (*25*). The highest rate of movement was into the kitchen from other downstairs areas, amounting to 40% per week. Male cockroaches were more mobile than females. To be effective, control procedures should be used throughout an apartment rather than confining them to the kitchen and bathroom, as is commonly done. Also, the entire apartment complex should be treated rather than only those apartments for which complaints have been received.

Changes in Habits

Some domiciliary species appear to have changed their habits. Regarding the brownbanded cockroach, Gould and Deay (*81*) stated in 1938 that "Except when in search of food, it seldom visits the kitchen, and confines its activities to other parts of the house." Concerning the German cockroach, they said, "In an infested house, it is confined to the kitchen and lavatories, where it hides behind baseboards, in cupboards, iceboxes, and dark corners, and around water pipes." Apparently the habits of the two species have changed, for now there appears to be no great difference in the way they are distributed in a dwelling, although the German cockroach generally becomes widely distributed throughout the building only after it has reached a high population in the kitchen.

Numerical data have been gathered that support this common observation (*210*). In apartments with low infestation levels, 79% of cockroaches were in kitchens and 16% in bathrooms and adjacent linen closets. In apartments with high infestations, 36% were in kitchens, 31% in bathrooms and/or linen closets, and 33% in bedrooms and/or linen closets (*210*).

In another study of 18 low-income houses, visual estimates of the numbers of German cockroaches in different rooms were recorded as follows: kitchen > dining room > hall > living room > bedroom > bath (*206*). The presence of German cockroaches in bathrooms cannot be satisfactorily explained solely on the basis of either food (possibly hairs, dermal scales, mold) or hiding places, but the bathroom often does provide water, an important factor in cockroach survival (*158*).

Since nonrepellent insecticides (boric acid and, more recently, microencapsulated diazinon) have been used for control, German cockroaches are again tending to be less widespread in buildings (A. Slater, personal communication). The brownbanded cockroach is still more widely distributed than the German cockroach in dwellings and has more of a tendency to seek resting places on high shelves, behind pictures on walls, and behind high moldings, where the highest temperatures in a room prevail (*82*).

Probably the increase in central heating of homes has been a factor in increasing the area of dispersal of the German cockroach. However, even if some rooms become cool at night, the insects can find warmth in or about such appliances as refrigerators, stoves, water heaters, television sets, radios, and electric clocks.

Many cockroaches can be sustained by food particles that fall to the floor where people eat their meals or snacks near the television set, which might be in the living room, bedroom, or den. Modern homes may have two or three bathrooms, as well as air conditioners and refrigerators with evaporating pans, to provide moisture. Beer and soft drink cans and bottles not completely emptied may be left in more locations than formerly, and these also provide food and water.

The oriental cockroach has also changed its habits (179). It can survive farther north than formerly reported, and, as previously mentioned, it may be found outdoors in almost continuous freezing weather. The German cockroach also can be found outdoors in temperate climates more often now than formerly reported (179).

Structural Preferences of Cockroaches

All pest cockroaches (except the Asian roach; see Chapter 9, this volume) tend to avoid light and seek dark areas in which to spend the daylight hours. These areas may be attics, wall voids, dropped ceilings, soffits, garbage chutes, pantries, cabinets, and closets, as well as the voids under the false floors of some of these structures; and in almost any crack or crevice. Such harborages may provide only darkness and seclusion, or may also provide warmth and moisture. In some buildings that are old or of loose or careless construction, numerous out-of-sight areas are readily accessible to cockroaches and provide ideal harborage and breeding locations and areas of escape from insecticide deposits. They also provide passageways for movement to uninfested areas.

Microhabitats

Cockroaches in infested areas are often found in cracks and crevices. The smallest German cockroach nymphs can enter cracks as narrow as 0.5 mm; adult males and females without oothecae require cracks at least 1.6 mm wide (201). However, when given the choice of a series of eight spaces ranging (in 1.6-mm increments) from 1.6 to 12.7 mm apart, 67% of the adults congregated in the 4.8-mm space when the spaces were arranged horizontally. When spaces were arranged vertically, the proportion of insects found in the 4.8-mm space increased to 85% (26).

Normally, all stages and instars are present under field conditions. German cockroaches tend to occupy all cracks, ranging in size down to those wide enough only for first-instar nymphs. Adults drive large nymphs into cracks of less width than the adults occupy, the large nymphs drive smaller nymphs into cracks of even less width, and so on until all cracks down to 0.5 mm in width are occupied by cockroaches of the appropriate size.

American cockroaches behave similarly, except that spaces suitable for the various stages and instars are wider (68). Thus, narrow spaces such as those encompassed by the corrugations of cardboard boxes can provide shelter for many young cockroaches. The habit that cockroaches have of occupying every available refuge em-

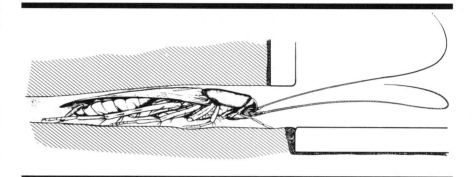

Figure 8.2
Characteristic position of a cockroach when at rest in a section of a narrow crevice.
SOURCE: Cornwell, P. B. 1968. *The Cockroach,* Vol. 1. Hutchinson, London.

phasizes the importance of examining paper grocery bags and any other containers of food, beverages, and laundry that may be brought into a home or a food facility to make sure they are free of these insects. If enough food and water are available, cockroach populations expand to the limit of the surface area of undisturbed dark harborage. Under these conditions, available harborage, rather than food and water, limits the size of a cockroach population. In most field situations, however, food and water are the common limiting factors (161).

In the food-storage room of a large "rest home," it was noted that thousands of German cockroaches had found harborage in a space of 2–3 mm between a shelf and a wall. The bodies of the insects were completely hidden from view except for a continuous band of waving antennae that extended from the aperture and could be seen by looking beneath the shelf (52). This is a common niche utilized by cockroaches (44) (Figure 8.2), a fact well known to PCOs.

Some areas where cockroaches harbor and breed may be inaccessible to insecticidal treatment or may be treated only through access holes drilled into voids. Most infested areas are accessible, yet represent only a small percentage of the total area inside the building. In general practice, insecticides are applied to a relatively small area of a building being treated for cockroaches.

Cockroaches moving from their harborage areas travel mainly along intersections, such as along the back edge of a shelf or the juncture of the floor or ceiling and wall. Insecticides or traps are most effective when placed at these intersections. The movements of cockroaches appear to be influenced more by the physical features of their environment than by such factors as food odors or pheromones (56–58, 62).

Effects of Structural Features on Cockroach Survival

In an experiment designed to study the distribution of German cockroaches in localized habitats, Ebeling et al. (58) constructed six mock-up closets, similar to those designed by Wagner et al. (197), each containing three shelves, in which cockroaches were confined by means of an electric barrier. The dimensions (in inches) of each closet were: height, 93; width, 31; depth, 39; and depth of shelves, 17. Distances (in inches) of the shelves from the floor were: lower shelf, 51; middle shelf, 65; and upper shelf, 79.

In one experiment, the mean percentages for the distribution of about 350 adult cockroaches in each of three empty closets, 24 hours after the insects were released, were 40.4 on the floor, 23.7 between the floor and the lower shelf, 3.7 between the lower and middle shelves, 4.6 between the middle and upper shelves, and 27.6 between the upper shelf and the ceiling (where the insects were released). The insects tended to congregate along intersections and in corners. Nearly two-thirds of them were below the lower shelf.

Location of Food and Water

To determine the effect of location of food and water on cockroach survival in large spaces, six empty closets were used. Dog food and water were placed on the floor in two closets, and on the top shelf in two closets. No food or water was provided in the other two closets. Fifty adult German cockroaches of each sex were released on the floor of one closet for each food-and-water arrangement and another 100 were released on the top shelf for each of the other closets.

The place where the cockroaches were released in these bare closets had no significant effect on their survival, but the location of the food and water was of critical importance. When food and water were on the top shelf, an average of only 3% of the insects survived for 18 days, whereas when food and water were on the floor, an average of 65.5% survived. There was no significant difference in the survival of males and females. The highest survival rate occurred when food and water were placed where the cockroaches preferred to congregate (58).

Complex Habitats

The addition of piles of clothing, newspapers, paper sacks, corrugated cardboard boxes, and the usual contents of closets increases cockroach survival. However, even under cluttered conditions, food and water must be close to the favored harborages to ensure survival. Cockroaches tend to congregate in areas that are physically attractive to them; the population declines if such areas are not also close to food and water. The success of the cockroach in human habitations probably depends on food scraps and water being located in dark, out-of-the-way places such as in cabinets, under the kitchen countertop, under the refrigerator and stove, along the back edges of shelves, and in cracks and crevices—places that are also structurally attractive to these insects. Favorable temperature and humidity would, of course, further enhance the attractiveness of such locations.

The mock-up closet experiments showed that German cockroaches are not able to maintain their numbers in an empty closet even when food and water are present. The presence of corrugated cardboard or the usual contents of a cluttered closet provide a necessary habitat for a thriving colony of cockroaches by adding the required degree of environmental complexity and safe refuges for small nymphs. A German cockroach population will decline, even when food and water are present, in an empty closet, box, or jar with nothing to increase the complexity of the environment (57; see also section on Influence of Light, below).

Considerations in Cockroach Management and Prevention

Crack-and-Crevice Treatment

The current tendency is to reduce the area treated and the quantity of insecticide applied as much as possible, particularly in areas where food is prepared or served. Labels for residual insecticide formulations now bear the wording, "Applications of this product in the food areas of food-handling establishments, other than crack-and-crevice treatment, are not permitted.'' This has resulted in the development of what has become known as crack-and-crevice (C&C) treatment, a technique in which small amounts of insecticide are introduced into cracks and crevices where cockroaches hide or through which they may enter a room or building. ''Such openings commonly occur at expansion joints between different elements of construction, and between equipment and floors. These openings may lead to voids such as hollow walls, equipment legs and bases, conduits, motor housings, junction or switch boxes'' (*117*).

Treatment Location Guides Cockroach location charts for restaurants, lounges, bars, homes, and apartments have been published to guide the PCO to the many locations where C&C treatment can be effectively applied. A toxicant usually has a longer residual life in a crack or crevice than when applied elsewhere, because of lower temperature, less moisture, and less air movement (*117*).

Blatticides Commonly used C&C insecticides in liquid formulations are acephate, bendiocarb, chlorpyrifos, diazinon, propoxur, and cypermethrin. Synergized pyrethrins or dichlorvos are often added for their flushing value. Sprays are usually applied with a compressed-air sprayer and a nozzle producing a pin-stream spray, so as to accurately confine the spray to the target area. In a laboratory test of C&C treatment, the highest mortality of German cockroaches occurred when the spray nozzle was inserted 4 mm inside the crack (*109*).

In a field test in an apartment complex in Virginia, there was approximately 30% higher mortality of German cockroaches when the insecticides (dilute emulsions of bendiocarb and acephate) were applied as a C&C treatment as compared to a fan-spray application (*211*). The C&C application required about three times longer (11 minutes compared to 3.5 minutes), but when one considers the time spent in preparation for fan-spray operations (entering the apartment, talking with the residents, and so on), the difference in application time is relatively small and is well justified by the great increase in control.

One formulation especially designed for C&C treatments contains a special solvent that evaporates almost instantly upon contact with air, leaving only the insecticide on the treated surface and virtually no runoff. C&C treatment is an important factor in cockroach management in food-processing establishments (*20, 117*).

Dusts and Baits Dusts may also be used in C&C treatments and in some situations may give better results because of their ability to penetrate deeply into cracks, crevices, and places inaccessible to liquid sprays. They are also usually registered at higher concentrations than liquids, so insects crawling over dusts are more likely to pick up a lethal dose of insecticide than when crawling over liquid spray deposits. This is especially true in the case of dusts that have, among their inherent physical characteristics, a positive electrostatic charge, such as boric acid and the original Dri-Die 67 formulation (silica gel with 4.7% ammonium fluosilicate). The electrostatic charge increases adherence to surfaces and to cockroaches that crawl through deposits of the dusts.

The nearly continuous monomolecular layer of ammonium fluosilicate on the Dri-Die 67 particles gives the dust a positive electrostatic charge. Dri-Die formulations are nonabrasive and are properly called sorptive dusts. They are insecticidal because they absorb the amorphous constituents of the insect epicuticle, leaving minute channels in an otherwise predominantly crystalline matrix and allowing water to evaporate from the water-bearing cuticulin substrate (*48, 51, 53*). The insecticidal efficacy of *abrasive* dusts is very limited and none has ever been proved to be of any practical value.

Dusts are less likely than sprays to be adversely influenced by the type of surface (wood, vinyl tile, asphalt tile, latex paint, and so on) on which they are deposited (*117, 156*). There is often an enormous difference in absorption rates of organic liquids into such surfaces (*58, 89, 131, 207*). As with liquids, specialized equipment is available for C&C application for dusts, such as the B&G Miniduster with adjustable tips, some fitting into cracks of very narrow width.

Toxic baits such as paste baits (in squeeze tubes) and pelletized formulations are also available. Like dusts, baits have longer residual life than liquid insecticides (*117*).

Microencapsulated Insecticides Microencapsulated insecticides would appear to be appropriate for C&C treatments. In fact, if this remarkably safe formulation continues to demonstrate the high degree of effectiveness noted by researchers to date, it may decrease the need for insecticides. The toxicant is covered with a thin chemical shell that retards its rate of volatilization and prolongs its residual life. For example, synergized pyrethrins, which would otherwise have no long-term insecticidal effectiveness against German cockroaches, were found to be about as effective as chlorpyrifos and diazinon one month after treatment when used in a 0.22% concentration as an encapsulated formulation (22).

Microencapsulated diazinon, for which the rat lethal dose (LD_{50}) is greater than 21,000 mg/kg of body weight, was applied as a 1% emulsion in apartment buildings. On the basis of percent reduction of German cockroaches 12 weeks after treatment (85.2%), it was approximately as effective as 0.5% chlorpyrifos (87.1%) and was surpassed in insecticidal effectiveness only by boric acid powder (95.6%) at one pound per apartment (172).

Sealing Cracks and Crevices In view of the importance of cracks and crevices as harborage for cockroaches and the various apertures that give cockroaches ingress and egress to wall voids and even adjoining rooms and apartments, it would appear to be good cockroach management to seal such cracks, crevices, and apertures. To test the feasibility of caulking as a technique to suppress cockroach populations, an experiment was designed in which the bathroom and kitchen sink cabinets were caulked in three apartments (only 41% of the cracks and crevices were sealed; the remainder were not accessible to the caulking gun and foam). No caulking was done in two other apartments. All five apartments were sprayed with 0.5% chlorpyrifos, with the nozzle set at "coarse fan." The apartments were trapped before treatment and at 1, 2, 6, and 20 weeks after treatment. The treatment was reapplied after 6 weeks.

Although after both the first and the second treatment there was initially a greater rate of decrease in cockroach populations in the caulked apartments, this initial advantage was later lost; in 2 weeks and again in 20 weeks, there was no significant difference in control between caulked and uncaulked apartments. The researchers attributed this lack of success to the large percentage of harborage sites that were not caulked and probably not reached by insecticide, and to the low level of sanitation. The insects that were denied harborage may have utilized clothes, soft drink containers, newspapers, and other clutter as alternative harborages.

It was concluded that German cockroach populations might be significantly reduced by a combination of caulking, sanitation, and improved construction, along with the careful application of a residual insecticide. However, there may also be situations in which it would be preferable to leave the cracks and crevices uncaulked and accessible to C&C treatments (70, 71).

Dry-Ice Fumigation

There are locations in which cockroaches cannot be reached via cracks and crevices. Of interest to the food industry are the food service carts used in hospitals. These carts have many areas where cockroaches can hide and avoid sprays and dusts. Experiments have been conducted to show that dry ice can be used to produce a safe and inexpensive fumigant gas (carbon dioxide) when applied according to a recommended dosage and time schedule (15, 189). At 26°, the minimum concentration of carbon dioxide required for the 24-hour period of the test was 21% in order to kill not only all the female German cockroaches with attached oothecae that were used as the test insects, but also to prevent hatching of their eggs over the 30-day period of observation. This method of fumigation has the advantage of being safe and not requiring specially trained personnel.

Exploratory Activity

Exploratory activity of German cockroaches was investigated in choice boxes, mock-up closets, and mock-up wall voids. Of these three devices designed to simulate typical cockroach harborage areas found in dwellings and other buildings, only the choice boxes (29, 61, 76, 79) were suited for simultaneous testing of a large number of insecticides.

Choice Box Tests Choice boxes (54) (Figure 8.3) were made of 0.25-inch white pine. They were 1 foot wide, 1 foot long, and 3.75

inches high, and had a partition wall in the middle that divided the box into two equal-sized compartments, one light and ventilated and one dark. A 0.5-inch hole was drilled as close as possible to the top center of the partition so that when the box was covered (with Plexiglass on

Figure 8.3
Wooden choice box with treated compartment kept dark by means of Masonite cover.
SOURCE: Ebeling, W., and D. A. Reierson. 1969. The cockroach learns to avoid insecticides. *California Agriculture* 23(2)12–15.

the light half and Masonite on the dark half), the hole served as the only entrance from one half of the box to the other. Insecticides were applied to the tempered Masonite floor of the dark halves. The boxes were later decontaminated by keeping them at 140° for 4 hours (62).

Adult male cockroaches (usually 20) were placed in the untreated part of each treated choice box, along with food and water. The partition holes were kept closed for 2 hours after the cockroaches were introduced. This gave the insects time to recover from being anesthetized (with carbon dioxide) and placed in the choice boxes. In daytime they gradually migrated to the dark half through the partition hole. If the hole was drilled even a short distance (e.g., 0.5 cm) from the top of the partition wall, the movement of cockroaches from the light to the dark half of the box was greatly delayed, demonstrating the strong tendency of the insects to crawl along intersections such as the juncture between the wall and the ceiling of the box (58, 62).

When German cockroaches are transferred to a new environment, they actively explore it. Exploratory activity of German cockroaches placed in new surroundings diminishes 65% within 30 minutes and eventually diminishes to about 20% of its initial intensity (49, 50). This low rate of activity is then maintained indefinitely. Unlike stimuli such as hunger and thirst, the exploratory drive is never satisfied and is inversely proportional to population density (55). Among the environmental factors that influence German cockroach movements in modified choice boxes (13), those that increased movement, based on the number of insects caught in sticky traps, were (a) environmental familiarity, (b) increased temperature, (c) increased number of holes between chambers, and (d) food deprivation. The factors that resulted in decreased catches were increased cockroach population density and increased fecal contamination.

Influence of Light Rauscher et al. (147) studied the effect of isolated factors (insecticide, food, water, and light) on the distribution of German cockroaches in choice boxes. They suggest that "test conditions should include only the factor being evaluated (i.e., choice box setup with both chambers dark) without food or water and with only one test surface (bottom panel) treated with insecticide." Current practice calls for treatment of only the removable bottom panels of choice boxes. The reason for light in the untreated chamber is to increase the severity of the test for repellency. Choice boxes generally

measure insecticidal efficacy and not just repellency alone (147). This, of course, is the reason they provide a good laboratory evaluation of the way an insecticide will perform in the field.

When the dark halves of choice boxes were treated with insecticide, mortality of German cockroaches increased with increase in light intensity in the untreated half, for they had greater difficulty habituating themselves to the lighted halves of the boxes. This also happened in empty mock-up closets in which toxicants were applied only at intersections where cockroaches normally crawl. But when the closets were cluttered with suitcases, boxes, paper sacks, clothes, and shoes to simulate normal conditions, increasing intensity of light *decreased* mortality, for the insects escaped the light, sought refuge in and under the clutter, and had less contact with treated surfaces. However, in continuous darkness, the rate of mortality also was lower in the cluttered closets (complex habitat) than in uncluttered closets, for the insects found harborage in the clutter and spent less time on the insecticide-treated surfaces. Thus, a complex environment decreases the efficacy of insecticidal treatments (58, 205).

Chemoreception

Cockroaches were represented among the ancient, primitive insects. They probably utilized sheltered, restricted habitats that were at least moderately damp, such as beneath stones and detritus and under bark. Food and mates were found by chance encounters. The waterproofing mechanism of the cuticle of such insects was relatively weak. It was improved cuticular waterproofing and the development of wings that freed insects from a restricted environment. Most cockroach species have wings but either do not fly about actively or, in some species, seldom, if ever, fly. [The Asian roach, *Blattella asahinai,* recently discovered in Florida, is a notable exception to this generalization (134; see also Chapter 9).] Improved chemoreception became advantageous for actively flying insects (125), but the primitive, cryptobiotic, nonflying or feebly flying insects such as the cockroaches continued to possess only relatively feeble powers of chemoreception, even among modern species.

Attraction to Food

It has long been known that cockroaches are attracted to certain food odors (199). Among the most attractive are bread, Coca Cola syrup, beer plus yeast, banana peels, and sliced apple or potato. But foods fail to attract cockroaches for more than short distances from their normal routes of travel or from the course of their random movements. German cockroaches can find foods that have been odor-neutralized as readily as foods in their normal, unaltered condition (*123*). A cockroach resting on a door hinge close to a potato-paste poison bait was observed for 2 weeks. The insect either did not readily accept the bait or did not wander in the right direction to find it (*123*).

When German cockroaches were placed near baits, they traveled in a random manner, continually sensing the area over which they were crawling, before contacting the bait (*123*). As might be expected, poison bait applied in a line across potential travel routes was more effective than bait placed on cards or jar lids distributed throughout the infested area. Mellanby (*120*) found that bait did not increase the catch of oriental cockroaches in his "Demon" cockroach traps.

Tests

A collaborator and I conducted an experiment to determine the extent to which German cockroaches and two other kinds of relatively primitive, cryptobiotic insects—firebrats (Thysanura) and earwigs (Dermaptera)—are attracted away from their normal routes of travel by food baits (*56*). We used three bait boxes, each 1.5 feet wide, 3 feet long, and 1 foot high (Figure 8.4). An electrified aluminum-foil barrier, as used in the mock-up closets, was installed inside the rim of each box to confine the test insects. In experiments with cockroaches, three wide-mouth, one-quart jars with bait and

three without bait were placed in each box. Each jar was provided with a film of finely divided, highly sorptive montmorillonite on its inner surface to serve as a barrier to trapped insects attempting to escape. Each jar also contained a tablespoonful of montmorillonite clay to kill the insects by desiccation.

The following quantities of bait were used per jar in three tests with cockroaches: (a) two half-slices of white bread; (b) a tablespoon of white flour; and (c) two slices of chipped beef. The jars were positioned 3 inches from the walls of the boxes so that the insects had to leave their regular pathway at the intersections of the floor and the walls of the box in order to encounter the jars. A watering jar and some dog chow were placed in the center of each box. Each box contained 100 cc of cockroaches of a size retained on a 4-mesh screen (mostly adult males and females), a total of approximately 2,000 insects for the three boxes. The bait boxes were covered with Celotex boards. The number of cockroaches caught in each jar in a 7-day period was determined. Bread attracted approximately one-half and beef attracted approximately one-third of the cockroaches into the trap jars. Flour attracted no significant number (*56*).

The cockroaches trapped in the control jars represented the degree of random exploratory activity of German cockroaches. The 7.6-fold greater number trapped in jars with bread indicates that cockroaches can be attracted, at least for short distances, to odors of dry foods. If bread is dampened, it is more attractive but it soon dries out or becomes moldy. It was advantageous to use a dry bait because all stages of the trapped cockroaches remained dry and could be retained on a screen of a mesh size that would also admit the clay (*59*).

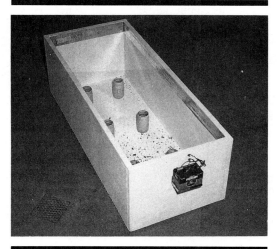

Figure 8.4

Research bait box with an electrified aluminum-foil barrier.
SOURCE: Ebeling, W., and D. A. Reierson. 1974. Bait trapping silverfish, cockroaches, and earwigs. *Pest Control* 42(4)24, 36–39.

Bread

Having determined the degree of attractancy of bread to cockroaches over a distance of 3 inches, it was then of interest to find out how far it could attract cockroaches within the dimensions of an 8-foot by 12-foot experimental kitchen. Four trap jars baited with bread, as described in the above experiment, were used in the test. Two were placed on the floor along its juncture with one of the walls and two were placed in the middle of the room, 6 feet from the wall. Dog food and water were available at two stations in the room where the insects normally traveled.

Trap yields were determined after alternate

3-day periods when (a) food was removed from the two food stations, and (b) food was returned. At the beginning of each period, 100 cc of German cockroaches retained on a 4-mesh screen (about 650 insects) were released in the room. The average yield per trap-day for the two traps placed along the floor/wall intersections was 13.5 cockroaches when food was absent from the two food stations, compared to 1.1 when the food was present. The average yield for the two traps in the middle of the room was 2.4 when food was absent compared to zero when food was present. Hunger greatly increased the exploratory activity of the insects. The experiments demonstrated the importance of placing traps where cockroaches normally travel in a room (i.e., at intersections, such as the juncture of wall and floor) and of removing from the room all other foods that may compete with the bait in the trap (i.e., practicing good housekeeping) (62).

Insecticide Baits

In insecticide baits, a toxicant is usually combined with an attractant, the latter usually a preferred food that serves as a carrier for the toxicant. Baits may be in the form of granules, pellets, pastes, or liquids. With a given quantity of a bait (e.g., boric acid in starch), small granules scattered widely are more effective than larger tablets not so widely distributed (192). (An important advantage of granules is that they are less likely to be found and eaten by small children and pets.) The addition of 0.5% lindane or fenitrothion to a boric acid bait resulted in an increased initial rate of mortality but, because of their repellency, these insecticides increased the period required for a 100% kill. On the other hand, methyl myristate at 0.02 to 0.2% or maltose at 10 to 15% concentration increased the amount of bait consumed (193).

In laboratory trials to compare the attractancy of various bait materials, Reierson and Rust (150, 151) used six bait boxes painted white, similar in size and shape to that shown in Figure 8.4. German cockroaches were confined to the boxes by an aluminum foil barrier (196). Traps (one-quart, wide-mouth glass jars) were placed in the boxes. Trapped cockroaches were retained by a 2-inch barrier of Teflon T-30 emulsion on the inside surface of each jar. Baited traps were placed in two corners of each box, and unbaited control traps were placed in the other two corners. The insects caught in the jars, along with hatched oothecae, were counted and removed daily.

The researchers started with volumetrically measured initial populations of a mixture of stages ranging from 1,020 to 2,930 insects per bait box. They allowed the insects to become accustomed to their new environment in the bait boxes for at least 4 days. The insects had access to dog chow and watering jars. The baits ranged in descending order of efficacy as attractants as follows: white bread > aggregation pheromone > ether extract of bread > acid extract of bread = banana > acid extract of Coca Cola > a proprietary 2% propoxur bait > five synthetic organic compounds reported to be highly attractive to some stages of the German cockroach (186).

Reierson and Rust (150, 151) attempted to draw conclusions as to the potential efficacy of trapping based on their counts of insects in the bait boxes when compared with counts of those in boxes without traps. (It should be borne in mind that during the 30-day period there would normally be a considerable increase in the cockroach population as the result of reproduction.) The average increase in cockroach population in the two boxes in which bread-bait traps were used was 15.2%, compared to 91.2% in the control boxes. The researchers concluded from these data that "inadequate control with traps baited with even very attractive substances would be expected as long as there is an abundance of food, water, and harborage. Additionally, the number of roaches trapped probably reflects the extent of an existing infestation rather than the degree of control being obtained."

Uses and Efficacy in Traps

Several firms sell cockroach traps. After experimenting with one of the available brands in which the attractant was a material described as having "a pleasant maple syrup odor," Barak et al. (14) recommended that traps be used for the following purposes:

(a) To detect low-level populations. The existence of a potential problem can be confirmed before a population explosion takes place.
(b) To locate problem areas or harborages. This can greatly enhance control efforts, allowing the PCO to intensify treatments in certain areas and perhaps solve continuing problems.
(c) To monitor population increases. Thus, the need for frequent and expensive applications or treatments can be minimized.
(d) To reduce or control infestations, as a primary method in certain instances but, more commonly, integrated with current methods of

chemical treatment to improve efficacy. Traps are more effective in reducing oriental cockroach populations than they are in reducing German cockroach populations.

Some investigators found that when attractants were used in the field to increase the effectiveness of insecticides against German cockroaches, they were ineffective. When the insects had adequate shelter and food, they were not attracted to purported lures over even short distances. Under field conditions, baits (blatticide plus attractant) did not provide satisfactory control, and their performance was not improved when the numbers of baits in kitchens were increased from 4 to 18 (150, 151, 174).

Comparison of Sprays and Baits

In tests made in an apartment complex, a spray of chlorpyrifos/dichlorvos (0.25%/0.25%) was compared with MaxForce (hydramethylnon) bait at 12 baits per apartment and Raid (chlorpyrifos) bait at 45 baits per apartment (25). Reduction in cockroach populations from the sprays peaked at 24 hours, followed by a steady decline, whereas reductions from bait treatments were initially low but were followed by a steady increase during 2 months of observation. At the end of that period, chlorpyrifos and hydramethylnon roach baits gave 88% and 86% reduction, respectively, in cockroach populations compared with 40% for chlorpyrifos/dichlorvos.

An average of 45 baits of chlorpyrifos per apartment gave a higher and more rapid initial kill compared to hydramethylnon, but after 2 weeks there was no difference in results between this treatment and hydramethylnon at 12 baits per apartment. Twelve hydramethylnon baits per apartment will control German cockroaches in severe infestations. Hydramethylnon would be the treatment of choice from the standpoint of ease of use and cost, but if a rapid reduction in the cockroach infestation is desired, the use of supplementary sprays is necessary (25).

Although the exploratory drive of German cockroaches is never satisfied, it does decline to about 20% of its maximum. Maximum activity occurs when the cockroaches are presented with a new environmental situation. Good performance of an effective bait might be maintained by periodically moving it to new strategic positions in order to renew exploratory activity of the cockroaches (152). Cockroach activity is also increased by hunger, thirst, and deprivation of accustomed harborage, pointing out again the importance of sanitation.

Traps

Monitoring Cockroach Populations

Traps have been used for many years as a means of estimating relative infestation levels of cockroaches in buildings before and after insecticidal treatment.

Mason Jar Traps

Traps used in houses and apartments are generally placed in the kitchen (59). These are wide-mouth, one-quart Mason jars with a film of sorptive clay on the inner surfaces, as previously described, and with two half-slices of fresh white bread as bait. Two traps are used per kitchen (one in back of the stove and one in back of the refrigerator); both are pushed up against the wall so as to be in the path most frequently traveled by cockroaches. They are left there for one week. In severe infestations, 1,500 or more cockroaches of all stages and instars have been trapped in a single jar. This represented only a small sample of the total population.

Roatel Versus Mason Jar Traps Ross (160) compared the Mason-jar trap with the Roatel, for trapping B. germanica. The Roatel had several one-way entrances at floor level. The two types of traps gave closely similar results except for small nymphs, which were less inclined, or less able, to climb the distance to the tops of the jars. The Roatels captured four times more small nymphs than the jars. In an area where water is scarce, cockroaches are more attracted to the traps containing water (in sponge-stoppered vials), but where water was plentiful, dog food attracted greater numbers.

Flushing Versus Mason Jar Traps Cockroaches can be flushed out from heavily infested areas, such as beneath and behind a refrigerator or stove, and counted as they come into view. Pyrethrins, usually at 0.25% concentration in an aerosol formulation, are used as the flushing agent. Piperonyl butoxide enhances the flushing and killing actions of pyrethrins. In experiments conducted in an apartment-house complex with low to moderate German cockroach infestations, trapping with jars baited with white bread was superior to flushing with an aerosol of 0.25% pyrethrins plus 1.25% piperonyl butoxide, and was approximately equal to light flushing with an aerosol of 3.34% pyrethrins normally used for monitoring cockroach populations. Visual examination unaided by flushing produced the lowest counts (150, 151).

Sticky traps

Cockroach population changes were monitored by Kuru Kuru traps (Tokiwa Chemical Industries, Ltd., Japan), 10 traps in each of 10 apartments (10). These are tent-shaped cardboard structures, open at both ends, with additional side openings, and with the inside surface coated with a nondrying adhesive containing a bait in which cockroaches become ensnared. For five consecutive nights, the sticky traps were removed, the captured cockroaches were counted, and new traps were set. The traps caught nearly 20,000 cockroaches, but the average number caught per night did not diminish. The data suggest that the traps had no effect on the population.

Subsequent, more extensive field trials were conducted with another brand of sticky trap, 15 replications of each, examined for 10 successive nights (11). A flat sticky surface with no sides was as effective as traps with sides, but the latter type was easier to handle. A sticky trap with a spacer that kept the trap 2 cm from the vertical surface against which traps are normally placed caught significantly fewer German cockroaches per trap: 21.4 compared to 49.5 for the standard traps (11). This confirmed the common observation that traps of any type must be placed flush against a vertical surface for maximum results, because cockroaches tend to travel along the intersection of vertical and horizontal surfaces.

Electrified Traps

In further tests, an electrified can trap was compared with a wide-mouth jar with a 2-cm-wide barrier of petroleum jelly inside the mouth of the jar and 1 cm down from the top (10). The unbaited electrified traps consisted of a 12-inch-square plastic bowl, 5 inches deep, with an electric barrier on the inside 1 inch below the rim, with a power range of 12–25 volts. They were far superior to the unbaited jars, trapping about 15 times more cockroaches. Laboratory tests had shown that some of the cockroaches backed away from a petroleum jelly barrier and were not captured. This difficulty can be avoided by dusting the inner surface of the jar with a sorptive clay which acts as a barrier and also kills the cockroaches by desiccation (59).

Laboratory Tests of Traps

Several experiments have been designed to compare the efficacy of various trapping methods (138). One study was set up in four mock-kitchen chambers (Figure 8.5) in a large, controlled-environment chamber at Purdue University. It was designed to simulate German cockroach harborage conditions commonly found in households. Electric barriers were installed at the perimeter of each chamber. Four population levels were used: 50, 100, 250, or 500. The population profile was balanced, with equal numbers of small nymphs (instars 1 to 3), large nymphs (instars 4 to 6), nongravid adult females, and adult males. After the insects had become acclimated to the chambers for one week, one population sampling technique was utilized in each chamber and trap yields were recorded. The chambers were then disassembled and all remaining insects were collected, counted, and destroyed. Various proprietary and experimental traps and baits were tested, as well as a visual counting method. An important finding was that the selected population sampling technique performed essentially the same with all four cockroach population densities.

Field Tests of Traps

On the basis of the above-mentioned laboratory study, coupled with the results of a field study in specific closet areas of urban apartments (136), five population sampling techniques were evaluated in low-income urban apartments. Of all the traps tested, the jar traps as used at the University of California (150) yielded the most accurate quantitative information while sampling the smallest portion of the population (138). In field trials, there was great random variability in numbers of German cockroaches in successive samplings of any particular trap and trap location. This was probably the result of changing spatial distribution and movement patterns of the German cockroach populations as affected by unspecified environmental changes. On the basis of these tests, it is recommended that 10–15 traps should be used in each apartment (75–140 m²) (138).

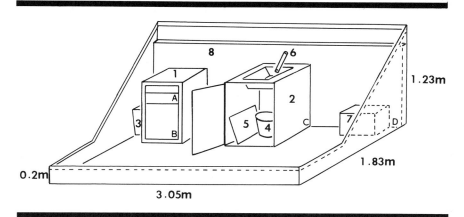

Figure 8.5
Diagram of a mock-kitchen chamber used in the laboratory study of population sampling techniques for German cockroaches. Numbered items are (1) cabinet with drawers, (2) sink unit, (3) empty wastebasket, (4) plastic bucket to catch dripping water, (5) two large grocery bags folded for storage, (6) plastic tubing carrying water into the sink, (7) case of empty beer bottles, and (8) wood paneling spaced 3 cm from back wall of chamber to create a wall void. The double line around the perimeter represents a double line of metal tape which, when charged by a 22-volt direct-current battery, served as an electric barrier to confine the cockroaches. Letters indicate locations where traps were positioned: (A) inside drawer, (B) inside cabinet, (C) inside sink unit on floor, and (D) in corner near beer case.

SOURCE: Owens, J. M., and G. W. Bennett. 1983. Comparative study of German cockroach (Dictyoptera: Blattellidae) population sampling techniques. *Environmental Entomology* 12(4)1040–1046.

Control of Cockroach Populations

Because of increasing environmental concerns and more regulatory restrictions on the use of insecticides, there will probably be an increased interest in environmentally acceptable concepts, including trapping, the latter not only for monitoring infestations, but also as an integral part of the cockroach management program (74, 143, 144). Trapping may be particularly useful for slow-breeding species such as American, smokybrown, and oriental cockroaches. Traps used for this purpose have usually consisted of pint jars that retain cockroaches either by means of a delicately balanced, hinged trap door or a band of petrolatum 3 or 4 cm wide applied to the upper interior surfaces of the jars. Such traps may be practical in certain household situations against species that breed indoors, such as the German and the brownbanded cockroaches (144), but in commercial usage a "convenient, disposable, long-lasting, inexpensive trap would be more practical" (14).

Sticky Traps

Although originally used principally to monitor cockroach populations (14), sticky traps with various configurations of adhesive have also been evaluated as to their potential as control devices. In one study using four brands of sticky traps, trap catches varied widely when the traps were placed in laboratory arenas and the cockroaches were confined by means of an electric barrier plus a petroleum jelly band (128). The species most likely to be trapped was *Blatta orientalis*; the least likely, *Supella longipalpa*. The relative effect of trapping on the various cockroach populations can be expressed as *B. orientalis* > *Periplaneta fuliginosa* = *P. americana* > *S. longipalpa* > *Blattella germanica*. During the 2-week trapping period, most cockroaches were caught during the first 24 hours. Although more German cockroaches were trapped than any other species, the decrease due to trapping was more than offset by the high reproductive potential of this species.

Electrified Traps

In a field test, an electrified trap (as described above) was placed on the floor in a corner of a hospital canteen with a bait of bread and water. Each week the cockroaches (*B. germanica*) were removed and the trap was reset with fresh bread and water. Weekly catches ranged from over 2,000 the second week the trap was in place to an average of less than 100 beginning with the fifth month of the test. The researchers doubted that complete eradication could be accomplished but noted that the same doubt could be raised in the case of insecticide treatments (31). In field trials of 2–4 months' duration with three unbaited, electrified traps in each of eight apartments in Kansas, cockroach population reductions ranged from 0 to 76% (10).

With electrified traps, the number of cockroaches captured was approximately proportional to the number of traps per apartment. With nine traps per apartment (the largest number used), 6,688 males, 6,182 females, and 55,475 nymphs were captured. In another experiment conducted in a heavily infested home, four types of traps were evaluated for 10 successive nights. One trap of each design was placed adjacent to the stove, refrigerator, and shower. The numbers trapped per night were recorded: electric traps, 1,082; sticky traps, 432; jar traps, 62.5; and jars wrapped with paper, 57.4. On the basis of sticky-trap catches on the first and tenth nights, the cockroach population was reduced 27%.

Baits

The most effective baits are fresh or dehydrated apple segments and/or pieces of sponge saturated with a commercially manufactured, oil-based, apple-flavored compound. These baits tend to lose their effectiveness after several weeks of use. Inexpensive, potent, synthetic attractants with long residual life would be much more suitable as baits. The feasibility of baits consisting of synthesized chemicals such as sex and/or aggregation pheromones should also be explored (74, 143, 144).

Outdoor Trapping

Outdoor trapping can reduce and might practically eliminate predominately outdoor species such as American and smokybrown cockroach populations and limit the number of cockroaches entering a home. The protection required for traps placed outdoors can be provided by cutting No. 10 cans open on the side and spreading them apart to produce a four-legged arrangement to fit loosely over the traps.

Repellency of Insecticides

The tendency of cockroaches to be repelled to at least some extent by all insecticides is one of the main factors limiting the effectiveness of chemical control of these insects (24, 27, 58, 59, 62, 94, 169–171, 183). Cockroaches that fail to pick up a lethal dose of toxicant after a number of brief contacts with insecticidal deposits may be successful in avoiding further contacts. Many may be driven into adjoining rooms or apartments. Thus, this behavorial characteristic, strongly exhibited by *Blattella germanica, Blatta orientalis, P. americana,* and *S. longipalpa,* is a major factor in the ecology of these cockroaches.

Even a repellent insecticide might be effective if it is lethal on initial contact and application is so thorough that it penetrates all refuges (*152*). But this uncommon contingency might be of no avail if the insect population has become resistant.

Effects on Interapartment Movement

Repellent insecticides can increase interapartment movement of cockroaches to levels as high as 30% per week (*12, 54, 59, 137*). Therefore, German cockroach control should be extended beyond the usual kitchen and bathroom treatments to include the entire apartment, the entire building, and ideally, the entire community (*137*). On one night following treatment, German cockroach populations increased in 37.5% of the apartments that were adjacent to units treated with chlorpyrifos plus dichlorvos and in 50% of the apartments adjacent to units treated with dichlorvos total-release bombs (*12*).

In one investigation involving the marking of approximately 6,000 adult German cockroaches of both sexes, 14% were recaptured (*167*). Of these, 88.3% were from the apartment in which the traps were placed and 11.7% were from an adjacent apartment. Of the cockroaches from the adjacent apartments, 75.1% came from those with common plumbing connections. Males moved about more actively than females.

Using laboratory-reared, mutant black male German cockroaches to avoid the need for labeling the insects in their interapartment migration studies, Akers and Robinson (*1*) found that apartments with poor sanitation can be significant sources of infestation for adjacent apartments with good sanitation.

Choice Tests

Various tests were developed for determining the behavior of German cockroaches when they are given the choice of entering or avoiding areas that would normally be attractive but which have residues of liquid insecticides or deposits of insecticidal dusts, dust diluents, or other powders. Areas such as voids under built-in cabinets (particularly common in dwellings) were sim-

ulated by devices that could be used under controlled conditions in the laboratory, namely mock-up closets, a simulated kitchen, choice boxes, and mock-up wall voids.

Twenty adult male nonresistant German cockroaches were placed in the light compartment of each of from three to five choice boxes (Figure 8.3), along with dog chow and water, for each insecticide used in an experiment. Generally, treatment consisted of painting the floor of the dark half of the box with 3 mL of insecticide solution or brushing on the floor 10 cc of insecticide dust. The insecticide was applied a day before the insects were placed in the light halves of the boxes. To determine the efficacy of the insecticide, the dead insects in both halves of each box were counted daily. To determine the degree of repellency of the insecticide, the live insects in the dark half were counted and compared with the number of live insects in the light half. On this basis, all insecticides were found to be repellent, but they differed in degree of repellency. Every powder tested was found to be repellent, even those that might be expected to be attractive, such as flour and sugar.

Boric Acid

The least repellent powder tested was boric acid. On the first 2 days, the percentage of live German cockroaches in the dark (and treated) halves of the boxes treated with boric acid was nearly as high as in the corresponding locations in the untreated controls. On the third day, all live insects were in the dark halves of the boxes treated with boric acid, but there were only a few that were still alive and they were too feeble to climb the partition wall and get back into the light halves of the boxes.

All tests showed that even boric acid is slightly repellent, but not sufficiently so to keep the cockroaches from repeatedly entering the dark halves of treated boxes and contacting the powder. Boric acid is less repellent and more toxic than borax and is therefore much more effective. Cockroaches remove dust from their antennae and legs by passing them through their mouthparts; thus, they consume significant quantities of boric acid. Boric acid also penetrates the cockroach cuticle (*64*), and has the advantage of having a positive electrostatic charge.

Although the tests showed flour and sugar to be repellent, cockroaches will feed on them when they are in the form of tablets. These powders did not reduce insecticidal effectiveness when mixed with boric acid in 50% concentrations, provided the cockroaches were continuously confined to the mixtures, as in covered petri dishes (62). But in choice boxes, as well as in tests made in apartment houses, the repellency of flour and sugar reduced the effectiveness of boric acid (59). Sugar also caused residues to become less fluffy when it deliquesced and subsequently dried up. The deposit could then no longer be picked up as readily by the cockroaches.

One percent concentrations of stabilizers, such as magnesium stearate or tricalcium phosphate, increase the intrinsic effectiveness of boric acid by causing it to be more finely divided, but they also cause it to be slightly more repellent (58). Stabilizers are useful in proprietary formulations because they prevent lumping and facilitate uniform distribution of fine films of powder. They allow for the maximum advantage to be derived from the inherent electrostatic charge of boric acid.

Repellent Versus Nonrepellent Formulations

In an experiment to show the basic difference in slope of the time-mortality curves for the German cockroach when treated with a particularly nonrepellent versus a highly repellent insecticide, 10 cc of screened boric acid was brushed uniformly onto the floor of the dark half of each of five choice boxes and 3 mL of 1% propoxur emulsion was applied in five others (54). Forty adult male German cockroaches were placed in the light halves of the boxes. The mortality curves in Figure 8.6 show the typical trend when boric

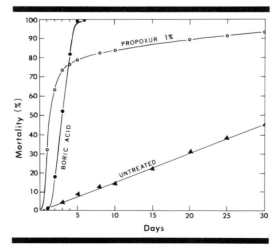

Figure 8.6
Time mortality curves for German cockroaches in treated choice boxes, showing characteristic curve for an insecticide of low toxicity but almost completely nonrepellent (boric acid) and one of high toxicity but highly repellent (propoxur).
SOURCE: Ebeling, W., and D. A. Reierson. 1969. The cockroach learns to avoid insecticides. *California Agriculture* 23(2)12–15.

acid, which is not sufficiently repellent to prevent repeated entries into the treated dark compartment, is compared with a repellent insecticide.

With boric acid, the rate of mortality follows the typical sigmoid time-mortality curve, whereas with the other insecticide the mortality rate may

initially follow this curve but eventually the curve becomes hyperbolic, showing the influence of insects that have "learned" to stay out of the dark compartment and have accustomed themselves to remaining continuously in the light halves of the choice boxes.

In an earlier experiment (62), the mortality curve for Dri-Die 67 in choice boxes was found to be no different than that of the untreated control. No tracks of German cockroaches were found on the floors of the choice boxes that were dusted with this fluorinated silica aerogel. It was avoided completely. The practical significance of repellency in field trials in which boric acid is compared with other insecticides has been repeatedly demonstrated (9, 59, 173, 176).

Behavior Modification

"Learning," as used above, is defined by Thorpe (188) as "that process which manifests itself by adaptive change in individual behavior as the result of experience." Thorpe believed that even under natural conditions the capacity of an insect for learned modifications of behavior is an important factor in its potential for survival. Learning by cockroaches has been stimulated, as with all other animals, by means of punishment or reward. In experiments conducted by Ebeling et al. (62), insecticide deposits took the place of the traditional punishment of electrical shock used by most animal behaviorists.

Predicting Performance

Devices such as choice boxes, which simulate places in buildings that provide harborage for cockroaches, give the researcher a much better idea of how an insecticide formulation will perform in actual field usage than does the conventional confinement of the insects with the insecticide deposit. In fact, the latter may be actually misleading, as was demonstrated in tests made with eight dust formulations of boric acid and with six emulsions of organic insecticides. With insecticide dusts or liquids, there is often an inverse relationship between their intrinsic toxicity and their effectiveness in choice boxes (58). Among the liquids, chlordane was the least toxic to nonresistant German cockroaches in a test in which the insects were confined to the residue of six insecticides in petri dishes. But it was the most effective of the six insecticides in choice boxes. Chlordane had relatively low toxicity and relatively low repellency. This may account for its excellent performance against German cockroaches before they became resistant. Two percent hydramethylnon (MaxForce, Combat) which has a slow action against cockroaches is also, like boric acid, practically non-repellent (*152*).

After initial trials with German cockroaches in choice boxes, experiments were extended to include comparative tests with the American, oriental, and brownbanded cockroaches. The descending order of repellency of insecticides tested was Drione > propoxur > diazinon > chlordane > sodium fluoride > boric acid. This was also the descending order of their toxicity to cockroaches when the insects were continuously confined to the deposits. In general, the effectiveness of these insecticides against the cockroaches in choice boxes was inversely proportional to their toxicity (and repellency) as determined in the petri dish tests (continuous exposure) (*60*).

A noteworthy exception was propoxur when used against the oriental cockroach. This is a relatively sluggish insect and apparently has less success than the other species in avoiding the extremely toxic propoxur deposits soon enough to avoid picking up a lethal dose. Propoxur at 0.5% concentration was found to be very effective against oriental cockroaches when applied in buildings and outdoors around the perimeters of the buildings. Treatment both outside and inside the building is particularly important with this species (*60*).

Bait Repellency

Repellency can also be demonstrated with baits. A bait of boric acid mixed 1:1 with oat flour was compared with 0.25% iodofenphos in a commercially produced, honey-based bait. The comparison was made in an apparatus that allowed adult male and female German cockroaches equal access to bait with or without toxicant (*16*). The test insects came from a field strain of German cockroaches resistant to dieldrin and bendiocarb but not to iodofenphos or boric acid. The toxic effects of boric acid plus oats became apparent more slowly than those of iodofenphos gel. But after exposure for 24 hours, all cockroaches with access to boric acid plus oats died within 9 days, whereas some of those exposed to iodofenphos gels survived after 21 days. No gel with iodofenphos was eaten after 4 days; the gel bait containing no insecticide was eaten throughout the duration of the experiment (*16*).

Wall Void Experiments

Laboratory experiments were conducted us-

Figure 8.7.
Setup for wall void experiments includes 5-gallon cans with a maze, food, and water, and an electrical barrier at the inside rim. Each can was connected with a wall void by a ⅞-inch pipe that allows the cockroaches to enter the void and return to the can.

SOURCE: Ebeling, W., R. E. Wagner, and D. A. Reierson. 1966. Influence of repellency on the efficacy of blatticides. I. Learned modification of behavior of the German cockroach. *Journal of Economic Entomology* 59(6)1374–1388.

ing insecticidal dusts in 24 mock-up wall voids designed to simulate those in standard hollow-wall construction—14 inches wide, 4 inches deep, and 8 feet high. Five-gallon cans were connected to the lower panels of the wall voids by means of ⅞-inch pipes that allowed for passage of insects between the cans and the voids (Figure 8.7). Each can was provided with a corrugated cardboard maze for harborage, along with food and water. An electric barrier was attached to the inner rim of each can. A cork was temporarily placed in each passage pipe. Enough dust (ca 40 cc) to cover the entire inner surface of the void with a film was blown into the void by means of a water-type fire extinguisher via half-inch holes. The bulk of the dust, however, settled onto the floor plate.

Fifty cc of German cockroaches (mostly adults) retained on a 4-mesh screen were placed in each can. The corks in the pipe passageways leading to the voids were removed 2 hours after the insects were released in the cans. By that time most of them had sought refuge in the mazes (*57, 58, 62*).

One might expect that the cockroaches would have little incentive to enter the wall voids as long as harborage, food, and water were provided in the cans where the cockroaches were released. But exploratory activity continues throughout the life of each cockroach, as mentioned previously. The insects entered untreated voids freely. If sufficient numbers were placed in the cans, they invaded the wall voids in such great numbers that they formed continuous masses along the intersections of the studs and the wall and even on parts of the plane surfaces. Within a month the crevices between studs and laths contained large numbers of first- and second-instar nymphs hatched after the adults had been placed in the cans.

As in the choice boxes, varying degrees of repellency were manifested by the different insecticides blown into wall voids. Repellency resulted in different numbers of cockroaches surviving the treatments, depending on the ability of the insects to learn to stay out of the treated wall voids before picking up a lethal dose of the insecticide.

In this experiment, the 50-cc quantity of cockroaches placed in each can was assumed to number 326, based on a count of one such quanity. Equating this number to 100, the relative numbers of live insects in the cans and connecting voids 30 days after they were released were determined: boric acid, 9; 2% diazinon, 63; borax, 89; Drione, 124; Dri-Die 67, 128; sodium fluoride, 139; 2% propoxur, 210; and untreated control, 330 (3.3-fold increase).

Boric acid caused the greatest reduction in the cockroach population. Borax and diazinon dust were the only other materials that reduced the number of cockroaches. Relative efficacy of the insecticides was not related to their toxicity to cockroaches as indicated by tests in which the insects were confined continuously with the insecticide and in which Dri-Die 67 (silica aerogel with 4.7% ammonium fluosilicate), sodium fluoride, diazinon, Drione, and propoxur were all far more potent than boric acid (58, 62). They were strikingly inferior in their ability to decimate a cockroach population when the insects had the choice of entering or avoiding the insecticide-treated but otherwise attractive wall voids.

Field Experiments

As in laboratory experiments, tests made in buildings showed that the least repellent insecticide, boric acid, resulted in the most effective control of cockroaches. Cockroaches continued to reenter the dusted areas until they succumbed to the relatively feeble toxic action of the powder. An added advantage of the boric acid is that it is inorganic and therefore has a long residual effectiveness when compared with organic insecticides (54, 59).

Most insecticides are at least to some degree repellent. When applying such insecticides, it is desirable to treat all potential harborages throughout a dwelling, even if some areas do not appear to be infested at the time of treatment. Cockroaches repelled from an infested area (e.g.,

kitchen or bathroom) may seek harborage in areas previously uninfested. In apartment buildings there is a tendency for cockroaches to migrate from treated to untreated apartments, suggesting the desirability of treating the latter also, if this can be arranged.

Simple Versus Complex Environments

Tests made in apartment buildings also showed that the results of insecticide treatments in "simple environments" (free of clutter) are far superior to those made in complex environments, with excessive clutter in pantries, cabinets, and closets (54). Clutter gives cockroaches ample opportunity to find harborage in locations free of insecticidal residues.

Liquid-Dust Combinations

Better results are obtained when liquids and dusts are used together in a control program than when either is used alone. Liquids can be applied in areas in which dusts would be unsightly. Liquids and dusts act in a complementary fashion (28, 30, 59, 61, 95, 126). The advantage of the liquid-dust combination is greatest when sanitation in a treated building is substandard (95).

Repellency Considerations

One of the most important aspects of ecology as it relates to cockroach control is to be aware of the role of repellency as it affects the efficacy of the various insecticides. Recognition of this fact has been an important factor in the striking success of the cockroach management program for student housing at the University of California, Berkeley, under the supervision of Arthur Slater of the university's Office of Environmental Health and Safety. This had been a particularly difficult program because of the objection by some of the students to the use of conventional insecticides.

According to Slater et al. (181),

> The use of built-in controls by treating voids with boric acid powder has turned these areas into traps and has resulted in a continuing two-thirds reduction in cockroach problems over a four year period. Spraying with Knox Out 2 FM® has enabled an additional reduction in numbers of treatments. Student response to the programs improved, because of the use of almost odorless pesticides. . . . This pest management (applied ecology) approach has been effectively utilized to resolve German cockroach problems in other private and institutional housing facilities.

Comparative Tests

In field experiments comparing (a) straight boric acid powder screened to eliminate lumps, (b) proprietary boric acid formulations, and (c) a wide array of organic insecticides that included those currently most commonly used by PCOs, one of the boric acid powder formulations (Roach Prufe) gave long-term results superior to those from other boric acid formulations, straight boric acid, and the organic insecticides (9, 176). Its superiority was evident one week after treatment, and in 3 months it was the only insecticide that had shown 100% control.

Previous field tests had confirmed that additives to give the original Roach Prufe formulation a light bluish color and a bitter taste (to reduce the chance of the powder being eaten by small children) had serendipitously increased its effectiveness as a blatticide (173). An increase in boric acid's inherent electrostatic charge, accomplished during the manufacturing process, probably added still more to the effectiveness of the Roach Prufe formulation.

Boric acid or silica aerogels are inorganic powders that can be dusted into enclosed areas such as attics, wall voids, dropped ceilings, soffits, and voids under cabinets and built-in appliances, at the time of construction of a building. The dust can be conveniently applied with a water-type fire extinguisher or a B&G Duster No. 2100 D through holes drilled in appropriate places. Such treatments eliminate many of the usual harborage and breeding areas for cockroaches for a very long period, possibly the life of the building, and should greatly facilitate control of cockroaches and several other kinds of insects that may enter the building (63, 127). Boric acid, being essentially nonrepellent, has the added advantage of acting as a trap to eliminate cockroaches attracted to the dusted harborage areas.

Pheromone Extracts to Reduce Repellency

Among natural products that might be added to insecticides to make them less repellent (or even attractive) to cockroaches, various pheromones are logical candidates for testing.

Sex Pheromones

The ability of cockroaches to respond to sex pheromones is relatively limited. Male German cockroaches have to make physical contact with females, which produce a nonvolatile sex attractant on their integument, before sexual discrimination takes place. This is known as contact chemoreception (162). The female sex pheromone of the American cockroach is more volatile and, when dispensed from traps, has attracted a significant number of adult males when compared with controls (19). However, attracting only adult males is an obvious weakness of such traps. Moreover, production of and response to sex pheromone by American cockroaches are strongly temperature dependent (4).

Aggregation Pheromones

Aggregation pheromones (APs), attracting both sexes and all stages, are expected to be more useful. Ebeling et al. (58) noted that sections of corrugated cardboard from mazes used for harborage of German cockroaches in rearing cans were very attractive to these insects. The cockroaches reached the immediate vicinity of the sections of feces-contaminated paper by random movements, then climbed onto and remained on the paper. But when the paper was dusted with either inert or toxic dusts, it was almost completely ignored. A chloroform extract of feces-contaminated, corrugated-paper mazes yielded a yellowish, viscous liquid that was also very attractive to German cockroaches.

Aggregation pheromones of cockroaches, present in the feces and on the bodies of the insects and spreading readily on the mobile epicuticular lipid, are responsible for the insects tending to aggregate (18, 33, 103–106, 113, 122, 140, 142, 155, 175, 184, 200). Not all cockroach species utilize APs, nor are APs strictly species specific when the insects are given a choice. Nymphs of *P. americana* and *P. fuliginosa* are most responsive to their own pheromones, but also demonstrate some degree of interspecific aggregation (142). But, in general, nymphs aggregate in response to their own pheromone (142, 168). This indicates that a species may mark its preferred harborages with its own AP (168). APs may also serve to mark forage sites and trails (113).

Wileyto et al. (200) made bioassays with German cockroaches from a natural population that had been subject to mortality due to control efforts. These cockroaches displayed a stronger aggregation response than the laboratory strain, a result that the researchers interpreted as being consistent with the hypothesis that "aggregation serves to ameliorate a harsh environment and promote individual survival."

Repellency Reduction Experimental evidence shows that cockroach fecal deposits and odor can reduce the repellency of many spray constituents (24). Although APs attract only over short distances, they might make the area of insecticidal activity more attractive (*155*). American cockroaches seldom inadvertently crawl over a surface sprayed with insecticide unless they are abruptly placed in a strange environment, but aggregation pheromones as well as sex pheromones show promise in decreasing the repellency of blatticides to this species (*17*).

Rust and Reierson (*170*) devised testing procedures to determine whether the AP of the German cockroach could be added to insecticides to reduce their repellency and thereby increase their insecticidal efficacy. They prepared a pheromone extract from cockroach feces collected from cans in which thousands of insects had been reared. They first determined the relative intrinsic toxicity of six insecticides when adult male German cockroaches were continuously confined in petri dishes. The order of toxicity, based on the time required for 50% knockdown (KT_{50}) was propoxur > Dri-Die 67 > chlorpyrifos > diazinon > Dri-Die 68 (silica aerogel without fluoride) > boric acid (*170*).

The next test involved a technique similar to the small-bowl assay (*105*) which gave the insects an opportunity to avoid contact with insecticide-treated surfaces. AP added to 21 cm² of filter paper at the rate of 2.5 and 25.0 mg of pheromone extract consistently increased the effectiveness of every insecticide except boric acid. The AP reduced repellency and increased the period of contact with the toxicant. Effectiveness was not greatly increased by using the higher of the two concentrations of AP.

Chemometer Tests Rust and Reierson (*170*) also experimented with a custom-made, 12-chamber transparent Plexiglas chemometer (Figure 8.8) to further test the degree of repellency of insecticides with and without AP. Twelve circular chambers, 5.6 (ID) × 2.8 cm, with removable lids, were connected by clear plastic tubing of 1.27 cm (ID) with a circular staging arena, 9.5 cm (ID) × 2.8 cm, also with a removable lid. Movement of air through the chemometer was facilitated by 1-mm-diameter holes in the lids of the chambers. One hundred anesthetized male German cockroaches were placed in the arena, along with a piece of dry dog chow. The number of dead and live insects was determined at 18 hours and at 7 days. The tests were replicated at least three times.

Relative Avoidance Index To facilitate comparison of insecticides, a Relative Avoidance Index (RAI) was used:

$$RAI = [kd\,(100 - \overline{X})]/KT_{50}$$

where $kd = 0.1238$ and represents the performance of 1% diazinon, $\overline{X}$ represents the mean number of dead insects at 7 days; and KT_{50} is the time period for 50% knockdown. The RAIs in decreasing magnitude of repellency were propoxur > Dri-Die 67 > Dri-Die 68 > chlorpyrifos > diazinon > boric acid. This order of repellency was almost the same as the decreasing order of intrinsic insecticidal activity. Except for Dri-Die 67 and boric acid, the addition of AP invariably reduced the RAIs.

Rust and Reierson (*170*) condensed and combined their data from the bowl assay device and the chemometer to show that with nymphs as well as adults, extracts of cockroach feces combined with toxicants increased mortality beyond that obtained with the toxicants alone (except for boric acid). The increase, however, was not statistically significant. The AP had its greatest effect in increasing cockroach mortality when used with diazinon, an insecticide of only moderate repellency, rather than propoxur, which the RAI test indicated had the highest degree of repellency among the insecticides tested.

In a later experiment, Rust and Reierson (*169*) showed that repellency of chlorpyrifos, diazinon, and propoxur was directly proportional to the concentration of toxicant in the insecticide residue contacted by the cockroaches. A given quantity of AP increased insecticidal effectiveness most when used with the lowest concentrations of toxicant applied. Thus, 0.25% chlorpyrifos + AP and 0.5% diazinon + AP produced significantly greater mortality than 0.75% chlorpyrifos and 1.0% diazinon, respectively, without AP. They also produced higher mortality than 1% propoxur.

Field Tests In field experiments conducted

Figure 8.8
Chemometer: From the central staging area cockroaches enter or avoid peripheral chambers in which treatments are applied.
SOURCE: Rust, M. K., and D. A. Reierson. 1977. Using pheromone extract to reduce repellency of blatticides. *Journal of Economic Entomology* 70(1)34–38.

in apartments, the ratio of AP to toxicant was only 1:1 or 1:2, because of limitations in the quantity of AP that could be obtained, whereas in laboratory tests the ratio was 42:1. Nevertheless, AP added to 0.25% chlorpyrifos, 0.5% diazinon, and 0.5% microencapsulated diazinon resulted in satisfactory reductions in German cockroach populations in a 2-week period, whereas when AP was not added, cockroach populations increased (169).

A crude AP extract on corrugated cardboard harborages confined in a "test pen" attracted German cockroaches (78). The number attracted increased almost linearly to a maximum of about 80% of the population at 1.4 grams of AP per harborage, but increased no further with a doubling of the amount of the pheromone. The addition of AP to fenitrothion wettable powder, known to be repellent to German cockroaches, increased its effectiveness against these insects. Its effectiveness increased with decreasing percentage of insecticide used in the sprays. The higher concentrations of conventional insecticides will continue to be used until AP can be produced synthetically on a large scale (78).

The chemical identity of German cockroach AP has not been determined, so the amount of actual pheromone required to make insecticides less repellent is not known. However, the data indicate that the effectiveness of residual insecticides or baits may be increased by increasing the period of time cockroaches will remain in contact with them; this can be accomplished with APs.

Insect Growth Regulators

Insect growth regulators (IGRs) are chemical compounds that alter insect growth and development. The most important are juvenile hormone (JH) analogs produced synthetically (121, 154). There are periods in the life cycles of insects when the presence or absence of JH plays a vital role in their normal development. For example, an IGR known as methoprene prevents immature fleas from becoming adults. It has been effectively used in flea control. IGRs are environmentally benign and are harmless to warm-blooded animals as well as to insect species other than the ones for which they were designed. Thus they cause no ill effects in the adult insect predators and parasites of the pest species against which they are applied.

Hydroprene, an IGR closely related to methoprene, has an effect on cockroaches similar to that of methoprene on fleas, resulting in abnormal nonreproductive adults. In one study of the potential role of hydroprene in the management of severe cockroach infestations in apartments, the effectiveness of this IGR applied as an aqueous spray was compared to that of a total-release aerosol formulation; the latter was more effective (25, 166).

Since hydroprene has no effect on cockroach populations until 3–6 months after treatment, it was obviously desirable to combine the hydroprene treatment with one that provides a quick knockdown. When hydroprene at 1.2% was used in combination with chlorpyrifos/dichlorvos at 0.5%, the cockroach population was immediately reduced to acceptable levels and the sterilizing effect of the hydroprene served to maintain a very low level for at least one year—6–9 months longer than chlorpyrifos/dichlorvos alone. It should be borne in mind, however, that during the winter months cockroaches declined in numbers even in the untreated apartments.

On the basis of their experiments with hydroprene, Bennett et al. (25) concluded, "A material which is capable of sustaining high reductions well beyond those seen with conventional control materials, plus having a high specificity to insects and low mammalian toxicity, is a welcome addition to the control arsenal against urban pests."

Resistance to Insecticides

When a strain of an insect species has a genetic resistance to an insecticide, that resistance may be the result of some physiological change within the insect's body, giving it the ability to "tolerate doses of toxicants which would prove lethal to the majority of individuals in a normal population of the same species." But the resistance may also be "behavioral," giving an insect "the ability to avoid a dose which would prove lethal" (69). In its early stages, behavioral, as contrasted with physiological, resistance might be confused with the innate tendency of cockroaches, whether resistant or not, to apparently avoid all insecticides, at least to some degree, as demonstrated in the choice-box experiments discussed previously.

Physiological resistance, however, can be detected, precisely defined, and statistically evaluated, at least in the laboratory, because suspected strains of cockroaches can be confined to known doses of insecticide and have no way of avoiding it. Their mortality rates can be compared with those of nonresistant strains available in the major entomological laboratories.

Organochlorine Compounds

German cockroaches first developed significant resistance in the 1950s; they became resistant to chlordane and other organochlorine insecticides (84, 86, 87, 100). During the next decade or so, organophosphorus and carbamate insecticides provided satisfactory control, but low to moderate resistance to these insecticides (85) and high-level resistance to malathion (23) soon became evident in field populations of the German cockroach. After subjecting adult female German cockroaches to selection pressure of diazinon and malathion for approximately 54 generations in the laboratory, Grayson (88) found that resistance to diazinon increased from 12 to 21 times and to malathion from 48 to 64 times. The German cockroach also demonstrated a high-level resistance to synergized pyrethrins (39).

Cross-Resistance

It was not long after resistance of the German cockroach to the various insecticide groups was documented that cross-resistance was discovered (43, 101). In one survey (132), a field-collected, "Baltimore" strain of the German cockroach was shown to be highly resistant (LT_{50} resistance ratio 90×) to bendiocarb even though this carbamate insecticide apparently had been applied only once against that strain. The Baltimore strain was also resistant to chlordane (8.2×), diazinon (3.7×), malathion (6.5×), and propoxur (13.3×), but was susceptible to chlorpyrifos, acephate, and fenitrothion.

Studies of resistance in which cockroaches are injected with a toxicant or confined to its residues are useful but do not give an adequate estimate of the potential of the toxicant under field conditions where its repellency adds an additional variable. Rust and Reierson (171) have taken both the levels of resistance and insecticide-avoidance behavior into account in their Potential for Effectiveness (PE) calculations.

Development of Resistance

Cochran (40) believes that changes in formulation, such as in the case of microencapsulated diazinon and lacquer-based chlorpyrifos, designed to produce long-lasting control, may result in higher levels of resistance to these toxicants as the result of a higher level of selection pressure. He concluded that a long-range need is for chemicals with modes of insecticidal activity differing from those currently utilized and that the impact of resistance problems may be lessened by the development and implementation of alternative cockroach management strategies. In the long run there is a need for new chemicals with modes of action different than those that were used at the time of his evaluation of the problem.

Delay of Resistance

Cochran (41) suggests a sequence of treatments to delay the development of resistance:

> In a regular sequential treatment regime use a synthetic pyrethroid material, followed by a boric acid product, followed by an Amdro-containing bait, followed by a less commonly used OP insecticide. Each material should be used only once in a given sequence. Obviously, this scheme could be modified both with respect to the order of material used and the number of toxicants employed, as long as the main concept is retained.

This short-term, sequential use of different insecticides with entirely different modes of action should combat or delay further development of resistance while allowing all these groups of insecticides to be retained for cockroach control. Wood (203) believes that contact by cockroaches with deposits of insecticide that are sublethal (because they are placed where the insects only occasionally contact them) is conducive to the development of resistance. Crack-and-crevice application would appear to be a method of treatment that would minimize this difficulty (see discussion above under Structural Preferences of Cockroaches). Wood also criticizes the treatment of apartments surrounding one known to be heavily infested without treating the latter. Cockroaches from the overcrowded "focus unit" then continually move into the treated units until some with the gene(s) for resistance come into contact with continuously degrading insecticides. Some will survive to produce resistant offspring. It may not be possible to avoid this difficulty if treatment of apartments is by request only.

No doubt any person with extensive experience in cockroach control in public housing will recall the many insecticide aerosol canisters he or she has encountered there, despite the pest control services provided by the housing management. Wood (203) notes that much of the cockroach population is repelled by this "persistent fogging" and is driven into untreated places of refuge from which the insects will gradually return as the insecticide degrades. As an aid in the effort to retard the development of resistance, Wood (203) suggests that people applying insecticides be given time to do a thorough job and then set up a monitoring schedule to ensure early detection and elimination of new infestations.

Natural Enemies

Natural enemies of domiciliary cockroaches are currently of relatively little importance in cockroach control, as indicated by the enormous populations of these insects that can develop when food, water, and suitable harborages are available. However, this may be in part because no effort has been made to search the regions of origin of the various pest species for possible predators and parasites that may not have been distributed throughout the world along with their hosts. Since the most important pest species are exotic invaders from the Ethiopian and Oriental biogeographical regions, exploration of those regions with the view of introducing to other areas of the world as many species of natural enemies as possible would appear to be a worthy and urgent objective. Rearing and distribution of these natural enemies under the best procedures of integrated pest management would then be the next order of business.

Predators

Roth and Willis (165) list as predators of cockroaches various species in 4 families of frogs, 3 families of fishes, 8 families of reptiles, 17 families of birds, and 18 families of mammals, including humans. Among predators that have been released in houses to control cockroaches, but about which most tenants might be expected to be somewhat less than enthusiastic, are frogs (118) and hedgehogs (178). Among arthropods, predators include dragonflies, mantids, reduviid bugs, carabids, rhipiphorid and dermestid beetles, wasps, ants, scorpions, and spiders. Roth and Willis (165) report two species of theridiid spiders as predators of cockroaches. Numerous specimens of a theridiid, Steatoda grossa, have been observed in cockroach colonies in mock-up closets at the University of California, Los Angeles. Many cockroach nymphs were caught in their webs (52). A sparassid spider, Heteropoda venatoria, has been observed feeding on American cockroaches beneath rock piles in Hawaii (209).

To protect themselves against capture by predators, cockroaches utilize evasive behavior, concealment, protective coloration, mimicry, or secretion of malodorous substances. The nocturnal activity of cockroaches may protect them from daytime predators. Cockroaches may also be protected by their swiftness and their habit of squeezing into narrow spaces.

Cannibalism

Cockroaches prey upon their own species, feeding on egg capsules, nymphs, and injured or weak adults, even when food appears to be abundant (165).

Pettit (141) observed cannibalism in a laboratory culture of B. germanica, but only molting insects were attacked. Adult cockroaches, and to a lesser degree nymphs, attacked molting insects, but only those that were older than third instar. The later the stadia, the more subject they were to attack, and molting adults suffered the greatest mortality. Pettit found no correlation between population density and cannibalism.

In the laboratory, container size limits cockroach population size by overcrowding. The high mortality of small nymphs under these conditions is particularly noteworthy (161).

Viruses

Until recently, viruses were not considered to be of great potential importance in cockroach management (96). However, Suto et al. (187) isolated a virus from Periplaneta fuliginosa that causes high mortality of nymphs and adults. Adults of P. australasiae, P. brunnea, and P. japonica were shown to be equally as susceptible. The newly isolated virus has many characteristics similar to those of a parvovirus isolated from the greater wax moth, Galleria mellonella.

Endoparasites and Ectoparasites

Generally, a large percentage of a cockroach population contains internal parasites. Many pathogens, as well as parasites and predators, of cockroaches have been listed by Jenkins (108). For example, two species of nematodes and seven protozoan parasites were found in 105 German cockroaches in New York City. One of the nematode species, Blatticola blattae, was present in 96.2% of the cockroaches collected. Among the protozoa, Nephridiophaga blattellae was found in 82.8% of the cockroaches, and three other species were also abundant (191).

Mites

The cockroach mite, Pimeliaphilus cunliffei (Pterygosomatidae), an obligatory parasite of cockroaches, must feed on the live insects to survive (it cannot subsist on feces, cast skins, or dead insects) (47). It has been accused of biting people. Its presence in homes is an indication of cockroach infestation (8).

Hymenopterous Egg Parasites

Encyrtidae It is the hymenopterous egg parasites that offer the most promise of being ef-

Figure 8.9
The hymenopterous parasite, *Comperia merceti*, ovipositing into an ootheca of a brownbanded cockroach.
SOURCE: Roth, L. M., and E. R. Willis. 1960. The biotic associations of cockroaches. *Smithsonian Miscellaneous Collections* 141:1–470.

fectively utilized in a cockroach management program. There are parasitic species in at least six families. The egg parasite *Comperia merceti* (Figure 8.9) is largely confined to the brownbanded cockroach (*80*). When it was inadvertently introduced into Hawaii, it practically eliminated the brownbanded cockroach there (*165*). This parasite has been recorded attacking brownbanded and German cockroaches in Florida and Texas (*130*).

Comperia merceti is being reared in an insectary at the University of California, Berkeley, and released in research facilities at the University (*180, 182*). Sometimes nymphs and adults of the brownbanded cockroach are suppressed with traps to accelerate the impact of the parasites. Slater et al. (*182*) concluded that *C. merceti* "shows promise as the most effective and efficient approach" for controlling the brownbanded cockroach in large buildings.

Coler et al. (*42*) assessed the impact of parasitism by *C. merceti* on a free-living population of *S. longipalpa* by means of trap hosts (cockroach oothecae). The researchers found little effect until they stimulated a great population increase among the cockroaches by giving them a supplemental feeding consisting of a prepared cockroach diet. Then, 3–4 months after the increase in the percentage of parasitized cockroaches, captures of cockroaches decreased abruptly, beginning with the smallest nymphs and followed by medium nymphs, large nymphs, and ultimately the adults. The researchers believe that this pattern of change in population age structure supports their conclusion that the population collapse was caused by *C. merceti*.

Eupelmidae Another parasite, *Anastatus blattidarum*, appeared to be effective in controlling the brownbanded cockroach in certain areas of Arizona (*75*). Some oothecae of this cockroach were parasitized in laboratory colonies in Ohio (*102*).

Eulophidae The systematics, biology, and ecology of *Tetrastichus hagenowii*, a eulophid egg parasite, have been well documented (*36, 67, 73, 163*). The oriental, German, American, and Australian (*Periplaneta australasiae*) cockroaches have been recorded as hosts of this parasite in Florida and Louisiana (*130*). Piper et al. (*145*) surveyed 17 eastern Texas and 4 Louisiana cities during 1974–75 and found *T. hagenowii* in 14 of the Texas and in all the Louisiana sites, accounting for 99.4% of the parasitism. All eggs within the parasitized oothecae were consumed, which is typical of this species. Even the eggs not consumed generally fail to hatch

(*37, 67*). The overall mean parasitization rates for each cockroach species were as follows: American, 46% (range 0–100%); smokybrown, 25% (range 0–77%); and oriental, 12.5% (range 0–100%). None of the 110 oothecae of the brown cockroach was parasitized. Exposure of 50 viable oothecae of this species to 100 mated female *T. hagenowii* did not result in successful parasitization.

Evaniidae Piper et al. (*145*) recorded *Evania appendigaster* (Evaniidae) parasitizing oothecae in Austin and Brownsville, Texas, with parasitization rates of 9% and 25%, respectively. This species also occurs in Houston and San Antonio. It has a circumtropical distribution. There are several useful accounts of its biology and distribution (*37, 46, 97, 112, 190*).

Unidentified evaniid parasites were so abundant in a home in Worthington, Ohio, that the occupants considered them to be a nuisance. They were abundant at windows as well as other places and flew about actively both day and night. The oriental cockroaches that were abundant in the basement of this house did not seem to annoy the occupants. Likewise, a barber in Columbus, Ohio, used a fly swatter to kill what he considered to be flies, but they were actually evaniids parasitizing cockroaches in the basement beneath the shop.

The most common evaniid found in Ohio was *Prosevania fuscipes* (*66*). Edmunds (*65*) obtained 1,213 specimens of hymenopterous parasites from 459 oothecae of *Parcoblatta* (woodroaches), mainly *P. pennsylvanica*: *Hyptia harpyoides* (Evaniidae), *H. thoracica*, *Systellogaster ovivora* (Pteromalidae), the last named also reported from the oriental cockroach (*130*), and *Syntomosphyrum blattae* (Eulophidae). The evaniids alone parasitized 6.7% of the overwintering woodroach oothecae in Ohio. There were also two dipterous parasites: *Coenosia basalis* (Muscidae) and *Megaselia* sp. (Phoridae).

Release of Egg Parasites in Buildings Piper and Frankie (*143*) were able to determine the density and effectiveness of the egg parasite, *Tetrastichus hagenowii*, by distributing laboratory-cultured, nonparasitized oothecae, each attached to an identifiable cardboard slip. The oothecae were later removed and dissected to determine the presence or absence of parasites.

Inundation of an area with *T. hagenowii* might be an effective control technique in closed systems such as residences, commercial buildings, and storehouses. In one test of this hypothesis, American cockroach oothecae were distributed from floor to ceiling in kitchens, bathrooms, and living rooms of several residences prior to release of the parasites. The parasites oviposited in oothecae located in all parts of the rooms. Maximum parasitization occurred when the ratio of wasps to oothecae was from 8:1 to 12:1. Not everyone would welcome indoor parasite releases to control cockroaches, just as many people do not use insecticides in their homes because of doubts as to their safety. Public understanding and acceptance of the use of biological control agents will contribute to the success of an urban cockroach management program (*143*).

Ecologically Sound Cockroach Management

Attitudes Toward Cockroaches

As stated previously, the cockroach is the animal "least liked by the American people" (*111*). The degree of revulsion toward cockroaches is indicated by the large proportion of the research effort that is directed against them. In recent years much of this research has consisted of studies of peoples' attitudes toward cockroaches and how it has affected control measures, particularly in public housing (*153, 157, 177, 204, 212*). Among the conclusions reached by these researchers was that peoples' attitudes were affected by the degree of cockroach infestation to which they had become accustomed. This indicated the possibility of utilizing an "aesthetic injury level" (AIL) as was suggested by Olkowski (*135*) in an urban pest management program for street trees. The smaller the number of cockroaches in a housing project, the lower the AIL (*212*).

Even in dwellings in which cockroaches have been eliminated, they can again enter via food packages. Among the packages on which they may be transported are corrugated cardboard cartons and burlap sacks of potatoes, onions, and so on. Infestation can originate anywhere along the line from the food plant to the dwelling, including the conveyances in which the food is transported. From the standpoint of the food industry, packaging and transportation are discussed in this volume in Chapters 23 and 24, respectively.

Integrated Pest Management

The terms "integrated control" and "integrated pest management" (IPM) originated among agricultural entomologists in relatively recent times (*185*), although many of the concepts embodied in the terms are as old as human struggle against agricultural and structural pests. It was the integration of these concepts into environmentally benign and maintainable pest management strategies that was new.

Newsom (*133*) defined IPM as "an ecologically based strategy for regulating populations of pest species at levels below economic-injury thresholds by use of the most appropriate control tactics available." In agriculture, IPM is sometimes highly complex and sophisticated, but the number of insects that might be tolerated in the field or orchard "at levels below economic-injury thresholds" may be far greater than would ordinarily be tolerated in a human dwelling. As Bennett (*21*) notes, "for most urban and industrial pest problems, there are no economic thresholds."

Role of Sanitation

Prime objectives in cockroach control are sanitation and the avoidance of clutter. Cockroaches would not be present if they found no food and water. Wright (*205*) conducted a survey in 612 single-family dwellings to determine the relationship of sanitation and clutter to German cockroach population density. In working with individual houses rather than multiple-family apartment buildings, he eliminated the effect of the migration of cockroaches from dwelling units that commonly occurs in the apartment buildings. Using six categories of sanitation and clutter ranging from "fairly clean–not cluttered" to "severely dirty–cluttered," Wright found a direct relationship (1% level) between the number of cockroaches visually counted and the sanitation category of the dwelling; the worse the sanitation, the greater the number of cockroaches counted.

The greater number of insecticide applications in the dwellings with poor sanitation had no significant effect on the relation of sanitation to cockroach populations. Even when large numbers of dead cockroaches were seen, the treated dwellings with poor sanitation still had the greatest number of survivors. Wright (*205*) concluded that satisfactory cockroach control may be almost impossible to achieve under such conditions.

Robinson et al. (*158*) recommend that after an entire building is thoroughly inspected, the places where numerous cockroaches are focused should be marked on a floor plan. Chemical control should be initiated only after a cleanup program is completed. They recommend that the minimum amount of insecticide be applied in such a way as to achieve the maximum effect after taking into consideration the resistance status and the repellent and residual qualities of the pesticide used.

Holistic Pest Control

Katz (*110*) points out that the terms "integrated pest management" and "pest control" imply that a certain level of infestation is expected and tolerated. However much one may agree with this concept as a guide to dealing with the German cockroach in multifamily apartment buildings, it receives no support among those whose tolerance level for these insects is zero and who are willing to pay the price and make the personal commitment to achieve their goal. For such individuals, Katz (*110*) suggests a newer concept of dealing with pests that he calls "holistic pest control" and which "implies *total* eradication of pest populations with a total commitment of the PCO and the customer. This is dedicated to removing causes rather than treating symptoms."

A requirement for PCOs would be a course in household ecology, which would include, besides the recognition of the characteristic habitats of pests in furniture and appliances of homes and commercial buildings, the following:

(a) sewerage systems—a repository for rich, nutritious effluent from garbage grinders, kept warm year round by wastewater and connected with miles of arteries below the frost line, sewers serve as a perfect haven for psychodid drain flies, American cockroaches, and rats;

(b) suspended ceilings and false walls;

(c) electrical conduit systems spanning structures in all directions;

(d) heating and air-conditioning ducts, utility pipe chases, and large wall openings for utility pipes;

(e) hollow modular concrete and steel units permitting pest movement from wall to wall; and

(f) retrofitted insulation with the vapor barrier left on the wrong side, creating superb conditions favoring insects and rats.

Katz (*110*) emphasizes that holistic pest control is expensive because it requires all the skills of highly trained people. Nevertheless, it is welcomed by homeowners greatly concerned about environmental pollution. And in commercial premises, once management understands the holistic concept and is willing to pay the price and commit resources for equipment and for training and motivating employees, progressive PCOs can be found who are able to provide the service. Nowadays they generally describe their service as "pest elimination" or "pest eradication" and claim the technology exists to enable them to succeed. Their service includes follow-up contact with the customer by supervisory personnel in person, or by telephone, and instruction of the customer regarding the urgency of good sanitation and the means by which it can be attained. Many PCOs believe that "sales of eradication services may be the fastest-growing segment of the industry today" (*195*).

Holistic pest control, or pest elimination, is admittedly very expensive. Generally, budget constraints require that less heroic measures be practiced. But even with merely "good control" as the goal, education and training of the customer are required. In the case of cockroach control in low-cost public or private housing, the task falls to the housing administrators. Cockroach management in the food industry requires an IPM strategy in which every employee of an establishment assumes a share of the responsibility for making the IPM approach successful.

Acknowledgments

I am deeply indebted to M. K. Rust and D. A. Reierson who reviewed the manuscript and made many suggestions for its improvement.

Cited References

1 Akers, R. C., and W. H Robinson. 1981. Spatial pattern and movement of German cockroaches in urban, low-income apartments (Dictyoptera: Blattellidae). *Proceedings of the Entomological Society of Washington* 83(1)168–172.

2 Alcamo, I. E., and A. M. Frishman. 1980. The microbial flora of field-collected cockroaches and other arthropods. *Journal of Environmental Health* 42(5)263–266.

3 Anonymous. 1967. Up a tree for American roach control. *Pest Control* 35(12)16.

4 Appel, A. G., and M. K. Rust. 1983. Temperature-mediated sex pheromone production and response of the American cockroach. *Journal of Insect Physiology* 29(4)301–305.

5 Appel, A. G., and M. K. Rust. 1985. Water distribution and loss in response to acclimation at constant humidity in the smokybrown cockroach, *Periplaneta fuliginosa* (Serville). *Comparative Biochemistry and Physiology* 80A(3)377–380.

6 Appel, A. G., and J. B. Tucker. 1984. Bionomics and control of the smokybrown cockroach. *Pest Management* 3(12)10–13.

7 Appel, A. G., D. A. Reierson, and M. K. Rust. 1983. Comparative water relations and temperature sensitivity of cockroaches. *Comparative*

Biochemistry and Physiology 74A(2)357–361.

8 Baker, E. W., et al. 1956. *A Manual of Parasitic Mites of Medical or Economic Importance.* National Pest Control Association, New York, NY.

9 Ballard, J. B., and R. E. Gold. 1983. The efficacy of selected home use insecticide products upon German cockroach populations, 1981. *Insecticide and Acaricide Tests* 8:53–54.

10 Ballard, J. B., and R. E. Gold. 1983. Field evaluation of two trap designs used for control of German cockroach populations. *Journal of the Kansas Entomological Society* 56(4)506–510.

11 Ballard, J. B., and R. E. Gold. 1984. Laboratory and field evaluations of German cockroach (Orthoptera: Blattellidae) traps. *Journal of Economic Entomology* 77(3)661–665.

12 Ballard, J. B., R. E. Gold, and J. D. Rauscher. 1984. Effectiveness of six insecticide treatment strategies in the reduction of German cockroach (Orthoptera: Blattellidae) populations in infested apartments. *Journal of Economic Entomology* 77(5)1092–1094.

13 Ballard, J. B., H. J. Ball, and R. E. Gold. 1984. Influence of selected environmental factors upon German cockroach (Orthoptera: Blattellidae) exploratory behavior in choice boxes. *Journal of Economic Entomology* 77(5)1206–1210.

14 Barak, A. V., M. Shinkle, and W. E. Burkholder. 1977. Using attractant traps to help detect and control cockroaches. *Pest Control* 45(10)14–20.

15 Barnhart, C. S., Jr. 1963. How to control cockroaches with dry ice fumigation. *Pest Control* 31(2)30.

16 Barson, G. 1982. Laboratory evaluation of boric acid plus porridge oats and iodofenphos gel as toxic baits against the German cockroach, *Blattella germanica* (L.) (Dictyoptera: Blattellidae). *Bulletin of Entomological Research* 72(2)229–237, 1 pl.

17 Bell, W. J. 1984. Bionomics and control of the American cockroach. *Pest Management* 3(3)12–19.

18 Bell, W. J., C. Parsons, and E. A. Martinko. 1972. Cockroach aggregation pheromones: Analysis of aggregation tendency and species specificity (Orthoptera: Blattidae). *Journal of the Kansas Entomological Society* 45(4)414–421.

19 Bell, W. J., et al. 1977. Attractancy of the American cockroach sex pheromone (Orthoptera: Blattidae). *Journal of the Kansas Entomological Society* 50(4)503–507.

20 Bennett, G. W. 1977. Cockroach manual. II. Control in food handling establishments. *Pest Control* 45(9)39–40.

21 Bennett, G. W. 1978. Technology transfer in urban pest management. In *Perspectives in Urban Entomology,* G. W. Frankie and L. S. Koehler,

eds. Academic Press, New York, NY.

22 Bennett, G. W., and R. D. Lund. 1977. Evaluation of encapsulated pyrethrins (Sectrol™) for German cockroach and cat flea control. *Pest Control* 45(9)44, 46, 48–50.

23 Bennett, G. W., and W. T. Spink. 1968. Insecticide resistance of German cockroaches from various areas of Louisiana. *Journal of Economic Entomology* 61(2)426–431.

24 Bennett, G. W., and C. G. Wright. 1971. Response of German cockroaches to spray constituents. *Journal of Economic Entomology* 64(5)1119–1124.

25 Bennett, G., E. Runstrom, and J. Bertholf. 1984. Examining the where, why and how of cockroach control. *Pest Control* 52(6)42–43, 46, 48.

26 Berthold, R., Jr., and B. R. Wilson. 1967. Resting behavior of the German cockroach, *Blattella germanica. Annals of the Entomological Society of America* 60(2)347–351.

27 Burden, G. S. 1975. Repellency of selected insecticides. *Pest Control* 43(6)16–18.

28 Burden, G. S., and J. L. Eastin. 1960. Field tests of dusts and sprays against German cockroaches. *Florida Entomologist* 43(1)15–18.

29 Burden, G. S., and J. L. Eastin. 1960. Laboratory evaluation of cockroach repellents. *Pest Control* 28(6)14.

30 Burden, G. S., and B. J. Smittle. 1961. New and current insecticides for German cockroach control. *Pest Control* 29(6)30.

31 Burgess, N. R. H., S. N. McDermott, and A. P. Blanch. 1974. An electrical trap for the control of cockroaches and other domestic pests. *Journal of the Royal Army Medical Corps* 120(3)173–175.

32 Burgess, N. R. H., S. N. McDermott, and J. Whiting. 1973. Aerobic bacteria occurring in the hind-gut of the cockroach, *Blatta orientalis. Journal of Hygiene* 71(1)1–7.

33 Burk, T., and W. J. Bell. 1973. Cockroach aggregation pheromone: Inhibition of locomotion (Orthoptera: Blattidae). *Journal of the Kansas Entomological Society* 46(1)36–41.

34 Busvine, J. 1971. *A Critical Review of the Techniques for Testing Insecticides.* Commonwealth Agricultural Bureaux, London.

35 Buxton, G. M., and T. J. Freeman. 1968. Positive separation of *Blattella vaga* and *Blattella germanica* (Orthoptera: Blattidae). *Pan-Pacific Entomologist* 44(2)168–169.

36 Cameron, E. 1955. On the parasites and predators of the cockroach. I. *Tetrastichus hagenowii* (Ratz.). *Bulletin of Entomological Research* 46(1)137–147.

37 Cameron, E. 1957. On the parasites and predators of the cockroach. II. *Evania appendigaster* (L.). *Bulletin of Entomological Research*

48(1)199–209.

38 Chamberlin, J. C. 1949. Insects of agricultural and household importance in Alaska with suggestions for their control. *Alaska Agricultural Experiment Stations, Circular* 9:1–49.

39 Cochran, D. G. 1973. Inheritance and linkage of pyrethrins resistance in the German cockroach. *Journal of Economic Entomology* 66(1)27–30.

40 Cochran, D. G. 1982. German cockroach resistance. *Pest Control* 50(8)16, 18, 20.

41 Cochran, D. G. 1984. Insecticide resistance in cockroaches—is it at a crossroad? *Pest Management* 3(8)26–31.

42 Coler, R. R., R. G. Van Driesche, and J. S. Elkinton. 1984. Effect of an oothecal parasitoid, *Comperia merceti* (Compere) (Hymenoptera: Encyrtidae), on a population of the brownbanded cockroach (Orthoptera: Blattellidae). *Environmental Entomology* 13(2)603–606.

43 Collins, W. J. 1973. German cockroach resistance. 1. Resistance to diazinon includes cross-resistance to DDT, pyrethrins, and propoxur in a laboratory colony. *Journal of Economic Entomology* 66(1)44–47.

44 Cornwell, P. B. 1968. *The Cockroach*, vol. 1. Hutchinson, London.

45 Cornwell, P. B. 1976. *The Cockroach*, vol. 2. Hutchinson, London.

46 Crosskey, R. W. 1951. Part II. The taxonomy and biology of the British Evanioidea. *Transactions of the Royal Entomological Society of London* 102(5)282–301.

47 Cunliffe, F. 1952. Biology of the cockroach parasite, *Pimeliaphilus podapolipophagus* Trägårdh, with a discussion of the genera *Pimeliaphilus* and *Hirstiella* (Acarina, Pterygosomidae). *Proceedings of the Entomological Society of Washington* 54(4)153–169.

48 Cunningham, B., and M. P. Kelly. 1989. Drione dust against *Blatta orientalis*. *International Pest Control* 31(4)90–92.

49 Darchen, R. 1952. Sur l'activité exploratrice de *Blattella germanica*. *Zeitschrift für Tierpsychologie* 9:362–372.

50 Darchen, R. 1955. Stimuli nouveaux et tendance exploratrice chez *Blattella germanica*. *Zeitschrift für Tierpsychologie* 12:1–11.

51 Ebeling, W. 1971. Sorptive dusts for pest control. *Annual Review of Entomology* 16:123–158.

52 Ebeling, W. 1975. *Urban Entomology*. Agriculture and Natural Resources, University of California, Oakland.

53 Ebeling, W. 1976. Insect integument: a vulnerable organ system. In *The Insect Integument*, H. R. Hepburn, ed. Elsevier Scientific, Oxford, England.

54 Ebeling, W., and D. A. Reierson. 1969. The cockroach learns to avoid insecticides. *California Agriculture* 23(2)12–15.

55 Ebeling, W., and D. A. Reierson. 1970. Effect of population density on exploratory activity and mortality rate of German cockroaches in choice boxes. *Journal of Economic Entomology* 63(2)350–355.

56 Ebeling, W., and D. A. Reierson. 1974. Bait trapping silverfish, cockroaches, and earwigs. *Pest Control* 42(4)24, 36–39.

57 Ebeling, W., and D. A. Reierson. 1977. Cockroach control as affected by building structure. In *Proceedings of the 1977 Seminar on Cockroach Control*, K. Westphal, ed. New York State Department of Health, [Albany].

58 Ebeling, W., D. A. Reierson, and R. E. Wagner. 1967. Influence of repellency on the efficacy of blatticides. II. Laboratory experiments with German cockroaches. *Journal of Economic Entomology* 60(5)1375–1390.

59 Ebeling, W., D. A. Reierson, and R. E. Wagner. 1968. The influence of repellency on the efficacy of blatticides. III. Field experiments with German cockroaches with notes on three other species. *Journal of Economic Entomology* 61(3)751–761.

60 Ebeling, W., D. A. Reierson, and R. E. Wagner. 1968. The influence of repellency on the efficacy of blatticides. IV. Comparison of four cockroach species. *Journal of Economic Entomology* 61(5)1213–1219.

61 Ebeling, W., R. E. Wagner, and D. A. Reierson. 1965. Cockroach control with Dri-Die and Drione. *Pest Control Operators News* 26(10)16–22; (11)16–19, 25.

62 Ebeling, W., R. E. Wagner, and D. A. Reierson. 1966. Influence of repellency on the efficacy of blatticides. I. Learned modification of behavior of the German cockroach. *Journal of Economic Entomology* 59(6)1374–1388.

63 Ebeling, W., R. E. Wagner, and D. A. Reierson. 1969. Insect-proofing during building construction. *California Agriculture* 23(5)4–7.

64 Ebeling, W., et al. 1975. Silica aerogel and boric acid against cockroaches: External and internal action. *Pesticide Biochemistry and Physiology* 5(1)81–89.

65 Edmunds, L. R. 1952. Some notes on the habits and parasites of native wood-roaches in Ohio (Orthoptera: Blattidae). *Entomological News* 63(6)141–145.

66 Edmunds, L. R. 1953. Some notes on the Evaniidae as household pests and as a factor in the control of roaches. *Ohio Journal of Science* 53(2)121–122.

67 Edmunds, L. R. 1955. Biological notes on *Tetrastichus hagenowii* (Ratzeburg), a chalcidoid parasite of cockroach eggs (Hymenoptera: Eulophidae; Orthoptera: Blattidae). *Annals of the*

Entomological Society of America 48(4)210–213.

68 Ehrlich, H. 1943. Verhaltensstudien an der Schabe *Periplaneta americana* L. *Zeitschrift für Tierpsychologie* 5:497–552.

69 Expert Committee on Insecticides. 1957. *Resistance of Insects to Insecticides*, 7th report. Technical Report Series 125. World Health Organization, Geneva, Switzerland.

70 Farmer, B. R., and W. H Robinson. 1984. Is caulking beneficial for cockroach control? *Pest Control* 52(6)28, 30, 32.

71 Farmer, B. R., and W. H Robinson. 1985. Caulking for roach control. Pest Control Technology 13(4)71, 73, 74.

72 Fleet, R. R., and G. W. Frankie. 1974. Habits of two household cockroaches in outdoor environments. *Texas Agricultural Experiment Station, Miscellaneous Publication* 1153:1–6.

73 Fleet, R. R., and G. W. Frankie. 1975. Behavioral and ecological characteristics of a eulophid egg parasite of two species of domiciliary cockroaches. *Environmental Entomology* 4(2)282–284.

74 Fleet, R. R., G. L. Piper, and G. W. Frankie. 1978. Studies on the population ecology of the smokybrown cockroach, *Periplaneta fuliginosa*, in a Texas outdoor urban environment. *Environmental Entomology* 7(6)807–814.

75 Flock, R. A. 1941. The field roach *Blattella vaga*. *Journal of Economic Entomology* 34(1)121.

76 Flynn, A. D., and H. F. Schoof. 1966. A simulated-field method of testing residual insecticide deposits against cockroaches. *Journal of Economic Entomology* 59(1)110–113.

77 Frishman, A. M., and I. E. Alcamo. 1977. Domestic cockroaches and human bacterial disease. *Pest Control* 45(6)16, 18, 20, 46.

78 Glaser, A. E. 1980. Use of aggregation pheromones in the control of the German cockroach (*Blattella germanica*). *International Pest Control* 22(1)7, 8, 21.

79 Goodhue, L. D., and C. Linnard. 1952. Determining the repellent action of chemicals to the American cockroach. *Journal of Economic Entomology* 45(1)133–134.

80 Gordh, G. 1973. Biological investigations on *Comperia merceti* (Compere), an encyrtid parasite of the cockroach *Supella longipalpa* (Serville). *Journal of Entomology* A47(2)115–123.

81 Gould, G. E., and H. O. Deay. 1938. Notes on the bionomics of roaches inhabiting houses. *Proceedings of the Indiana Academy of Science* 47(1937)281–284.

82 Gould, G. E., and H. O. Deay. 1940. The biology of six species of cockroaches which inhabit buildings. *Purdue University, Agricultural Experiment Station, Bulletin* 451:1–31.

83 Graffar, M., and S. Mertens. 1950. Le rôle des blattes dans la transmission de salmonelloses. *Annales d'Institut Pasteur* 79(5)654–660.

84 Grayson, J. M. 1953. Effects on the German cockroach of 12 generations of selection for survival to treatment with DDT and benzene hexachloride. *Journal of Economic Entomology* 46(1)124–127.

85 Grayson, J. M. 1965. Resistance to three organophosphorus insecticides in strains of the German cockroach from Texas. *Journal of Economic Entomology* 58(5)956–958.

86 Grayson, J. M. 1966. Recent developments in the control of some arthropods of public health and veterinary importance. Cockroaches. *Bulletin of the Entomological Society of America* 12(3)333–338.

87 Grayson, J. M. 1966. Results of 1965 NPCA-sponsored research at VPI on cockroach control. *Pest Control* 34(6)12–15.

88 Grayson, J. M. 1973. 1972 Roach control research. *Pest Control* 41(3)30, 32, 34.

89 Grayson, J. M. 1980. Pydrin™ on different surfaces for control of resistant German cockroaches. *Pest Control* 48(6)19, 20, 22.

90 Grothaus, R. H., et al. 1981. Superroach challenged by microcomputer. *Pest Control* 49(1)16–18; (2)19–23; (3)24–30.

91 Gunn, D. L. 1931. Temperature and humidity relations of the cockroach. *Nature* 128(3222)186–187.

92 Gunn, D. L. 1934. The temperature and humidity relations of the cockroach (*Blatta orientalis*). II. Temperature preference. *Zeitschrift für Vergleichende Physiologie* 20(5)617–625.

93 Gunn, D. L. 1935. The temperature and humidity relations of the cockroach. III. A comparison of temperature preference, and rates of desiccation and respiration of *Periplaneta americana*, *Blatta orientalis*, and *Blattella germanica*. *Journal of Experimental Biology* 12(2)185–190.

94 Gupta, A. R., and Y. S. Das. 1974. Orthene against German and American cockroaches. *Pest Control* 42(1)15–16.

95 Gupta, A. R., et al. 1973. Effectiveness of spray-dust-bait combination. *Pest Control* 41(9)20, 22, 24.

96 Guthrie, D. M., and A. R. Tindall. 1968. *The Biology of the Cockroach*. St. Martin's, New York, NY.

97 Haber, V. R. 1920. Oviposition by an evaniid, *Evania appendigaster* Linn. *Canadian Entomologist* 52(9)248.

98 Hafez, M., and A. M. Afifi. 1956. Biological studies on the furniture cockroach *Supella supellectilium* Serv., in Egypt (Orthoptera: Blattidae). *Bulletin de la Société Entomologique d'Egypte* 40:366–396.

99 Hayes, W. J., Jr. 1975. *Toxicology of Pesticides*. Williams & Wilkins, Baltimore, MD.

100 Heal, R. E., K. B. Nash, and M. Williams. 1953. An insecticide-resistant strain of the German cockroach from Corpus Christi, Texas. *Journal of Economic Entomology* 46(2)385–386.

101 Heuvel, W. J. van den, and D. G. Cochran. 1965. Cross resistance to organophosphorus compounds in malathion- and diazinon-resistant strains of *Blattella germanica*. *Journal of Economic Entomology* 58(5)872–874.

102 Hull, G., Jr., and R. H. Davidson. 1958. The biology of the brown-banded cockroach and its relative susceptibility to five organic insecticides. *Journal of Economic Entomology* 51(5)608–610.

103 Ishii, S. 1970. Aggregation of the German cockroach, *Blattella germanica* (L.). In *Control of Insect Behavior by Natural Products*, D. L. Wood, R. M. Silverstein, and M. Nakajima, eds. Academic Press, New York, NY.

104 Ishii, S. 1970. An aggregation pheromone of the German cockroach, *Blattella germanica* (L.). 2. Species specificity of the pheromone. *Applied Entomology and Zoology* 5(1)33–41.

105 Ishii, S., and Y. Kuwahara. 1967. An aggregation pheromone of the German cockroach *Blattella germanica* L. (Orthoptera: Blattellidae). I. Site of the pheromone production. *Applied Entomology and Zoology* 2(4)203–217.

106 Ishii, S., and Y. Kuwahara. 1968. Aggregation of German cockroach (*Blattella germanica*) nymphs. *Experientia* 24(1)88–89.

107 Iwao, S. 1963. On a method for estimating the rate of population interchange between two areas. *Researches on Population Ecology* 5(6)44–50.

108 Jenkins, D. W. 1964. Pathogens, parasites and predators of medically important arthropods. *Bulletin of the World Health Organization* 30(suppl.)1–150.

109 Karner, M., R. G. Price, and L. A. Roth. 1978. Laboratory evaluation of crack and crevice treatment for control of *Blattella germanica* by using various nozzle types, nozzle heights, and crack widths. *Journal of Economic Entomology* 71(1)105–106.

110 Katz, H. L. 1982. Holistic pest control. In *Handbook of Pest Control*, 6th ed. Franzak & Foster, Cleveland, OH.

111 Kellert, S. R. 1980. American attitudes toward and knowledge of animals: An update. *International Journal for the Study of Animal Problems* 1(2)87–119.

112 Kieffer, J. J. 1920. Hymenoptera. Family Evaniidae. *Genera Insectorum*, fasc. 2, pp. 1–13.

113 Kitamura, C., H.-S. Koh, and S. Ishii. 1974. Possible role of feces for directional orientation of the German cockroach, *Blattella germanica*

L. (Orthoptera: Blattellidae). *Applied Entomology and Zoology* 9(4)271–272.

114 Knipling, E. B., and W. N. Sullivan. 1967. Insect mortality at low temperatures. *Journal of Economic Entomology* 50(3)368–369.

115 Mallis, A. (ed.). 1982. *Handbook of Pest Control*, 6th ed. Franzak & Foster, Cleveland, OH.

116 Mampe, C. D. 1972. The relative importance of household insects in the continental United States. *Pest Control* 40(12)24, 26–28.

117 Mampe, C. D. 1976. Roach control in the food handling business. *Pest Control* 44(6)14–16, 18, 54, 55; (7)27–30, 32, 34; (8)44, 46, 49, 50.

118 Marlatt, C. L. 1908. Cockroaches. *USDA Bureau of Entomology Circular* 51:1–14.

119 Mellanby, K. 1939. Low temperature and insect activity. *Proceedings of the Royal Society of London* B127(849)473–487.

120 Mellanby, K. 1940. The daily rhythm of activity of the cockroach, *Blatta orientalis* L. 11. Observations and experiments on a natural infestation. *Journal of Experimental Biology* 17(3)278–285.

121 Menn, J. J., and M. Beroza (eds.). 1972. *Insect Juvenile Hormones, Chemistry and Action*. Academic Press, New York, NY.

122 Metzger, R., and K.-H. Trier. 1975. Zur Bedeutung der Aggregationspheromone von *Blattella germanica* und *Blatta orientalis*. *Angewandte Parasitologie* 16(1)1627.

123 Miesch, M. D., and D. E. Howell. 1967. An evaluation of baits for cockroaches. *Pest Control* 35(6)1620.

124 Mizukubo, T. 1981. A revision of the genus *Blattella* (Blattaria: Blattellidae) of Japan. I. Terminology of the male genitalia and description of a new species from Okinawa Island. *Esakia* 17(Nov.)149–159.

125 Moore, B. P. 1967. Chemical communication in insects. *Science Journal* 3(9)44–49.

126 Moore, R. C. 1972. Boric acid-silica dusts for control of German cockroaches. *Journal of Economic Entomology* 65(2)458–461.

127 Moore, R. C. 1973. *Cockroach Proofing. Preventive Treatments for Control of Cockroaches in Urban Housing and Food Service Carts*. Bulletin 740. Connecticut Agricultural Experiment Station, New Haven, CT.

128 Moore, W. S., and T. A. Granovsky. 1983. Laboratory comparisons of sticky traps to detect and control five species of cockroaches (Orthoptera: Blattidae and Blattellidae). *Journal of Economic Entomology* 76(4)845–849.

129 Moreland, D. 1985. Pest control in food plants. *Pest Control Technology* 13(8)40–42, 44, 46, 48.

130 Muesebeck, C. F. W., K. V. Krombein, and H. K. Townes. 1951. *Hymenoptera of America*

North of Mexico. Agriculture Monograph 2. U.S. Department of Agriculture, Washington, DC.

131 National Pest Control Association. 1966. *Cockroaches and Their Control.* NPCA Technical Release 11–65, pp. 1–10.

132 Nelson, J. O., and F. E. Wood. 1982. Multiple and cross-resistance in a field-collected strain of the German cockroach (Orthoptera: Blattellidae). *Journal of Economic Entomology* 75(6)1052–1054.

133 Newsom, L. D. 1975. Pest management: Concept to practice. In *Insects, Science and Society*, D. Pimentel, ed. Academic Press, New York, NY.

134 November, J. 1986. Asian roaches invade America. *Pest Control Technology* 14(11)72–74, 76.

135 Olkowski, W. 1973. A model ecosystem management program. *Proceedings of the Tall Timbers Conference on Ecological Animal Control by Habitat Management* 5:103–113.

136 Owens, J. M. 1980. Some aspects of German cockroach population ecology in urban apartments. Dissertation, Purdue University, West Lafayette, IN.

137 Owens, J. M., and G. W. Bennett. 1982. German cockroach movement within and between urban apartments. *Journal of Economic Entomology* 75(4)570–573.

138 Owens, J. M., and G. W. Bennett. 1983. Comparative study of German cockroach (Dictyoptera: Blattellidae) population sampling techniques. *Environmental Entomology* 12(4)1040–1046.

139 Palermo, M. T. 1960. The cockroach twins. *Pest Control* 28(6)12.

140 Persoons, C. J., and F. J. Ritter. 1979. Pheromones of cockroaches. In *Chemical Ecology: Odor Communication in Animals*, F. J. Ritter, ed. Elsevier, Amsterdam.

141 Pettit, L. C. 1940. The roach *Blattella germanica* (Linn.): Its embryogeny, life history, and importance. Dissertation, Cornell University, Ithaca, NY.

142 Piper, G. L. 1977. Aggregation tendency in two domiciliary cockroaches. *Southern Entomologist* 2(2)8893.

143 Piper, G. L., and G. W. Frankie. 1978. Integrated management of urban cockroach populations. In *Perspectives in Urban Entomology*, G. W. Frankie and L. S. Koehler, eds. Academic Press, New York, NY.

144 Piper, G. L., et al. 1975. *Controlling Cockroaches Without Synthetic Organic Insecticides.* Texas Agricultural Experiment Station and Extension Service L-1373. Texas A&M University, College Station.

145 Piper, G. L., G. W. Frankie, and J. Loehr. 1978. Incidence of cockroach egg parasites in urban environments in Texas and Louisiana. *Environmental Entomology* 7(2)289–293.

146 Raffy, A. 1930. Réaction des blattes aux variations de témperature. *Comptes Rendus des Séances de la Société de Biologie* 104(21)657–658.

147 Rauscher, J. D., R. E. Gold, and W. W. Stroup. 1985. Effects of chlorpyrifos and environmental factors on the distribution of German cockroaches (Orthoptera: Blattellidae) in Ebeling choice boxes. *Journal of Economic Entomology* 78(3)607–612.

148 Rehn, J. A. G. 1945. Man's uninvited fellow traveller—the cockroach. *Scientific Monthly* 61:265–276.

149 Reichenbach, N. C., and W. J. Collins. 1984. Multiple logit analyses of the effect of temperature and humidity on the toxicity of propoxur to German cockroaches (Orthoptera: Blattellidae) and western spruce budworm larvae (Lepidoptera: Tortricidae). *Journal of Economic Entomology* 77(1)31–35.

150 Reierson, D. A., and M. K. Rust. 1977. Trapping, flushing, counting German roaches. *Pest Control* 45(10)40, 42–44.

151 Reierson, D. A., and M. K. Rust. 1978. The effect of food, harborage and trapping in a population of the German cockroach. *Folia Entomologica Mexicana* 39–40:151.

152 Reierson, D. A., and M. K. Rust. 1984. Insecticidal baits and repellency in relation to control of the German cockroach, *Blattella germanica* (L.). *Pest Management* 3(6)26–32.

153 Reierson, D. A., M. K. Rust, and R. E. Wagner. 1978. The current status of German cockroach control and methodology for evaluating insecticides as blatticides. *Pest Control* 47(3)14–16, 18, 19, 78.

154 Riddiford, L. M., A. M. Ajami, and C. Boake. 1975. Effectiveness of insect growth regulators in the control of populations of the German cockroach. *Journal of Economic Entomology* 68(1)46–48.

155 Ritter, F. J., and C. J. Persoons. 1975. Recent developments in insect pheromone research in particular in the Netherlands. *Netherlands Journal of Zoology* 25(3)261–275.

156 Robinson, W. H. 1985. New look at an old tool. *Pest Control* 53(5)36–37.

157 Robinson, W. H, and P. A. Zungoli. 1985. Integrated control program for German cockroaches (Dictyoptera: Blattellidae) in multiple-unit dwellings. *Journal of Economic Entomology* 78(3)595–598.

158 Robinson, W. H, R. C. Akers, and P. K. Powell. 1980. German cockroaches in urban apartment buildings. *Pest Control* 48(7)18–20.

159 Rosen, A. 1962. One million roaches. *Pest Control* 30(5)92–93.

160 Ross, M. H. 1981. Trapping experiments with the German cockroach, *Blattella germanica* (L.) (Dictyoptera: Blattellidae), showing differential effects from the type of trap and the environmental resources. *Proceedings of the Entomological Society of Washington* 83(1)160–163.

161 Ross, M. H., and C. G. Wright. 1977. Characteristics of field-collected populations of the German cockroach (Dictyoptera: Blattellidae). *Proceedings of the Entomological Society of Washington* 79(3)411–416.

162 Roth, L. M., and E. R. Willis. 1952. A study of cockroach behavior. *American Midland Naturalist* 47(1)66–129.

163 Roth, L. M., and E. R. Willis. 1954. The biology of the cockroach egg parasite, *Tetrastichus hagenowii* (Hymenoptera: Eulophidae). *Transactions of the American Entomological Society* 80(2)53–72, 3 pl.

164 Roth, L. M., and E. R. Willis. 1957. The medical and veterinary importance of cockroaches. *Smithsonian Miscellaneous Collections* 134(10)1–147, 7 pl.

165 Roth, L. M., and E. R. Willis. 1960. The biotic associations of cockroaches. *Smithsonian Miscellaneous Collections* 141:1–470.

166 Runstrom, E. S., and G. W. Bennett. 1984. Efficacy of hydroprene on field populations of German cockroaches, 1982. *Insecticide and Acaricide Tests* 9:409.

167 Runstrom, E. S., and G. W. Bennett. 1984. Movement of German cockroaches (Orthoptera: Blattellidae) as influenced by structural features of low-income apartments. *Journal of Economic Entomology* 77(2)407–411.

168 Rust, M. K., and A. G. Appel. 1985. Intra- and interspecific aggregation in some nymphal blattellid cockroaches (Dictyoptera: Blattellidae). *Annals of the Entomological Society of America* 78(1)107–110.

169 Rust, M. K., and D. A. Reierson. 1977. Increasing blatticidal efficacy with aggregation pheromone. *Journal of Economic Entomology* 70(6)693–696.

170 Rust, M. K., and D. A. Reierson. 1977. Using pheromone extract to reduce repellency of blatticides. *Journal of Economic Entomology* 70(1)34–38.

171 Rust, M. K., and D. A. Reierson. 1978. Comparison of the laboratory and field efficacy of insecticides used for German cockroach control. *Journal of Economic Entomology* 71(4)704–708.

172 Rust, M. K., and D. A. Reierson. 1979. Insecticide candidates for German cockroach control in apartments. *Pest Control* 47(5)14–16.

173 Rust, M. K., and D. A. Reierson. 1980. Performance of insecticides against cockroaches in field trials, 1979. *Insecticide and Acaricide Tests* 5:212–213.

174 Rust, M. K., and D. A. Reierson. 1981. Attraction and performance of insecticidal baits for German cockroach control. *International Pest Control* 23(4)106–109.

175 Rust, M. K., T. Burk, and W. J. Bell. 1976. Pheromone-stimulated locomotory and orientation responses in the American cockroach. *Journal of Animal Behaviour* 24(1)52–67.

176 Rust, M. K., D. A. Reierson, and A. M. Van Dyke. 1983. Performance of insecticides for German cockroach in apartments, 1982. *Insecticide and Acaricide Tests* 8:55.

177 Sawyer, A. J., and R. A. Casagrande. 1983. Urban pest management: A conceptual framework. *Urban Ecology* 7(2)145–157.

178 Shipley, A. E. 1916. *More Minor Horrors*. Smith Elder, London.

179 Shuyler, H. R. 1956. Are German and oriental roaches changing their habits? *Pest Control* 24(9)9–10.

180 Slater, A. J. 1984. Biological control of the brownbanded cockroach, *Supella longipalpa* (Serville), with an encyrtid wasp, *Comperia merceti* (Compere). *Pest Management* 3(4)14–17.

181 Slater, A. J., et al. 1979. German cockroach management in student housing. *Journal of Environmental Health* 42(1)21–24.

182 Slater, A. J., M. G. Hulbert, and V. R. Lewis. 1980. Biological control of brownbanded cockroaches. *California Agriculture* 34(8–9)16–18.

183 Smittle, B. J., G. S. Burden, and W. A. Banks. 1968. Cockroach insecticides. How repellent are they? *Pest Control* 36(11)9–10.

184 Sommer, S. H. 1975. Zum Aggregationsverhalten bei *Blattella germanica*. Angewandte Parasitologie 16(3)135–141 (English summary).

185 Stern, V. M., et al. 1959. The integrated control concept. *Hilgardia* 29(2)81–101.

186 Sugawara, R., S. Kurihara, and T. Muto. 1975. Attraction of the German cockroach to cyclohexyl alkanoates and *n*-alkyl cyclohexaneacetates. *Journal of Insect Physiology* 21(5)958–964.

187 Suto, C., F. Kawamoto, and N. Kumada. 1979. A new virus isolated from the cockroach, *Periplaneta fuliginosa* (Serville). *Microbiology and Immunology* 23(3)207–211.

188 Thorpe, W. H. 1956. *Learning and Instinct in Animals*. Methuen, London.

189 Tompkins, G. J., and G. E. Cantwell. 1973. The use of dry ice to control German cockroaches in hospital food service carts. *Pest Control* 41(11)24–26.

190 Townes, H. 1949. The Nearctic species of Evaniidae (Hymenoptera). *Proceedings of the U.S. National Museum* 99(3253)525–539.

191 Tsai, Y.-H., and K. M. Cahill. 1970. Parasites

of the German cockroach (*Blattella germanica* L.) in New York City. *Journal of Parasitology* 56(2)375–377.

192 Tsuji, H., and S. Ono. 1969. Laboratory evaluation of several bait factors against the German cockroach, *Blattella germanica* (L.). *Japanese Journal of Sanitary Zoology* 20(4)240–247.

193 Tsuji, H., and S. Ono. 1970. Glycerol and related compounds as feeding stimulants to cockroaches. *Japanese Journal of Sanitary Zoology* 21(3)149–156.

194 Twomey, N. R. 1966. A review of the biology and control of German cockroach, *Blattella germanica* (L.), in California. *California Vector Views* 13(4)27–37.

195 Vega, R. 1986. Pest eradication: Is it attainable? *Pest Control Technology* 14(6)46–48.

196 Wagner, R. E., W. Ebeling, and W. R. Clark. 1964. An electric barrier for confining cockroaches in large rearing or field collecting cans. *Journal of Economic Entomology* 57(6)1007–1009.

197 Wagner, R. E., W. Ebeling, and D. A. Reierson. 1966. Control of cockroaches in sewers. *Public Works* 97(1)82–84.

198 Walter, V. 1968. A roach by any other name. *Pest Control* 36(2)32.

199 Wileyto, E. P., and G. M. Boush. 1983. Attraction of the German cockroach, *Blattella germanica* (Orthoptera: Blattellidae), to some volatile food components. *Journal of Economic Entomology* 76(4)752–756.

200 Wileyto, E. P., G. M. Boush, and L. M. Gawin. 1984. Function of cockroach (Orthoptera: Blattidae) aggregation behavior. *Environmental Entomology* 13(6)1557–1560.

201 Wille, J. 1920. Biologie und Bekämpfung der deutschen Schabe (*Phyllodromia germanica* L.). *Monographien zur Angewandten Entomologie*, suppl. 1 of *Zeitscrift für Angewandte Entomo-logie* 7(5)1–140.

202 Willis, E. R., and N. Lewis. 1957. The longevity of starved cockroaches. *Journal of Economic Entomology* 50(4)438–440.

203 Wood, F. E. 1980. Cockroach control in public housing. *Pest Control* 48(6)14–16, 18.

204 Wood, F. E., et al. 1981. Survey of attitudes and knowledge of public housing residents toward cockroaches. *Bulletin of the Entomological Society of America* 27(1)9–13.

205 Wright, C. G. 1979. Survey confirms correlation between sanitation and cockroach populations. *Pest Control* 48(9)28.

206 Wright, C. G., and R. C. Hillmann. 1973. German cockroaches: Efficacy of chlorpyrifos spray and dust and boric acid powder. *Journal of Economic Entomology* 66(5)1075–1076.

207 Wright, C. G., and R. B. Leidy. 1980. Cardboard affects cockroach control chemicals. *Pest Control* 48(8)42–43.

208 Wright, C. G., and H. C. McDaniel. 1973. Further evaluation of the abundance and habitat of five species of cockroaches on a permanent military base. *Florida Entomologist* 56(3)251–254.

209 Zimmerman, E. C. 1948. *Insects of Hawaii, Vol. 2. Apterygota to Thysanoptera*. University of Hawaii, Honolulu.

210 Zungoli, P. A. 1982. Aspects of dispersal and population structure of *Blattella germanica* (L.) in field habitats and attitudes concerning aesthetic injury level. Dissertation, Virginia Polytechnic Institute and State University, Blacksburg.

211 Zungoli, P. A., and W. H Robinson. 1982. Crack and crevice outshines fan-spray treatment. *Pest Control* 50(6)20, 22.

212 Zungoli, P. A., and W. H Robinson. 1984. Feasibility of establishing an aesthetic injury level for German cockroach pest management programs. *Environmental Entomology* 13(6)1453–1458.

9
The Asian Cockroach: Implications for the Food Industry and Complexities of Management Strategies

Richard J. Brenner

Of the nearly 4,000 described species of cockroaches in the world, most are feral, rarely seen in association with humans (*11*). A recent checklist shows that 66 species and subspecies of cockroaches have been collected in North America (*23*). Pest species are relatively few in number; some are peridomestic, such as those in the genera *Periplaneta* and *Blatta*, living around and frequently in the domestic environment, but their survival is not obligatorily tied to humans. These species are characterized by their size (relatively large), lengthy developmental times (months to 1+ years), low reproductive potential, and minimal mobility. They often have significant reservoir populations near the domestic environment. Members of the smallest group of pestiferous species are strictly domestic—best exemplified by the German cockroach, *Blattella germanica*, or the brownbanded cockroach, *Supella longipalpa*. This group is intimately associated with humans and their structures. Typically, these smaller domestic species exhibit quick maturation from egg to adult (weeks to 1+ months), thereby allowing high reproductive potential, and limited mobility that is restricted by their virtual

Brenner: Medical and Veterinary Entomology Research Laboratory, USDA Agricultural Research Service, P.O. Box 14565, Gainesville, FL 32604.

confinement to buildings. Prior to the mid-1980s, no major pestiferous species occurred in the United States (USA) that could be characterized as having substantial feral reservoirs, rapid development, and enhanced mobility.

Surprisingly, virtually all pestiferous species of cockroaches in the USA have been introduced inadvertently within the past 200 years (*24*). Although not all introduced species are pests, the appearance of any new species is cause for concern. For example, brownbanded cockroaches presumably entered at Miami in 1903 and had spread to virtually all of the 48 contiguous states by the early 1970s (*11*).

The Asian cockroach, *Blattella asahinai*, probably represents the most recent arrival (Figure 9.1). The species was described from specimens taken in sugarcane fields on Okinawa (*22*). Roth (*28*) independently described this cockroach as *B. beybienkoi* but later declared that *B. asahinai* had priority (*29*). This is one of 45 described species of *Blattella* worldwide and is most closely related morphologically and phyletically to the German cockroach (*28*). Consequently, the geographic distribution of the Asian cockroach in its native range (Figure 9.2) has been clouded by misidentifications of specimens collected as far back as the late 1800s, and likely is much more extensive in Southeast Asia and the Indian Ocean region than the few reports in the literature would lead us to believe.

Figure 9.1
Adult (A) and nymph (B) of the Asian cockroach, *Blattella asahinai*.
 SOURCE: B. Bjork, USDA Agricultural Research Service Information Service.

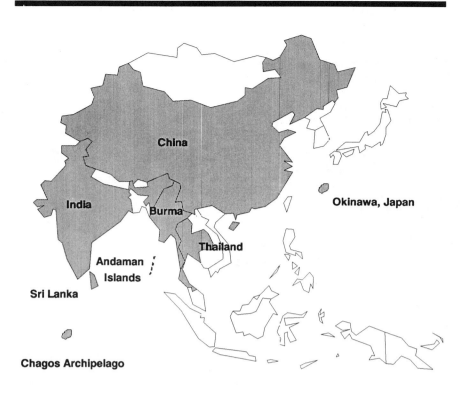

Figure 9.2
Geographic distribution of Asian cockroaches in their native range, based on listings of specimens by Mizukubo (22) and Roth (28). Countries in which Asian cockroaches have been identified are indicated, but this should not be construed as uniform distribution within each country.

Figure 9.3
Gas chromatographs of Asian, German, and Asian-German hybrid cockroaches.
SOURCE: D. Carlson, USDA Agricultural Research Service, Gainesville, FL.

U.S. Distribution and Identification

Asian cockroaches were first found in the Western Hemisphere in the winter of 1985–86 near Lakeland, Florida (6, 8, 17, 29). Roth identified the specimens on the basis of male genitalia [variation in morphological traits of *Blattella* females worldwide makes identification of females unreliable (28)]. However, Mizukubo (22) differentiates female Asian cockroaches from other Japanese *Blattella* on the basis of wing characteristics, and Ross and Mullins (27) have found some differences in nymphs and oothecae for Asian and German cockroaches. Laboratory studies have demonstrated that Asian and German cockroaches are capable of some hybridization, further obfuscating identification (10, 26). Because control measures for Asian cockroaches differ markedly from those of German cockroaches (see below), early identification of the infesting species is imperative.

These difficulties in identifying closely related species and hybrids, and the complete absence of nymphal keys to *Blattella*, necessitated the development of alternative methods of taxonomy based on qualitative and quantitative analyses of the cuticular hydrocarbons (10). The technique involves a 10-minute soaking of the specimen in 1 mL of hexane, followed by injection of 1 μL of the wash into a gas chromatography apparatus. The resultant curves represent the diversity of the hydrocarbon molecules and the relative amount of each. Detailed analyses have demonstrated that profiles are species-specific for the closely related *Blattella* cockroaches, thereby providing an easily recognizable "hydrocarbon speciesprint." Hybrids from Asian and German parents are also distinguishable (Figure 9.3). The accuracy of the technique is such that a single leg, wing, or fragment of a cast skin is sufficient to determine species (10).

Soon after the initial confirmation of Asian cockroaches in Florida, a cursory survey revealed an ovoid pattern of distribution extending across ca 50 km, from the port of Tampa northeast to Lakeland. Consequently, only three counties were considered infested (Figure 9.4A). By early 1989, this number had increased to at least 10, and infestations are probable in several other counties (Figure 9.4B). Asian cockroaches exhibit a diurnal period of intense activity when they are likely to infest crates, potted plants, and other vegetable material (see below). Therefore, this pest is viewed as having potential for passive transport in tourists' travel trailers (J. Haslem, personal communication).

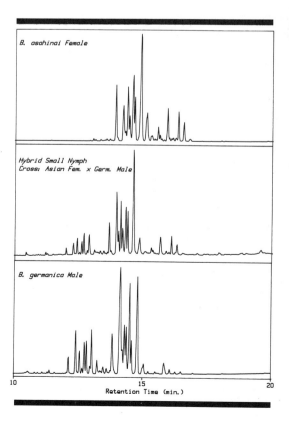

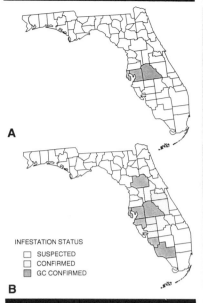

A

INFESTATION STATUS
☐ SUSPECTED
☐ CONFIRMED
■ GC CONFIRMED

B

Bionomics and Behavior

Even though Asian and German cockroaches have similar morphological traits, their habitat preferences, activity patterns, flight ability, and the effect of light on their behavior are diagnostic. The unique combination of these traits makes Asian cockroaches an unprecedented pest. Principal habitats can be either peridomestic or feral but typically are characterized by an accumulation of leaf litter or abundant ground cover in an area partially or wholly shaded (8) (Figure 9.5).

Laboratory studies comparing the net reproductive rates of Asian and German cockroaches have yielded mixed results, depending on whether the German strains are field-caught or laboratory-adapted (M. Ross, personal communication). In general, developmental rates are similar to those of German cockroaches. Using mark-release-recapture rates, field populations of Asian cockroaches have often been estimated to be in excess of 100,000 per acre (8). Historically, residents of central Florida first become aware of Asian cockroaches when populations approach about 30,000 per acre; within 18 months, these populations often increase to about 100,000 per acre and generally fluctuate near that level, although some populations have approached 200,000 per acre.

Daily activity profiles of Asian cockroaches within these habitats begin within minutes of sunset and peak within the next hour. Activity continues at a less frenzied level until ca 90 minutes before sunrise, at which time adults and nymphs move back into the substrate where they remain during the daylight hours (Figure 9.6).

In part, activity levels and flight behavior fluctuate in response to normal diel changes in light. During the day, Asian cockroaches tend to shun light, scurrying deep into the leaf litter or vegetation when disturbed. Flight rarely covers more than a few meters and is directed downward or toward adjacent plants.

As the sun sets (i.e., when both light quality and quantity change), adults and nymphs move from beneath the leaf litter onto plants or other objects in contact with the ground. In feral environments, this behavior continues for several hours, and may include a period of inactivity followed by foraging. Asian cockroaches have been observed feeding on dead insects (oftentimes trapped in spider webs), detritus and/or associated microbes, pollen and nectar in flowers, and honeydew on aphid-infested plants. However, in peridomestic environments, adult Asian cockroaches are attracted to reflected light

Figure 9.4
Distribution of Asian cockroaches in Florida in (A) 1986 and (B) 1989. GC-confirmed infestations have been confirmed by gas chromatography; "confirmed" infestations are based on firsthand reports from professional pest control operators or county extension personnel specifically trained to identify Asian cockroaches; "suspected" infestations are based on reports from professional pest control operators and the public, and are also based on known distributions in nearby counties.

Figure 9.5
Typical principal habitat of Asian cockroaches characterized by leaf litter and shade.

Figure 9.6
Time line of typical activity profile of Asian cockroaches.

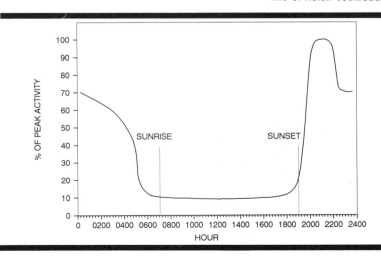

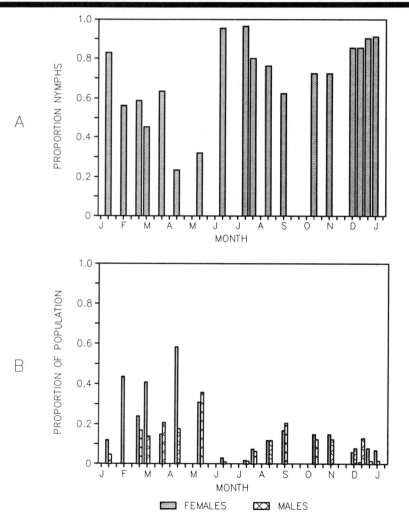

Figure 9.7
Asian cockroaches clustered on the ceiling in the areas of brightest reflected light at an apartment complex in Hillsborough County, FL.

Figure 9.8
Seasonality of age structure of Asian cockroach populations in Florida. Note abundance of nymphs (A) during June and July, and changes in abundance and proportions of adult sexes (B), presumably due to seasonal effects on survival and inherent characteristics of longevity by sex.

such as illuminated areas and light-colored walls (Figure 9.7), or to weak light sources such as television screens. An examination of the spectral sensitivity of Asian and German optic nerves indicated significant differences in the wavelenghts of light that they can perceive. During this time, Asian cockroaches demonstrate their proficiencies of mobility. They are capable of covering tens to hundreds of meters per flight in controlled and often circuitous paths from dark, wooded areas toward illuminated buildings. Flight occurs if temperatures at sunset are at least 21°, winds are light, and the area is not experiencing drought (unpublished data). Window panes and glass doors also are common landing sites when rooms are illuminated in the evening.

Upon gaining entrance, the adults fly to the brightest rooms and alight on the brightest walls. As occupants change lighting levels throughout the building, Asian cockroaches move accordingly. These adults are often attracted to light-colored clothing. Those that land are usually quickly brushed off, but often recover their balance in flight and return to the clothing, giving one the perception that Asian cockroaches have little fear of humans. Once all lights are extinguished, Asian adults and nymphs seek moisture. During subsequent days, they are found most commonly in potted plants, among moist objects in bathrooms and kitchens, and in crack-and-crevice habitats common to German cockroach infestations. Laboratory research has demonstrated that Asian cockroach nymphs are attracted to the German cockroach aggregation pheromone (*26*).

Unlike German cockroach populations that show a consistent age structure over time (*18*), Asian cockroach populations in Florida exhibit seasonality of age structure (Figure 9.8). Although this results in the presence of a substantial proportion of mobile Asian adults during all or parts of 9 months, two distinct periods occur when control strategies can be targeted toward the spatially concentrated and relatively immobile nymphal stages. It is not known whether indoor populations show a similar seasonality in age structure.

Implications for the Food Industry

The presence of any species of cockroach in a food-handling establishment has attendant health implications which include

(a) psychological stressors on employees and clientele;

(b) the possible transfer of pathogenic microorganisms to food or utensils and subsequent likelihood of disease;

(c) production of offensive odors or chemical substances that result in nausea or vertigo when inadvertently ingested;

(d) production and accumulation of allergenic materials; and

(e) risks associated with the use of pesticides to combat infestations (7).

Concerning Asian cockroaches, their unusual attraction to light immediately after sunset, accompanied by adept flight abilities and relatively high rates of reproduction (comparable to other *Blattella* spp.), firmly qualifies this species as a strong psychological stressor.

An additional concern arises from the production of allergens. German cockroaches have been identified as the most common source of cockroach allergy (hypersensitization) in the USA. Allergists and immunologists have determined that at least 11 proteins arising from cast skins, frass, or whole bodies can cause allergy in humans (25); these allergens are heat stable and persistent (2, 3); that 40–60% of asthmatics have serious allergy to cockroaches (15); and allergy to cockroaches tends to be more prevalent in women than men (16). As might be expected, Asian cockroaches produce allergens similar to or the same as allergens produced by German cockroaches (13), suggesting that the incidence of cockroach allergy may become greater in areas with widespread Asian cockroach populations.

Hypersensitization may occur by inhalation of airborne allergens, through dermal contact with allergen-contaminated surfaces, or possibly through ingestion of allergen-contaminated food. This latter mechanism has not been evaluated scientifically but may be of concern when cockroaches infest agricultural commodities and food-handling establishments. There is a published account of a woman who appeared to have an allergy to chocolate and wheat; her symptoms resolved with desensitization therapy based on *Periplaneta* extracts, reflecting American cockroach infestations in cacao bean warehouses and wheat bins (20).

Within the past year, several anecdotal accounts and direct observations have been recorded associating Asian cockroaches with agricultural commodities. These have included

(a) their consumption of tassels on sweet corn to the extent of complete detasseling;

(b) sporadic feeding on guava fruit (adults were serendipitously captured in Medfly sticky traps);

Figure 9.9
Adult Asian cockroach feeding on ripe strawberries before harvest.

(c) infestation of cabbage plants, with primary or secondary damage from feeding;

(d) feeding on strawberries in the field;

(e) consumption of rose and petunia flowers;

(f) heavy infestations in organically maintained gardens (turnips, corn, tomatoes); and

(g) infestations of wholesale and retail nursery plants (readily apparent at the first watering).

Perhaps most disconcerting was the report of Asian cockroaches in strawberry fields (9). An inspection of a field during daytime (personal observation, March 1989) revealed approximately 60% of the plants with Asian cockroaches at the base and under the black plastic ground cover. Observations at night revealed several adults and nymphs on virtually every plant, perching on leaves and berries; much of the feeding was secondary to previous damage, but adults were also seen eating through the skin and consuming pulp of umblemished fruit (Figure 9.9). Although the economic implications have not been adequately assessed in any of these instances, the potential to cause serious problems seems plausible. Large populations, high energy requirements attendant to flight, and suspected high cuticular permeability predisposes this species to seek biotic resources rich in carbohydrates and liquid water, such as fruits, flowers, and succulent plants.

Infestation and Reinfestation

Unprecedented Problems

Historically, cockroach infestations have been viewed as a problem of the tenant or business owner; German, American, and oriental cockroach populations are relatively immobile and typically arise within the property bounds—he

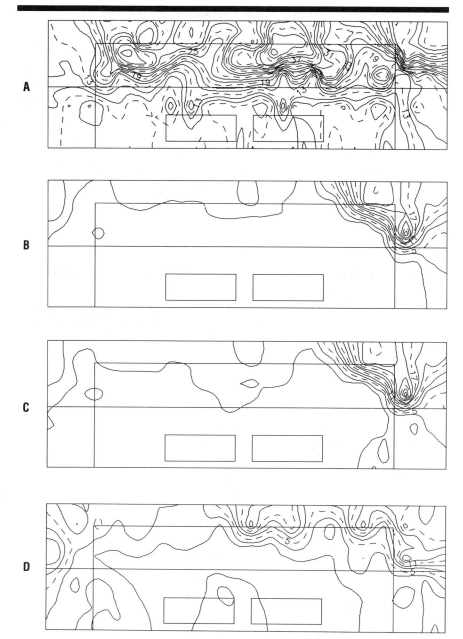

Figure 9.10

Contour maps showing (A) distribution of adult and nymphal Asian cockroaches at pretreatment, (B) nymphal cockroaches 2 weeks posttreatment, (C) nymphal and adult cockroaches at 2 weeks posttreatment, and (D) nymphal and adult Asian cockroaches at 2 months posttreatment. The two smaller rectangles (lower center) represent buildings; the larger rectangle (length, 32 m) delineates the treated area (1 acre) to the north (top), west, and east of the buildings (south side of buildings bounded by extensive parking lots). The horizontal line through center of each plot represents a property line between a heavily wooded area (above line) and well-kept lawn (below). The area was treated with a pelletized bait (Dursban Asian Roach Bait, Southern Mill Creek, Tampa) at 150 lb/acre; the upper right-hand corner of the area shown did not receive bait. Contours (minimum = 1, intervals of 2) reflect the number of cockroaches per sticky trap per night based on data from 150 traps. Differences in contours between B and C reflect adults moving toward security lights between the buildings and at the lower right of the treated area (location of security lights). Note that by the end of 2 months, posttreatment populations (adults) were becoming reestablished near these light sources and near a third light source just to the left of the treated area. See Brenner et al. (8) for a general description of contouring procedures.

who produces the problem is responsible for securing remedies. But, obtaining adequate control of Asian cockroaches is exacerbated by their attraction to light, diffuse feral/peridomestic habitat, and flight abilities. Problems occurring at one establishment repeatedly arise from other properties—thus, the owner has no jurisdiction over the areas where control measures can be effective. Moreover, activity profiles of Asian cockroaches not only favor passive transport from infested regions, but virtually guarantee frequent reinfestation (Figure 9.10). To prevent or minimize problems associated with Asian cockroaches, diligent surveillance of areas likely to become infested is required.

High-Risk Areas

As Asian cockroach infestations spread, certain areas will be at higher risk for infestation and reinfestation. Factors that favor infestation include

(a) adjacent undeveloped land or heavily wooded property;

(b) light-colored buildings or structures with large windows through which inside lights shine during the evening;

(c) outside security lights that remain on during the night, thereby providing a diffuse area of brightness on the ground;

(d) use of natural mulches in landscaping and frequent watering of verdant lawns;

(e) well-illuminated shipping/receiving bays with loose-fitting doors; and

(f) businesses with extensive indoor plantings (Figure 9.11).

Given the recent reports involving agricultural commodities, grain storage facilities and food distribution warehouses also may be at elevated risk for Asian cockroach infestation.

Surveillance

Sticky traps are used successfully to monitor German cockroach infestations indoors when they are placed near principal habitats (*1*). Successful application of similar techniques with Asian cockroaches, however, will require placement of traps in various locations to identify (a) primary foci (outdoor reservoir populations characterized by a high proportion of nymphs); (b) dispersive (expansion) populations (flying adults); and (c) passively transported populations (all stages). Even these procedures must be undertaken with the knowledge that life stages may be seasonally restricted (Figure 9.12).

Virtually any baited sticky trap will suffice; these should be emplaced in late afternoon and retrieved early the following morning to reduce likelihood of damage from moisture or scavenging by ants and mice. Surveillance for adults expanding from primary foci (often unrecognized) should focus on the distribution of outdoor lighting, especially lights close to buildings near grasses and mulches. An occasional inspection for resting adults on light-colored or brightly lit walls, glass doors, and windows in rooms commonly lighted after sunset also would be prudent. Primary outdoor foci can be discovered by means of sticky traps placed in shaded areas where leaf litter accumulates, beneath bushes and trees, and along fence lines where leaves, grasses, and debris are usually undisturbed by property maintenance procedures. Be especially careful with new mulches and peat because spread of Asian cockroaches into previously uninfested neighborhoods, nurseries, and communities has occurred in central Florida in this manner (J. Hinton, Hillsborough County Cooperative Extension, Florida, personal communication).

Populations resulting from passive transport of infested goods likely will assume characteristics of German cockroaches, seeking harborage sites near sources of moisture and food. What appear to be German cockroaches may not be; Asian adults as well as Asian-German hybrids take flight when challenged with insecticide sprays (R. S. Patterson, USDA Agricultural Research Service, Gainesville, FL, personal communication), as was seen when a Tampa warehouse was treated for a presumed German cockroach infestation (D. Carlson, Young Pest Control, Tampa, FL, personal communication).

Strategies for Suppression of Asian Cockroaches

Although laboratory studies have shown that Asian cockroaches are highly susceptible to virtually all insecticides (30), there is a vast difference between killing and controlling Asian cockroaches. Because this species is so active during our leisure time and is attracted to our lighting, virtually every invading adult will be seen, thereby establishing the aesthetic injury level (31) at one cockroach. Populations that may become established indoors probably can be controlled (i.e., eradicated) with a combination of materials, as is done commonly with German cockroach infestations. However, in addition to standard applications of residual sprays and growth regulators, toxic baits should be used in areas attractive specifically to Asian cock-

Figure 9.11
A typical scenario that favors Asian cockroach infestation. This lawn and garden store has become uncontrollably infested with Asian cockroaches that are attracted by the bright lights in the parking lot. Adults fly from adjacent wooded area where their population levels are in excess of 50,000 per acre. Sod, potted plants, and loose mulch are heavily infested. From here, adult cockroaches move into the store and into the public cafeteria. The adjacent shopping plaza (9 shops, 1 restaurant) has similar problems.

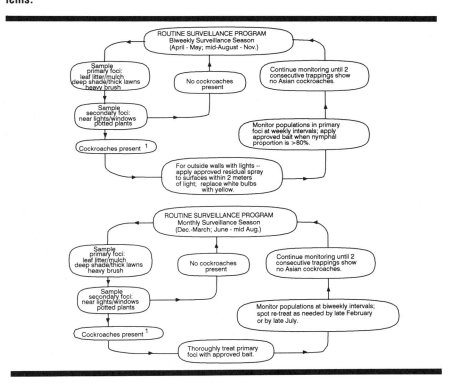

Figure 9.12
Flow diagram detailing suggested methods for routine surveillance for primary and secondary Asian cockroach populations, and a strategy for control.
[1]In primary and secondary foci: Suggests that populations are well established. If populations can be controlled now, be aware that reinfestations likely will occur if building is within the range of feral Asian cockroach populations. Be vigilant for reinfestation in secondary foci.

In primary only: Suggests that infestation is recent and possibly the result of infested commodities being transported to your property. Attempt to locate commodity and determine point of origin. Notify provider.

In secondary only: Suggests presence of new infestation resulting from adults flying from nearby primary foci which likely arise from neighboring property. With permission of land owners, check neighboring properties in effort to locate principal (reservoir) populations; if these cannot be eliminated, infestation at secondary foci will occur by adults, and encroachment of nymphs may occur at your property line as populations expand.

Asian Cockroach **127**

Figure 9.13
Spatial and temporal changes in the distribution of adult stages (A, C) and all stages (B, D) of Asian cockroaches in May (A, B) and June (C, D) on a property in central Florida. See Figure 9.5 for view toward lower right corner of property. Contours (minimum = 1, intervals of 3) reflect number of cockroaches per sticky trap per night based on data from 55 traps. Small rectangles reflect location of buildings on 1.5 acres; largest rectangle (upper right corner) delineates a neighboring property.

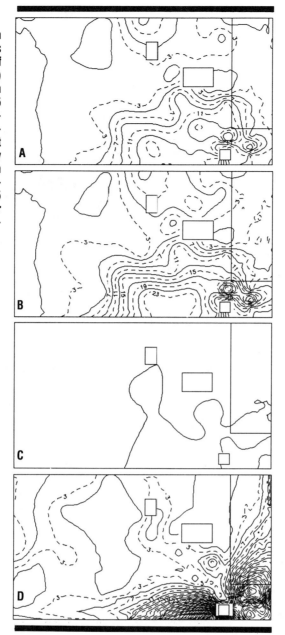

roaches, e.g., near potted plants, greenhouses, and terrariums.

In spite of these efforts, Asian cockroach adults will continue to be seen indoors so long as the principal outdoor foci remain untreated. Given a low tolerance threshold, clients of food stores, restaurants, and department stores will notice these persistent invaders; typically, adults are seen in the aisles and food-serving areas on a daily basis (J. Bradley, Arab Pest Control, Tampa, FL, personal communication).

Consequently, in latitudes where principal foci exist outdoors, the strategy becomes complex. The entomologically and environmentally sound tactic, based on suppressing the most concentrated, immobile, and accessible life stages (CIA concept; *4, 5, 14*), is difficult to implement. In spatial terms, nymphal stages are most concentrated, but the extensive distribution of suitable habitats can hardly be described as concentrated. Furthermore, in most instances, many of the foci within the flight range of adults are legally inaccessible, in that these feral reservoirs fall outside the jurisdiction of the owner of the affected property who cannot treat beyond his or her boundaries.

Use of electrocutor grid traps may seem a plausible approach on first inspection, but there are several serious shortcomings. Foremost, Asian cockroach adults are not attracted to ultraviolet wavelengths, but to reflected light from incandescent and fluorescent lamps. Conceivably, ultraviolet lamps could be replaced with bulbs producing attractive wavelengths, and electrified reflective surfaces could be used in lieu of wire grids. But the actual percentage of adults flying to lights is extremely small, given the enormity of a typical population, and it can be argued that the addition of a "landing beacon" may serve only to attract more fertile adults to the premises. Thus, efforts to control populations on the basis of adult suppression appear tantamount to treating the symptoms and not the disease. It should be mentioned, however, that, like German cockroaches (*19*), Asian cockroaches do not see yellow light. Replacement of white bulbs with yellow might reduce the level of attraction; alternatively, it is possible that this may only change the landing pattern, shifting distribution from outside lights toward windows of illuminated rooms.

Legal issues aside, there are, however, two temporal concentrations when (a) highly mobile adults are exceedingly rare and the relatively immobile nymphs are predictably concentrated in recognizable habitats (June and July); and (b) there is a preponderance of nymphs and when temperatures below 21° at sunset render adults relatively immobile (December, January, February) (Figure 9.8). Application of toxic baits to principal foci during these times will provide the greatest probability of success. Thus, recognition of principal foci, diligent monitoring of same, and careful timing of treatments are requisite to managing Asian cockroach populations. This requires client education because the windows of opportunity coincide with the periods in which the client perceives little or no problem.

This is best illustrated by Figure 9.13, depicting a typical situation in eastern Hillsborough County, Florida. In 1986, this particular region had no known Asian cockroach populations. A property owner recalled one week in late April 1987 when the number of Asian adults on light-colored walls rose from zero one night to dozens the next (in retrospect, coincident with the first warm evening). Use of toxic granules on mulches at the perimeter of the house, boric acid powders near doorways and in the carport, and sprays directed at the outside walls appeared effective as judged by the absence of adults in early June. But the perceived security was false; nymphal populations in shaded areas of accumulated leaf litter were enormous, and the resultant adult population in mid-August was uncontrollable. The solution that is obvious to entomologists (but not clients) requires application of a toxic bait in June or July, and a second treatment in December or January, if trap catches document reinfestation. Industry has begun developing toxic baits (and some sprays) specifically for Asian cockroaches and is seeking EPA registration for use in principal habitats.

Projected Strategies

Perhaps the greatest need lies in developing regional control strategies. Unlike mosquitoes, which are perceived both as a health threat and of regional administrative concern (i.e., neighborhood, municipality, county), cockroaches continue to be considered only as nonagricultural nuisances and a localized problem (home- or business owner). With no county, state, or federal regulatory agencies having legal jurisdiction, the solutions have been addressable only by private industry. Safeguarding some crops, warehouses, food-processing plants, food-serving industries, and health care facilities ultimately may require the implementation of regional control strategies against Asian cockroaches. The likelihood of this occurring at any governmental level seems improbable, given financial constraints, public consciousness of pesticides, and the unprecedented legislation necessary to enact such a program. However, perhaps such a program may be undertaken by a consortium of pest control operators working in concert with a chamber of commerce, a tourist bureau, land developers, or a homeowners' association.

Given that clients and landowners object primarily to the adult Asian cockroaches, direction of at least some efforts toward this stage seems unavoidable. The recent development of potent, light-stable insect growth regulators for nymphs offers some promise because the morphogenic side effects produce nonfunctional wings in the resultant adults. Asian cockroaches that do not fly obviously will be less likely to arrive at brightly lit walls. Cockroach repellents (not yet commercially available), applied to outside walls, door and window frames, and internal surfaces of food-processing or food-delivery equipment, also may reduce the likelihood that infestations will be established (*12, 21*).

Acknowledgments

This chapter represents a compilation of observations and studies conducted independently and in cooperation with many individuals and organizations, including the Asian Cockroach Task Force (R. Pierce, Chairman) of the Florida Pest Control Association; J. Bradley, Arab Pest Control, Tampa; T. Himelberger, formerly with Southern Mill Creek, Tampa; J. Hinton, Hillsborough County Cooperative Extension; P. G. Koehler, University of Florida, Gainesville; Lady Bugs Pest Control (K. Everett, J. Gibson), Tampa; R. S. Patterson and L. Pickens, USDA Agricultural Research Service (ARS). Special thanks to B. Bjork, USDA-ARS Information Service, for photographs (Figure 9.1) and to D. Carlson, USDA-ARS, Gainesville, FL, for providing Figure 9.3. Thanks to peer reviewers R. S. Patterson, D. Focks, and D. Williams.

Cited References

1 Ballard, J. B., and R. E. Gold. 1982. The effect of selected baits on the efficiency of a sticky trap in the evaluation of German cockroach populations. *Journal of the Kansas Entomological Society* 55(1)86–90.

2 Bernton, H. S., and H. Brown. 1964. Insect allergy—preliminary studies of the cockroach. *Journal of Allergy* 35(6)506–513.

3 Bernton, H. S., and H. Brown. 1967. Cockroach allergy II: The relation of infestation to sensitization. *Southern Medical Journal* 60(8)852–855.

4 Brenner, R. J. 1988. Activity and distribution of cockroaches with implications to developing control strategies. *Proceedings of the Second National Conference on Urban Entomology* (College Park, MD, Feb. 21–24), pp. 33–40.

5 Brenner, R. J. 1988. Focality and mobility of some peridomestic cockroaches in Florida (Dictyoptera: Blattaria). *Annals of the Entomological Society of America* 81(4)581–592.

6 Brenner, R. J., P. G. Koehler, and R. S. Patterson. 1986. A profile of America's newest import, the Asian cockroach. *Pest Management* 5(Nov/Dec)17–19.

7 Brenner, R. J., P. G. Koehler, and R. S. Patterson. 1987. Implications of cockroach infestations to human health. *Infections in Medicine* 4(8)349–355, 358–359, 393.

8 Brenner, R. J., R. S. Patterson, and P. G. Koehler. 1988. Ecology, behavior, and distribution of *Blattella asahinai* (Orthoptera: Blattellidae) in central Florida. *Annals of the Entomological Society of America* 81(3)432–436.

9 Carey, S. 1989. Asian cockroach sighted in Florida strawberry fields. *Citrus & Vegetable Magazine* 52(7)35, 38–39.

10 Carlson, D. A., and R. J. Brenner. 1988. Hydrocarbon-based discrimination of three North American *Blattella* cockroach species (Orthoptera: Blattellidae) using gas chromatography. *Annals of the Entomological Society of America* 81(5)711–723.

11 Cornwell, P. B. 1968. *The Cockroach*, vol. 1. Hutchinson, London.

12 Hagenbuch, B. E., et al. 1987. Two chemical repellents for control of German (Orthoptera: Blattellidae) and American cockroaches (Orthoptera: Blattidae). *Journal of Economic Entomology* 80(5)1022–1024.

13 Helm, R. M., et al. 1988. Cross-reactivity of cockroach (CR) allergens: RAST inhibition and immunoblot analysis of Asian cockroach (ACR) (*Blattella asahinai*) and German cockroach (GCR) (*Blattella germanica*) extracts. *Journal of Allergy and Clinical Immunology* 81(1)269 (Abstr. No. 404).

14 Horsfall, W. R. 1985. Mosquito abatement in a changing world. *Journal of the American Mosquito Control Association* 1(2)135–138.

15 Kang, B. C., and J. L. Chang. 1985. Allergenic impact of inhaled arthropod material. *Clinical Reviews in Allergy* 3:363–375.

16 Kang, B. C., et al. 1988. The cockroach asthma (BACR): A variant of severe environmental bronchial asthma. *Journal of Allergy and Clinical Immunology* 81(1)272 (Abstr. No. 416).

17 Koehler, P. G., and R. S. Patterson. 1987. The Asian roach invasion. *Natural History* 96(11)28, 30, 32, 34, 35.

18 Koehler, P. G., R. S. Patterson, and R. J. Brenner. 1987. German cockroach (Orthoptera: Blattellidae) infestations in low-income apartments. *Journal of Economic Entomology* 80(2)446–450.

19 Koehler, P. G., et al. 1987. Spectral sensitivity and behavioral response to light quality in the German cockroach (Dictyoptera: Blattellidae). *Annals of the Entomological Society of America* 80(6)820–822.

20 Marchand, A. M. 1966. Allergy to cockroaches. *Boletín de la Asociación Médica de Puerto Rico* 58(1)49–52.

21 McGovern, T. P., and G. S. Burden. 1985. Carboxamides of 1,2,3,6-tetrahydropyridine as repellents for the German cockroach, *Blattella germanica* [Orthoptera (Dictyoptera): Blattellidae]. *Journal of Medical Entomology* 22(4)381–384.

22 Mizukubo, T. 1981. A revision of the genus *Blattella* (Blattaria: Blattellidae) of Japan. I. Terminology of the male genitalia and description of a new species from Okinawa Island. *Esakia* 17(Nov)149–159.

23 Pratt, H. D. 1988. Annotated checklist of the cockroaches (Dictyoptera) of North America. *Annals of the Entomological Society of America* 81(6)882–885.

24 Rehn, J. A. 1945. Man's uninvited fellow traveler—the cockroach. *Scientific Monthly* 61:265–276.

25 Richman, P. G., et al. 1984. The important sources of German cockroach allergens as determined by RAST analyses. *Journal of Allergy and Clinical Immunology* 73(5)590–595.

26 Ross, M. H. 1989. Asian cockroach research. *Pest Management* 8(2)20–22.

27 Ross, M. H., and D. E. Mullins. 1988. Nymphal and oothecal comparisons of *Blattella asahinai* and *Blattella germanica* (Dictyoptera: Blattellidae). *Journal of Economic Entomology* 81(6)1645–1647.

28 Roth, L. M. 1985. A taxonomic revision of genus *Blattella* Caudell (Dictyoptera: Blattaria: Blattellidae). *Entomologica Scandinavica* 22(suppl.):1–221.

29 Roth, L. M. 1986. *Blattella asahinai* introduced into Florida (Blattaria: Blattellidae). *Psyche* 93(3–4)371–374.

30 Wadleigh, R. W., P. G. Koehler, and R. S. Patterson. 1989. Comparative susceptibility of North American *Blattella* (Orthoptera: Blattellidae) species to insecticides. *Journal of Economic Entomology* 82(4) 1130–1133.

31 Zungoli, P. A., and W. H Robinson. 1984. Feasibility of establishing an aesthetic injury level for German cockroach pest management programs. *Environmental Entomology* 13(6)1453–1458.

10
Beetles: Coleoptera

Richard T. Arbogast

Beetles constitute the most important and largest order of insects attacking stored products. According to Hinton (*64*), 600 species of beetles, representing 34 families, have been found associated with stored products in various parts of the world. This number includes species that consume stored products, that feed on molds growing on stored products, and that occur in the storage habitat only as predators. The most serious pest species, as well as some less serious but commonly encountered species, have been selected for treatment here.

The amount of damage inflicted by stored-product insects in a particular storage situation is largely a function of their feeding habits and the population levels they are able to attain. Population level depends upon the number initially present, the time available for population growth, and its biotic potential modified by the combined action of predators, parasites, diseases, and other adverse factors.

Biotic potential is usually expressed as the innate capacity for increase in numbers (r_m) or as its antilog (λ), the rate of self-multiplication per unit time or the finite rate of increase (*3*). Another parameter, known as the capacity for increase (r_c) (*85*), is sometimes used. Both r_m and r_c are determined by life-history features (rate of development, age-specific fecundity, age-specific mortality) and are characteristic of a particular species or strain, but their expression

is influenced by the prevailing physical conditions (e.g., temperature, humidity) and by the nutritional situation. The innate capacity for increase represents the growth rate of a population with a stable age distribution on a particular diet under a given set of physical conditions when the population is free from the suppressive effects of crowding, predators, disease, and so on. The capacity for increase is an approximation of r_m but also represents the actual rate of increase of a population with no overlapping of generations or the actual rate of increase of a continuously breeding population in the first few generations arising from a group of animals all of the same age.

Food Habits, Life History, and Biotic Potential

Anobiidae

Lasioderma serricorne, cigarette beetle

The biology of the cigarette beetle as reported here is based largely on a paper by Howe (*70*). The cigarette beetle is perhaps the most ubiquitous of all stored-product pests. It occurs throughout the tropical and subtropical regions of the world, and although it is restricted by low temperature and humidity, it occurs commonly in warm buildings throughout the temperate regions. It breeds in a wide variety of commodities including plant and animal materials. These include aniseed (*Pimpinella anisum*), areca nuts (betel nuts, *Areca catechu*), bamboo, beans, biscuits, cassava (*Manihot esculenta*), chick-peas, cocoa beans, coffee beans, copra, coriander,

Arbogast: Stored-Product Insects Research and Development Laboratory, U.S. Department of Agriculture, P.O. Box 22909, Savannah, GA 31403.

cottonseed, cottonseed meal, cumin, dates, dried banana, dried cabbage, dried carrots, dried fruit, drugs, flax tow, wheat flour (including atta), ginger, grain, peanuts, herbs, herbarium specimens, insecticides containing pyrethrum, juniper seed, licorice root, nutmeg, raisins, rhubarb, rice, seeds of trees and other plants, spices, yeasts, tobacco, dried insects, dried fish and fish meal, and leather.

The eggs are laid singly in crevices, folds, or depressions in the food. The incubation period decreases from 20–22 days at 20° to 5–6 days at 35°, and then increases again to about 7 days at 37.5° (70). The minimum temperature for hatching lies between 17.5° and 20°; the maximum, between 37.5° and 40°. Low humidity inhibits hatching, but at 30° some eggs hatch at relative humidities (rh) as low as 20%. The effect of low humidity becomes more rigorous above or below this temperature, and higher humidities are required for hatching. Humidity affects the duration of the egg stage only at the lower limit of humidity tolerance, but at extremely low humidities, hatching is noticeably delayed.

Newly hatched larvae are very active, negatively phototactic, and often enter very small holes. Thus they can infest a variety of packaged goods, including spices in closed cans, apparently by entering through minute openings in the seams or around the lids (90). Older, scarabeiform larvae are also negatively phototactic but are less active.

The number of larval instars ranges from four to six; Howe (70) observed only four at 30° and 70% rh. The lower threshold for larval development is between 17.5° and 20° and the upper threshold is between 37.5° and 40°. Larvae become inactive at temperatures between 15.5° and 19.5° and enter dormancy at temperatures slightly below this. They can remain dormant for several months and may be able to overwinter if the temperature does not drop too low.

The optimum relative humidity for development is 70–80%. At higher humidities, development is slower, with heavy mortality near the limiting temperatures. The minimum humidity for development depends upon the temperature and the commodity. Thus, the lower limit for development on wheatfeed (fine bran plus a small quantity of endosperm) is 50% rh at 20° or 37.5°, and 25% rh at 30°, but on peanuts the lower limit is 40% rh at 30°. The duration of the larval period on wheatfeed at 70% rh decreases from about 70 days at 20° to 16 days at 32.5°, and then increases to about 26 days at 37.5°. Mor-

tality reaches 55% at the upper temperature limit. At 30°, the duration of the larval stage on wheatfeed decreases from about 38 days at 25% rh to 18 days at 70–80% rh and then increases to about 20 days at 100% rh. Mortality is highest near the humidity extremes.

Upon reaching maturity, cigarette beetle larvae construct fragile cells of food and waste material cemented together with a secretion from the midgut. Pupal development requires from 4 days at 30–37.5° to 12 days at 20°, and is not affected by humidity. The adults remain within the pupal chamber until the cuticle hardens, a period that depends upon temperature (but not upon humidity) and which lasts 4 days at 32.5°, 5 days at 37.5°, and 12 days at 20°.

The length of the developmental period and

Table 10.1

Developmental period and percentage survival to the adult stage of *Lasioderma serricorne* on various foods

Product	Developmental period[a] (days)	Survival rate (%)
Howe (70)		
Wheatfeed	26.1	88
Wholemeal flour	26.5	100
Cottonseed meal	26.9	93
Cowpeas		
crushed	27.5	86
damaged	28.2	90
Yeast powder	28.4	100
Coconut meal	31.2	93
Cowpeas, whole	31.6	63
Peas, whole	33.0	3
Locust beans, crushed	33.6	83
Peanuts, whole	34.4	50
Cocoa beans, crushed	55.2	86
Locust beans, whole	68.5	33
Cocoa beans, whole	105.6	15
LeCato (90)		
Standard medium[b]	35	90
Paprika	70	72
Cayenne pepper	71	85
Curry powder	122	2
Chili powder	135	2

[a] Howe (70) used the mean number of days from hatching to adult emergence from the pupal chamber at 30°, 70% rh. LeCato (90) used the median number of days from egg (within 12 hours of hatching) to adult emergence (presumably from the pupal cell) at 28°; humidity was not reported.
[b] White flour, white cornmeal, and brewers' yeast (10:10:1.5 by volume).
SOURCES: Howe, R. W. 1957. A laboratory study of the cigarette beetle, *Lasioderma serricorne* (F.) (Col., Anobiidae), with a critical review of the literature on its biology. *Bulletin of Entomological Research* 48(1)9–56, 2 pl.; LeCato, G. L. 1978. Infestation and development by the cigarette beetle in spices. *Journal of the Georgia Entomological Society* 13(2)100–105.

the mortality of the immature stages vary considerably with the commodity upon which the larvae develop (Table 10.1). Products such as whole peas, curry powder, and chili powder are relatively poor diets and do not support rapid population growth. Furthermore, although infestations of *L. serricorne* have been reported in a wide variety of spices, many spices including black pepper, cinnamon, cloves, East Indian mace, mustard, nutmeg, sage, thyme, and turmeric cannot support development (*90*).

Howe (*70*) did not observe feeding by the adults, but Lefkovitch and Currie (*99*) demonstrated that adults feed readily on the same food as the larvae and that mated females lay few eggs unless provided with food. Longevity is apparently reduced by oviposition, since mated females deprived of food live longer than mated females that are fed (*99*). Adults are short-lived, the lifespan depending upon temperature and humidity, with the shortest lifespans at high temperatures and low humidities (*70*). On average, at 70% rh, males live 21 days at 35° to 43 days at 20°, and females live 18–46 days.

Adult cigarette beetles probably reach sexual maturity while they are still in the pupal chamber; they mate several times after emergence. The shortest preoviposition period observed, following emergence from the pupal chamber, ranged from 4 days at 30° and 35° to 12 days at 20°. The oviposition period ranges from 6–9 days at 35° to 14–20 days at 20°. Oviposition reaches an early maximum and then declines sharply (Figure 10.1). The oviposition period is prolonged, and the rate of oviposition falls off less rapidly when females have access to water.

Mean fecundity on wheatfeed at 70% rh increases rapidly with temperature between 20° and 25° (Figure 10.2A). Fecundity is maximum between 25° and 35° and declines rapidly between 35° and 37.5°. At 30°, females not provided with water show a marked decrease in fecundity. Percentage egg hatch is greatest (76–78%) at 22.5–25°. Eggs fail to hatch at 37.5°, and only 50% hatch at 20°. Egg mortality also increases sharply near the lower humidity limits, which vary with temperature. In general, percentage hatch varies between 60% and 90% at favorable humidities, being greater at higher humidities.

The relationship between temperature and λ on wheatfeed at 70% rh is illustrated in Figure 10.2B. The minimal temperature for population increase lies just below 20°, and the rate of population growth increases rapidly with temperature up to 35°. Numerous studies have been

conducted on the lethal effects of high and low temperatures on various stages of *L. serricorne*. At 75% rh, third and fourth instars are killed by a 3-week exposure to 4.4°. Third instars are also killed in 3 weeks by exposure to 7.2°, but a 5-week exposure is required to kill fourth instars. At 10°, 60% of third instars and 20% of fourth instars are killed in 11 weeks. Mortality at 10° is higher at lower humidities (45% or 55%) (*31*). Mullen and Arbogast (*112*) reported median lethal exposures (LT_{50}) of 91, 60.8, 25.5, 2.4, 0.4, and 0.3 hours for eggs at 5, 0, −5, −10, −15, and −20°, respectively. Adults are killed by a 2-hour exposure to −13°, a 4-hour exposure to −12.5°, an 8-hour exposure to −9°, or a 4-day exposure to −5.5°. More than 6 days are required to kill all adults at 4°. Exposure for 24 hours at 50° or for a few minutes at 60° kills all stages (*70*).

Stegobium paniceum, drugstore beetle

The drugstore beetle infests nearly any dry plant or animal material, but it has been suggested that this wide range of food preferences may reflect the existence of different races (*98*). It has been recorded from black and red pepper, drugs, stored grain, spices, tobacco, leather, books, wood, and textiles. It is a cosmopolitan insect but is characteristically more temperate than tropical.

There are four to six larval instars, the last of which constructs a cocoon or pupal chamber

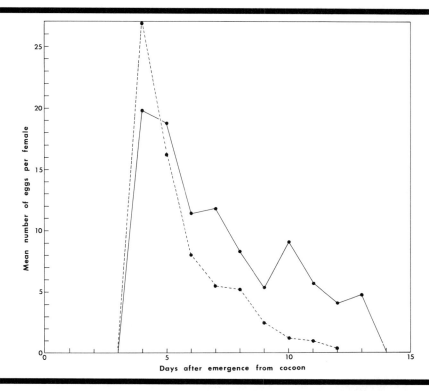

Figure 10.1
Oviposition of *Lasioderma serricorne* on wheatfeed at 30° and 70% rh when the beetles were supplied with water (solid line) and when no water was supplied (dashed line).
SOURCE: Howe, R. W. 1957. A laboratory study of the cigarette beetle, *Lasioderma serricorne* (F.) (Col., Anobiidae), with a critical review of the literature on its biology. *Bulletin of Entomological Research* 48(1)9–56, 2 pl.

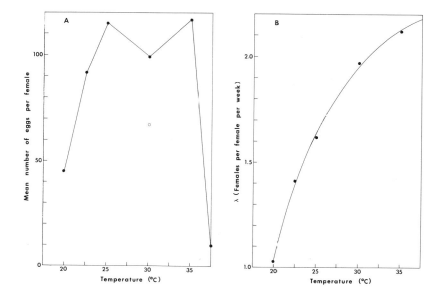

Anthribidae
Araecerus fasciculatus, coffee bean weevil

The coffee bean weevil or cacao weevil is essentially worldwide in distribution but is more common in tropical and subtropical countries where is attacks a wide variety of crops, both in the field and in storage. Its reported hosts include coffee, corn, Brazil nuts, Chinese ginger, palm kernels, cacao, cassava, peanuts, nutmeg, Bambarra groundnuts (*Vigna subterranea*), monkeypod (*Pithecellobium saman*), sweet potato, avocado, sorghum, pawpaw, dried plantain (*Plantago* spp.), and yams (*111*). It is abundant in the southern USA where it breeds on dried fruit, coffee, cornstalks, corn, and the seeds and seed pods of a wide variety of plants (*1*).

A. fasciculatus prefers food material with a relatively high moisture content and breeds well on material that has reached a state of decay. Thus, 90% rh is very favorable for oviposition and development on corn, nutmeg, or cacao at 27°, and even 100% rh is favorable as long as the food does not become moldy or covered with condensation (*47*). The susceptibility of "robusta" (*Coffea canephora*) and "arabica" (*C. arabica*) coffees varies considerably with humidity (*86*). Both are susceptible to infestation at 100% rh, and in either variety the weevil completes development in 30–35 days at 30° and 40–60 days at 25°. Both varieties become less susceptible at lower humidities, and varietal differences appear. Thus, *A. fasciculatus* completes development in 65–90 days on arabica at 80% rh (25°), but cannot develop on robusta. The effects of humidity on the minimum developmental period at 27° are shown in Table 10.2 (*47*).

The incubation period of the egg is 5–8 days, and the duration of the pupal period is 6–7 days at 27° and humidities between 50% and 100% (*47*). On corn, adults live for 27–28 days at 50% rh and 86–134 days at 90% rh. Few live more than 20 days on cacao at relative humidities below 80%. Males reach sexual maturity in 3 days and females in 6 days after emergence. Adults mate more than once, but one mating is sufficient for normal oviposition. Females usually begin laying eggs within half an hour after mating, although oviposition is sometimes delayed as much as 6 hours. Oviposition reaches a maximum during the second and third week after emergence, when the highest rate is 2.4 eggs/female per day.

An ovipositing female forces the ovipositor

Figure 10.2
Effect of temperature on (A) the mean fecundity, and (B) the finite rate of increase (λ) of *Lasioderma serricorne* on wheatfeed at 70% relative humidity when the beetles were supplied with water. The effect of withholding water at 30° is indicated by the open circles.

SOURCE: Howe, R. W. 1957. A laboratory study of the cigarette beetle, *Lasioderma serricorne* (F.) (Col., Anobiidae), with a critical review of the literature on its biology. *Bulletin of Entomological Research* 48(1)9–56, 2 pl.

in which it pupates. However, cocoon formation depends upon the conditions under which the insect develops; no cocoons are formed under some temperature and humidity conditions. The incubation period of the egg and the duration of the pupal period decrease with increasing humidity as well as with increasing temperature. The adults remain within the cocoon for variable periods after eclosion. Unlike adult *L. serricorne*, adult *S. paniceum* do not feed (*98*).

The temperature limits for development are 15° and 34°; the lowest humidity for development is 35% rh. The developmental period on wheatfeed is shortest (about 40 days) at 30° and 60–90% rh. The developmental period at 70% rh increases to about 150 days at 17.5° and to about 60 days at 32.5°; at 30°, it increases to about 90 days at 40% rh. Mortality of immature stages is lowest at 20–30° and 95% rh and at 27.5° and 50% rh. Adults live longest (about 85 days) at 17.5° and 50–70% rh. The adult lifespan is shortest (about 13 days) at 32.5° and 70% rh. Fecundity is highest (about 75 eggs/female) between 20° and 25° and 60–70% rh. *S. paniceum* does not survive as well as *L. serricorne* at high temperatures, and at its maximum rate it develops more slowly. *L. serricorne* is better able to cope with dry conditions and appears to lay more eggs, but the adults live for a shorter period than those of *S. paniceum.*

Table 10.2

Effect of humidity on the minimum developmental period (in days) of *Araecerus fasciculatus* reared on various commodities at 27°

Food	Relative humidity (%)				
	60	**70**	**80**	**90**	**100**
Corn	57	51	43	35	29
Nutmeg	69	59	51	43	38
Cacao	—	—	66	45	37

SOURCE: El-Sayed, M. T. 1935. On the biology of *Araecerus fasciculatus* de Geer (Col., Anthribidae), with special reference to the effects of variations in the nature and water content of the food. *Annals of Applied Biology* 22(3)557–577.

into the food material with a series of thrusts and then deposits an egg (*47*). Most eggs, usually more than 98%, are thus concealed within the food. Females oviposit in coffee berries, and the larva feeds at first on the pulp and then within the seed (*10, 11*). Eggs are also laid on soft corn after harvest, but damage to stored corn is minimal because the kernels become too hard (*1*).

Oviposition and egg hatchability vary directly with the moisture content of the food and are maximum at 90–100% rh (*47*). Females lay eggs readily on corn at 8.4% mc (60% rh), but oviposition is much lower at 6.7% mc (50% rh), and they do not usually lay eggs on cacao at less than 7.5% mc (80% rh) or on nutmegs in equilibrium with less than 60% rh. Hatchability on corn at 27° ranges from about 40% at 50% rh to nearly 100% at 100% rh.

Bostrichidae

Prostephanus truncatus, larger grain borer

The larger grain borer is an established pest of corn in Central America and has occasionally been observed in other parts of the world (*141*). This pest is now well established in Africa. It often attacks corn before harvest and continues to breed in it throughout storage, especially if the corn remains unshelled (*32*). It is capable of attacking a wide range of commodities, but among the commodities tested [corn, wheat, rice, cowpeas, butter beans (*Phaseolus lunatus*), haricot beans (*P. vulgaris*), peanuts, cocoa beans, coffee beans], it is able to breed successfully only on corn (*140*). The combined duration of the larval and pupal stages is shortest (27 days) at 32° and 80% rh and longest (78 days) at 22° and 50% rh on finely ground corn (*141*). At 70% rh, the combined larval and pupal period increases from 30 days at 30° or 32° to 40 days at 35° and to 61 days at 22°. At all temperatures, the developmental period is much longer when the relative humidity is lower than 70%. Mortality is generally low in the zone bounded by 27–32° and 70–80% rh.

Rhyzopertha dominica, lesser grain borer

The lesser grain borer is a cosmopolitan pest of a wide variety of foods, chiefly cereals, but including grain of all kinds, flour, macaroni, beans, chick-peas, pumpkin seeds, dried potato, edible bulbs, wood, and many other materials (*124*).

Oilseeds and spices (except greater cardamom, *Amomum subulatum*) are unsatisfactory for larval development (*78*). Larvae are also unable to develop on whole pulses but are able to develop on some of the same pulses when the seeds are broken. Presumably, the larvae are unable to penetrate intact seed coats. Larvae are able to develop quite well on whole kernels of wheat, rough rice, brown rice, and pearl millet; sorghum and barley are less satisfactory, and milled rice is unsatisfactory. Development is better on broken kernels of wheat, rough rice, sorghum, and barley than on whole kernels; broken kernels of wheat and rough rice were the best of the diets tested by Kapoor (*78*). The mean period required for development from hatching to adult emergence (34°, 75% rh) ranges from 28 to 33 days on cereal grains and from 30 to 38 days on pulses.

Population growth is most rapid on wheat, followed by pearl barley, rye, and oats; populations decline on rice (*131*). In another study (*108*), there was an increase in numbers of *R. dominica* on milled rice, although the increase was much less than on brown rice. Production of progeny was greater on brown rice of several varieties than on either rough or milled rice of the same varieties. With brown rice there is little difference among varieties, but with rough rice or milled rice varietal differences are apparent (*108*).

Eggs are laid either on kernels of grain (in which case they are usually laid in batches) or loose in the frass produced by the insect (in which case they are usually laid singly) (*124*). On corn, the egg is most often laid beneath the palea at the embryo end of the kernel. The duration of the egg stage ranges from about 32 days at 18.1° to about 5 days at 36° or 38°; egg mortality is lowest (3–10%) at 29.1–32° (*28*). The mean incubation period at 70% rh is 7–8 days at 28° and 11 days at 25° (66).

Larval development is more rapid on whole grain than on meal made from the same grain. Young larvae cannot penetrate undamaged kernels (66). The campodeiform first instar is active and burrows into any convenient rift in the seed coat. The second instar is less active than the first but is still capable of locomotion. The third instar is scarabeiform, apparently incapable of locomotion on a flat surface. The fourth and fifth instars, if they occur, are similar to the third (124).

When larvae develop in whole grain, they remain within a kernel during their development through four or five molts (124). On whole meal, the larvae molt two to seven times, usually three or four (66, 124).

The number of molts is correlated with the suitability of conditions for normal development. Thus, there is a reduction in number of instars only at 28°, which is near the optimum for rapid development, and there is an increase in number of instars at 25°. On average, larval development on whole meal at 70% rh requires 27–31 days at 28° and 46 days at 25° (66).

When R. dominica develops in whole grain, pupation takes place in an enlarged cell at the end of the larval tunnel (135). The pupa lies free wherever it is formed (124). At 70% rh, pupal development requires an average of 5–6 days at 28° and 8 days at 25° (66).

The limiting temperatures for development on wheat of 14% moisture content (mc) (in equilibrium with air at 70% rh) are 18.2° and 38.6° (25). Males and females develop at the same rate. The developmental period from egg to adult is shortest (25 days) at 34°; also, mortality among immature stages is lowest (22%) at this temperature. The developmental period increases to 84 days at 22° and to 36 days at 38.2°. Mortality reaches 53% at 22° and 86% at 38.2°; aside from failure of the eggs to hatch, most mortality occurs among first instars. Development is slower on grain with a moisture content below 14%. Thus, the developmental period is increased 12 days at 12% mc (38.2°), 4 days at 11% mc (38°), 10 days at 11% mc (26°), 10 days at 10% mc (36°), and 14 days at 10% mc (30°). Callow adults remain in the pupal cell for 3–5 days after eclosion (28.9–30°) before beginning to feed and tunnel out of the kernel. They begin ovipositing 15 days later (135).

Oviposition increases markedly when the adults have damaged kernels on which to feed (26). More than 95% of the eggs in one series of tests were deposited on the grain, mostly in scalpel cuts made in the kernels, or under the loose testa covering the embryo. Total fecundity is greatest between 26° and 34° on wheat of 9–14% mc. The highest fecundity was 415 eggs/female on wheat of 14% mc at 34° (25). The oviposition period lasted 112 days under these conditions. The rate of oviposition is very low at 18.3°; only 38 eggs/female were laid in 14% mc wheat during a 4-month test period. The highest temperature at which eggs are laid is 39°, but the rate is only 2 eggs/female per week. Moisture content between 11% and 14% has no effect on oviposition, which is still high at 9% mc. However, no eggs are laid on wheat with a moisture content below 8%. At 34° and above, at all moisture contents except 8%, the rate of oviposition peaks within 2 weeks and then gradually declines. Under all other conditions, the rate of oviposition fluctuates widely about an average over most of the adult lifespan and never reaches a definite peak. Most females survive for several days after oviposition ceases.

Adult lesser grain borers fly readily, at least during certain times of the year and under certain light conditions (124). Their flight is not well oriented, and they may be caught and carried by air currents (135).

The tolerance of the adults to cold in relation to starvation was investigated by Swain (155). At 30°, the first death in a group of beetles deprived of food occurred after 116 hours, and all had died within 18 days. The time required to reach 100% mortality in groups of beetles exposed to cold ranges from 12 minutes at −25° to 570 minutes at −5°. This time is reduced to 2 minutes at −25° and to 420 minutes at −5° by 36 hours of starvation. Further starvation produces an additional reduction in survival time. It is clear that the stress of starvation mitigates against survival at low temperatures. The harmful effects of low temperature tend to disappear when the adults are returned to favorable temperatures after exposure for nonlethal periods to low temperatures above −18°. However, starved adults exposed to −18° or lower for 20 minutes or more enter a "stress-induced lethargy" from which they do not recover. Such adults are unable to feed and their longevity is reduced significantly. This phenomenon is not apparent in normally fed beetles.

The rate of increase is maximum ($\lambda = 2.14$ females/female per week) at 34° on 14% mc wheat (27). The temperature range for population growth ($\lambda > 1$), which is about 19–38° on 14% mc wheat, becomes narrower as moisture content decreases. Thus, on wheat of 11%,

10%, or 9% mc, the ranges are about 22–38°, 25–37°, or 31–35°, respectively. *R. dominica* is incapable of population increase at any temperature on wheat of 8% mc.

Bruchidae

Several beetles of the family Bruchidae are serious pests of cultivated leguminous seeds (pulses). Those that occur in storages lay eggs on developing pods and seeds in the field, and because development may not be completed until after the seeds mature, they are often brought into storehouses with the harvested crop. Some species, such as *Bruchus pisorum, B. rufimanus,* and *Bruchidius atrolineatus,* are unable to breed on dried pulses and thus die out under storage conditions. Others, however, such as *Acanthoscelides obtectus, Zabrotes subfasciatus,* and various species of *Callosobruchus,* breed readily under storage conditions, and their numbers may increase rapidly. Among the latter group are the most serious storage pests of this family.

Most stored-product bruchids are capable of infesting a wide range of leguminous seeds, but each species is characteristically associated with certain host plants (*151*). Numerous studies have been made to determine the susceptibility of various seeds to bruchid attack. The results are inconsistent; varieties of a plant from different parts of the world often appear to differ more in susceptibility than do some species (*74*). Single host plant species appear in the literature under an array of common names, some of which are local names and others of which represent distinct varieties. Inconsistency in the association of common names with particular Latin names adds to the confusion.

The life histories of six species—*A. obtectus, Z. subfasciatus, Callosobruchus maculatus, C. chinensis, C. analis,* and *C. rhodesianus*—are quite similar (*74*). With the exception of *A. obtectus,* the eggs of all species are firmly attached to the seeds or pods, and the hatching larvae bore directly through the egg shell into the host. The eggs of *A. obtectus,* on the other hand, are deposited loosely among stored seeds or are inserted into cuts or cracks in growing pods, and the newly hatched larva wanders freely for a time before penetrating a seed. Posthatch development occurs entirely within the seed.

The larva burrows by feeding on the cotyledon and molts four times before reaching the pupal stage. As it feeds and grows, the larva forms a chamber that becomes perfected during the final larval stadium. Frass, which at this stage is produced in long translucent coils, is pushed against the walls of the chamber to form a smooth, hard lining. The entire chamber is lined except for a small circular area adjacent to the seed coat. The larva reduces this area to a thin window by eating away the inner portion of the seed coat. It then pupates with its mandibles facing the window.

The newly formed adult may remain in the chamber for several days before escaping by pushing out the window. The adults are usually sexually mature when they emerge from the seed and mate shortly after emergence. Preoviposition and oviposition periods are short, and most eggs are laid within a week after adult emergence. Oviposition is reduced by low humidity in all six species, so the highest fecundity is achieved at 70–100% rh. Adults are short-lived and take little or no solid food but will drink water and nectar.

Acanthoscelides obtectus, bean weevil

The bean weevil apparently originated in South and Central America and has now become established throughout most of the world. It is a well-known pest of *Phaseolus vulgaris* and is rarely seen on other seeds (*151*). Nonetheless, it also occurs on *Vicia faba* and has recently been found infesting Bambarra groundnut, *Vigna* (=*Voandzeia*) *subterranea,* from Malawi, so it is quite capable of infesting crops other than species of *Phaseolus.* This bruchid is a moderately successful fungivore, capable of feeding and reproducing on 4 of 23 species of fungi isolated from insect-infested grain (*147*).

The temperature range for development extends from just below 15° to between 32.5° and 35°; development is inhibited by very high as well as low humidities (*74*). The developmental period on haricot beans is shortest (about 28 days) at 30° and 70–80% rh. Fecundity is maximum (about 70 eggs/female) at 25° (70% rh).

The performance of the adult bean weevils produced on several varieties of *P. vulgaris* is shown in Table 10.3. On each variety, the preoviposition period is one day, and about two-thirds of the total egg complement is laid during the first 4 days following adult emergence from the seed (*82*). On yellow beans at 75% rh and various temperatures, the fecundity of *A. obtectus* is maximum (about 55 eggs/female) at 20–29°. Mortality of the immature stages is high; the minimum, which occurs at 25°, is about 58%. The mean developmental period is shortest (about 32 days) at 29°, increases to about 92 days at 18°, and to 36 days at 32°. No beetles complete development at 11°. The finite rate of increase is maximum (1.30 females/female per day) at 29° and decreases to 1.10 at 18° and 1.19 at 32° (*83*).

Callosobruchus analis

This species, associated with *Vigna u. unguiculata* and *V. r. radiata,* probably originated in Southeast Asia or India (*151*). Until 1957, it was considered synonymous with *C. maculatus.* The minimum temperature for development lies between 17.5° and 20° and the maximum, at high humidity, is probably close to 37° (*74*). Optimum conditions for rapid development are 32.5° and about 70% rh; under these conditions development on cowpeas requires about 27 days. The greatest numbers of eggs (about 70 per female) are laid at 35° (70% rh). The mean developmental period is about 29 days, and the mean fecundity is 96 eggs/female for *C. analis* reared on mung beans at 30° and 70% rh (*127*).

Callosobruchus chinensis, southern cowpea weevil

The southern cowpea weevil has essentially the same distribution as *C. maculatus* but has not become established in Australia. Its hosts are *Vigna u. unguiculata* (=*Dolichos biflorus*), *V. r. radiata, V. angularis* (=*Phaseolus angularis*), *Cicer arietinum, Lens culinaris, Pisum sativum,* and possibly *Lablab purpureus* (*151*).

The lower temperature limit for development is a little under 17.5° and the upper limit, at high humidity, is probably close to 37° (*74*). The optimum conditions for rapid development are 32.5° and 90% rh; under these conditions development requires an average of 22 days. The greatest number of eggs (about 50 per female) are laid at 22.5° (70% rh), and oviposition decreases little with temperature up to 37.5°.

Table 10.3

Oviposition period and fecundity of *Acanthoscelides obtectus* reared at 29° and 75% rh on varieties of *Phaseolus vulgaris*

Variety	Mean oviposition period (days)	Mean total fecundity (eggs/female)
Yellow	13	57
Kidney	18	51
Golden Grain Snap	10	29

SOURCE: Krnjaić, S. 1971. The effect of diet type on fecundity and life table of *Acanthoscelides obtectus* Say. *Zastita Bilja* 22(114)45–51 (English summary).

The mean developmental period is about 22 days and the mean fecundity is 78 eggs/female when these weevils are reared at 30° and 70% rh on mung beans (*127*). The mean developmental periods are about 43–68 days at 27.5° and 22–24 days at 30° for *C. chinensis* collected from 15 localities on Honshu, Japan, and reared on adzuki beans (*V. angularis*) at 75% rh (*116*).

When *C. chinensis* was reared on chick-peas at various combinations of temperature and humidity ranging from 25° to 35° and from 50% to 90% rh, the greatest average number of eggs (about 70 eggs/female) was laid at 30° and 90% rh. Mean fecundity was nearly as high (about 60 eggs/female) at 35° and 90% rh, 30° and 70% rh, and 25° and 70% or 90% rh. Females reared at other combinations of temperature and humidity laid an average of about 50 eggs each. Hatchability was about 90% under the most favorable conditions (*8*).

Callosobruchus maculatus, cowpea weevil

The cowpea weevil occurs throughout the tropical and subtropical regions of the world. It is associated with these species of legumes: *Vigna unguiculata, V. radiata, V. subterranea, Cajanus cajan, Cicer arietinum, Lens culinaris, Glycine max,* and possibly *Lablab purpureus* (=*Dolichos lablab*) (*151*). When this and three other species—*Callosobruchus chinensis, C. analis,* and *C. rhodesianus*—were reared on garden peas (*Pisum sativum*) and on cowpeas (*V. unguiculata*), it was found that although all four species can complete development on garden peas, their performance was poor, i.e., development was slower, mortality was higher, and developmental limits were more restricted than when they were reared on cowpeas (*74*).

There are two morphologically distinct forms of *C. maculatus* (*158*): One, the flightless form, breeds on stored seeds; the other, the active (flying) form, breeds on growing seeds. In addition to differences in morphology, the two forms differ in physiology, chemical composition of the body, behavior, and response to crowding.

The flying form oviposits on developing beans in the field, and the resulting larvae are brought into storage when the beans are harvested. Adults produced from these larvae are the flightless form. They remain in the storage facility and oviposit on the stored beans. Eventually, some adults of the flying form are again produced, and these fly to the field where they oviposit on the growing crop. Production of the flying form is apparently governed by environmental factors that operate during preimaginal development. Larval crowding, high temperature (which often results from crowding), low moisture content, continuous darkness, or continuous light stimulate production of the flying form. The first four of these factors are commonly associated with stored beans.

Adult longevity is about twice as long in the flying form as in the flightless form. Adults of the flightless form mate and oviposit soon after emergence, but females of the flying form emerge with immature ovaries so that mating and oviposition are delayed for 3–4 days. The flying form lays fewer eggs, requires a higher humidity, and in general is better able to withstand low temperature than the flightless form (158) (Tables 10.4 and 10.5). In the flying form, fecundity is maximum at 25° or 30°; hatchability is maximum at 25°. In the flightless form, on the other hand, fecundity is maximum at 35° or higher, and hatchability, which is poor at low temperatures, reaches a maximum at 35°. The optimum relative humidity for fecundity (at 30°) is 100% in the flying form compared to 75% in the flightless form, and the optimum humidity for hatchability is also higher in the flying form (75% rh) than in the flightless form (50% rh).

The influence of population density on fecundity is also somewhat different in the two forms. Very low densities are most favorable for oviposition of the flightless form, and the number of eggs laid falls exponentially as density increases. In contrast, very low densities apparently have an adverse effect on oviposition of the flying form. The number of eggs laid increases with density to a maximum level and then falls with further population increases.

The developmental minimum of *C. maculatus* is between 17.5° and 20° and the maximum is probably near 37° at high humidity (74). Surprisingly, the eggs are quite tolerant of low temperatures, the median lethal exposure (LT_{50}) being about 110, 35, 16, 2.7, 1.3, and 0.3 hours at 5°, 0°, −5°, −10°, −15°, and −20°, respectively (112).

The developmental period on cowpeas is shortest (21 days) at 30° (70% rh) and the greatest numbers of eggs (about 100) are laid at 35° (74). When reared on mung beans (*Vigna radiata*), the developmental period is 24 days and the fecundity rate is 128 eggs/female at 30° and 70% rh (127). The rates of development and fecundity of *C. maculatus* on various pulses are shown in Table 10.6 (46, 127).

Table 10.4

Effect of temperature on fecundity and egg hatchability of the flying and the flightless forms of Callosobruchus maculatus

Temp. (°C/°F)	Mean fecundity (eggs/female)		Egg hatchability (%)	
	Flying	Flightless	Flying	Flightless
15/59	20.0	56.2	45.9	64.1
20/68	39.8	54.2	45.8	50.4
25/77	41.4	64.3	57.2	43.2
30/86	44.3	61.1	40.2	29.1
35/95	36.6	77.1	22.5	1.8

SOURCE: Utida, S. 1972. Density dependent polymorphism in the adult of *Callosobruchus maculatus* (Coleoptera, Bruchidae). *Journal of Stored Products Research* 8(2)111–126.

Table 10.5

Effect of humidity at 30° on fecundity and egg hatchability of flying and flightless forms of Callosobruchus maculatus

Relative humidity (%)	Mean fecundity (eggs/female)		Egg hatchability (%)	
	Flying	Flightless	Flying	Flightless
20	18.7	43.1	54.5	78.8
50	31.4	54.4	76.8	90.9
75	29.9	62.4	87.6	92.5
100	43.7	61.2	82.2	67.7

SOURCE: Utida, S. 1972. Density dependent polymorphism in the adult of *Callosobruchus maculatus* (Coleoptera, Bruchidae). *Journal of Stored Products Research* 8(2)111–126.

Table 10.6

Development and fecundity of Callosobruchus maculatus reared on various pulses at 28° and 75% rh

Food	Mean developmental period (days)	Mean fecundity (eggs/female)
Vigna unguiculata		
Black-eyed cowpeas	26	62
Cowpeas (local variety)	29	39
Vigna faba		
Broad beans	28	110
Phaseolus sativum		
Garden peas	39	93
Lens culinaris		
Lentils	66	23
Astragalus cicer		
Mountain chick-peas	28	72

SOURCE: El-Halfawy, M. A., J. M. Nakhla, and N. H. Isa. 1972. Effect of food on the fecundity, longevity and development of the southern cowpea weevil, *Callosobruchus maculatus* F. *Agricultural Research Review* 50:67–70.

Callosobruchus rhodesianus

This is an African species associated with *Vigna u. unguiculata* (*151*). The minimum temperature for development is below 17.5°, probably about 16°, and the maximum is between 35° and 37.5°. The developmental period on cowpeas is shortest (about 25 days) at 30° (70% rh) (*74*).

Zabrotes subfasciatus, Mexican bean weevil

The Mexican bean weevil is another species that apparently originated in South and Central America and is now widely distributed over the world. *Phaseolus vulgaris*, *P. lunatus*, *Vigna subterranea*, *V. radiata*, and *V. unguiculata* are its hosts (*151*). The three latter species represent recent host range expansions.

The lower developmental limit for *Z. subfasciatus* is probably a little below 20° and the upper limit is between 35° and 37.5° (*74*). However, few adults emerge at 35° when the humidity is low. Optimum conditions for development on haricot beans are about 32.5° and 70% rh; under these conditions, development requires about 24 days. Fecundity is maximum (about 40 eggs/female) at 25° (70% rh).

Z. subfasciatus is unusual among the Bruchidae in that it deposits 30–40% of its eggs in clumps of two or more eggs, a habit that may be an adaptation to life in beans of low moisture content (*157*). Mortality is lower among larvae hatching from clumped eggs than among those hatching from scattered eggs, even when the total number of eggs on a bean is the same. This effect is most pronounced on drier beans. Utida (*157*) suggested that water produced by the aggregated larvae might soften the bean, thereby reducing mortality.

Cleridae

Necrobia rufipes, redlegged ham beetle

The cosmopolitan redlegged ham beetle or copra beetle is the most important of a group of insects that infest meats that have become dried to some extent by evaporation during long storage or as a result of prolonged smoking or both (*142*). It is one of the most destructive pests of copra (*54*). In addition, it has been found feeding on cheese, ham, bacon, fish and salt fish, dried egg yolk, bones and bone meal, drying carrion, dried figs, palm nut kernels, and guano.

The adults and larvae are markedly predatory and cannibalistic (*142*). Sometimes they eliminate infestations of the cheese skipper, *Piophila*

casei, by preying on the maggots. Adult ham beetles eat adults of their own species, as well as eggs and larvae, even when other prey is available. Newly hatched larvae feed on unhatched eggs of their own species as well as the eggs of *Dermestes maculatus*.

The eggs are usually deposited in clusters and are cemented in place. During warm weather (mean daily temperature between 21.7° and 31.7°), the minumum incubation period ranges from 4 to 6 days.

Larvae molt two or three times and complete development in about 17 days. Full-grown larvae infesting smoked meat migrate from the greasy material in which they feed to seek dark, dry locations such as crevices in which to build their pupal chambers. These chambers are constructed with a white substance emitted from the mouth of the larva in frothy droplets that harden as soon as they are deposited. Pupation sometimes takes place without the protection of a cell, but exposed pupae are readily eaten by adults. In fact, adults sometimes break into pupal cells and eat the pupae.

About 13 days are spent in the pupal cell, including portions of the adult and larval periods as well as the pupal period. Adults may remain within the cell for a day or two after they become fully pigmented. They emerge by chewing an irregular hole in the wall of the cell. Adults mate shortly after emergence and at frequent intervals during the oviposition period. The preoviposition period may be as brief as 2 days. The eggs are laid in batches, often with periods of a few days to 6 weeks or more between batches. Information on the longevity and oviposition of adult *N. rufipes* reared on stale bacon or on larvae of *P. casei* is presented in Table 10.7. The data for the two diets were collected at slightly different temperatures, but it is nonetheless clear that longevity and reproduction are increased markedly by predation (*142*).

Cucujidae

Ahasverus advena, foreign grain beetle

The foreign grain beetle is a cosmopolitan insect that has been recorded from a wide variety of stored products including grain and cereal products, copra, cocoa, peanuts and most other oilseeds and oilseed products, dried fruit, herbs, spices, and roots (*164*). It usually occurs in large numbers only when the produce is moldy. It is capable of breeding on pure cultures of some molds commonly associated with stored grain (Table 10.8; *39*). It can, however, breed successfully in the absence of any visible mold on wheat germ or on rolled oats or whole wheat flour to which either yeast or wheat germ has been added, but it cannot breed on either rolled oats or whole wheat flour alone. Apparently, most stored foodstuffs are deficient in some nutritional factor required by *A. advena*; yeast, wheat germ, and molds provide adequate amounts of this unknown factor (*164*).

The entire life cycle from egg to adult on rolled oats and yeast requires an average of 22 days at 27° and 75% rh (*38*). The eggs, which are usually laid singly but sometimes in clusters of two or three, hatch in 4–5 days. There are four or five larval instars, and larval development is completed in 11–19 days. When fully grown, the larva constructs a chamber of food particles cemented together and then attaches itself to the substrate with a brownish substance secreted from the anal aperture. Pupation occurs after a prepupal period of 1–2 days, and adults emerge 3–5 days later. Females begin laying eggs 3–4 days after emergence, and in most females, periods of oviposition (20–30 days) alternate with periods during which no eggs are laid (5–23 days). Each female normally goes through two to four of these nonovipositional periods. During oviposition cycles, females generally lay 1–4 eggs/day but occasionally lay 8–12 eggs/day. The rate of oviposition peaks during the first 15 days after emergence and again after 90–105 days. Mated males and females have average lifespans of 159 and 208 days, respectively. Unmated beetles live considerably longer—males 275 days, females 301 days.

The developmental period, mortality among immature stages, and oviposition rate of *A. advena* reared on various molds are shown in Table 10.8. This beetle cannot develop on some molds, possibly because they are toxic or lack essential nutrients. When given a choice between several species of molds, females oviposit preferentially on those that are suitable for larval development.

The foreign grain beetle requires rather moist conditions. Low humidities are unsuitable both for larval development and for oviposition. Survival from larva to adult on rolled oats and yeast is 95% or more at 66–92% rh; none survives at 58% rh (*38*). Humidity has no effect on the duration of the egg, prepupal, or pupal stages, but the rate of larval development decreases with decreasing humidities so that the mean period from egg to adult increases from 19 days at 92% rh to 24 days at 66% rh. Most larvae have four instars at high humidities, but as humidity de-

Table 10.7

Average longevity and oviposition of adult *Necrobia rufipes* reared on stale bacon or on larvae of *Piophila casei*

Fecundity and longevity	Bacon[a]	Larvae[b]
Total fecundity (eggs/female)	137	906
Male longevity (days)	133	255
Female longevity (days)	136	280
Preoviposition period (days)	17	25
Oviposition period (days)	78	187

[a] At mean daily temperatures of 14.4–30.6°.
[b] At mean daily temperatures of 15.6–32.2°.
SOURCE: Simmons, P., and G. W. Ellington. 1925. The ham beetle, *Necrobia rufipes* DeGeer. *Journal of Agricultural Research* 30(9)845–863.

creases, the proportion of larvae with five instars increases. *A. advena* is unable to breed on peanuts or copra at humidities below 70% and is unable to breed on cocoa below 80% rh, the lowest humidity at which visible mold develops (*164*).

Each female produces an average of 1.5 eggs/day at 75% rh (27°), but the rate of oviposition diminishes to 0.1 egg/female per day at 58% rh (*38*). The preoviposition period at 58% rh ranges from 4 to 20 days; 65% of the females never oviposit.

Cathartus quadricollis, squarenecked grain beetle

The squarenecked grain beetle is cosmopolitan in distribution and has been found in wheat, rolled barley, rice, ripe or dried fruits, cacao, tobacco, and oil palms (*166*). In the USA, it attacks corn in the field, especially in the South (*1*). Its bionomics are poorly known. Water-

Table 10.8

Developmental period (egg to adult), percentage mortality among immature stages, and oviposition rate of *Ahasverus advena* reared on pure cultures of various molds at 27° and 85% rh

Mold	Mean developmental period (days)	Mortality (%)	Mean oviposition rate (eggs/female per day)
Aspergillus amstelodami[a]	18	13	6
Penicillium citrinum[b]	19	9	5
Cladosporium spp.	21	11	7
Aspergillus candidus[c]	26	67	—

[a] *A. amstelodami* (Mangin) Thom *ex* Church.
[b] *P. citrinum* Thom.
[c] *A. candidus* Link *ex* Fries.
SOURCE: David, M. H., R. B. Mills, and D. B. Sauer. 1974. Development and oviposition of *Ahasverus advena* (Waltl) (Coleoptera, Silvanidae) on seven species of fungi. *Journal of Stored Products Research* 10(1)17–22.

house and Nowosielski-Slepowron (*159*) compared the rate of development and mortality of a field strain and a laboratory strain of *C. quadricollis* at various combinations of temperature and humidity (food medium not noted). They found that at 80% rh, the developmental period ranges from 24 days (field strain) or 26 days (laboratory strain) at 25° to 20 days or less at 27.5–30°. At 27.5°, the developmental period ranges from 20 days or less at 80–85% rh to about 22 days at 70% rh. Development is most rapid (egg to adult in less than 20 days) and mortality lowest (< 20%) at 27.5–28° and 80–85% rh. Viability is negligible above 30° or below 20° and also at any relative humidity below 65%.

Yoshida (*166*) determined the effect of density on the development and reproduction of *C. quadricollis* in whole wheat flour plus yeast at 30° and 70% rh. Development from egg to adult required about 6 weeks, and the minimum preoviposition period was 3 days. Production of progeny was maximum (6.8 per female) at a density of two pairs of beetles in 2 grams of food. The beetles proved quite sensitive to crowding, and production of progeny declined to 0.2 per female as density increased to 32 pairs.

Cryptolestes Species

Six species of the genus *Cryptolestes* are known to infest stored products (*75*). Of these, *C. ferrugineus* and *C. pusillus* are cosmopolitan. The distribution of *C. pusillus* is restricted by low temperature and by low humidity, but in the wet tropics it is more abundant than *C. ferrugineus*. *C. turcicus* is distributed throughout the temperate regions of the world except for New Zealand and Australia. *C. pusilloides* occurs in eastern Africa, Australia, and South America but may be extending its range. *C. capensis* occurs in Europe and South Africa, and *C. ugandae* is apparently limited to central Africa.

Cryptolestes species feed on grain and cereal products as well as a variety of other materials. *C. ferrugineus* has been recorded from wheat, flour, oilseeds, cassava root, dried fruits, chilies, bean cakes, and gum dammar (*Agathis dammara*); *C. pusillus* from wheat, corn, rice, barley, cottonseed, copra, coffee berries, bulbs, nutmegs, and gum dammar; and *C. turcicus*, from copra, cacao, chilies, and nutmegs (*40*). *C. ugandae* has been found to occur naturally in corn, millet, peanuts, cassava, cowpeas, and cottonseed (*93*). *C. pusilloides* occurs on a wide variety of foods including wheat and wheat products, sorghum, corn, rice, barley, oilseeds (occasionally), almonds, and kibbled locust beans (*Ceratonia siliqua*) (*96*). *Cryptolestes turcicus* and *C. capensis* occur commonly in milling machinery in flour and provender mills (*94, 95*).

The best larval food for *C. ferrugineus* is wheat germ (*129*). Mortality on this diet is only 5% and the mean developmental period is 22 days (32.2°, 75% rh). Bran is second with 11% mortality and a developmental period of 50 days, followed by flour with 61% mortality and a developmental period of 56 days. The larvae feed preferentially on the germ of whole kernels, but they also feed on the endosperm and sometimes hollow out the entire kernel leaving only the outer seed coat. Among rye, wheat, corn, rice, oats, barley, sunflower seeds, flax, and soybeans, rye is the most susceptible to attack, followed by wheat, corn, and rice in that order; the remaining commodities were not attacked (*129*).

Growth of mold in the endosperm renders it more suitable as larval food. In general, molds in the diet also favor development and survival of *C. turcicus* (*30*). *C. ferrugineus* is able to complete development at 33° and 75% rh on 10 of 33 species of seed-borne fungi tested (*145*). The developmental period is shortest (22 days) on *Trichothecium roseum* (Persoon) Link *ex* S. F. Gray and longest (34 days) on *Fusarium moniliforme* Sheldon.

Cryptolestes species are apparently unable to feed on sound grain, but they can feed on kernels with very slight imperfections or injuries, and such kernels are always present, even in grain considered undamaged. First-stage larvae of *C. ferrugineus* cannot feed on perfectly intact wheat but readily attack kernels with even slight breaks in the bran layer (*129*). Neither adults nor larvae of *C. pusillus* can penetrate perfectly intact grain, but small defects in the seed coat of an appreciable number of kernels, caused by normal handling of the grain, renders the seeds vulnerable to attack (*40*).

The life cycles of the stored-product *Cryptolestes* species are similar (*37, 40, 93-95, 96, 129, 148, 149, 161*). Eggs are deposited singly in crevices or furrows in kernels of grain, in spaces between kernels, or in debris. At 25° and 75% rh, *C. pusillus* begins laying eggs within 4 days after emergence, and oviposition sometimes continues for more than 34 weeks (*40*). The rate of oviposition varies considerably during the egg-laying cycle, but there is no evidence of a marked peak, and there is only a slight decline toward the end of the cycle. Females usually continue to lay eggs until they die. Davies (40) recorded a mean fecundity of 242 eggs/female and noted that repeated mating results in higher levels of egg production.

Larvae of *C. ferrugineus* burrow into kernels of grain but may leave their burrows in search of more favorable food (129); the other species probably exhibit the same habits when feeding on whole grain. There are four larval instars, the last of which forms a cocoon with silk produced by prothoracic glands. Larvae infesting whole grain often form cocoons in chambers within the kernels. The robustness of the cocoon differs among species. In *C. pusillus* and *C. turcicus,* the cocoon is a tough silken envelope with food particles and frass, among other things, incorporated in the outer layer. In the other species, the cocoon is flimsy, consisting of a cell in the food material supported by silk. To some extent, cocoon formation is influenced by temperature and humidity, at least in *C. turcicus* and *C. ugandae,* and under some conditions, few cocoons, if any, are formed. Newly emerged adults remain within the cocoon for a few days before chewing out.

In at least some species, and probably in all of them, adults as well as larvae are cannibalistic and will destroy eggs, prepupae, and pupae of their own kind. Pupae and prepupae of *C. ferrugineus* are frequently attacked and eaten by wandering larvae (129).

Statistics summarizing the development and population growth of stored-product *Cryptolestes* are presented in Table 10.9. The optimum humidity for rapid development and population growth of all six species is near 90%. *C. ferrugineus* and *C. capensis* can withstand dryness, but the other species cannot breed at humidities much below 50% (96).

Crowding of *C. ferrugineus* causes a reduction in oviposition and rate of development and an increase in mortality, so that r_m declines as density increases (149). Thus, in whole wheat flour at 30° and 70% rh, reported values for r_m are 0.87, 0.76, and 0.42 female/female per week at densities of 1, 4, and 16 beetles/gram of flour, respectively. The corresponding values of λ are 2.39, 2.14, and 1.52 females/female per week. The greatest change in rate of increase results from a reduced rate of development, increased mortality being the second most important factor.

Oryzaephilus Species

The sawtoothed grain beetle, *Oryzaephilus surinamensis,* and the merchant grain beetle, *O. mercator,* are cosmopolitan pests of stored grain, cereal products, dried fruit, oilseeds, and other foodstuffs (68, 105). *O. surinamensis* is more frequently associated with cereal grains and cereal products; *O. mercator,* with oilseeds and their derivatives. However, *O. mercator* has become firmly established in Canada as a household pest of processed cereals, particularly those with a high oil content (105). *O. mercator* also feeds on seed-borne fungi and is able to complete development at 33° and 75% rh on 18 of 23 species tested (145). Development is fastest (21 days) on *Mucor sphaerosporus* Hagen and the longest (40 days) on *Chaetomium funicola* Cooke.

O. surinamensis cannot complete development on whole polished rice, but it develops well on rice polish (136). Rice bran, ground brown rice, ground rough rice, and whole brown rice are progressively less favorable in the order named. *O. surinamensis* cannot develop on food that contains few or no carbohydrates. It is unable to attack perfectly sound grain but can attack grain with even small lesions in the bran

Table 10.9

Development and population growth of stored-product *Cryptolestes* on wheatfeed at 90% rh

| Species | Reference | Temperature limits for development | | Minimum developmental period | | Maximum population growth rate | |
		Minimum	Maximum	Temperature	Duration (days)	Temperature	λ (females/female per week)
C. ferrugineus[a]	(148)	17.5–20.0	40.0–42.5	35.0	21	35.0	2.72
C. pusillus	(37)	15.0–17.5	37.5–40.0	32.5–35.0	22–23	35.0	2.0
C. pusilloides	(96)	<15.0	35.0–37.5	30.0–35.0	21–23	30.0	1.72
C. turcicus	(95)	15.0–17.5	35.0–37.5	35.0	30	27.5	1.54
C. capensis	(55)	10.0–15.0	32.5–35.0	32.5	31	30.0	1.72
C. ugandae	(93)	<17.5	>35	30.0–32.5	22	30.0	1.87

[a] Whole wheat flour at 70% rh.

layer over the germ. It often feeds only on the germ but can complete development on endosperm alone. It also grows well on dried fruits such as currants and figs (52).

O. surinamensis may supplement its vegetable diet by feeding on eggs and dead adults of stored-product moths. This supplementation has little effect on the rate of population increase when the insect feeds on favorable diets such as corn, wheat, or rice, but it has a marked effect when the insect feeds on less favorable diets, such as peanuts (87).

The life histories of O. surinamensis and O. mercator are similar. The eggs are laid singly or in small clusters. In coarse food, such as grain, they are usually deposited in crevices, but in finely ground material, such as flour, they are laid loosely. There are typically three larval instars although there are sometimes two or four, and the average number is slightly higher in O. mercator than in O. surinamensis (68). When the larva reaches maturity, it usually constructs a crude pupal cell by cementing together particles of food material. Before pupating, the larva fastens its caudal extremity to a solid object, and the pupa remains thus attached.

O. surinamensis and O. mercator have been reared on wheatfeed and a number of other diets (68). The upper limit of temperature for development of both species lies between 37.5° and 40°, but egg and larval mortality are high at 37.5°. Both species develop successfully at 20°, but at 17.5°, only a few O. mercator complete development, and in O. surinamensis adult eclosion is unsuccessful. The optimum temperature range for development is about 30–35° for O. surinamensis and 30–32.5° for O. mercator. At 70% rh, the mean developmental period (egg to adult) of O. surinamensis ranges from less than 20 days at 32.5° to about 80 days at 20°, and that of O. mercator ranges from less than 25 days at 32.5° to about 100 days at 20°.

Eggs of both species hatch at temperatures between 17.5° and 40°, but egg mortality is high near the upper and lower limits. The mean incubation period ranges from about 3 days at 40° to about 16 days at 17.5° and is not influenced by humidity. At 70% rh, egg mortality is about 10% between 22.5° and 35°, 50% at 37.5°, and 75–90% at 40°.

At 30° and 70% rh, O. surinamensis larvae developing on wheatfeed pupate in an average of 12 days; O. mercator larvae pupate in 16 days. Larval mortality is less than 10% in both species. On coconut meal, the larval period of O. surinamensis is extended to 32 days; O. mer-

cator, to 23 days. Larval mortality of O. surinamemsis increases to 40% on a medium of coconut meal. The duration of the pupal period, which is the same in both species, ranges from about 4 days at 37.5° to about 16 days (20°) and is not influenced by humidity. There is little pupal mortality, but at 17.5°, emerging adults of O. surinamensis are unable to free themselves from the pupal skin. Development of O. surinamensis at humidities between 70% and 84% requires from 20 days at 35° to 69 days at 20° on rolled oats, 25 days at 35° to 71 days at 20° on English walnuts, and 36 days at 30° to 81 days at 20° on raisins (156).

Low humidity does not prevent development of O. surinamensis but does prevent development of O. mercator at the same temperatures (68). Thus, O. mercator is able to complete development at 10% rh only at temperatures between 25° and 32.5°. In a study of the influence of humidity on the development, oviposition, and adult lifespan of these species on rolled oats (supplemented with brewers' yeast) at 30° (5), it was found that the rate of development increases with humidity in both species, but at all humidities O. surinamensis develops faster. The mean developmental period (egg to adult) of O. surinamensis ranges from 19 days at 74% rh to 24 days at 12% rh, and that of O. mercator ranges from 20 to 25 days. Neither species is able to complete development at 96% rh, and mortality is heavy (56–58%) at 12% rh. At intermediate humidities, mortality is light to moderate (12–20%) except for a 39% mortality rate in O. surinamensis at 33% rh.

The fecundity of both species is maximum at 56% and 74% rh. Oviposition usually begins during the first week of adult life and reaches a maximum rate during the second or third week (Figure 10.3). The rate of egg laying remains high for about 10 weeks and then declines steadily. The average total fecundity is about 280 eggs/female in O. surinamensis and about 260 in O. mercator; the maximum numbers of eggs laid by a single female were 432 and 360, respectively. The length of the oviposition period and the total number of eggs laid decreases with humidity.

At 12% rh, most females do not begin laying eggs until the second week after eclosion, and the oviposition period is shortened to 3 weeks. Mean total fecundity is reduced to 52 eggs/female in O. surinamensis and to 31 eggs/female in O. mercator. Average adult lifespan of mated female O. surinamensis ranges from 4 weeks at 12% rh to 19 weeks at 74% rh, but three females

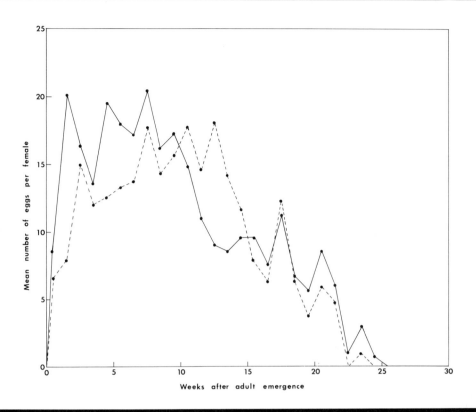

Figure 10.3
**Oviposition of *Oryzaephilus sur-*
inamensis (solid line) and *O.*
mercator (dashed line) on rolled
oats and brewers' yeast at 30°
and 74% rh**
SOURCE: Arbogast, R. T. 1976.
Population parameters for *Oryzaephi-*
lus surinamensis and *O. mercator*: Effect of relative humidity. *Environmental*
Entomology 5(4)738–742.

lived for more than 29 weeks; the average adult lifespan of mated female *O. mercator* ranges from 5 to 18 weeks.

The rate of population growth of *Oryzaephilus* on rolled oats at 30° increases rapidly with humidity between 12% and 56% and then levels off. The values of λ obtained at 12% rh are 1.35 and 1.39 females/female per week for *O. surinamensis* and *O. mercator*, respectively. The maximum λ for *O. surinamensis* is 2.08 at 56% and 74% rh; the maximum for *O. mercator* is 2.01 at 56% rh. Neither species is capable of population growth at 96% rh. At some critical humidity, probably between 90% and 95%, the rate of population growth begins to decline rapidly with increasing humidity. Howe (*68*) reported values of 2.16 and 1.91 for λ in *O. surinamensis* and *O. mercator*, respectively, on wheatfeed at 30° and 70% rh, and values of 2.64 and 2.09 at 33°.

Adults and larvae of *O. surinamensis* are somewhat resistant to cold and are able to withstand 3 weeks of exposure to temperatures between $-1.1°$ and $1.1°$, but all stages are killed by a one-day exposure to temperatures between $-15.6°$ and $-17.8°$ (*14*). The median lethal exposure for the eggs of this species ranges from about 25 hours at 5° to less than one hour at $-15°$ (*112*).

Curculionidae

Sitophilus Species

The grain weevils of the genus *Sitophilus* (*S. granarius*, *S. oryzae*, and *S. zeamais*) are worldwide in distribution, but *S. granarius* is clearly a species of temperate climates. Although occasionally found in tropical and subtropical regions, it does not thrive there. All three species attack grain and cereal products, but they are primarily pests of whole grain. A description of the feeding habits of *S. granarius* (*16*) probably applies just as well to the other species. *S. granarius* breeds in chick-peas and all the common grains such as corn, oats, barley, rye, wheat, kafir, buckwheat, and millet. It cannot breed in finely divided farinaceous material such as flour but breeds readily in manufactured cereal products such as macaroni and noodles and in milled cereals that have become badly caked from excess moisture. The adults feed on a variety of seeds and cereal products including flour, but they do not oviposit on particles too small to serve as food for the development of a single larva.

Female weevils usually deposit their eggs within the endosperm of grain kernels but sometimes oviposit in the germ. A narrow cavity equal in depth to the length of the proboscis is chewed in the kernel, the ovipositor is inserted into this cavity, and an egg is deposited in the bottom. As the ovipositor is withdrawn, a gelatinous material is deposited on top of the egg and packed down even with the surface of the kernel. This material quickly hardens into a protective plug. More than one egg is often laid in a single kernel, but the larvae are cannibalistic and only rarely does more than one adult emerge. Upon hatching, the larva immediately begins to burrow in the seed, forming a winding tunnel that increases in size as the larva grows. There are four larval instars, the last of which forms a pupal cell at the end of its burrow, using a mixture of frass, borings, and larval secretions to wall in the open end of the burrow. Newly emerged adults remain within the kernel at least until the cuticle darkens, and some may remain to feed for a considerable time.

An understanding of the nomenclature that has been applied to members of the *Sitophilus* complex is necessary for sorting out bionomic information on *S. oryzae* and *S. zeamais*. Halstead (*59*) presented a brief outline of the synonymy as follows: *Sitophilus oryzae* (L.) = *Calandra sasakii* Takahashi = *Sitophilus sasakii* Floyd and Newsom = small strain of Birch and Richards. *Sitophilus zeamais* Motschulsky = *Sitophilus oryzae* Floyd and Newsom = large strain of Birch and Richards. Howe (*67*) described his work on the biology of *S. zeamais* under the name *S. oryzae* (large strain).

Sitophilus granarius Development of the granary weevil on wheat at 21° requires from 57 to 71 days, depending on the moisture content of the grain (Table 10.10) (*128*). At 25° (70% rh), development is completed in 45 days; females live for 174 days and lay an average of about 200 eggs. There are apparently no periodic fluctuations in oviposition rate, but sudden temporary changes do occur for unknown reasons. In addition, oviposition varies according to population density (Figure 10.4), age, temperature, and humidity. Reduced oviposition due to aging obscures the density effect in females older than about 100 days. The numbers of eggs laid at 17°, 21°, and 25° is probably in the ratio of 43:100:268, and oviposition ceases at about 9.5°. Oviposition at 21° increases rapidly with relative humidity between 50% (0.2 egg/female per day) and 70% (1.4 eggs/female per day) and then increases more gradually up to 100% (2.0 eggs/female per day). In general, hatchability of the eggs under favorable conditions is about 80% but is slightly lower in eggs laid by older females.

The estimated values of λ for *S. granarius* reared at 70% rh range from 1.12 females/female per week at 17° to 1.54 females/female per week at 25° (*2*). The capacity for increase (r_c) and the weekly rate of self-multiplication (R_c = antilog r_c) of one laboratory strain and seven field strains of *S. granarius* at 15° in wheat of 14% mc have been reported by Evans (*49*). Values of R_c ranged from 1.05 to 1.09 females/female per week, with a mean of 1.07. Differences in r_c at 15° were caused largely by differences in population vigor (expressed in terms of fertility at an optimum temperature, 27°) rather than by differences in cold tolerance related to previous temperature history at the site of collection. *S. granarius* is the most cold-hardy of the grain weevils (Table 10.11).

Sitophilus oryzae The life cycle of the rice weevil from oviposition to adult emergence from the grain averages 35 days at 27° and 69% rh (*137*). The mean incubation period of the egg is 5 days, and the hatchability is about 75%. Each larval stadium averages 5 days in duration, so that larval development is completed in about 20 days. The pupal period lasts for an average of 5 days, and adults remain within the kernels for an additional 5 days.

The temperature limits for development of *S. oryzae* on wheat (14% mc, in equilibrium with 70% rh) are 15.2° and 34° (*25*). The mean developmental period from egg to adult is shortest (25 days) at 29.1°. Development requires an

Figure 10.4
The relationship between oviposition rate and population density (logarithmic scale) of adult female *Sitophilus granarius* on wheat at 25° and 70% rh.
SOURCE: Richards, O. W. 1947. Observations on grain-weevils, *Calandra* (Col., Curculionidae). I. General biology and oviposition. *Proceedings of the Zoological Society of London* 117(1)1–43.

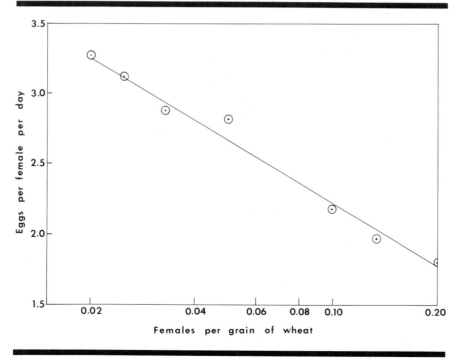

average of 94, 31, or 26 days at 18.2°, 25.4°, or 32.3°, respectively. Reduction of moisture content to 11% adds 4.7 days to the developmental period at 18.2° and 2.4 days at 25.4° or 29.1°. Development is also slower when more than one larva infests a single grain of wheat. Birch (25) observed developmental periods of 110, 44, or 36 days at 18.2°, 25.4°, 29.1°, respectively, when there were three larvae/kernel.

At a density of 1 weevil/50 kernels of wheat, oviposition is maximum (344–384 eggs/female) at 25.5–29.1° and 14–20% mc (26). No eggs are laid at 13° or above 35°. Also, no eggs are laid on grain with a moisture content below 10% and very few are laid when the moisture content is below 11%. Oviposition is influenced by population density, but this effect is proportional to temperature and is dependent on the age of the weevils. Thus, when density is increased from 1 weevil/50 kernels to 1 weevil/10 kernels, the total number of eggs laid at 25.5–29.1° (14% mc) decreases to 265–296 eggs/female, primarily due to a reduction in the rate of oviposition during the first 8 weeks. A similar increase in density at 23° causes no loss in egg production. The lifespan of adult weevils and the age at which oviposition reaches a maximum rate varies little with temperature. The highest rate is usually achieved 1–2 weeks after emergence; oviposition then declines gradually until the females die at an average age of 3 months.

The optimum temperature for population increase in wheat of 14% mc (70% rh) is 29.1, at which temperature λ = 2.15 females/female per week, and the temperature limits for population increase ($\lambda > 1$) are near 15° and 34° (27). The favorable temperature range narrows as moisture decreases, so that there is little population increase at any temperature on wheat of 10.5% mc and none on wheat of 10% mc. The rate of increase on corn of 13% mc (70% rh) at 29.1° (λ = 1.52 females/female per week) is much lower than on wheat. The finite rate of increase at 30° and 75% rh (15–15.5% mc) depends on wheat variety used as the food medium and ranges from 1.57 to 1.73 females/female per week (144).

Evans (48) determined the capacity for increase (r_c) and the weekly rate of self-multiplication (R_c) of one laboratory population and seven field populations of S. oryzae at 15° in wheat of 14% mc. Values of R_c ranged from 1.07 to 1.10 females/female per week, with a mean of 1.09, which is somewhat higher than the value (λ = 1.01 females/female per week) reported by Birch (27) for the same conditions of tem-

Table 10.10

Duration of developmental stages of *Sitophilus granarius* in wheat at 21°

Relative humidity (%)	Moisture content (%)	Duration of developmental stage (days)			
		Egg	Larva	Pupa	Adult remains within grain
50	11.1	7	43	11	10
60	12.6	7	37	10	10
70	14.2	6	35	8	9
80	15.8	6	34	8	9

SOURCE: Richards, O. W. 1947. Observations on grain-weevils, *Calandra* (Col., Curculionidae). I. General biology and oviposition. *Proceedings of the Zoological Society of London* 117(1)1–43.

Table 10.11

Resistance of *Sitophilus* weevils to low temperature

Species and stage	Survival time at selected temperatures (°C/°F)				
	Days				Hours
	5/41	0/32	−5/23	−10/14	−15/5
S. granarius					
Egg & adult	152	67	26	14	19
Larva	32	19	18	2	11
Pupa	133–147	39–47	23–25	6	13–16
S. oryzae					
Egg & adult	21	7.5	4	15	3.75
Larva	—	7	12	90	8
Pupa	—	16–17	3–4.5	24–97	2
S. zeamais					
Egg	30	18	13	—	24
Larva & pupa	130	37	22	—	16
Adult	100	37	23	13	16

SOURCES: Maceljski, M., and Z. Korunic. 1973. Contribution to the morphology and ecology of *Sitophilus zeamais* Motsch. in Yugoslavia. *Journal of Stored Products Research* 9(4)225–234; Rumyantsev, P. D. 1959. *The Biology of the Pests of Stored Grains*. Publishers for Technical and Economic Literature on Questions Relating to Grain, the Animal Feed Industry, and the Storage of Agricultural Products, Moscow (in Russian); Zakladnoi, G. A., and V. F. Ratanova. 1986 [1973]. *Stored-Grain Pests and Their Control*. Kolos, Moscow. Translated from the Russian by Amerind Publishing Company, New Delhi. Available in the USA from National Technical Information Service, Springfield, VA.

perature, moisture, and population density. As in S. granarius, differences in r_c between strains were caused by differences in population vigor rather than by differences in cold tolerances related to previous temperature history. S. oryzae is the least cold-hardy of the grain weevils (Table 10.11).

Sitophilus zeamais The life cycle of the maize weevil averages 35 days in hard red winter wheat at 27° and 69% rh (138). The mean incubation period of the egg is 5 days, and hatchability is 75%. The first larval stadium lasts 4 days and the last, 6 days, including one day as a prepupa. The second and third stadia are each passed in 5 days, so larval development is completed in 20 days. Pupal development is completed in 5 days, and the adults remain within the kernel for an additional 5 days (*138*).

Development on English wheat at 70% rh requires from 40 days at 25° to 110 days at 18° (*67*). At 21°, the developmental period increases from 48 days at 80% rh to 74 days at 50% rh. Egg hatchability is estimated to be about 90% and natural mortality of first instars ranges from 3.5% at 80% rh to 30% at 50% rh. Natural mortality among older larvae and pupae is low but cannibalism caused by overcrowding markedly increases total mortality.

Oviposition rates at 100% rh range from 1.3 eggs/female per day at 17° to 3.4 eggs/female per day at 25° (*67*). Egg production decreases sharply at relative humidities of 60% and below, and maximum egg laying is apparently not attained unless many kernels are available for oviposition. Thus, in Howe's experiments (*67*), each female laid an average of 4.5 eggs/day at 21° and 80% rh when there were one or two kernels/female, compared to 6.8 eggs/day when there were many kernels per female.

At 29.1° and 70% rh, males live an average of 20 weeks on wheat (14% mc) and 25 weeks on corn (13% mc), whereas females live about 18 weeks on both commodities. The finite rate of increase at the same temperature and humidity is 1.76 and 1.55 females/female per week on wheat and corn, respectively (*27*). In its tolerance to cold, *S. zeamais* is intermediate between *S. granarius* and *S. oryzae* (*106, 130, 167*) (Table 10.11).

Other Curculionidae

Two other curculionid species should be mentioned, although neither is as important economically as the three species just discussed. These are the broadnosed grain weevil, *Caulophilus oryzae*, and the sweetpotato weevil, *Cylas formicarius elegantulus*. The former species attacks soft or damaged grain but is unable to breed in dry, hard, uninjured grain (*1*). It is widespread in Florida and is occasionally found in Georgia and South Carolina. The sweetpotato weevil breeds in the crowns and fleshy roots of the sweet potato and closely related plants and is widely distributed in tropical and subtropical regions (*45*). It continues to breed in harvested roots during curing and storage when temperature is favorable, and it is capable of developing large populations under storage conditions.

Dermestidae

Hinton (*64*) notes that 52 species of Dermestidae have been recorded from stored prod-

ucts. These can be divided into three groups according to the type of food eaten: (a) species that breed only on materials containing animal protein; (b) species that normally breed on animal matter but which can breed successfully on vegetable matter; and (c) one species (*Trogoderma granarium*) that is apparently restricted to grain and cereal products. Most losses to stored products are caused by the larvae. Adults of many genera do not eat the same food as the larvae but instead feed on pollen or nectar. Adults of other genera feed only to a slight extent, if at all, on the larval food. Only in *Dermestes* do the adults feed to a significant degree on stored commodities, and even in this genus the larvae are more important as pests.

Attagenus unicolor, black carpet beetle

The larvae of the black carpet beetle, formerly also known as *A. megatoma* and *A. piceus*, feed on materials of animal origin such as wool, feathers, hair, and fur, as well as on cereal products (*56*). They are often serious pests in flour mills. The mean duration of the egg stage is 22, 10, or 6 days at 18.3°, 23.9°, or 29.4°, respectively. At 23.9°, relative humidity ranging from 20 to 93% has little if any effect on the duration of the egg stage.

The time required for larval development on rat fur and fish meal at 25–26.7° (humidity uncontrolled) ranges from 269 to 639 days in males and from 258 to 545 days in females. Larvae can be divided into two groups according to development. One group, consisting of more than half the larvae, pupate in less than 375 days. Larvae in the other group begin pupating 3 or 4 months later. Larvae reared at room temperature on a diet of rat fur, fish meal, and precooked oat flakes complete development in 283–572 days (males) or 259–569 days (females). Males reared on feathers and oat flakes complete development in 455–577 days and females in 509–580 days.

In general, beetles with a short larval life have fewer instars than those with a long larval life. At 25–26.7°, the number of instars ranges from 7 to 17. At room temperature, the number of instars ranges from 8 to 15 when larvae are fed a diet of fur, fish meal, and cereal, and from 11 to 16 when they are fed a diet of feathers and cereal. Larval mortality is 56% on the former diet and 29% on the latter. Larval mortality at 25–26.7° on a diet of fish meal and woolen cloth is 21%.

When black carpet beetles are reared at 30°

and 45–55% rh on a diet of ground Purina Laboratory Chow mixed with 5% (wt/wt) of brewers' yeast (18), the males undergo 4–11 molts and complete larval development in an average of 254 days. Females undergo 5–9 molts and complete development in 260 days. Males pass through an average of seven instars; females, eight. Larval mortality is only 3%. Larvae attain about 80% of their final weight during the first third of the larval period and then enter an obligatory period of decelerated or arrested development [diapause in the sense established by Nair and Desai (114)] that occupies nearly half the larval period (20). During this period the larvae feed intermittently and undergo occasional molts, but respiration is reduced. Larvae that are isolated without food after developing for 16–24 weeks usually undergo one or two starvation-induced molts but pupate at the same age as fed larvae (19). Pupation is somewhat delayed when larvae are isolated after 28 or 33 weeks. Pupal weight depends upon the age at which the larvae are isolated and upon the number of times they molt.

After entering the pupal stage, some pupae remain for several days, or even for the entire pupal period, with the caudal extremity covered by the last larval skin (56). Other pupae free themselves shortly after pupation. The mean duration of the pupal stage is 17, 9, or 5 days in males, and 18, 9, or 6 days in females at 18.3°, 23.9°, or 29.4°, respectively. The duration of the pupal stage at 23.9° is not influenced by relative humidity between 20% and 100%.

Newly emerged adults usually remain enclosed in the last larval skin for several hours and do not become very active until the cuticle completely darkens. The mean preoviposition period is 9, 6, or 3 days, the mean oviposition period is 13, 11, or 8 days, and the mean total fecundity is 61, 83, or 74 eggs/female at 18.3°, 23.9°, or 29.4°, respectively. Females kept at 12.8° lay no eggs. Females do not oviposit at a regular rate throughout the oviposition period; eggs are laid on an average of only 5 or 6 days with days of inactivity intervening.

The adults are apparently short-lived. Mated males live an average of 17 days at 18.3° to 40 days at 29.4°; mated females live an average of 15–38 days. Unmated males live 18–61 days and unmated females 22–76 days (56). However, the beetles in these tests were not provided with food or water, provision of which might have increased their longevity. Several studies indicate that in nature the adults feed on floral nectar and possibly pollen (64).

Dermestes lardarius, larder beetle

The larder beetle, or bacon beetle, is a well-established, cosmopolitan pest of stored animal products. It feeds on almost any dry or decomposing animal material and sometimes on plant material as well (64). It is known to attack such materials as hair, fur, feathers, bones, bone and fish meal, bacon, ham, sausage, cheese, dried fish, potatoes, and rye bread.

The eggs are laid singly on the larval food, and the larvae pass through a variable number of instars. Females usually molt one more time than males and require a longer time to develop. When suitable material is available, mature larvae bore tunnels in it in which to pupate; otherwise they pupate in the open. Pupae formed in the open usually remain encased in the last larval skin, but those formed in tunnels are usually naked, with the last larval skin used along with other debris to close the tunnel entrance. Females require a protein-rich diet for egg production. Although they usually mate more than once, only one mating is required for them to attain maximum fecundity.

When larder beetles were reared on a moistened mixture of beefburger bits, ham bits, and ground dog biscuits at 22° and 80% relative humidity, the eggs hatched in 4–6 days (mean incubation period, 4 days). Larval development of males required 40–44 days (mean = 41); females, 46–48 days (mean = 47). Males molted eight times and females nine times. The duration of the pupal period was 20–22 days (mean = 21) (63).

In another study (33), the beetles were reared on a mixture of fish meal, wheat germ, and cholesterol at various conditions of temperature and humidity. At 65% rh, the beetles were able to complete development at temperatures between 15° and 32.5°, but mortality was very high at the two extremes. There was little difference between males and females in rate of development. At temperatures between 25° and 32.5°, the combined duration of the larval and pupal stages was between 45 and 50 days on the average and increased to about 145 days at 15°. At 25°, the mean developmental period increased from 42–43 days at 80% rh to 69–70 days at 40% rh. Longevity of mated adults that were given an opportunity at weekly intervals to drink a 30% sugar solution ranged from 23 days at 32.5° to 250 days at 20° (65% rh). At 25°, mean longevity ranged from 221 days at 40% rh to 290 days at 80% rh.

No eggs were produced at 32.5°. At 20°, there was a long preoviposition period (85–146 days) followed by an oviposition period of 55–189 days. The durations of the preoviposition and oviposition periods were influenced little by temperature or humidity within the range of 20–25° and 40–80% rh, but at 25° there was sometimes an early period of egg laying (6–31 days in duration) that was completely separated in time from the main oviposition period. Daily egg production averaged 0.4–1.8 per female during the early period and 0.5–1.1 per female during the main period. Total average fecundity ranged from 41 to 81 eggs/female. Egg hatchability was extremely variable but averaged about 50% and was affected little by temperature or humidity.

Dermestes maculatus, hide beetle

The hide beetle is a cosmopolitan pest that attacks a wide variety of substances with high protein content (64). It has been recorded from horn, feathers, fur, bristles and glue of brushes, hog bristles, bone, carcasses, skin and hides of all kinds, dog biscuits, cacao beans, dried fish, ham, bacon, cheese, and so on.

The incubation period of the egg varies from 2 days at 35° to 6 days at 21° but is not affected by humidity. The number of larval instars and the time required for larval development increase as temperature and the quantity or quality of food decrease. At 75% rh on a diet of dried meat, larvae molt seven to nine times at 21° and complete development in an average of 50 days. At 35°, they molt six to seven times and complete development in 19 days. Combined larval and pupal mortality is about 6%, 10%, or 25% at 20°, 27°, or 35°, respectively. Larval development is also influenced by humidity. At 55% rh and 27°, larvae molt six to nine times and pupate 34 days after hatching. At 75% rh, they molt five to seven times and pupate after 28 days. Dried meat and dried bolti (a cichlid fish of Africa and Asia Minor) provide the best diets in terms of developmental rate and survival, whereas dried sardines, dried rabbit skin, and dried cheese are less satisfactory (Table 10.12) (12).

Upon reaching maturity, the larvae construct a pupal chamber by boring into any compact substance that may be available, a habit that often causes a great deal of damage to commodities on which they do not feed (64). The list of materials known to be damaged in this manner includes cork, fiberboard, books, tobacco, tea, linen, cotton cloth, woolens, salt, sal ammoniac (NH_4Cl), plaster molds, flexible asbestos, and lead. The opening of the pupal

Table 10.12

Larval development of *Dermestes maculatus* on various commodities at 27° and 75% rh

Dried food	Number of molts	Mean larval period (days)	Mortality (%)
Meat	5–7	27	10
Fish			
Boltis	6–8	29	10
Sardines	7–11	43	15
Rabbit skin	8–10	49	40
Cheese	10–14	91	76

SOURCE: Azab, A. K., M. F. S. Tawfik, and N. A. Abouzeid. 1972. Factors affecting development and adult longevity of *Dermestes maculatus* De Geer (Coleoptera: Dermestidae). *Bulletin de la Société Entomologique d'Egypte* 56:21–32.

chamber is usually closed by debris made during the excavation, or sometimes by the last larval skin. Pupal development is not influenced by humidity and requires from 5 days at 35° to 12 days at 21° (12).

Adults mate soon after emergence, and mating is followed by a preoviposition period of from 6 days at 35° to 13 days at 21° (13). Repeated mating increases fecundity, and females that fail to mate until long after emergence lay few eggs. The eggs are deposited singly or in batches of 2–25, usually in crevices or other sheltered places (11). The rate of oviposition is influenced strongly by temperature but only slightly by relative humidity between 55% and 75% (12). The duration of the oviposition period at 75% rh ranges from 28 days at 35° to 130 days at 21°, and egg production fluctuates considerably from week to week. Fecundity increases with temperature from 214 eggs/female at 21° to 362 eggs/female at 27° and then declines to 83 eggs/female at 35°.

Females require a continuous supply of food and water to achieve maximum fecundity. Those deprived of water lay few eggs, and those deprived of food lay none at all. Females provided with dried meat and water lay an average of 318 eggs each (at 29.6° and 61.2% rh), whereas those provided only with dried meat lay an average of only 17 eggs each. The type of food consumed also has a pronounced effect on oviposition. Females supplied with dried meat and a sugar solution produced an average of 432 eggs each compared with 362 eggs each produced by females supplied with dried meat and water. Females fed dried sardines, dried rabbit skin, or dried cheese produce about 43, 23, or

70 eggs, respectively. Adult longevity is maximum at 21° and 75% rh and is decreased by higher temperature or lower humidities. At 21° and 75% rh, males live an average of 169 days and females live 173 days (12).

Trogoderma glabrum

T. glabrum is capable of infesting a wide variety of materials of both plant and animal origin but develops best on animal feeds, especially on mixed feeds (laying mash, cat food, dog food) and rolled barley; poultry laying mash supports the largest populations (7). Increase in numbers is slow on cornmeal, egg noodles, oatmeal, spaghetti, wheat germ, and whole wheat flour. Dry beans, dried fruit, and nuts are poor diets, and although larvae survive on these commodities, they do not pupate. Larvae also survive but fail to pupate on fish meal and dead moths. Populations increase faster on processed cereals (rolled barley, cracked corn, and rolled oats) than on the corresponding whole grains. Nonetheless, the ability of this species to build and maintain large populations on whole grain was demonstrated when serious infestations developed in stored wheat and shelled corn in Kansas (160).

At 65–70% rh, the total developmental period from oviposition to adult eclosion on whole wheat flour is 140, 50, 41, or 45 days at 21.1°, 26.7°, 32.2°, or 37.8°, respectively (male and female data pooled). The corresponding survival from egg to adult is 9, 44, 38, or 42% (77). The eggs are deposited randomly among wheat kernels and hatch in an average of 4 days at 37.8° to 12 days at 21.1°. Percentage hatch ranges from 97 at 21.1° to 72 at 37.8°. No eggs hatch at 15.6°, although some may show partial development.

At 32.2°, male larvae reared on a mixture of whole wheat flour and wheat germ molt four or five times and require an average of 30 days to reach the pupal stage. Female larvae molt five or six times and require an average of 35 days to complete development. Larval development on a complex diet of cornmeal, wheat shorts, dog food, rolled oats, brewers' yeast, honey, glycerol, and whole wheat requires from 24 days at 32.2° to 132 days at 21.9° in males and from 27 to 146 days in females. Temperatures below the developmental threshold are lethal if exposure is prolonged, but surviving larvae mature and pupate normally when returned to higher temperatures. Mortality among larvae exposed to 2.2° reaches 80% after 40 days and 100% after 50 days; 70% survive a 150-day exposure

to 15.6°, a temperature well below optimum.

Larvae do not penetrate deeply into masses of stored grain but feed in the upper 2 or 3 feet. They feed mainly on the germ of whole grain but also eat the endosperm and bran after the germ has been consumed. All larval instars are capable of feeding on wheat with the bran layer intact, but the rate of development is slower than on whole wheat flour; 34 days and 38 days are required for larval development of males and females, respectively, compared to 30 days and 34 days on flour (77). There is a difference in survival between larvae reared on whole kernels (50%) and those reared on flour (65%), but this difference is not statistically significant (77). Larvae apparently seek out and feed on cracked grain, so the proportion of cracked kernels in a mass of wheat must have a marked effect on population growth. Larval development is slower and mortality higher on wheat or shelled corn containing 1% cracked kernels than on the same grains containing 10% or 15% cracked kernels.

The average duration of the pupal stage ranges from 7 days (males) or 9 days (females) at 21° to 3 days (both sexes) at 37.8°. Pupal mortality is 3% or less at all temperatures; 40% of the pupae are killed by a 6-day exposure to 2.2°, and all are killed by a 14-day exposure. The pupae are covered by the last larval skin throughout development, and newly formed adults remain quiescent within this skin for a period of 2–7 days before emerging and mating.

Mating is followed by a preoviposition period of 1–3 days and an oviposition period of 2–7 days, depending upon temperature. Total fecundity is 64, 92, 68, or 37 eggs/female at 21.1°, 26.7°, 32.2°, or 37.8°, respectively. Adults do not feed. At 60–70% rh, longevity of mated females averages from 7 days at 37.8° to 26 days at 21.1°; unmated females live longer, 16–42 days. Mated males live 5–24 days.

The influence of various constant temperatures and humidities and the effect of various programmed climates on the growth, development, and reproduction of T. glabrum have been described (7). Humidity has little effect on the duration of the egg stage, but low humidity and high temperature or high humidity and low temperature generally interact to reduce percentage hatch. Optimal conditions for larval development (in a mixture of wheat germ and dry skim milk) are 32.2° and 70% rh. At 32.2°, development from egg hatch to adult maturation (elytra fully darkened) requires 28, 35, or 42 days in males and 32, 40, or 48 days in females at 70, 50, or 30% rh, respectively.

Males molt five times and females six times at 30° and 60–70% rh when reared on a medium of Purina Laboratory Chow, nonfat dry milk, wheat germ, meat and bone meal, and brewers' yeast (22). The larval period lasts from 26 to 30 days in males and from 32 to 36 days in females; female larvae sometimes enter a supernumerary seventh larval stadium, which delays pupation. When sixth-instar females are isolated without food, they molt within 2 weeks and at irregular intervals thereafter, assuming progressively smaller larval dimensions with each molt. Normal progressive development is restored when they are provided with food. However, pupation among these "regrown" larvae is erratic, and many undergo additional larval molts after they should have pupated (23).

Trogoderma granarium, khapra beetle

The khapra beetle is among the world's most serious pests of stored grain. With the possible exception of South America, it has been recorded on every continent. It was once established in California, Arizona, and New Mexico (102) but has apparently been eradicated from those states. It has steadily extended its range during the present century, demonstrating a capacity to colonize new areas. With a few exceptions, its distribution outside artificially heated habitats coincides with the area of the world having at least 4 months with a mean temperature above 20° and a relative humidity below 50%. Although high humidity has adversely influenced its general spread, the species thrives at high humidity in the laboratory. The difference between its performance in the laboratory and that in the field may be caused by competitive interaction with other stored-product pests under field conditions (21).

The khapra beetle feeds on a wide variety of stored products. Unlike most other dermestids, it prefers whole grain and cereal products to foods of animal origin, but it will feed on such materials as dried blood, dried milk, and fish meal (102). It has been recorded from wheat, barley, oats, rye, corn, rice, peanuts, flour, bran, malt, flaxseed, alfalfa seed, tomato seed, pinto beans, black-eyed peas, sorghum, grain straw, alfalfa hay, various nut meats, spaghetti, noodles, cottonseed meal, dried fruits, and dried lima beans.

The larvae are unable to penetrate sound grain until after the third molt, but before then they are able to take advantage of even small lesions in the bran layer (57). The larvae feed on the germ and endosperm of malted barley and some-times hollow out the grain completely. On wheat they often eat only the germ and overlying bran but sometimes eat the adjacent endosperm as well.

The larvae are unable to penetrate undamaged pulses, but they develop on the flours of green gram, *Vigna r. radiata* (= *V. aureus, Phaseolus aureus*), black gram, *V. mungo* (= *P. mungo*), chick-peas (= Bengal gram, Kabuli gram), *Cicer arietinum*, cowpeas, *V. u. unguiculata*, and pigeon peas, *Cajanus cajan*, as well as they do on wheat flour (24). Development on soybean (*Glycine max*) and lentil (*Lens culinaris*) flour is much slower. The larvae are unable to complete development on French bean (*P. vulgaris*) flour.

The facts of the life history of *T. granarium* are available from several sources (17, 57, 102). Hadaway (57) reported *T. granarium* to be a serious pest of malted barley in England and gave an account of its growth, development, and reproduction on this product. Eggs are laid singly and are either deposited loosely among the grain kernels or tucked away in the grooves or cracks left at the end of the kernels by removal of the dried shoots. Duration of the egg stage is affected by temperature (but not by humidity) and ranges from 4 days at 40° to 10 days at 25°. Eggs hatch at all relative humidities between 2% and 73% and at all temperatures between 25° and 40°.

The number of larval instars varies under any given set of conditions, but there are usually more instars at lower than at higher temperatures. Thus, there are four or five instars at 35° (50% rh), but there are between five and eight at lower temperatures. Also, females usually molt one more time than males and require more time to complete development (Table 10.13). Optimal conditions for larval development are apparently 35° at 50–73% rh. Development is quickest under those conditions and at 40° and 50% rh; mortality is lowest at 35° and 25–73% rh. Although the larvae are resistant to low humidity, they develop faster at higher humidities. They become quiescent at 20° and, in Hadaway's tests (57), none had pupated after 12 months.

The pupae remain covered, except for a small portion of the dorsum, by the last larval skin. Pupal development requires from 3 days at 40° to 6 days at 25°, and the teneral adult remains within the last larval skin for a period of 1–3 days depending upon temperature. Pupal development is not affected by humidity.

Adult khapra beetles are short-lived. They do

not normally feed, and they require neither food nor water to attain full longevity and fecundity. Mating takes place shortly after emergence and is followed by a preoviposition period that ranges from a negligible amount of time at 40° to 2–3 days at 25°. Females require only one mating to produce a full complement of eggs and to achieve normal fertility. Most of the eggs are laid during the first few days of the oviposition period, after which the daily number decreases rapidly. Egg production varies little with temperatures between 25° and 40°, but adults become inactive at 20° and no eggs are produced (Table 10.14). Oviposition occurs at all relative humidities between 2% and 73%, but more eggs are laid at higher than at lower humidities. Eggs hatch at all temperatures and humidities in the range of 25–40° and 2–73% rh. Females live for only a few days beyond the end of the oviposition period. The total lifespan of mated females at 50–73% rh ranges from 5 days at 40° to 13 days at 25°. At 2–25% rh, females live for 4 days at 40° and 11 days at 25°. Longevity of adult males is about the same as that of females at 40°, but at lower temperatures males usually live longer than females. At 25°, males live an average of 12 days at 2% rh to 16 days at 73% rh.

After they have attained full growth, larvae may enter a form of diapause in which they continue to feed and molt intermittently but fail to pupate, and they may remain in diapause for over 6 years when food is present (115). This diapause is distinct from retrogressive molting, which is induced by starvation in *T. granarium* and other dermestids. Diapausing larvae are usually heavier than other larvae and often hide in crevices away from food. They may emerge from hiding at irregular intervals to feed, and having fed, they may return to hiding and revert to dormancy, or less frequently, they may pupate. The incidence of diapause is influenced by temperature, food, population density, and the genetic makeup of the population, but it is induced primarily by the combined effects of crowding and exposure to temperatures of 30° or less.

"Density-independent" diapause (DID) occurs in about 30% of larvae reared in isolation on wheat or corn flour (114). No diapause occurs at 35°, but at diapause-inducing temperatures the incidence of diapause is increased by a suboptimal diet. In crowded cultures, about 90% of the larvae enter a density-dependent diapause (DDD) on unconditioned flour. Neither accumulation of feces nor the presence of old

Table 10.13

Effects of temperature and humidity on larval development of *Trogoderma granarium*

| Temp. (°C/°F) | Relative humidity (%) | Mean duration of larval stage (days) | | Mean survival (%) |
		Males	Females	
25/77	2	60	68	10
	25	44	48	38
	50	39	47	49
	73	33	36	45
30/86	2	28	35	73
	25	27	33	83
	50	24	31	89
	73	22	27	82
35/95	2	32	33	65
	25	23	25	91
	50	19	21	90
	73	17	20	92
40/104	2	35	38	61
	25	22	24	79
	50	18	19	86
	73	22	24	85

SOURCE: Hadaway, A. B. 1956. The biology of the dermestid beetles, *Trogoderma granarium* Everts and *Trogoderma versicolor* (Creutz.). Bulletin of Entomological Research 46(4)781–796.

Table 10.14

Effects of temperature and humidity on the oviposition period, fecundity, and fertility of *Trogoderma granarium*

Temp. (°C/°F)	Relative humidity (%)	Mean oviposition period (days)	Mean number of eggs/ female	Mean hatch (%)
20/68	←——————No oviposition——————→			
5/77	2	4.4	25	49
	25	5.7	40	52
	50	7.5	39	50
	73	6.8	43	58
30/86	2	2.9	29	51
	25	3.9	30	53
	50	4.8	41	60
	73	5.0	41	58
35/95	2	3.0	27	55
	25	3.4	29	50
	50	3.4	46	75
	73	3.6	44	66
40/104	2	2.5	23	49
	25	3.0	34	60
	50	3.3	39	54
	73	3.3	38	58

SOURCE: Hadaway, A. B. 1956. The biology of the dermestid beetles, *Trogoderma granarium* Everts and *Trogoderma versicolor* (Creutz.). Bulletin of Entomological Research 46(4)781–796.

food is necessary, and although supplementation of the food with yeast improves the growth rate, it does not prevent diapause. A temperature of 35° prevents diapause to some extent, but pupation of some larvae is still delayed considerably.

Both DDD and DID can be terminated in about 10 days by raising the temperature from 30 to 37° (115). However, when the larvae are severely crowded, a few still take as long as 100 days to pupate. At 30°, isolated larvae that enter DID sometimes pupate when they are only 3 months old, but more commonly when they are about 5 months old. Food renewal is ineffective in terminating this diapause. About 25% of DDD larvae that are kept crowded on old flour pupate at about the same age. Moreover, an immediate burst of pupation occurs and subsequent pupation is enhanced when DDD larvae are isolated and provided with fresh food.

Diapausing larvae may weigh twice as much as nondiapausing larvae and have greater reserves of fat, protein, and glycogen (41). In general, they produce females that lay more eggs, because if food is available, their reserves are not depleted and can be utilized for egg production. For example, females produced from diapausing larvae reared on crushed wheat at 30° and 52% rh lay an average of 119 eggs compared to 61 eggs laid by nondiapause females.

Trogoderma inclusum

T. inclusum exhibits diverse feeding habits and has been found on a wide range of raw agricultural products, processed foods, mixed animal feeds, and natural hosts (detritus and cast skins of bees and beetles, for example). The rate of population growth is highest on pollen, followed by poultry laying mash, dry dog food, and dry cat food, and is lowest on whole barley and dry beans (153). Most whole cereal grains, peanuts, raisins, macaroni, spaghetti, and white flour are also poor diets.

T. inclusum is a pest of malt in England. Hadaway (57) gave an account of its growth, development, and reproduction on this commodity under the name *Trogoderma versicolor*. Young larvae are unable to feed on kernels that are perfectly sound, but they are capable of attacking kernels with only small lesions in the bran layer. The larvae often feed only on the germ and overlying bran, but they sometimes consume the adjacent endosperm as well, and occasionally feed on the endosperm so extensively that the grain is completely hollowed out,

leaving only the husk. Adults do not normally feed and require no food or water to attain normal fecundity and longevity. The eggs are laid singly and are either deposited loosely or inserted in grooves or cracks in the grain. The duration of the egg stage is affected by temperature, but not humidity, and ranges from 4 days at 40° to 20 days at 20°.

The number of larval instars varies under any given set of temperature and humidity conditions, even among individuals from eggs laid by the same female. At 35° and 50% rh, larvae molt four or five times; at lower temperatures, between five and eight times. Male larvae usually have one less molt than females and require a few days less to complete development.

Optimum conditions for larval development are 30–35° and 50–73% rh (Table 10.15). The most rapid development in both sexes occurs at 35° and 50% rh; the lowest larval mortality occurs at 30° and 73% rh. Larvae cannot complete development at 15°, and they are adversely affected by low humidity, especially at low or high temperatures. Under such conditions, development is prolonged and mortality is high. Larvae are unable to complete development at 2% rh at either 25° or 40°.

The pupa remains within the last larval skin, with only a part of the dorsal surface exposed, as does the adult until the integument has hardened and darkened. Pupal development and the amount of time spent by the adults within the last larval skin is not appreciably influenced by humidity, but is a function of temperature. The mean duration of the pupal stage ranges from 3 days at 40° to 13 days at 20°. The adult remains within the larval skin from one day at 40° to 6 days at 20°.

Adults sometimes mate immediately after emergence, and a single mating is sufficient for attainment of full fecundity and fertility. Mating is followed by a preoviposition period of one day or less at 35° and 40° to as long as 3 days at 25°. The preoviposition period is greatly extended and oviposition irregular at 20°. Most eggs are laid during the first few days of oviposition, after which the daily number decreases rapidly. Oviposition and percentage hatch are influenced by temperature and humidity (Table 10.16). Egg production is low at 40°, but increases with decreasing temperature to a maximum at 30° and then decreases again until the adults become inactive at 20°.

No oviposition occurs at 2% rh and 20° or 40°, but does occur at all humidities at the intermediate temperatures. At any given temperature, egg production increases with humidity. Fertility is low at 20° and 40°. When eggs are held at 2%, 25%, 50%, and 73% rh at either 20° or 40°, they hatch only at 50% and 73% rh (57). At intermediate temperatures, some eggs hatch at all humidities, but at any given temperature, fertility generally increases with humidity. Adults are short-lived. At 73% rh, males live for 10 days at 40° to 38 days at 20° on the average, and mated females live for 8–38 days. Adult lifespan at any temperature decreases with humidity.

Trogoderma variabile, warehouse beetle

The warehouse beetle occurs on a wide range of food materials. Although many diets sustain populations of the beetle, population increase is greatest on animal feeds (mixed feeds and processed grains), whole kernels of barley and wheat, pollen, and various grocery commodities such as egg noodles, oatmeal, wheat germ, and whole wheat flour (122). Performance is generally poor on dry lima and kidney beans.

Loschiavo (103) reared T. variabile at 32° and 70% rh on a 1:1 mixture of finely ground wheat and bran. Eggs are laid singly and are deposited loosely in crushed wheat or tucked into crevices of whole kernels. The mean duration of the egg stage is 7 days. There are normally six larval instars, but many fully developed larvae enter diapause (used here in the same sense as with T. granarium) and fail to pupate even though molting continues at irregular intervals. The proportion of larvae entering diapause increases from 32 to 67% when they are exposed daily to room temperature (21–28°) and increases to 80% when they are handled daily as well as exposed to room temperature.

Development of nondiapausing larvae requires an average of 34 days. Most larvae pupate near the surface of the food. The larval skin splits dorsally during the last larval molt but is not cast off, so that the pupa remains enclosed in the skin. Pupal development is completed in 4 days. Newly formed adults spend an average of 2 days within the last larval skin, but some spend as long as 7 days there. Adults mate shortly after emergence.

When warehouse beetles are reared at 30° and 60–70% rh on wheatfeed (29), the mean duration of the egg, larval, and pupal stages is 6, 26, and 5 days, respectively. Newly emerged adults remain within the last larval skin for an average of 2 days. Nondiapausing larvae pupate within 7 weeks after hatching, but diapausing larvae delay pupation and remain the same size

Table 10.15

Effects of temperature and humidity on larval development of *Trogoderma inclusum*

| Temperature (°C/°F) | Relative humidity (%) | Mean duration of larval stage (days) | | Mean survival (%) |
		Males	Females	
20/68	25	263	262	13
	50	197	—	2
	73	191	238	32
25/77	2	—	—	0
	25	51	56	86
	50	48	55	76
	73	41	50	78
30/86	2	68	81	38
	25	43	47	69
	50	34	38	91
	73	31	38	93
35/95	2	98	95	28
	25	40	51	91
	50	27	35	87
	73	31	38	71
40/104	2	—	—	0
	25	96	104	19
	50	68	75	33
	73	53	84	47

SOURCE: Hadaway, A. B. 1956. The biology of the dermestid beetles, *Trogoderma granarium* Everts and *Trogoderma versicolor* (Creutz.). *Bulletin of Entomological Research* 46(4)781–796.

Table 10.16

Effects of temperature and humidity on the oviposition period, fecundity, and fertility of *Trogoderma inclusum*

Temperature (°C/°F)	Relative humidity (%)	Mean oviposition period (days)	Mean number of eggs/ female	Mean hatch (%)
15/59	←————No oviposition————→			
20/68	2	2.0	0	0
	25	3.2	18	0
	50	3.4	20	30
	73	8.4	66	50
25/77	2	4.1	45	45
	25	4.8	52	53
	50	7.5	84	78
	73	7.7	113	76
30/86	2	3.8	54	51
	25	4.6	64	58
	50	5.6	100	76
	73	7.0	109	82
35/95	2	2.2	30	62
	25	2.6	32	73
	50	3.6	55	79
	73	4.5	50	74
40/104	2	0	0	0
	25	2.5	17	0
	50	2.5	31	18
	73	1.8	20	13

SOURCE: Hadaway, A. B. 1956. The biology of the dermestid beetles, *Trogoderma granarium* Everts and *Trogoderma versicolor* (Creutz.). *Bulletin of Entomological Research* 46(4)781–796.

for as long as 2 years. The proportion of larvae entering diapause increases as the volume of food decreases but is not related to the amount of food per larva.

Partida and Strong (*122*) observed the development of *T. variabile* under a variety of constant and programmed conditions of temperature and humidity. At 70% rh, the duration of the egg stage ranges from 6 days at 37.8° to 18 days at 21.1°. Development from egg hatch to adult maturation (elytra fully darkened) on a 1:1 mixture of dry skim milk and wheat germ at 50% rh requires from 29 days at 37.8° to 63 days at 23.9° in males and from 33 to 68 days in females. Males generally pupate after one less molt than females, but there is overlap. The minimum number of molts required before pupation ranges from five to nine and varies considerably for a given set of climatic conditions. At 32.2°, development requires from 27 days at 70% rh to 33 days at 30% rh in males and from 29 to 38 days in females.

Loschiavo (*103*) noted that all life stages were present in an unused mill and warehouse in Canada that had not been heated the previous winter, an indication that *T. variabile* can survive winter temperatures in some parts of that country. He found that mature larvae are not killed by a 6-day exposure to 1°, 10°, or 20°. Adult survival is also high at low temperatures, and although oviposition is inhibited, females exposed for a short period lay a full complement of eggs when returned to favorable conditions.

Loschiavo (*104*) described the influence of temperature at 70% rh on adult longevity and ovipostion. The mean lifespan of adult males ranges from 9 days at 40° to 50 days at 17.5°. Males live for 30, 62, or 80 days at 4.4°, 10°, or 12.8°, respectively. The relationship between temperature and female longevity is illustrated in Figure 10.5. Lifespan is maximum in both sexes at 12.8°. Loschiavo (*104*) suggested that below this temperature, life is shortened by the increasing lethal effects of cold, and above it by the increasing rate of metabolic breakdown.

The preoviposition period decreases from about 16 days at 17.5° to less than one day at 32.5° and then increases again to about 2 days at 37.5°. The oviposition period (Figure 10.5) decreases from about 10 days at 17.5° to 4 days at 32.5° and above. Oviposition peaks during the first 3 days at temperatures between 27.5° and 35° and then rapidly declines. At 37.5°, oviposition is maximum on the 4th, 5th, and 6th days. The percentage of fertile females increases with temperature from 62% at 17.5° to 94% at 35°, and then decreases at higher temperatures.

Fecundity is maximum between 27.5° and 30° (Figure 10.5). A few eggs are laid at 15.6° but these are deformed and do not hatch. Eggs of normal appearance are laid at 17.5°, but hatchability is low (<2%) and oviposition is sporadic. Females do not oviposit at temperatures below 15.6° or at 40°. Apparently, the lower threshold for egg laying is near 16° and the upper threshold is between 38° and 40°. The lower threshold of hatchability appears to be near 18°.

Alternately exposing females to 32.5° and 15.6° at 24-hour intervals does not reduce total egg production below that observed at a constant temperature of 32.5°, but the number of females ovipositing and the number of eggs laid per female is lower during the cooler periods. Also, the oviposition period is longer than at a constant temperature of 32.5°, but no eggs are laid after the 14th day. The rate of self-multiplication (λ) has been calculated to be 1.5 females/female per week at 32.2° and 70% rh (*104*) and 1.7 females/female per week at 30° (*29*).

Nitidulidae

Twenty-two species of Nitidulidae, most of which belong to the genus *Carpophilus*, have been recorded from stored products (*64*). Sixteen species of *Carpophilus* have been associ-

Figure 10.5

Effect of temperature on the longevity (◐), oviposition period (●), and total fecundity (◑) of *Trogoderma variabile* at 70% rh.
SOURCE: Loschiavo, S. R. 1967. Adult longevity and oviposition of *Trogoderma parabile* Beal (Coleoptera, Dermestidae) at different temperatures. *Journal of Stored Products Research* 3(4)273–282.

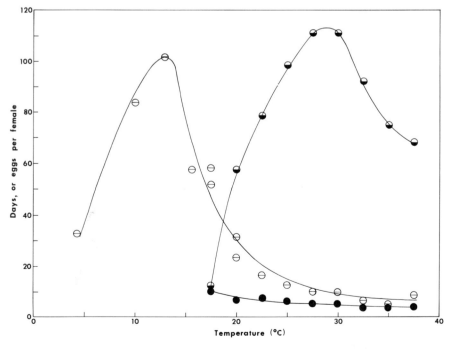

ated with stored products in different parts of the world, and some have been widely distributed by commerce (142). These beetles are most often found in association with fruit or moldy food.

Carpophilus dimidiatus, corn sap beetle

The corn sap beetle is a cosmopolitan pest of a wide variety of stored foods, especially dried fruit and cereals (42, 43, 64, 154). However, it rarely damages cereal products that are in good condition. In the field it normally feeds on decaying fruit and vegetation and on sap exuding from injured plants.

When C. dimidiatus was reared on dates at various temperatures at a high humidity maintained by moist sand, development from egg to adult, on average, ranged from 15 days at 32.2° to 49 days at 18.3°. The mean durations of the various stages were as follows: egg, 2–4 days; larva, 5–16 days; prepupa, 3–12 days; pupa, 5–18 days (101). When C. dimidiatus was reared on dried onions at 30° and 70% rh, the mean developmental period was 36 days, and adults lived an average of 13 days. The mean durations of the egg, larval, and pupal stages were 5, 24, and 7 days, respectively.

When reared on wheat bran at various temperatures and humidities, C. dimidiatus completed development at all temperatures between 17.5° and 32.5° at humidities above 50% (97). However, the rate of development at any given set of conditions varied considerably. The performance of C. dimidiatus on wheat bran was relatively poor, apparently due to the absence of appropriate molds in the diet (97). The total developmental period was shortest (26 days) at 32.5° and 90% rh, but survival was low under these conditions. Survival was greatest at 22.5–27.5° and 70% rh, conditions at which development required a range of 38–42 days. Mortality was greater at higher or lower temperatures and humidities. Mean longevity of adult males and females was about the same. Longevity at each temperature was greatest at the highest humidities. Thus at 90% rh, longevity was greatest (134 days) at 20°. At 70% rh, longevity was greatest (75 days) at temperatures between 22.5° and 25°. The most favorable conditions for population growth on wheat bran are probably 22.5–25° at 70% rh (97).

Carpophilus hemipterus, driedfruit beetle

The driedfruit beetle (64) is found throughout the temperate and tropical regions of the world. It has been found in a wide variety of ripe and decomposing fruit in the field and is a serious pest of dried fruit. It has also been recorded from stored corn, cornmeal, wheat, oats, rice, beans, nuts, peanuts, cottonseed, copra, spices, drugs, bread, sugar, and honey, among others (42, 43, 64, 118, 154).

Eggs are laid singly on ripe or fermenting fruit in the field or in stored products. The larvae probably molt three times. Pupation in the field usually takes place in heavy soil, but in stored products, pupation takes place within the infested commodity. In warm regions and in heated buildings, breeding is continuous, and there are several generations per year. When winter temperatures are too low to permit breeding, the beetles hibernate as mature larvae, pupae, or adults in the soil, in stored fruit, or in fruit left on the ground or on the trees. The adults sometimes live for more than a year. They are active during the day as well as at night and have been shown to migrate as far as 2 miles in 4 days.

In date palm groves, populations of C. hemipterus and C. dimidiatus build up in fermenting dates that have dropped to the ground prematurely. Later they infest ripened dates on the trees and are carried into storage (79, 101).

When C. hemipterus was reared on dates at various temperatures at a high humidity maintained by moist sand, development from egg to adult required an average of 12 days at 32.2° to 42 days at 18.3°. The mean durations of the various stages were: egg, 1–4 days; larva, 4–14 days; prepupa, 3–8 days; pupa, 4–16 days (101). When C. hemipterus was reared on dried onions at 30° and 70% rh, the mean developmental period was 32 days, and adults lived an average of 11 days. The mean durations of the egg, larval, and pupal stages were 4, 22, and 6 days, respectively (76).

When C. hemipterus was reared on dried fermented peaches under variable temperature and humidity conditions (64, 143), the greatest number of eggs laid by one female was 2,134 at a mean temperature of 28.1°, and the lowest number was 461 at a mean temperature of 14.7°. The overall mean number of eggs per female was 1,071 at mean temperatures ranging from 14.6° to 30°. Mean longevity was 146 days in males and 103 days in females. The preoviposition, oviposition, and postoviposition periods averaged 3, 61, and 39 days, respectively.

Insects reared at mean temperatures between 26.1° and 30.3° completed development from egg to adult in an average of 19 days with an average of 2, 10, and 7 days being spent in the egg, larval, and pupal stages, respectively.

Tenebrionidae

Alphitobius diaperinus, **lesser mealworm**

The lesser mealworm is a cosmopolitan insect commonly associated with damp or moldy grain and cereal products. It also occurs commonly in broiler houses and is suspected of transmitting avian leucosis (*84*). Both larvae and adults have been observed feeding on the carcasses of dead birds, and at least the larvae may feed on moribund birds as well.

The eggs are cemented to the substrate with a clear sticky secretion, and there is a tendency for them to be deposited in groups of 12 (*163*). Adults held for 25 days on whole wheat kernels of about 12% mc show the highest rate of oviposition (6 eggs/female per day) at 32.2°. The oviposition rate declines to 4 eggs/female per day at 38° and to 1 egg/female per day at 15.6°; no eggs are laid at 10°. Egg hatch is maximum (89%) at 32.2°, declining to 77% at 38° and to 80% at 15.6°.

When *A. diaperinus* is reared on a diet of ground wheat at high humidity (over water), the mean developmental period (egg to adult) ranges from 42 days at 38° to 97 days at 15.6° (*163*). The incubation period ranges from 3 to 10 days. There are 6–11 larval instars, the average number increasing with declining temperature. Mature larvae form a chamber in the food material in which they pupate after a prepupal period of 4–7 days. Pupal development is completed in 6–10 days. The preoviposition period is 10–11 days and shows little relationship to temperature.

No larvae develop beyond the first instar at 10°. Otherwise, survival is maximum (60%) at 32.2° and declines to 52% at 38° or 27% at 15.6°. There is a positive correlation between the best larval development and the presence of a fungus (*Aspergillus*).

The estimated mean adult longevity at room temperature (21–24°) on a semisynthetic diet based on chicken starter mash is over 400 days (an adult in one test lived at least 703 days) (*125*). The preoviposition period ranges from 4 to 30 days (mean = 13 days). Oviposition continues at a more or less steady rate (mean of 4 eggs/female per day) throughout most of the adult lifespan (*125*).

Cynaeus angustus, *larger black flour beetle*

Prior to 1938, when it was discovered infesting stored grain in Iowa, Kansas, and Washington, the larger black flour beetle was known only from debris at the base of yucca plants in California (*81*). In 1939, it was found in a flour mill in Seattle; by 1941, it had been found in stored grain in Minnesota, South Dakota, Missouri, Nebraska, and Illinois. It is now fairly common in farm-stored grain throughout the North Central states (*1*).

The eggs are usually laid singly and generally are placed in protected places. They hatch in about 3 or 4 days at 30°. There are 9–11 larval instars, and larval development is completed in 22–92 days at 30°. Overcrowding slows the rate of larval development considerably and sometimes prevents pupation. Mature larvae usually select or construct a sheltered location for pupation. In wheat, pupation generally occurs in a cell constructed among the kernels. At 30°, the duration of the pupal stage ranges from 4 to 6 days with an average of 5 days. Females generally begin laying eggs 5–7 days after emergence, and on corn of 17–18% mc, they lay from 4 to 5 eggs/day. If an oviposition period of 80–100 days and an average oviposition rate of 4.5 eggs/day is assumed, each female has a potential fecundity of 360–450 eggs. Adults probably live for a year or longer under storage conditions (*81*).

In general, *C. angustus* develops equally well on corn, oats, and barley, but wheat is less satisfactory. When a moisture gradient was provided by a layer of relatively dry grain over a layer of damp and moldy grain, development from egg to adult at 30° required an average of 35, 37, 38, or 40 days on barley, wheat, oats, or corn, respectively. Mortality ranged from 34 to 39% on barley, oats, and corn, but was about 51% on wheat.

Corn with a high moisture content is generally very favorable for development until the action of bacteria and fungi cause an accumulation of free water and possibly fermentation gases. The mean developmental period at 30° is 52, 55, or 61 days, and mortality is 27, 21, or 29% on corn of 16.7, 14.1, or 10.7% mc, respectively. Also, larvae as well as adults are cannibalistic, and the rate of cannibalism is apparently greatest on food of low moisture content.

The larger black flour beetle is able to feed on undamaged grain, but development is favored by the presence of damaged kernels. Thus, at 30° and 11.2% mc, development requires an average of 62 days or 78 days and mortality is 28% or 54% on cracked or whole corn, respectively. When feeding on undamaged kernels, the larvae attack the germ by chewing a more or less circular hole through the pericarp. When it is given a choice of several grains, *C. angustus* shows a marked preference for corn and barley over wheat and oats, and a 2:1 preference for corn over barley. This apparent preference for corn is borne out by field observations. *C. angustus* feeds voraciously on several species of fungi isolated from insect-infested grain, but among these it is able to reproduce only on *Cladosporium cladosporioides* (Fresenius) de Vries (*147*).

Gnatocerus cornutus, broadhorned flour beetle

The broadhorned flour beetle is a cosmopolitan pest of cereal and animal products. Shepherd (*139*) found it feeding on flour, cornmeal, dog biscuits, corn, pancake flour, yeast cakes, bran, and farina. He noted that it shows a preference for flaky material. Morison (*110*) observed that it is usually associated with infestations of moths and stated that although it can breed on farinaceous materials alone, it fares better when it has access to insect prey. He showed experimentally that larvae and adults of *G. cornutus* prey on eggs of *Anagasta kuehniella* and that they will also eat dead or injured adults, larvae, and pupae of that species.

Pimentel (*123*) reared *G. cornutus* on meat-bone meal plus Haydak's formulation (yeast, cereal, and milk products) at a variety of temperatures and humidities (expressed as saturation deficits in millimeters of mercury). This beetle is able to complete development at temperatures between 15° and 32° and at saturation deficits from nearly 0 to 12.7 mm. Optimum conditions range from 24° to 30° and from 2.5- to 7.5-mm saturation deficit. Eggs hatch in 5 days at 30° to 8 days at 20° under favorable humidity conditions, but low humidities make it difficult for larvae to emerge from the eggs.

Larvae show a broad tolerance for temperature and especially for humidity. They develop well at most conditions suitable for egg hatch except at very high humidities (0.9- and 1.0-mm saturation deficit) at 27° and 30°; such failures may be caused by mold growth. Larval development requires 39 days, including 2 days as a prepupa, and pupal development requires 10 days at 27° and 6.4-mm saturation deficit (about 76% rh). The pupal and prepupal stages are very sensitive to low moisture levels so that survival decreases sharply above 8-mm saturation deficit and none survive at or above 19 mm. Survival is not affected, however, by temperatures between 20° and 32°. Adults live about 8 months under optimum conditions. The largest number of eggs laid (estimated from the number of larvae produced) by a single female was 360, produced over a period of 5 months (*123*).

Gnatocerus maxillosus, slenderhorned flour beetle

The slenderhorned flour beetle has been reported from a variety of commodities including wheat and other cereal grains, pumpkins, tamarind seeds, pulses, nutmegs, and peanuts. The mean incubation period of the egg is about 4 days at 27.5° and 30° (75% rh). Egg hatch is 92% at 27.5° but is only 64% at 30°. The number of larval instars varies. At 27.5°, there are seven or eight instars. Larval development on a 10:1 mixture of wheat flour and yeast is completed in an average of 30 days. At 30°, there are eight or nine instars and development is completed in 34 days. Larval mortality is much higher (31%) at 30° than at 27.5° (1%). Pupal development is completed in 6 days at 27.5° and in 5 days at 30° (*117*).

At favorable humidities, the temperature range for development extends from 17.5° to about 35°. The favorable humidity range varies with temperature, but successful development can occur at relative humidities as low at 7% and as high as 100%. At 20°, the mean developmental period from egg to adult ranges from 115 to 132 days, depending upon humidity. The mean developmental period is shortest (35–40 days) at 30°. Mortality is lowest at temperatures between 25° and 30°, and is 20% or less at some humidities within this range. At 20°, 22.5°, and 35°, mortality usually exceeds 50% (*117*).

Latheticus oryzae, longheaded flour beetle

The longheaded flour beetle has been reported from most parts of the world as a pest of wheat, rice, corn, barley, rye, flour, and similar products (*1*). Hafeez and Chapman (*58*) determined its rates of development and mortality on whole wheat flour supplemented with yeast at various combinations of temperature and humidity. *L. oryzae* requires a rather high temperature and humidity for successful development, but excessive humidity is detrimental. The minimum temperature for completion of development is near 25°. A combination of 35° and 85% rh is apparently optimum; under these conditions, mortality is very low and development is completed in an average of 22 days. The incubation period of the egg ranges from 3 days at 40° to 10 days at 25°. Humidity has little effect on the length of the incubation period, but high humidity reduces hatchability considerably. Thus, egg hatch at 35° is 94% at 75% rh or 98% at

85% rh but is only 8% at 95% rh.

Under favorable conditions, there are six or seven larval instars, but there are more instars in larvae reared on poor diets and at low temperatures. The mean larval period is 15 days (minimum) at 35° and 85% rh. At 85% rh, it increases to 17 days at 40° and to 95 days at 25°; at 35° it increases to 16 days at 75% rh and to 20 days at 95% rh. Larval mortality is very high (>95%) at 95% rh, regardless of temperature; and it is also high (32–100%) at 75% and 85% rh when the temperature is 25° or 40°, respectively. The duration of the pupal stage ranges from 3 days at 40° to 10 days at 25° and is not affected by humidity. There is little or no pupal mortality at any combination of temperature and humidity.

Palorus Species

Six species of *Palorus* and three species belonging to the closely related genera *Coelopalorus* and *Palorinus* have been recorded from stored products (*61*): *Palorus subdepressus, P. ratzeburgii, P. laesicollis, P. genalis, P. cerylonoides, P. ficicola, Coelopalorus foveicollis, C. carinatus,* and *Palorinus humeralis.* A seventh species of *Palorus*—*P. shikhae*—was found in old rice among other storage pests in India (*132*), but Halstead (*60*) believes that this taxon cannot correctly be assigned to the genus *Palorus* or to any other genus of the *Palorus* group.

In the storage habitat, *Palorus* species are usually associated with primary pests, particularly *Sitophilus.* They develop large populations when temperature and humidity are high, as in old grain residues. In such situations, the foods available to these beetles include cereal grains (often mold- or insect-damaged), cereal products (e.g., flour, meal), molds, yeasts, and all stages of primary insect pests (living and dead) and their fecal matter. In nutrition experiments, Halstead (*61*) showed that intact wheat kernels, flour (with associated molds and yeasts), and frass can be utilized as food by *Palorus,* but that only the germ of intact kernels is eaten. The frass of *Sitophilus* proved to be the best food, better than whole wheat flour plus yeast. *Palorus* species probably prey on the inactive stages of other insect species, and there may be some cannibalism, but dead insects are probably of little value as food. Two species of *Palorus* are more widespread and more frequently encountered in stored products than the others. *P. ratzeburgii* is cosmopolitan and *P. subdepressus* is nearly so, being distributed throughout the world except for the northern temperate region.

Palorus ratzeburgii The smalleyed flour beetle (*61*) infests granaries, bakeries, and feed and flour mills. It has been recorded from grain, flour, meal, barley in British maltings, corn, peanuts, and cassava meal. The mean duration of the egg stage at 70% rh is about 4 days at temperatures between 30° and 40° and increases to 33 days at 17.5°. Hatchability is high (85–100%) at temperatures between 20 and 40°, but is only 57% at 17.5°; no eggs hatch at 15° or 42.5°.

The mean duration of the larval period on whole wheat flour at 70% rh ranges from 18 days at 32.5° to 55 days at 20°. Larvae fail to complete development at 17.5° and 40°; the combined mortality of eggs and larvae is minimal (14–18%) between 30° and 35°. The lower humidity threshold for development is between 40% and 50% rh at 20° and between 10% and 20% rh at 25° or 30°. The mean duration of the larval period at 30° ranges from 20 days at 80% rh to 34 days at 20% rh. The combined mortality of eggs and larvae is minimal (14–18%) between 60% and 70% rh.

Pupal development is not affected by humidity and, on average, is completed in 4 days at 37.5° to 14 days at 20°. Pupal mortality is negligible except at 20°, where at 60% and 70% rh a third or more of the pupae die. However, near the upper temperature limit for development (37.5°), pupae produce a large percentage of abnormal adults which soon die. Production of abnormal adults is also high at 20° and 60% rh. At 30° and 80% rh, oviposition peaks about 9–36 days after pairing, and the mean daily egg production is 3 eggs/female per day.

Palorus subdepressus The depressed flour beetle (*61*) has been found in wheat, rice, corn, sorghum (guinea corn, durra) (*Sorghum bicolor*), sago (*Metroxylon sagu*) flour, peanuts, cacao, copra, illipe nuts (*Madhuca* spp.), pollards, pepper, and ginger. It has been reported to infest warehouses, granaries, and feed and flour mills. The eggs, as in many tenebrionids, have a sticky surface to which food particles adhere. At 70% rh, eggs hatch at temperatures between 17.5° and 37.5° but not at 15° or 40°. Hatchability is over 80% (usually over 90%) at temperatures between 20° and 35° but is somewhat lower near the limits for hatching. The mean incubation period at 70% rh ranges from 3 days at 32.5° to 20 days at 17.5°. At 25°, neither hatchability nor length of the incubation period is affected by humidity.

Typically, there are seven larval instars, although there are sometimes eight or nine. The mean duration of the larval period on whole wheat flour at 70% rh ranges from 28 days at 35° to 96 days at 20°. At 17.5° and 37.5°, larvae fail to survive beyond the first or second instar. The optimum temperature for development (minimum developmental period and mortality) is between 30° and 32.5°; larval development at this temperature is completed in 29–31 days, and the combined mortality of eggs and larvae is 8–16%. The lower humidity threshold for development is between 60% and 70% rh at 20° and between 40% and 50% rh at 25 or 30°. The mean duration of the larval period at 30° ranges from 21 days at 90% rh to 47 days at 50% rh. The optimum humidity for development is about 80% rh; larval development at this humidity and 30° is completed in an average of 27 days, and the combined mortality of eggs and larvae is 8%.

Mature larvae construct a pupal cell in the food material and enter a prepupal stage that lasts for 1–2 days at 30° or 3 days at 25°. On average, pupal development is completed in a range from 4 days at 35° to 14 days at 20°. Humidity has little effect on the pupal period at 25° or 30°, but at 20°, the pupal period is extended by low humidity. Mortality among pupae is unusual, but at extreme temperatures, and also to some extent at extreme humidities, emerging adults sometimes fail to shed the pupal skin completely or to expand the elytra and thus die in a short time.

The mean oviposition rate at 70% rh is highest (2 eggs/female per day) at 30° and oviposition peaks 20 days after eclosion. Viable eggs are laid at 20° and 35°, but these temperatures are near the lower and upper limits for oviposition, and performance is poor. The oviposition rate at 30° is increased to 2.5 eggs/female per day by an increase in relative humidity from 70% to 80%. Maximum individual egg totals observed at 30° were 303 and 635 at 70% and 80% rh, respectively; the corresponding oviposition periods were 171 and 237 days.

Tenebrio Species

The dark mealworm, *T. obscurus*, and the yellow mealworm, *T. molitor*, are cosmopolitan, but in the USA, *T. molitor* is essentially restricted to the northern states whereas *T. obscurus* occurs throughout the country (*34*). Both species show a preference for dark, moist situations and breed in refuse grain, coarse cereal, and mill products that accumulate in dark corners, under bags, in bins, and in similar places. In addition to cereal products, they feed on materials of animal origin such as meat scraps, dead insects, and feathers, and they are sometimes found in the litter of chicken houses where feathers and waste grain are mixed with excrement.

Under storehouse conditions in Washington, DC, both species of mealworms usually require at least one year to complete development (*34*). Fifty *T. obscurus* reared from egg to adult on graham flour and meat scraps reached the adult stage in 105–675 days [mean ± standard deviation (SD) = 351 ± 125], and 41 *T. molitor* reached the adult stage in 301–649 days (mean ± SD = 521 ± 125). The mean durations (±SD) of the developmental stages (in days) were as follows:

	Egg	Larva	Pupa
T. obscurus	13 ± 2	323 ± 126	15 ± 3
T. molitor	6 ± 1	501 ± 124	14 ± 2

The number of larval molts varies, ranging from 12 to 24 in *T. obscurus* and from 9 to 20 in *T. molitor*. However, about half of the *T. obscurus* molt 14–15 times, and about half of the *T. molitor* molt 17–19 times. Larvae may pupate as soon as they become full grown or they may remain in the larval stage for many months with little change in size or outward appearance. Pupation occurs on the surface of the food material after a short prepupal period, during which the larvae are inactive. The pupa is naked, not enclosed in a pupal chamber.

At Washington, DC, both species normally produce one generation a year, although under favorable conditions *T. obscurus* may have a partial second generation, and some individuals of both species may require 2 years to complete development. Both species overwinter in the larval stage. Adults begin to appear in the spring and are present throughout the summer. Adults of *T. obscurus* begin to emerge a month or more before those of *T. molitor*. Adults mate within a few days after emergence and at intervals throughout their lives. At room temperature, *T. obscurus* begins laying eggs 9–20 days after emergence and *T. molitor*, 5–18 days after emergence. The beetles live for 2–3 months, and the oviposition period extends over most of the lifespan. Oviposition in the laboratory is irregular, and there are frequent intervals between periods of egg laying. The number of eggs laid in one day during periods of oviposition ranges from 1 to 62 in *T. obscurus* and from 1 to 40 in *T. molitor*. The average total fecundity of *T. obscurus* ranges from 73 to 970 (mean = 463); that of *T. molitor* ranges from 77 to 576 (mean = 276).

When reared on whole wheat flour at 25° and 72% rh, *T. molitor* undergoes 11 larval instars, and the mean durations of the egg, larval, and pupal stages are 8, 192, and 7 days, respectively. The adult beetles live for an average of 47 days. The mean incubation period of the egg ranges from 6 days at 30–35° to 17 days at 15°; hatchability is highest (71%) at 30°. Pupal development requires from 7 days at 25–35° to 48 days at 15° (*80*).

Mealworm larvae can survive for long periods without food or moisture (*34*). Of 50 *T. obscurus* larvae held at room temperature without food or water, 25 lived for 6 months, 8 lived for 8 months, and 1 lived for 9 months. As long as the food contains adequate amounts of digestible material, moisture is of little consequence to larval development (*113*). Larvae of *T. molitor* develop successfully on whole meal at humidities between 9% and 80% rh, and little or no improvement is achieved by providing drinking water. However, moisture is critical when the food contains large amounts of roughage, and metabolic water per unit weight of food is consequently low. Thus, drinking water improves development of *T. molitor* larvae on bran at high humidities and is essential for development at low humidities.

Tribolium Species

In his comprehensive review of the biology of *Tribolium*, Sokoloff (*150*) mentioned nine species as pests or potential pests of stored products: *T. castaneum, T. confusum, T. madens, T. audax, T. destructor, T. anaphe, T. thusa, T. brevicorne*, and *T. parallelum. T. castaneum* and *T. confusum* are the most common and widespread of these species as well as the most important economically. Both are cosmopolitan, but *T. castaneum* is essentially an insect of warmer climates, whereas *T. confusum* occurs in cooler climates (*55*). The effect of temperature on distribution is obscured, however, by the fact that either species may become established in heated buildings in climates much colder than they could otherwise tolerate.

T. castaneum and *T. confusum* infest a wide variety of plant and animal products. They have been recorded from grain, flour and other cereal products, peas, beans, cacao, cottonseed, nuts, dried fruits, forest products, vegetables, drugs, spices, milk chocolate, dried milk, and hides (*55*). Flour, especially whole wheat flour, and other milled products of grain are favored diets. The beetles do not attack sound grain, but feed on the broken kernels that are always present. *T. castaneum* is a major pest of stored peanuts in many parts of the world. It feeds preferentially on the germ and only after the germ is destroyed does it feed to any extent on the cotyledons (*4*).

The eggs of *T. confusum* and *T. castaneum* are laid directly in flour or other food material in which the insects are living. They may be laid free in the flour or attached to the surface of the container, but the adults apparently show little preference for cracks or crevices as oviposition sites. The eggshell is coated with a sticky substance that aids in attaching the eggs to surfaces and causes small particles to adhere to them. Under favorable conditions, about 90% of the eggs laid by young females are viable, but very old females may lay infertile eggs for considerable periods before oviposition finally ceases.

The number of larval instars is not fixed but ranges from 5 to 11 or more, usually 7 or 8. This variation results from individual characteristics of the insects as well as from external conditions such as food, temperature, and humidity. The larvae are fairly active but are usually negatively phototactic and live concealed in the food material. In flour mills and warehouses, they may be found in cracks and crevices, but more often they occur in a quantity of food, feeding near the surface or under a piece of wood, paper, or other similar object on the surface. Fully grown larvae come to the surface of the food material or seek a sheltered place for pupation. The pupa is naked, without protection of any kind.

Adults of both species have well-developed wings, but only *T. castaneum* has been observed to fly, and it is not a strong flier. Essentially, the adults have the same habits as the larvae except that they do not always remain concealed, and numbers of them may be observed crawling on flour, grain, or other surfaces in mills and storehouses (*55*).

Howe (*69, 71*) observed the development of *T. castaneum* and *T. confusum* on wheatfeed, a high-grade cereal product. He found the optimum conditions for rapid development of *T. castaneum* to be between 35° and 37.5° at a relative humidity over 70%. Under these conditions, development is completed in 19–20 days (egg, 3 days; larva, 12–13 days; pupa, 4 days). Development can be completed in a month or less at temperatures as low as 30° and at a relative humidity as low as 30%. The limits of development are imposed mostly by larval mortality, especially among early instars. The minimum temperature for development is between 20° and 22.5°; the maximum is between 37.5° and 40° when the relative humidity is either low (10–30%) or high (90%), but it is over 40° at intermediate humidities (50–70%). The durations of the egg and pupal periods are not affected by humidity, nor is the percentage egg hatch, except that no eggs hatch at 10% rh and 40°. Hatchability is usually about 80%. The rate of larval development increases with humidity at any temperature. Larval mortality is less than 20% except at 40° and at combinations of low temperature and low humidity.

The optimum, maximum, and minimum temperatures for development of *T. confusum* are all about 2.5° lower than for *T. castaneum*. The developmental periods of the two species are about equal between 23° and 27°, but *T. confusum* develops more quickly than *T. castaneum* at lower temperatures and more slowly at higher temperatures.

The optimum conditions for rapid development of *T. confusum* are 32.5° with the relative humidity above 70%. Under these conditions, development is completed in 25 days (egg, 4 days; larva, 16 days; pupa, 6 days). Development can be completed in 30 days or less with low mortality at temperatures between 30° and 35° with relative humidity as low as 30%. Above 35°, mortality exceeds 20% at all humidities and exceeds 60% at humidities below 50%. The minimum temperature for development is between 17.5° and 20°, but it is higher at low humidities (10%). The maximum is just above 37.5°, except that it is lower at very low and very high humidities (10% and 90%). The durations of the egg and pupal periods are not affected by humidity, nor is the percentage egg hatch. No eggs hatch at 15° or 40°, and only about 60% hatch at 37.5°. Under all other conditions, hatchability is about 90%. Larval development is fastest at high humidities, and larval mortality is usually less than 16%, except under extreme conditions (20° and 90% rh, 37.5° and 10% rh or less).

Many commodities, such as peanuts, corn, wheat, or rice, are less suitable diets for *Tribolium* than wheatfeed or other high-quality diets used for maintaining laboratory cultures. Development of *T. castaneum* on peanuts is slower and more variable and mortality higher than on wheatfeed (69). Also, the rate of development of *T. castaneum* on peanuts increases less rap-

idly with temperature, and the effect of humidity is more pronounced, possibly because at any given relative humidity, the moisture content of peanuts is much lower than that of wheatfeed. Development from egg to adult at 35° and 70% rh requires 46 days on peanuts compared to 20 days on wheatfeed. The increased duration results entirely from slower larval development.

Development and reproduction of *Tribolium* on suboptimal diets may be greatly stimulated by predation and saprophagous feeding on other insects (91). Larval development of *T. castaneum* is slower, and fewer progeny are produced on cracked corn than on a high-quality diet of white flour, white cornmeal, and yeast used to maintain cultures of *Tribolium* in the laboratory. However, when cracked corn is supplemented with eggs or dead adults of the Indianmeal moth, *Plodia interpunctella,* the rate of development and the production of progeny increase markedly, even though eggs or adult moths alone constitute a less satisfactory diet than cracked corn. Dead eggs or adults of *P. interpunctella* added to shelled peanuts or whole kernels of corn, wheat, or polished rice greatly increase the population growth rate of *T. castaneum* on these commodities by increasing the rates of natality and survival (89).

Fungi may also play a significant role in the nutrition of *Tribolium*. Both *T. castaneum* and *T. confusum* are capable of breeding on certain species of seed-borne fungi (146). The beetles fed to at least some extent on all 24 kinds of fungi offered to them in one series of tests; *T. castaneum* fed well on 4 species and oviposited on 16; *T. confusum* fed well on 3 species and oviposited on 10. *T. castaneum* was able to complete development on 8 species of fungi and *T. confusum* on 7 species (146).

Adult *Tribolium* may live for longer than 3 years, and females may lay eggs for more than a year (55). Males of *T. castaneum* held at 27° or at room temperature live an average of 547 days; females, 226 days. The average lifespan of adult *T. confusum* held under the same conditions is 634 days for males and 447 days for females. The oviposition period of *T. castaneum* averages 148 days at 27° and 174 days at room temperature (maximum, 308 days). The oviposition period of *T. confusum* averages 235 days at 27° and 277 days at room temperature (maximum, 432 days). The greatest number of eggs laid during one series of tests was 956 for *T. castaneum* and 976 for *T. confusum*; averages were 327 and 458, respectively (55).

With wheatfeed as food, the shortest preoviposition period for *T. castaneum* ranges from 2 days at 37.5° to 10 days at 22.5° and is affected little by humidity (*72*). The average oviposition rate under most conditions of temperature and humidity reaches a maximum during the first week of egg laying and then slowly declines. At 70% rh, fecundity (average for the first 7 weeks of oviposition) is highest (539 eggs/female) at 32.5° and declines to 355 eggs/female at 37.5° and 119 eggs/female at 22.5°. Mean fecundity at 30% rh ranges from 42 eggs/female at 25° to 413 eggs/female at 35°; at 30° and 2% rh, it is 209 eggs/female. Hatchability ranges from 75 to 96% and is not consistently affected by either temperature or humidity (*72*).

At 70% rh, λ reaches a maximum of 2.71 females/female per week at 35° and declines to 2.58 at 37.5° and to 1.29 at 22.5°. At 30% rh, λ ranges from 1.35 at 25° to 2.28 at 35°. At 30° and 2% rh, λ is 1.38. The maximum rate of increase of *T. confusum* is somewhat lower (λ = 60 females/female per lunar month or 2.64 females/female per week) than that of *T. castaneum* (*73*).

Trogositidae

Tenebroides mauritanicus, cadelle

Larvae and adults of the cadelle feed readily on many kinds of nuts, seeds, and dried fruits and vegetables as well as on grains and their milled products. In addition, the adults may attack and eat the larvae of any other insect that they encounter, including larvae and newly emerged adults of their own kind. Back and Cotton (*15*) fed the beetle in the laboratory with larvae of the cheese skipper (*Piophila casei*) and larvae of various flour beetles (*Tribolium*, among others). Adult cadelles readily eat the adults of

Rhyzopertha dominica, Sitophilus oryzae, and *Oryzaephilus surinamensis* (*165*). The type of food available has a marked influence on the rate of larval development; oviposition seems to be stimulated by predation on other insects.

Larval development is completed in about 69 days (May-July in Washington, DC) on corn, wheat, or graham flour, but development on barley flour takes about 2 weeks longer; development on rice is not completed until the following summer. Larvae are unable to complete development on refined white flour (*15*).

The eggs are either deposited loosely in food materials or inserted in crevices (crevices are preferred). The eggs are placed side by side in batches that usually contain 10–60 eggs each, but as few as 4 or as many as 91 have been found in a single batch (mean = 25). The minimum incubation period observed was 7 days at a mean temperature of 26.7–29.4°. At a mean temperature of 21.1–21.7°, the incubation period ranged from 15 to 17 days.

The larvae usually molt three or four times, but five, six, or seven times is not unusual when the larval period is long. Aside from any influence of changing temperature, the duration of the larval stage depends upon the time of year at which eggs hatch. The observed relationships among month of hatch, temperature, percentage pupating the same year, and maximum and minimum larval period are shown in Table 10.17. The temperatures listed in the table are means for the minimum larval period and provide an indication of the influence of temperature on the rate of development. The minimum developmental period does not extend over winter except in larvae that hatch in August.

Mature larvae burrow into soft wood and hollow out a small chamber before pupating. The

---

Table 10.17

Relationships among month of egg hatch, temperature, percentage pupating the same year, and maximum and minimum larval period of *Tenebroides mauritanicus*

Month of egg hatch	Temperature (°C/°F)	Percentage pupating same year	Minimum larval period (days)	Maximum larval period (days)
March	22.7/72.8	100	93	138
April	26.1/79	100	80	104
May	26.7/80	90	60	376
June	27.8/82	40	48	414
July	26.7/80	10	62	346
August	21.7/71	0	233	318

SOURCE: Back, E. A., and R. T. Cotton. 1926. *The Cadelle.* Department Bulletin 1428. U.S. Department of Agriculture, Washington, DC.

open end of the tunnel is closed by a mixture of borings and a larval secretion. Larvae sometimes crawl between boards and form a cell there, especially if the boards are of hardwood. They may also pupate in a hollowed-out kernel of corn, or if no suitable material is available, they pupate without forming a cell. Larvae that hibernate usually construct their pupal cells in the fall but do not pupate until spring. The duration of the pupal period ranges from 8 days at a mean temperature of 28.9° to 25 days at 21.1°. When development is completed within a single season, the duration of the total developmental period ranges from 67 to 134 days. When development is interrupted by hibernation, the developmental period ranges from 271 to 410 days.

The adults remain in the pupal chamber for a week or more before emerging. They mate soon after leaving the pupal cell and at irregular intervals throughout life. The preoviposition period (following transformation to the adult) varies from 15 to 210 days. Under summer conditions, it usually lasts about 2 weeks. The preoviposition period is longer during the spring, and adults emerging in late summer or early fall do not oviposit until the following spring. Many adults live for over a year and have been known to live for as long as 22 months; they may lay eggs for a period of more than a year. The longest oviposition period observed was 14 months; the shortest, 2 months.

Females that emerge during the summer usually lay a portion of their eggs the same year and cease laying with the approach of winter. They resume oviposition in February or March in warm buildings, or later in unheated warehouses, and continue laying eggs until the full complement has been produced. Females that emerge early in the season complete oviposition and die before winter. The pattern of oviposition is irregular. Eggs may be laid daily or every other day during the period of peak oviposition, but intervals of 10–14 days between batches of eggs are not uncommon, even when weather conditions are favorable. The number of eggs laid by a single female ranges from 436 to 1,319 (mean = 910).

There are two generations and possibly a partial third per year in the vicinity of Washington, DC, and perhaps three generations/year in tropical climates when favorable food is available. Overwintering occurs in the adult or larval stage. The number of adults overwintering in warehouses and granaries is augmented in the spring and early summer by newly emerged adults that have passed the winter as larvae. Overwintering adults begin ovipositing early in the spring and continue to oviposit all summer. Eggs laid in early spring develop to adults by the end of June or early July. The midsummer adults produce larvae that usually overwinter in various stages of development. Many reach maturity by fall and a few complete transformation to the adult stage at that time. Larvae that overwinter transform to adults in the spring. These adults lay eggs all summer and hibernate during the following winter. Development is more or less continuous in tropical and subtropical climates, with much overlapping of generations.

Factors Affecting Population Growth

Interactions Within and Among Species

Interactions among individuals of the same or different species of stored-product insects can have a positive or negative effect on population growth rates. Some population processes, such as mutual interference and cannibalism, act to hold growth rates below the maximum for a particular commodity and set of physical conditions. Other processes, such as facultative predation, cause suppression of the prey population while often simultaneously stimulating growth of the predator population. This effect is especially dramatic when the predator infests a commodity that is deficient in essential nutrients that are provided by the prey.

Effects of Crowding

Reproductive Rates The adverse effects of crowding on the reproductive rates of several species (e.g., *Sitophilus granarius*, Figure 10.4) have already been noted, but underpopulation as well as crowding can reduce the actual rate of population increase below that predicted by r_m. When a few individuals are present in a large space, the probability of mating is lowered and the average fecundity of the population is consequently reduced. But aside from the effect on mating, low population density has a suppressive effect on the fecundity of at least some beetles. In these species, egg production per female increases with the size of the population up to an optimum density and then declines as the population continues to grow. Maclagan (*107*) demonstrated that such an optimum density for oviposition exists in both *Tribolium confusum* and *S. granarius*. When he maintained cultures of *T. confusum* in small quantities of whole wheat

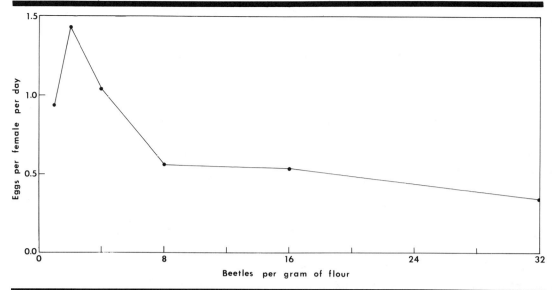

Figure 10.6

Relationship between oviposition rate and population density of adult *Tribolium confusum* in flour at 30° and 90% rh.

SOURCE: Maclagan, D. S. 1932. The effect of population density upon rate of reproduction with special reference to insects. *Proceedings of the Royal Society* B111(B773)437–454.

flour, which he renewed at frequent intervals, oviposition peaked at a density of two beetles/gram of flour (Figure 10.6). When he maintained *S. granarius* in small quantities of wheat, the females never oviposited in all the kernels available to them, nor did egg production reach a maximum rate until there were at least eight times as many kernels available as were actually used. When the number of kernels was increased above this level, however, oviposition declined.

Population Growth There have been numerous studies on the effects of crowding on population growth. Crombie (*35*), for example, showed that populations of *Rhyzopertha dominica* and *Oryzaephilus surinamensis* living in wheat rise to a maximum and then decline as food becomes exhausted and conditioned by frass. The eventual extinction of the insects results from larval mortality rather than reduced fecundity. When the wheat is maintained at a constant level by periodic removal of frass and addition of fresh grain, populations reach an equilibrium density and fluctuate about this level indefinitely. Further population growth is apparently checked by heavy mortality (>90%) among the immature stages.

Crombie (*36*) obtained similar results with populations of *T. confusum* and *O. surinamensis* living in flour. On unrenewed flour, populations of both species rise to a maximum and then decline. The eventual extinction of *T. confusum* results from failure of the larvae to develop and pupate; extinction of *O. surinamensis* also results from this cause and from cessation of oviposition. When the flour is periodically renewed, populations increase to a maximum level and then remain steady. The rate of population growth

under these conditions is determined by the rates of oviposition and development on one hand and by cannibalistic predation on the other.

Predation Many stored-product beetles show some tendency for predation, both intraspecific (cannibalistic) and interspecific. They are facultative predators; that is, their prey does not constitute an essential part of their diet, even though it may stimulate their population growth markedly under certain circumstances. Examples of facultative predation by stored-product beetles have already been cited in the discussions of life histories and habits. Perhaps the most thoroughly documented case is that involving *Tribolium castaneum* and *T. confusum* (*121, 150*). The adults and larvae of both of these beetles are voracious predators that feed on the eggs and pupae of their own and other species. The intensity of their predatory behavior depends upon species, sex, and strain. *T. castaneum* is characteristically more predatory than *T. confusum*, and females are more voracious than males. *T. castaneum* shows a selective preference for pupae of *T. confusum* over its own pupae, and *T. confusum* shows a preference for pupae of *T. castaneum*.

Predation is also influenced by commodity, the life stage of the predator, the quantity of prey available, and the density of the predator. Cannibalism is a major cause of death in *Tribolium* populations, and since the destruction of eggs and pupae increases as adult and larval populations increase, it serves as an effective mechanism for regulating population numbers. Thus, adult populations of *Tribolium* are stabilized at levels well below the limits set by available food.

Essentially the same processes (mutual interference, predation, conditioning of medium, and so on) operate in mixed-species populations as operate in single-species populations. However, these processes may be more detrimental to one species than to another, or they may actually work to the advantage of one species while hindering or having little effect on the other. As a result, when two or more species share the same habitat, the changes in their population densities are often quite different from those observed when each species occurs alone. In fact, if the requirements of two species are essentially the same, one usually eliminates the other. The outcome of interactions among species depends upon the characteristics of the environment and upon the competitive abilities of the species involved. Competitive ability is a complex characteristic that incorporates factors such as predatory behavior, tolerance of crowding, reproductive rate, and so on.

Competition Crombie (*35*), for example, found that competition between *T. confusum* and *O. surinamensis* living in flour depends upon mutual predation, and that *T. confusum* eliminates *O. surinamensis* when all stages of both species are accessible, because *T. confusum* is the more voracious predator. However, when the flour contains small-bore glass tubing that excludes *T. confusum* and provides a refuge for *O. surinamensis* pupae, both species survive.

Lefkovitch and Milnes (*100*) found that at least four interacting factors influence competing populations of *Cryptolestes ferrugineus* and *C. turcicus*:

(a) At moderately high population densities, *C. turcicus* has higher survival and developmental rates on finely divided media than does *C. ferrugineus*.

(b) The pupae of *C. ferrugineus* are more vulnerable to predation, because their fragile cocoons afford little protection compared with that afforded by the tough silken cocoons of *C. turcicus*.

(c) *C. ferrugineus* develops as rapidly on coarse material as on fine, but *C. turcicus* does not.

(d) Availability of refuges for pupation increases survival of *C. ferrugineus*.

When these factors are considered along with the moisture requirements of the two species, it becomes evident that dry conditions, coarse food, low larval density, and availability of pupal refuges would favor dominance of *C. ferrugineus*, whereas moist conditions, fine food, moderate or high larval density, and lack of pupal refuges would favor dominance of *C. turcicus*.

The complex nature of interactions among species is illustrated by the results of a multifactorial experiment reported by LeCato (*88*). He determined the change in numbers of four species of stored-product insects (*T. castaneum* = Tc, *O. surinamensis* = Os, *Cryptolestes pusillus* = Cp, and *Plodia interpunctella* = Pi) on whole or cracked corn during a 10-week period. Each species was studied alone and in all combinations with the others. His diagrammatic models for two-species cultures (Figure 10.7) portray the significant interactions (suppression or stimulation) that occurred between species and indicate the influence of commodity on these interactions. However, the interaction between two species was sometimes modified by the presence of a third or fourth species. In two-species cultures on whole corn, for example, both Tc and Pi significantly reduced the population growth of Cp. Yet when all three species occurred together in association with Os, population growth of Cp increased. LeCato attributed this change to two factors: (a) mutual interference among Tc, Os, and Pi reduced predation on Cp by Tc and Pi; and (b) damage to the corn caused by the feeding of Tc, Os, and Pi provided more food for Cp, which has difficulty feeding on whole corn.

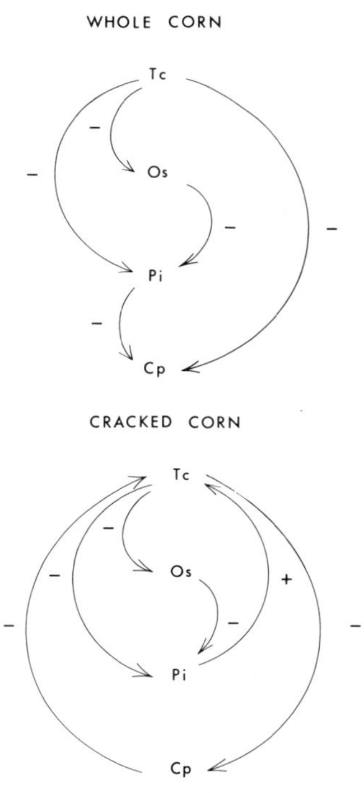

Predators, Parasites, and Pathogens

Coleopterous pests of stored products are attacked by many different predators including mice, spiders, mites, other beetles, bugs, and fly larvae. They are parasitized by various Hymenoptera including Trichogrammatidae, Bethylidae, and Pteromalidae and are subject to infection by a number of pathogens. Any of these agents can act to prevent beetle populations from realizing their full biotic potential, but the degree of population regulation achieved and the ecological processes involved are poorly known. As parameters in population dynamics, the actions of predatory and parasitic insects are essentially the same, i.e., the prey (host) is destroyed, often before it reproduces. Parasitic insects are, in fact, specialized predators often referred to as parasitoids. Pathogens, on the other hand, are true parasites. A portion of their hosts may survive to reproductive age, but development is often prolonged, and fecundity and fertility are frequently reduced.

Predators

Beetles Predatory beetles of several families including Carabidae, Staphylinidae, and His-

Figure 10.7
Models of interactions among four species of stored-product insects in two-species cultures. Suppression or stimulation of population growth are indicated by − or +, respectively. Tc = *Tribolium castaneum*, Os = *Oryzaephilus surinamensis*, Pi = *Plodia interpunctella*, and Cp = *Cryptolestes pusillus*.
SOURCE: LeCato, G. L. 1975. Interactions among four species of stored-product insects in corn: A multifactorial study. *Annals of the Entomological Society of America* 68(4)677–679.

teridae enter the storage habitat and may prey upon stored-product insects. Hinton (65) noted that 14 histerid species have been found in stored products or storage structures in various parts of the world. Most of these breed only where the local relative humidity is about 90–95%, and some have been found in considerable numbers in damp and heated waste grain. These beetles are exclusively predatory, but their impact upon pest populations has not been assessed. Among the histerid beetles, *Saprinus semistriatus* and *S. semipunctatus* have been found to prey upon larvae of *Dermestes lardarius* and *D. frischii* in stores of air-dried and smoked fish, and the former species has also been reported from a culture of *Tenebrio* on bran.

Bugs Anthocorid bugs of the subfamily Lyctocorinae are among the predatory insects most commonly encountered in storage ecosystems. The effectiveness of one of these bugs, *Xylocoris flavipes,* in regulating beetle populations has been carefully evaluated. The results of this evaluation serve to illustrate the impact that predators can have on populations of these pests.

The impact of *X. flavipes* on populations of *T. castaneum* was demonstrated in an experiment with infested lots of in-shell peanuts (126). Each lot (about 210 liters) was placed in a plywood bin with a capacity of about 1,800 liters and then adult beetles were added at the rate of 2.3 pairs/liter of peanuts. One week later, adult predators were added to all but one bin at rates of 0.3–2.3 pairs/liter. Samples taken 15 weeks after the peanuts were infested showed that the predator had suppressed the population growth of the beetles and had reduced the number of damaged peanut kernels by 66%. The beetle population had increased by over eightfold in the bin that received no predators but had decreased by 20–80% in the others, depending upon the initial density of the predators.

Xylocoris flavipes also had a pronounced impact on populations of *O. surinamensis* in small lots (about 32 liters) of shelled corn contained in fiber drums (about 98-liter capacity) (6). Suppression of the beetles ranged from 97% to 99%. Populations that were free of predation increased more than 1,900-fold in 15 weeks from an initial density of 0.6 pair/liter of corn. When predators were added to the drums (one week after the beetles) at a rate of 0.2 pair/liter, increase in the beetle population was reduced to 65–fold, and when predators were added at rates of 0.3, 0.6, or 0.9 pair/liter, increase was reduced to about 20-fold.

X. flavipes can effectively regulate residual beetle populations in empty storage facilities. LeCato et al. (92) introduced 15 pairs each of *O. surinamensis* and *T. castaneum* into a room in a warehouse and then estimated their population density at regular intervals. The room was empty except for a small quantity of rolled oats scattered on the floor to simulate grain debris and two boards placed on the floor to provide hiding places. In the absence of predators, the beetle populations increased nearly 100-fold in 100 days, but when 30 pairs of *X. flavipes* were introduced one week after the beetles, growth of the beetle population was suppressed (Figure 10.8).

Parasites

Wasps The impact of hymenopteran parasites on populations of stored-product beetles has been illustrated by results of a number of studies. Finlayson (51), for example, presented evidence that *Cephalonomia waterstoni,* a bethylid parasite of *Cryptolestes,* can effectively suppress the population growth of its host. He compared a population of *Cryptolestes* that he had studied, and which was under attack by the parasite, with a parasite-free population that had been studied earlier by Oxley and Howe (119). Both populations occurred in wheat and were concentrated in extensive hot spots that were in an advanced stage of development, central temperature being well over 40°. The two populations differed markedly in the numbers of beetles at approximately corresponding temperatures, especially at the lower temperatures. At 30°, for example, Finlayson (51) found a population density of 48 beetles/kg of wheat compared to a density of over 2,800 beetles/kg reported by Oxley and Howe (119). Finlayson concluded that the parasite had reduced the numbers of *Cryptolestes* considerably and had slowed its spread into the cooler regions of the grain bulk.

Williams and Floyd (162) showed that the pteromalid wasps, *Anisopteromalus calandrae* and *Choetospila elegans,* can reduce population growth of *Sitophilus zeamais* in stored corn. They conducted experiments on 250-gram replicates of shelled corn, each contained in a 1.9-liter round fiber carton and infested by the addition of 40 adult weevils. The cartons were held either in the laboratory or in a farm-type storage bin exposed to the elements. Five female parasites were added to each carton 21 days after the weevils, and the number of weevils was determined after 2 months and again after 4

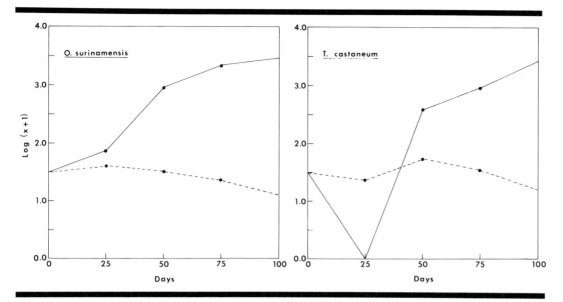

Figure 10.8

Changes in population density of *Oryzaephilus surinamensis* and *Tribolium castaneum* in an empty room of a warehouse, as indicated by the numbers of adult beetles (*x*) in samples of rolled oats swept from the floor. The predatory bug, *Xylocoris flavipes*, was either present in the room (dashed lines) or absent (solid lines).

SOURCE: LeCato, G. L., J. M. Collins, and R. T. Arbogast. 1977. Reduction of residual populations of stored-product insects by *Xylocoris flavipes* (Hemiptera: Anthocoridae). *Journal of the Kansas Entomological Society* 50(1)84–88.

months. Population growth of the weevil was reduced 25–50% by *C. elegans* and more than 50% by *A. calandrae*.

Some insight into the effectiveness of the bethylid wasp, *Holepyris sylvanidis,* in regulating populations of *Oryzaephilus* can be gained from information reported by Spitler and Hartsell (*152*). In order to evaluate pirimiphos-methyl as a protectant for in-shell almonds, they placed treated and untreated almonds in drums and stored them in a room where they were subjected to infestation by several species of stored-product insects, including *Oryzaephilus mercator.* When the almonds were first placed in storage and at monthly intervals thereafter, 10,000 *O. mercator* were released in the room. The beetle population in the untreated nuts reached a peak of about 1,000 adults in 10 samples after 6 months of storage. At that time, the control drums were invaded by *H. sylvanidis,* and the number of living beetles fell abruptly. The beetle population reached a low of 200 adults in 10 samples after 8 months, at which time no parasites were found. The beetle populations then began to increase again, but leveled off following a resurgence of the parasite. The test was terminated after 12 months, but it appears that after overtaking the host population, the parasite was beginning to hold it at an equilibrium density of about 250-300 beetles in 10 samples, even though a large source of immigrant beetles was provided at weekly intervals.

Pathogens

Stored-product beetles are subject to infection by a number of pathogenic organisms including bacteria, fungi, and protozoa, but the most important are protozoa of the Class Sporozoa (Orders Gregarinida, Coccidia, and Microsporida). These organisms generally produce chronic debilitative diseases that reduce population growth by increasing mortality while slowing development and lowering fecundity. At least some of them are widespread and relatively common among natural populations of stored-product beetles, and they undoubtedly play an important role in the regulation of these populations.

Among the gregarines, most members of the Suborder Eugregarina do little damage to host tissue and can be considered commensals, although they may become mildly pathogenic under conditions of dietary or other environmental stress. Harry (*62*), for example, showed that when larvae of *Tenebrio molitor* infected with the eugregarine *Gregarina polymorpha* are reared under optimal conditions, they complete larval development as rapidly and achieve the same pupal weight as noninfected larvae. When infected larvae are grown on a suboptimal diet, however, their ability to complete development and the pupal weight they achieve are reduced. Another example is provided by Dunkel and Boush (*44*) who reported that larvae of *Attagenus unicolor* infected with the eugregarine *Pyxinia frenzeli* lose weight almost twice as rapidly during starvation as do gregarine-free larvae. They suggested that since feral populations of *A. unicolor* are frequently infected with *P. frenzeli* and commonly subjected to partial starvation, *P. frenzeli* may serve as a continuous check on population growth. Schwalbe and Baker (*133*), however, found little difference in weight loss during starvation between *A. unicolor* larvae infected with this organism and uninfected larvae. They suggested that the inconsistency between their results and those of Dunkel and Boush may have resulted from an improved nutritional state of the larvae in their tests, or from a difference in the degree of infection.

In contrast to eugregarines, gregarines of the Suborder Neogregarina (Schizogregarina) are virulent pathogens. Finlayson (50) reported that when adult female *Cryptolestes pusillus* are placed in a culture medium containing spores of *Mattesia dispora,* all of their offspring die in the larval stage, so that the adult population declines rather than increases. He also found that this organism infects and kills the larvae of *C. ferrugineus* and a third, unidentified species of *Cryptolestes,* but that the larvae of *C. turcicus* are apparently immune. Species of *Mattesia* also infect various species of *Trogoderma* and are thought to be responsible for erratic population trends in these beetles. Schwalbe et al. (134) showed that the virulence of one of these pathogens, *Mattesia trogodermae,* infecting larvae of *T. glabrum,* is influenced by dosage and temperature. When they exposed larvae to 100 mg of spore powder per gram of culture medium for 24 hours, the median survival time ranged from 20 days at 35° to 29 days at 25°. Ashford (9) found that another neogregarine, *Lymphotropha tribolii,* prevents normal larval development and increases larval mortality in *T. castaneum.* Furthermore, although the longevity and fecundity of 75% of the survivors in his tests were normal, the remainder of the survivors were weakened and sterilized.

The Orders Coccidia and Microsporida also include virulent pathogens of stored-product beetles. For example, an unidentified species of the coccidian genus *Adelina* limits growth of *T. castaneum* populations by increasing mortality, especially among the immature stages, and may also act by reducing fecundity (120). A microsporidan, *Nosema whitei,* drastically reduces the rate of development of *T. castaneum* and delays molting after the second molt. Infected adults lay few eggs, although egg viability is unaffected. The pathogenic effect of *N. whitei* on *T. castaneum* is increased by dietary stress (109). Mortality is higher and death occurs sooner among larvae on diets deficient in protein and cholesterol than among larvae reared on optimal diets (53).

Acknowledgment

The author is indebted to Sheryll S. Shelton, formerly Library Technician at this laboratory, for assistance in locating literature on stored-product beetles and for obtaining copies of references.

Cited References

1 Agricultural Research Service. 1986. *Stored-grain Insects.* Agriculture Handbook 500. U.S. Department of Agriculture, Washington, DC.

2 Andersen, F. S. 1963. Intrinsic rate of increase of *Calandra granaria. Årsbereting Statens Skadedyrlaboratorium,* 1959 & 1960, pp. 78–81.

3 Andrewartha, H. G., and L. C. Birch. 1954. *The Distribution and Abundance of Animals.* University of Chicago, IL.

4 Appelbaum, S. W. 1969. The suitability of groundnuts for the development of *Tribolium castaneum* (Herbst) (Coleoptera, Tenebrionidae). *Journal of Stored Products Research* 5(1)305–310.

5 Arbogast, R. T. 1976. Population parameters for *Oryzaephilus surinamensis* and *O. mercator*: Effect of relative humidity. *Environmental Entomology* 5(4)738–742.

6 Arbogast, R. T. 1976. Suppression of *Oryzaephilus surinamensis* (L.) (Coleoptera, Cucujidae) on shelled corn by the predator *Xylocoris flavipes* (Reuter) (Hemiptera: Anthocoridae). *Journal of the Georgia Entomological Society* 11(1)67–71.

7 Archer, T. L., and R. G. Strong. 1975. Comparative studies on the biologies of six species of *Trogoderma*: *T. glabrum. Annals of the Entomological Society of America* 68(1)105–114.

8 Arora, G. L., and T. Singh. 1970. The biology of *Callosobruchus chinensis* (L.) (Bruchidae: Coleoptera). *Research Bulletin of the Punjab University* (n.s.) 21(12)55–66.

9 Ashford, R. W. 1970. Some relationships between the red flour beetle, *Tribolium castaneum* (Herbst) (Coleoptera, Tenebrionidae), and *Lymphotropha tribolii* Ashford (Neogregarinida, Shizocystidae). *Acta Protozoologica* 7(36)513–529.

10 Autuori, M. 1931. Dados biologicos sobre o *Araecerus fasciculatus* De Geer (Col. Anthribidae). *Revista de Entomologia* 1(1)52–62.

11 Azab, A. K., M. F. S. Tawfik, and N. A. Abouzeid. 1972. The biology of *Dermestes maculatus* De Geer (Coleoptera: Dermestidae). *Bulletin de la Société Entomologique d'Egypte* 56:1–14.

12 Azab, A. K., M. F. S. Tawfik, and N. A. Abouzeid. 1972. Factors affecting development and adult longevity of *Dermestes maculatus* De Geer (Coleoptera: Dermestidae). *Bulletin de la Société Entomologique d'Egypte* 56:21–32.

13 Azab, A. K., M. F. S. Tawfik, and N. A. Abouzeid. 1972. Factors affecting the rate of oviposition in *Dermestes maculatus* De Geer

(Coleoptera: Dermestidae). *Bulletin de la Société Entomologique d'Egypte* 56:49–59.

14 Back, E. A., and R. T. Cotton. 1926. Biology of the saw-toothed grain beetle, *Oryzaephilus surinamensis* Linné. *Journal of Agricultural Research* 33(5)435–452.

15 Back, E. A., and R. T. Cotton. 1926. *The Cadelle*. Department Bulletin 1428. U.S. Department of Agriculture, Washington, DC.

16 Back, E. A., and R. T. Cotton. 1926. *The Granary Weevil*. Department Bulletin 1393. U.S. Department of Agriculture, Washington, DC.

17 Badawi, A. 1973. The biology of two species of *Trogoderma* existing in Egypt (Coleoptera—Dermestidae). *Bulletin de la Société Entomologique d'Egypte* 57:239–246.

18 Baker, J. E. 1977. Growth and development of the black carpet beetle on the laboratory diet. *Annals of the Entomological Society of America* 70(3)296–298.

19 Baker, J. E. 1977. Synchronized pupation in starved and fed larvae of the black carpet beetle. *Annals of the Entomological Society of America* 70(3)299–302.

20 Baker, J. E., S. M. Woo, and P. T. M. Lum. 1979. Respiratory metabolism during postembryonic development in the black carpet beetle. *Annals of the Entomological Society of America* 72(5)676–680.

21 Banks, H. J. 1977. Distribution and establishment of *Trogoderma granarium* Everts (Coleoptera: Dermestidae): Climatic and other influences. *Journal of Stored Products Research* 13(4)183–202.

22 Beck, S. D. 1971. Growth and retrogression in larvae of *Trogoderma glabrum* (Coleoptera: Dermestidae). 1. Characteristics under feeding and starvation conditions. *Annals of the Entomological Society of America* 64(1)149–155.

23 Beck, S. D. 1971. Growth and retrogression in larvae of *Trogoderma glabrum* (Coleoptera: Dermestidae). 2. Factors influencing pupation. *Annals of the Entomological Society of America* 64(4)946–949.

24 Bhattacharya, A. K., and N. C. Pant. 1969. Growth and development of khapra beetle, *Trogoderma granarium* Everts (Col., Dermestidae) on pulses. *Bulletin of Entomological Research* 59(3)383–388.

25 Birch, L. C. 1945. The influence of temperature on the development of the different stages of *Calandra oryzae* L. and *Rhizopertha dominica* Fab. (Coleoptera). *Australian Journal of Experimental Biology and Medical Science* 23(1)29–35.

26 Birch, L. C. 1945. The influence of temperature, humidity, and density on the oviposition of the small strain of *Calandra oryzae* L. and *Rhizopertha dominica* Fab. (Coleoptera). *Australian Journal of Experimental Biology and Medical Science* 23(2)197–203.

27 Birch, L. C. 1953. Experimental background to the study of the distribution and abundance of insects. I. The influence of temperature, moisture and food on the innate capacity for increase of three grain beetles. *Ecology* 34(4)698–711.

28 Birch, L. C., and J. G. Snowball. 1945. The development of eggs of *Rhizopertha dominica* Fab. (Coleoptera) at constant temperatures. *Australian Journal of Experimental Biology and Medical Science* 23(1)37–40.

29 Burges, H. D. 1961. The effect of temperature, humidity and quantity of food on the development and diapause of *Trogoderma parabile* Beal. *Bulletin of Entomological Research* 51(4)685–696.

30 Chang, S. S., and S. R. Loschiavo. 1971. The influence of some fungi in flour, and humidity on the survival and development of *Cryptolestes turcicus* (Coleoptera: Cucujidae). *Canadian Entomologist* 103(2)261–266.

31 Childs, D. P., et al. 1970. Low temperature effect upon third- and fourth-instar cigarette beetle larvae. *Journal of Economic Entomology* 63(6)1860–1864.

32 Chittenden, F. H. 1911. The lesser grain borer. The larger grain borer. *USDA Bureau of Entomology Bulletin* 96(3)29–52.

33 Coombs, C. W. 1978. The effect of temperature and relative humidity upon the development and fecundity of *Dermestes lardarius* L. (Coleoptera, Dermestidae). *Journal of Stored Products Research* 14(23)111–119.

34 Cotton, R. T., and R. A. St. George. 1929. *The Meal Worms*. Technical Bulletin 95. U.S. Department of Agriculture, Washington, DC.

35 Crombie, A. C. 1945. On competition between different species of graminivorous insects. *Proceedings of the Royal Society* B132(869)362–395.

36 Crombie, A. C. 1946. Further experiments on insect competition. *Proceedings of the Royal Society* B133(870)76–109.

37 Currie, J. E. 1967. Some effects of temperature and humidity on the rates of development, mortality and oviposition of *Cryptolestes pusillus* (Schönherr) (Coleoptera, Cucujidae). *Journal of Stored Products Research* 3(2)97–108.

38 David, M. H., and R. B. Mills. 1975. Development, oviposition, and longevity of *Ahasverus advena*. *Journal of Economic Entomology* 68(3)341–345.

39 David, M. H., R. B. Mills, and D. B. Sauer. 1974. Development and oviposition of *Ahasverus advena* (Waltl) (Coleoptera, Silvanidae) on seven species of fungi. *Journal of Stored Products Research* 10(1)17–22.

40 Davies, R. G. 1949. The biology of *Laemophloeus minutus* Oliv. (Col. Cucujidae). *Bulletin of Entomological Research* 40(1)63–82.

41 Desai, A. K., G. K. Kapnavar, and K. S. S. Nair. 1973. Influence of diapause on fecundity of *Trogoderma granarium* Everts (Dermestidae, Coleoptera). *Journal of Animal Morphology and Physiology* 20(2)197–199.

42 Dobson, R. M. 1954. The species of *Carpophilus* Stephens (Col. Nitidulidae) associated with stored products. *Bulletin of Entomological Research* 45(2)389–402.

43 Dobson, R. M. 1959. Notes on the taxonomy and occurrence of *Carpophilus* Stephens (Col. Nutidulidae) associated with stored products. *Entomologist's Monthly Magazine* 95(1142)156–158.

44 Dunkel, F., and G. M. Boush. 1969. Effect of starvation on the black carpet beetle, *Attagenus megatoma*, infected with the eugregarine *Pyxinia frenzeli*. *Journal of Invertebrate Pathology* 14(1)49–52.

45 Edmond, J. B. 1971. *Sweet Potatoes: Production, Processing, Marketing.* AVI, Westport, CT.

46 El-Halfawy, M. A., J. M. Nakhla, and N. H. Isa. 1972. Effect of food on the fecundity, longevity and development of the southern cowpea weevil, *Callosobruchus maculatus* F. *Agricultural Research Review* 50:67–70.

47 El-Sayed, M. T. 1935. On the biology of *Araecerus fasciculatus* De Geer (Col., Anthribidae), with special reference to the effects of variations in the nature and water content of the food. *Annals of Applied Biology* 22(3)557–577.

48 Evans, D. E. 1977. The capacity for increase at a low temperature of several Australian populations of *Sitophilus oryzae* (L.). *Australian Journal of Ecology* 2(1)55–67.

49 Evans, D. E. 1977. The capacity for increase at a low temperature of some Australian populations of the granary weevil, *Sitophilus granarius* (L.). *Australian Journal of Ecology* 2(1)69–79.

50 Finlayson, L. H. 1950. Mortality of *Laemophloeus* (Coleoptera, Cucujidae) infected with *Mattesia dispora* Naville (Protozoa, Schizogregarinaria). *Parasitology* 40(34)261–264.

51 Finlayson, L. H. 1950. The biology of *Cephalonomia waterstoni* Gahan (Hym., Bethylidae), a parasite of *Laemophloeus* (Col., Cucujidae). *Bulletin of Entomological Research* 41(1)79–97.

52 Fraenkel, G., and M. Blewett. 1943. The natural foods and the food requirements of several species of stored products insects. *Transactions of the Royal Entomological Society of London* 93(2)457–490.

53 George, C. R. 1971. The effects of malnutrition on growth and mortality of the red rust flour beetle *Tribolium castaneum* (Coleoptera: Tenebrionidae) parasitized by *Nosema whitei* (Microsporidia: Nosematidae). *Journal of Invertebrate Pathology* 18(3)383–388.

54 Gonzalez, J., J. P. Quilae, and N. C. Manalo. 1957. Study of the biology of the copra beetle. *Araneta Journal of Agriculture* 4(1)69–83.

55 Good, N. E. 1936. *The Flour Beetles of the Genus* Tribolium. Technical Bulletin 498. U.S. Department of Agriculture, Washington, DC.

56 Griswold, G. H., and M. Greenwald. 1941. *Studies on the Biology of Four Common Carpet Beetles.* Agricultural Experiment Station Memoir 240. Cornell University, Ithaca, NY.

57 Hadaway, A. B. 1956. The biology of the dermestid beetles, *Trogoderma granarium* Everts and *Trogoderma versicolor* (Cruetz.). *Bulletin of Entomological Research* 46(4)781–796.

58 Hafeez, M. A. M. A., and G. Chapman. 1966. Effects of temperature and high relative humidity on the rate of development and mortality of *Latheticus oryzae* Waterhouse (Coleoptera, Tenebrionidae). *Journal of Stored Products Research* 1(3)235–242.

59 Halstead, D. G. H. 1962. The rice weevils, *Sitophilus oryzae* (L.) and *Sitophilus zeamais* Mots.; identification and synonymy. *Tropical Stored Products Information* 5:177–179.

60 Halstead, D. G. H. 1967. A revision of the genus *Palorus* (*sens. lat.*) (Coleoptera: Tenebrionidae). *Bulletin of the British Museum (Natural History), Entomology* 19(2)61–148.

61 Halstead, D. G. H. 1967. Biological studies on species of *Palorus* and *Coelopalorus* with comparative notes on *Tribolium* and *Latheticus* (Coleoptera: Tenebrionidae). *Journal of Stored Products Research* 2(4)273–313.

62 Harry, O. G. 1967. The effect of a eugregarine *Gregarina polymorpha* (Hammerschmidt) on the mealworm larva of *Tenebrio molitor* (L.). *Journal of Protozoology* 14(4)539–547.

63 Hilali, M., H. K. Dahle, and K. Aurstad. 1972. Life history and food-spoiling enzymes of *Dermestes lardarius* (L.). *Norsk Entomologisk Tidsskrift* 19(1)25–32.

64 Hinton, H. E. 1945. *A Monograph of the Beetles Associated with Stored Products,* vol. 1. British Museum (Natural History), London.

65 Hinton, H. E. 1945. The Histeridae associated with stored products. *Bulletin of Entomological Research* 35(4)309–340.

66 Howe, R. W. 1950. The development of *Rhizopertha dominica* (F.) (Col., Bostrichidae) under constant conditions. *Entomologist's Monthly Magazine* 86(1028)1–5.

67 Howe, R. W. 1952. The biology of the rice weevil, *Calandra oryzae* (L.). *Annals of Applied Biology* 39(2)168–180.

68 Howe, R. W. 1956. The biology of the two

common storage species of *Oryzaephilus* (Coleoptera, Cucujidae). *Annals of Applied Biology* 44(2)341–355.

69 Howe, R. W. 1956. The effect of temperature and humidity on the rate of development and mortality of *Tribolium castaneum* (Herbst) (Coleoptera, Tenebrionidae). *Annals of Applied Biology* 44(2)356–368.

70 Howe, R. W. 1957. A laboratory study of the cigarette beetle, *Lasioderma serricorne* (F.) (Col., Anobiidae), with a critical review of the literature on its biology. *Bulletin of Entomological Research* 48(1)9–56, 2 pl.

71 Howe, R. W. 1960. The effects of temperature and humidity on the rate of development and the mortality of *Tribolium confusum* Duval (Coleoptera, Tenebrionidae). *Annals of Applied Biology* 48(2)363–376.

72 Howe, R. W. 1962. The effects of temperature and humidity on the oviposition rate of *Tribolium castaneum* (Hbst.) (Coleoptera, Tenebrionidae). *Bulletin of Entomological Research* 53(2)301–310.

73 Howe, R. W. 1965. A summary of estimates of optimal and minimal conditions for population increase of some stored products insects. *Journal of Stored Products Research* 1(2)177–184.

74 Howe, R. W., and J. E. Currie. 1964. Some laboratory observations on the rates of development, mortality and oviposition of several species of Bruchidae breeding in stored pulses. *Bulletin of Entomological Research* 55(3)437–477.

75 Howe, R. W., and L. P. Lefkovitch. 1957. The distribution of the storage species of *Cryptolestes* (Col., Cucujidae). *Bulletin of Entomological Research* 48(4)795–809.

76 Ibrahim, M. M., A. Koura, and M. El-Halfawy. 1970. Ecological and biological studies in some insects infesting dried onions in U.A.R. *Agriculture Research Review* 48:59–63.

77 Kantack, B. H., and R. Staples. 1969. *The Biology and Ecology of* Trogoderma glabrum *(Herbst) in Stored Grains*. Nebraska Agricultural Experiment Station Bulletin 232. University of Nebraska, Lincoln.

78 Kapoor, S. 1964. Nutritional studies on *Rhizopertha dominica* F. (Bostrichidae: Coleoptera). I. Effects of various natural foods on larval development. *Indian Journal of Entomology* 26(3)289–295.

79 Kehat, M., et al. 1966. Development of population and control of *Carpophilus dimidiatus* F., *C. hemipterus* L. and *Coccotrypes dactyliperda* F. in dates. *Israel Journal of Agricultural Research* 16(4)173–176.

80 Koura, A., M. A. El-Halfawy, and S. I. Bishara. 1972. On the biology of *Tenebrio molitor* L. in Egypt (Coleoptera: Tenebrionidae). *Bulletin de la Société Entomologique d'Egypte* 56:297–300.

81 Krall, J. L., and G. C. Decker. 1946. The biology of *Cynaeus angustus* Lec. *Iowa State College Journal of Science* 20(4)385–402.

82 Krnjaić, S. 1971. The effect of diet type on fecundity and life table of *Acanthoscelides obtectus* Say. *Zastita Bilja* 22(114)45–51 (English summary).

83 Krnjaić, S. 1971. The influence of temperature on the population growth of *Acanthoscelides obtectus* Say. *Zastita Bilja* 22(114)53–58 (English summary).

84 Lancaster, J. L., Jr., and J. S. Simco. 1967. Biology of the lesser mealworm, a suspected reservoir of avian leucosis. *University of Arkansas Agricultural Experiment Station Report Series* 159:1–11.

85 Laughlin, R. 1965. Capacity for increase: A useful population statistic. *Journal of Animal Ecology* 34(1)77–91.

86 Lavabre, E. M., and B. Decazy. 1968. Contribution a l'étude des problèmes posés par le stockage des cafés dans les pays de production. Premières données sur le comportement de l'*Araecerus fasciculatus* à température et humidité contrôlées. *Café, Cacao, Thé* 12(4)321–342.

87 LeCato, G. L. 1973. Sawtoothed grain beetle: Population growth on peanuts stimulated by eating eggs or adults of the Indian meal moth. *Annals of the Entomological Society of America* 66(6)13–65.

88 LeCato, G. L. 1975. Interactions among four species of stored-product insects in corn: A multifactorial study. *Annals of the Entomological Society of America* 68(4)677–679.

89 LeCato, G. L. 1975. Red flour beetle: Population growth on diets of corn, wheat, rice or shelled peanuts supplemented with eggs and adults of the Indian meal moth. *Journal of Economic Entomology* 68(6)763–765.

90 LeCato, G. L. 1978. Infestation and development by the cigarette beetle in spices. *Journal of the Georgia Entomological Society* 13(2)100–105.

91 LeCato, G. L., and B. R. Flaherty. 1973. *Tribolium castaneum* progeny production and development on diets supplemented with eggs or adults of *Plodia interpunctella*. *Journal of Stored Products Research* 9(3)199–203.

92 LeCato, G. L., J. M. Collins, and R. T. Arbogast. 1977. Reduction of residual populations of stored-product insects by *Xylocoris flavipes* (Hemiptera: Anthocoridae). *Journal of the Kansas Entomological Society* 50(1)84–88.

93 Lefkovitch, L. P. 1957. The biology of *Cryptolestes ugandae* Steel and Howe (Coleoptera, Cucujidae), a pest of stored products in Africa.

Proceedings of the Zoological Society of London 128(3)419–429.

94 Lefkovitch, L. P. 1962. The biology of *Cryptolestes capensis* (Waltl) (Coleoptera, Cucujidae). *Bulletin of Entomological Research* 53(3)529–535.

95 Lefkovitch, L. P. 1962. The biology of *Cryptolestes turcicus* (Grouvelle) (Coleoptera: Cucujidae), a pest of stored and processed cereals. *Proceedings of the Zoological Society of London* 138(1)23–35.

96 Lefkovitch, L. P. 1964. The biology of *Cryptolestes pusilloides* (Steel and Howe) (Coleoptera, Cucujidae), a pest of stored cereals in the Southern Hemisphere. *Bulletin of Entomological Research* 54(4)649–657.

97 Lefkovitch, L. P. 1966. Some observations on the life cycle of *Carpophilus dimidiatus* (F.) (Coleoptera, Nitidulidae) on wheat bran. *Journal of Stored Products Research* 2(2)163–165.

98 Lefkovitch, L. P. 1967. A laboratory study of *Stegobium paniceum* (L.) (Coleoptera: Anobiidae). *Journal of Stored Products Research* 3(3)235–249.

99 Lefkovitch, L. P., and J. E. Currie. 1967. Factors affecting adult survival and fecundity in *Lasioderma serricorne* (F.) (Coleoptera, Anobiidae). *Journal of Stored Products Research* 3(3)199–212.

100 Lefkovitch, L. P., and R. H. Milnes. 1963. Interaction of two species of *Cryptolestes* (Coleoptera, Cucujidae). *Bulletin of Entomological Research* 54(1)107–112.

101 Lindgren, D. L., and L. E. Vincent. 1953. Nitidulid beetles infesting California dates. *Hilgardia* 22(2)97–118.

102 Lindgren, D. L., L. E. Vincent, and M. E. Krohne. 1955. The khapra beetle, *Trogoderma granarium*. *Hilgardia* 24(1)1–36.

103 Loschiavo, S. R. 1960. Life-history and behavior of *Trogoderma parabile* Beal (Coleoptera: Dermestidae). *Canadian Entomologist* 92(8)611–618.

104 Loschiavo, S. R. 1967. Adult longevity and oviposition of *Trogoderma parabile* Beal (Coleoptera, Dermestidae) at different temperatures. *Journal of Stored Products Research* 3(4)273–282.

105 Loschiavo, S. R., and L. B. Smith. 1970. Distribution of the merchant grain beetle, *Oryzaephilus mercator* (Silvanidae: Coleoptera) in Canada. *Canadian Entomologist* 102(8)1041–1047.

106 Maceljski, M., and Z. Korunic. 1973. Contribution to the morphology and ecology of *Sitophilus zeamais* Motsch. in Yugoslavia. *Journal of Stored Products Research* 9(4)225–234.

107 Maclagan, D. S. 1932. The effect of population density upon rate of reproduction with special reference to insects. *Proceedings of the Royal Society* B111(B773)437–454.

108 McGaughey, W. H. 1973. Resistance to the lesser grain borer in "Dawn" and "Labelle" varieties of rice. *Journal of Economic Entomology* 66(4)1005.

109 Milner, R. J. 1972. *Nosema whitei*, a microsporidian pathogen of some species of *Tribolium*. III. Effect on *T. castaneum*. *Journal of Invertebrate Pathology* 19(2)248–255.

110 Morison, G. D. 1925. Notes on the broad-horned flour beetle [*Gnathocerus* (*Echocerus*) *cornutus*, Fabr.]. *Proceedings of the Royal Physical Society of Edinburgh* 21(1)14–18.

111 Mphuru, A. N. 1974. *Araecerus fasciculatus* De Geer (Coleoptera: Anthribidae): A review. *Tropical Stored Products Information* 26:7–15.

112 Mullen, M. A., and R. T. Arbogast. 1979. Time-temperature-mortality relationships for various stored-product insect eggs and chilling times for selected commodities. *Journal of Economic Entomology* 72(4)476–478.

113 Murray, D. R. P. 1968. The importance of water in the normal growth of larvae of *Tenebrio molitor*. *Entomologia Experimentalis et Applicata* 11(2)149–168.

114 Nair, K. S. S., and A. K. Desai. 1972. Some new findings on factors inducing diapause in *Trogoderma granarium* Everts (Coleoptera, Dermestidae). *Journal of Stored Products Research* 8(1)27–54.

115 Nair, K. S. S., and A. K. Desai. 1973. The termination of diapause in *Trogoderma granarium* Everts (Coleoptera, Dermestidae). *Journal of Stored Products Research* 8(4)275–290.

116 Nakamura, H. 1969. Geographic variation in the ecological characters in *Callosobruchus chinensis* L. *Japanese Journal of Ecology* 19(4)127–131.

117 Ntifo, S. E. A., and B. J. A. Nowosielski-Slepowron. 1973. Developmental period and mortality of *Gnathocerus maxillosus* (F.) (Coleoptera, Tenebrionidae) under various conditions of temperature and humidity. *Journal of Stored Products Research* 9(1)51–59.

118 Okumura, G. T., and R. G. Strong. 1965. Insects and mites associated with stored foods and seeds in California. Part II. *Bulletin of the California Department of Agriculture* 54(1)13–23.

119 Oxley, T. A., and R. W. Howe. 1944. Factors influencing the course of an insect infestation in bulk wheat. *Annals of Applied Biology* 31(1)76–80.

120 Park, T. 1948. Experimental studies of interspecies competition. 1. Competition between populations of the flour beetles, *Tribolium con-*

fusum Duval and *Tribolium castaneum* Herbst. *Ecological Monographs* 18(2)265–307.

121 Park, T., et al. 1970. Cannibalism of pupae by mixed-species populations of adult *Tribolium*. *Physiological Zoology* 43(3)166–184.

122 Partida, G. J., and R. G. Strong. 1975. Comparative studies on the biologies of six species of *Trogoderma*: *T. variabile*. *Annals of the Entomological Society of America* 68(1)115–125.

123 Pimentel, D. 1949. Biology of *Gnathocerus cornutus*. *Journal of Economic Entomology* 42(2)229–231.

124 Potter, C. 1935. The biology and distribution of *Rhizopertha dominica* (Fab.). *Transactions of the Royal Entomological Society of London* 83(4)449–482.

125 Preiss, F. J., and J. A. Davidson. 1971. Adult longevity, preoviposition period and fecundity of *Alphitobius diaperinus* in the laboratory (Coleoptera: Tenebrionidae). *Journal of the Georgia Entomological Society* 6(2)105–109.

126 Press, J. W., B. R. Flaherty, and R. T. Arbogast. 1975. Control of the red flour beetle, *Tribolium castaneum*, in a warehouse by a predaceous bug, *Xylocoris flavipes*. *Journal of the Georgia Entomological Society* 10(1)76–78.

127 Raina, A. K. 1970. *Callosobruchus* spp. infesting stored pulses (grain legumes) in India and a comparative study of their biology. *Indian Journal of Entomology* 32(4)303–310.

128 Richards, O. W. 1947. Observations on grain-weevils, *Calandra* (Col., Curculionidae). I. General biology and oviposition. *Proceedings of the Zoological Society of London* 117(1)1–43.

129 Rilett, R. O. 1949. The biology of *Laemophloeus ferrugineus* (Steph.). *Canadian Journal of Research* D27(3)112–148.

130 Rumyantsev, P. D. 1959. *The Biology of the Pests of Stored Grains*. Publishers for Technical and Economic Literature on Questions Relating to Grain, the Animal Feed Industry, and the Storage of Agricultural Products, Moscow (in Russian).

131 Sandner, H. K. 1959. Investigation of the ecology of the lesser grain borer, *Rhizopertha dominica* (F.) (Coleoptera: Bostrichidae). *Ekologia Polska* B5(2)181–185.

132 Sarup, P., S. M. Chatterji, and M. G. R. Menon. 1960. Taxonomic studies on Indian Tenebrionidae (Coleoptera). II. A new species of *Palorus* Mulsant, predaceous on *Latheticus oryzae* Waterhouse, a pest of stored products. *Indian Journal of Entomology* 22(4)239–243.

133 Schwalbe, C. P., and J. E. Baker. 1976. Nutrient reserves in starving black carpet beetle larvae infected with the eugregarine *Pyxinia frenzeli*. *Journal of Invertebrate Pathology* 28(1)11–15.

134 Schwalbe, C. P., G. M. Boush, and W. E. Burkholder. 1973. Factors influencing the pathogenicity and development of *Mattesia trogodermae* infecting *Trogoderma glabrum* larvae. *Journal of Invertebrate Pathology* 21(2)176–182.

135 Schwardt, H. H. 1933. Life history of the lesser grain borer. *Journal of the Kansas Entomological Society* 6(2)61–66.

136 Schwardt, H. H. 1934. *The Saw-Toothed Grain Beetle as a Ricemill Pest*. Arkansas Agricultural Experiment Station Bulletin 309. University of Arkansas, Fayetteville.

137 Sharifi, S., and R. B. Mills. 1971. Developmental activities and behavior of the rice weevil inside wheat kernels. *Journal of Economic Entomology* 64(5)1114–1118.

138 Sharifi, S., and R. B. Mills. 1971. Radiographic studies of *Sitophilus zeamais* Mots. in wheat kernels. *Journal of Stored Products Research* 7(3)195–206.

139 Shepherd, D. 1924. Life history and biology of *Echocerus cornutus* (Fab.). *Journal of Economic Entomology* 17(5)572–577.

140 Shires, S. W. 1977. Ability of *Prostephanus truncatus* (Horn) (Coleoptera: Bostrichidae) to damage and breed on several stored food commodities. *Journal of Stored Products Research* 13(4)205–208.

141 Shires, S. W. 1979. Influence of temperature and humidity on survival, development period and adult sex ratio in *Prostephanus truncatus* (Horn) (Coleoptera, Bostrichidae). *Journal of Stored Products Research* 15(1)5–10.

142 Simmons, P., and G. W. Ellington. 1925. The ham beetle, *Necrobia rufipes* DeGeer. Journal of Agricultural Research 30(9)845–863.

143 Simmons, P., W. D. Reed, and E. A. McGregor. 1931. *Fig Insects in California*. USDA Circular 157, U.S. Department of Agriculture, Washington, DC, pp. 1–71.

144 Singh, K., N. S. Agrawal, and G. K. Girish. 1973. Studies on the population of *Sitophilus oryzae* Linn. in high yielding varieties of wheat under different ecological conditions. *Bulletin of Grain Technology* 11(1)50–58.

145 Sinha, R. N. 1965. Development of *Cryptolestes ferrugineus* (Stephens) and *Oryzaephilus mercator* (Fauvel) on seed-borne fungi. *Entomologia Experimentalis et Applicata* 8(4)309–313.

146 Sinha, R. N. 1966. Development and mortality of *Tribolium castaneum* and *T. confusum* (Coleoptera: Tenebrionidae) on seed-borne fungi. *Annals of the Entomological Society of America* 59(1)192–201.

147 Sinha, R. N. 1971. Fungus as food for some stored-product insects. *Journal of Economic Entomology* 64(1)3–6.

148 Smith, L. B. 1965. The intrinsic rate of natural

increase of *Cryptolestes ferrugineus* (Stephens) (Coleoptera, Cucujidae). *Journal of Stored Products Research* 1(1)35–49.

149 Smith, L. B. 1966. Effect of crowding on oviposition, development and mortality of *Cryptolestes ferrugineus* (Stephens) (Coleoptera, Cucujidae). *Journal of Stored Products Research* 2(2)91–104.

150 Sokoloff, A. 1974. *The Biology of* Tribolium *with Special Emphasis on Genetic Aspects,* vol. 2. University of Oxford, England.

151 Southgate, B. J. 1978. The importance of the Bruchidae as pests of grain legumes, their distribution & control. In *Pests of Grain Legumes: Ecology and Control,* S. R. Singh, H. F. Van Emden, and T. A. Taylor. eds. Academic Press, London.

152 Spitler, G. H., and P. L. Hartsell. 1975. Pirimiphos-methyl as a protectant for stored inshell almonds. *Journal of Economic Entomology* 68(6)777–780.

153 Strong, R. G. 1975. Comparative studies on the biologies of six species of *Trogoderma: T. inclusum. Annals of the Entomological Society of America* 68(1)91–104.

154 Strong, R. G., and G. T. Okumura. 1958. Insects and mites associated with stored foods and seeds in California. *Bulletin of the California Department of Agriculture* 47(3)233–249.

155 Swain, W. R. 1975. Cold tolerance in relation to starvation of adult *Rhyzopertha dominica* (Coleoptera: Bostrichidae). *Canadian Entomologist* 107(10)1057–1061.

156 Thomas, E. L., and H. H. Shepard. 1940. The influence of temperature, moisture, and food upon the development and survival of the saw-toothed grain beetle. *Journal of Agricultural Research* 60(9)605–615.

157 Utida, S. 1967. Collective oviposition and larval aggregation in *Zabrotes subfasciatus* (Boh.) (Coleoptera, Bruchidae). *Journal of Stored Products Research* 2(4)315–322.

158 Utida, S. 1972. Density dependent polymorphism in the adult of *Callosobruchus maculatus* (Coleoptera, Bruchidae). *Journal of Stored Products Research* 8(2)111–126.

159 Waterhouse, F. L., and B. J. A. Nowosielski-Slepowron. 1965. Differences between identically reared laboratory and field stocks of the flour beetle *Cathartus quadricollis* Guer. *Proceedings, Twelfth International Congress of Entomology* (London), pp. 664–665.

160 White, G. D., and H. E. McGregor. 1957. Epidemic infestations of wheat by a dermestid, *Trogoderma glabrum* (Herbst). *Journal of Economic Entomology* 50(4)382–385.

161 Williams, G. C. 1954. Observations on the life history of *Laemophloeus minutus* (Ol.) (Col. Cucujidae) when bred on various stored cereals and cereal products. *Bulletin of Entomological Research* 45(2)341–349.

162 Williams, R. N., and E. H. Floyd. 1971. Effect of two parasites, *Anisopteromalus calandrae* and *Choetospila elegans,* upon populations of the maize weevil under laboratory and natural conditions. *Journal of Economic Entomology* 64(6)1407–1408.

163 Wilson, T. H., and F. D. Miner. 1969. Influence of temperature on development of the lesser mealworm, *Alphitobius diaperinus* (Coleoptera: Tenebrionidae). *Journal of the Kansas Entomological Society* 42(3)294–303.

164 Woodroffe, G. E. 1962. The status of the foreign grain beetle, *Ahasverus advena* (Waltl) (Col., Silvanidae), as a pest of stored products. *Bulletin of Entomological Research* 53(3)537–540.

165 Yoshida, T. 1975. Predation by the cadelle *Tenebroides mauritanicus* (L.) (Coleoptera, Ostomidae) on three species of stored-product insects. *Scientific Reports of the Faculty of Agriculture, Okayama University* 45(8)10–16.

166 Yoshida, T. 1976. The effect of crowding on the rate of reproduction in the square-necked grain beetle, *Cathartus quadricollis* (Guer.) (Coleoptera, Silvanidae). *Scientific Reports of the Faculty of Agriculture, Okayama University* 47(2)1–5.

167 Zakladnoi, G. A., and V. F. Ratanova. 1986 [1973]. *Stored-Grain Pests and Their Control.* Kolos, Moscow. Translated from the Russian by Amerind Publishing Company, New Delhi. Available in the USA from National Technical Information Service, Springfield, VA.

11
Spider Beetles: Ptinidae

Robert W. Howe

Spider beetles are typically found under bark, in rotting logs, or in animal nests, including those in caves. Only one species, *Ptinus tectus*, is a pest of economic importance, and this is serious only in northwestern Europe. Because of their habitat preferences, almost all kinds of spider beetles could invade storage facilities and become minor nuisances.

Identification

The vast literature on spider beetles is so rife with misidentifications that information on even common species can be accepted only with caution. Hisamatsu (*7*) considers that specimens of *Gibbium* from Asia and the New World are *G. aequinoctiale* rather than *G. psylloides*, which is European. Recently, Bellés and Halstead (*1*) have given extensive lists of the distribution of reliably identified specimens of both species. They show that *G. aequinoctiale* is the more widespread and was the species used in laboratory studies by Howe and Burges (*18*).

It is likely that many specimens of *Mezium* have been wrongly named. *Ptinus fur* has been an omnibus name covering many species, even including *P. tectus*. One species of *Ptinus*, *P. clavipes*, has a gynogenetic triploid form, *mobilis*, that has no male. Early keys in which the name *P. latro* is used for this form key out to the nonexistent male, thus emphazing the difficulty of identification in this group.

Howe: Slough Laboratory, Ministry of Agriculture, Fisheries, and Food, Slough SL3 7HJ, England. Current address: 28, Crossbush Road, Bognor Regis PO22 7LT, West Sussex, England.

There are useful keys to the economic (*6*) and American (*2, 23*) species. Seven species are known from Japan (*7*). Manton (*21*) produced a useful key to larvae; it was slightly amended by Hall and Howe (*4*). There is some disagreement about the correct Latin name for *Ptinus tectus*; Brown (*2*) believes that *P. ocellus* should be used. Since the species is a significant pest only in Europe and has a restricted distribution in North America (and is not an important pest there), I believe that the name widely used in Europe should be given precedence.

Habitats and Feeding Habits

Spider beetles are scavengers. Most storage species are able to thrive on both animal and vegetable remains (*15*), but a few in the *Niptus* group died out when confined on fishmeal (*18*) (Table 11.1). Coombs and Woodroffe (*3*) placed *Ptinus tectus* in the food chain of the breakdown of cereal grain residues between the weevil, *Sitophilus granarius,* and the dermestid, *Attagenus pellio.* Ptinids are able to feed on dry corpses of insects and vertebrates and on excreta. Animal-nest habitats are well suited to their needs. Larvae as well as adults drink from wet droppings of birds and rodents, from condensation, and from plumbing and roof leaks in structures. *Niptus hololeucus* has often been found in bathrooms. *N. hololeucus* and many other species have been found on mortar and in other places where there was no obvious source of food. Those on mortar were probably feeding on algae or lichens.

In general, spider beetles are associated with premises rather than with stored commodities.

177

Table 11.1
Developmental periods (oviposition to adult emergence in days) on fish meal and wheatfeed at 70% relative humidity

Species	Fish meal at 23°	Wheatfeed at various temperatures (°C)										
		13	15	17.5	20	22.5	23	27	28	30	33	35
Gibbium aequinoctiale	100	—	—	—	120	—	—	—	—	52	45	61
Mezium affine	—	—	—	—	135	—	—	—	—	62	62	—
Niptus hololeucus	—	—	—	—	121	—	—	—	—	—	—	—
Pseudeurostus hilleri	—	—	—	—	67	—	—	—	—	—	—	—
Ptinus tectus	78	171	—	—	78	—	63	61	70	—	—	—
Stethomezium squamosum	105	—	—	—	107	—	—	—	—	—	—	—
Tipnus unicolor	—	—	196	120	126	127	—	—	—	—	—	—
Trigonogenius globulum	119	—	—	—	98	—	—	—	—	—	—	—

SOURCES: Howe, R. W. 1953. The intrinsic rate of increase of some ptinid beetles. *Annals of Applied Biology* 40(1)121–133; Howe, R. W. 1955. The biology of *Tipnus unicolor* Pill. and Mitt. *Entomologist's Monthly Magazine* 91(1097)253–257; Howe, R. W., and H. D. Burges. 1952. The biology of five ptinid species found in stored products. *Bulletin of Entomological Research* 43(1)153–186; Howe, R. W., and H. D. Burges. 1953. A laboratory study of the biology of *Ptinus tectus* Boield. *Bulletin of Entomological Research* 44(3)461–516; Howe, R. W., and H. D. Burges. 1953. The biology of *Mezium affine* Boieldieu. *Entomologist's Monthly Magazine* 91(1090)73–75.

They secrete themselves in dark, moist crevices and refuges where spilled food materials accumulate. They lay eggs through sacking or scatter them on the surface of bulked commodities. The larvae develop mainly near the surface of the bulk where they spin mucous feeding shelters and eventually tougher silky cocoons in which they pupate. The adults have a daily activity rhythm determined by various combinations of light intensity, temperature, and humidity; they are most active around dusk. When dawn comes they seek shelter in dark crevices.

Life Cycle

The developmental periods of all spider beetles (*10, 11, 16, 18–20*) (Tables 11.1, 11.2) are longer than those of many other important storage pests and, consequently, their rates of increase are comparatively slow. *Tribolium castaneum*, for instance, can increase in number by as much as 70% per week, in contrast to 39% by *Ptinus tectus*. The latter and all storage genera other than *Ptinus s. l.* have uncomplicated life cycles with only three larval instars; their rates of multiplication depend fairly simply on

Table 11.2
Developmental periods (oviposition to adult emergence), percentage of eggs becoming adults, and fecundity on three foods at 70% relative humidity and 25°

Species	Fish meal		Flour		Wheatfeed		Egg yield	Weekly increase (%)	
	Developmental time (days)	% Eggs becoming adults	Developmental time (days)	% Eggs becoming adults	Developmental time (days)	% Eggs becoming adults		20°	25°
Gibbium aequinoctiale	87	40–70	76	40–70	72	40–70	280	—	26
Mezium affine	—	—	—	—	78	20–40	460	—	17
Niptus hololeucus	—	0	—	0	96	<10	50	19	4
Pseudeurostus hilleri	—	0	58	40–70	53	40–70	15	27	7
Ptinus tectus	67	>70	64	>70	51	>70	370	29	39
Stethomezium squamosum	125	40–70	91	40–70	80	40–70	170	—	19
Tipnus unicolor	—	—	—	—	159	<10	0	12	0
Trigonogenius globulum	139	20–40	82	40–70	74	40–70	160	—	25

SOURCES: Howe, R. W. 1953. The intrinsic rate of increase of some ptinid beetles. *Annals of Applied Biology* 40(1)121–133; Howe, R. W. 1955. The biology of *Tipnus unicolor* Pill. and Mitt. *Entomologist's Monthly Magazine* 91(1097)253–257; Howe, R. W., and J. O. Bull. 1956. The oviposition rate of *Pseudeurostus hilleri* (Reitt.). *Entomologist's Monthly Magazine* 92(1103)113–115; Howe, R. W., and H. D. Burges. 1952. The biology of five ptinid species found in stored products. *Bulletin of Entomological Research* 43(1)153–186; Howe, R. W., and H. D. Burges. 1953. A laboratory study of the biology of *Ptinus tectus* Boield. *Bulletin of Entomological Research* 44(3)461–516; Howe, R. W., and H. D. Burges. 1953. The biology of *Mezium affine* Boieldieu. *Entomologist's Monthly Magazine* 91(1090)73–75.

the rates of development, oviposition, and survival.

Species of *Stethomezium, Mezium,* and *Gibbium* are apparently subtropical in origin; they are capable of completing two or three generations a year in warm regions (*10, 11, 18–20*) (Table 11.3). The genera *Trigonogenius, Pseudeurostus, Tipnus,* and *Niptus* include temperate species that are able to survive cool winters and lay more eggs at 20° than at 25° (Table 11.2). Thus, *Niptus* and *Pseudeurostus* species lay about 150 eggs at 20°; *Tipnus* species lay about 100 eggs at 20° and 150 eggs at 17.5°. However, of these, only *Pseudeurostus hilleri* seems able to complete more than one generation per year in northwestern Europe. That species, like *Ptinus tectus,* is able to complete three or more generations a year. Around 1939, *Pseudeurostus hilleri* appeared to be spreading rapidly in northwestern Europe (*9, 25*). Possibly because it lays too few eggs, it did not repeat the record of range expansion that *P. tectus* achieved in Europe around 1900 after arrival from, apparently, New Zealand or Tasmania.

The true *Ptinus* species differ in many ways from species in other genera. The males are almost always strong fliers and, in some species such as *P. sexpunctatus,* both sexes fly. Where flight is normally confined to the male, the sexes are dimorphic. All of these species are frequently associated with insect nests. Although, like all spider beetles, they need a high relative humidity to develop and they benefit from the availability of drinking water, many are associated with insects in semidesert areas. Others are associated with old or rotten wood (*22*).

All species seem to have a resting stage in the life cycle that restricts them to one generation a year. Some, like *P. villiger* (*24*), survive very cold winters. The resting stage usually extends the last larval instar, as has been observed for *P. exulans,* a species that lives in spider nests in Tasmania (*5*), as well as for *P. fur* and *P. sexpunctatus* (*17*), *P. pusillus* (*12*), *P. clavipes* (*13*), and the *mobilis* form of *P. clavipes* (*14*). At higher temperatures (usually above 22.5°), the resting stage is seldom as a larva but instead the adult form remains within the cocoon so that the total life cycle under constant conditions always approximates one year. This dormancy is probably a facultative diapause, but this conclusion is speculative since light duration was not controlled in any of the associated experimental work.

Spider beetles are unlikely to become serious pests except in cool damp climates. In these they

Table 11.3

Maximum and optimum temperatures for development of various spider beetles

Species	Temperature (°C)[a]	
	Optimum	**Maximum**
Tipnus unicolor	18	25
Niptus hololeucus	20	25
Ptinus tectus	23	28
Pseudeurostus hilleri	24	28
Trigonogenius globulum	24	28
Stethomezium squamosum	25	29
Mezium affine	30	34
Gibbium aequinoctiale	33	36

[a] These temperatures are valid if they remain constant. Comparable values for the means of varying temperatures would be about 3° lower. The minimum for all species is about 10°.

SOURCES: Howe, R. W. 1953. The intrinsic rate of increase of some ptinid beetles. *Annals of Applied Biology* 40(1)121–133; Howe, R. W. 1955. The biology of *Tipnus unicolor* Pill. and Mitt. *Entomologist's Monthly Magazine* 91(1097)253–257; Howe, R. W., and J. O. Bull. 1956. The oviposition rate of *Pseudeurostus hilleri* (Reitt.). *Entomologist's Monthly Magazine* 92(1103)113–115; Howe, R. W., and H. D. Burges. 1952. The biology of five ptinid species found in stored products. *Bulletin of Entomological Research* 43(1)153–186; Howe, R. W., and H. D. Burges. 1953. A laboratory study of the biology of *Ptinus tectus* Boield. *Bulletin of Entomological Research* 44(3)461–516; Howe, R. W., and H. D. Burges. 1953. The biology of *Mezium affine* Boieldieu. *Entomologist's Monthly Magazine* 91(1090)73–75.

thrive best where food residues are allowed to accumulate and harborages are available, where rodents are numerous, or where there are bat roosts or bird nests in storage buildings or on window ledges. Cleanliness and prevention are better than control measures because spider beetles are not very susceptible to insecticides. *P. tectus* is one of the species most tolerant of phosphine at 10–15° (*8*). Oil-based films of pyrethroids synergized with piperonyl butoxide can be applied to structural surfaces to control this species.

Cited References

1 Bellés, X., and D. G. Halstead. 1985. Identification and geographical distribution of *Gibbium aequinoctiale* Boieldieu and *Gibbium psylloides* (Czenpinski) (Coleoptera: Ptinidae). *Journal of Stored Products Research* 21(3)151–155.

2 Brown, W. J. 1940. A key to the species of Ptinidae occurring in dwellings and warehouses in Canada (Coleoptera). *Canadian Entomologist* 72(6)115–122.

3 Coombs, C. W., and G. E. Woodroffe. 1973. Evaluation of some of the factors in ecological succession in an insect population breeding in stored wheat. *Journal of Animal Ecology* 42(2)305–322.

4 Hall, D. W., and R. W. Howe. 1953. A revised key to the larvae of the Ptinidae associated with stored products. *Bulletin of Entomological Research* 44(1)85–96.

5 Hickman, V. V. 1974. Notes on the biology of *Ptinus exulans. Journal of the Entomological Society of Australia* (N.S.W.) 8:7–14.

6 Hinton, H. E. 1941. The Ptinidae of economic importance. *Bulletin of Entomological Research* 31(4)331–381.

7 Hisamatsu, S. 1970. The Ptinidae of Japan (Coleoptera). *Ageha* 11:14–20 (in Japanese, English summary).

8 Hole, B. D., et al. 1976. The toxicity of phosphine to all developmental stages of thirteen species of stored product beetles. *Journal of Stored Products Research* 12(4)235–244.

9 Howe, R. W. 1940. New records of insects in grain stores. *Entomologist's Monthly Magazine* 76(911)73–75.

10 Howe, R. W. 1953. The intrinsic rate of increase of some ptinid beetles. *Annals of Applied Biology* 40(1)121–133.

11 Howe, R. W. 1955. The biology of *Tipnus unicolor* Pill. and Mitt. *Entomologist's Monthly Magazine* 91(1097)253–257.

12 Howe, R. W. 1956. The biology of *Ptinus pusillus* Sturm. *Entomologist's Monthly Magazine* 92(1108)331–333.

13 Howe, R. W. 1956. The biology of *Ptinus hirtellus* Sturm and some notes on *P. latro* F. *Entomologist's Monthly Magazine* 92(1110)369–373.

14 Howe, R. W. 1958. The developmental period of *Ptinus clavipes* Panz. form *mobilis* Moore (=*P. latro* auct.). *Entomologist's Monthly Magazine* 94(1133)236–237.

15 Howe, R. W. 1959. Conclusions and additional remarks. *Bulletin of Entomological Research* 50(2)287–326. (The work of Slough Laboratory on the biology of 14 species of spider beetles is summarized in this paper.)

16 Howe, R. W., and J. O. Bull. 1956. The oviposition rate of *Pseudeurostus hilleri* (Reitt.). *Entomologist's Monthly Magazine* 92(1103)113–115.

17 Howe, R. W., and H. D. Burges. 1951. The biology of *Ptinus fur* (L.) and *P. sexpunctatus* Panzer. *Bulletin of Entomological Research* 42(3)488–511.

18 Howe, R. W., and H. D. Burges. 1952. The biology of five ptinid species found in stored products. *Bulletin of Entomological Research* 43(1)153–186.

19 Howe, R. W., and H. D. Burges. 1953. A laboratory study of the biology of *Ptinus tectus* Boield. *Bulletin of Entomological Research* 44(3)461–516.

20 Howe, R. W., and H. D. Burges. 1953. The biology of *Mezium affine* Boieldieu. *Entomologist's Monthly Magazine* 91(1090)73–75.

21 Manton, S. M. 1945. The larvae of the Ptinidae associated with stored products. *Bulletin of Entomological Research* 35(4)341–365.

22 Morrison, B. 1970. Notes and observations on Scottish ptinids with particular reference to the south-east region. *Proceedings of the British Entomological and Natural History Society* 3:3–8.

23 Papp, C. S. 1962. An illustrated and descriptive catalogue of the Ptinidae of North America. *Deutsches Entomologische Zeitschrift* 9(5)367–423.

24 Watters, F. L. 1964. Locomotor activity of the hairy spider beetle at the surface of stored wheat. *Journal of Economic Entomology* 57(6)889–891.

25 Zacher, F. 1939. Verschleppung und Einbürgerung von Vorratsschädlingen [Introduction and establishment of stored-product pests]. In *Proceedings of the 7th International Congress of Entomology* (Berlin, 1938), vol. 4, pp. 2919–2926.

12
Biology and Ecology of Moth Pests of Stored Foods

Patrick D. Cox and Christopher H. Bell

About 70 species of moths, mainly from the families Pyralidae, Tineidae, Oecophoridae, and Gelechiidae, have been associated with infestations of stored products. However, only seven species can be considered as widely distributed, major pests of stored foods: *Ephestia* (*Cadra*) *cautella*, *E.* (*Anagasta*) *kuehniella*, *E. elutella*, *E.* (*Cadra*) *figulilella*, *Plodia interpunctella*, *Corcyra cephalonica*, and *Sitotroga cerealella*. In addition, the following species sometimes reach pest status: *Paralipsa gularis* on nuts (9); *Pyralis farinalis* and *Nemapogon granella* on grain; *Ephestia* (*Cadra*) *calidella* on carobs and dried fruit (33, 35); *Tineola bisselliella* on foodstuffs such as soybean meal in warehouses (76); *Niditinea fuscipunctella* in poultry houses from which it can spread to private dwellings (61); *Endrosis sarcitrella* mainly on stored cereal residues (105); and *Hofmannophila pseudospretella* on stored foodstuffs, clothing, and furnishings (104). The last two species sometimes infest warehouses, mills, and domestic premises.

Moths attacking stored products are typically inconspicuous, the larvae being whitish and usually well hidden in their habitat, and the adults dull brown and/or grey with a wingspan of less than 2.5 cm. Many can thrive in environments lacking access to free water. *Ephestia kuehniella* and *Tineola bisselliella* can develop at humidities approaching zero (41); apparently they are able to use their own metabolic water. In com-

mon with all Lepidoptera, storage moth larvae produce varying quantities of silk, and this can create serious problems, apart from the actual destruction of the commodity, by matting food particles together and clogging machinery used in the food industry. Particularly dense webbing can be produced by *Ephestia* spp., *Plodia interpunctella*, and *Corcyra cephalonica*.

Origins and Feeding Habits

Since human beings have stored foods only comparatively recently, it is interesting to speculate on what might have been the earlier natural habitats of modern storage pests (see also Chapter 3). It seems likely that some moths were originally associated with food sources provided by other animals. For example, *Hofmannophila pseudospretella* and *Endrosis sarcitrella* are still found regularly in birds' nests, and the tineid clothes moths are scavengers on dry animal materials. The wax moths, *Galleria mellonella* and *Achroia grisella*, occur in bees' nests, feeding on the pollen and honeycomb. Storage moths have been found in the food stores and nests of mammals, particularly rodents.

Some species, such as *Sitotroga cerealella* on wheat and corn and *Mussidia nigrivenella* and *Pyroderces rileyi* on corn, infest grain in the standing crop. This is probably their original habitat, and the more sheltered storage environment was colonized when people unwittingly brought the pests into stores at harvest time. Similarly, modern pests of nuts (such as *Paralipsa gularis* and *Ectomyelois ceratoniae*) and of dried fruits (such as *Ephestia calidella* and *E. figulilella*) were probably associated with the natural products either on the tree or when they

Cox and Bell: Slough Laboratory, Ministry of Agriculture, Fisheries, and Food, Slough SL3 7HJ, England.

had ripened and fallen to the ground, long before people took an interest in storing them.

The origins of other moths, such as *Ephestia cautella, E. elutella, E. kuehniella, Plodia interpunctella,* and *Corcyra cephalonica,* which now attack a wide variety of dry plant materials, are less clear. Perhaps they were general scavengers on seeds and other plant materials wherever such materials accumulated in cracks and crevices sheltered from the weather. The present-day food preferences of the seven major pest species are shown in Table 12.1.

In bulk storages, moths tend to infest the surface layer only, and even where bags of produce are stacked on pallets, most infestation is located in the outer bags. Depth of penetration varies with the commodity, length of infestation, and species. In bulk wheat, larvae of *Ephestia elutella* and *Sitotroga cerealella* have been recorded at depths of 30 cm after lengthy periods of infestation (*89, 102*). The larger grain size of corn permits deeper penetration than with other cereals. In laboratory studies with *S. cerealella,* larvae penetrated to at least 12 cm in corn columns within 10 weeks but only to half this depth in wheat or sorghum (*72*). Similarly, the depth of egg deposition by *P. interpunctella* was up to 8 cm in corn but only 4.5 cm in rye (*88*).

Distribution and Life Histories

Each of the major pest species has a wide developmental range and can complete its life cycle in 3–5 weeks under ideal conditions (Table 12.2). Although they have become virtually cosmopolitan in distribution through international trade, these moths still exhibit preference for their original climates. Many species came from tropical or warm-temperate regions and moved into protected storage situations at latitudes higher than that of their original range. *Ephestia kuehniella* and *E. elutella,* however, are apparently of Northern Hemisphere, temperate origin and are excluded from the tropics because they cannot tolerate long exposures to high temperatures. They have, however, been successful in spreading to temperate zones in the Southern Hemisphere.

All seven major pest species are at least moderately cold-hardy; all except *Corcyra cephalonica* can survive most winters in temperate climates (Table 12.2 summarizes information on their bionomics). Under optimal conditions, single pairs of *Plodia interpunctella, Ephestia cautella,* and *Sitotroga cerealella* can each produce up to 125,000 offspring in 3 months. The life cycle of the common pests and some other species are considered below in more detail (in alphabetical order by family and genus).

Gelechiidae

Sitotroga cerealella, Angoumois grain moth

This gelechiid species requires whole grain in which to develop. Eggs are laid mainly at night on the outside of grains in cracks, grooves, or holes made by other insects (*51*). A single female lays 80–200 eggs. The larval and pupal stages are spent inside the grain. Two or three larvae may develop in one kernel of corn, but from other grains only one adult can be produced.

Although the eggs can hatch at temperatures down to 12° and the species is able to survive winters in the British Isles (*94*), the minimum temperature for population increase is 16° (*55*). Similarly, although eggs hatch at 36°, development is not completed above 34° (*19*). The optimum relative humidity (rh) for development is 75–80%. The egg stage lasts about 10 days at 20°, 6–7 days at 25°, and 5–6 days at 30° (*19, 26*). The pupal stage lasts about 20 days at 20°, 10–12 days at 24–27°, and 8 days at 30°. Total developmental time at 70–90% rh is

---

Table 12.1

Frequency of occurrence of moth species on stored commodities

Species	High	Medium	Low
Corcyra cephalonica	Rice; cocoa; peanuts; oilseed products	Corn; cereal products; seeds; pulses and products	Sorghum; millet; nuts; copra
Ephestia cautella	Dried fruits and vegetables; nuts, oilseeds, and products; cereals and products; cocoa	Pulses and products; copra; carobs	Fallen fruits and nuts; spices; citrus pulp
Ephestia elutella	Cereals; dried fruits and vegetables; tobacco; cocoa	Oilseeds and products; nuts; cereal products	Carobs; pulses and products
Ephestia figulilella	Dried fruits, especially dates, grapes, and figs	Nuts; carobs	Cereal products; oilseed products; cocoa
Ephestia kuehniella	Cereal products, especially flour	Cereals; nuts; dried fruits and vegetables	Oilseeds and products; dried citrus pulp
Plodia interpunctella	Dried fruits; nuts; cereals; oilseeds and products	Cereal products; cocoa; carobs; confections	Dried vegetables and products; pulses; seeds; citrus pulp; fallen fruits
Sitotroga cerealella	Wheat; corn	Rice; sorghum	Cereals; seeds; peanuts

Table 12.2

Bionomics of moth pests of stored products

Species	Temperature (°C)			Relative Humidity (%)		Duration of life cycle (days)[a]	Maximum rate of increase per lunar month[b]	References
	Minimum for population increase	Optimum for development	Maximum for development	Minimum for population increase	Optimum for development			
Corcyra cephalonica	18	30	35–37	30	80	25	15	*(38, 55)*
Ephestia cautella	17	30	37	20	75	28	60	*(9, 22, 55, 63)*
Ephestia elutella	10–12	25	30	20	70	40	15	*(9, 55)*
Ephestia figulilella	15	30	36	30	70	40	20	*(34)*
Ephestia kuehniella	8–12	25	28	0	70	40	50	*(9, 37, 55, 56)*
Plodia interpunctella	18	30	35	20	75	25	60	*(9, 55, 63)*
Sitotroga cerealella	16	30	35–37	30	75	30	50	*(19, 55)*

[a] At optimum temperature and humidity on good food.
[b] Maximum rate of increase in numbers that can be expected over a 28-day period under ideal conditions, starting from a population with all life stages present.

about 30 days at 30° and 40 days at 25°.

S. cerealella can be responsible for considerable heating of grain; cultures held at constant temperature can heat to more than 5° above ambient, accelerating development accordingly. Larval development also causes an increase in the moisture content of grain (*70*).

Like many other moths, the peak time for flight activity is at dusk. Females alighting on grain are stimulated to oviposit. Air passed through grains with some mold growth acts as an attractant to adults (*67*). Wheat germ acts as an arrestant for the female but stimulates biting and feeding behavior in larvae hatching from eggs laid on grain (*27*).

In mixed cultures, *S. cerealella* can coexist with *Oryzaephilus surinamensis,* an external grain feeder, but is eventually suppressed totally by the internal feeders, *Sitophilus zeamais* or *Rhyzopertha dominica* (*6, 77*). The presence of wing scales from the moth inhibits the development of *S. zeamais* (*5*).

Oecophoridae

Endrosis sarcitrella, whiteshouldered house moth

E. sarcitrella occurs widely as a scavenger, particularly in premises with a long history of food storage and infestation where vegetable debris and moldy residues are available. It is rarely a pest of clean commodities. The female lays eggs deep in crevices by extending her ovipositor. Eggs hatch at 10–29°, requiring 42 days at 10°, 15 days at 15°, and 6–7 days at 25° (*105*). Larvae require a minimum of 80% rh. They take about 10–11 weeks at 15° and about 5–6 weeks at 25° to develop. The pupal stage lasts 25 days at 15° and 10–11 days at 25°. The optimum temperature range for development is 24–26° (*55*). Development continues throughout the winter in sheltered situations. No diapausing stages have been found (*54, 105*).

Under favorable conditions, explosive bursts of population growth occur. Up to four generations may be completed each year. A population of this species was studied in an unheated Scottish grain silo. After the winter, adults began to emerge in March, whereas other moth species in the same building did not appear until June (*83*). The sex ratio of emerging adults is not constant and has been reported as 23:1 in favor of females (*105*) or 4:1 in favor of males (*2*). Andersen (*4*) reported that the ratio worsens for females as the mother ages, indicating a preferential reabsorption of female embryos, but he could not relate the sex ratio to population density.

Hofmannophila pseudospretella, brown house moth

Commonly occurring in buildings, this oecophorid moth feeds on dried and decaying organic matter in cool humid situations. It is more commonly encountered in domestic premises than other storage moths. In industry it can attack bulk wheat and flour, but its pest status is reduced by its long life cycle (usually univoltine) and need for relative humidities higher than 80% (55). It can metabolize keratin, but apart from occasional damage to carpet backings, it is rarely more than just a nuisance in the textile industry (30). The larvae do cause problems by boring into corks of wine bottles (104).

Eggs hatch at 10–29°, requiring about 25 days for development at 15° and 10 days at 25°. The larval feeding period is 70–80 days at 20–25° and 70–90% rh, but is over 180 days at 10° (104).

Generally, H. pseudospretella overwinters as a fully grown larva in diapause. In laboratory studies, diapause ensued in all larvae reared under natural or controlled temperatures in darkness and lasted longest in those reared and held at high temperatures (104; see also section on Diapause, below). Diapause was avoided in larvae reared at temperatures increasing throughout development from about 10° to 25°. After overwintering in an unheated silo in Scotland, adults began to emerge in June (83). The pupal stage lasts about 14 weeks at 10°, 7 weeks at 15°, 25 days at 20°, and 15 days at 25° (104). Larger females can lay 400–500 eggs.

Pyralidae

Corcyra cephalonica, rice moth

Limits for complete development are 17° and 35° at 70% rh (38). At 15°, all larvae die early in development but at 37.5°, although a few manage to pupate, no adults are produced. On a diet of wheatfeed, yeast, and glycerol, highest survival (70–80%) and most rapid development (26 days) from egg hatch to adult emergence occurs at 30–32.5° and 70% rh. Development is completed in the range 15–80% rh but few adults emerge at 15% and none at 90% unless a mold inhibitor is present in the food.

The rice moth often takes over from Ephestia kuehniella in hot damp climates as the major moth pest of flour mills (44). For example, specimens of C. cephalonica (but not E. kuehniella) were found in dry as well as damp hot areas of India (97), and an apparent gradation from C. cephalonica to E. kuehniella was noted down the eastern coast of South America from the hot climate of Rio de Janeiro to cooler Buenos Aires (44). Low moisture content of food may also be a factor here, since E. kuehniella can complete development at a humidity near zero whereas the limit for C. cephalonica is in the range 10–15% rh.

Analyses of the distribution records of C. cephalonica in the literature and the records of its occurrence on imports to Britain confirm that this species is a major pest, mainly in areas of tropical climate (38). It is unlikely that C. cephalonica can multiply fast enough to become a pest at temperatures below 18° (55). This species may be able to survive mild winters in temperate climates but is unlikely to breed there except in unusual circumstances. For example, it has been observed breeding in midwinter in a warehouse in England where a heavy infestation caused peanuts to heat (43).

The egg period is influenced by temperature but not by humidity in the range 20–80%. Incubation takes about 10 days at 20° and 4 days at 30°. Eggs hatch at temperatures from 17.5° to 32.5°. Hatch is adversely affected by low humidity; only about 1% of the larvae emerge at 20% rh. All eggs die after 7 days' exposure at 10°. At 15°, although a few newly laid to one-day-old eggs exposed for 14 days hatch, none completes development to adult. Development at 28° and 70% rh is faster on millet (33–41 days) than on sorghum (46–55 days) (86). The final instar larva spins a tough, closely woven, double-layer cocoon covered by food particles and debris (7, 25). Adult males tend to emerge 1–2 days earlier than females (38).

Ephestia calidella, carob moth

The carob moth is confined mainly to the Mediterranean region, although a larval diapause increases its potential as a pest in more temperate regions (34; see also section on Diapause, below). Limits for complete development are similar to those for E. figulilella (29). At 30° and 70% rh, development is completed in 27 days on a diet of wheatfeed, yeast, glycerol, and glucose, 41 days on ground carobs, and 59 days on almonds (35). Only about 23% of the eggs hatch at 15°; none hatch at lower temperatures. At 15°, survival of larvae from egg hatch to the adult stage is less than 5%; at 37.5°, all larvae die soon after hatching (33).

Ephestia cautella, almond moth

At 70% relative humidity, the temperature range over which E. cautella can complete development is from about 15° to 36° (22). At 30°, development occurs between 20% and 90% rh. Development is quickest at 30–32° and 70–80% rh, taking a mean of 30 days from oviposition to adult emergence, when this species is reared on a mixture of wheatfeed, wheat germ, and yeast. This period increases to 145 days at 15.5°

and 70% rh, and 51 days at 30° and 20% rh. Near the optimum conditions for rapid development, the egg period occupies about 10% of the developmental cycle, the larval period 70%, and the pupal period 20% (9).

At 26° and 85% rh, larval development takes a mean of 41 days on pearl barley, 47 days on wheat germ, 49 days on wheat bran, 55 days on soybeans, 65 days on rolled rice, 67 days on peanuts or almonds, 68 days on cocoa beans, and 87 days on sesame seeds (75). On peanuts, development is completed over the range of 16–34° and 10–95% rh (74). In another study at 27° and 60% rh, the most suitable natural products for speed of development and yield of adults were cornmeal, cracked corn, and cracked runner peanuts; whole grains were least suitable (60).

Eggs hatch between 15° and 37.5° (22). Eggs laid at 25° and 70% rh take 17 days to hatch at 15°, 7–8 days at 20°, 4–5 days at 25°, and 3 days at 30°. The pupal stage lasts about 18 days at 20°, 9 days at 25°, and 7 days at 30°. Egg and pupal periods are not greatly influenced by humidity.

Larval periods are generally longer at lower humidities; near 100% rh, the larvae die young unless a mold inhibitor is added to the food. In some strains, larval diapause may prolong development (17, 50). In the absence of diapause, *E. cautella* is moderately susceptible to cold, only surviving winters in which temperatures do not fall below 0° (94). Large larvae and pupae are the forms most resistant to cold.

The shortest exposures necessary to prevent large larvae from completing development are one day at −1°, 5 days at 0°, 32 days at 5°, or 83 days at 10° (21). Eggs laid at 25° and allowed to develop to various ages survive well when transferred to 15° but fail to hatch if exposed for only 2 weeks at 10° (9). Ninety-five percent of the eggs are killed by a 9-hour exposure to −10° (73).

Ephestia elutella, tobacco moth

On most foods, *E. elutella* takes longer to develop than other phycitine species; from egg to adult on an optimal food (wheatings, yeast, and glycerol) takes about 6–7 weeks at 25° and 70% rh and 11–12 weeks at 20° (9). In addition, a strong tendency to diapause at lower temperatures or in short photoperiods greatly prolongs the developmental period and renders the species univoltine in cool climates (10, 11, 98; see section on Diapause, below). In the United Kingdom, emergence of adults begins in late May and reaches a peak at the end of June, with a small second generation in September and October (83).

The temperature range for development is 10–30°. At 15° and 30°, development can be completed at 25% rh (9), indicating that the minimum relative humidity may be much lower. Eggs laid at 25° require about 20 days to hatch at 15°, 10 days at 20°, 6–7 days at 25°, and 4–5 days at 30°. Diapause in larvae can last up to 450 days at constant temperatures in the laboratory. Pupae require 6–7 weeks at 15°, 18–23 days at 20°, 12–15 days at 25°, and 10–11 days at 30° to complete development. Adults emerging from pupae incubated at 30° can be infertile (9).

In warehouses, adults live up to 3 weeks. Most eggs are laid at dusk or during the night through the first week of adulthood (102). At 25°, fertilized females deprived of water live only about a week and lay most of their eggs during the first 4 days (103). Fecundity and longevity are increased by 30% or more when water is available. Up to twice as many eggs are laid by females reared on corn or on wheatings and glycerol than on whole wheat (103). Under optimal conditions, the normal fecundity is 150–200 eggs/female, but some individuals have produced up to 327 eggs.

Ephestia figulilella, raisin moth

Limits for complete development of *E. figulilella* at 70% rh are about 15° and 36° (33). The lower humidity limit lies between 30% and 50%, except above 35° and below 22.5° where it rises to between 50% and 70%. The optimum conditions for survival and speed of development are around 30° and 70% rh. Under these conditions on a diet of wheatfeed, yeast, and glucose, development from egg hatch to adult emergence takes a mean of 36 days, increasing to 122 days at 17.5°. Only 10% of eggs hatch at 15°; none hatch below this temperature. Larvae die within a few days after hatching at 15° and 37.5°. Development is completed in 34 days on ground carobs and in 56 days on almonds at 30° and 70% rh (35).

Ephestia kuehniella, Mediterranean flour moth

The lower limit for complete development is about 12°, although a few strains can develop from egg to adult at 10° (56). Development can be completed at 28° but not at 31°. Males become sterile when reared at 30° (79). Rearing in continuous light also induces sterility in males (84), as it also does in *Plodia interpunctella*.

The effect may be partially reversed if males emerging from pupae reared in continuous light are transferred within the first few days to a 12-hr-light/12-hr-dark cycle (83).

Generally, development is quicker and survival is better at 70% rh than at lower humidities. However, at 20° and 25°, survival is good (80–90%) even at 15% rh, and at 25°, about 35% survival to the adult stage has been recorded at humidities approaching zero. Development at such low humidities has been attributed to the ability of larvae to utilize their own metabolic water (41). Development is fastest at 25° and 75% rh, taking about 10 weeks on white flour (56). Whole-meal flour is an optimal diet for larvae of E. kuehniella (41).

Oviposition occurs at 7.5° but not at 5°. Eggs hatch at temperatures between 10° and 31°, regardless of the ambient humidity. When laid at 25°, eggs require about 16 days to hatch at 15°, 8 days at 20°, 5 days at 25°, and 4 days at 30°. At all temperatures, the pupal stage lasts rather more than twice as long as the egg stage.

Larvae of the Mediterranean flour moth have been classed as cold-hardy; they can survive temperatures down to −10° for short periods (3–4 days) (94). Eggs have survived exposure at −9 to −10° for 5 days but not for 7 days. All developmental stages are killed by 24 hours at −18° (67). In a laboratory strain, some (up to 22%) eggs of various age groups laid at 25° developed into adults at 25° after a 14-day exposure at 10° but none did so after 21 days of exposure (9). No newly laid to one-day-old eggs survived 14 days' exposure at 7.5° but 4% of the older eggs did survive. In one strain acquired from a mill, survivals of 16–27% were recorded from eggs aged 0–1, 1–2, and 2–3 days at the start of a 14-day exposure at 7.5°; and of 50–70% after a 21-day exposure at 10°. Eggs over one day old showed a small survival rate (up to 3%) after 28 days at 10° or 21 days at 7.5°. Generally, eggs held at 25° become more tolerant of cold when aged over one day. Newly hatched larvae become less resistant to cold after they start to feed (87).

During a population study of E. kuehniella in a large unheated grain store in Scotland, it was observed that about 30% of last-instar larvae survived the winter. Most of the larvae that entered the prepupal stage before temperatures rose in spring failed to pupate. Adults from larvae that pupated later began to emerge in June (29). Some fully fed, last-instar larvae entered diapause during the winter. In laboratory tests, nearly 50% of the larvae entered diapause when reared in continuous darkness, and up to 30% did so in short photoperiods (37; see also section on Diapause, below). Eighty percent of diapausing larvae survived an exposure of 4 weeks at −2.5°, whereas only 10% of nondiapausing larvae did so (36).

Paralipsa gularis, stored-nut moth

A native of Southeast Asia and still commonly seen infesting tamarind fruit on trees in southern India, this pest has been imported into Europe and North America on a range of commodities including nuts, peanuts, cocoa, currants, dried plums, soybeans, and some cereals. Development is completed at temperatures in the range of 15–33°, requiring 12–15 weeks at 24° and about 6 months at 18–20° (91). Females lay 150–250 eggs which require 4–5 days to hatch at 30°. There are normally one or two generations per year. Mature larvae that enter diapause in response to temperatures below 22° survive the winter or cool season. Most diapausing larvae reared at 31° pupate after 3 or 4 months' exposure at 15–18°, but only 25% do so even after 14 months at 25° (92). Eggs do not survive exposure to 0°; 100% mortality of eggs occurs after 25 days at 2°, and after 30 days at 5° (62).

Plodia interpunctella, Indianmeal moth

A cosmopolitan feeder on dried fruits, cereals, and many other products, such as nuts and oilseeds, P. interpunctella can develop very rapidly under favorable conditions. Development proceeds at temperatures in the range of 18–35°, and can be completed in about 60 days at 20°, 30 days at 25°, and 25 days at 30° (9, 99). The sex ratio is approximately 1:1 but the emergence of males and females of a particular generation can be displaced by food shortage, the females tending to emerge later (78).

The fertility of males can be lowered by rearing them in continuous light (65) or at temperatures above 33° (63). On average, the female lays 150–200 eggs during the first few days after mating (32, 64, 100). Eggs do not hatch at 15° and require 7–8 days for development at 20°, 4–5 days at 25°, and 3–4 days at 30° (9, 100). Young larvae can survive at temperatures down to 10° (96). The number of larval instars varies from five to seven (100). The pupal stage lasts 15–20 days at 20°, 8–11 days at 25°, and 7–8 days at 30° (9).

Diapause induced in response to short photoperiods (10), low temperature, or high population pressure (99) may greatly extend the developmental periods. Diapause provides a means for the species to overwinter at higher latitudes in unheated situations. The extent to which different strains diapause varies greatly, those from the tropics or long reared in laboratories showing a reduced capacity (*16*; see also section on Diapause, below). Larvae entering diapause spin a dense hibernaculum. At the onset of pupation, the larva chews a hole in the silk before spinning a flimsy pupal cocoon. At the limits of its range, *P. interpunctella* may have only one or two generations per year, but as many as eight generations per year may occur in warmer climates (*100*).

Pyralis farinalis, meal moth

P. farinalis has a wide distribution in temperate regions of Europe and North America, occurring on farm-stored grain, in poultry houses, mills, domestic premises, and cellars. It occurs mostly in the damper situations of infested premises where some mold growth is evident. Like other storage moths, it attracts attention to itself by matting together grains on the surface of a bulk. It prefers hard wheat to soft, infests coarse brans and meals rather than flour, and also occurs on sesame cake, peanuts, straw, vegetable refuse, and moldy bread crumbs in houses. A single female can produce 200–500 eggs (*32*). In summer the life cycle can be completed within 2 months and a larval diapause enables the species to overwinter.

Tineidae

Nemapogon granella, European grain moth

Comparatively little is known about the bionomics of this species. It has been recorded in Europe, North and South America, Japan, and parts of Africa. It has been classified as moderately hardy or hardy since a few larvae survived an English winter with temperatures at or below 0° on 23 days (*94*). It is often associated with moldy and damp grain residues. It is believed to have a facultative diapause, and the number of generations per year varies from one to four depending mainly on the climate.

Niditinea fuscipunctella, poultry house moth

This species has not been studied in much detail in the laboratory so far. It is commonly found in poultry houses (*1*) and birds' nests. In southern California, larvae showed annual spring and early summer peaks followed by smaller mid- and late-summer surges (*61*).

Tineola bisselliella, webbing clothes moth

This moth is distributed throughout the temperate regions of the world. Mean development from egg to adult on impregnated flannel is completed in 63 days at 23.5° and 50% rh, and 55 days at 31° and 66% rh (*76*). At 25°, the optimum relative humidity for speed of development and survival is 75%, although most larvae (about 90%) can complete development down to 30% rh (*49*).

This species is considered to be hardy and has survived a winter during which there were 37 days with temperatures at or below 0° (*94*). Larvae have been reported to survive 67 days at −6.7 to −3.9° and 21 days at −15 to −12.2° (*8*). Development and reproduction can occur at 33°, and eggs hatch at 35° (*80*). However, adult males become sterile in 2 days at 35°, and all stages are killed within 4 hours at 41°. Eggs hatch in 37 days at 13° and in 7 days at 33° (*48, 76*).

Diapause

A character common among stored-product moths is a cold-tolerant, photoperiodically induced diapause. Species shown to possess a diapause so far include *H. pseudospretella*, *N. granella*, and *P. gularis* (all discussed briefly under Distribution and Life Histories, above), but most work has been done on the diapause of species in the genera *Ephestia* and *Plodia*. These phycitine pests are essentially long-day species, diapausing as last-instar larvae. Within this group, the tendency to diapause ranges from very low to very high. Only one of the two temperate-zone species mentioned in Table 12.2, *E. elutella*, regularly relies on diapause to overwinter. However, during laboratory studies on the other species (*E. kuehniella*), some larvae in 7 out of 10 English stocks reared in isolation entered diapause as did larvae in stocks from Scotland, Finland, Japan, and Egypt (*39*).

In *E. kuehniella*, diapause is influenced by nutrition as well as by temperature and photoperiod. For example, in a strain from Scotland, the incidence of diapause was about 50% on diets of whole maize flour or wheat flour and about 12% on an artificial laboratory diet containing glycerol, yeast, and wheatfeed (*39*). Nearly 50% of the larvae of this strain entered diapause when reared at 25° in continuous darkness, and up to 30% did so in short photoperiods (*34*). Diapause lasted 2–3 months in most photoperiods at 20° and 25°, but in continuous darkness at 25° it lasted about 7 months. Termination of diapause was hastened (by about 3 months) by chilling at 7.5° for 6 weeks.

Recent work has indicated that diapause in *E. kuehniella* may be linked with the stimuli controlling the production of ecdysteroids (*46*). Larval interaction in the presence of light provided a strong stimulus for larvae to spin the outer envelope of the pupal cocoon which in turn led to the production of ecdysteroids and pupal ecdysis. Constant darkness at 26–27° and isolation of larvae in a limited space inhibited production of the cocoon and extended the wandering stage by several days and, in a few individuals, by up to 2 weeks.

In *E. elutella*, diapause usually lasts 6–8 months (*11*) and is stimulated by response of the newly molted, last-instar larvae to photoperiods of 14 hours or less (*10*) or to temperatures of 20° and below. The sensitivity of larvae to light at this time is high, intensities down to 0.1 lux eliciting a response (*12*). The termination of diapause is best synchronized when an extended period at moderately low temperatures (5–10°) in winter is followed by a warm (>20°) spring (*14*).

The warm-temperate species, *E. figulilella* and *E. calidella*, diapause in photoperiods of 13 hours or less (*15*) but rearing at moderately low temperatures (20°) may be less effective in initiating diapause than in *E. elutella*, most larvae continuing morphogenesis in long photoperiods. Although at constant temperatures some individuals remain in diapause for as long as one year, in nature most larvae pupate after 3–5 months in diapause.

Most strains of the cosmopolitan *Plodia interpunctella* possess a diapause broadly similar to that of *E. figulilella* and *E. calidella*, with a critical photoperiod ranging from 12.5 to 13.5 hours (*16*), but the intensity of diapause in different strains varies considerably. Thus, in some, diapause may be induced only at about 20° and may last no longer than 2–4 months, whereas in others it can be induced at temperatures above 25° and can last for up to 9 months. Studies on the photoperiodic clock controlling diapause in *P. interpunctella* have indicated that the response is triggered by an hourglass-type measurement of the duration of the scotophase (*57*).

In the predominantly tropical/subtropical *E. cautella*, the most important moth pest of stored products, diapause occurred in some individuals from 19 of 20 samples collected from different parts of the world (*17*). More larvae entered diapause in continuous darkness than in short photoperiods or continuous light (*15, 17*). Rearing at 25° rather than 20° prevented diapause in all but one strain. Generally, diapause was quite short, rarely lasting longer than 4 months even under constant conditions. Exposure to temperatures just below the developmental threshold broke the diapause within 30 days.

Diapause renders insects harder to control by fumigants and other insecticides (*13, 40*). In the warehouse, heating and artificial lighting tend to reduce the incidence of diapause but at the same time cause an increase in the growth rates of pest populations. Continuous lighting is not effective in reducing diapause but if lighting in heated premises is controlled to give a summer day-length, diapause can be reduced. However, since diapause is also induced by low temperatures, such a lighting system may not prevent the diapause of pests in the structural framework of the buildings.

Pest populations have the potential to adjust quickly to prevailing warehouse conditions; therefore, the incidence of diapause in a population can be rapidly selected for or against. Thus, manipulation of warehouse environmental conditions can support control measures only in the short term.

Larval and Adult Behavior

Moth larvae are strongly negatively phototactic until they are about to pupate. They are also sensitive to crowding, a factor that can delay development and result in smaller larvae and adults. Larvae of some phycitines secrete kairomones from their mandibular glands (*71*). The quantity of these chemicals increases as larval density increases. Presence of too much or too little of these secretions causes increased wandering by the larvae and also affects adult oviposition (*31*).

The application of juvenile hormones to some species, such as *Ephestia kuehniella*, has produced high mortality when given to migratory larvae and prepupae (*53*). However, when applied early in the last instar, larval appetite increases, resulting in superlarvae and delayed pupation.

Adults are mainly nocturnal, emerging in the late afternoon and evening, apparently in response to changes in light intensity and temperature; they mate and lay eggs after the subsequent dusk. *E. cautella* has a peak flight activity in late evening and a second smaller peak just before dawn (*47*). Sleep behavior of *E. kuehniella* has been described, deepest sleep being characterized by the antennae crossed over the back and the abdomen covered by the wings (*3*). The attractiveness of light of different wavelengths has been studied in some detail. For example, more adults of *E. kuehniella* are attracted to green electroluminescent lamps than to ultraviolet ones (*93*).

All the major kinds of storage moths have been shown to produce sex pheromones. The chief component of the female pheromone of several phycitine pest species has been identified as *cis*-9,*trans*-12-tetradecadien-1-yl acetate (*20, 59*). Within 24 hours after emergence, the female becomes stationary on a suitable surface and assumes the calling posture in which the abdomen is lifted between the wings and the plumes of the scent glands are extruded. Males locate the females probably by flying against air currents while working along the interface between the presence or absence of pheromone (*66*). Sight is probably important only in the last stage of the approach.

Recently, sex pheromones have been extracted from the wing glands of male *E. elutella* and *Plodia interpunctella* (*58, 69*). The characteristic rapid wing beating of males prior to coupling no doubt facilitates the dispersal of these pheromones which appear to be important in the final stage of courtship. The pheromones released from male *P. interpunctella* elicit a turning response in receptive females (*69*). Although most work has been done on the pheromones of phycitines, in recent years certain pheromones from *Corcyra cephalonica* and *Sitotroga cerealella* have also been isolated (*90, 101*).

Synthetic pheromone traps are used for detecting infestations and estimating populations (*28, 81*). Attempts have been made to use such traps to lure insects to a sterilant, pathogen source, or contact insecticide; some success along these lines has been achieved (*24*). Synthetic pheromones have been used in laboratory tests to disrupt mating in *P. interpunctella* and *E. cautella* (*95*).

Predators, Parasites, and Pathogens

Mites such as *Cheyletus eruditus* and *Blattisocius tarsalis* prey upon the eggs and early instars of storage moths (*82, 104, 105*). Larger larvae and adults may be parasitized by *Pyemotes* species. Some spiders such as the European argiopid, *Nuctenea umbraticus*, often trap moths in their loosely constructed and very sticky webs, but the readily detached wing scales allow most moths to escape from webs of other spiders. Other predators include bugs such as *Lyctocoris campestris*, ants, and even other storage

pests such as the beetles, *Tribolium confusum* and *Tenebroides mauritanicus*, in mixed infestations. The most important parasites are members of the Hymenoptera, especially the braconid, ichneumonid, and chalcid wasps which often heavily parasitize larval populations (see also Chapter 15) and may also act as vectors of pathogenic microorganisms (*18, 52, 78*).

As in many other Lepidoptera, the larval gut of pyralids operates at a relatively high pH and so is susceptible to attack by spore-forming bacteria, particularly *Bacillus thuringiensis*. Highly virulent strains of this organism, available as commercial formulations, have been used for control with varying degrees of success, the main problem in bulk storage being effective distribution (*23*). The gelechiid, *Sitotroga cerealella*, is less susceptible to *B. thuringiensis* because the larvae spend most of their life protected inside kernels of grain (*68*).

Other pathogenic organisms include polyhedrosis and granulosis viruses, fungi of the genera *Entomophthora* and *Beauveria*, and protozoa such as the microsporidan, *Nosema plodiae*, and the schizogregarine, *Mattesia dispora*.

Pest Status

The moth species described above have adapted with great success to the storage environment, making full use of the protected habitat with its plentiful food supplies. *Ephestia cautella* is surpassed in importance as a postharvest loss source only by *Tribolium castaneum* (*45*). Although seldom so great a source of loss as *E. cautella*, *Plodia interpunctella* has the widest distribution of all moths generally infesting stored foods and is truly a global pest. In tropical and subtropical regions of the Old World, the worst moth pest in mills is *Corcyra cephalonica*. In temperate zones, the principal pest of flour mills is *E. kuehniella*. Whenever there is a policy to store grain for several seasons, such as today in Berlin or in Great Britain during the 1940s, the most successful moth pest is *E. elutella*, also a major pest of tobacco. The most serious moth pest of grain in the New World and Africa is *Sitotroga cerealella*.

The more important storage moths, especially *E. cautella*, can attack an extremely wide range of foodstuffs and related commodities. This potential for adaptation is a major factor in their remarkable success as storage pests.

Cited References

1 Agricultural Development and Advisory Service. 1980. Insects in Poultry Houses. Leaflet

537. Ministry of Agriculture, Fisheries and Foods, Alnwick, United Kingdom.

2 Andersen, F. S. 1956. Effects of crowding in *Endrosis sarcitrella*. *Oikos* 7(11)215–226.

3 Andersen, F. S. 1968. Sleep in moths and its dependence on the frequency of stimulation in *Anagasta kuehniella*. *Opuscula Entomologica* 33(1–2)15–24.

4 Andersen, F. S. 1972. Biology of grain pests. (a) Effect of "batch number" on sex ratio and number of eggs in *Endrosis*. (b) Natural occurrence of *Bacillus thuringiensis* in a Danish strain of *Ephestia kuehniella*. *Årsbereting Statens Skadedyrlaboratorium* 1971:50.

5 Ayertey, J. N. 1979. An unusual mortality of *Sitophilus zeamais* caused by *Sitotroga cerealella* in mixed laboratory cultures. *Proceedings of the Second International Working Conference on Stored Product Entomology* (Ibadan, Nigeria, 1978), pp. 328–337.

6 Ayertey, J. N. 1979. The growth of single and mixed laboratory populations of *Sitophilus zeamais* and *Sitotroga cerealella* on stored maize. *Researches on Population Ecology* 21(1)1–11.

7 Ayyar, P. N. 1934. A very destructive pest of stored products in South India. *Bulletin of Entomological Research* 25(2)155–169.

8 Back, E. A., and R. T. Cotton. 1926. Insect control in upholstered furniture. *Furniture Warehouseman* 6(5) (original pagination unknown; reprint consists of 7 pp.).

9 Bell, C. H. 1975. Effects of temperature and humidity on development of four pyralid moth pests of stored products. *Journal of Stored Products Research* 11(3)167–175.

10 Bell, C. H. 1976. Factors governing the induction of diapause in *Ephestia elutella* and *Plodia interpunctella*. *Physiological Entomology* 1(2)83–91.

11 Bell, C. H. 1976. Factors influencing the duration and termination of diapause in the warehouse moth, *Ephestia elutella*. *Physiological Entomology* 1(3)169–178.

12 Bell, C. H. 1977. The sensitivity of larval *Plodia interpunctella* and *Ephestia elutella* to light during the photoperiodic induction of diapause. *Physiological Entomology* 2(3)167–172.

13 Bell, C. H. 1977. Toxicity of phosphine to the diapausing stages of *Ephestia elutella*, *Plodia interpunctella* and other Lepidoptera. *Journal of Stored Products Research* 13(4)149–158.

14 Bell, C. H. 1983. The regulation of development during diapause in *Ephestia elutella* (Hübner) by temperature and photoperiod. *Journal of Insect Physiology* 29(6)485–490.

15 Bell, C. H., and C. R. Bowley. 1980. Effect of photoperiod and temperature on diapause in a Florida strain of the tropical warehouse moth, *Ephestia cautella*. *Journal of Insect Physiology* 26(8)533–538.

16 Bell, C. H., et al. 1979. Diapause in twenty-three populations of *Plodia interpunctella* (Hübner) (Lep., Pyralidae) from different parts of the world. *Ecological Entomology* 4(3)193–197.

17 Bell, C. H., et al. 1983. Diapause in twenty populations of *Ephestia cautella* (Walker) (Lepidoptera: Pyralidae) from different parts of the world. *Journal of Stored Products Research* 19(3)117–123.

18 Benson, J. F. 1973. The biology of the Lepidoptera infesting stored products, with special reference to population dynamics. *Biological Review* 48(1)1–26.

19 Boldt, P. E. 1974. Effects of temperature and humidity on development and oviposition of *Sitotroga cerealella*. *Journal of the Kansas Entomological Society* 41(1)30–35.

20 Brady, U. E., and D. A. Nordlund. 1971. *cis*-9,*trans*-12 Tetradecadien-1-yl acetate in the female tobacco moth *Ephestia elutella* (Hübner) and evidence for an additional component of the sex pheromone. *Life Sciences* 10(pt. 2)(14)797–801.

21 Burges, H. D. 1956. Some effects of the British climate and constant temperature on the life-cycle of *Ephestia cautella*. *Bulletin of Entomological Research* 46(4)813–835.

22 Burges, H. D., and K. P. F. Haskins. 1965. Life-cycle of the tropical warehouse moth, *Cadra cautella*, at controlled temperatures and humidities. *Bulletin of Entomological Research* 55(4)775–789.

23 Burges, H. D., and J. A. Hurst. 1977. Ecology of *Bacillus thuringiensis* in storage moths. *Journal of Invertebrate Pathology* 30(3)131–139.

24 Burkholder, W. E. 1979. Application of pheromones and behaviour-modifying techniques in detection and control of stored product insects. *Proceedings of the Second International Working Conference on Stored-Product Entomology* (Ibadan, Nigeria, 1978), pp. 56–65.

25 Carmona, M. M. 1958. A Entomofauna dos Produtos Armazenados. *Corcyra cephalonica*. Estudos, Ensaios e Documentos 55. Junta de Investigação do Ultramar, Ministerio do Ultramar, Lisbon.

26 Castro, J. de J. 1951. Ecological studies on the Angoumois grain moth, *Sitotroga cerealella*, and the rice weevil, *Sitophilus oryzae*. Thesis, Iowa State College, Ames.

27 Chippendale, G. M., and R. A. Mann. 1972. Feeding behavior of Angoumois grain moth larvae. *Journal of Insect Physiology* 18(1)87–94.

28 Cogan, P. M., and D. Hartley. 1984. The effective monitoring of stored product moths using a funnel pheromone trap. *Proceedings of the*

Third International Working Conference on Stored-Product Entomology (Manhattan, KS, 1983), pp. 631–639.

29 Cole, D. B., and P. D. Cox. 1981. Studies on three moth species in a Scottish port silo, with special reference to overwintering *Ephestia kuehniella*. *Journal of Stored Products Research* 17(4)163–181.

30 Cole, J. H. 1962. *Hofmannophila pseudospretella* (Stnt.) (Lep., Oecophoridae), its status as a pest of woolen textiles, its laboratory culture and susceptibility to mothproofers. *Bulletin of Entomological Research* 53(1)83–89.

31 Corbet, S. A. 1973. Oviposition pheromone in larval mandibular glands of *Ephestia kuehniella*. *Nature* 243(5043)537–538.

32 Cotton, R. T. 1956. *Pests of Stored Grain and Grain Products*. Burgess, Minneapolis, MN.

33 Cox, P. D. 1974. The influence of temperature and humidity on the life-cycles of *Ephestia figulilella* Gregson and *Ephestia calidella* (Guenée) (Lepidoptera: Phycitidae). *Journal of Stored Products Research* 10(1)43–55.

34 Cox, P. D. 1975. The influence of photoperiod on the life-cycles of *Ephestia calidella* (Guenée) and *Ephestia figulilella* Gregson (Lepidoptera: Phycitidae). *Journal of Stored Products Research* 11(2)75–85.

35 Cox, P. D. 1975. The suitability of dried fruits, almonds, and carobs for the development of *Ephestia figulilella*, *E. calidella* and *E. cautella*. *Journal of Stored Products Research* 11(3–4)229–233.

36 Cox, P. D. 1987. Cold tolerance and factors affecting the duration of diapause in *Ephestia kuehniella* Zeller (Lepidoptera: Pryalidae). *Journal of Stored Products Research* 23(3)163–168.

37 Cox, P. D., et al. 1981. Diapause in a Glasgow strain of the flour moth, *Ephestia kuehniella*. *Physiological Entomology* 6(4)349–356.

38 Cox, P. D., et al. 1981. The influence of temperature and humidity on the life-cycle of *Corcyra cephalonica*. *Bulletin of Entomological Research* 71(2)171–181.

39 Cox, P. D., et al. 1984. The incidence of diapause in seventeen populations of the flour moth, *Ephestia kuehniella* Zeller (Lepidoptera: Pyralidae). *Journal of Stored Products Research* 20(3)139–143.

40 Cox, P. D., et al. 1984. The effect of diapause on the tolerance of larvae of *Ephestia kuehniella* to methyl bromide and phosphine. *Journal of Stored Products Research* 20(4)215–219.

41 Fraenkel, G., and M. Blewett. 1944. The utilization of metabolic water in insects. *Bulletin of Entomological Research* 35(1)127–139.

42 Fraenkel, G., and M. Blewett. 1946. The die-tetics of the caterpillars of three *Ephestia* species, *E. kühniella*, *E. elutella* and *E. cautella*, and of a closely related species, *Plodia interpunctella*. *Journal of Experimental Biology* 22(3–4)162–171.

43 Freeman, J. A. 1950. Methods of spread of stored products insects and origin of infestation in stored products. *Proceedings of the Eighth International Congress of Entomology* (Stockholm, Sweden, 1948), pp. 815–825.

44 Freeman, J. A. 1962. The influence of climate on insect populations of flour mills. *Proceedings of the Eleventh International Congress of Entomology* (Vienna, Austria, 1960) 2:301–308.

45 Freeman, J. A. 1977. Prediction of new storage pest problems. In *Origins of Pest, Parasite, Disease and Weed Problems*, J. M. Cherret and G. R. Sagar, eds. Blackwell, Oxford, England.

46 Giebultowicz, J. M., B. Cymborowski, and J. P. Delbecque. 1984. Environmental control of larval behaviour and its consequences for ecdysteroid content and pupation in *Ephestia kuehniella*. *Physiological Entomology* 9(4)409–416.

47 Graham, W. M. 1970. Warehouse ecology studies of bagged maize in Kenya, I–IV. *Journal of Stored Products Research* 6(2)147–180.

48 Griswold, G. H. 1944. Studies on the Biology of the Webbing Clothes Moth (*Tineola bisselliella*). Memoir 262. Agricultural Experiment Station, Cornell University, Ithaca, NY.

49 Griswold, G. H., and M. F. Crowell. 1936. The effect of humidity on the development of the webbing clothes moth (*Tineola bisselliella*). *Ecology* 17(2)241–250.

50 Hagstrum, D. W., and J. E. Sharp. 1975. Population studies on *Cadra cautella* in a citrus pulp warehouse with particular reference to diapause. *Journal of Economic Entomology* 68(1)11–14.

51 Hammad, S. M., M. G. Shenouda, and A. L. El-Deeb. 1967. Studies on the biology of *Sitotroga cerealella* Oliv. *Bulletin de la Société Entomologique d'Egypte* 51:257–268.

52 Hassell, M. P., J. H. Lawton, and R. M. May. 1976. Patterns of dynamical behaviour in single-species populations. II. Predator rate of increase. *Journal of Animal Ecology* 45(2)471–486.

53 Hong, T. K. 1975. Effects of a synthetic juvenile hormone and some analogues on *Ephestia* spp. *Annals of Applied Biology* 80(2)137–145.

54 Howe, R. W. 1962. The influence of diapause on the status as pests of insects found in houses and warehouses. *Annals of Applied Biology* 50(3)611–614.

55 Howe, R. W. 1965. A summary of estimates of optimal and minimal conditions for population increase of some stored products insects. *Journal of Stored Products Research* 1(2)177–184.

56 Jacob, T. A., and P. D. Cox. 1976. The influence of temperature and humidity on the life-cycle of *Ephestia kuehniella* Zeller (Lepidoptera: Pyralidae). *Journal of Stored Products Research* 13(3)107–118.

57 Kikukawa, S., and S. Masaki. 1984. Interacting effects of photophase and scotophase on the diapause response of the Indian meal moth, *Plodia interpunctella*. *Journal of Insect Physiology* 30(12)919–925.

58 Krasnoff, S. B., and K. W. Vick. 1984. Male wing-gland pheromone of *Ephestia elutella*. *Journal of Chemical Ecology* 10(4)667–679.

59 Kuwahara, Y., et al. 1971. Sex pheromone of the almond moth and the Indian meal moth: *cis*-9,*trans*-12-tetradecadienyl acetate. *Science* 171(3973)801–802.

60 Le Cato, L. G. 1976. Yield, development and weight of *Cadra cautelia* and *Plodia interpunctella* on twenty-one diets derived from natural products. *Journal of Stored Products Research* 12(1)43–47.

61 Legner, E. F., and R. E. Eastwood. 1973. Seasonal and spatial distribution of *Tinea fuscipunctella* on poultry ranches. *Journal of Economic Entomology* 66(3)685–687.

62 Le Torc'h, J. M. 1977. Cold storage as a means of protection against stored products insects: Laboratory tests with insects of dried French plums. *Revue de Zoologie Agricole et de Pathologie Végétale* 76(4)109–117.

63 Lum, P. T. M. 1977. High temperature inhibition of development of eupyrene sperm and of reproduction in *Plodia interpunctella* and *Ephestia cautella*. *Journal of the Georgia Entomological Society* 12(3)199–203.

64 Lum, P. T. M., and B. R. Flaherty. 1969. Effect of mating with males reared in continuous light or in light-dark cycles on fecundity in *Plodia interpunctella*. *Journal of Stored Products Research* 5(2)80–94.

65 Lum, P. T. B., and B. R. Flaherty. 1970. Effects of continuous light on the potency of *Plodia interpunctella* males. *Annals of the Entomological Society of America* 63(5)1470–1471.

66 Marsh, D., J. S. Kennedy, and A. R. Ludlow. 1978. An analysis of anemotactic zigzagging flight in male moths stimulated by pheromone. *Physiological Entomology* 3(3)221–240.

67 Mathlein, R. 1961. Studies on some major storage pests in Sweden, with special reference to their cold resistance. *Statens Växtskyddsanstalt Meddelanden* 12(83)1–49.

68 McGaughey, W. H., and R. A. Kissinger. 1978. Susceptibility of Angoumois grain moths to *Bacillus thuringiensis*. *Journal of Economic Entomology* 71(3)435–436.

69 McLaughlin, J. R. 1982. Behavioral effect of a sex pheromone extracted from forewings of male *Plodia interpunctella*. *Environmental Entomology* 11(2)378–380.

70 Misra, C. P., C. M. Christensen, and A. C. Hodson. 1961. The Angoumois grain moth, *Sitotroga cerealella*, and storage fungi. *Journal of Economic Entomology* 54(5)1032–1033.

71 Mudd, A. 1983. Further novel 2-acylcyclohexane-1,3-diones from lepidopteran larvae. *Journal of the Chemical Society, Perkin Transactions* 1(9)2161–2164.

72 Muhihu, S. K. 1984. Depth of infestation by *Sitotroga cerealella* into grain layers of wheat, maize and sorghum. *Tropical Stored Products Information* 47:34–38.

73 Mullen, M. A., and R. T. Arbogast. 1979. Time-temperature-mortality relationships for various stored-product insect eggs and chilling times for selected commodities. *Journal of Economic Entomology* 72(4)476–478.

74 Nawrot, J. 1979. Effect of temperature and relative humidity on population parameters for almond moth (*Cadra cautella*). *Prace Naukowe Instytutu Ochrony Roślin* 21(2)41–52.

75 Nawrot, J. 1979. Population parameters for almond moth (*Cadra cautella*) reared on natural products. *Prace Naukowe Instytutu Ochrony Roślin* 21(2)53–60.

76 Pereira, M. 1960. Contribuição para o Estudo da *Tineola bisselliella* e seu Combate. Estudos, Ensaios e Documentos 78. Junta de Investigação do Ultramar, Ministerio do Ultramar, Lisbon, Spain.

77 Pingale, S. V., and G. K. Girish. 1967. Role of density on the multiplication of stored grain insect pests. *Bulletin of Grain Technology* 5(1)12–20.

78 Podoler, H. 1974. Analysis of life-tables for a host and parasite (*Plodia-Nemeritis*) ecosystem. *Journal of Animal Ecology* 43(3)653–670.

79 Raichoudhury, D. P., and S. E. Jacobs. 1937. Experiments on the sterility of *Ephestia kuehniella* in relation to high temperature (30°C). *Proceedings of the Zoological Society of London* A107(3)283–288.

80 Rawle, S. G. 1951. The effects of high temperature on the common clothes moth, *Tineola bisselliella*. *Bulletin of Entomological Research* 42(1)21–40.

81 Reichmuth, C., et al. 1978. Die Fängigkeit pheromonbeköderter Klebefallen für Speichermotten (*Ephestia elutella* Hbn.) in unterschiedlich dicht befallenen Getreidelägern. *Zeitschrift für Angewandte Entomologie* 86(2)205–212.

82 Richards, O. W., and W. S. Thomson. 1932. A contribution to the study of the genera *Ephestia*, Gn. (including *Strymax*, Dyar), and *Plodia*, Gn. (Lepidoptera, Phycitidae), with notes on parasites of the larvae. *Transactions of the En-*

tomological Society of London 80(2)169–250, 8 pl.

83 Richards, O. W., and N. Waloff. 1946. The study of a population of *Ephestia elutella* living on bulk grain. *Transactions of the Royal Entomological Society* 97(11)253–298.

84 Riemann, J. G., and R. L. Rudd. 1974. Mediterranean flour moth: Effects of continuous light on the reproductive capacity. *Annals of the Entomological Society of America* 67(6)857–860.

85 Riemann, J. G., M. Johnson, and B. Thorson. 1981. Recovery of fertility by Mediterranean flour moths transferred from continuous light to light:dark. *Annals of the Entomological Society of America* 74(3)274–278.

86 Russell, V. M., G. G. M. Schulten, and F. A. Roorda. 1980. Laboratory observations on the development of the rice moth, *Corcyra cephalonica,* on millet and sorghum at 28°C and different relative humidities. *Zeitschrift für Angewandte Entomologie* 89(5)488–498.

87 Salt, R. W. 1936. *Studies on the Freezing Process in Insects.* Technical Bulletin 116. Agricultural Experimental Station, University of Minnesota, [St. Paul].

88 Schmidt, H.-U. 1982. The depth of egg deposition of the Indian meal moth, *Plodia interpunctella* Hbn., in rye and maize. *Anzeiger für Schädlingskunde, Pflanzenschutz and Umweltschutz* 55(1)1–4 (in German, English abstract).

89 Simwat, G. S., and B. S. Chahal. 1980. Effect of storage period and depth of stored grains on the insect population and resultant loss of stored wheat with the farmers. *Bulletin of Grain Technology* 18(1)35–41.

90 Singh, D., H. S. Sidhu, and P. S. Kalsi. 1973. Isolierung und Chemie der Geschlechts-Pheromone der Reismotte *Corcyra cephalonica* (Stainton), Galleriidae: Lepidoptera. *Reichstoffe, Aromen, Körperpflegemittel* 23(7)213–216 (English summary).

91 Smith, K. G. 1956. The occurrence and distribution of *Aphomia gularis* (Zell.) (Lep., Galleriidae), a pest of stored products. *Bulletin of Entomological Research* 47(4)655–667, 1 pl.

92 Smith, K. G. 1965. Some aspects of the biology of *Paralipsa (Aphomia) gularis* (Zell.) (Lepidoptera) in relation to its distribution. *Proceedings of the Twelfth International Congress of Entomology* (London, 1964), p. 626.

93 Soderstrom, E. L. 1970. Effectiveness of green electroluminescent lamps for attracting stored-product insects. *Journal of Economic Entomology* 63(3)726–731.

94 Solomon, M. E., and B. E. Adamson. 1955. The powers of survival of storage and domestic pests under winter conditions in Britain. *Bulletin of Entomological Research* 46(2)311–355, 2 pl.

95 Sower, L. L., and G. P. Whitmer. 1977. Population growth and mating success of Indian meal moths and almond moths in the presence of synthetic sex pheromone. *Environmental Entomology* 6(1)17–20.

96 Stratil, H. H., and C. Reichmuth. 1984. Development and longevity of young larvae of the stored product moths *Ephestia cautella, E. elutella,* and *Plodia interpunctella* at low temperatures. *Anzeiger für Schädlingskunde, Pflanzenschutz und Umweltschutz* 57(2)30–33.

97 Stroyan, H. L. G. 1946. Biological notes on Indian stored products insects. *Entomologist* 997(79)135–139.

98 Strümpel, J. 1969. Entwicklungszyklen einiger an Rohkakao schädlichen Insekten. *Anzeiger für Schädlingskunde und Pflanzenschutz* 42(11)161–165 (English summary).

99 Tsuji, H. 1963. Experimental studies on the larval diapause of the Indian meal moth *Plodia interpunctella.* Thesis, Kyushu University, Fukuoka, Japan.

100 Tzanakakis, M. E. 1959. An ecological study of the Indian meal moth, *Plodia interpunctella,* with emphasis on diapause. *Hilgardia* 29(5)205–246.

101 Vick, K. W., J. A. Coffelt, and M. A. Sullivan. 1978. Disruption of pheromone communication in the Angoumois grain moth with synthetic female sex pheromone. *Environmental Entomology* 7(4)528–531.

102 Waloff, N., and O. W. Richards. 1946. Observations on the behaviour of *Ephestia elutella* living on bulk grain. *Transactions of the Royal Entomological Society* 97(12)299–355.

103 Waloff, N., M. J. Morris, and E. C. Broadhead. 1948. Fecundity and longevity of *Ephestia elutella. Transactions of the Royal Entomological Society* 99(6)245–268.

104 Woodroffe, G. E. 1951. A life-history study of the brown house moth, *Hofmannophila pseudospretella. Bulletin of Entomological Research* 41(3)529–533.

105 Woodroffe, G. E. 1951. A life-history study of *Endrosis lactella. Bulletin of Entomological Research* 41(4)749–760.

13
Synanthropic Flies: Diptera

Mir S. Mulla and Lal S. Mian

Many species of flies are found in human biocoenoses such as households, food-processing plants, and slaughterhouses, and in association with domestic animals. The degree of association of flies with humans and animals ranges from a complete dependence on human-controlled environments (eusynanthropy) to partial dependence on man-made habitats (hemisynanthropy). In eusynanthropy, the requirements for larval development such as food and other biotic and abiotic needs are invariably met in man-made habitats. In hemisynanthropy, however, flies can exist independently of human habitats, although the influence of these habitats can be advantageous in the production of large populations. There is yet a third and much looser association (zoophily) in which certain flies come into contact with humans through domesticated animals in barns, stables, or pastures.

Synanthropic flies can become an actual or potential menace to human health by invading human dwellings and making contact with human food after breeding in or visiting and feeding on excrement, dead animal matter, or other contaminated materials. The role of flies can be significant in transmitting pathogenic organisms, or their mere presence can simply constitute a nuisance, interfering with human comfort. Because of the involvement of flies in the spread of disease organisms and their economic importance in urban and rural areas, a substantial literature exists on their taxonomy, biology, ecology, and medical importance. Sev-

eral reference sources provide comprehensive information on synanthropic flies (5, 7–9, 13, 21, 24, 29, 32, 33, 38). In this chapter, ecological aspects of various pest flies in the families Calliphoridae, Chloropidae, Drosophilidae, Muscidae, Piophilidae, Psychodidae, and Sarcophagidae are analyzed in relation to problems in the food industry and food service establishments. Doubtless many other species and families that are not treated here could also be considered as pests of food.

Synanthropy

Synanthropy, as an ecological phenomenon, refers to the coexistence of flies with humans, i.e., their occurrence in human habitations (anthropobiocoenoses) over some period of time. Ecologically speaking, the groupings of flies can be determined by such requirements as food, abiotic factors [temperature, relative humidity (rh)], and biotic factors (population density, reproductive potential, and competition). The food of adult synanthropic flies includes human and animal feces (coprophagy), decaying organic matter (saprophagy and necrophagy), fresh animal protein (carnivory), and nectar. The larvae of these flies thrive in and depend on for food almost the same substrates as the adults. Besides searching for food, the larvae and/or adults of certain synanthropic flies seek shelter in human establishments (e.g., farm buildings, barns, homes, restaurants).

Eusynanthropes

Flies in this category maintain a close association with anthropobiocoenoses where their

Mulla and Mian: Department of Entomology, University of California, Riverside, CA 92521.

195

entire development takes place (*31*). Hence, these insects are found in households, food-processing plants, slaughterhouses, sanitary landfills, garbage containers, and rendering plants. Some eusynanthropes are largely endophilous, i.e., restricted to human biocoenoses for both food and shelter, as are *Musca domestica* and certain *Drosophila* species.

Exophilous eusynanthropes are not strictly tied to anthropobiocoenoses for their abiotic and biotic needs, nor are they closely associated with humans trophically. This group includes *Calliphora vicina, Muscina stabulans, Phaenicia sericata,* and *Fannia canicularis.* Although this last species remains outdoors, it usually attains high population levels in anthropobiocoenotic milieux. *F. canicularis* has two population cycles, one synanthropic, the other independent of humans. Moreover, this species exhibits some seasonal endophilous behavior, especially during the hot summer months.

Hemisynanthropes

Basically, hemisynanthropic flies live independently of humans. Hemisynanthropy is apparent only during an occasional or temporary interference of humans with natural areas, such as the use of latrines associated with camping, fieldwork, and military camps. Numerous species could be considered hemisynanthropes: *Lucilia caesar, L. illustris, Calliphora vomitoria, Pyrellia cyanicolor, Ophyra leucostoma, Piophila vulgaris, Hylemya* spp., *Muscina pabulorum,* and *Mydaea urbana.*

It is difficult to draw a definite line between hemisynanthropes and exophilous eusynanthropes. A species that acts as a hemisynanthrope, such as *Calliphora vicina* in central Europe, may behave as an exophilous eusynanthrope in northern Europe. A related species, *C. vomitoria,* which is shade-loving and lives in forests at high elevations, can build large populations in the vicinity of mountain chalets, hotels, and health resorts, and can actively invade human dwellings. Synanthropy, therefore, appears to be a dynamic phenomenon in which hemisynanthropy may alter or coincide with exophilous eusynanthropy.

Zoophily

Zoophilous flies are associated with the human biocoenosis through the excreta of domestic animals, especially ruminants and poultry. Symbovine flies associated with ruminants may take the pasture form or the stable form. In the former case, symbovines, such as species of *Haematobia, Hydrotaea, Mesembrina, Morellia, Musca,* and *Orthelia,* build up large populations in cattle pastures without any immediate dependence on human biocoenoses. In the stable form, symbovine flies are associated with the feces of stabled animals directly in and around animal shelters. This group includes *Haematobia irritans, Stomoxys calcitrans,* and stable populations of *Musca domestica* and *Fannia* spp. Zoophiles that breed in chicken manure in and around poultry houses include species of several genera: *Fannia, Musca, Muscina, Ophyra,* and *Stomoxys.*

Ecology

Calliphoridae

The members of this family, generally called blow flies, bluebottles, and greenbottles, are found almost everywhere. They are metallic blue or green in color. They may become abundant around houses, slaughterhouses, meat-processing plants, and garbage dumps (*5*). They are strong fliers, often traveling great distances. For example, the black blow fly, *Phormia regina,* has been recovered 4–28 miles distant from the point of release (*34*). Blow flies are most active on warm sunny days; they are less active and generally rest on vegetation when the weather is cool and cloudy.

Certain blow flies such as *Phaenicia sericata* and *Eucalliphora lilaea* freely enter houses for feeding and oviposition. Blow flies that enter houses, food-processing plants, and industrial establishments can be more annoying than the common house fly (*Musca domestica,* discussed below) because their large size, buzzing sound, and indiscriminate flight throughout the building. Normally, blow flies breed in carrion, meat, and excrement; they may also oviposit on decaying vegetable matter.

The distribution and habits of blow flies are summarized in Table 13.1. They are multivoltine; the winter is passed as full-grown larvae in the soil. Four species of *Chrysomya,* formerly unknown in the New World, have recently become established in the Neotropics (*6*).

Calliphora erythrocephala, Large Blow Fly

The large blow fly or bluebottle fly is widely distributed and common in Europe and North America. This fly has a bluish-black thorax with a dark, metallic-blue abdomen. It is basically an outdoor insect, but frequently enters houses for oviposition and shelter. The female deposits her eggs on fresh, decaying, or cooked meat, and on dead insects. Bluebottles commonly alight on human feces wherever these are exposed. The larvae are necrophagous. Hewitt (15) reared the larvae on fresh rabbit meat at an average temperature of 23°. Under these conditions, the incubation period was 10–20 hours, the larval period, 7–8 days, and the pupal period, 14 days. The entire development was completed in 22–23 days.

Lucilia illustris, Sheep Maggot

The sheep maggot or greenbottle fly derives one of its common names from its habit of feeding on flesh on the backs of sheep. Generally, it is an outdoor insect, but under adverse weather conditions it seeks shelter in houses, especially in rural areas. About the size of *Musca domestica*, it is brilliantly colored a burnished gold, sometimes with a bluish or shiny green hue. The adults frequent dead animals and the excreta of humans and animals. The larvae feed on and develop in these substrates, although they have also been reported to feed on the wool and flesh of sheep (15, 21). Human feces also support larval development (18, 21).

Pollenia rudis, Cluster Fly

The cluster fly resembles the house fly but is larger, more robust, and slower in movement. Cluster flies derive their name from the fact that they enter houses in the fall for hibernation and remain in clusters in closets and unused rooms. They may concentrate under open ceilings or walls or may crawl behind window casings, moldings, loose wallpaper, pictures, or furniture. In the spring, cluster flies migrate from their hibernation places to the living areas where they readily become a nuisance. However, in contrast to other flies, cluster flies are not of direct public health importance. The female cluster fly oviposits in the soil. The larvae are parasitic in earthworms. Teneral adults must have a soft, moist soil through which they are able to burrow from their subterranean puparia to the soil surface. Rainfall seems to facilitate large-scale emergence of adults by softening the soil (32).

Chloropidae

Chloropids are small flies sometimes brightly colored with yellow or shining black. Most common in meadows and on pasture grasses, they occur in a wide variety of habitats. The larvae of many kinds of chloropids are phytophagous, feeding on grasses and cereals. Some species are scavengers and a few are parasitic or predatory.

Table 13.1
Distribution and habits of calliphorid flies

Species	Distribution[a]	Larval habitats	Synanthropy
Calliphora terraenovae	Nearctic	Carrion, decomposing organic matter	Hemisynanthrope
C. uralensis	Palaearctic	Moist excreta, decaying meat	Hemi- to eusynanthrope
C. vicina	Holarctic	Decomposing meat, feces	Eusynanthrope (endophilous)
C. vomitoria	Holarctic	Saprophagous, decaying matter	Hemi- to eusynanthrope (exophilous)
Chrysomya albiceps	Palaeotropic to Palaeosubtropic, Neotropic	Predatory on other synanthropes in carrion	Hemisynanthropes
C. chloropyga	Ethiopian, Neotropic	Carrion, sheep flesh	Hemisynanthrope
C. marginalis	Palaearctic, Palaeotropic	Carrion, meat, cow dung, feces	Hemisynanthrope to symbovine
C. megacephala	Palaearctic to Oriental, Neotropic	Carrion, decaying organic matter	Hemi- to eusynanthrope (exophilous)
C. rufifacies	Australasian to Oriental, Neotropic	Carrion, predatory on maggots of other synanthropes	Hemisynanthrope
Cochliomyia macellaria	Nearctic to Neotropic	Carrion, garbage	Hemisynanthrope
Lucilia caesar	Palaearctic	Saprophagous, decaying organic matter	Hemisynanthrope
L. illustris	Holarctic	Carrion, garbage	Hemisynanthrope
Phaenicia cuprina	Australasian, Ethiopian	Carrion, garbage, decaying organic matter	Hemisynanthrope
P. pallescens	Holarctic	Decaying fruits and meat	Hemi- to eusynanthrope (exophilous)
P. sericata	Holarctic	Decaying meat and plant matter	Eusynanthrope (exophilous)
Phormia regina	Holarctic	Carrion and decaying plant matter	Hemi- to eusynanthrope (exophilous)
Pollenia rudis	Palaearctic to Nearctic	Parasitic on earthworms	Hemisynanthrope

[a] *Australasian* = Biogeographic region including the lands of the central and southern Pacific Ocean; *Ethiopian* = Africa south of the Sahara, southern Arabia, and sometimes Madagascar and the adjacent islands; *Holarctic* = northern parts of the Old and New World and comprising the Palaearctic and Nearctic regions; *Nearctic* = Greenland, arctic America, and northern and mountainous parts of North America; *Neotropic* = South America, West Indies, and tropical North America; *Oriental* = Asia south and southeast of the Himalayas and the Malay archipelago west of Wallace's line; *Palaearctic* = region including Europe, Asia north of the Himalayas, northern Arabia, and Africa north of the Sahara; *Palaeosubtropic* = the littoral of the Mediterranean Sea in Europe, Turkey, Syria, Lebanon, Israel, and Africa north of the Atlas Mountains; *Palaeotropic* = Oriental and Ethiopian regions.
SOURCE: Greenberg, B., and D. Povolny. 1971. Bionomics of flies. In *Flies and Disease, vol. 1. Ecology, Classification and Biotic Associations*. Princeton University, Princeton, NJ.

Hippelates Species, Eye Gnats

Species of the genus *Hippelates* are different from other chloropids in their habits and behavior. They breed in soil rich in organic matter. Adults are attracted to animals, where they feed on pus, blood, and mucous secretions. Their attraction to the eyes of humans and animals gives them the name of eye gnats. They are vectors of yaws and pinkeye. Adult eye gnats can be annoying in the early morning and late afternoon, or anytime during the day in the shade cast by dense shrubbery, date gardens, or houses (5, 13).

Species that occur in lower elevations in California include *Hippelates collusor*, *H. dorsalis*, *H. impressus*, *H. pusio*, *H. robertsoni*, and *H. hermsi*; all except *H. hermsi* are pests. *H. collusor* has been incriminated in the transmission of bacterial conjunctivitis or pinkeye common in children in the Coachella Valley of southern California. Eye gnat problems have been reported in the Imperial and San Joaquin valleys of California. The potential for eye gnat problems exists in the Mojave Desert if and when suitable conditions associated with cultivation and irrigation are realized. *H. collusor* has also been closely associated with human annoyance and probable transmission of the pinkeye in the southwestern United States and Mexico; and *H. pusio* is implicated in the southern United States (13).

Eye gnat breeding areas are generally situated no more than 5 miles from residential areas. *H. collusor* has been observed to fly about 4 miles in the direction of the wind (27). *H. impressus*, known to occur in California, Texas, and Mexico, is pestiferous in foothill and mountain regions. In summer, heavy rains in the mountains result in outbreaks of this species in the adjacent desert regions of California (26).

H. collusor, the most important species of eye gnats in California, oviposits in moist, friable soils. Such conditions in tilled soils can support large populations (25). The larvae feed on decaying organic matter in soil. Other species breed in alfalfa fields, golf courses, lawns, ditch banks, river basins, and lake shores. Farmland enriched with organic matter reliably produces large populations of eye gnats. The larvae migrate to the surface of media that are too moist or they burrow deep into media that are too dry. Depending on the nature of food medium, temperature, and moisture, the larval period averages 11.4 days on human feces, 8.7 days on dog dung, and 17 days on decaying oranges (12). Under temperatures in the range of 27–41°, the life cycle takes about 3 weeks (14).

Drosophilidae

Members of the family Drosophilidae are commonly called pomace flies, small fruit flies, or vinegar flies. These small, often yellowish-colored flies are found in almost all kinds of decaying and fermenting vegetation and fruits.

The larvae of most species feed on decaying fruits and fungi. Some species occur as ectoparasites on caterpillars or predators on mealybugs.

Vinegar flies are common pests in homes, restaurants, fruit markets, canneries, wineries, cellars, and other places that contain attractive fermenting food materials (e.g., vinegar, cider, wine, beer). In Arizona, grapefruit peels dumped about a mile away from Tempe or half a mile from Yuma produced large populations of vinegar flies. Swarms of *Drosophila repleta* and *D. melanogaster* entered houses in these cities and caused severe annoyance (4).

Female flies lay their eggs on the surface of fermenting or rotting fruits and vegetables, or in dirty garbage cans. Vinegar flies have a very high biotic potential; they take only 8–10 days to complete the life cycle in the warmer months. The most important and widely known species is *D. melanogaster*; other species that can become pests are *D. funebris*, *D. repleta*, *D. busckii*, *D. affinis*, *D. falleni*, *D. tripuncta*, and *D. hydei*.

Muscidae

In this large group of flies are many important pests of humans and animals. Muscids are found in almost every biotope. The most widely known member of this family is the common house fly, *Musca domestica*, a nonbiting species that breeds abundantly in a variety of organic waste substrates such as feces and decaying materials of plant and animal origin. The house fly has been incriminated in the transmission of typhoid fever, dysentery, yaws, anthrax, some forms of conjunctivitis, and other diseases of humans and animals.

A close relative of the house fly is the face fly, *Musca autumnalis*, a pest that congregates primarily on the faces of animals. The biting members of the family are the stable flies (*Stomoxys calcitrans*), horn flies (*Haematobia irritans*), and tsetse flies (*Glossina* spp.). The stable fly breeds chiefly in piles of decaying straw, lawn clippings, and kelp on beaches. The horn fly breeds in fresh cow dung. The tsetse flies are vectors of sleeping sickness and similar diseases of humans and animals in Africa.

Fannia Species

The cosmopolitan little house fly (*F. canicularis*) is more abundant than *Musca domestica* (discussed below) for a short time in the spring months, March through June, in temperate zones. With the onset of hotter weather, it is replaced by *Musca domestica*. Little house flies retreat in small numbers to rooms of the house not used for cooking where they may be seen hovering in a jerky and zigzag fashion around chandeliers and other hanging fixtures (*15, 21*). They also hover in entryways, garages, and shaded areas.

The breeding habits of the little house fly are almost the same as those of *Musca domestica*. Favorable breeding materials include decaying and fermenting vegetable matter and the excrement of chickens, humans, horses, and cows (*13*). The larvae have been found in the nests of bumblebees (*Bombus terrestris*) and pigeons and in snail shells and old cheese (*15, 21*). The life cycle of this insect takes 24–29 days to complete (*21*). The durations of different stages at 27° and 65% rh were 1.5–2 days for the egg stage, 8–10 days for the larval stage, and 9–10 days for the pupal stage. The period from egg to adult was 18.5–22 days (*36*).

The latrine fly (*F. scalaris*) resembles *F. canicularis* in general appearance and life cycle except for its larger size and bluish-black thorax in contrast to blackish-gray of the little house fly. *F. scalaris* prefers human excrement to other manures. In addition to its breeding in semiliquid human excrement, the latrine fly also breeds in mushrooms, rotting fungi, and decaying vegetable matter (*7, 13, 15, 21*).

The coastal fly, *F. femoralis*, is just over half the size of *F. canicularis* and is of less importance as a pest than the little house fly (discussed above). *F. benjamini* is becoming a pest in California, especially in suburban expansions near the foothills and in foothill recreational areas. Most common along the streams, these flies annoy people by entering their ears, nostrils, and eyes. The flies are attracted to perspiration and mucous secretions (*5*).

Musca autumnalis, Face Fly

The face fly resembles the house fly (*M. domestica*) but is larger. The abdomen of the female is black and that of the male is orange or cinnamon-buff, whereas the abdomen of the house fly is yellow. The face fly is common in Africa, Asia, and Europe. It reached North America in 1952. This species has gained considerable importance as a pest since it has spread over Canada and all parts of the United States except the southern regions. The face fly is primarily a pest of animals but it can also be annoying to humans. In Europe, it has been incriminated as vector of the eye worm, *Thelazia rhodesii*. It is also known to transmit the bacterium, *Moraxella bovis*, causing pinkeye in cattle (*1, 37*).

During the summer months, the adult flies feed on secretions around the head of cattle and other animals. Male flies rest on fence posts, tree leaves, or similar objects, but are rarely found on animals. At night, both sexes rest on vegetation. Hibernation by the face fly takes place as unmated adults.

The adult flies often enter houses, stables, and barns in large numbers and can cause considerable annoyance to people. The females lay eggs beneath the surface of fresh cow dung. The incubation period of the eggs is about one day, and the yellow larvae feed for 2–4 days before they pupate. Unlike the reddish-brown puparium of the house fly, the puparium of the face fly is dirty white. The life cycle is complete in about 2 weeks (*13*).

Musca domestica, House Fly

The house fly is the most widely distributed insect and is commonly associated with humans over almost the entire globe. It is most abundant during the warmer months of the year (*15, 21*). In North America and Europe, the house fly is most abundant during July through September; in South America, it is abundant from October to February; in Australia, the house fly season stretches from October to March; and in the more tropical areas, the fly season is much longer.

Dispersal Considering the seasonal variations in fly abundance in different parts of the world, it is obvious that climatic factors have a great impact on the distribution and abundance of this insect. The house fly can disperse readily and is able to travel either by its own efforts or possibly by the aid of winds. The distances traveled in urban or suburban areas range from 90 to 570 feet (*15, 21*), 900 to 3,600 feet (*19*), and 0.25 mile (*28*). Under rural conditions, the range of fly dispersal is greater—900 to 5,100 feet (*2*). Hodge (*17*) recovered flies 1.25–6 miles off the shore of Lake Erie, suggesting that these insects were carried by the wind.

The relatively short distances traveled under the city conditions could be due to such factors as the availability of food and shelter. These factors could both inhibit flight activity and stimulate alighting after short distances in search of food and breeding places. Almost any reasonable distance could be covered by a house fly under compulsion to find food and shelter (*30*). When these requirements are not available close by, the insect is compelled to go farther. In localities having numerous houses, the usual maximum distance traveled by this insect seems to be about a quarter of a mile.

Factors that affect the dispersal of the house fly may be physical, physiographical, and physiological (30). Among the physical factors (wind, temperature, humidity, rain, light, and so on), wind appears to most greatly affect the dispersal of flies. The house fly can fly with the wind (negative anemotaxis) (15) or against or across the wind (positive anemotaxis) (16). In the case of positive anemotaxis, the stimulus could be odors borne by the wind; the actual reaction might have been chemotaxic, the wind acting only as an agent to transport the stimulus.

Dispersal of the house fly is temporarily halted due to decreased flight activity when it rains. Low temperatures retard locomotion, but optimum temperatures promote increased activity, resulting in rapid breeding and greater distribution. House flies become inactive at temperatures below 7.2°; temperatures below 0° are lethal.

Human activity may result in the transfer of fly populations from one place to another—in garbage transported from residential areas to dumps, in fruit-hauling trucks traveling from farms to packing sheds and processing plants, and on grapes carried from vineyards to wineries. Adult flies transported in this manner cause serious threats to the sanitary standards of food-processing establishments and wineries.

Flight activity becomes possible at an air temperature of 11.6° and reaches maximum intensity at 32.2°. Above that point it declines rapidly and then ceases at the thermal death point of 44.4° (7, 38). Relative humidity affects the response of flies to temperature. High as well as low temperatures are lethal to house flies when the humidity is high. The flies live longest at a temperature of 15.5° and at a relative humidity between 42% and 55%. Flies reach optimum physiological activity at a high temperature and low relative humidity (32). Activity is diurnal, reaching a peak between 2 p.m. and 4 p.m., the hottest and driest hours of the day (3). Adult flies are generally inactive at night, but they do respond to artificial illumination (positive phototaxy).

Physiographic factors such as topography, bodies of water, and woodlands may affect fly dispersal. The house fly travels along depressions and avoids crossing elevations (40). In the city, congested areas with high buildings could offer conditions that would restrict fly dispersal. However, under these circumstances, it seems likely that house flies would also respond to other stimuli such as food and breeding sites. Small wooded areas seem to offer no barrier to the dispersion of the house fly. In one study in Montana (30), house flies were able to reach the city from the dump either by traversing a wooded area 600 feet wide or by taking a roundabout course (2,400 feet) up an old channel of a river. That report (30) suggests that the presence of bodies of water such as streams under city conditions does not constitute a barrier to fly dispersal (17).

Of the several physiological factors affecting dispersal, two are especially important—the presence of feeding stimuli and the availability of breeding sites. Adult feeding areas may not serve as suitable breeding sites for the larvae. The attraction of adult flies to sources of food and breeding material has been demonstrated (20). Marked flies were released at a dairy barn about 2,100 feet distant from a stable and a kitchen. Marked flies were recovered at the kitchen but not at the stable, thus showing the preference of adult flies for the feeding area. However, gravid flies in a feeding area respond to stimuli from breeding sites when they are ready to lay eggs (30).

In spite of their seemingly incessant activity, adult house flies actually rest much of the time, with a preference for edges. This resting behavior makes flypaper a useful adjunct to other control measures. Indoors, flies rest on light strings, electrical wires, light fixtures, exposed pipes, walls, ceilings, and other similar places. Outdoors, they rest mainly on fences, electrical wires, edges of buildings, and vegetation (32).

Adult Food House flies search actively for food during the daylight hours. They require two or three feedings per day. Since adult house flies have sponging mouthparts, their food must be either in the liquid state or easily soluble in the salivary and crop secretions. Common foods are milk, sugar, blood, and many other food substances commonly available in human biocoenoses. Water is essential; adult flies cannot live for more than 48 hours without water. In addition to foods for human consumption, house flies are also strongly attracted to feces and other types of decaying organic materials.

Breeding Habits The breeding areas of the house fly are varied. The flies are attracted to warm, moist organic substrates that can provide adequate nourishment to the larvae. Human feces and animal manures appear to be excellent media, with horse manure being the most suitable of all, supporting a very high number of larvae per unit weight (e.g., often about 1,200 larvae/pound of manure). Manures of other animals such as cows, pigs, rabbits, and birds, especially chickens, are suitable breeding substrates. Breeding of the house fly occurs in a variety of habitats such as privies, exposed feces, and partially digested sludges from sewage treatment plants. In urban areas, improper management of garbage dumps and pet feces can result in severe infestations (22).

Female flies deposit their eggs in cracks and crevices in the breeding medium. Upon hatching, the larvae move to those regions in the breeding medium that meet their temperature requirements. Feeding larvae prefer a temperature range of 30–35°; full-grown larvae ready to pupate prefer lower temperatures. In tropical and subtropical regions, fly breeding continues throughout the year; in more temperate areas, it is interrupted during winter.

In the eastern half of the USA, populations of flies build up gradually during the spring and summer, reaching a maximum in late summer or early fall. In the southcentral and southwestern areas, population levels may be high during the spring, drop off markedly during the hot, dry, midsummer months, then peak again during the early fall.

In temperate zones, house flies pass the winter in hibernation and/or in intermittent breeding in protected areas. The eggs and larvae show little tolerance of cold. Adult flies do not emerge if the pupae have been exposed to temperatures below 11° for 20–25 days, or 8.8° for 24 hours. Adult flies can be kept alive for long periods at temperatures in the range of 10–26.6° (32).

Musca vicina, Egyptian House Fly

The Egyptian house fly, a species closely related to *M. domestica*, is the most predominant synanthrope in urban and rural areas of Egypt (10, 11). Although found throughout the year, the fly season really begins in March, with this species becoming most abundant in early summer (April–June) and autumn (September) (11). Temperatures in these months (28–35°) are very favorable for the breeding and development of this fly. In midsummer (July–August), high temperatures (50° or more) prevent oviposition and the development of the early stages. During winter, especially in January, this fly becomes rare, primarily because of low temperatures.

Breeding Habits The local distribution of *M. vicina* is mainly dependent on the availability of favorable breeding media, the most important of which are horse dung in towns and donkey dung in the countryside. Since the stables of these animals are often cleaned at irregular in-

tervals of weeks or months, the floor becomes covered with a thick, compact mass of litter, and food remains mixed with manure and urine. The surface of this material remains slightly dry. The deeper layers offer ideal larval habitats where sufficient humidity and lower temperatures prevail. The temperature of the bedding, generally 30–35°, seems very favorable for the ovipositing flies.

In rural areas, because of poor sanitation and the lack of sewage disposal systems, human feces play an important role in the production of flies. Sheep and goat dung are also important as sources of fly production, but swine manure is of minimal importance in Egypt. *M. vicina* does not breed in buffalo, cow, or calf dung (10).

Factors Affecting Development The development of this fly in its breeding media is affected by factors such as temperature, humidity, and larval food. At about 30°, the duration of the life cycle is 6.7 days; at 25°, the life cycle lasts about 9 days (11). During the summer months (July and August), for example, horse manure, being loose and coarse, dries up rapidly in the hot, dry weather and becomes unsuitable for fly breeding.

The nature and kind of food medium can affect fly breeding significantly. In one study involving different media—horse, pig, donkey, sheep, and human dung and kitchen refuse—the duration of development of *M. vicina* was about 8 days in horse and pig dung, about 10 days in sheep dung, 14 days in human excrement, and about 24 days in kitchen refuse (11). Under natural conditions, however, the physical state (loose or compact) and the chemical and biological nature of the breeding media influence the development of flies. Excessive dryness and the heat of fermentation of the breeding media strictly limit fly production.

Muscina Species, False Stable Flies

False stable flies are larger than the common house fly. They breed in decaying vegetable matter, human excrement, poultry and other manures, and occasionally in garbage. Full-grown larvae can feed on maggots of other flies. Larval development takes 15–25 days. Adults frequently enter houses in search of food such as meat, fruits, and vegetables. *M. stabulans* is more robust than the common house fly. Its general coloration is dark gray. Adult false stable flies are nonbiting, but they can carry pathogens in and on their bodies; the larvae are able to cause intestinal myiasis in humans. This is a eusynanthropic and endophilous species, especially in rural areas. The adults alight on walls of privies and on fruits in orchards and vineyards. Eggs are laid on decaying organic matter and excrement. The larvae breed in human excrement, rotting cow dung, poultry manure, and decaying fruits. The first-instar larvae are mostly saprophagous; older instars are predatory on larvae of other flies (7, 13). Other species such as *M. assimilis* and *M. pabulorum* have almost the same habits as those of *M. stabulans*.

Ophyra Species, Dump Flies

Dump flies, shiny black and smaller than house flies, are widely distributed in the USA. They are found around garbage disposal sites and are abundant in urban areas. The adult flies can cause great annoyance by entering houses, privies, and slaughterhouses. The larvae develop in a variety of substrates such as human and animal excrement, kitchen wastes, and animal carcasses.

The bronze dump fly (*O. aenescens*) is found in the Neotropical and Oceanian regions as well as in the USA. The adults are attracted to carrion and to animal and human excrement, all of which serve as suitable media for larval development. In Central America, the flies enter houses and alight on foods causing annoyance and health hazards.

O. anthrax is widely distributed in the Caucasus and Central Asia. Since it is a thermophilous species, it does not reach the colder regions. The adults are frequently associated with toilets, garbage dumps, animal carcasses, and dwellings. Suitable substrates for larval development include feces of humans and animals, kitchen wastes, and animal carcasses.

O. capensis is distributed in the southern areas of the Palaearctic region. It closely resembles *O. leucostoma* in its biology and ecology.

The adults of *O. leucostoma*, a common Holarctic, heliophilous species, alight on vegetation, especially the undersides of leaves. The larvae require breeding media similar to those of *O. anthrax*. Second- and third-instar larvae prey on other larvae, especially those of the house fly. The adults commonly enter privies and slaughterhouses where their presence becomes a nuisance and a health hazard.

O. nigra is distributed in the Australasian region and is associated closely with humans in the tropics. In Queensland, it is a carrion feeder. The larvae develop in manure, and pupation takes place in very moist manure which ensures successful emergence of adults.

Stomoxys calcitrans, Stable Fly

The stable fly closely resembles the house fly, but it is slightly larger and more robust. Its association with cow sheds and stables has given it the name of stable fly. It is widely distributed, being common in Europe, Canada, and the United States from July to October. It is frequently found outdoors, but occasionally enters houses. During the winter months, it remains inside barns and stables. Both sexes of the stable fly are bloodsucking, feeding chiefly on cattle, horses, dogs, and occasionally humans.

The stable fly breeds in a wide variety of substrates: horse, cow, poultry, and sheep manure; human excrement; the straw of oats, rice, barley, and wheat; fermenting grass clippings; kelp on beaches; decaying vegetable and animal matter; and rotting fungi. The life cycle from oviposition to adult emergence may take from 13 days to 8 weeks, depending on the temperature and other factors. The longest developmental period of about 10 weeks was observed during the fall from October 14 to December 15 (15, 21). In temperate climates, larvae as well as pupae can survive in warm situations during the winter months.

Piophilidae

The members of this family, called skipper flies, are small insects. Adults are metallic black or bluish in color. The larvae of some species are scavengers; others feed on cheese and meat. A good example of the latter is the cheese skipper, *Piophila casei*, often considered a serious pest in cheese, salted beef, smoked fish, brine-cured fish, bones, animal carcasses, human cadavers, and various other substances high in animal protein.

The female lays eggs on the surface of the food medium. The full-grown larva is capable of "skipping" as much as 25 cm horizontally and 15 cm vertically. The entire life cycle may be as short as 12 days. The adult fly feeds on the liquid fractions of the larval food. The adults live 3 or 4 days. They usually appear in warmer seasons (5, 35) when they can become a problem in the food industry by infesting cheese or cheese products, salted beef, and cured fish.

Psychodidae

Psychodids are small, hairy, mothlike insects. The larvae of the terrestrial psychodids (Phlebotominae) are found in decaying plant and animal matter, mud, and moss; those of the aquatic forms (Psychodinae) occur in water. Adults of the aquatic breeders rest in moist, shady places outdoors; indoors they are found around sinks, drains, and in sewage treatment plants. Phlebotomine sand flies are bloodsucking and are found in the southern USA and in the temperate and tropical regions of the world. Phlebotomine sand flies are known vectors of pathogens of several diseases. Most psychodine sand flies are not involved in the transmission of human diseases, but they can be annoying to humans in outdoor and indoor situations.

Moth Flies

Moth flies (Psychodinae) develop in the muck or gelatinous deposits in sewage disposal beds, septic tanks, wastewater lagoons, compost, and dirty garbage containers. They may also breed in drains of sinks and bathtubs, especially the overflow drains, in tree holes, rain barrels, in most any moist organic solids, and in birds' nests with accumulations of moist excreta. Moth flies are commonly associated with sewage filter beds where the larvae and pupae live in the gelatinous coating of the filter materials or on other objects bearing decomposing organic matter.

The larvae of moth flies are probably beneficial because they eat the fungi, bacteria, and sludge in sewage disposal beds and break down the gelatinous materials into small fecal pellets that are easily washed away. Adult moth flies become pestiferous when they enter homes in large numbers. They cause annoyance by getting into the eyes, nose, and ears of people. Some cases of bronchial asthma from allergic reactions to *Psychoda alternata* have been reported among workers at a sewage plant in South Africa (5).

Several species of moth flies are found in the USA. The Pacific drain fly, *Psychoda pacifica,* is a pest along the Pacific coast from Alaska to California. *P. alternata* is distributed from Florida to Massachusetts and from Washington to California. The adults of this species are weak fliers and prefer shade in the daytime. At night they are attracted to lights. Adults may enter houses in large numbers and can be annoying to people.

Other species of *Psychoda* distributed in North America are *P. satchelli* and *P. cinerea.* Another species widely distributed in the USA is *Telmatoscopus albipunctatus.* It is found breeding in sinks and drains; its larvae have been found in mud, tree holes, rain barrels, and shallow pools with organic debris. This species also breeds in the rinds of candlenut fruits, *Aleurites moluccana* (Euphorbiaceae) (39).

Table 13.2
Distribution and habits of sarcophagid flies

Species	Distribution[a]	Larval habitats	Synanthropy
Sarcophaga argyrostoma	Holarctic	Carrion, human feces	Hemisynanthrope
S. bullata	Nearctic	Decaying organic matter	Hemisynanthrope
S. carinata	Palaearctic	Earthworms, decaying meat	Hemisynanthrope
S. haemorrhoidalis	Cosmopolitan	Carrion, excrement, exposed meat	Hemisynanthrope
S. hirtipes	Palaearctic	Animal and human feces, decomposing animal and plant materials	Hemisynanthrope
S. melanura	Holarctic	Animal excrement, garbage, fruits and sweets in markets	Hemisynanthrope
S. misera	Palaearctic, Oriental, Australasian	Carrion, horse manure	Hemisynanthrope
S. peregrina	Palaearctic, Oriental, Australasian	Decomposing organic matter, human feces	Hemisynanthrope
S. stercoraria	Holarctic	Animal and human feces	Hemisynanthrope or eusynanthrope (in rural areas)

[a] *Australasian* = Biogeographic region including the lands of the central and southern Pacific Ocean; *Ethiopian* = Africa south of the Sahara, southern Arabia, and sometimes Madagascar and the adjacent islands; *Holarctic* = northern parts of the Old and New World and comprising the Palaearctic and Nearctic regions; *Nearctic* = Greenland, arctic America, and northern and mountainous parts of North America; *Neotropic* = South America, West Indies, and tropical North America; *Oriental* = Asia south and southeast of the Himalayas and the Malay archipelago west of Wallace's line; *Palaearctic* = region including Europe, Asia north of the Himalayas, northern Arabia, and Africa north of the Sahara; *Palaeosubtropic* = the littoral of the Mediterranean Sea in Europe, Turkey, Syria, Lebanon, Israel, and Africa north of the Atlas Mountains; *Palaeotropic* = Oriental and Ethiopian regions.

SOURCE: Greenberg, B., and D. Povolny. 1971. Bionomics of flies. In *Flies and Disease, vol. 1. Ecology, Classification and Biotic Associations*. Princeton University, Princeton, NJ.

Sarcophagidae

Sarcophagids or flesh flies resemble some blow flies but are generally darker in color and have gray stripes on the thorax. Adults feed on sugar-containing materials such as nectar, sap, fruit juices, or honeydew. The larvae of some species breed in carrion, excrement, and decaying vegetable matter; other kinds of larvae parasitize certain invertebrates (e.g., grasshoppers, lepidopterans, and snails) and several kinds of vertebrates including humans (13). Sarcophagids are not as likely to enter houses as the calliphorids, but some sarcophagids are considered to be synanthropic from spring through fall. The distribution and habits of several species of sarcophagids are summarized in Table 13.2.

Sarcophaga haemorrhoidalis, Redtailed Flesh Fly

S. haemorrhoidalis is found everywhere but in the Oriental and Australasian regions. A thermophilous species, it is not found in the cooler areas of Europe and Asia. Basically a hemisynanthropic fly, it often tends to become eusynanthropic in the warmer areas, especially in the Ethiopian region where it is common in and around human habitations. In such situations in Africa, the adults are often found indoors. They often alight on fresh feces. The larvae may be found in carrion and exposed meats, but they generally breed in fecal matter.

Sarcophaga melanura

S. melanura, a Holarctic species, frequents fruits, sweets, and uncovered produce in markets in Russia. The adults are generally found around human, horse, and cattle excrement; they are also found in human dwellings. The larvae breed in the feces of pigs and other animals (7).

Fly Problems in Food Industries

In this discussion of the ecological aspects of synanthropic flies, we have indicated that many kinds of flies can become problems in food industries such as food-processing plants, slaughterhouses, and restaurants. The fly problem in the food industry has three aspects. The first is the annoyance caused to food factory workers by the presence of flies. In order of severity, these are muscoids, calliphorids, sarcophagids, chloropids, and drosophilids. Owing to the degree of synanthropy of these insects, the efficiency of the workforce in food industries could be greatly decreased by the nuisance caused by adult flies. Also, moth flies could be annoying to people using the restrooms in food plants.

A second aspect is the role of flies in causing health problems or disease conditions in food-industry personnel. Although there are no reported cases in the literature per se, the potential exists, in particular for disease transmission by muscoids (especially the house fly) and chloropids (*Hippelates* spp.), and myiasis production by calliphorids and sarcophagids.

The third and most important aspect is the infestation or contamination of food materials by flies. Such infestation could result in serious aesthetic, hygenic, and economic problems. For example, the occasional presence of dead house flies in bottled soft drinks as a result of invasion of the beverage factories by flies justifiably arouses aesthetic and hygenic concerns among consumers. The infestation of fruits by *Drosophila* species, and of cheese, salted beef, and cured fish by *Piophila casei*, definitely constitutes an economic problem.

Cited References

1 Cheng, T.-H. 1967. Freqency of pinkeye incidence in cattle in relation to face fly abundance. *Journal of Economic Entomology* 60(2)598–599.

2 Copeman, S. M., F. M. Howlett, and G. Merriman. 1911. An experimental investigation on the range of flight of flies. Reports of Local Government Board on Public Health and Medical Subjects, N.S. No. 53. Further Reports (No. 4) on Flies as Carriers of Infection, pp. 1–10.

3 Dakshinamurty, S. 1948. The common housefly, *Musca domestica* L., and its behaviour to temperature and humidity. *Bulletin of Entomological Research* 39(3)339–357.

4 Deonier, C. C. 1942. Insect pests breeding in vegetable refuse in Arizona. *Journal of Economic Entomology* 35(3)457–458.

5 Ebeling, W. 1975. *Urban Entomology*. Agriculture & Natural Resources, University of California, Oakland.

6 Gagné, R. J. 1981. *Chrysomya* spp., Old World blow flies (Diptera: Calliphoridae), recently established in the Americas. *Bulletin of the Entomological Society of America* 27(1)21–22.

7 Greenberg, B. 1971. *Flies and Disease, vol. 1. Ecology, Classification and Biotic Associations*. Princeton University, Princeton, NJ.

8 Greenberg, B. 1973. *Flies and Disease, vol. 2. Biology and Disease Transmission*. Princeton University, Princeton, NJ.

9 Greenberg, B., and D. Povolny. 1971. Bionomics of flies. In *Flies and Disease, vol. 1. Ecology, Classification and Biotic Associations*. Princeton University, Princeton, NJ.

10 Hafez, M. 1941. Investigations into the problem of fly control in Egypt. *Bulletin de la Société Fouad Ier d'Entomologie* 25:99–144.

11 Hafez, M. 1941. A study of the biology of the Egyptian common house-fly: *Musca vicina* Macq. (Diptera: Muscidae). *Bulletin de la Société Fouad Ier d'Entomologie* 25:163–189.

12 Hall, D. G. 1932. Some studies on the breeding media, development, and stages of the eye gnat, *Hippelates pusio* Loew (Diptera: Chloropidae). *American Journal of Hygiene* 16(3)854–864.

13 Harwood, R. F., and M. T. James. 1979. *Entomology in Human and Animal Health*, 7th ed. Macmillan, New York, NY.

14 Herms, W. B., and R. W. Burgess. 1930. A description of the immature stages of *Hippelates pusio* Loew and a brief account of its life history. *Journal of Economic Entomology* 23(3)600–603.

15 Hewitt, C. G. 1914. *The Housefly*, Musca domestica *Linn.—Its Structure, Habits, Development, Relation to Disease and Control*. Cambridge University, London.

16 Hindle, E. 1914. The flight of the house fly.

Proceedings of the Cambridge Philosophical Society 17(4)310–313.

17 Hodge, C. F. 1913. The distance house flies, blue bottles and stable flies may travel over water. *Science* 38(980)512–513.

18 Howard, L. O. 1900. A contribution to the study of the insect fauna of human excrement (with special reference to the spread of typhoid fever by flies). *Proceedings of the Washington Academy of Science* 2(32)541–604, 2 pl.

19 Howard, L. O. 1911. *House-flies.* Farmers Bulletin No. 459. U.S. Department of Agriculture, Washington, DC.

20 Hutchinson, R. H. 1915. A maggot trap in practical use; an experiment in house-fly control. *Bulletin of the U.S. Department of Agriculture* 200:1–15.

21 Keiding, J. 1986. VII. The House Fly—Biology and Control, Report No. WHO/VBC/86.937, World Health Organization, Geneva, Switzerland. 63 pp.

22 Lewallen, L. L. 1954. Biological and toxicological studies of the little house fly. *Journal of Economic Entomology* 47(6)1137–1141.

23 Lindsay, D. R., and H. I. Scudder. 1956. Nonbiting flies and disease. *Annual Review of Entomology* 1:323–346.

24 Metcalf, C. L., W. P. Flint, and R. L. Metcalf. 1962. *Destructive and Useful Insects*, 4th ed. McGraw-Hill, New York.

25 Mulla, M. S. 1963. An ecological basis of suppression of *Hippelates* eye gnats. *Journal of Economic Entomology* 56(6)768–770.

26 Mulla, M. S. 1971. *Hippelates impressus,* an eye gnat new to California (Diptera: Chloropidae). *California Vector Views* 18(1)1–4.

27 Mulla, M. S., and R. B. March. 1959. Flight range, dispersal patterns and population density of the eye gnat, *Hippelates collusor. Annals of the Entomological Society of America* 52(6)641–646.

28 Nuttall, G. H. F., E. Hindle, and G. Merriman. 1913. The range of flight of *Musca domestica.* Reports of Local Government Board on Public Health and Medical Subjects, N.S. No. 85. Further Reports (No. 6) on Flies as Carriers of Infection, pp. 20–41.

29 Oldroyd, H. 1964. *The Natural History of Flies.* Norton, New York, NY.

30 Parker, R. R. 1916. Dispersion of *Musca domestica* Linnaeus under city conditions in Montana. *Journal of Economic Entomology* 9(3)325–354, 3 pl.

31 Povolny, D. 1971. Synanthropy. In *Flies and Disease, vol. 1. Ecology, Classification and Biotic Associations.* Princeton University, Princeton, NJ.

32 Pratt, H. D., K. S. Littig, and H. G. Scott. 1975. *Household and Stored-Food Insects of Public Health Importance and Their Control.* Center for Disease Control, Atlanta, GA.

33 Pratt, H. D., K. S. Littig, and H. G. Scott. 1976. *Flies of Public Health Importance and Their Control.* Center for Disease Control, Atlanta, GA.

34 Schoof, H. F. 1959. How far do flies fly and what effect does flight pattern have on their control? *Pest Control* 27(4)16–24, 66.

35 Simmons, P. 1927. *The Cheese Skipper as a Pest in Cured Meats.* Bulletin 1453. U.S. Department of Agriculture, Washington, DC.

36 Steve, P. C. 1960. Biology and control of the little house fly, *Fannia canicularis,* in Massachusetts. *Journal of Economic Entomology* 53(6)999–1004.

37 Steve, P. C., and H. H. Lilly. 1965. Investigations on transmissibility of *Moraxella bovis* by the face fly. *Journal of Economic Entomology* 58(3)444–446.

38 West, L. S. 1951. *The House Fly. Its Natural History, Medical Importance, and Control.* Comstock, Ithaca, NY.

39 Williams, F. X. 1943. Biological studies in Hawaiian water-loving insects—Part III. Diptera or flies—C. Tipulidae and Psychodidae. *Proceedings of the Hawaiian Entomological Society* 11(3)313–338.

40 Zetek, J. 1914. Dispersal of *Musca domestica* Linné. *Annals of the Entomological Society of America* 7(1)70–72.

14
Ants: Formicidae, Hymenoptera

Susan H. Beatson Campbell

Ants are social insects. They live in colonies usually composed of a large number of individuals. The population consists of various castes, each caste performing a specific function within the colony. Many species of ants may be encountered foraging in warehouses, in catering and food-manufacturing establishments, and in domestic dwellings.

Ants have existed for many millions of years, the earliest record being from the Magothy Formation (Upper Cretaceous) at Cliffwood, New Jersey (22). The two ants found embedded in amber there in 1966 are probably some 100 million years old. Since these specimens are very similar in appearance to modern ants, it is likely that ants existed much earlier, possibly even as far back as the Jurassic. By the Eocene Period (some 70 million years ago), many species of ants had developed and are commonly found as fossils.

Ants, in common with all insects, have a rigid external covering, the exoskeleton. This consists of several different layers and, apart from providing a basis for muscle attachment, it screens out dangerous solar rays and reduces the amount of water vapor lost, always a problem for insects living on dry land. The ant body is divided into three major regions: head, thorax, and gaster. An unusual feature of ants and most Hymenoptera is the development of the most anterior section of the abdomen into a petiole composed of

Campbell: Welsh Office, Agriculture Department, 66, Ty Glas Road, Llanishen, Cardiff CF4 52B, South Wales.

one or two segments. This is a narrow tube hinged in such a way that the gaster can be brought forward under the thorax until its tip reaches the jaws of the ant. This adaptation is very useful to ants, enabling them to grasp prey or an enemy with the jaws and immediately sting it (some kinds of ants, however, do not have stingers).

The antennae bear important sense organs. They are fully hinged and may be carried either well forward of the jaws or folded back against the head when the ant finds itself in a potentially dangerous situation. Many chemical stimuli, probably including olfactory ones, can be detected by the antennae. The antennae undoubtedly help an ant to perceive the size and shape of objects in its immediate area. They are probably sensitive to sound and to vibrations in the substrate. The antennae and the forelegs may also be used for communicating with other workers in the nest by means of a conspicuous mechanical action.

The eyes of most ants are of the usual compound variety but are not particularly well developed except in some of the males. Sexual males and females have light-sensitive ocelli on top of the head.

Ants are surprisingly strong. Some are able to move twice or more their own weight in food material when out foraging. They learn their way around the vicinity of the nest in many ways; gravity, chemical patterns, and the shape of objects are important, as are light patterns from trees and bushes which help with visual orientation. The intensity of gravity is measured by ants through organs in the joints of their legs; this provides them with an estimate of their angle with respect to vertical.

Some ants leave chemical trails on the ground. These are important navigational clues. Workers can be seen running along with their antennae close to the ground. Foragers returning to the nest use the antennae to communicate information about food supplies. In other ants, the scent trail alone is enough to rouse workers to leave the nest and start foraging for food. The trail-making chemical usually originates from the tip of the gaster, but some species lay trails with their feet. In some ants, the sting is extruded and used rather like the nib of a pen to spread the chemical along the ground. In other ants, the trail chemical is spread from the anus or from a gland that opens directly onto the tip of the gaster.

The Colony

Ants live in colonies made up of different castes. The colonies themselves vary greatly in size and in the number of queens that each colony may contain. There are normally three distinct castes in the ant society: sexual females, sexual males, and workers. Workers may vary in size, some being considerably larger than others and performing special duties.

Sexual Females

Fertile females (queens) are often the largest ants in the caste system. Their primary function is reproduction (the egg production process is centered in the enlarged gaster). The mating flight is usually short-lived, but this mass flight of newly formed sexuals from a colony is a well-known cause of complaint from owners of dwellings, hotels, and catering establishments. Not all species swarm in this way; some ants mate within the nest or on the ground. During mating, the male transfers enough sperm to the queen to last her a lifetime (queens of some species live for many years).

After mating, the queen discards her wings and searches for a suitable nesting site. On finding one, she digs out the first nest cell. Egg laying then begins, and the production of a new colony is under way. The number of queens present in a colony varies widely from species to species. Colonies of *Lasius niger* contain only one queen; colonies of *Monomorium pharaonis* may contain as many as 100 queens. On occasion, the queens of some species leave the nest to forage; those of some other species never do so.

Sexual Males

The only function of male ants is to mate with the sexual females. The males die after mating. Most males are winged and are intermediate in size between workers and queens. They often have extremely large eyes. In species that mate in the air, it is not unusual to see the queens carrying the males in flight. The male ants do not long remain within the nest and many, of course, die without mating. Only large or very old colonies produce male ants.

Workers

Workers (females that remain sexually immature) develop from eggs laid in the new cells, and these are first cared for by the queen as the new colony gradually enlarges. When suffecent workers have emerged, the duties of caring for the eggs and the young and of enlarging the nest and protecting the colony are taken over by them. There are great differences in size and structure among workers in a colony. In some species there are major and minor workers, the major workers being very much larger. Major workers, which often have greatly enlarged jaws, defend the colony. Minor workers may specialize in nest building and repair, foraging for food, or caring for the young.

Life Cycle

There are four stages of development in the life history of an ant: egg, larva, pupa, and adult. The eggs laid by the queen vary in size and shape according to species. On hatching, a small legless larva begins feeding, molts several times, and then pupates. The pupae in some species are naked; in others they are enclosed within a silk cocoon. The latter, often referred to as "ant eggs," are sometimes sold in pet shops. After the adult ant emerges from the pupal case, a day or two may be required for the body to harden and darken in color. The length of time between egg laying and the emergence of the adult varies greatly depending on the species of ant, the temperature and, of course, the time of year.

Important Pests

Anoplolepis longipes, Longlegged Ant

This species probably originated in Southeast Asia. It has been spread along trade routes to many areas of the Old World tropics and to Chile and Mexico. *A. longipes* workers measure some 4 mm long and are light brown. The species is in the interesting position of being both a beneficial insect and a pest, depending on the situation in which it is found. Workers are considered to be beneficial when they attack the beetles *Melittomma insulare* Fairmaire (Lymexylidae) and *Oryctes monoceros* (Olivier) (Scarabaeidae), both serious pests of coconut palms. It has also been recorded that centipedes, cockroaches, and larger species of ants have been greatly reduced in number where this ant is abundant.

On the debit side, *A. longipes* tends various honeydew-producing insects on breadfruit (*Artocarpus altilis*), jackfruit (*A. heterophyllus*), and coffee (*Coffea arabica*). Heavy outbreaks of sooty mold (*Capnodium* spp., Capnodiaceae) are nearly always attributed to the activities of *A. longipes*. In the Seychelles, where it is known as the "crazy ant" (not to be confused with *Paratrechina longicornis*) (*13*), this newly imported ant (which is thought to have arrived around 1962) has become rapidly established and has developed into a serious pest. It has caused problems in a hospital, in hotels, and in food- and drink-processing firms (*16*).

Nesting occurs in a variety of places—in trees, in soil (if it is of the light, well-drained type), in fallen tree trunks, and in leaf litter. Around human habitations, practically any moist crevice may be utilized, such as cracks in walls, spaces under sheet metal, and under the cover of rubbish and the earth around trees. The colonies contain many queens and the extent of each individual colony seems to be somewhat ill-defined. No aggressive tendencies are shown by members of one colony toward those of another. Mating flights have not been observed, but males and sexual females both apparently have functional wings. The most common method of spread appears to be by colonial budding. Production of sexual stages appears to vary from country to country; in the Seychelles, it is related to the wet season (November to March).

Foraging also appears to be somewhat variable. In the Seychelles, peaks of foraging activity are reached in the late afternoon and early evening. This species shows a distinct preference for sugary foods, but insects, alive or dead, are also collected. Centipedes and large cockroaches are not immune to attack. Problems may also occur with domestic animals, especially poultry, where the young may be killed and older birds may be disturbed.

Inside houses, hotels, or factories, the foragers gain access to foodstuffs, containers, and machinery of all types. The ants cause considerable irritation to the eyes, mouth, and nose when they crawl over sleeping people; they are, therefore, a serious problem for hoteliers. Chemicals (formic acid and other organics) released from crushed ants attract other ants to the site. Foraging workers readily collect crushed specimens of their own species and carry them back to the nests.

Camponotus Species, Carpenter Ants

The genus *Camponotus* consists of a large number of species distributed widely around the world (*19*). Many carpenter ants are generally rather large, varying in size from 6 mm for a minor worker up to 12 mm for a queen. Their coloration varies according to the species but many are black or red and black. Other kinds of ants also utilize cavities in wood as nesting sites, and care should be taken not to confuse such species with carpenter ants. Further, not all species of carpenter ants nest in wood.

The life cycles of the major pest species of carpenter ants are similar to those of most other ants. Swarming occurs in the spring and summer after the colony has been established for some 3–6 years. Only mature colonies produce the sexual stages. By that time the colony will have grown to several thousand workers. After mating, the males die and the newly fertilized females look for suitable sites to start new nests. This is usually in moist and partially damaged wood which may be outside or inside a building. Rotten window frames are common nest sites, as is the void between the roof and ceiling of porches; hollow porch posts and columns are often used. Some species may nest occasionally in hollow veneer doors and other small cavities; in such cases the wood does not show tunneling since the ants are utilizing only the existing cavities.

The queen does not venture out of the nest. She feeds the first larvae with secretions from her mouth. During the first year, the colony is very small, consisting only of the queen and some dozen or so workers and various immature stages. Colony size rapidly increases after the first year if conditions are favorable. Workers are constantly engaged in cutting new galleries in the wood where the colony is living. These tunnels are not regular in shape or direction but are generally excavated along the grain following the softer parts of the wood. The occupied galleries are immaculately kept; the walls have a smooth, sand-papered appearance unlike the frass-clogged tunnels produced by subterranean

Figure 14.1
Pharaoh ants (*Monomorium pharaonis*) (Formicidae). Top, male; center, queen; bottom, worker.

termites (but similar to the tunnels of drywood and powderpost termites). Food fragments and other debris are collected, carried from the nest, and deposited outside. These piles of frass (insect fragments and sawdust-like materials) are often a useful clue that carpenter ants are present and active.

Carpenter ants are omnivorous. They are particularly fond of sweet substances and collect vast amounts of honeydew. In some instances, the ants protect aphids, and large numbers of ants may be seen congregating around them. Fruit and other sweet items such as jelly, sugar, honey, and syrup are readily taken; meat grease and fat are equally acceptable.

Iridomyrmex humilis, Argentine Ant

This ant is a native of Brazil and Argentina. Because it is spread by trade, it is now a very well known tramp species which may be encountered the world over. It first appeared in the USA in New Orleans in 1891; it reached Northern Ireland in 1900. The workers vary in color from light to dark brown and measure 2.2–2.6 mm long. The queens are about 6 mm long and are darker brown in color. Swarming is seldom seen and mating usually takes place within the nest early in the summer (*19*).

In tropical areas, nests are established in rotten wood, in soil, and in any damp place that is well shaded. The ants are found in a wide variety of buildings, including greenhouses, and also in basements and around manholes. In temperate areas, these ants have exploited the increased use of central heating and live in situations similar to those utilized by *Monomorium pharaonis* (discussed below). Hospitals, restaurants, and food manufacturing premises have all suffered severe infestations.

The colony, which can become very large and may contain many queens, never stays in one place for very long. The worker ants are very aggressive and tend to eliminate other species of ants in the vicinity of the colony. They attack cockroaches, dismembering them and carrying the pieces to the nest. Like many other species of ants, the workers move in well-defined trails. Their food consists of a wide variety of items but sweets seem to be much preferred. Meat, insects, seeds, and fruits are also attractive to Argentine ants.

Lasius niger, Black Garden Ant

L. niger, the black garden ant, is probably the best-known member of this genus in Asia and Europe. The brownish-black workers and queens measure about 5 mm and 15 mm long, respectively. The black garden ant is the only formicid indigenous to Britain that regularly enters houses, hotels, hospitals, warehouses, and factories. Nesting occurs mainly outdoors under concrete driveways or patios, but there have been records of nests being found in buildings where temperatures were kept continually high throughout the year. Indoors, any suitable cavity is exploited for nesting purposes.

L. niger is as common in towns and cities as it is in rural areas. Mating flights occur in warm weather during August and September. Huge swarms are often encountered but these are normally short-lived. After mating, the fertilized queen selects a site and starts to produce a single-queen colony. Sweet commodities are particularly attractive to this species, but many other products are routinely investigated by foraging workers. One instance has been reported of *L. niger* nesting in urea-formaldehyde foam insulation in an external wall of a domestic kitchen. The ants easily excavated living space in the foam and deposited the resulting debris outside the nest.

Monomorium pharaonis, Pharaoh Ant

This small tropical ant (Figure 14.1) has spread all over the world with trade and has established itself in many temperate regions. The name pharaoh ant is thought to have originated from a belief that this ant was one of the plagues of ancient Egypt. The geographical origin of this species was probably tropical Africa. This species is a serious pest of the catering and food-processing industries. So far, this ant has been generally unable to adapt to temperate climates, but it has been observed nesting outdoors in California. It is important, however, in temperate areas because it takes advantage of centrally heated buildings of all kinds. In Britain, it was found in the absence of central heating in an apartment block where the ants had nested near small gas-fired water heaters.

The yellowish-brown workers are only about 2 mm long, queens are about 4 mm long, and sexual males are somewhat smaller than the queens and are dark brownish-black. The queens forage occasionally. Colonies may contain many thousands of workers and numerous queens. Although queens as well as males are winged, mating occurs within the nest, and then the males die. Mating takes place throughout the year.

Queens may leave the nest and set up new colonies but the many colonies that infest some buildings probably consist of only a single supercolony. There is no antagonism or aggression between these different coexisting colonies. There appears to be a very complex relationship between all stages within the colony. Only some 30–50 workers, together with some preadult stages, are required to start a new nest. Very often, small colonies of this type progress far better than similar groups containing only a queen and a number of workers.

Budding following mass migrations of very large colonies is rarely observed. An instance of this was observed once in a hospital. A mass of ants measuring approximately 1 meter by 2 meters was moving across a storeroom floor. Many queens were present as were thousands of workers, many of which were carrying eggs, larvae, and pupae. Such mass migrations cause practical problems, because within a very short period one large colony splits up into a vast number of smaller colonies scattered all over a building. This spreading within a suitable building rapidly leads to very heavy infestations particularly when humidity, temperature, and food supplies are favorable. Humidity is of particular importance; the availability of water appears to be more important in the choice of a new site than proximity to food.

Pharaoh ants utilize many niches within buildings as nest sites, such as behind ceramic wall tiles and light fixtures, within false ceilings, walls, door lintels, window frames, electric clocks, and baseboards, and in the linings of commercial and laboratory refrigerators. Cookers, heated food carts, sinks and wash basins, the backs of drainboards, and dishwashers may also be infested. Automatic vending machines dispensing food and drink are also admirable nesting sites, as are the many machines used in food manufacturing in various factory premises. The list is almost endless. However, the point of entry into a wall or ceiling cavity does not mean that the nest is in the immediate vicinity. I have never actually seen a nest in situ in spite of stripping walls of tiles and investigating numerous wall and ceiling cavities.

Almost any food suitable for animals or humans attracts pharaoh ants. Dead insects, mice, and other similar items also attract large numbers of foraging workers; meat and meat products are particularly sought after, as are cakes and other types of confectionary. Food is stored within the nest in the form of piled up granules that probably have been chewed and partially digested by the ants. Once a trail to a food supply is laid, recruitment of foraging ants follows rapidly. Prominent trails of ants may be seen moving back and forth from the nest entrances to the food source (Figure 14.2).

Foraging ants readily become trapped in sticky sweets. Pharaoh ant workers have been observed to cross hot stoves to reach a supply of minced beef, hamburger, or other similar foods. These ants can be more than just a nuisance; they are capable of carrying food-poisoning organisms (3). There have also been many reports in Britain of these ants biting people, particularly nursing staff, at night (4). Slight rashes have been observed in the area of the alleged bites but severe reactions have not been encountered.

Pharaoh ants have been observed in warehouses in tropical countries where they feed on stored grain and a wide range of other commodities including dried herbs. Although there have been no published reports of these ants infesting factories and catering establishments in tropical areas, such infestations probably occur.

Other Monomorium Species

Other species of *Monomorium* occur from time to time in dwellings, warehouses, and factories. *M. minimum* regularly invades houses in areas of the USA where temperatures are warm enough (19). *M. destructor,* a species dispersed worldwide through commerce, has been responsible

Figure 14.2
Trail of pharaoh ants, *Monomorium pharaonis* (Formicidae), between nest entrances and food source.

for damaging clothes and fabrics and destroying the rubber insulation around the electrical cables in shops in Indonesia. In two cases, ants chewed holes through 0.8-mm lead sheeting around the wiring. *M. latinode* has attacked and damaged rubber articles in Java. *M. barbatulum* has been recorded in granaries in Egypt.

Ochetomyrmex auropunctatus, *Little Fire Ant*

This Neotropical pest has become established in Florida (*19*). Its color ranges from pale brown to golden brown. It is extremely small, measuring some 2 mm long. The little fire ant is very susceptible to cold and does not appear outside the nest until the temperature has risen sufficiently. Nesting occurs from time to time in houses. Meat and a variety of dairy products are favored as food. The colonies become very large, and more than one queen may be present at any given time. When the nest is disturbed or damaged in any way, these ants attack and sting in much the same way as their larger cousins (*Solenopsis invicta* and *S. richteri*) mentioned below.

Paratrechina longicornis, *Crazy Ant*

P. longicornis is now circumtropical in distribution and has spread widely elsewhere by way of world commerce. The dark brown to black workers measure about 2 to 4 mm long. In Britain, crazy ants breed quite successfully in centrally heated buildings and greenhouses. They nest in wall cavities, behind plasterboard, under baseboards, and in many other similar locations. They probably cannot survive outdoors in temperate areas. In warmer areas, outdoor nests are situated in soil that is afforded some protection by rocks or plants, or in rotten wood or garden walls. The workers of this species travel great distances from the nest when foraging. Finding the nest may therefore be difficult. Crazy ants feed readily on a wide variety of food materials including meats, meat products, seeds, fruits, cakes, and biscuits.

P. vividula, a smaller species, is a pest in Britain and in various other places around the world.

Solenopsis Species, *Fire Ants*

There are many species of fire ants ranging through tropical and subtropical regions. Some are also found in more temperate areas such as Britain where *Solenopsis fugax* is endemic. Revisionary work continues on the taxonomy, hybridization, and range of imported fire ants.

Imported Fire Ants

The black imported fire ant, *Solenopsis richteri*, arrived in the USA around 1918 at Mobile, Alabama; now it is found only in a small area in the northeastern corner of Mississippi and the northwestern corner of Alabama. Around 1930, the red imported fire ant, *S. invicta,* appeared in the same area, probably carried there from the Mato Grosso of Brazil. Since then, *S. invicta* (or, more probably, its hybrid with *S. richteri*) has spread over a large area of southeastern USA (*8, 14*).

The biology and morphology of these two imported fire ant species and their hybrids are very similar. They measure some 3–6 mm long. Swarming usually occurs from May to September. Large numbers of winged males and females are seen mostly in the evening following a particularly warm day. After mating, the females shed their wings and search for suitable nesting sites, such as under stones, bricks, or boards on the ground. Some of the new queens may find their way directly into the walls of buildings through various cracks and holes. The queen lays eggs and tends the young until the workers are able to take over these duties. Usually only one queen is found in each colony, but occasionally several may be present.

The workers may be divided into two groups: majors and minors. Major workers rarely sting but they do bite. Minor workers are extremely pugnacious and sting as well as bite. They swarm out in vast numbers when the nest is disturbed and run frantically about vibrating their raised gasters. On finding the intruder, they attack vigorously and hold on by their mandibles; they pull the skin of the enemy outward, swing the gaster forward between their legs, and then sting repeatedly.

Ant venoms are mostly of the proteinaceous type, but the venom of the red imported fire ant is unusual in that it contains a necrotizing toxin, solenamine. Human deaths have been attributed to stings by these ants but none of these reports has been substantiated. Potentially fatal allergic reactions can occur when certain susceptible individuals are stung (*5*).

Imported fire ants prefer food with a high oil content; foods containing sugars are less favored. Stores and warehouses may be invaded by large numbers of foraging workers, and products such as barley, wheat, maize, and rice, particularly of the cracked or ground variety, are carried away and consumed.

The spread of *S. invicta* appears to have been facilitated by man-made disturbances that have apparently removed the natural hazards this species otherwise faces at critical stages of its life history. Red imported fire ants are apparently able to take advantage of their inadvertent carriage in ships, planes, cars, and trucks. They have been observed floating down rivers in vast numbers. Imported fire ants associated with humans seem to occur in far greater numbers and are more persistent than those found in more natural situations where human interference has been minimal (7).

Solenopsis molesta, Thief Ant

This tiny ant, measuring some 2 mm long, is found throughout most of the USA (19). Its color ranges from yellowish-brown to dark brown. The cuticle is shiny. Swarming occurs from the late summer months to early autumn. Nests are found in rotten wood, in soil and, on occasion, within buildings. Colonies can become extremely large. Food with a high protein content is much favored. These ants do not eat sugar. However, many foods used in catering are subject to infestation.

Minor Pests

Several other kinds of ants have been found foraging in houses and in establishments where food is stored or manufactured. By and large, these ants nest mainly outside such buildings, but the foraging workers may enter buildings under certain circumstances.

Lasius brunneus

L. brunneus was first recorded as a domestic pest in Britain in 1958 (12). It was found nesting in a private dwelling where it was raiding the kitchen for cooked meat, jam, and cake. *L. brunneus* is normally a tree-dwelling species, particularly favoring old oak trees. However, the species has been recorded infesting a canteen and nesting in the insulation of a refrigerator. In the instances where nesting occurred in dwellings, the nests were usually located in rotting timbers within the structure of the building. This minor pest is now found in Europe and from North Africa east to Pakistan.

Pheidole *Species*

This is a vast genus containing species from most regions of the world (19). These ants are commonly called bigheaded ants because the head of the major worker is very large in proportion to the rest of the body. Minor workers are rather small, about 2–3 mm long. Colonies are usually found in exposed soils, in rotten wood, or under various objects that give sufficient cover. Colonies are usually small and, as far as is known, nesting occurs only very rarely in buildings (two instances are known of nesting in heated buildings in Britain). Malawi peanuts have been damaged by members of this genus. Bigheaded ants are especially attracted by high-protein foods and honeydew. Sugary foods are also taken. The workers generally forage wherever food for human consumption is handled.

Prenolepis imparis, *False Honey Ant*

False honey ants have a shiny cuticle and a reddish-brown to almost black color (19). Workers are about 2–3 mm long. They are among the earliest ants to swarm, usually in early spring. The nests are only very rarely found in buildings; typically, the colonies are small and are found in shaded soils containing clay. Workers seek sweet foods by preference, but also take live or dead insects and decaying or overripe fruit. Feeding to repletion causes the gaster to become enlarged. Buildings where sugary commodities are manufactured or stored are prime candidates for infestations by *P. imparis.*

Tapinoma sessile, *Ordorous House Ant*

T. sessile is encountered in buildings. Swarming rarely occurs; more typically, mating takes place within the nest in the early part of the summer. Most nests are situated in the walls of buildings, beneath flooring, and at a shallow depth beneath stones and boards. Virtually any kind of food, even photographic film (6), is acceptable to these ants, but sugary foods such as honeydew seem to be preferred. Some colonies may become extremely large and may contain many queens. The workers form well-defined trails to foraging areas.

Control of Pest Ants

Control of any species of insect depends on correct identification, and this is certainly the case with ants. Unless the ant is positively identified and something is known of its life history, a great deal of unnecessary and wasteful insecticidal treatment may take place. It is advisable to catch several individuals if at all possible and dispatch them in a securely sealed, rigid container to a government agency, pest control company, museum, or other resource that can assist in identification. Searching for ant nests is often difficult and time-consuming, because the nest may well be some distance from the point at which workers are seen entering the soil, wall, or other cover.

Pyrethrins

Foraging workers may usually be dealt with safely by applying pyrethrins. These chemicals have a quick knockdown effect and are safe to use in areas where food is present. However, their length of effectiveness is limited, and several treatments may be needed to alleviate the situation. Label instructions should be followed carefully. However, because most ant colonies contain many thousands of workers, this type of treatment cannot readily exterminate the nest. It is purely a measure to temporarily relieve the infestation until more definitive control measures can be exercised.

Residuals

Contact insecticides have their uses in certain situations, but retreatment is likely to be necessary. The chemicals that can be used for such treatment vary greatly all over the world, and local requirements—legal as well as practical—should be first ascertained and then followed. Some worker ants appear to be able to detect and avoid treated areas. In other instances, workers have been observed to suspend foraging activity for several days after treatment. It is most important to use a chemical with a long residual life and to ensure that a sufficiently large area is treated. Spot treatment of only the areas where ants are actually seen is ineffective since the source of the trouble has not been reached.

Baits

Baiting appears to be the ideal method of ant control. It lends itself readily to the ant's habit of carrying food back to the nest for the young and the queens. Unfortunately, the results achieved are sometimes far from what is desired. In some species, an alarm mechanism may be triggered causing nest breakup; the colony rapidly decamps or splits up into several colonies and begins to search for a more suitable site. Even if only a few ants of all stages survive and are able to relocate, the colony may rapidly reestablish itself, especially if several queens are present.

It will appear initially that such treatments have been successful because there will be a rapid drop in the number of visible ants and all will remain quiet for a period of several weeks or months; but then the infestation will suddenly flare up, often in several different places. Ants in desperate situations commonly devour their own larvae, eggs, and other workers. For the ant, survival of the colony is always the overriding factor.

The most effective control to date for *Monomorium pharaonis* is a bait, MaxForce Pharaoh Ant Killer, containing hydramethylnon. Using only this bait (i.e., no other ant control measures such as premises sanitation were used), R. E. Wagner (personal communication, 1989) achieved 80% control in 48 hours and eradication within one week in an apartment house in southern California.

It has been estimated that if only 10% of colonies of imported fire ants survive after an area has been treated, the whole area will rapidly be reinfested by the survivors (17). However, it has been shown that queens of the red imported fire ant from established colonies are unable to reestablish a colony after all the workers have been killed. This is probably due to the fact that an established queen produces and tends eggs in a manner totally different from that of a newly mated queen establishing her first nest cell (20). With *M. pharaonis*, only some 30–50 workers with various preadult stages are needed to restart a small colony.

Biological Control

Control of some ant species, especially *M. pharaonis*, has been achieved by the use of juvenile hormone analogues (JHA). JHA permanently sterilizes the queens of *M. pharaonis* (9–11). Field trials in buildings, where strict sanitation could be enforced, demonstrated that infestations can be completely eradicated within about 18 weeks. Use of JHA has not been quite so successful against the red imported fire ant. The queens were able to recover after feeding on bait containing JHA (21). Later work showed that JHA affected only larvae during the first instar (18).

Investigations into biological control of ants using specific fungal diseases remain in a preliminary stage. Some kinds of ants are known to be susceptible to a host-specific fungal pathogen (15). *Thelohania* species (Nosematidae, Microsporida) have been isolated from black imported fire ant workers in Uruguay and Argentina, and from *Solenopsis* species in Uruguay (2). *S. invicta* is susceptible to *Thelohania* species (1). Many of the sampled colonies of *S. invicta* in Uruguay and Argentina showed low levels of infection in March, but by May these nests had been abandoned or had become drastically reduced in size. The infected colonies were less vigorous and the workers were less aggressive.

Bacillus thuringiensis has been tried with some success against *Monomorium pharaonis* (*23*). It is, of course, far too early to speculate on what role *B. thuringiensis* and *Thelohania* species may play in control of pest species of ants, but these and other pathogens may well become important management tools in the future.

Cited References

1 Allen, G. E., and W. F. M. Buren. 1974. Microsporidian and fungal diseases of *Solenopsis invicta* in Brazil. *Journal of the New York Entomological Society* 82(2)125–130.

2 Allen, G. E., and A. Silveira-Guido. 1974. Occurrence of Microsporida in *Solenopsis richteri* and *Solenopsis* sp. in Uruguay and Argentina. *Florida Entomologist* 57(3)327–329.

3 Beatson, S. H. 1972. Pharaoh's ants as pathogen vectors in hospitals. *Lancet* i(7747)425–427.

4 Beatson, S. H. 1975. Problems of hospital pest control. *Proceedings of the Fourth British Pest Control Association Conference* (St. Helier) 4(7)1–3.

5 Brown, L. L. 1972. Fire ant allergy. *Southern Medical Journal* 65(3)273–277.

6 Brown, L. R. 1969. Odorous house ant feeds on photographic film. *Journal of Economic Entomology* 62(4)955–956.

7 Buren, W. F., G. E. Allen, and R. N. Williams. 1978. Approaches toward possible pest management of the imported fire ants. *Bulletin of Entomological Research* 24(4)418–421.

8 Diffie, S., R. K. Vander Meer, and M. H. Bass. 1988. Discovery of hybrid fire ant populations in Georgia and Alabama. *Journal of Entomological Science* 23(2)187–191.

9 Edwards, J. P. 1975. The effects of juvenile hormone analogues on laboratory colonies of Pharaoh's ants, *Monomorium pharaonis* (L.) (Hymenoptera, Formicidae). *Bulletin of Entomological Research* 65(1)75–80, 1 pl.

10 Edwards, J. P. 1975. The use of juvenile hormone analogues for the control of some domestic insect pests. *Proceedings of the Eighth British Insecticide and Fungicide Conference* 1:267–276.

11 Edwards, J. P., and B. Clarke. 1978. Eradication of Pharaoh's ants with bait containing the insect juvenile hormone analogue methoprene. *International Pest Control* 20(1)5, 6, 8–10.

12 Green, A. A., and J. Kane. 1958. *Lasius brunneus* (Latr.) (Hym., Formicidae) as a domestic pest. *Entomologist's Monthly Magazine* 94(1131)181.

13 Haines, I. H., and J. B. Haines. 1978. Colony structure, seasonality and food requirements of the crazy ant *Anoplolepis longipes* (Jerd.) in Seychelles. *Ecological Entomology* 3(2)109–118.

14 Hedges, S. 1987. Imported fire ant: Nemesis of the South. *Pest Control Technology* 15(5)58–60, 84, 88, 90.

15 Leston, D. 1973. The ant mosaic—tropical tree crops and the limiting of pests and diseases. *Pans* 19(3)311–341.

16 Lewis, T., et al. 1976. The crazy ant (*Anoplolepis longipes* Jerd.) (Hymenoptera, Formicidae) in Seychelles, and its chemical control. *Bulletin of Entomological Research* 66(1)97–111.

17 Morrill, W. L. and J. A. Bass. 1976. Flight and survival of alate red imported fire ants after Mirex treatment. *Journal of the Georgia Entomological Society* 11(3)203–208.

18 Robeau, R. M., and S. B. Vinson. 1976. Effects of juvenile hormone analogues on caste differentiation in the imported fire ant. *Journal of the Georgia Entomological Society* 11(3)198–203.

19 Smith, M. R. 1965. *House-Infesting Ants of the Eastern United States*. Technical Bulletin 1326. U.S. Department of Agriculture, Washington, DC.

20 Stringer, C. E., et al. 1976. Red imported fire ants: Capacity of queens of established colonies and of newly-mated queens to establish colonies in the laboratory. *Annals of the Entomological Society of America* 69(6)1004–1006.

21 Troisi, S. J., and L. M. Riddiford. 1974. Juvenile hormone effects on metamorphosis and reproduction in the fire ant, *Solenopsis invicta*. *Environmental Entomology* 3(1)112–116.

22 Wilson, E. O., F. M. Carpenter, and W. L. Brown, Jr. 1967. The first Mesozoic ants, with the description of a new subfamily. *Psyche* 74(1)1–19.

23 Wisniewski, J. 1975. Control of Pharaoh's ants in zoological gardens with *Bacillus thuringiensis*. *Angewandte Parasitologie* 16(1)43–49.

15

Hymenopterous Parasites of Stored-Food Insect Pests

Gordon Gordh and Hollister Hartman

Parasitism is a very common phenomonen in the animal kingdom. It should come as no surprise, then, that many of the common kinds of moths and beetles that infest stored food have parasites, some of which belong to the Order Hymenoptera. In other words, parasitic wasps are natural components of storage ecosystems. There are two common species of minute, nonparasitic wasps associated with food. The fig wasp, *Blastophaga psenes* (L.) (Agaonidae), is intimately involved in the life of domestic (California) and imported (southern Europe) figs. The fennel seed wasp, *Systole geniculata* Foerster (Eurytomidae), is typically found wherever fennel is stored. In this chapter, the more common kinds of parasitic wasps that the food-pest manager may expect to find in storage premises are pointed out.

Although general reviews of the parasitoid phenomenon have been published (*38, 93*) and there is an identification key to the more common species of parasitoid wasps often found in the storage environment (*43*), there has been no comprehensive review of the parasitic wasps associated with stored foods. The nearest approaches to such a study were papers by Waterston (*94*) and Cotton and Good (*17*). The frequent failure of researchers to identify accurately, even in published reports, the insects associated with stored products is in part responsible for this lack of a comprehensive analysis. Also, the problem is complicated by the fact that pesticides are frequently used, killing the parasites along with the pests, or just killing

the parasites and leaving the insecticide-resistant hosts free to propagate ad libitum.

Nevertheless, there is a large body of information regarding parasites of stored-product pests. An enormous fauna of parasitic Hymenoptera is potentially capable of being associated with insect pests of stored products, but a full appreciation of the extent of this association is hampered by our poor nomenclatural and taxonomic knowledge of the parasitic taxa under consideration. Parasites of stored-product pests are found among the superfamilies Chalcidoidea, Chrysidoidea (= Bethyloidea), Evanioidea, and Ichneumonoidea.

Chalcidoid families with known parasites of storage pests include the chalcidids, encyrtids, eulophids, eupelmids, eurytomids, pteromalids, and trichogrammatids.

The only chrysidoid family represented among parasites of food pests is Bethylidae. Of the other families, sclerogibbids are ectoparasites of nymphal Embioptera; chrysidids attack sawfly larvae, phasmatid eggs, and solitary bee or wasp progeny; and dryinids are endoparasites of auchenorrhynchous Homoptera.

Among the evanioid families, only the Evaniidae are parasites of food pests (cockroaches); aulacids attack only wood-boring Coleoptera and Hymenoptera, whereas gasteruptiids prey upon wood-nesting solitary bees and sphecid wasps.

Of the ichneumonoid families, only the Braconidae and Ichneumonidae are represented. Inappropriate host ranges exclude the remaining families from consideration as parasites of stored-food insects: Aphidiids attack only aphids, and stephanids, only wood-boring Coleoptera and Hymenoptera. Paxylommatid host preferences are poorly known.

Gordh: Division of Biological Control, Department of Entomology, University of California, Riverside, CA 92521.

Hartman: Population Biology Program, University of California, Riverside, CA 92521.

217

Chalcidoidea

The superfamily Chalcidoidea is exceptionally large, geographically cosmopolitan, and biologically exceedingly diverse (41). Numerically, among the superfamilies mentioned here, the Chalcidoidea are most abundantly represented in stored products. Representatives of seven families have been recovered from stored products—Chalcididae, Encyrtidae, Eulophidae, Eupelmidae, Eurytomidae, Pteromalidae, and Trichogrammatidae. Mymaridae have been poorly studied but are also described below.

Chalcididae

About 100 genera and fewer than 1,500 species comprise the family Chalcididae. The family is cosmopolitan in distribution (80), but not all higher taxonomic groups are found in various zoogeographic regions. Chalcidids are characterized by relatively large bodies (among the largest in the Chalcidoidea), robustness, 13-segment antenna, and an enormously enlarged hind femur. In terms of body structure, chalcidids are most similar to the Leucospididae.

Taxonomically, chalcidids have not been studied comprehensively. The Japanese species have been revised (50) and the North American (56) and South American (22–25) faunas have been cataloged. The Lepidoptera is the largest group of hosts attacked, but Diptera, other Hymenoptera, and Coleoptera are also hosts for Chalcididae. Two species of chalcidids have been reported as parasites of *Corcyra cephalonica*: *Antrocephalus aethiopicus* Masi and *A. mahensis* Masi.

Encyrtidae

This cosmopolitan family is among the largest families of the chalcidoids, with about 500 genera and 2,900 species described. Encyrtids are among the most morphologically plastic chalcidoids and thus it is difficult to characterize them. Usually, encyrtids have forward-advanced pygostyli (sometimes located near the propodeum). There is considerable taxonomic activity in the Encyrtidae, owing in part to the importance of this group in applied biological control. There is a host catalog (84) and catalogs of North American (56) and Neotropical species (20–25). The Japanese species (83), the Palaearctic fauna (89, 90), and the Argentine fauna (21) have been reviewed. Keys have been prepared to the genera found in the Nearctic (91) and Ethiopian (72) regions.

Encyrtids are predominantly parasites or hyperparasites of homopterous insects, especially scale insects. Encyrtids attack all immature stages of their hosts and are invariably internal parasites. Although encyrtids have focused on Homoptera, the actual host spectrum of the family is exceedingly broad, ranging from larval and nymphal ticks (*Hunterellus* Howard, *Ixodiphagus* Howard), spider egg sacs (*Amira* Girault), the eggs of other insects (*Ooencyrtus* Ashmead, *Avetianella* Trjapitzin), and aculeate Hymenoptera (*Coelopencyrtus* Timberlake) to Orthoptera (*Leefmansia* Waterston) and cockroach oothecae (*Comperia* Gomes)(63). Encyrtids can be solitary or gregarious; at least two genera are polyembryonic (*Copidosoma* Ratzeburg, *Paralitomastix* Mercet).

Although their host associations are broad and the stages of host attacked are extensive, encyrtids have rarely been associated with stored food. Apparently all species of *Cerchysiella* Girault (= *Zeteticontus* Silvestri) are parasites of larval Coleoptera, and perhaps occasionally attack Diptera (81).

The genus *Comperia* has undergone recent revision (71). Presently, seven species are known, six exclusively from Africa, but the host associations have been confirmed only for *Comperia merceti* (Compere) and *C. alfierri* (Mercet). *C. merceti*, cosmopolitan in distribution, has been recovered from warehouses where it attacks only oothecae of the brownbanded cockroach, *Supella longipalpa* (40). It is commonly taken in buildings where it is frequently found at the windows.

These are the only encyrtids that we have been able to identify as parasites of stored-food pests, but unquestionably more will come to light as host associations for encyrtids become better known and the parasite fauna of the storage habitat is systematically and intensively studied. Candidate genera that we predict will be found in association with insect pests of stored food include *Cerchysius* Westwood, *Tachinaephagus* Ashmead, *Oobius* Trjapitzin, and *Homalotylus* Mayr.

Eulophidae

The Eulophidae, a large family with more than 300 genera and 3,000 species, are cosmopolitan. Most of the known species are parasitic or hyperparasitic (a few are phytophagous). Eulophids are characterized by four-segmented tarsi, the antenna with nine or fewer segments, and frequently with the body lightly sclerotized and therefore shriveled in dried, point-mounted specimens. The Palaearctic (20), Nearctic (56), and Neotropical (22–25) species have been cataloged.

Biologically, eulophids are exceedingly diverse in host associations and stages attacked. Eulophids are egg, larval, and pupal parasites. They develop externally or internally and are solitary or gregarious. The family appears to focus generally on holometabolous hosts, and many species are found in association with leafminers or gall-formers. Although only a few eulophid species have been used extensively in applied biological control, other members of the family certainly merit more careful scrutiny; probably some species are at least potentially important.

The genus *Entedon* is exceptionally large and at

least one species, *E. longiventris* Ratzeburg, has been associated with *Stegobium paniceum*.

The genus *Aprostocetus* Westwood, now including parts of the genus *Tetrastichus* Haliday, is enormous (*47*). It is cosmopolitan in distribution and the host range is incredibly large. *A.* (*Tetrastichodes*) *hagenowii* (Ratzeburg) is worldwide in distribution and attacks the oothecae of several species of cockroaches, most of which are domiciliary (*75*). *Tetrastichus periplanetae* Crawford has been reported from the American cockroach, *Periplaneta americana. T. coerulescens* Ashmead (a senior synonym of *T. doteni* Crawford) is a notorious hyperparasite of other Hymenoptera that are primary parasites on a variety of Lepidoptera and Coleoptera associated with stored products. *T. australiae* Gahan has been reported from the oothecae of the Australian cockroach, *P. australasiae,* in Sumatra. Other records of *Tetrastichus* species remain to be confirmed.

Paraolinx is a New World genus with five species. *P. typica* (Howard) [= *P. nigriventris* (Girault)] has been reported as a parasite of stored-product insects, but this record needs confirmation.

The genus *Melittobia* Westwood is rather small, but is worldwide in distribution. Members of this genus are parasites of aculeate Hymenoptera. However, in the laboratory, these wasps can be induced to parasitize many species of insects. One species, *M. chalybii* Ashmead, has been reported on from pests of stored products, but this identification is questionable because a sibling species complex exists under this name.

Eupelmidae

The Eupelmidae, a moderate-sized family of about 50 genera and 700 species, are cosmopolitan in distribution and morphologically similar to the Encyrtidae. Members of the family attack a variety of hosts in the orders Coleoptera, Diptera, Lepidoptera, Hemiptera, Homoptera, Orthoptera, and Hymenoptera. *Eupelmus cushmani* (Crawford) has been reared from several species of Coleoptera of the families Anthribidae (*17*), Curculionidae, and Bruchidae. It seems likely that several other species of eupelmids could be associated with insect pests of stored food.

Eurytomidae

This highly distinctive family of chalcidoids is cosmopolitan in distribution and moderately large with about 70 genera and 1,000 species. Morphologically, eurytomids are highly distinctive, characterized by a quadrate pronotum, nonmetallic coloration (except *Chrysidea* Spinola), small coxae, robust body, and slender hind femora. Eurytomids have been rather poorly studied, but the world genera have been listed (*10*), the western Palaearctic fauna has been revised (*46*), and the genera of two subfamilies in the Soviet Union have been reviewed (*97*).

Eurytomids have proportionately more phytophagous species than other families of chalcidoids (excluding the Agaonidae). Eurytomids form a biological transformation series; some species are phytophagous and parasitic, whereas others are exclusively parasitic. All of these biological attributes are apparently found in the genus *Eurytoma* Illiger.

The genus *Aximopsis* Ashmead appears to be Neotropical and Oriental in distribution, perhaps tropicopolitan. Two species of *Aximopsis* have been associated with *Araecerus fasciculatus*: *A. javensis* Girault and *A. tephrosiae* Girault (*17*).

One species of the genus *Eurytoma*—*E. tylodermatis* Ashmead—is widespread in the USA and has been recovered from numerous species of bruchids, curculionids, and *A. fasciculatus* (Anthribidae) (*17*). The host list given by Peck (*66*) is far broader than one can imagine and perhaps represents several species that have been misidentified. This is most likely considering the difficulty of identifying species in this genus.

Mymaridae

The Mymaridae, another cosmopolitan family, has about 90 genera and 1,200 described species. Morphologically, mymarids are characterized by their small-to-minute body size, strongly reduced forewing venation, disproportionately long marginal cilia, disproportionately long antenna with a well-developed club in the female, hindwing long and narrow with anterior and posterior margins parallel, and tarsi always four- or five-segmented (never three-segmented).

Taxonomically, mymarids have been poorly studied (owing in part to their size and infrequent occurrence in museum collections). There is one paper on the world genera (*1*); other taxomonic publications on this family appear to be limited to specific geographical regions. It seems that with more intensive investigation this group will be shown to be larger than presently recognized.

Biologically, mymarids are all parasites within the eggs of other insects. There are records about other host stages, but these reports are questionable. Several species have been used in applied biological control, with success ranging from moderate to spectacular (*19*). Mymarids are poorly represented as parasites of stored-food pests, although this is almost certainly a function of sampling techniques and/or the minute size of these parasites.

Pteromalidae

This large, cosmopolitan family of chalcidoids is one of the most difficult to identify because it is structurally diverse and identification is based primarily on negative morphological characteristics.

The world fauna has not been studied comprehensively, and the literature is diverse (5, 45). Presently, there are more than 550 genera and about 3,000 species. The Nearctic fauna has been characterized through numerous isolated descriptions, but there has been no comprehensive study of this group and there are no published keys to the genera. The Nearctic (56) and Neotropical (22–25) species have been cataloged.

Pteromalids are quite diverse biologically. One tribe, the Brachyscelidiphagini (predominantly Australian in distribution), has some species that are gall-formers. However, most species are probably primary parasites or are occasional facultative hyperparasites.

About 25 species of Pteromalidae have been associated with insect pests of stored products, but many of the names reported in the literature are probably misidentifications or junior synonyms of older names. Moreover, there are several pteromalid taxa that have been reported as parasites of stored-product pests, but not in storage situations. A complete taxonomic revision of the family is absolutely necessary before the importance of this group can be accurately assessed. This is especially true of the species in the genera *Pteromalus* Swederus and *Dibrachys* Foerster.

Several pteromalid species attack pests of stored tobacco (2), *Pteromalus* (= *Habrocytus*) *cerealellae* (Ashmead) attacks the Angoumois grain moth, *Sitotroga cerealella* (36), and *Aplastomorpha vandinei* Tucker [= *Anisopteromalus calandrae* (Howard)] parasitizes the rice weevil, *Sitophilus oryzae* (17).

All species of the Cerocephalinae (52) for which the biology is known are parasites of anobiid, scolytid, bostrichid, and curculionid beetles. The group is cosmopolitan in distribution, and one species—*Choetospila elegans* Westwood—is exceptionally common and has been reared from several species of beetles (e.g., *Sitophilus* spp.) found in storage situations.

The genus *Lariophagus* Crawford includes one species, *L. distinguendus* (Foerster), that has been frequently recovered from several species of Coleoptera in storage situations (92). The references to *Meraporus graminicola* in the older literature in all probability refer to *L. distinguendus*.

Meraporus Walker is a small genus found in North America and Europe. *M. requisitus* Tucker is a parasite of *S. oryzae* in North America (17). There is also a problem with *M. requisitus*. Although this parasite may attack *S. oryzae*, it is probably not a member of *Meraporus* as that genus is understood by European workers.

The European species of *Dibrachys* have been revised (45) and there is a key to the North American species (48). *D. cavus* Walker has been reported attacking *Plodia interpunctella* and several species of Coleoptera. The host list of this parasite is unusually long and includes several orders of insects—Lepidoptera, Coleoptera, other Hymenoptera—and spiders (56). It may function in a primary, secondary, or tertiary parasitic role. The species is Holarctic in distribution, but there are records from other zoogeographic regions that must be confirmed.

Most species of the genus *Zatropus* Crawford (8) prefer larvae of Coleoptera (curculionids, bruchids), but other species attack cecidomyiids and microlepidoptera. *Z. incertus* (Ashmead) parasitizes *S. oryzae* and several other species of beetles. Another *Zatropus* species attacks *Caulophilus oryzae* (17).

Dimachus Thomson is a small genus, presumably restricted to the Palaearctic. *D. discolor* (Walker) has been recovered from *Stegobium paniceum* (45) and from *Ptinus tectus* (= *P. ocellus*) (17).

One member of the genus *Ptinobius* Ashmead—*P. texanus* Crawford—has been reared from *Araecerus fasciculatus* (17).

Norbanus Walker is a small but widespread genus in need of taxonomic revision. The biology for most species is unknown, but fragmentary records suggest that members of this genus are mostly parasites of Coleoptera, and occasionally attack sawflies and Lepidoptera. A *Norbanus* species was reported as a parasite of *Lasioderma serricorne* in the Philippines (17).

Pteromalus, a cosmopolitan genus, now also includes species formerly assigned to *Habrocytus* Thompson (6). *Pteromalus puparum* (L.) is a worldwide species for which innumerable hosts have been reported (in our opinion, many of these reports are misidentifications). It is a pupal parasite of butterflies, but it has also been reported as a hyperparasite of ichneumonids and braconids attacking pupae of Lepidoptera (66). Although it has not been reported as a parasite of insect pests in storage, its biological versatility suggests that it may occur adventively in these hosts. Species collected in storage situations include *P. pyrophilus* Kollar on *Sitotroga cerealella*, *P. tritici* Gourlay on *Sitophilis granarius* and *S. oryzae*, and a *Pteromalus* species on *L. serricorne* (17).

Pteromalus cerealellae appears well adapted to the Angoumois grain moth. The parasite is apparently rather host-specific and attacks only the larval stage of the moth, but cannot penetrate the seed until the host larva has expanded its cell to the seed coat. The parasite develops externally and is often cannibalistic (36).

Anisopteromalus calandrae (39, 96) is a cosmopolitan species of Pteromalinae which has been reared from several species of Coleoptera in grain and tobacco and from *Ephestia elutella*. *A. calandrae*, a primary external parasite on the host, is usually solitary, although not to the extreme extent noted for *P. cerealellae*.

Spalangia Latreille (9) is a cosmopolitan genus consisting of about 100 species. Its form is highly distinctive compared to the other subfamilies of Pteromalidae. Most of the host records indicate that *Spalangia* is parasitic on pupae of Muscidae and related families of filth- and carrion-inhabiting flies.

Species of *Spalangia* that attack food pests include *S. cameroni* Perkins, *S. nigroaenea* Curtis, *S. nigripes* Curtis, *S. endius* Walker, and *S. drosophilae* Ashmead (as parasites of *Musca domestica*); *S. cameroni*, and *S. nigra* (on *Dacus curcurbitae*); *S. gemina* Bouček and *S. endius* (on *D. dorsalis*); and *S. endius* (on *Ceratitis capitata*). Host specificity in *Spalangia* has not been studied.

Tritneptis Girault (= *Systellogaster* Gahan) includes only two species (*49*). *T. ovivora* (Gahan) is a parasite of cockroach oothecae, especially those of *Blatta orientalis* (*75*).

Members of the genus *Hemitrichus* Thomson are found in North America and Europe. *H. seniculus* (Nees) has been reported from *Ptinus tectus* (= *P. ocellus*) (*45*).

Trichogrammatidae

This family presently consists of about 70 genera and nearly 500 species. The family is cosmopolitan in distribution and many of the larger genera (*Trichogramma* Westwood, *Aphelinoidea* Girault, *Oligosita* Walker, and *Paracentrobia* Howard) are also found throughout the world. Morphologically, trichogrammatids may be recognized by their minute body size, three-segmented tarsi, and distinctively clubbed antennae with highly reduced number of funicular segments.

Some basic taxonomic work has been done (*26, 62, 68*), but more extensive taxonomic study of this family will reveal that it is composed of substantially greater numbers of species than our present knowledge would indicate.

Biologically, trichogrammatids form a cohesive group in that all species for which the biology is known are exclusively egg parasites of other insects. As such, they are among the smallest insects (*Megaphragma mymaripenne* Timberlake, 0.18 mm long, is apparently the world's smallest insect). The small size of the trichogrammatids makes passive dispersal by wind seem like a probable explanation for their widespread distribution.

Several species of *Trichogramma* Westwood are propagated for the biological control of various agricultural pests throughout the world. *Trichogramma pretiosum* (Riley) attacks *Cadra cautella* and *Plodia interpunctella* (*70*). *T. minutum* Riley parasitizes various stored-product insects. Given our poor knowledge of the taxomony of this genus, and the fact that many species have been repeatedly misidentified, it seems quite likely that several species of *Tricho-*

gramma, in addition to *T. minutum* and *T. pretiosum*, are involved.

Chrysidoidea

The Chrysidoidea (= Bethyloidea) is a primitive group of aculeate Hymenoptera in which the ovipositor is modified into a sting and no longer serves a role in oviposition. The superfamily contains seven families and is cosmopolitan in distribution (*12*). Several families, however, are numerically small and restricted in geographical distribution. All species are apparently parasitic on other insects, but the hosts for many species are unknown. Only Bethylidae are associated with storage pests.

Bethylidae

This is a large family with about 100 genera and 2,000 species (*31*). The family is cosmopolitan in distribution, but it is exceptionally well represented in tropical areas. Some of the taxonomic studies of regional fauna have been undertaken in the New World (*28–33*), France (*4*), England (*67, 72*), and in Oriental (*58*) and Ethiopian (*3*) regions.

Bethylids are distinctive, easily recognized by their antlike habitus, small body size (usually less than 5 mm and nearly always less than 10 mm), the antenna 12- or 13-segmented, the body usually dorsoventrally depressed, and the front legs fossorial; many species are brachypterous or apterous. When wings are present, the venation is typically reduced.

Bethylids have strongly fossorial habits, relatively long lifespans, and high reproductive potentials. Typically, bethylids search out their hosts in concealed situations (e.g., galleries, within seeds, rolled leaves). When they find the host larva, they paralyze it with their sting, and oviposit upon the surface of the body. The larvae develop externally. The parasites are gregarious, with a few to several larvae developing on a single host. After feeding, the larvae spin silken cocoons and pupate adjacent to the shriveled host.

The importance of bethylids in the control of storage pests lies partly in the known host spectrum of these wasps. They are exclusively primary external parasites on the larvae and pupae of Coleoptera (*95*) and microlepidoptera. Bethylid species that attack caterpillars seem restricted to the microlepidoptera (*42, 44, 57*). Most of the bethylid species studied thus far are not host-specific; rather they appear to be habitat-specific, i.e., they attack hosts in concealed situations.

At least 17 species of bethylids in eight genera have been associated with insect pests of stored products. *Cephalonomia* Westwood appears to have the greatest number of species associated with storage insect species. All species of *Cephalonomia* parasitize beetle larvae. Other genera include *Holepyris* Kieffer (often on caterpillars); *Laelius* Ashmead (dermestid larvae);

Goniozus Foerster (caterpillars); *Plastanoxus* Kieffer (ciid, cucujid, anobiid larvae); *Rhabdepyris* Kieffer (beetle larvae); and *Scleroderma* Latreille (beetle larvae).

Evanioidea

The relationship of this rather small superfamily to other apocritous Hymenoptera is uncertain. Three families are currently recognized as constituting the Evanioidea, namely the Evaniidae, Gasteruptiidae, and the Aulacidae. In older classifications, these were frequently assigned to the Proctotrupoidea. The aulacids (about 150 species, worldwide distribution) are parasites of wood-boring Hymenoptera and Coleoptera. The gasteruptiids (about 300 species, cosmopolitan except not reported in Micronesia) are parasites in nests of solitary aculeate Hymenoptera. Only one family, Evaniidae, has been associated with food pests.

Evaniidae

The evaniids represent an isolated group of very poorly known (biologically and systematically) parasitic Hymenoptera. The family is cosmopolitan in distribution and contains 14 genera and about 400 species (*18, 51, 85*). Morphologically, evaniids are characterized by a long, cylindrical petiole placed high on the propodeum (not near the coxae) and a small laterally compressed gaster. Evaniids are unique among the parasitic Hymenoptera in that they have a long anal lobe at the base of the hind wing.

Evaniids are interesting in that all species are parasitic in the oothecae of cockroaches. The family is cosmopolitan, but most species are tropical or subtropical. Although published distributional records indicate the contrary, we are of the impression that evaniids strongly prefer humid habitats. In the United States, evaniids are more common in the eastern part of the country, but some species occur in xeric habitats. In Australia, evaniids are typically coastal. Possibly, evaniids can survive in areas with low humidity provided that water is continually available. Evaniids are frequently collected at drains and windows in domestic situations.

Five species of evaniids have been associated with roaches. These species were, at one time or another, all placed in the genus *Evania* Fabricius. However, taxonomic study of these species shows that at least two have been generically misplaced. The correct names are *Evania appendigaster* (L.), *E. dimidiata* Spinola, *E. erythraspis* Cameron, *Prosevania fuscipes* (Illiger), and *Szepligetella sericea* (Cameron).

Evania appendigaster is the best-known evaniid (*11*). Although it is cosmopolitan in distribution, doubtless having been spread by commerce, it is probably of Oriental origin. The host list for this species is extensive (and probably reflects many misidentifications

of both parasite and host), but includes *Blatta orientalis* and *Periplaneta* spp. (*63*). Cameron (*11*) observed three to four generations of parasites per year; the parasites were more rapid breeders than their hosts. This parasite is solitary within the roach ootheca; all eggs within the ootheca are consumed by the developing larva.

The biology of *Prosevania fuscipes* [= *P. punctata* (Brullé)] is similar to that of *E. appendigaster* (*27*). *P. fuscipes* has been recovered from several species of domiciliary cockroaches.

Szepligetella Bradley includes at least five species in Australia. *S. sericea* reportedly attacks oothecae of the cockroaches *P. americana* and *P. australasiae* in Hawaii (*82*) and is widely distributed in Melanesia, Polynesia, and Micronesia.

Ichneumonoidea

This large group of parasitic Hymenoptera is considered the most primitive superfamily in the Division Parasitica. Most of the species of Ichneumonoidea are rather large-bodied compared to other Parasitica. Characters used to separate the Ichneumonoidea from other superfamilies of parasitic Hymenoptera include the long antenna (14 or more segments), the costal and subcostal veins fused in the forewing, and the (usually) two-segmented trochanter. The Ichneumonoidea of the Holarctic region have been best studied taxonomically.

Five families are presently recognized in the Ichneumonoidea: Aphidiidae, Braconidae, Ichneumonidae, Paxylommatidae (= Hybrizontidae) and Stephanidae. Two families, Braconidae and Ichneumonidae, have been associated frequently with insect pests in stored products; the others have not.

Braconidae

This family is cosmopolitan in distribution and contains about 2,000 species in the Nearctic realm and 10,000 worldwide. Morphologically, it is most closely related to the Aphidiidae and Ichneumonidae and separated from the former on biology and from the latter by having only one recurrent vein in the forewing. Taxonomically and biologically, the Braconidae are relatively well studied (*20, 59, 61, 77–79*).

At least 12 species have been associated with insect pests of stored products. Among these are *Habrobracon crassicornis* (Thomson), *H. kitcheneri* (Dudgeon & Gough), *H. hebetor* (Say), *H. brevicornis* (Wesmael), *Chremylus rubiginosus* (Nees), *Hecabolus sulcatus* Curtis, *Spathius exarator* (L.), *Apanteles araeceri* Wilkinson, *A. carpatus* (Say), *A. nephoptericis* (Packard), *Meteorus ictericus* (Nees), and *Bassus* (= *Agathis*) *hawaiicola* (Ashmead).

Although the host associations are unknown for several species in some subfamilies (e.g., Telengainae, Gnathobraconinae), the host spectrum for the family is not particularly broad when compared with other families of parasitic Hymenoptera. Braconids are predominantly parasitic on larval Lepidoptera, but representatives do attack other Holometabola except Siphonaptera, Trichoptera, and Mecoptera. Typically, braconids attack the immature stages of their hosts, but some (several genera in the Euphorinae) parasitize adult hosts.

Habrobracon hebetor attacks several species of Lepidoptera associated with stored grain and flour (*64, 65*).

Ichneumonidae

The family is cosmopolitan in distribution and is composed of an estimated 60,000 species (*86*). Morphologically, most ichneumonids can be distinguished from other members of the superfamily on the basis of a second recurrent vein in the forewing (a few lack this character). Most ichneumonids have more than 16 segments in the antenna, but this character is not absolute.

Several species have been associated with stored-product insects, but some of these names are apparently *nomina nuda*, and others appear to be misidentifications. The species of ichneumonids known to be parasites of stored-product insects include *Trathala flavoorbitalis* (Cameron), *Venturia canescens* (Gravenhorst), *Mesostenus gracilis* (Cresson), *Hypsicera curvator* (Fabricius), *H. femoralis* (Geoffroy) and, possibly, *Diadegma chrysostictum* (Gmelin).

The genus *Venturia* Schrottky is cosmopolitan and numerically large. The European species have been revised (*54*) and the Ethiopian (*88*) and Nearctic (*56*) species have been cataloged. *Venturia canescens* is a parasite commonly found in many temperate and tropical areas of the world (*14–16, 37, 74*). It has frequently been recovered from buildings where grains and flour are stored. The species is an internal larval parasite, commonly in *Anagasta kuehniella*, *Plodia interpunctella*, and *Cadra cautella* (*70*), but numerous hosts have been demonstrated experimentally (*76*).

The Palaearctic species of *Diadegma* Foerster (*86*) have been revised (*53*). The genus is large and cosmopolitan, with more than 90 Palaearctic species and about 45 Nearctic species. *D. chrysostictum* has been recovered from *A. kuehniella*, *Galleria mellonella*, and *P. interpunctella*. Detailed accounts of the biology and ecology of this parasite with an extensive host list have been published under the name *Horogenes chrysostictos* (*34, 35*). The parasite has been commonly recovered from flour mills in England, often along with *Venturia canescens*. *D. chrysostictum* is strongly attracted to the odors of flour and oatmeal.

Trathala Cameron is another large, cosmopolitan genus, species of which have been recovered from a variety of lepidopterous hosts. *T. flavoorbitalis*, the species associated with stored products, has been taken from several areas in the world including Japan, Australia, India, and Hawaii. It was purposely introduced into the United States in 1933 but has not been recovered subsequently. Hosts include *Ostrinia nubilalis* and *Grapholita molesta*. The parasite is of Oriental origin. Its host spectrum is broad and includes several families of Lepidoptera (*7*). This parasite is probably not an effective natural enemy of stored-food insects because it does not prefer confined spaces and it is readily attracted to light. Laboratory rearing of the parasite may be difficult because it does not copulate under confined conditions.

Mesostenus Gravenhorst is "moderately large and nearly worldwide in distribution" (*56*). The Neotropical species have been revised (*69*). At least one species, *M. gracilis*, has been reported as a parasite of the pupae of the tobacco moth, *Ephestia elutella*, in stored tobacco. This parasite might be an effective natural enemy of the tobacco moth (*2*).

Hypsicera Latreille is a large genus of the Holarctic, Ethiopian, and Oriental realms. *H. femoralis*, one of the four Nearctic species, is distributed throughout the world and is frequently encountered in buildings where it is presumably a parasite of larval Lepidoptera that feed on stored products (*56*). A second species, *H. curvator*, also widely distributed, attacks *Tinea pellionella*.

Proctotrupoidea

The Proctotrupoidea, a large, cosmopolitan superfamily consisting of about 10 families, is most closely related to the Chalcidoidea. Morphologically, proctotrupoids are similar to chalcidoids in their small-to-minute body size and frequently reduced wing venation. They can be distinguished from chalcidoids on the basis of structure of the prothorax (posterolateral margin extending to the tegula). The body is usually more heavily sclerotized than is the case with chalcidoids. The middle tibia usually has two spurs (except Platygasteridae and Scelionidae).

Biologically, all species of Proctotrupoidea are parasitic (a few species are hyperparasitic). No proctotrupoids are phytophagous, gall-formers, or parasitic on leafmining insects or lepidopterous larvae. Some proctotrupoids are more abundant in the tropics, but there are several families that appear to be more cool-adapted and circumboreal in distribution.

Scelionidae

This family, cosmopolitan in distribution, contains nearly 80 genera (*60*). All species for which the biology is known are insect egg parasites or parasitic in the egg sacs of spiders (*13*).

Final Comments

Storing grains, legumes, and so forth, creates a unique habitat for pest and parasite alike. The pest is confronted with a concentrated resource, and those that can adapt to the low-humidity conditions are apt to attain high population densities. This situation imposes several ecological constraints on potential parasites: Parasites will be favored

(a) that are host- rather than habitat-specific or that can quickly evolve specificity for the new habitat;
(b) that are not moisture-dependent or that have mechanisms that expedite survival in low-humidity environments;
(c) that do not require highly dispersed, low-density host availability; and
(d) that effectively exploit a resource continuously available in time.

Members of the family Ichneumonidae are habitat-specific, and evidence is scarce that speciation to new niches proceeds at a pace adequate to include recent man-made habitats. They are very moisture dependent; hatching is delayed in an arid atmosphere (*13*), high-humidity microhabitats are sought, and water must be imbibed daily (*87*). They are very active, strong fliers, tending to disperse away from sites of intended biological control, and they oviposit in widely spaced hosts rather than fully utilizing localized concentrations. Finally, their life cycles are often univoltine, thus limiting their potential reproductive capacity and ability to utilize a dense, continuously available host.

The Braconidae differ ecologically from the Ichneumonidae in several ways, which may account for their apparently greater success as parasites of insect pests of stored food. They exploit niches that the Ichneumonidae have not invaded, such as parasitizing the imaginal stage and insects with incomplete metamorphosis (*61*). The females generally do not insist on widely dispersed hosts for oviposition and thus may be more effective at high pest densities. Life cycles are often multivoltine and need not be completely correlated with host development (especially among the ectoparasitic Cyclostomi).

Juillet (*55*) found that ichneumonid activity was maximum under conditions of moderate temperatures and high humidity, whereas braconids were most active in a high-temperature, low-humidity environment. Adaptations in some subfamilies that may facilitate survival in dry habitats include host-feeding with a tube, embryonic enlargement within the host body, and enclosure of the first-instar larva by the embryonic membrane.

Some chalcidoids also host-feed with a tube, which may allow access to moisture otherwise unavailable in the artificial storage environment. Many are also previously dry-adapted to xeric habitats, which might facilitate a shift to pests of stored products, especially if the parasite's orientation were host- rather than habitat-specific. The factor that may contribute the greatest weight to the success of chalcidoids as parasites is their behavioral plasticity. The presence of numerous sibling species complexes in the Chalcidoidea suggests that this group is in a active state of evolution and speciating rapidly. Rapid speciation would allow them to quickly invade new niches via host-switching and microgeographic isolation which results from their small size and sib-mating behavior.

The Chrysidoidea represent the opposite end of the spectrum from the Ichneumonidae. Females, rather than being active dispersers, are frequently wingless; host-feeding is obligatory rather than occasional; gregariousness is the rule and superparasitism even by several species is tolerated rather than shunned; and development is always external (often with maternal care) rather than internal as is usual for ichneumonids. Thus, many of the factors listed above as disadvantageous to reproductive success for the Ichneumonidae work in favor of the Bethylidae. Low dispersability and tolerance of the presence of others are especially useful for adapting to storage situations where host resources are often highly concentrated. Many bethylids prefer the kinds of hosts that are most detrimental to stored products (wood- and seed-infesting Lepidoptera and Coleoptera).

Cited References

1 Annecke, D. P., and R. L. Doutt. 1961. The genera of the Mymaridae (Hymenoptera: Chalcidoidea). *South Africa Department of Agriculture, Technical Services, Entomology Memoir* 5:1–71.

2 Bare, C. O. 1942. Some natural enemies of stored-tobacco insects, with biological notes. *Journal of Economic Entomology* 35(2)185–189.

3 Benoit, P. L. G. 1963. Monographie des Bethylidae d'Afrique Noire (Hymenoptera). *Musée Royal de l'Afrique Centrale* 119:1–95.

4 Berland, L. 1928. *Faune de France 19. Hymenoptera Vespiformes. II. (Eumenidae, Vespidae, Masaridae, Bethylidae, Dryinidae, Embolemidae).* Lechevalier, Paris.

5 Bouček, Z. 1976. African Pteromalidae (Hymenoptera): New taxa, synonymies and combinations. *Journal of the Entomological Society of Southern Africa* 39(1)9–31.

6 Bouček, Z., and M. W. R. de V. Graham. 1978. British check-list of Chalcidoidea (Hymenoptera): Taxonomic notes and additions. *Entomologist's Gazette* 29(4)225–235.

7 Bradley, W. G., and E. D. Burgess. 1934. *The Biology of* Cremastus flavoorbitalis *(Cameron), an Ichneumonid Parasite of the European Corn Borer.* Technical Bulletin 441. U.S. Department of Agriculture, Washington, DC.

8 Burks, B. D. 1955. A redefinition of the genus *Zatropis,* with descriptions of three new species (Hymenoptera,

Pteromalidae). *Proceedings of the Entomological Society of Washington* 57(1)31–37.

9 Burks, B. D. 1969. Species of *Spalangia* Latreille in the United States National Museum Collection (Hymenoptera: Pteromalidae). *Smithsonian Contributions to Zoology* 2:1–7.

10 Burks, B. D. 1971. A synopsis of the genera of the family Eurytomidae (Hymenoptera: Chalcidoidea). *Transactions of the American Entomological Society* 97(1)1–89.

11 Cameron, E. 1957. On the parasites and predators of the cockroach. II. *Evania appendigaster* (L.). *Bulletin of Entomological Research* 48(1)199–209.

12 Carpenter, J. M. 1986. Cladistics of the Chrysidoidea (Hymenoptera). *Journal of the New York Entomological Society* 94(3)303–330.

13 Clausen, C. P. 1940. *Entomophagous Insects.* McGraw-Hill, New York, NY.

14 Corbet, S. A. 1968. The influence of *Ephestia kuehniella* on the development of its parasite *Nemeritis canescens. Journal of Experimental Biology* 48(2)291–304.

15 Corbet, S. A. 1971. Mandibular gland secretion of larvae of the flour moth, *Anagasta kuehniella,* contains an epideictic pheromone and elicits oviposition movements in a hymenopteran parasite. *Nature* 232(5311)481–484.

16 Corbet, S. A., and S. Rotheram. 1965. The life history of the ichneumonid *Nemeritis* (*Devorgilla*) *canescens* (Gravenhorst) as a parasite of the Mediterranean flour moth *Ephestia* (*Anagasta*) *kuehniella* Zeller, under laboratory conditions. *Proceedings of the Royal Society of London* A40(4–6)67–72.

17 Cotton, R. T., and N. E. Good. 1937. Annotated list of the insects and mites associated with stored grain and cereal products, and of their arthropod parasites and predators. *USDA Miscellaneous Publication* 258:1–81.

18 Crosskey, R. W. 1951. The morphology, taxonomy, and biology of the British Evanioidea (Hymenoptera). *Transactions of the Royal Entomological Society of London* 102(5)247–301.

19 De Bach, P. 1964. *Biological Control of Insect Pests and Weeds.* Reinhold, New York.

20 Delucchi, V., and G. Remandiere (eds.). 1971. *Index of Entomophagous Insects.* Le François, Paris.

21 De Santis, L. 1963. Encírtidos de la República Argentina (Hymenoptera: Chalcidoidea). *Anales de la Comisión de Investigaciones Científicas* 4:9–422.

22 De Santis, L. 1967. *Catálogo de los Himenópteros Argentinos de la Serie Parasítica, Incluyendo Bethyloidea.* Comisión de Investigaciones Científicas, Provincia de Buenos Aires Gobernación, La Plata.

23 De Santis, L. 1979. *Catálogo de los Himenópteros Calcidoideos de América al Sur de los Estados Unidos.* Publicación Especial. Provincia de Buenos Aires Comisión de Investigaciones Científicas, La Plata.

24 De Santis, L. 1980. *Catálogo de los Himenópteros Brasileños de la Serie Parasítica Incluyendo Bethyloidea.* Universidade Federal do Paraná, Curitiba.

25 De Santis, L. 1981 [1983]. Catálogo de los himenópteros calcidoideos de América al sur de los Estados Unidos. Primer supplemento. *Revista Peruana de Entomología* 24(1)1–38.

26 Doutt, R. L., and G. Viggiani. 1968. The classification of the Trichogrammatidae (Hymenoptera: Chalcidoidea). *Proceedings of the California Academy of Science* (4th series) 35(20)477–586.

27 Edmunds, L. R. 1954. A study of the biology and life history of *Prosevania punctata* (Brullé) with notes on additional species (Hymenoptera: Evaniidae). *Annals of the Entomological Society of America* 47(4)575–592.

28 Evans, H. E. 1964. A synopsis of the American Bethylidae (Hymenoptera, Aculeata). *Bulletin of the Museum of Comparative Zoology* 132(1)1–222.

29 Evans, H. E. 1970. West Indian wasps of the subfamilies Epyrinae and Bethylinae (Hymenoptera: Bethylidae). *Proceedings of the Entomological Society of Washington* 72(3)340–356.

30 Evans, H. E. 1977. A revision of the genus *Holepyris* in the Americas (Hymenoptera: Bethylidae). *Transactions of the American Entomological Society* 103(3)531–579.

31 Evans, H. E. 1978. The Bethylidae of America North of Mexico. *Memoirs of the American Entomological Institute* 27:1–332.

32 Evans, H. E. 1979. The genus *Dissomphalus* in northwestern South America (Hymenoptera: Bethylidae). *Proceedings of the Entomological Society of Washington* 81(2)276–284.

33 Evans, H. E. 1979. Additions to knowledge of the bethylid fauna of Hispaniola (Hymenoptera: Bethylidae). *Proceedings of the Entomological Society of Washington* 81(3)456–459.

34 Fisher, R. C. 1959. Life history and ecology of *Horogenes chrysostictos* Gmelin (Hymenoptera, Ichneumonidae), a parasite of *Ephestia sericarium* Scott (Lepidoptera, Phycitidae). *Canadian Journal of Zoology* 37(4)429–446.

35 Fisher, R. C. 1962. The effect of multiparasitism on populations of two parasites and their host. *Ecology* 43(2)314–316.

36 Fulton, B. B. 1933. Notes on *Habrocytus cerealellae,* parasite of the Angoumois grain moth. *Annals of the Entomological Society of America* 26(4)536–553.

37 Ganesalingam, V. K. 1974. Mechanism of discrimination between parasitized and unparasitized hosts by *Venturia canescens* (Hymenoptera: Ichneumonidae). *Entomologia Experimentalis et Applicata* 17(1)36–44.

38 Gauld, I., and B. Bolton (eds.). 1988. *The Hymenoptera,* Oxford University, Cambridge, England.

39 Ghani, M. A., and H. L. Sweetman. 1955. Ecological studies on the granary weevil parasite, *Aplastomorpha calandrae* (Howard). *Biologia* 1(2)115–139.

40 Gordh, G. 1973. Biological investigations on *Comperia merceti* (Compere), an encyrtid parasite of the cockroach *Supella longipalpa* (Serville). *Journal of Entomology* A47(2)115–123.

41 Gordh, G. 1975. Some evolutionary trends in the Chalcidoidea (Hymenoptera) with particular reference to host preference. *Journal of the New York Entomological Society* 83(4)279–280.

42 Gordh, G. 1976. Goniozus gallicola *Fouts, a Parasite of Moth Larvae with Notes on other Bethylids (Hymenoptera: Bethylidae, Lepidoptera: Gelechiidae).* Technical Bulletin 1524. U.S. Department of Agriculture, Washington, DC.

43 Gordh, G. 1991. Parasitic wasps (Apocrita, Hymenoptera). In *Insect and Mite Pests in Food: An Illustrated Key,* J. R. Gorham, ed. Agriculture Handbook 655. U.S. Department of Agriculture, Washington, DC.

44 Gordh, G., and H. E. Evans. 1976. A new species of *Goniozus* imported into California from Ethiopia for the biological control of pink bollworm and some notes on the taxonomic status of *Parasierola* and *Goniozus* (Hymenoptera: Bethylidae). *Proceedings of the Entomological Society of Washington* 78(4)479–489.

45 Graham, M. W. R. de V. 1969. The Pteromalidae of north-western Europe (Hymenoptera: Chalcidoidea). *Bulletin of the British Museum (Natural History) Entomology, Supplement* 16:1–908.

46 Graham, M. W. R. de V. 1970. Taxonomic notes on some western Palearctic Eurytomidae (Hymenoptera: Chalcidoidea). *Proceedings of the Royal Entomological Society of London* B39(10)139–152.

47 Graham, M. W. R. de V. 1987. A reclassification of the European Tetrastichinae (Hymenoptera: Eulophidae), with a revision of certain genera. *Bulletin of the British Museum of Natural History (Entomology)* 55(1)1–392.

48 Grissell, E. E. 1974. A new *Dibrachys* with a key to the Nearctic species (Hymenoptera: Pteromalidae). *Florida Entomologist* 57(3)313–320.

49 Grissell, E. E. 1985. Some nomenclatural changes in the Chalcidoidea (Hymenoptera). *Proceedings of the Entomological Society of Washington* 87(2)350–355.

50 Habu, A. 1960. A revision of the Chalcididae (Hymenoptera) of Japan, with descriptions of sixteen new species. *Bulletin of the National Institute of Agricultural Science* C11:131–363.

51 Hedicke, H. 1939. Evaniidae, Part 9. In *Hymenopterorum Catalogus.* Dr. W. Junk, The Hague.

52 Hedqvist, K.-J. 1969. Notes on Cerocephalini with descriptions of new genera and species (Hymenoptera: Chalcidoidea: Pteromalidae). *Proceedings of the Entomological Society of Washington* 71(3)449–467.

53 Horstmann, K. 1973. Nachtrag zur Revision der europäischen *Diadegma*-Arten. *Beiträge zur Entomologie* 23(1–4)131–150.

54 Horstmann, K. 1973. Übersichtüber die europäischen Arten der Gattung *Venturia* Schrottky (Hymenoptera, Ichneumonidae). *Mitteilungen der Deutschen Entomologischen Gesellschaft* 32(1)7–12.

55 Juillet, J. A. 1964. Influence of weather on flight activity of parasitic Hymenoptera. *Canadian Journal of Zoology* 42(6)1133–1141.

56 Krombein, K. V., P. D. Hurd, Jr., and D. R. Smith (eds.). 1979. *Catalog of Hymenoptera in America North of Mexico,* 3 vols. Smithsonian Institution, Washington, DC.

57 Kühne, H., and G. Becker. 1974. Zur Biologie und Ökologie von *Scleroderma domesticum* Latreille (Bethylidae, Hymenoptera), einem Parasiten holzzerstörender Insektenlarven. *Zeitschrift für Angewandte Entomologie* 76(3)278–303.

58 Kurian, C. 1955. Bethyloidea (Hymenoptera) from India. *Agra University Journal of Research* 4:67–156.

59 Marsh, P. M., S. R. Shaw, and R. A. Wharton. 1987. An identification manual for the North American genera of the family Braconidae (Hymenoptera). *Memoirs of the Entomological Society of Washington* 13:1–99.

60 Masner, L. 1980. Key to the genera of Scelionidae of the Holarctic Region, with descriptions of new genera and species (Hymenoptera: Proctotrupoidea). *Memoirs of the Entomological Society of Canada* 113:1–54.

61 Matthews, R. W. 1974. Biology of Braconidae. *Annual Review of Entomology* 19:15–32.

62 Nagaraja, H., and S. Nagarkatti. 1973. A key to some New World species of *Trichogramma* (Hymenoptera: Trichogrammatidae), with descriptions of four new species. *Proceedings of the Entomological Society of Washington* 75(3)288–297.

63 Narasimham, A. U., and T. Sankaran. 1979. Domiciliary cockroaches and their oothecal parasites in India. *Entomophaga* 24(3)273–279.

64 Payne, N. M. 1933. The differential effect of environmental factors upon *Microbracon hebetor* Say (Hymenoptera: Braconidae) and its host *Ephestia kühniella* Zeller (Lepidoptera: Pyralidae). I. *Biological Bulletin* 65(2)187–205.

65 Payne, N. M. 1934. The differential effect of environmental factors upon *Microbracon hebetor* Say (Hymenoptera: Braconidae) and its host, *Ephestia kühniella* Zeller (Lepidoptera: Pyralidae). II. *Ecological Monographs* 4(1)1–46.

66 Peck, O. 1963. A catalog of the Nearctic Chalcidoidea (Insecta: Hymenoptera). *Canadian Entomologist,* Supplement 30:1–1092.

67 Perkins, J. F. 1976. Hymenoptera. Superfamily Bethyloidea (excluding Chrysidae). *Handbooks for the Identification of British Insects* 6(3a)1–38.

68 Pinto, J. D., G. R. Platner, and E. R. Oatman. 1978. Clarification of the identity of several common species of North American *Trichogramma* (Hymenoptera: Trichogrammatidae). *Annals of the Entomological Society of America* 71(2)169–180.

69 Porter, C. 1973. A revision of the South American species of *Mesostenus* (Hymenoptera, Ichneumonidae). *Acta Zoológica Lilloana* 30:227–267.

70 Press, J. W. 1989. Compatability of *Xylocoris flavipes* (Hempitera: Anthocoridae) and *Venturia canescens* (Hymenoptera: Ichneumonidae) for suppression of the almond moth, *Cadra cautella* (Lepidoptera: Pyralidae). *Journal of Entomological Science* 24(1)156–160.

71 Prinsloo, G. L., and D. P. Annecke. 1978. On some new and described Encyrtidae (Hymenoptera: Chalcidoidea) from the Ethiopian Region. *Journal of the Entomological Society of Southern Africa* 41(2)311–331.

72 Prinsloo, G. L., and D. P. Annecke. 1979. A key to the genera of Encyrtidae from the Ethiopian Region, with descriptions of three new genera (Hymenoptera: Chalcidoidea). *Journal of the Entomological Society of Southern Africa* 42(2)349–382.

73 Richards, O. W. 1939. The British Bethylidae (*s.s.*). *Transactions of the Royal Entomological Society of London* 89(8)297–305.

74 Rogers, D. 1972. The ichneumon wasp *Venturia canescens*: Oviposition and avoidance of superparasitism. *Entomologia Experimentalis et Applicata* 15(3)190–194.

75 Roth, L. M., and E. R. Willis. 1960. The biotic associations of cockroaches. *Smithsonian Miscellaneous Collections* 141:1–470.

76 Salt, G. 1975. The fate of an internal parasitoid, *Nemeritis canescens,* in a variety of insects. *Transactions of the Royal Entomological Society of London* 127(2)141–161.

77 Shenefelt, R. D. 1965. A contribution towards knowledge of the world literature regarding Braconidae (Hymenoptera: Braconidae). *Beiträge zur Entomologie* 15(3–4)243–500.

78 Shenefelt, R. D. 1969–1978. Braconidae 1–8, 10. In *Hymenopterorum Catalogus.* Dr. W. Junk, The Hague.

79 Shenefelt, R. D., and P. Marsh. 1976. Braconidae 9. In *Annals of the Entomological Society of America* 71(2)169–180.

80 Steffan, J.-R. 1958. La distribution geographique des Chalcididae (Hymenoptera). *Proceedings of the Tenth International Congress of Entomology* (Montreal, 1956) 1:799–804.

81 Subba Rao, B. R. 1972. On *Zeteticontus* Silvestri, with descriptions of three new species (Hymenoptera: Encyrtidae). *Entomophaga* 17(2)179–195.

82 Swezey, O. H. 1929. Notes on the egg-parasites of insects in Hawaii. *Proceedings of the Hawaiian Entomological Society* 7(2)282–292.

83 Tachikawa, T. 1963. Revisional studies on the Encyrtidae of Japan (Hymenoptera: Chalcidoidea). *Memoirs of Ehime University (section VI, Agriculture)* 9:1–264.

84 Tachikawa, T. 1978. Hosts of the Encyrtidae of the world (Hymenoptera: Chalcidoidea). *Transactions of Shikoku Entomological Society* 14:43–63.

85 Townes, H. K. 1949. The Nearctic species of Evaniidae (Hymenoptera). *Proceedings of the U.S. National Museum* 99(3253)525–539.

86 Townes, H. K. 1969. The genera of Ichneumonidae, part 3. *Memoirs of the American Entomological Institute* 13:1–307.

87 Townes, H. K. 1972. Ichneumonidae as biological control agents. In *Proceedings, Tall Timbers Conference on Ecological Animal Control by Habitat Management* (Tallahassee, FL, 1971) 3:235–248.

88 Townes, H. K., and M. Townes. 1973. A catalog and reclassification of the Ethiopian Ichneumonidae. *Memoirs of the American Entomological Institute* 19:1–416.

89 Trjapitzin, V. A. 1971. Review of genera of Palearctic encyrtids (Hymenoptera: Encyrtidae). *Transactions of the All-Union Entomological Society* 54:68–119 (in Russian).

90 Trjapitzin, V. A. 1978. *Handbook of Insects of the European Part of the USSR* (part 2) 3:236–328 (in Russian).

91 Trjapitzin, V. A., and G. Gordh. 1978. A review of the Nearctic Encyrtidae (Hymenoptera, Chalcidoidea). *Entomological Review* 57:257–270, 437–448.

92 Van den Assem, J. 1971. Some experiments on sex ratio and sex regulation in the pteromalid *Lariophagus distinguendus. Netherlands Journal of Zoology* 21:373–402.

93 Waage, J., and D. Greathead (eds.). 1986. *Insect Parasitoids.* Academic Press, Orlando, FL.

94 Waterston, J. 1921. Report on parasitic Hymenoptera bred from pests of stored grain. *Royal Society of London Grain Pests (War) Commission Report* 9:8–32.

95 Yamada, Y. 1955. Studies on the natural enemy of the woollen pest, *Anthrenus verbasci* Linné (*Allepyris microneurus* Kieffer) (Hymenoptera, Bethylidae). *Mushi* 28(3)13–29.

96 Yoshida, S. 1978. Behaviour of males in relation to the female sex pheromone in the parasitoid wasp, *Anisopteromalus calandrae* (Hymenoptera: Pteromalidae). *Entomologia Experimentalis et Applicata* 23(2)152–162.

97 Zerova, M. D. 1976. Chalcid family Eurytomidae, subfamilies Rileyinae and Harmolitinae. Fauna USSR, Hymenoptera. *Zoological Institute of the Academy of Sciences of the USSR* 7:1–229 (in Russian).

16
Pest Bird Ecology and Management

William B. Jackson

Infestation or invasion of food production or storage facilities by birds may result in contamination problems. Too often the band-aid approach of attempting to deal with the situation without evaluating the environment or understanding the pest species leads to frustration, lack of success, and further contamination problems.

Birds are a very different kind of pest; simply regarding them as flying rodents, as many are prone to do, will result in an inept management program. Successful bird management requires knowledge of the biology, ecology, and behavior of the species, as well as an understanding of population dynamics (3). Although such an endeavor has its difficulties, positive results are possible, and the professional and financial rewards can be gratifying (27).

Avian Physiology and Behavior

Since birds as well as their products can be contaminants, examination of avian structures and knowledge of behavioral patterns are necessary. Feathers are unique to birds, but there are other numerous, but less obvious, structural and physiological modifications that have been essential to the evolution of the bird as a flying machine. Flight enables birds to travel vast distances, to transport diverse microorganisms, to invade many habitats and niches, and to move

Jackson: Department of Biological Sciences, Bowling Green State University, Bowling Green, OH 43403.

daily between feeding and nesting or roosting sites.

Feathers

Feathers are produced by follicles in the skin. When fully erupted, feathers contain no living tissues. Typically, feathers are molted twice a year, but one of these plumage changes may be only partial. Old feathers, pushed out by new ones, flutter earthward, possibly landing on food products or containers or otherwise contaminating the environment of the food facility.

Any feather fragment in food constitutes contamination, but being able to identify the source may be necessary to eliminate the problem. Keys for feather identification have been published (5, 9), and although specific identification of whole feathers and sometimes of fragments is possible, few laboratories have the expertise.

Feathers give birds a smooth contour and provide aerodynamic qualities. They number more than 2,000 on a small bird and insulate the body against heat loss, thus conserving energy needed for flight. The normal bird, in preening, keeps the feathers oiled and in place. When a surfactant is applied to birds, this barrier is broken and, thus, the insulating properties of the feathers are diminished or lost. This leads to the development of hypothermia with potentially fatal results. Such a strategy is involved with the use of Tergitol (PA-14) for the control of roosting blackbirds (34).

Feathers also provide an environment for ectoparasites such as mites or lice. These arthropods may get separated from their hosts and become incorporated into a product, find harborage within a nest or crevice, or attach themselves to a human or domestic animal substitute. Most of these ectoparasites are merely nuisances, but bite irritation and disease transmission are possible in certain situations.

Leg Structure

Reflection over a Thanksgiving drumstick will bring the realization that the bird leg has an extra segment when compared to a human leg. This has occurred with the separation and fusion of several bones and the elongation of the ankle. As a result, the hunching down of a bird on its perch causes the Achilles tendon, which controls toe flexion (curling), to be pulled over a joint. The more the bird settles down, the tighter its grip becomes on the perch. This explains why sleeping birds do not fall off their perches.

Although the typical songbird utilizes such a mechanism, not all birds have this kind of a grasping foot. Ducks are flat-footed; pigeons do not readily perch on twigs but prefer larger branches or flat ledges. Consequently, toxic wick perches (Rid-A-Bird) are manufactured in two basic styles: a round configuration for perching birds (sparrows, starlings) and a flat style for pigeons (17).

Even though the skin on the feet of birds is scaly and highly cornified, some sensitivity remains. Sticky repellents (e.g., Tanglefoot, Roost-No-More, 4 the Birds, Hot Foot) are effective because birds react adversely to the feel of the material. Toxic substances from perches can be absorbed through the feet despite their protective sheathing.

Disease Associations

Pathogens and other microorganisms can be carried on the feet and legs of birds, just as birds can transport algae to isolated bodies of water (25, 32). That salmonella bacteria (and other pathogens) might be moved mechanically from fecal-contaminated roosts to grain stores or food-processing operations seems likely, but the practical importance of this probable transfer has not been studied or documented. Similarly, birds may be involved in moving pathogens from one animal feeding operation to another. For example, it has been suggested that starlings are responsible for transmission of gastroenteritis to swine (16, 39).

Birds have been implicated as reservoirs, vectors, hosts, or mechanical carriers for many other pathogens and parasites. Of these, St. Louis encephalitis is the most notable; it is carried by urban birds and transmitted to humans by mosquitoes (28). Deer ticks, vectors of Lyme disease, may be distributed by migrating birds, perhaps accounting for the apparent rapid spread of the disease in the USA (1).

Fungal diseases (especially histoplasmosis) have been associated with established bird roosts where droppings accumulate, mix with the soil, and provide a rich culture medium for fungus development. Subsequent environmental disturbances (e.g., raking, construction) disperse spores to be inhaled by workers or nearby observers. A serious tuberculosis-like infection can result. A detailed account of the diseases associated with three common pest species has been published (40), but that reference lacks both a bibliography and a critical evaluation of the epidemiological importance of birds as reservoirs of human disease.

Uric Acid Production

Birds, like their reptilian ancestors, do not produce soluble urea as a metabolic waste product. Instead, insoluble uric acid is formed. It is the white color of this material that gives accumulations of bird droppings their characteristic appearance. Such materials, collecting on metal surfaces, are corrosive, unsightly, and difficult to wash away.

Alimentation

Mastication occurs not in the mouth but in the gizzard, a highly muscular portion of the stomach, where gravel or sand is required as "grist for the mill." Consequently, food items can be swallowed whole, cracked or broken by birds with heavy bills, or ripped out in chunks by raptors. Feeding habit studies based on stomach content analysis are made difficult because of the rapid and thorough grinding of food in the gizzard. Although not always possible, analysis of crop (an outpocketing of the esophagus) contents provides a better basis for assessing feeding preferences. In some species, the crop is utilized for temporary food storage. It is the crop that permits rapid feeding on an abundant food source and, at nesting time, the transport and subsequent regurgitation of food for nestlings. In pigeons, the lining cells of the crop, under control of the same hormone that induces lactation in mammals, slough off, forming "pigeon milk." Waste products from the digestive tract and those from the kidneys comingle in the cloaca and are then expelled from the body.

Brain Structure and Behavior

The development and specialization of the

bird brain has several implications for pest management.

Cerebrum

The cerebrum in birds is smooth and small in contrast to that of mammals which is large and convoluted. Since a convoluted cerebrum is associated with complex learning pathways, it can be concluded from the simpler morphology of the avian cerebrum that birds are largely dependent upon innate (instinctive) behavior and relatively simple learning configurations. Thus, basic flight, song, courtship, nesting, and migration behaviors are not learned. However, trial-and-error experience or imitation (especially early in life) may modify some of the basic, unlearned behaviors.

Odor Perception

The part of the brain responsible for odor perception is relatively small, causing many anatomists to assume that all chemical sensation capabilities are poorly developed in birds. This conclusion seems to be supported by the bird's lack of oral mastication and subsequently its lack of opportunity for tasting. However, in recent years, more critical studies have demonstrated that discrimination among many substances does in fact occur (11, 19, 30, 31, 38). This attribute can be utilized in the protection of seeds and maturing crops with repellent chemicals such as Mesurol (6, 8, 18, 20).

Feeding Behavior

Feeding patterns are part of the daily routine. This is why birds usually must be baited over a period of weeks at a given site or open trap prior to the use of toxicants or closing the trap door. Selectivity for control efforts can be achieved by bait size or location of the baiting site. Pigeons accept a bait of whole-kernel corn, but most nontarget species do not. Sparrows feed most readily on the ground. Pigeons accept elevated sites. In addition, limiting the potential for exposure to toxic baits to the November-March period in northern states further reduces accidental poisoning of nontarget species, since most of these have migrated south for the winter. (*Note*: The use of strychnine-treated grain baits, often used in the past to effect rapid population reductions, currently is prohibited under a federal court order.)

Auditory Responses

Call notes are part of a bird's inborn behavioral repertoire, with individuals responding automatically to specific calls. In early control efforts, the squawks of birds held by their feet in front of a microphone were recorded, amplified, and played back. Birds responded to these distress calls and sometimes left the area. Later, alarm and distress calls were distinguished, the former being found to be more effective. However, recordings tended to be species specific. More recently, synthetic alarm tapes have been developed (e.g., Megastress, AvAlarm), considerably expanding the versatility of this approach and making commercialization possible (*14*).

The hearing range of birds is similar to that of humans. Loud noises that frighten or stress us are likely to frighten birds, hence the success of acetylene or propane exploders in moving blackbirds out of corn fields. Recent claims for ultrasonic repellers, which emit sounds beyond the hearing ranges of birds as well as humans, should be regarded with caution. No experimental or test data documenting efficacy have been provided, but this does not deny the possible value of such devices.

Patterns of Movement

In its daily routine, a bird may roost (or nest) in one area, feed in another area, and loaf in still a third area. Movement patterns between these areas become established and difficult to change. As a result, routinized behavior makes up the predominant part of a bird's behavioral repertoire. Birds, indeed, are creatures of habit.

Sparrows, for example, often attempt repeatedly to build a nest on a loading dock even though their work is removed or destroyed daily. Pigeons often persist in roosting on ledges even in the presence of repellents. Starlings and blackbirds daily return to the same roost trees. Repeated efforts, over successive days (or even weeks), may be required to dislodge birds from their established patterns. (See discussion, below, on Management of Movement.)

In addition to daily movement routines, most bird species have an innate, twice-annual migration rhythm. This means that some protected species are problems only on wintering grounds or when assembled in premigratory or migratory flocks early in the spring or late in the summer. Others are destructive only when feeding their young on ripe cherries or other fruits. Thus, the definition of *pest* may have temporal or geographic components.

The principal pest species (house sparrows, pigeons, and starlings), however, are not migratory, though some seasonal movements of starlings are known (*21*); local, seasonal shifts, with the formation of winter aggregations, occur commonly. Some pest control operators (PCOs) have utilized such knowledge in securing a removal contract just prior to the spring breakup of these flocks and have then taken credit (and pay) for the natural event.

Roosting and Nesting

Winter flocks use shrubbery and trees in shopping malls, ivy on building walls, overhangs on shipping docks or open sheds, or clumps of decorative evergreens in a landscaping design for night roosts. Protected ledges on buildings or electric signs also may be utilized. Whether or not this behavior is considered to be a pest problem is determined by the numbers and species of birds as well as by the proximity of the roosts to food-processing, storage, and shipment operations, or to publicly used areas.

Most birds have a brief nesting period coinciding with a season of favorable weather and abundant food supplies. When a nest or even the clutch of eggs is destroyed, most female birds make renesting attempts. Many species begin a second nesting even when the first one has been successful. Pigeons, unlike other species, can reproduce during much of the year, though most nestings are made in the spring and the summer.

Figure 16.1
Survival curves for starlings, grackles, and robins banded as nestlings. (See also Table 16.1 for references.)

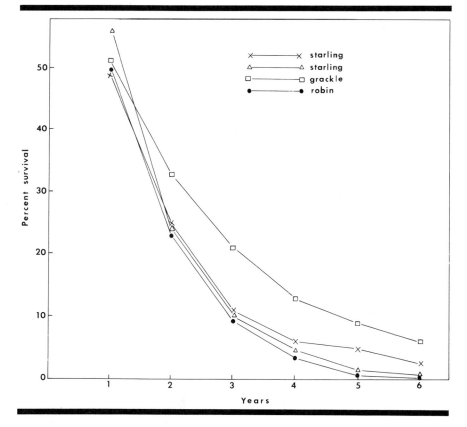

Nest Location

Pest birds build nests on or in structures; repeated removal often does not discourage them. House sparrows (a kind of weaver finch) build loosely woven nests within cavities or behind protective surfaces; they may chip away at styrofoam insulation and enlarge existing openings. Sometimes they build the nest in a tree or a bush, a reversion to ancestral nesting patterns. Starlings, similarly, are cavity nesters, usually being restricted to hollow trees or to cavities (e.g., light fixtures, vents) in the exterior surfaces of buildings. Pigeons utilize ledges or other flat surfaces under many different conditions.

Birds that nest in cavities or protected sites, and even those with exposed nests, often utilize a nearby perch as an integral part of their approach pattern. This behavioral characteristic makes toxic perch installations (Rid-A-Bird) successful.

Fire Hazards

Nests constitute a fire hazard when they are built in switch boxes, sporadically used heaters, or electric motors. Sparrows sometimes carry still-glowing cigarette butts to their grass-fiber nests, with disastrous consequences to buildings. Bird-related fire losses have been estimated at $8.5 million annually (*33*).

Management Strategies

Our concern is less with numbers of nests built (though these and their associated inquilines constitute important sources of contamination in themselves) than with the number of birds fledged, for these are the individuals making up the bulk of the next season's breeding population. Normal mortality rates for birds are generally high (Figure 16.1, Table 16.1); appropriate management techniques for pest birds should be designed to push these rates even higher. Krakowski and Payne (*23*) point out that nests are highly vulnerable, and that concentrating control efforts on nest removal (or prevention) is critical for pigeon management programs.

Ideally, chemosterilants (e.g., Ornitrol) could be used to limit populations by reducing successful reproduction. However, the production of infertile eggs demonstrated in the laboratory has not yet been transferable to feral populations, perhaps because of deficient formulations (*12, 42*). Improved effectiveness is claimed for more recent formulations.

Seasonal Stress

The most stressful season for birds is winter

when limited food supplies, low temperatures, and wind result in greater thermoregulatory stress and increased mortality. Reductional programs in the fall are inefficient, since normal winter mortality removes many of these birds anyway. Early spring trapping or poisoning has a higher benefit/cost ratio. On the other hand, it may be the fall flocks that constitute the greatest threat to food storage or processing operations; this threat may require an appropriate and more immediate preventive or corrective campaign against the birds.

Because birds are so mobile, a biological vacuum created at one facility by a control program may be relatively short-lived. Often, by the next breeding season, young birds will have begun to use the site unless they have been inhibited by sanitation (reduction of the available food supply) and birdproofing (closing of access to roosting and nesting areas) (see Chapter 22).

Localized Control

Measures that involve managing the movement of birds can be used for local control. Because small things can spook birds, a pigeon exhibiting the signs of strychnine toxicosis may cause the rest of the flock to veer away from feeding at that site. This principle is utilized in the case of Avitrol when only a few individuals in a blackbird or pigeon flock eat treated grain; their abnormal flight displays and calls result in the flock leaving the area (26). Similarly, the nontoxic sticky repellents (e.g., Tanglefoot, Roost-No-More) have both a visual (spook) effect and a tactile repellency when a ribbon or bead is laid down on a roosting surface.

Recent studies (4, 10) have evaluated Bird Scaring Reflective Tape in a variety of environments and with diverse avian species. This synthetic resin film tape (11-mm wide), with a metallic coating that reflects sunlight, produces a flashing effect, and emits a loud humming noise in the wind. Initially utilized with agricultural crops, its use has been extended successfully to structural environments. An essential feature is that tape installation must provide for the pulsating response to wind.

Rotating orange lights are advertised and sold as bird repellers. These lights are often mounted around loading docks. I know of no experimental data or other documentation to validate the advertised claims for their efficacy. Such lights, however, do tend to impress human visitors (e.g., sanitation inspectors). Similarly, mounted wooden or plastic owls do little to discourage bird roosting; often such objects simply

Table 16.1

Survival rates for starlings, grackles, and robins banded as nestlings

Species	Average Annual Survival Rate (%)	Reference
Starling	48	(36)
Starling	53	(6)
Starling	50	(12, 23)
Grackle	51	(36)
Red-winged blackbird	53	(36)
Robin	47	(12)

provide additional perches. Some observations suggest that if such models can be kept in constant motion, repellency is more likely to be achieved. Such devices that are now being advertised need to be carefully evaluated for efficacy.

An important limitation of such localized measures was exhibited at Bowling Green State University, which had a resident population of about 200 pigeons. Reduction by nonchemical means was mandated. In several months of trapping, 652 pigeons were removed. Obviously, the adjacent town's abundant pigeon population had expanded into the biological vacuum created on campus.

Communitywide Control

If bird problems can be addressed on an areawide or communitywide basis, then the feeding, loafing, nesting, and roosting sites can be addressed as part of an integrated program. Unless the garbage, spilled grain, deserted buildings, and other conditions attractive to pest birds can be reduced, any population reduction achieved in just one location is likely to be only temporary. However, such areawide programs that permanently reduce the carrying capacity of the community environment for pest species are the most difficult to implement (36).

A study by Krakowski and Payne (23) in Stevens Point, Wisconsin, one of very few community studies of pigeons, documents distinctive roost and movement patterns. Food and water were available year-round, and breeding was distributed over much of the year. Despite an annual increment of 43%, a high dispersal rate allowed the population to remain relatively constant.

Building Design

Another aspect of management strategy is to design buildings and associated landscaping so that birds do not become problems at all (*15*, *22*). This kind of design awareness has been forced upon airport architects (*35*), but industrial and commercial planners seemingly have given little thought to these problems. Quality control and environmental sanitation personnel are thus forced into a catch-up operation in an effort to maintain facilities that are bird-free.

Protected Status of Birds

It is necessary to be able to identify birds and to have some knowledge of the laws and regulations affecting them. Unlike rats, who have few friends, a bird that is a pest to one person may be a beloved songbird to another. Laws and regulations may originate at national, state, or local levels; a species not protected at one level may be protected at another.

Federal Laws

Fortunately, most pest birds are afforded little protection. The Federal Migratory Bird Act mandates protected status for most native, migratory species; other actions of Congress have extended this protection to most species initially excluded. However, blackbirds are exempted when they are damaging or are about to damage crops. House sparrows, starlings, and pigeons, being nonnative and nonmigratory (except for some starling populations), are automatically exempted. Thus, there is no federal prohibition against killing or removing any of these three species at any time. However, all toxic chemicals used for bird control are designated by the Environmental Protection Agency as restricted-use materials; people using them must be licensed by appropriate state agencies. Such chemicals are not available to the general public.

Local Laws

State laws that may (at least technically) protect these pest birds are not likely to be enforced. On the other hand, local ordinances may protect pigeons within a city. Enforcement of these laws has made life difficult for some PCOs and city officials. Discharge of firearms and outdoor use of toxic baits within a city usually are prohibited, but permits often are obtainable by PCOs upon application to carry out a specific control program.

In some situations, federally protected species may be a serious problem. Gulls along the eastern seaboard roost and loaf on many buildings, and their presence provides the potential for fecal contamination of incoming and outgoing food shipments. In addition, they may damage roofing. Waterfowl utilizing warehousing grounds adjacent to waterways could constitute a similar hazard (*2*). Swallows, robins, and other songbirds may build nests or perch over critical areas or within structures. Birds may depredate and contaminate raisins and other fruits drying outdoors (*29*). Woodpeckers can seriously damage wood siding on buildings. Such birds, even though they are endangering food storage or products or causing significant damage, may not be killed without a specific permit (application fee, $25.00) from the U.S. Fish and Wildlife Service. Usually, nonlethal approaches are prescribed initially. In some states, once eggs have been laid, the nests may not be disturbed.

Birds that accidentally enter structures must be dealt with on an individual basis at the time. This might involve opening windows or doors, using nets or tennis rackets, or using other appropriate techniques (e.g., shotgun, air rifle, slingshot). Technically, pest birds that are protected may be harassed, but they may not be killed or injured without appropriate permits.

Conclusion

In order to undertake bird management projects, pest management specialists need to think "like a bird"—to be a "bird brain." They need to view the environment as a bird does. They need to spend many hours, especially in the early morning and late afternoon, observing birds—their roosting, feeding, and loafing areas and movement patterns. Unless the behavior of the pest birds is understood, any management program is likely to be only partially successful (*41*). Management of pest birds is a unique field requiring approaches different from those used with cockroaches, mosquitoes, or rats. It requires intense, creative, devoted personnel—and lots of time.

Cited References

1 Battaly, G. R., D. Fish, and R. C. Dowler. 1987. Seasonal occurrence of *Ixodes dammini* and *Ixodes dentatus* (Acari: Ixodidae) on birds in a Lyme disease endemic area of southeastern New York state. *Journal of the New York Entomological Society* 95(4)461–468.

2 Baur, F. 1976. Bird problems and food storage and processing facilities. Viewpoint from industry. In *Proceedings. Seventh Bird Control Seminar* (Bowling Green State University, Bowling Green, OH), pp. 112–115.

3 Baur, F. J., and W. B. Jackson (eds.). 1982. *Bird Control in Food Plants—It's a Flying Shame.* American Association of Cereal Chemists, St. Paul, MN.

4 Bruggers, R. L., et al. 1986. Responses of pest birds to reflecting tape in agriculture. *Wildlife Society Bulletin* 14(2)161–170.

5 Chandler, A. C. 1916. A study of the structure of feathers, with reference to their taxonomic significance. *University of California Publications in Zoology* 13:243–246.

6 Conover, M. 1983. Using conditioned taste aversions to protect blueberries from birds: Comparison of two carbamate repellents. In *Proceedings. Ninth Bird Control Seminar* (Bowling Green State University, Bowling Green, OH), pp. 205–206.

7 Coulson, J. C. 1960. A study of the mortality of the starling based on ringing recoveries. *Journal of Animal Ecology* 29(2)251–271.

8 Crase, F. T., and R. W. deHaven. 1976. Methiocarb: Its current status as a bird repellent. In *Proceedings. Seventh Vertebrate Pest Conference* (Monterey, CA), pp. 46–50.

9 Day, M. G. 1966. Identification of hair and feather remains in the gut and faeces of stoats and weasels. *Journal of Zoology* 148(2)201–217.

10 Dolbeer, R. A., P. P. Woronecki, and R. L. Bruggers. 1986. Reflecting tapes repel blackbirds from millet, sunflowers, and sweet corn. *Wildlife Society Bulletin* 14(4)418–425.

11 Dolbeer, R. A., P. P. Woronecki, and R. A. Stehn. 1986. Resistance of sweet corn to damage by blackbirds and starlings. *Journal of the American Society of Horticultural Science* 111(2)306–311.

12 Erickson, W. A., and W. B. Jackson. 1983. Use of the chemosterilant Ornitrol in feral pigeon (*Columba livia*) control. In *Proceedings. Ninth Bird Control Seminar* (Bowling Green State University, Bowling Green, OH), pp. 261–269.

13 Farner, D. S. 1945. Age groups and longevity in the American robin. *Wilson Bulletin* 57(1)56–74.

14 Fitzwater, D. 1970. Sonic systems for controlling bird depredations. In *Proceedings. Fifth Bird Control Seminar* (Bowling Green State University, Bowling Green, OH), pp. 110–119.

15 Geis, A. D. 1976. The effects of building design and quality on nuisance bird problems. In *Proceedings. Seventh Vertebrate Pest Conference* (Monterey, CA), pp. 51–53.

16 Gough, P. M., J. W. Beyer, and R. D. Jorgenson. 1979. Public health problems: TGE. In *Proceedings. Eighth Bird Control Seminar* (Bowling Green State University, Bowling Green, OH), pp. 137–142.

17 Jackson, W. B. 1978. Rid-A-Bird perches to control bird damage. In *Proceedings. Eighth Vertebrate Pest Conference* (Sacramento, CA), pp. 47–50.

18 Jackson, W. B., A. Teklehaimanot, and B. Ali. 1977. Use of methiocarb as an avian repellent on blueberries. *Plant Health Newsletter: Colloquium on Crop Protection Against Starlings, Pigeons, and Sparrows.* EPPO Publ. B84, pp 181–189.

19 Kare, M. R., and J. G. Rogers, Jr. 1976. Sense organs. In *Avian Physiology,* P. D. Sturkie, ed. Springer-Verlag, New York, NY.

20 Kassa, H., and W. B. Jackson. 1979. Mesurol as a bird repellent on grapes in Ohio. In *Proceedings. Eighth Bird Control Seminar* (Bowling Green State University, Bowling Green, OH), pp. 59–64.

21 Kessel, B. 1953. Distribution and migration of the European starling in North America. *Condor* 55(2)49–68.

22 King, C. B. 1982. Facility design for bird exclusion. In *Bird Control in Food Plants—It's a Flying Shame.* American Association of Cereal Chemists, St. Paul, MN.

23 Krakowski, J., and N. F. Payne. 1986. Population ecology of rock doves in a small city. *Wisconsin Society of Sciences, Arts and Letters* 74:50–57.

24 Lack, D. 1943. The age of some British birds. *British Birds* 36(9)193–197.

25 Maguire, B. 1963. The passive dispersal of small aquatic organisms and their colonization of isolated bodies of water. *Ecological Monographs* 33(2)161–185.

26 Mampe, C. D. 1976. Current status report: Pigeon control. In *Proceedings. Seventh Bird Control Seminar* (Bowling Green State University, Bowling Green, OH), pp. 95–101.

27 Martin, D. F., and G. G. Benson. 1976. Bird control as a profit source. *Pest Control Technology* 4(2)8, 10, 12–13; 4(3)26, 28–31, 40–42.

28 McLean, R. G., and T. W. Scott. 1979. Avian hosts of St. Louis encephalitis virus. In *Proceedings. Eighth Bird Control Seminar* (Bowling Green State University, Bowling Green, OH), pp. 143–155.

29 Palmer, T. K. 1973. Bird management on the farm and at the processing plant. In *Proceedings. Sixth Bird Control Seminar* (Bowling Green State University, Bowling Green, OH), pp. 15–16.

30 Rogers, J. G., Jr. 1974. Responses of caged red-winged blackbirds to two types of repellents. *Journal of Wildlife Management* 38(3)418–423.

31 Rogers, J. G., Jr. 1978. Some characteristics of conditioned aversion in red-winged blackbirds. *Auk* 95(2)362–369.

32 Schlichting, H. E., Jr. 1960. The role of waterfowl in the dispersal of algae. *Transactions of the American Microscopical Society* 79(2)160–166.

33 Schneider, D. E., and M. W. Fall. 1970. The role of bird management in fire protection. In *Proceedings. Fifth Bird Control Seminar* (Bowl-

ing Green State University, Bowling Green, OH), pp. 53–55.

34 Smith, R. N. 1970. The use of detergent spraying in bird control. In *Proceedings. Fifth Bird Control Seminar* (Bowling Green State University, Bowling Green, OH), pp. 138–140.

35 Solman, V. E. F. 1976. Aircraft and birds. In *Proceedings. Seventh Bird Control Seminar* (Bowling Green State University, Bowling Green, OH), pp. 83–88.

36 Steckel, J. 1970. A municipal bird control project. In *Proceedings. Fifth Bird Control Seminar* (Bowling Green State University, Bowling Green, OH), pp. 120–123.

37 Stewart, P. A. 1978. Survival tables for starlings, red-winged blackbirds, and common grackles. *North American Bird Bander* 3(3)93–94.

38 Stone, R. 1979. Behavioral and physiological problems associated with development of CURB. In *Proceedings. Eighth Bird Control Seminar* (Bowling Green State University, Bowling Green, OH), pp. 90–95.

39 Summers, R. W., G. C. Pritchard, and H. B. L. Brookes. 1983. The possible role of starlings in the spread of TGE in pigs. In *Proceedings. Ninth Bird Control Seminar* (Bowling Green State University, Bowling Green, OH), pp. 301–306.

40 Weber, W. J. 1979. *Health Hazards from Pigeons, Starlings, and English Sparrows.* Thomson, Fresno, CA.

41 Yeager, R. C. 1971. The day the starlings came. *Pest Control* 39(8)30, 32, 53.

42 Yusuf, H. M. 1985. Efficacy of Ornitrol as a pigeon chemosterilant. Thesis, Bowling Green State University, Bowling Green, OH.

17
Ecology and Management of Bats as Food-Industry Pests

Denny G. Constantine

Bats nearly always have positive effects on the food industry (3). Insect-eating bats consume agricultural pests in flight. Tropical, fruit-eating bats play an indispensable role in pollination and dispersion of seeds of many tropical fruits. They eat only unwanted ripe native fruit and thus are not a threat to orchard crops, which are picked prior to ripening. On the negative side, however, since the discovery of America by Europeans, tropical American vampire bats have limited livestock production in Latin America by spreading rabies. And, bats occasionally live in buildings where their wastes may foul food stores. The present discussion, which concerns stored-food pests, accordingly is meant to assist in preventing access of bats to food-storage buildings.

Species and Habits

There are about 850 species of bats. They vary in size from those with a wingspread of several inches to species with a wingspan in excess of 5 feet. The various species are specialized in eating habits. They may be categorized as insect-eating, fruit-eating, nectar-drinking, fish-eating, meat-eating, or blood-drinking; some are omnivorous. Bats are found throughout the world to the limits of tree growth, and they increase in numbers and kinds as the tropics are approached. They are active throughout the year in the tropics, but hibernate during winter in temperate zones or migrate to warm areas. Many bats are essentially cold-blooded when at rest, assuming the temperature of their environment. Most species are gregarious, living in colonies of a few to hundreds or even thousands in dark, confined areas such as caves, barns, attics, and warehouses. Some aggregations in caves number in the millions. Other species live as solitary individuals, usually in open places, such as tree foliage.

Bats generally have one or two young, but some species have as many as four. Gregarious bats usually leave their young clinging in groups to the ceilings or walls of their roost while they fly out to forage for food. Others, particularly solitary bats, take very young bats with them. Young bats are equipped with recurved teeth that maintain a firm hold in nipple tissue. Among some gregarious bats, suckling the young has been observed to be largely a community effort, any female providing milk to any baby, whereas other species seek out their own young.

Banding studies have shown that some bats can live as long as 30 years.

Health Precautions

Although numerous and varied microbial agents of disease have been associated with bats (3), only rabies and histoplasmosis pose serious problems to humans.

Rabies

Bat rabies is known throughout the Americas. A few cases have been reported in Europe, and similar deadly, rabies-related viruses have been reported from several kinds of bats in Africa

Constantine: Veterinary Public Health Unit, Department of Health Services, 2151 Berkeley Way, Berkeley, CA 94704.

and Europe. Of the various microbial infections, only rabies and rabies-related viruses are known to produce overt disease and death in the bats. Usually, the rabies-infected bat ceases to eat or fly or does so ineffectually. Eventually, the bat hangs separately, perhaps trembling, and the wings may hang partially open. Finally, it may die in that position or death may come after it falls from the roost. Young, normal bats often fall from roosts and are mistakenly suspected of being infected with rabies. Unlike other animals, bats do not experience rabies outbreaks. The infected bat is rare or uncommon despite crowded conditions in colonies, the prevalence of infection in colonies being 0.1–0.5%.

Transmission of rabies occurs via bite, so bites must be avoided. Airborne transmission of the virus is known to occur only in the presence of a million or more bats in caves. House bats do not attack or bite unless handled, even when the bat is infected with rabies. Preexposure immunization against rabies has been recommended for people who run a high risk of repeated accidental exposure to infection (2, 6).

Histoplasmosis

Histoplasma capsulatum, the causative agent of histoplasmosis, occurs naturally in the soil of many warm, humid areas of the world. It can also be shed from minute lesions in the intestines of some asymptomatic infected bats in certain geographic areas. Whatever its source, it may multiply to dangerous levels in fecal deposits such as those of birds or bats. Infection occurs when the organism's spores are inhaled or ingested.

Most cases in humans are subclinical, but miners of bat guano have succumbed following inhalation of overwhelming doses. Regular cleanup of fecal wastes should eliminate the opportunity for a buildup of infective doses. The fungus has been isolated from bat guano in the eastern United States, Mexico, Central America, South America, Africa, Israel, Romania, Malaya, and Australia.

Other Pathogens and Ectoparasites

The spectrum of microbes associated with bats is similar to that seen in other mammalian groups. Common-sense sanitation should be used to avoid health problems. For example, *Leptospira* may be present in urine, and *Salmonella* may be present in feces. Bats may support various parasites, including protozoans, helminths, mites, ticks, bat ''bed bugs,'' streblid and nycteribiid flies,

and fleas. The arthropods can become abundant in populous bat roosts, and they often adhere to bats on their foraging flights. Although some of these arthropods are known to bite people who enter populous roosts, consequent disease has rarely been suspected and never proved. Except for some ticks, certain fleas, and streblid flies, the mouthparts of these ectoparasites are usually inadequate to penetrate human skin; thus, the parasites quickly die in the absence of bats. Only mites are occasionally abundant enough to be annoying when they crawl over sensitive skin areas. Dermatitis may result from scratching.

Bat Management

The only lasting method of ridding a building of bats is to physically bar their entry to the building (4, 7). Previously, this was done after the bats had left on their evening flight to forage for insects, or after they had departed on their fall migration. Now, however, the use of a simple one-way valvelike mechanism in the bat entryway makes it possible to do the exclusion job during daytime (5) (Figures 17.1 and 17.2). Bats do not gnaw, so it is relatively easy to block their entryways in most instances. Attempts to kill the bats or to repel them are usually ineffective or only temporarily effective at best; moreover, these methods have often created formidable health hazards. These objectionable approaches are discussed below because of the necessity to discourage demands for a quick chemical solution.

Seasonal Behavior

Attempts to rid premises of bats must take into account the seasonal behavior patterns of the bats and the effect of that behavior on the success of the eradication effort and on the public health. Bats generally live in buildings only during the warm months, hibernating elsewhere during winter. Females aggregate in buildings in spring to bear and rear their young, whereas males tend to be solitary, scattering to higher altitudes or otherwise cooler places at that time. By midsummer, the young bats start flying, and afterward the males may join the group. Bats usually depart as the weather cools and insect activity decreases. Sometimes bats are present in a building during winter as either constant or intermittent hibernators, perhaps flying and feeding whenever the weather warms. In especially warm geographic areas, bats may be present and active all year. Superimposed on this pattern, migrating groups of bats often stop over at buildings during spring and fall. Some

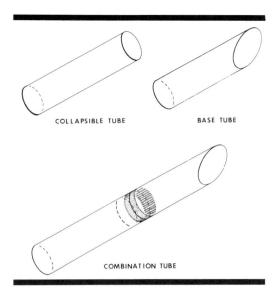

COLLAPSIBLE TUBE   BASE TUBE

COMBINATION TUBE

buildings are attractive to bats primarily or exclusively at night, and few or none may be found there during daytime.

Batproofing

It is apparent from the foregoing that batproofing of buildings or efforts to expel the bats should not be undertaken from spring to midsummer because young bats, incapable of efficient flight, might be present in the roosts. These young mammals would either be trapped inside or scattered outside. Trapped bats may crawl from a sealed attic or wall to the building interior, and grounded young bats could bite inquisitive people and pets. Therefore, batproofing should be done anytime from late summer to spring.

One or more one-way valvelike devices should be installed in primary exits to permit bats to depart but to prevent their reentry. These devices allow any hibernating bats to leave as well, should exclusion be undertaken after the commencement of hibernation. Otherwise, the hibernating bats would be trapped inside the building and eventually die and decay, or they might work their way through cracks into building spaces occupied by people or stored goods.

The specific approach to physically exclude bats from a building can vary markedly, depending on the situation. A building in good repair often has only one or a few bat entry sites, usually identifiable by dark soiling about the holes and by the presence of bat droppings beneath them. If careful inspection indicates that bats can depart from only one or a few holes, all except the primary entryway(s) should be sealed with wood, wire mesh, caulking compound, or other materials. Then, a one-way

valvelike device(s) should be installed in the main exit(s), and the bats permitted to depart at night through the valve(s). The building should be checked at dusk for one or more evenings to confirm that no bats are entering or leaving through previously undetected, unsealed holes. It should be kept in mind that some bats can squeeze through a quarter-inch slot. Eventually, the valvelike device can be removed and the final entryway sealed.

Some buildings are seemingly impractical to batproof because of peculiarities of construction or general delapidation, entryways being exceedingly numerous. The interiors of such buildings usually offer little shelter to bats, which often retire to slots or crevices in the walls or ceiling for protection from light and drafts. In instances of this kind, batproofing can be done by covering or filling the slots or crevices after the bats have left or been evicted (a blast of compressed air will suffice to accomplish this).

Where bats hang in the dark, open spaces of large buildings, entering through many holes, such as under the edge of a corrugated metal roof, the open space may be made undesirable for bats by lighting, accomplished by opening doors or windows (also creating undesirable drafts in the process) or by the installation of electric lights. In an effort to escape the light or drafts, the bats will leave or attempt to retreat to slots or crevices, which will have to be covered or filled, as mentioned above.

Problems of Tile Roofs

An especially difficult challenge has been that of excluding bats from harborage under Spanish tile roofing from which they sometimes enter attics through holes or cracks. Bats often enter the tile spaces through the open (lowermost) tile ends and also where the sides of the tiles overlap the sides of the building. Individual variation in tile shape and other factors usually render manufactured plugs (used to block the openings presented by the lowermost row of tiles) ineffective in blocking bats and actually make the spaces more desirable for bats in some instances by lessening drafts. Expansion and contraction of the tiles, resulting from temperature changes, loosen or break many blocking materials. However, a solution has been found to this problem.

It has been observed that bats do not enter the end tile openings where the presence of a rain gutter has blocked ready access. The bats usually enter unguarded tiles by swooping upward in an arc, the terminal part of which is about vertical, and landing on or just under the

Figure 17.1
One-way valvelike bat excluder and components. Each segment is 2 inches wide and 6 inches long. The lightweight, readily pliable, collapsible tube overlaps the heavier, relatively rigid, plastic base tube to which it is attached with tape. Bats freely exit through the apparatus, but the collapsible tube closes behind them, barring reentry.

Figure 17.2
Examples of tube installations: (A) attached in slot or hole with metal strap and screws; (B) attached at slot or hole with duct tape; (C) attached at vertical slot with duct tape, which is also used to block slot; (D) attached at attic entry-exit with metal strap and screws.

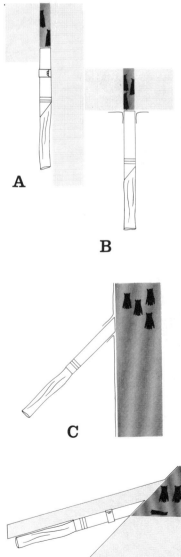

A

B

C

D

ledge that is directly under the open tiles. Where bats enter the sides of tiles which curve over (overlap) the building walls, they usually have to land on the wall before crawling inside.

Rain gutters should be installed directly under the open tile ends and under the overlapped tiles as well. The upper edge of the gutter should extend outward some 8 inches from that point. Furthermore, the gutter should be tightly fastened flush with the wall. Some gutter installations leave an open space behind the gutter, where bats can crawl to the tiles after alighting on the wall. Other gutter installations leave a deep, dark slot between the gutter and wall, providing a shelter for bats in the slots.

Negative Effects of Destruction of Bats

Destruction or attempted destruction of bats is contraindicated for public health and other reasons. Therefore, the U.S. Environmental Protection Agency (EPA) has registered no toxicants for this purpose. DDT, small quantities of which once were used for bat destruction, is no longer available. The reasons that bat destruction is contraindicated are many, but the foremost are discussed below. First, the many toxicants and other related control techniques actually killed relatively few bats; reproduction or recruitment from surrounding areas kept pace with the mortality (1, 6, 8, 9). Second, the continuous production of disabled, grounded bats for at least 6 years following poisoning of bat roosts results in a continuous series of bat-bite incidents involving curious people and pets who investigate the fallen animals. Bat bites must be regarded as rabies exposures and are to be managed accordingly until proved otherwise (2), which often is impossible because of the escape or destruction of the bat. Therefore, bitten people should receive partial or complete antirabies treatment, and bitten pets should be destroyed or quarantine up to 6 months. Third, attempted destruction of bats displaces the only permanent solution, which is exclusion.

Impracticality of Repellents

No compounds have been registered recently with the EPA for use as bat repellents, but it is anticipated that certain chemicals will be registered eventually. Repellents, especially sprays, are generally best avoided in attempting to rid buildings of bat colonies that are resident during daytime hours, but some repellents are of limited usefulness in other special situations. Both paradichlorobenzene and naphthalene (as flakes or mothballs) have been used to repel bats from enclosed places such as attics, where the fumes concentrate as the crystals evaporate. For bat roosts, 2–5 pounds are usually recommended. However, experience shows that the repellent effect is usually disappointing and temporary at best. Pet repellent sprays containing oil of mustard (allyl isothiocyanate), sprayed heavily on roosting surfaces when bats are absent, may repel bats for about 2 months.

My experience has shown a repelling effect on bats by contact with fiberglass batting (or padding, available in rolls) made of coarse fibers, such as some of the products used in previous years for insulation. However, recent products are far less repellent and even may be a comfort to bats. Lights and strategically directed breezes produced by electric fans have been used successfully to repel bats. In contrast, tests indicate that ultrasound is ineffective and causes fretfulness in nearby people.

The reasons that chemical repellents are contraindicated in day roosts are similar to objections to the use of poisons: The repellent effect is temporary; repelling the bats tends to displace the permanent solution—physical exclusion; and repellents are almost certain to be misapplied, scattering disabled bats or flushing flightless immature bats.

In theory, a repellent such as a spray would be applied to the roost interior when the bats are absent. That objective is not often sought, however, because it would require waiting until the bats leave at night. And young, flightless bats might be in the roost constantly during spring and the first half of summer. Lack of visibility into and within bat harborages may prevent suitable inspection. Often the worker prefers not to enter an attic, electing instead to apply the spray blindly through exterior openings. The result is to drench the bats with a chemical that burns the eyes, mucous membranes, and skin. The consequences are grounded, distressed bats scattered throughout town, creating the appearance of a bat rabies epizootic.

Repellents, especially nonchemical ones, find some legitimate application if the worker is familiar with the many aspects of the bat harborage problem and is diligent in avoiding misapplication. One justifiable application would be at night roosts in open porches, carports, and similar places, where foraging bats hang up at night to rest. Light is sometimes an effective deterrent in these instances.

People who undertake debatting operations when bats are in the roost or who expect to come into contact with bats are advised to wear leather work gloves to prevent bites if they handle the animals. In addition, they should consider undergoing preexposure immunization against rabies, as suggested above, especially if this kind of activity is undertaken repeatedly in localities where bats are infected with rabies.

Cited References

1 Barclay, R. M. R., D. W. Thomas, and M. B. Fenton. 1980. Comparison of methods used for controlling bats in buildings. *Journal of Wildlife Management* 44(2)502–506.

2 Centers for Disease Control. 1984. Rabies prevention—United States, 1984. *Morbidity and Mortality Weekly Report* 33(28)393–402, 407–408.

3 Constantine, D. G. 1970. Bats in relation to the health, welfare, and economy of man. In *Biology of Bats,* vol. 2, W. A. Wimsatt, ed. Academic Press, New York, NY.

4 Constantine, D. G. 1980. Bat rabies and bat management. *Bulletin of the Society of Vector Ecologists* 4(1979)1–9.

5 Constantine, D. G. 1982. Batproofing of buildings by installation of valvelike devices in entryways. *Journal of Wildlife Management* 46(2)507–513.

6 Expert Committee on Rabies. 1984. *Seventh Report.* World Health Organization Technical Report Series No. 709, 104 pp.

7 Greenhall, A. M. 1982. *House Bat Management.* Resource Publ. 143. Fish and Wildlife Service, U.S. Department of the Interior, Washington, DC.

8 Hurley, S., and M. B. Fenton. 1980. Ineffectiveness of fenthion, zinc phosphide, DDT and two ultrasonic rodent repellers for control of populations of little brown bats (*Myotis lucifugus*). *Bulletin of Environmental Contamination and Toxicology* 25(3)503–507.

9 Kunz, T. H., E. L. P. Anthony, and W. T. Rumage III. 1977. Mortality of little brown bats following multiple pesticide applications. *Journal of Wildlife Management* 41(3)476–483.

18
Bionomics and Integrated Pest Management of Commensal Rodents

Stephen C. Frantz and David E. Davis

Modern approaches to solutions of problems besetting mankind are turning more and more toward investigating the causes of these problems. Controlling causes means shutting off end results at their source, thereby forming the basis for prevention. So it is with the control of rodents. Lasting success can best be accomplished only when the reasons for infestation are controlled. To paraphrase this, [rodent] control means environmental control [J. C. Jones (175)].

This quotation implies that the basic concepts to be presented here were certainly well known more than three decades ago. Somehow, over the intervening years, pest control has often gotten off base. It commonly exhibits a preprogrammed rigidity where treatment is given automatically in accordance with a predetermined schedule, that is, regardless of the biological realities of the situation. Every few years a new arsenal of control materials appears in the marketplace as we nuture our predominant approach to abatement of pest infestations through attempts to eradicate the animals themselves with techniques that do not adequately consider the whole environment (304).

Frantz: Wadsworth Center for Laboratories and Research, New York State Department of Health, Albany, NY 12201.

Davis: 777 Picacho Lane, Santa Barbara, CA 93108.

All photographs are by S. C. Frantz and all drawings are by C. Feller, unless otherwise indicated.

Holistic Pest Management

We must recognize the fact that there is no magic pill; no single, simple remedy can be relied upon to solve rodent pest problems in a lasting way. Rodent control must be considered in terms of the environment in which the pest is active; it must have, as an overriding principle, the bionomics of the rodent in concert with its whole environment and the periodic changes that occur therein. Integrated pest management (IPM) is such a systems (holistic) approach, a relatively new term for earlier concepts of rodent control espoused by Crouch, Davis, Emlen, Silver, and others (66, 68, 103, 267). In the 1930s, Hugo Hartnack (a Chicago pest control operator and early "maverick" of modern pest control) advocated a holistic-environmental approach to pest control and suggested that preventive pest control needed to be practiced rather than treating the pest problem after the fact (273).

The actual toll of rodent depredation to the food industry is not known. Estimates vary considerably with country surveyed, rodent species present, kinds of commodities attacked, packaging techniques utilized, time in storage, and so on, as well as methods of assessing loss (41, 152, 158, 293). Although the variables are numerous, there is general agreement that postharvest losses are enormous (especially in developing countries), not only in terms of food items and packaging, but in terms of human health and toil, environmental damage, and losses

243

Figure 18.1

Lesser bandicoot rats on bagged rice in a food warehouse (Calcutta, India); losses are caused by consumption, contamination (feces, urine, hair, ectoparasites), and damage to bags.

of fertilizer, energy resources, time, and money.

In September 1975, the U.S. Secretary of State told the United Nations General Assembly that better storage and better pesticides could prevent enough losses to match the total of all food assistance worldwide (*56*). Most authorities agree that prevention of postharvest losses caused by vertebrate and invertebrate pests needs to be given high priority in food production systems worldwide. Stored foods at all stages—raw materials through finished products—are often subject to repeated attacks by rodents, attacks that result in significant economic, health, and aesthetic damages (Figure 18.1).

IPM for commensal rodents, as presented here, includes relevant information on bionomics and intervention techniques. The main focus is on the food industry (including warehousing raw materials, food processing, product storage, retailing, and catering), but the concepts presented will be useful to a wider audience of pest management professionals.

Concepts of Integrated Pest Management

Definition of IPM

Integrated pest management is a decision-making process in which all interventions (treatments, actions) are brought to bear on a pest problem with the goal of providing the most effective, economical, and safest remedy possible (*162, 221, 222, 280*). Simply stated, IPM is a process for determining *If* intervention is needed or justified; *When* intervention is needed; *Where* intervention is needed; and *What* intervention is needed.

The basic problems are that commensal rodents live in and around man-made structures, causing damage to people and property, and that rodents are difficult to eradicate once they become established. An apparently logical approach in addressing rodent-caused problems is first to learn more about the animals to be managed and then to intervene to reduce their impact on the quality of human lives. Knowledge of the animals' biology and typical habits is necessary to determine if pest suppression is needed at all. A client's report of a "rat" running inside a warehouse loading area is not necessarily an immediate cause for alarm. Only after a thorough investigation will one know if a rat or a mouse was actually involved, what particular species, and if an infestation is present or if the observed animal was simply passing through. The presence of damage or potential for damage, including legal implications, must also be considered when justifying the need for intervention.

Intervention Strategies

Interventions should be undertaken when the probability of maximum impact on the pest population is greatest, thus reducing the problem. Essentially, the strategy is to reduce the pest population to non-economic, non-injury densities. If possible, management procedures should be initiated when rodent numbers are at a seasonal low, the time when the greatest proportion of the population might be affected; expenditures on labor and materials can also be minimized at this time. However, if the presence of the pest represents a serious threat or is causing considerable damage, the intervention(s) should be applied promptly regardless of seasonal fluctuations in population size.

The efficacy of any intervention will be greatly enhanced by matching the correct tactic to the specific problem location rather than by applying a general treatment to a larger area. This moderate approach may also result in the use of smaller quantities of rodenticides, reduction of accidental poisoning hazards, and a less expensive program. The corollary is that the boundaries of the target area must be drawn sufficiently large to include the possibility of solving the problems within it. In the example of the "rat" in the loading area of a warehouse, the concern is the origin of the rat—from a recent shipment of goods, from within the warehouse, or from a nearby infestation. The appropriate interven-

244   Stephen C. Frantz and David E. Davis

tion would be different for each potential source. However, the boundaries of the program would usually include the entire premises indoors and at least the building perimeter outdoors; some areas might require active treatments, whereas monitoring would suffice in others.

Basically, the four steps—*if, when, where,* and *what*—define a particular rodent pest problem. Instead of immediately responding with a predetermined treatment, IPM considers all possible interventions and uses whatever steps are necessary when and where they will be most effective in reducing a problem while maintaining a high standard of safety for nontarget species. With IPM, application of chemicals is often not the first line of attack; however, the judicious use of rodenticides does play an important role in most management programs. The overall strategy necessary for managing a particular rodent problem may involve various interventions including educational, physical, legal, biological, mechanical, and chemical techniques. Note that various components of the interventions could be included under more than one category.

Basic Philosophy for Program Development

Any pest management program must fit the specific needs of the pest problem under consideration. One must be aware that within complex ecological settings, there may be effects or outcomes (e.g., indirect, interactive, or cumulative) that are not necessarily predictable. Hence, one should endeavor to think through all identifiable effects during the planning phase so as to prevent consequent environmental costs.

Interventions should be chosen that are least disruptive to naturally occurring control actions. In the Caribbean, undiluted endrin was sometimes dumped into irrigation ditches to control field rats (J. O. Williams, personal communication). This unwarranted action not only endangered human health, but also killed many wild vertebrates, some of which would normally prey on rodent and insect pests. The point is that one should not make work for oneself but should allow naturally occurring controls to work together with IPM interventions. "When you kill off the natural enemies of the pests, you inherit their work" (*221*).

Choosing Interventions

Interventions that are most in harmony with both short-term and long-term human and environmental health should be chosen. In the above example, the benefits from the short-term, fast-kill of rodents with endrin did not outweigh the long-term, negative consequences of broad-spectrum killing of nontarget species, the possibility of human toxicosis, and the persistence of endrin in the environment. It is important to clarify that interventions should be the outcome of responsible decisionmaking, a process that can only be expected from well-informed individuals. An operational definition of responsibility is provided by the philosopher Charles Frankel [cited by Hardin (*151*)]: "a decision is responsible when the person (or group) that makes it has to answer for it to those who are directly or indirectly affected by it." Anything less would certainly create unnecessary risks for clients of the pest control industry.

Recent changes in public attitudes and government regulations have increased the pressure to minimize the use of pesticides on food at all stages from crop to table (*59*). Clients and workers at all levels are demanding, and rightly so, details regarding health and environmental hazards of the materials used by service companies. Without question, we do want to utilize techniques that relieve the problem, but we do not want our interventions to come back to haunt us at a later date in the form of nontarget poisonings or rodenticide resistance. To the fullest extent possible, techniques used should be those most likely to have a lasting effect. For example, reducing the causative conditions of a rodent infestation will surely provide more lasting relief from rodent depredation than will rodent removal (e.g., killing) techniques alone; further, the rodents cannot develop resistance to a lack of food, water, and harborage.

Ease of Application

A major consideration in selecting interventions is ease of application; that is, they should be reasonably easy to perform effectively (and correctly). Listing all possible approaches, with the advantages and disadvantages of each, will help the pest manager to make logical choices. The attitudes of the human population involved must also be given ample weight in this selection process. On one occasion in Nepal, one of us (SCF) was required to remove all rat traps from a small grist mill in a mountain village. The mill owner, a devout Buddhist, did not, however, object to the rodent exclusion measures (rodentproofing) suggested as an alternative to trapping.

Interventions should be chosen that conserve nonrenewable energy resources. Since much of urban rodent control involves the problem of solid waste management, it makes little sense to use pest management techniques that add to this problem. Nonbiodegradable (plastic) place packs, bait trays, and bait stations marketed as "disposable" or "throw away" are examples. "There is no away to throw to" (*151*); sooner or later we will encounter our wastes again. These items unnecessarily add to the many millions of tons of solid wastes produced annually in the United States (USA) and elsewhere (*1, 230*). In rodent pest management, alternative durable or biodegradable fabrications are available and need to be used on a broader scale.

Cost/Benefit Analysis

Practically speaking, cost-effectiveness is a major consideration that knits together IPM program development and must be assessed in the long term as well as the short term. However, cost/benefit analyses of pest management involving complex natural systems are difficult and should probably be kept modest. Economic valuation relies critically on understanding and measuring (quantitatively, if at all possible) the physical, chemical, and biological effects of our interventions (*163*). Economic valuation is, generally, the last step in analysis.

Available methods for placing monetary values on nonmarket goods and services are not well developed. For example, it is difficult to place a value on life (human or nonhuman) and damages to such life. When most future costs and benefits are discounted, and especially those of nonhumans are disregarded, cost/benefit analysis becomes a specious exercise (*18*). This section and the remainder of the chapter are intended to help in planning and evaluating IPM interventions, with due consideration being given to environmental quality.

Some factors to consider in food-industry plants are (a) current rodent damage, (b) lag time to initiate the control program, and (c) legal implications. For example, consider the case where one wants to use a pelletized rodenticide for a relatively quick kill of rats in a bread factory. The quick kill may well save the client money otherwise lost to rat damage and product contamination. However, rats may also carry away some of the rodenticide pellets and, in the process, drop some that could end up in a baked loaf of bread. Legal problems may ensue, a substantial quantity of product may need to be recalled and destroyed, and production time could be lost to cleaning and inspection. In the long run, a rodenticide that rats could not easily carry away or some other treatment (e.g., glueboards,

traps) would be a more appropriate intervention. Certainly, *prevention* of an indoor infestation would be the best overall strategy.

Clients must also realize that technical competence, not the low price of a quick fix, should be the major consideration in choosing pest management services. In a recent case, a pest control firm was reported to be utilizing one technician for one-half day per week to service four floors in a large office/commercial building in New York City (*132*). That is, in a period of about 4 hours/week, one technician was to adequately monitor, evaluate, provide interventions, and make recommendations regarding 100,000 ft^2 of complex floor space utilized by five independent tenants with a wide variety of functions (flea market, pizza parlor, sporting goods store, offices, and so on). The cost for this "professional" service was only $6,000 per annum. Needless to say, the pest problems were not abated by this paltry effort which was unprofessional and neither holistic nor realistic.

Components

The major components of a systems approach to pest management remain basically the same regardless of the pest species involved. These components, in the order in which they appear programmatically, are (a) monitoring, (b) establishing an injury level, (c) interventions, and (d) evaluation. In any pest control program, the pest manager should address each of these components in developing a clear, overall strategy for presentation to the client (e.g., food-plant operator, premises owner, tenant). If the project is small, the statement could be verbal, but usually it should be written.

Minimally, the statement should provide

(a) the name of the species to be controlled,
(b) the population and/or damage reduction to be expected (100%, 50%, nuisance levels, and so on),
(c) the methods to be used,
(d) why such actions were selected,
(e) the duration of control activities,
(f) who will do it,
(g) possible hazards to nontarget species,
(h) situations that might prevent or reduce success, and
(i) when the extent of program achievement will be evaluated.

The four major program components are summarized below.

Monitoring

Monitoring is essentially a systematic survey

at regular intervals which keeps one continually aware of all aspects of the pest situation and establishes baseline data that enable later evaluations as well as application to future projects of a similar nature. A monitoring system functions to:

(a) locate and identify the pest species;

(b) regularly estimate the rodent population directly or through signs of the infestation (this process may be expanded to include surveillance for pesticide resistance, disease, poisoning of nontarget species, and so on);

(c) identify natural enemies (owls, snakes, cats, dogs) and potential pests (mice may become important once the major pests, rats, have been suppressed);

(d) investigate causative conditions (particularly food, water, and shelter resources) and modes of entry that might be altered; human behavior is often a major factor (e.g., unsanitary waste disposal practices);

(e) be aware of other management decisions and practices that might affect the pest population and the efficacy of the program (nearby construction, landscaping, and so on);

(f) note weather or seasonal changes that may affect program efficacy, size of the pest population, and so on (e.g., the amount of food and harborage available is more critical during cold weather than under less stressful conditions).

Monitoring is an ongoing activity throughout an IPM program, with its purpose changing to meet specific needs at particular points in time. The initial survey is often called an "inspection" or "baseline survey" whereas a follow-up survey might be called an "evaluation." The literature provides several useful monitoring formats (*35, 78, 134, 145, 157, 222, 277, 292, 311*). Whatever recordkeeping system is devised, it must be specific and easy to use, but adequately reflect the variables being studied. In order to meet the high standards of the food industry, extremely detailed inspection and/or monitoring systems are needed to provide the practical working base for IPM programs (*59, 137, 160, 188*).

Thresholds and Tolerances

Injury level (tolerance limit, action threshold) is the size of the rodent population that can be correlated with injury sufficient to warrant intervention (i.e., injury that has become intolerable). This threshold for action may be determined by economic, medical, and/or aesthetic considerations, depending upon the particular pest situation, institutions involved,

Figure 18.2
This mouse in a box of commercial breakfast cereal prompted a consumer to bring suit against the manufacturer (New York).

sanitary and legal standards, and individual tolerances for rodents and their depredating effects. Obviously, there can be no set standard for all situations. A dairy farmer may well be able to tolerate a few rats in barns where forage crops are stored; however, the dairy products stored and sold on the same farm need to be rodent-free in accordance with government regulations. The consequences of a single rat or mouse in the cleanroom of a flour mill or in a computer center would undoubtedly be more severe than in a warehouse for storing concrete drain tiles. Rodent contamination of food products (Figure 18.2) may precipitate regulatory action.

Each situation must be considered as a separate entity with its own factors for determining what course of action needs to be taken. A practical approach to determining an initial "working injury level" is to correlate successive quantitative observations of the rodent population, as obtained from monitoring records, with evidence of rodent damage (e.g., loss of product, contamination, worker complaints). At later evaluations, the injury level can be revised up or down as experience is gained on the specific target site.

In the now-defunct Federal Urban Rat Control Program, action was required when exterior monitoring records showed that a city block had (a) greater than 2% of premises infested with rats; (b) greater than 30% of premises with unapproved refuse (garbage and rubbish) storage; and (c) greater than 15% of premises with exposed garbage available to rats (*78, 84*). A sim-

Figure 18.3
Skull of *Rattus norvegicus*, a typical murid rodent, showing dental characteristics.

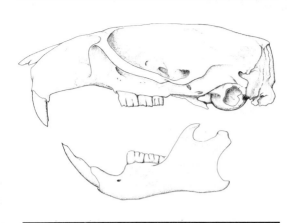

Figure 18.4
Self-sharpening action of rodent incisors.
SOURCE: Adapted from Brooks, J. E., and F. P. Rowe. 1987. *Commensal Rodent Control*. Rept. No. WHO/VBC/87.949, World Health Organization, Geneva, Switzerland, 107 pp.

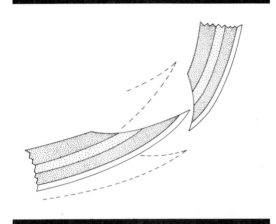

A

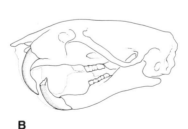

B

C

Figure 18.5
Comparison of normal incisor occlusion (A) with a malocclusion (splayed bottom incisors) (B, lateral view; C, anterior view).

ilar approach for establishing action thresholds according to predetermined levels of urban rodent infestation (adapted from industrial quality control methodology) was proposed by Rennison (*243*). Using an injury level based only on economic losses due to house mouse damage to livestock building insulation, Timm and Fisher (*291*) concluded that most livestock producers would find it cost-effective to exclude rodents from buildings whenever possible.

For the U.S. food industry, the Food and Drug Administration (FDA) has set limits at or above which they will take legal action against a product to remove it from the market (*47*). For example, in peanut butter, the defect action level (DAL) is reached when a sample averages or exceeds one rodent hair per 100 g; in wheat the limit is 9 mg of rodent excreta pellets and/or pellet fragments per kilogram. Such allowances are made by the FDA in recognition of the fact that food totally free of contaminants is not a practical objective; however, the DALs are subject to adjustments as analytical techniques improve.

Interventions

Interventions are measures taken by a pest manager to reduce the size of the pest population and/or the concomitant damage. Interventions must fit into a well-balanced plan that is based upon such factors as the bionomic particulars of the pest species, size of the pest population, location and extent of the problem, and climatic conditions. When possible, interventions should be well coordinated so as to enhance their individual effects. For example, rodentproofing supplemented by poisoning will often have a greater and more prolonged effect than either action alone. Poisoning (to reduce rodent numbers) during habitat modification will also help to ensure against possible short-term detrimental effects (e.g., increased human/rodent contacts) as the environment is disrupted.

Evaluation

Evaluation is the necessary final step to determine the outcome of any intervention and to discover whether or not the program objectives have been achieved. This step is actually a continuation of ''monitoring'' used to determine what will occur next in the management program; this could involve additional treatments, a change of treatments, only monitoring, and so on. The decision for future actions is based on *progress to date*, not on a *calendar date*. Without documented, quantitative/temporal assessments of infestation, pest control becomes guesswork.

Rodent Biology and Behavior

A closer examination of IPM components and an explanation of the basic features of rodent biology and behavior show why IPM is ''biological,'' safe, and effective.

Recognizing Rodents

The first item on the agenda for investigating an alleged rodent infestation is to determine if rodents are indeed the problem. If the pest animal can be captured or otherwise closely examined, it is relatively easy to determine if it is a rodent.

Rodent Teeth

About half of all known mammal species are rodents. A key physical feature for recognizing rodents is their well-developed front cutting teeth, the incisors: one pair in the upper jaw and one pair in the lower jaw. The incisors are separated from the chewing teeth (molars) by a gap (diastema) (Figure 18.3). This important feature allows the rodent to draw in the sides of the lips (to close its mouth behind its incisors) while

248   Stephen C. Frantz and David E. Davis

continuing to gnaw. It can discard unwanted fragments without taking them into the mouth (92). Thus, to exclude rodents from a structure, it is generally more important that structural materials be hard rather than merely distasteful.

A second important feature of rodent incisors is that they grow continuously from the root and are self-wearing (self-sharpening). This latter effect occurs because the points of the upper and lower incisors rub against one another like two opposing chisels. The softer layers on the posterior tooth surface are worn away, leaving the hard enamel layer to form a point on the front surface (Figure 18.4) (35, 284). Thus, rodents need not gnaw to keep their incisors worn down; however, gnawing on hard objects contributes to their wear. Rate of growth varies with extent of the use of the teeth; thus, the lower incisors, which do most of the work during gnawing, grow faster than the upper pair (92).

Continuously growing incisors can produce an abnormal occlusion (malocclusion). When an incisor is misaligned or broken and no longer opposes its counterpart in the other jaw, the unopposed incisor continues to grow (Figure 18.5). This excessive incisor growth may eventually force the mouth open, partially blocking the mouth cavity, and result in reduced food intake or starvation. Maloccluded incisors sometimes curve around and penetrate the brain. Growth of incisors exterior to the jaw sometimes produces a grotesque anomaly.

Murid Rodents

Of all the rodents, Old World rats and mice (Family Muridae, comprised of more than 500 species) are best known. Commensal rats and mice are those that live with or near human beings and depend on them for at least part of their food and shelter. The most common of these in the USA (and probably worldwide) are the Norway rat (*Rattus norvegicus*), the roof rat (*Rattus rattus*), and the house mouse (*Mus musculus*). These three species are reasonably easy to distinguish from one another (see Table 18.1 and Figure 18.6) (101, 207, 219, 258).

Rattus norvegicus

The Norway rat (brown, grey, wharf, or sewer rat) is the most common urban and rural rat in much of the USA and is the largest of the commensal species (Figure 18.7). The Norway rat is distinguished from the roof rat by its blunt nose, relatively small eyes and ears, and a tail that is shorter than head and body length together. It is lighter in color on the ventral surface

(compared to the dorsum). Its foot pads are smooth and flush with the sole; its ears are covered with short hairs (Table 18.1). In temperate countries, Norway rats occupy urban and rural habitats, but in warmer countries they tend to be more restricted to larger cities and ports.

Rattus rattus

The roof rat (black or ship rat) is the most common ocean-going rat (Figure 18.8). It is often abundant in ports and their environs and may also be found in inland urban and agricultural areas in the southern USA, California, Hawaii, and in some tropical countries. Roof rats occur in three relatively distinct color types, each of which was once thought to be a separate species until it was learned that they interbreed freely. The large protruding eyes, large mobile ears (which are virtually hairless), long, uniformly dark tail (longer than head and body), and lamellated (in a fine fingerprint pattern) foot pads and toe tips are key physical features (Table 18.1). Roof rats are streamlined in overall appearance and are extremely adept as climbers.

Bandicota bengalensis

Another common pest of food-storage and food-processing facilities is the lesser bandicoot rat, *B. bengalensis*, also called the bandicoot or the Indian mole-rat (Figure 18.9). Lesser bandicoots, found only in South and Southeast Asia, are also important agricultural pests. This rat resembles the Norway rat in coloration, but is smaller and has a more blunted snout. When excited or disturbed, the lesser bandicoot becomes very vocal, emitting a harsh nasal grunting or snorting, and erects its long dorsal guard hairs.

Mus musculus

The adult house mouse (Figure 18.10) is similar in appearance to a young roof rat of the same size; however, the latter can be differentiated because it has a relatively larger head, hind feet, and body (Figure 18.6). In addition, a young, mouse-sized rat would usually not be found outside the immediate vicinity of its burrow or nest until it grows to be larger than an adult mouse. House mice are more widely distributed worldwide than the above rat species and are more likely to be found living independently of humans.

Cricetid Rodents

Species of the genera *Peromyscus* and *Microtus* belong to the family Cricetidae (=Cri-

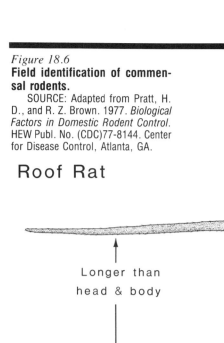

Figure 18.6
Field identification of commensal rodents.
SOURCE: Adapted from Pratt, H. D., and R. Z. Brown. 1977. *Biological Factors in Domestic Rodent Control.* HEW Publ. No. (CDC)77-8144. Center for Disease Control, Atlanta, GA.

Roof Rat

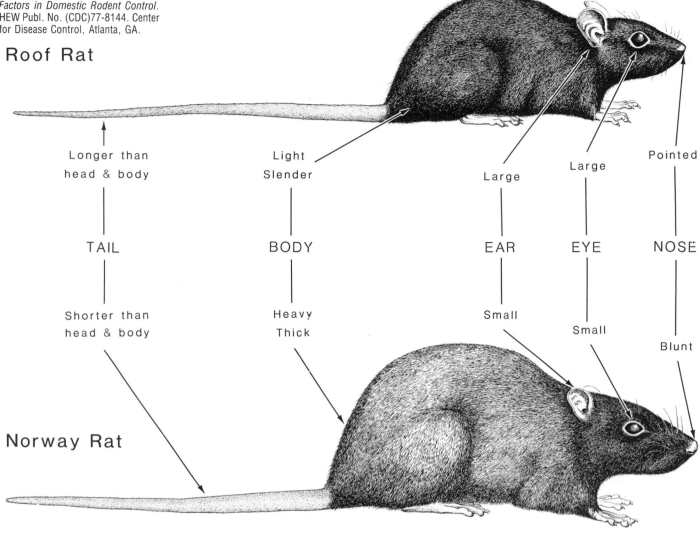

Longer than head & body	Light Slender		Large	Large	Pointed
TAIL	BODY		EAR	EYE	NOSE
Shorter than head & body	Heavy Thick		Small	Small	Blunt

Norway Rat

House Mouse

Small — HEAD

Small — FEET

Table 18.1

Field characteristics and measurements of adult commensal rodents

Characteristics	Norway rat, *Rattus norvegicus*	Lesser bandicoot rat, *Bandicota bengalensis*	Roof rat, *Rattus rattus*	House mouse, *Mus musculus*
Weight	105–600 g medium to large size	90–360 g medium size	80–350 g medium size	12–25 g small size
Head and body	18–26 cm nose blunt heavy, stocky body	13–23 cm rounded head, nose blunt stocky body	16–23 cm nose pointed light, slender body	6–11 cm nose pointed slender body

Tail	15–21 cm
	shorter than head and
	  body
	thick, blunt, scaly
	bicolored (dark upper,
	  lighter below)
	sparse, short, stiff hairs
	often drags on ground
		11–20 cm
		shorter than head and
		  body
		thick, blunt, scaly
		unicolored (medium dark)
		
		very sparsely haired
		often drags on ground
			18–25 cm
			longer than head and
			  body
			long and thin, fine scales
			unicolored (uniformly dark)
			
			essentially hairless
			carried high with movement
				6–11 cm
				equal to or a little longer than
				  head and body
				long and thin
				unicolored (uniformly dark)
				
				essentially hairless

Ears[a]	18–23 mm thick, furry relatively small appear half-buried in pelage	18–25 mm thick, sparse-haired relatively small	21–28 mm thin, hairless large, prominent stand out from pelage	15 mm or less thin, sparse-haired large, prominent stand out from pelage
Eyes	relatively small	relatively small	relatively large	relatively large
Hind foot[b]	40–45 mm, broad	21–39 mm	20–40 mm, narrow	<20 mm
Teats	12	12–16 (sometimes 20)	10–12	10

Fur/pelage	coarse
	reddish-brown to greyish-
	  brown back
	greyish to yellow-white
	  belly
		coarse, rough
		greyish-brown to reddish
		  brown back
		greyish-white belly
		prominent, long guard
		  hairs on back
			finer (than Norway rat)
			tawny brown, brownish-grey
			  or blackish back
			white to yellow-white,
			  brownish-grey or grey belly
				fine
				brownish-grey to greyish-back
				
				greyish to greyish-white belly

[a] Length from inside notch.
[b] Length from heel to tip of longest toe.

cetinae of Muridae) (about 500 species). House mice should not be confused with the similar-sized, white-footed mice and deer mice (*Peromyscus* spp.) that sometimes invade structures from nearby woodlots or fields (46). The adult deer mouse has a distinctly bicolored (darker on the upper side than below) tail; the soft, full pelage of the upperside is sandy to brown or grey and the underside is white. Additionally, deer mice have white feet and relatively larger eyes than house mice.

Meadow mice and voles (*Microtus* spp.) are also occasionally found indoors. The adults are roughly twice the body size (17.5 cm long) of house mice and have, proportionately, a much shorter tail (3.8 cm) (204). The ears of meadow mice are furred and do not project much above the pelage of the head.

Signs

When first investigating an alleged infestation, we rarely get an opportunity to directly observe the rodents. Therefore, we must rely on evidence or signs to determine if rodents are present, which species, relative numbers, and where they are active. When a food-processing

Figure 18.7
Norway rat, *Rattus norvegicus*.

Figure 18.8
Roof rat, *Rattus rattus*;
note "halo" rubmarks on wall.

Figure 18.9
Lesser bandicoot rat, *Bandicota bengalensis*.

plant is inspected (a legislated form of monitoring) by a representative of a regulatory agency, the presence of rodent signs may be interpreted in such a way as to constitute the violation of laws or regulations. For example, the following is a partial list of violations quoted from one "Notice of Hearing" by the New York State Department of Agriculture and Markets regarding one grain-milling facility:

> Rat warrens were noted in the dirt floor of the basement, especially under a loose pile of lumber.
>
> Three distinct areas in the basement, which has dirt floors, had active rodent holes to the extent that the dirt floor crumbled and an inspector dropped approximately four to six inches when stepping over a series of three entry holes.
>
> Over 100 rat droppings per square yard were noted alongside the cross conveyor and under the wheat storage bin.
>
> Mouse droppings were noted along all the wall and corner areas throughout the establishment.
>
> Rat tracks (over 30 prints) were noted in an area of approximately one square yard between two elevator legs and in spilled flour on the second floor.
>
> Corncobs carried by rodents appeared over and under lumber piled in the basement.
>
> Six dead mice, some old and dried up and others plump and fresh, were found in a room above the third floor by the top of the grain bins; one mouse skeleton was found in a nest made from cloth bag fibers on a wall beam on the second floor.

Some signs of rodent presence (particularly droppings, burrows, tracks) are useful for pro-

viding indirect, relative estimates of population size. During monitoring procedures, a quantified accounting of signs is made and the signs are then removed or obliterated before beginning a lag period followed by subsequent counts. Thus, over the course of an IPM program, one can determine if a population is increasing or decreasing (i.e., evaluate whether the interventions are having success). Such relative estimates will be sufficiently reliable and cost- and time-efficient for most rodent pest management programs.

Sounds

Rodents make noise, especially at night, when crawling and climbing in walls, above dropped ceilings, under cabinets, and in machinery or other enclosed spaces. Sounds made by rodents gnawing on solid objects may also be detected. Squeaks and other vocalizations can often be heard when rodents are fighting, mating, and caring for their young. Note that other household pests such as chipmunks, squirrels, and bats make some sounds similar to those of rats and mice. It requires practice and/or the combination of other signs to distinguish among the various intruders.

Pet Excitement

Pet excitement may occur when cats and dogs hear or smell rodents in a wall, under a cabinet or kitchen appliances, and so on. Pets may hear rodent sounds better than humans do because their auditory sensitivity crosses into the ultrasonic range of rodents. Cats and dogs sometimes sniff or scratch excitedly at the floor or wall indicating the general location of rodents.

Fecal Pellets

Fecal droppings are found wherever rodents are active—along their movement paths, near walls, behind objects, and near food and shelter resources (Figure 18.1). Different sizes of droppings indicate animals of different ages or species. Of the three common commensal species, Norway rat droppings are the largest (Figure 18.11), but may be confused with those of roof rats which tend to be smaller and curved. Empirical discriminant functions (*164*) have been developed to aid in identifying single droppings (with 80% accuracy) or samples of two or more droppings (with 95% accuracy) of Norway and roof rats. Some cockroach droppings appear similar to those of house mice, but the latter are more pointed and elongated and usually contain hairs (visible under magnification). One rat

Figure 18.10
House mouse, *Mus musculus*.
SOURCE: Photo by J. K. Clark.

dropping may contain as many as 200 hair fragments (41).

Numbers of droppings are not very useful in estimating the size of an infestation. The number of droppings depends on the type of food, moisture content, availability of water, extent of coprophagy, and so on. In addition, estimates based on caged-animal studies may be lower than for free-ranging rats with higher daily food intake. One study at the Rodent Control Evaluation Laboratory (New York State Department of Health) showed that adult Norway rats fed a standard laboratory diet and water *ad libitum* excreted an average of 37 droppings per animal per day (range = 16–55). Coprophagy was minimal because most of the pellets fell through the wire-mesh floors of the cages. For roof rats, the average was 59 droppings/animal per day (range = 31–126). In studies at the Haffkine Institute in Bombay, rats fed mixed cereal grains (in granular form) with water available had daily excretions of fecal pellets as follows: Norway rats, 38; roof rats, 37–38; and lesser bandicoot rats, 45. On a diet of wet cereal mash, roof rats consumed no water, and the fecal output increased to 50 pellets/rat per day. Barnett (20) allowed four enclosed Norway rat populations (10–26 rats each) access to one ton of bagged wheat for 84–196 days. The rats produced 172 droppings (whole and fragments) per kilogram of wheat; up to 320 rat hairs per kilogram of wheat were recovered. In all, 70.4% of the wheat had been fouled (spilled from bags, mixed with feces, and so on) over the course of the study.

The house mouse produces 50 or more droppings per day (250). In a simulated bulk wheat storage study of 150-day duration, it was estimated that a mouse population (averaging 12 mice) deposited 434 droppings and 1,204 hairs per kilogram of wheat (295). Further, 65% of the surface seeds and 55% of all seeds were contaminated with urine.

Although the number of rodents is difficult to correlate with the number of droppings, the rate of accumulation of droppings in specified areas or in bait stations can provide a relative census tool. After counts are made at a prescribed interval, the area under study must be swept clean and the droppings removed. This method is particularly useful where disturbance of the study area by nontarget species is likely to occur.

Close examination of droppings can disclose whether or not an infestation is active. Fresh droppings are moist, soft, and shiny or dark. Old ones are dry, hard, and dull or whitish; they

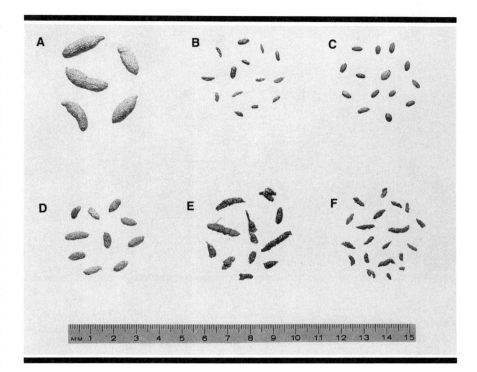

crumble under pressure. The color of the droppings may be a useful indicator of the diet of rodents. When rodents feed on dyed materials, including many rodenticide baits, their feces may become similarly colored. For example, a waxy diet such as soap can give the droppings a waxy, light-colored appearance. Specific locations of droppings can tell how rodents are gaining access to a structure. Accumulations of droppings indicate where rodents spend considerable time and may be effective locations for placement of traps and poisons.

By adding a fluorescent pigment (e.g., Helecon No. 1953, United States Radium Corp., Morristown, NJ) to bait, one can determine the daily range of rodents feeding at a specific location (121, 131). The resultant droppings (when fresh or dry) fluoresce when activated by long-wave ultraviolet illumination (black light). The technique works especially well indoors or in areas with little ground cover or clutter. A more recent technique uses metallic flake particles in baits to provide several marker colors for determining percentages of animals feeding at numerous bait placements simultaneously (113).

Urine Stains

Urine stains (wet or dry) of mice and rats naturally fluoresce under ultraviolet illumination and can be used to determine areas of activity. Recording the extent of staining and then removing such stains can be used with subsequent observations to indicate relative changes

Figure 18.11
Comparison of fecal pellets of selected commensal vertebrates. (A) *Rattus norvegicus*, Norway rat (adult); (B) *Mus musculus*, house mouse; (C) *Peromyscus* sp., deer mouse; (D) *R. norvegicus*, Norway rat (4 weeks old); (E) *Eptesicus fuscus*, big brown bat; (F) *Myotis lucifugus*, little brown bat.

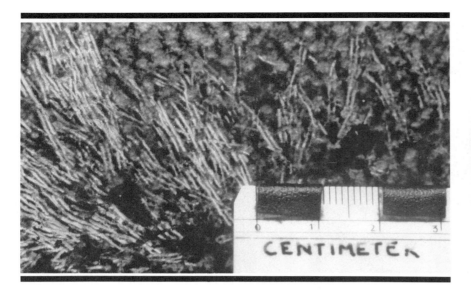

in activity. Rodents dribble small amounts of urine and other urogenital secretions wherever they move; such stains may be found along frequented runways and on stored goods. One adult rat roaming over the surface of bulk stored wheat can contaminate as much as 10,000 kernels/day (*41*). A pair of rats can void 5.7 liters of urine per year. Dried urine fluoresces blue/white if fresh and yellow/white if old (*138*). The presence of urine strongly suggests that the goods have been contaminated by rodents. To verify that the material observed is truly rodent urine, it is necessary to use one or more standarized tests (*306*). In some cases, rats and mice consistently urinate and defecate in particular spots that become conspicuous even without use of black-light illumination.

Odors

Odors resulting from deposits of fermenting urine and feces and from body oils are discernable in closed spaces even when only a few rodents are involved. With practice, one can distinguish between the odors of the Norway rat, the roof rat, and the house mouse.

Gnawmarks

Gnawmarks may be present as actual toothmarks (Figure 18.12) or as small particles resulting from gnawing (e.g., woodchips about the size of coarse sawdust for rats), and as holes around pipes passing through walls, in baseboards, window and door casings, doors, cabinets, packaged goods, or wherever rodents enlarge an opening in order to pass through a solid object. The woodchips and gnawmarks of recently attacked wood are light-colored, whereas older ones are dark. The size of the tooth-

marks—3.5 mm wide for adult Norway rats (*207*)—and the diameter of these holes can be used to distinguish between rats and mice. When young, rats can enter openings as small as 1.25 cm (0.5 inch) in diameter; young house mice can pass through openings about half that size. Adult-size rat holes (especially burrows) are more typically around 75 mm (3 inches) in diameter (Figure 18.13); those of mice are around 20 mm (0.75 inch) in diameter.

Rodents gnaw almost anything they can get between their upper and lower incisors and, theoretically, can cut through any material softer than the enamel of their incisors. Norway rat tooth enamel (lower incisors) is rated at 5.5 on Moh's geological hardness scale; that is, the

enamel's hardness lies roughly between that of iron and steel (*92*). Comparable values have been found for the roof rat and the house mouse. Theoretically, they could gnaw iron with a value of 4 (Moh's scale); however, in practice they effectively gnaw only those items with a hardness of 3.5 or less (e.g., aluminum, lead, copper). With their chisel-like incisors exerting a pressure of 1.7 metric tons/cm^2 (*41*), they can gnaw through most building woods, sheetrock, aluminum siding, plastic garbage cans, soft mortar, asphalt, lead pipe, telephone cables, and electric utility wires.

A significant percentage of urban fires of unknown origin have been attributed to rodents gnawing electrical cables (*267, 309*). Rodents easily cut into fiberglass insulation, make tunnels through it, and use it as nesting material. Rats and mice also gnaw soap, clothing, jute grain bags, leather goods, and food. The appearance of damaged grain can be used to distinguish rats from mice: rats often leave half grains mixed with smaller pieces; mice tend to nibble around the outside of a grain, leaving behind a bitten core and many smaller pieces (*292*).

In a study using an indirect census method based on rat-gnawing of objects placed in the environment (*262*), stakes were driven into the ground of the study area, with 16–21 cm projecting aboveground to provide a gnawing surface; evidence of gnawing was recorded at preset time intervals. This technique correlated well with census baiting results, though the space utilization pattern was somewhat different with each method. Bite marks on human and animal victims are useful rodent signs in public health work.

Burrows

Burrows 7–10 cm (2.75–4 inches) in diameter in soil are most commonly made by Norway rats. They may be found in any soft ground, particularly stream banks or where burrow entrances may be concealed—in low, dense vegetation (Figure 18.14); under concrete slabs, foundation walls, lumber (Figure 18.15), and piles of rubbish; and among the rocks of rip-rap shorelines. Garbage piles, refuse dumps, and sewers, providing both food and harborage, are also favored. Indoors, Norway rats burrow in the earthen floors of basements, warehouses, and animal quarters. They also utilize bales of hay or straw and other suitable materials (Figure 18.16). The roof rat burrows only occasionally, as does the house mouse [2- to 3-cm (0.75- to 1.25-inch) diameter burrows] when it lives outdoors. If it is suspected that a burrow connects the ground surface with a sewer, dye can be flushed down the burrow while nearby manholes are observed for traces of dye.

The burrow systems of Norway rats may be simple or very complex, with many entrances, tunnels, and nesting chambers (*44, 232*). Burrowing is sometimes so extensive under a poured concrete slab (without reinforcement) that the floor cracks and collapses into the burrows. A burrow entrance is usually surrounded by a small mound of excavated soil (fresh soil indicates current activity); if this mound is not present, as in an earth-floored warehouse, it may indicate that the rat exited at that point, having burrowed from the other side of a nearby wall or fence. Figure 18.17 shows extensive damage to a lawn by the burrows (excavated) of the lesser bandicoot rat.

Figure 18.15
Typical burrow/harborage locations for Norway rats.
SOURCE: Rodent Control Evaluation Laboratory, New York State Health Department, Albany, NY, file photo).

Figure 18.16
Norway rat "burrow" in a bale of straw.
SOURCE: Rodent Control Evaluation Laboratory, New York State Health Department, Albany, NY, file photo).

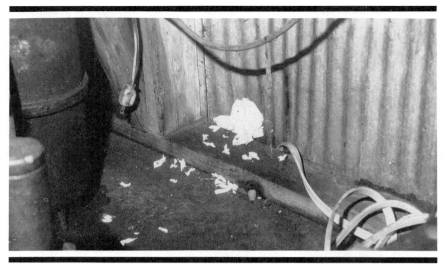

Burrow systems sometimes have bolt-holes (escape holes) which are entrances containing loosely packed soil, debris, or vegetation that can be pushed out easily. In Calcutta, closed bolt-holes of lesser bandicoot rats were found along unpaved roads in the daytime whereas at night many of the same entrances were open (*122*). Because of this complexity of bolt-hole use, burrow counts may not be a good indicator of the number of rodents present. However, if counts are made consistently at about the same time of day, burrow reopenings should provide a relative measure of changes in population size for evaluation purposes before and after some intervention is made.

In such census work, a preintervention closing of all burrows with soil should be completed on the initial day, with a return a few days later to count only those burrows reopened. This provides a baseline count and would be the time to fumigate or place bait into burrows in a rodenticide program (*130, 170, 277*). After each intervention is completed, burrows are again closed and are later reinspected at the same interval until activity ceases or the treatments are terminated. After the last routine inspection, burrows are once again closed and then revisited after at least 10 days for a final evaluative inspection. An alternative to filling burrow entrances with soil is to stuff them with a wad of paper (e.g., newspaper); the same procedure can be used with rodent holes in walls (Figure 18.18). The burrow activity method is useful in study areas otherwise disturbed by nontarget species and is not hampered by inclement weather.

Figure 18.18
Chewed paper (stuffed into hole through wall, overnight) indicates presence of rats, but low importance of opening because rats did not totally remove the paper from the hole (Slingerlands, NY).

Figure 18.19
Food cache (nut hulls, cherry pits, snail shells) of *R. rattus* under a woodpile at a residence in San Jose, CA.

Figure 18.17
Excavated burrow system of *B. bengalensis* in a residential lawn (Kathmandu, Nepal).

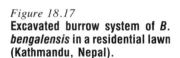

Nests, Food Caches

Food caches and nests are found indoors and outdoors in areas offering concealment and access to food resources. They may be found in warehouses, garages, basements, attics, storage areas, closets, under cabinets, between double walls, in false ceilings, and so on. Common outdoor locations include woodpiles (Figure 18.19), dense vegetation, and under the hood and in the seats of unused automobiles. Finely shredded paper, cloth, insulation (even fiberglass), and vegetation or other fibrous materials are often used for nest components.

The size of the nest indicates whether rats or mice are the builders and the location may clarify which rat species is involved. Norway rats prefer (though do not restrict themselves to) basements and lower stories indoors (including the closed spaces beneath cabinets) and ground-level areas outdoors. Roof rats prefer rafter areas, upper stories, and attics indoors, but will also nest in stuffed furniture (e.g., mattresses, chairs; see Figure 18.20) and beneath cabinets. When outdoors, they prefer elevated areas, as in thick ornamental shrubs, but they occasionally nest under woodpiles or plant debris. House mouse nests may be found anywhere suitable, but house mice prefer ground areas when outdoors. Their nests are quite small [12.5 cm (5 inches) in diameter] compared to those of Norway rats [ca 20 cm (8 inches)].

Tracks

Tracks in the form of footprints or marks made by the tail may be found on dusty surfaces, sand, soft soil, and in snow. The tracks of the five-toed rear foot are more commonly found than the four-toed forefoot, although both may be present (35). Fresh tracks are clearer and sharper than older, dust-covered tracks. Various-sized tracks indicate rodents of different species, sizes, or ages. Tailmarks are more likely to be made by adult Norway rats than by roof rats or young Norway rats, both of which carry their tails high. Holding a flashlight at a low angle helps one to see tracks on dusty surfaces; track direction often helps the inspector to locate rodent entry points as well as food and harborage resources.

Nontoxic tracking patches (chalk powder, unscented talc, lampblack), approximately 30 cm long by 10 cm wide, can be used to determine areas of rodent activity, especially when infestations are new. Dri-Die 67 (Fairfield American Corporation, Medina, NY) covers old rodent droppings and tracks and reveals new signs; it also kills ectoparasites that leave the rodents'

bodies. Fluorescent pigments have been used in tracking patches (176) and in sticky baits (121, 122, 131) to provide details of movement routes including food, water, and harborage resources and entry points to buildings. Grain flour should not be used for tracking patches because it could attract insect pests or the rodents themselves may consume it. The powders should be lightly dusted or painted onto a smooth surface or on pieces of some weighty material (e.g., an asphalt/plastic floor tile; see Figure 18.21) that can be conveniently prepared in advance of their onsite use (179, 264). When used outdoors, tracking patches must be sheltered from weather and nontarget animals; various bait stations are large enough for this purpose, as are boards leaned against walls.

To allow comparisons of data from various studies, Kaukeinen (179) developed a quantifiable track rating system for Norway rats, using

Figure 18.21
Norway rat on chalk-dust-covered tracking board is about to pass through a hole monitored by an Actimeter (a passive infrared motion detector).
SOURCE: Rodent Control Evaluation Laboratory, New York State Health Department, Albany, NY, file photo.

a standardized board size of 7.6 cm × 15.2 cm. Boards, especially those on which individual tracks are not readily distinguished, are rated in five categories: 0 tracks, 1–5 tracks, 6–10 tracks, 10–20 tracks, and over 20 tracks. A board wiped "clean" by rodent traffic would be categorized at "over 20." It is recommended that the same investigators read the boards throughout a study in order to reduce possible variation in interpretation. The rating system can be adapted as necessary for other species; Spaulding and Jackson (277) used six categories and found their technique reliable for monitoring Norway rat and house mouse activity in most situations.

Runs

Runs (trails, runways) are smooth, worn areas that the feet of rodents have beaten as a path on the ground or other substratum. Runs may be found in lawns and near buildings (Figure 18.22),

Figure 18.22
Run (trail) of *R. norvegicus* on soil next to foundation of a building (Milwaukee, WI).

Figure 18.20
Overstuffed chair in Miami, FL, home served as nesting site for roof rats; note extensive rubmarks.
SOURCE: Photo by A. Ros.

Figure 18.23
Rubmarks of *R. norvegicus* on wall.
SOURCE: Rodent Control Evaluation Laboratory, New York State Health Department, Albany, NY, file photo).

Figure 18.24
Characteristic "halo" rubmark of *R. rattus* in a kitchen (Miami, FL); note other horizontal and vertical rubmarks.
SOURCE: Photo by A. Ros.

walls, and fences. Indoors, they are dust-free trails along walls and behind boxes and other stored goods. The runs of rats are larger and more conspicuous than those of mice. These trails are particularly useful in deciding where to site traps, glueboards, and baits so they will be encountered by rodents.

A passive infrared motion detector/counter (Actimeter System) was developed by Kaukeinen (*179*) to monitor rodent activity; it is especially useful where the runs pass through narrow openings (e.g., a hole in a warehouse door). Both body heat and motion are required to activate a count; thus, an animal sitting in the

viewing path (20-degree angle) of the Actimeter is not counted until it leaves the field of view. The small (3.8 × 10.8 × 4.5 cm), portable detector/counter units (Figure 18.21) are each powered by a 9-volt transistor battery which allows at least one week of continuous operation. The units endure transportation, handling, high temperatures, and rainfall, although it is best to protect the units from precipitation or at least shelter the viewing tube. Undoubtedly, miscounts are possible, but the system provides a useful relative measure of activity (*136*).

Rub Marks

Rub marks consist of a mixture of dust, dirt, and oils from the fur of rodents that is deposited where rodents regularly rub against walls (Figure 18.23), baseboards, beams, pipes, and on outside edges of holes. The rub marks of roof rats are often found on elevated areas and may be characterized by a "halo effect" (Figure 18.24). That is, where they sit with their body in contact with a surface leaves one type of mark; another separate, semicircular mark is deposited farther away by their front feet as they stand up to investigate higher climbing routes. Another type of rub mark or trail peculiar to roof rats is discussed below in the section on physical abilities. Norway and roof rat rub marks under ceiling beams (Figure 18.25) are different in that the former's marks are broken in the middle whereas the latter's are continuous (*284*). Mouse rub marks usually cover rather small areas and are much more difficult to see than those of rats.

Live Sightings

Sightings of live rodents confirm the presence of an active infestation in the immediate area. Both rats and mice are capable climbers; thus, it is important to think three-dimensionally and to look up high as well as on the floor and ground areas when inspecting for rodents and their signs. It is also important to reverse one's direction of walk in order to discern signs visible only from a different angle. Daytime observations of rats usually indicate that the population is at least moderately high because rats are nocturnal (except under special circumstances) and only subordinate animals are likely to be diurnal (and only due to the behavioral pressures associated with high densities). Less predictable than rats, mice may be seen during the day or night.

For nightime observations, a powerful flashlight can be used (e.g., the rechargeable

SL-15 Streamlight, 15,000 cp; Streamlight, Inc., Norristown, PA). The eyes of rodents reflect light. More discrete observations can be made by placing a red filter over the light source to produce softer shadows and to be less disturbing to the rodents. A pop-off red plastic filter (approximately equivalent to Kodak Wratten 89B), an accessory from a Justrite electric headlamp, fits well over the lens of the SL-15. If one's hands must be free, the Justrite headlamp with filter is an excellent alternative to a handheld flashlight.

Visual sightings can be quantified by systematically observing a limited area of high rodent activity at regular intervals over the course of an intervention program. Though less accurate, an accounting of complaints of sightings by clients may be useful for some purposes. Similarly, the "call-back" rate may be the best indication of the effectiveness of a pest control technician's interventions (76). Where it is difficult to verify mouse locations, as in wall recesses, Frishman (139) suggests using a ULV spray of synergized pyrethrum which reportedly irritates the mice and often causes them to move to places where they can be observed.

Census

The overall purpose of monitoring is to determine the number of rodents present in the designated target area or in some other way to measure the severity of the problem. Precise estimates of commensal rodent populations are very difficult and usually impractical to obtain, and are not necessary for most IPM programs. Basically, it is important to know if more rodents are present in one area or building than in another, or to understand whether a population is increasing or decreasing over the course of one's interventions.

Direct and indirect census methods are available (97, 116, 146, 170, 179, 199, 209, 212, 277). The techniques will vary with the species under study, purpose of the study, local conditions, available manpower, and so on. With regard to conducting field efficacy studies of rodenticides, the U.S. Environmental Protection Agency (EPA) requires two independent census evaluations. Such efficacy studies may require somewhat different and more detailed information than would be necessary in evaluating large-scale commensal rodent problems.

Sampling by Traps

Trapping indices such as capture-mark-recapture (CMR) are widely used direct census pro-

cedures for detailed study of rodent populations (69); sophisticated schemes exist for analysis of individual and population data (191, 212). With CMR, a sample of animals is captured, marked, and returned to the population. Another sample is taken later, and the ratio of marked-to-unmarked individuals in the population is determined. This determination allows mathematical estimation of the size of the population. In

Figure 18.25
Common house rat (*Rattus rattus brunneus*) crosses under roof beam in a grain mill (Kathmandu, Nepal).

another direct mathematical method, the number of animals removed each day from an area (e.g., by trapping) is plotted against the total number previously removed. A projection of this rate of removal over time allows estimation of population size.

Unfortunately, the validity and practicality of trapping techniques are hampered in habitats where buildings and streets provide obstacles to rodent movements and where people vandalize and steal traps (106, 170, 179). Although such techniques are most informative regarding population age structure, sex ratio, birth, death, and movement characteristics, they require considerable effort and expense and are best reserved for special research purposes. In most rodent IPM programs and in some rodenticide efficacy work, evaluations of rodent activity through indirect census methods will be sufficient. CMR results were found comparable to data obtained from counts of fecal droppings, burrow activity, tracking activity, and placebo bait consumption (277). Other indirect census techniques were discussed in the Signs section, above.

Census Baits

Placebo bait consumption involves presenting a highly acceptable bait over a fixed time period both before and after interventions and then comparing the changes in consumption. Wherever possible, census baits should be in solid block form or finely ground in order to prevent hoarding; baits must be well-protected from nontarget species and weather (e.g., in tamper-resistant bait stations; see Appendix 18.1). Allowance is made for moisture changes in the exposed bait by use of a control bait point(s) not accessible to rodents. Rodents must be allowed an adjustment period of a few to several days to begin feeding on placebo baits (87), and to then stabilize (double the amount taken until it levels off for two consecutive nights) their daily consumption before data collection begins.

Waterers

Through use of graduated water containers, water consumption can also be used as a rodent census method in dry areas or where water resources can be restricted. As with census baits, waterers must be protected from weather and nontarget species (e.g., in a high-profile bait station; see Appendix 18.1), and a control waterer must be maintained to account for evaporation.

Measures of Infestation

With some effort, it may be possible to convert indirect census figures such as placebo bait consumption into rodent numbers. Food consumption by Norway rats was reported to vary with body weight; mean daily intakes were 7.8–21.1% of body weight (49). A figure of 24–25 g/rat per day is often used as the base consumption figure for Norway rats. However, 15 g/rat per day may be a more useful figure if the rats have access to alternative food resources (which is often the case). An enclosed roof rat population (50 rats) was offered mixed grains and free access to water, kitchen wastes, and other refuse; average consumption per rat per day was reported to be 9.9 g of grains and 1.42 g of kitchen wastes and other refuse (184). In another study where only pearl millet and water were offered without the garbage alternative, the average consumption was 20.0 g/rat per day. In a one-year study of caged roof rats offered mixed cereal grains and water, average daily consumption was nearly 10 g/rat (82). In the confined circumstances of ship infestations, 50 g may be used as a base consumption figure to estimate the minimum number of rats present (172).

For urban rodent control programs, it is probably rarely worthwhile to calculate the average number of rodents in each infested premises (97), as has been done, for example, for New York (67) and Folkestone (99). A more practical measure for urban areas is the prevalence of rodent-infested premises. This proportion of premises can be estimated simply by inspecting a random sample of premises (78).

Apart from the relative ease with which it can be obtained, there are three reasons for using prevalence of rodent-infested premises as a measure of the severity of the urban rodent problem (97):

(a) This measurement is readily understood by those who must conduct the program and by the people who benefit from it.

(b) This measurement can be directly associated with observations of causative conditions that largely form the basis for IPM interventions.

(c) Premises form the most readily identifiable and smallest units of land of which an urban area is composed. Therefore, premises provide a good basis for systematic surveys and subsequent interventions that can be seen to have a direct relationship to the measure of success of the IPM program, i.e., the reduction of the proportion of premises infested.

While premises are being inspected for the presence or absence of rodents via indirect census methods, simultaneous parallel observations of structural problems, human behavior, and other causative conditions will help in formulating the details of subsequent IPM strategies (35, 78, 134, 208). Though designed for insect control, the series of Prescription Treatment System pest management manuals from Whitmire Research Laboratories, Inc. (St. Louis, MO) provides excellent floor-plan charts (in three-dimensional perspective) and inspection information useful for rodent IPM in production kitchens, fast-food establishments, supermarkets, hotels, motels, resorts, and so on.

Daily Activity Period

Both rats and mice are mainly nocturnal, especially if the infested area has significant human activity during the day (22, 44, 112, 126, 180, 235, 240, 250). Generally, two peaks of activity occur, one at dusk and another just before dawn (Figure 18.26). However, activity period can be greatly influenced by factors such as high population density, food shortages, and human activity, which can result in the rodents

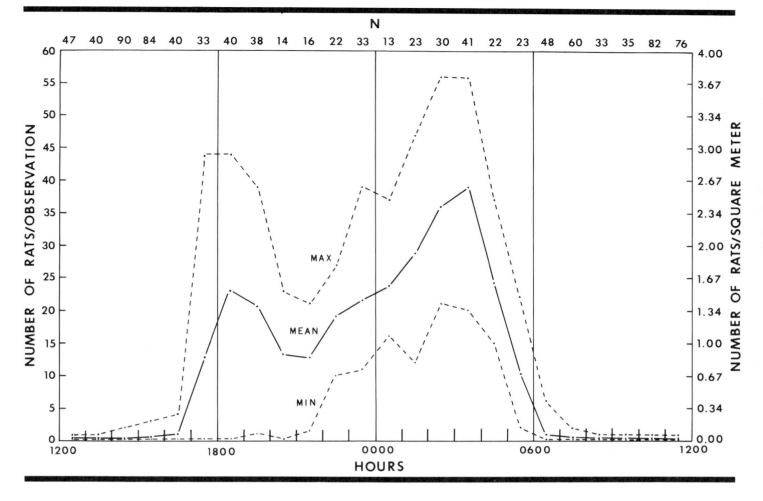

adjusting to other schedules, even becoming diurnal. In two cactus groves of Curzon Park in downtown Calcutta, a population of a few hundred bandicoot rats was totally diurnal (Figure 18.27). The rats' activity period coincided with peak human activity when the rats were fed by office workers, students, and other passersby, much as squirrels and pigeons are fed elsewhere.

Mice are intermittent, erratic feeders and, of course, are small and easily concealed; they frequently exhibit more daytime activity than rats, especially when left undisturbed (as in a closed warehouse).

Rodent management specialists must consider both night- and daytime activity depending on the particular situation being investigated. Nocturnal activity can be advantageous for outdoor fumigation. Burrows can be fumigated during the daytime when most rodents are inside and thereby maximize efficacy of the procedure. On the other hand, if some treatment is to be used that has long-term exposure hazards, it should be used only at night (and picked up each morning), thereby maximizing exposure to rodents

Figure 18.26
Daily activity pattern of *B. bengalensis* in a grain warehouse in Calcutta, India, when grain quantity was held relatively constant during the winter season. N is the number of observations upon which each corresponding point in the graph is based.
SOURCE: Frantz, S. C. 1973. Behavioral ecology of the lesser bandicoot rat, *Bandicota bengalensis* (Gray), in Calcutta. Dissertation, School of Hygiene and Public Health, Johns Hopkins University, Baltimore, MD.

Figure 18.27
Colony of *B. bengalensis* in downtown Calcutta park is totally diurnal and takes advantage of the food provided by people.
SOURCE: Photo by S. R. Frantz.

Bionomics and Integrated Pest Management of Commensal Rodents   **261**

and minimizing exposure to humans or other nontarget species.

Orientation and the Senses

Tracks, runs, and rub marks are interrelated; when rodents regularly move over an area, they make at least one of these signs. Once rodents leave the nest, they use pathways offering the most concealment, best routes of escape, and shortest distances to necessary resources. These movements are highly influenced by information picked up by the senses of touch, taste, smell, vision, and hearing.

Touch

Touch is highly developed in rats and mice, even in newborns. It enhances the ability of rats and mice to rapidly run pathways in the dark. Sensitive vibrissae extending out from the face (Figures 18.7 through 18.9) and long guard hairs on their backs aid rodent movements that are commonly, and preferably, made while in contact with vertical surfaces such as walls, baseboards, or fences. This habitual "in-touch" (thigmotactic) behavior results in the formation of rub marks and trails. Its predictability of expression makes rodents vulnerable to our management efforts because it indicates where we should provide interventions so as to intercept the rodents and thus maximize the probability of success.

Olfaction

Rats and mice have a keenly developed olfactory sense that also influences their movement patterns. When active, a rat or mouse continually moves its head about while sniffing and often raises up on its haunches to sample the air. Rodents leave odor trails of urine and other urogenital secretions which mark trails (and are cues for movement), delineate territories (the limits of movement range), and make it possible to detect sexually active mates (orients movement). Other odors in their habitat may provide positive cues (e.g., stored food) or negative cues (e.g., a predator) for orienting movement. Since commensal rodents essentially live with humans, the odor of humans is not a deterrent to rodent movements.

Taste

The sense of taste is highly developed. Rats and mice repeatedly nibble and taste substances in their environment. In various studies, rats have been observed to detect minute quantities (less than 1 ppm) of impurities, poisons, or other substances (29, 33, 242, 245). Rodents may reject on the basis of taste or smell alone foods previously recognized as either distasteful or as having been associated with illness. This remarkable ability has significant implications for rodenticide applications since it can lead to bait refusal and sublethal intoxication. Norway and bandicoot rats generally feed more heavily on one food at one time than do roof rats and house mice (both tend to be erratic feeders). Because the quantity of any one material consumed may be low, sublethal intoxication may be more common among the latter two species in interventions involving poison baits.

Hearing

Both rats and mice emit and respond to two levels of acoustics: a sonic range, audible to humans; and an ultrasonic range beyond the typical human threshold of around 20 kHz. The two levels of hearing responses are functional in that sounds such as humans talking or walking, a barking dog, and so on, may be signals for a rodent to hide (rats and mice are especially sensitive to any sudden noise). On the other hand, the sounds of rodents eating grain in a room may attract other passing rodents to a potential site for food and social interaction.

Ultrasonics are especially suited to nocturnal behavior and are used in echolocation for orientation [but are not nearly as well developed as in bats (246)] and in social or other communications (2, 144, 218, 220, 249, 270). Rat pups emit ultrasonic distress signals that function to recall the mother to the nest.

The ultrasounds of house mice are of greatest intensity in the 70-kHz range, just beyond the upper limit of hearing of the domestic cat (305). Ultrasounds made by mice at frequencies heard by cats would be rapidly attenuated in the air and scattered or absorbed by objects in the environment (e.g., bags of grain, blades of grass); thus, although a cat may be able to detect some signals at close range (within a few meters), ones farther away would be difficult for a cat to hear and to locate.

Sight

Vision in rats and mice is not as well developed as in humans, but it is specialized for nocturnal situations. Although their visual acuity is poor, rats and mice have a high sensitivity to light, can detect motion in very dim light, and can recognize simple patterns and objects of different sizes. With this combination of abilities, rodents typically move from lighted to darkened areas, a behavioral trait that has obvious benefits for survival. Rats detect movement at distances up to 10 m, have good depth perception to at least 1 m, and are able to gauge the effort needed for varying jumps; echolocation may also be used in the latter effort. Mice can identify objects at least 15 m away. Rats and mice apparently lack color vision, most colors presumably appearing to them as various shades of grey. Yellow and green, common colors of vegetable food items, would be revealed as light grey and probably would be attractive. This fact is utilized in dyeing poisonous baits bright yellow or green in order to repel birds and people, but not rodents; however, many dyes have an adverse taste or odor, and EPA may have restrictions on their use (*109, 223*).

Movement and the Senses

Commensal rats and mice memorize the details of their habitat—the pathways, obstacles, hiding places, and water and food resources. The actual muscular movements necessary, for example, to move down a pathway and to take shelter are also learned (*35, 180, 274*). This type of orientation uses a kinesthetic sense, a memory of muscular/physical coordination, and greatly aids survival especially when orienting in the dark. However, when a commonly used pathway is blocked, rodents repeatedly try to negotiate the route that their sense of orientation has informed them should be there. The obstruction might be a blocked hallway between grain warehouses or a covered hole leading to shelter or food (Figure 18.28). The associated disorientation may be momentary or even last a few hours during which time the animal may be exposed to adverse circumstances such as predators. Exploration or reexploration soon enable the rodent to reorient and learn any new aspects of its environment.

Avoidance and Exploration

The use of well-established runs can be interrupted or prevented merely by placing some unfamiliar object (e.g., a box, a brick) on or near the pathway (*22, 60, 61*). The same new-object reaction (neophobia) can be evoked by placing a familiar object in a new position within a familiar environment. Hence, neophobia occurs when there is a change in an otherwise familiar situation. The important factor seems to be the contrast between what a rodent expects (based on previous experience) and what is actually observed or encountered. The existence of neophobia is not universal and may vary with

Figure 18.28
Bandicoot rats gather at closed door (note chain latch at base) of grain warehouse; the door was usually left ajar at night (Calcutta, India).

species and habitat, as discussed below.

Two important characteristics of neophobia are that it is primarily seen among Norway rats and roof rats settled in stable areas, and that it is temporary (for a few days or even hours in some cases). Rats will habituate to the new situation if all else remains unchanged. This behavior may be significant when placing traps, baits, and bait stations. Such materials must be placed where the rats and mice will come in contact with them, but since neophobia may be elicited (at least in rats), it is often useful to place such objects unbaited, or baited and unset, for at least a few days before a poisoning or trapping campaign begins. On the other hand, mice tend to intensify exploration when their environment is disturbed (e.g., moved pallets or bags of grain), a habit that may make it easier to catch them in traps or attract them to baits. Mice have been observed to feed heavily from newly placed feeding stations even when there was already an abundant food supply available (*62*).

In a totally new situation, as when rats or mice have just been introduced into a warehouse, or because of habitat disruption (e.g., demolition of an infested building), exploration dominates behavior. Avoidance and neophobic behavior are minimal as the rodent learns about the resources (food, shelter, escape routes) of the new environment. Similarly, neophobia may be attenuated or absent in an unstable or changing environment (e.g., in a warehouse with rapid turnover of goods) where constant relearning of available resources may be critical for survival.

In new situations, exploration dominates, giving rodents a maximum amount of information about their surroundings in a short period of time. Once settled, they run about rapidly using their rote, kinesthetic sense of orientation; however, they continue to explore and reexplore

Table 18.2

Comparison of selected studies on home range sizes of urban rats

Species	Location	Estimated home range diameter (meters)[a]		Reference
		Average	Maximum	
B. bengalensis	Calcutta[b]	50	146	(121, 131)
B. bengalensis	Calcutta[b]	—	60	(278)
B. bengalensis and R. rattus	Calcutta[c]	—	73	(260)
R. norvegicus	Los Angeles[d]	—	120	MAR[e]
R. norvegicus	Baltimore[c]	24	60	(68)
R. norvegicus	Baltimore[c]	—	46	(73)
R. rattus	San Francisco[c]	—	92–152	(32)
R. rattus	Calcutta[f]	—	82	(260)
R. rattus	Los Angeles[c]	—	37	MAR[e]

[a] Different criteria for home range estimates are used by different authors; see references for details. Single linear movements by rats often exceed the population average.
[b] Howrah, an industrial/grain warehouse area.
[c] Residential area.
[d] College campus.
[e] M. A. Recht, California State University, Dominguez Hills, Carson, CA, 1988, personal communication.
[f] Industrial/warehouse area.

which keeps them abreast of changes. New objects are regarded with suspicion, cautiously investigated over time and, if all else remains unchanged, adapted to by the rodents. The strength of the various behaviors appears to depend largely on the contrast between typical habitat conditions and what is observed at a particular point in time. Although there is some species variation, the overall behavioral balance (the sum total of the various exploratory and avoidance behaviors) maximizes survival of rodents (24, 35, 122, 180).

It is important to adapt pest management procedures to take advantage of predictable "weaknesses" in rodent behavior. For example, we know that newly introduced animals tend to be exploratory though cautious, thigmotactic, and food- and shelter-seeking. This combination of behaviors makes it "bio-logical" to place bait stations, sheltered traps, glueboards, and so on, along entrance walls or other stationary objects to more easily intercept the newcomers. This type of ongoing monitoring is a key program element for finding new infestations.

Range of Movement

Range of movement depends largely upon distances between locations of food and suitable shelter and therefore can vary considerably between different rodent species and environment/climate combinations (Table 18.2). Range (or range diameter) should be interpreted not as a continuous area of equal utilization by rodents, but as a network of paths or runs. In essence, the movement range equals at least the minimum area required to meet daily nutritional and behavioral needs. Under stable environmental conditions (i.e., food and shelter conditions that remain relatively the same over time), rodent movements tend to be limited. Norway rats in a stable urban environment have been found to live for long periods within an area roughly 30–46 m (100–150 ft) in diameter (73). Greater daily movements may occur when the food and harborage are widely separated (240). Roof rats may have a larger, more three-dimensional range of movement due to their arboreal/scansorial habits, but data on this subject are quite limited (32, 241, 260).

House mice apparently stay within about 10 m (33 ft) of their harborage sites (38, 104) and may be restricted to only a few meters when food and harborage coincide (310). The movement range of mice may also be viewed three-dimensionally; for example, mice may live within a single stack of food materials stored on pallets in a warehouse and consisting of only a few cubic meters. Range tends to be greater when food and harborage are scarce.

Under unstable, changing conditions where food and harborage may be frequently altered, as with grain warehouses and refuse dumps, movements may be greatly expanded. This condition may explain some of the diversity of figures given for rodent movement ranges in Table 18.2 where stability of environment was not always reported.

In a study of *B. bengalensis* in Calcutta grain warehouses, most daily movements (standard range diameter = 68% of all movement) occurred within a 50-m diameter; wider daily range of movements occurred within 146 m (maximum range diameter = 98% of all movement) (*131*). Particularly long movements may occur if rodents have access to protected travel routes including narrow passages between buildings (as they did in the Calcutta study), in sewerage systems, or in dense, low-lying ground cover such as the vast stretches of ice plant (*Carpobrotus edulis*) along southern California freeways (Figure 18.29). Thus, in restricting rodent movements, it is important to think of "barriers" whether they are created by adding objects (e.g., a wall across a passage between buildings) or by taking something away (e.g., closely mowing the weeds and grass that may surround a food-processing plant).

Climatic conditions or marked seasonal changes may cause rats and mice to move greater distances. Mass migrations occur infrequently and are usually associated with sudden and drastic changes in environmental conditions, such as floods or crop failures (*284*). With the onset of cold weather in autumn, rats and mice may move significant distances from fields to farm buildings or other shelters. Drought may have a similar effect. Comparable movements may occur as a population increases in size and as excess, subordinate animals are forced out of a particular colony.

Physical Abilities

Norway and lesser bandicoot rats are far more active burrowers, but less efficient (although capable) climbers than are roof rats and house mice. Norway rats commonly gain entry to buildings through gnawed holes or via their burrows which have been known to reach a depth of 2 m in loose, soft soil. Generally, their burrows are less than 0.5-m deep (*44, 232*). The lesser bandicoot rat is similar to the Norway rat in being mostly a ground-dwelling species; all three rat species also inhabit sewers.

Harborage

The construction of below-grade, L-shaped curtain walls to prevent rodents from gaining access under ground floors of buildings takes full advantage of thigmotactic behavior (*259, 267, 297*; see also Chapter 22). Rodents begin burrowing a short distance from a building; they dig down at a steep angle until they contact the foundation wall, and then they burrow down parallel to the wall. Having made contact with the wall, they will not dig away from the wall to circumvent an obstruction. Of course, a con-

crete floor is more effective at excluding rodents from a building than a curtain wall and should be used whenever possible. In Calcutta, earth-floored grain warehouses are sometimes covered with a mixture of tar and sand to deter burrowing by lesser bandicoot rats.

Roof rats are more adapted to a climbing or arboreal habit; their foot pads give them a good grip (*207*). In buildings they are mainly found at roof height on beams, window ledges, and so on, and almost invariably so when Norway or lesser bandicoot rats are competitors (*22, 122, 284*). In the absence of more competitive species, roof rats sometimes burrow beneath rocks or the walls of buildings and invade sewers. Outdoors, depending on the kinds of vegetation and other conditions, roof rats live and nest in the tops of or around the bases of trees, in rotting vegetation, and in thick undergrowth that sometimes provides food as well as harborage (Figures 18.29, 18.30). For harborage, rats as well as mice are quick to exploit wall cavities, voids above dropped ceilings, spaces under and behind cabinets and machinery, dead spaces in machinery, roof spaces, and service shafts and ducts of buildings. In grain warehouses, they nest and travel in crevices between the bags and even burrow into the bags themselves (Figure 18.31).

Swimming

Rats and mice are generally good swimmers. Norway and roof rats readily swim through the water trap of a toilet bowl or floor drain, thus gaining access to buildings from sewers (*35, 235, 292*). Mice are less likely than rats to dive below the surface of water (*174*).

Figure 18.29
Vast stretches of ice plant (*Carpobrotus edulis*) along California highways provide rats with protected movement routes, food, and harborage (San Bernadino, CA).

Figure 18.30
Dense foliage (*Hedera canariensis*) covering a fence provides roof rats with food, harborage, and protected movement routes (Ontario, CA).

that a rat might grip in order to climb over it. Even a horizontal edge of sheet metal (<1 mm) is sufficient for roof rats to grip hold of with the claws of their forefeet.

From a standstill, rats can jump to a height of about 60 cm and mice to at least 30 cm (*235, 284*). With a running start or using a vertical surface as a spring board, Norway rats can clear 90 cm (*235*), roof rats, about 1.25 m, and house mice, 60 cm (*204*). Roof rats are capable of bouncing against a vertical surface to jump further vertically or diagonally to reach objects, harborage sites, and so on. Repeated use of such routes leads to soiled marks (from their feet) which are definitive signs for roof rats.

Roof rats and house mice are notoriously good climbers, whereas Norway rats are generally less capable (*235*). The Norway rat can climb if necessary (and when under pressure to escape or obtain food) and may exhibit an agility approximating that of roof rats operating under normal conditions. Rats and mice can climb or traverse almost anything they can grip or hold onto with their claws; this includes climbing vertical brick walls, pipes, ropes, and crossing horizontal wires. Smooth surfaces can be climbed if there is some object against which the rat or mouse can brace its back.

Reaching and Climbing

Rats can reach about 33 cm high along vertical walls and can easily hop up onto objects at this height. In fabricating sheet metal guards for vertical surfaces, a height of about 46 cm should be used to provide some margin of safety and ensure that rats will not be able to reach over it. Such guards must lack holding points (e.g., rough seams, closely spaced nail heads)

A major interspecies difference in abilities would appear to be the ease and speed with which roof rats can cross narrow-gauge wires and loose ropes and remain upright (Figure 18.32). Norway rats tend to move more slowly, the tail waving from side to side to help maintain balance, and they cannot remain upright while

Figure 18.31
B. bengalensis burrowing directly into a bag of rice in a warehouse (Calcutta, India).

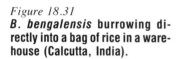

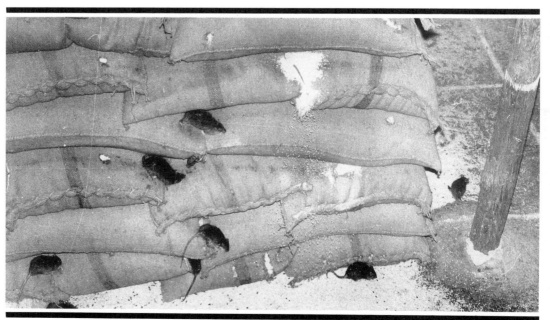

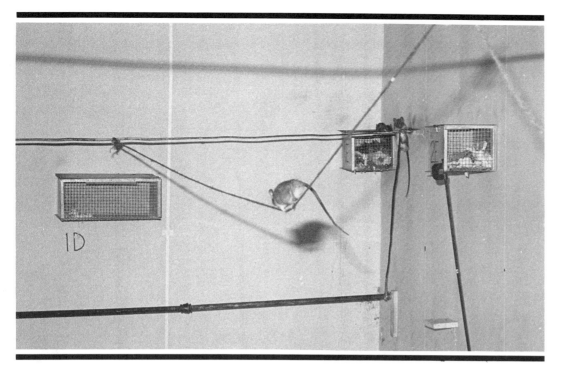

Figure 18.32
R. rattus readily crosses a loose rope (tied between two wires) while remaining upright and balanced.
SOURCE: Rodent Control Evaluation Laboratory, New York State Health Department, Albany, NY, file photo).

crossing loose, narrow-gauge ropes or wires. Norway rats often use their tails in a prehensile fashion to "grip" a pipe or other object being climbed (Figure 18.33). Similarly, lesser bandicoot rats have been observed climbing service pipes on building exteriors in Bombay to reach elevated food and harborage sites (*124*).

Roof rats have been observed to climb a 90-degree corner of a room wall covered with smooth sheet metal without obvious imperfections or other clawholds such as nail- or screwheads. Generally, it appears that rats and mice can climb any vertical surface on which they can get a clawhold, but many of these remarkable feats would only be attempted under highly stressful conditions. It must be remembered that rats and mice probably exploit their habitat only to the degree "necessary." Thus, in rodentproofing work, some arbitrary line must be drawn between what is probable for rodent behavior and what is actually possible. This approach should result in exclusion practices that are effective and not prohibitively expensive.

Feeding Habits

Preferences

Commensal rats and mice are omnivorous: Rodents are opportunists in feeding habits, generally following seasonal or localized abundance of food resources (*23, 82, 91, 100, 122, 165, 177, 187, 282, 288*). Parental food preferences are transmitted to the young (at least in rats), but various other factors can influence diet. The mother rat's diet apparently influences the taste and odor of her milk, and her offspring are sufficiently sensitive to these cues for them to later prefer solid foods with the same taste and odor (*45, 141*). Undoubtedly, food residues on the mother's face, fur, and forefeet are also available and influencial to young still in the nest (*24*).

Later, the young regularly follow adults on ventures outside the maternal nest; therefore, they come in contact with the food resources of adults. Also, young animals tend to explore and feed in an area that contains the residual olfactory cues left by the adults (e.g., odor of fermented fecal matter and urine). Hence, this mechanism that conditions food preferences, coupled with the tendency of rodents to avoid new foods, tends to provide some protection to the colony from ingestion of potentially dangerous foods. The acute sense of taste of rodents further enhances their survival.

Norway Rats

Norway rats are steady eaters, preferring cereal grains, seeds, nuts, meats, fish, and some types of fruit (*257, 266*). Given a free choice, they can select a nutritionally balanced diet, with a preference for fresh items over rancid, stale, or contaminated foods. Rats seem to compensate for qualitative and quantitative variations in their food resources (*23, 29, 82, 195, 257*). City garbage provides a more balanced diet than most rats can obtain on farms; as a result, city rats tend to mature more rapidly and are larger

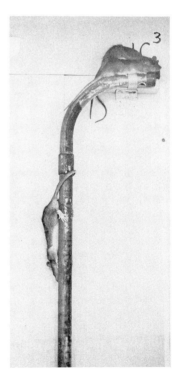

Figure 18.33
R. norvegicus uses its tail in a prehensile fashion while climbing down a pipe.
SOURCE: Rodent Control Evaluation Laboratory, New York State Health Department, Albany, NY, file photo).

Figure 18.34
Roof rats gnawed a hole in the lid of a plastic bucket used to store dog food (San Jose, CA).

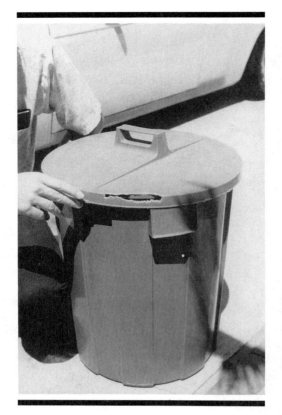

(65). Overall, their diet is similar to that of roof rats, but Norway rats more frequently forage in sewers and garbage (including landfills).

Roof Rats

The roof rat is an erratic feeder with preferences for fruits and nuts. Roof rats tend not to depend on household garbage but will take advantage of available dog food (Figure 18.34). In much of suburban California, roof rats have a rich and varied diet that includes fruits, nuts, avocado seeds, carob pods, garden snails (Figure 18.19), and a variety of ornamental [e.g., Algerian ivy (*Hedera canariensis*)] and native plants [e.g., Himalayan blackberry (*Rubus discolor*)]. In parts of Florida, roof rats are serious pests in citrus groves and sugarcane fields, whereas in some other areas they avoid citrus groves—obviously, local resources significantly affect dietary choice *(235)*. Both Norway and roof rats supplement their diets with available, nutritious materials (e.g., insects, worms, crabs, undigested food particles in dog feces) and occasionally ingest unusual items (e.g., glue, soap, plaster, putty).

Bandicoots

The lesser bandicoot rat's food preferences have been studied in cultivated areas *(25, 202)*, urban grain warehouses *(126, 131)*, and the laboratory *(82, 122, 253, 278)*. It prefers a diet of rice, wheat, pearl millet, various lentils, and other grains, but the actual diet is varied, including cockroaches, dead birds, young roof rats, and sugarcane. It hoards much more food grain than it eats *(226)*. Lesser bandicoots may be serious pests in mustard oil mills where they gnaw into seed bags to obtain other grains that accidentally get mixed with mustard seeds during harvest; mustard seed itself is not a preferred food. They apparently need free water in the grain warehouse habitat. They have been observed drinking undiluted human urine in latrines.

House Mice

The house mouse prefers cereal grains and seeds but has been known to exist on a wide variety of food including insects, frozen meat, and flour. Mice generally prefer foods high in fat or protein such as lard, butter, nuts, bacon, meat, and sweets such as chocolate candy as dietary supplements *(204)*. They often develop strong preferences; for example, they may choose a particular brand of candy bar in a supermarket and bypass anything else *(138)*. In testing caged

and free-living mice, Klimstra *(180)* found some of their favored "household" foods (in the USA) to be salted peanuts, "Cheez-it" crackers, cornmeal, and wheat germ. Prunes and pineapples, in liquid or solid form, were also highly favored.

Food preferences of a given rodent colony may be determined by closely scrutinizing spilled, partially eaten stored goods or garbage and by carefully examining the harborage area and runs leading to the harborage. Where rodents (especially rats) have been cutting into stored foods, the damage is quite evident; particles are commonly dropped while a rodent carries food to haborage—thus, the selection by the rodents becomes obvious. For baiting purposes, test-baiting (simultaneous offerings of a variety of food materials) will quickly help to determine an appropriate bait base for a particular control situation.

Daily Requirements for Food and Water

Food Losses to Rodents

In general, the daily food requirement of rodents equals about 10% of their body weight. Thus, a 250-g adult Norway rat would eat about 25 g/day or about 9 kg/year; an average house mouse would consume about one-tenth as much. In laboratory studies, adult lesser bandicoots ate about 12 g/day. This amount was estimated to be about one-third the daily ration required by free-living rats (278). Deoras (82) reported a daily consumption of about 14 g of mixed cereal grains by caged bandicoots. An estimated 200 lesser bandicoot rats regularly fed in one relatively small food-grain warehouse in Calcutta; they consumed nearly 2 metric tons of grain annually (125). This would be the equivalent daily rice ration for the average Indian for about 11 years! This estimate may be considerably less than the *total* amount of grain lost, since the average lesser bandicoot has been reported to eat or destroy nearly 70 g of grain daily (225, 278).

In a study in Nepal, it was conservatively estimated that 70% of premises surveyed were infested indoors with rats, mice, and shrews (123), with an average estimated food loss per infested dwelling of 65 kg/year. The estimated population of rats, mice, and shrews per Bangladesh farm family could cause a potential food (paddy) loss per farm family of 50 kg/year (295).

Not all of these various estimates have accounted for the additional losses due to waste, hoarding, and partially eaten and contaminated food items that may be several times greater than that eaten. After enclosed Norway rat populations (10–26 rats) had each been given free access to one ton of bagged wheat for 12–28 weeks (diet was supplemented with cabbage and horse liver; water was supplied *ad libitum*), it was reported that 25.1% of the grain remained in the sacks, 70.4% was fouled (spilled on the floor, contaminated with urine, fecal droppings, and hair), and 4.4% was otherwise lost (20).

Mice and roof rats tend to be nibblers and thereby damage a greater number of food items per unit of time than do steadier eaters such as Norway rats or lesser bandicoot rats. In a simulated bulk grain storage study, wheat was exposed to house mice for 150 days in a container 6.4 cm deep $\times$ 1.97 m^2. The mouse population (averaging about 12 throughout the study) consumed an average of 35 g/day, and about 0.33 kg of grain waste was found on the floor surrounding the container (295).

Utilization of Water

Water requirements vary greatly with diet, but Norway rats tend to need about 15–30 mL free water per day when on dry foods. Lesser bandicoot rats also appear to need free water, but roof rats evidently have less of a free-water requirement. All three species can probably subsist without free water when they have access to succulent foods. On a grain diet, all three species regularly drink available water (82). Water intake for *R. rattus* is highest when the rats are on a dry, high-protein diet.

Mice are generally able to live without free water, relying on metabolic water alone, unless the diet is high in protein or roughage (115). The metabolic processing of large amounts of protein increases the excretion of urine, and processing roughage increases water loss by way of feces. Thus, if mice are depending on a high-protein, dry diet, control should be enhanced by eliminating available free water and, if allowable, use of water-based rodenticides.

Actually, in most cases, mice and rats will be attracted to water if it is available. They may obtain water from dripping taps, leaking roofs, drains, wash buckets, early morning dew on roofs or grass, and so on. The Norway rat is more prone to suffer from water shortage, but under extreme conditions of deprivation the fertility of roof rats and house mice also declines (284). Wherever possible, abolition of freely available water should be part of a rodent pest management program.

The necessity of food and water (depending on species) for rodent survival has obvious implications for IPM interventions by sanitation, exclusion, and poisoning. Where food rather than harborage is the limiting factor for population growth, it may be practical or feasible to reduce available food resources and thereby reduce the rat population in direct proportion to the reduction in food. Applying poison baits *immediately* after an appreciable reduction in available food or water will enhance the rate of population attrition.

Indoors, in food-handling establishments, preventing rodent access to foods is a much more desirable intervention than poisoning. If poisoning is necessary, scrupulous care must be taken to ensure that no food contamination will result from poison application. Consumption by the rodents of a nontoxic prebait or census bait will help to roughly estimate the number of rodents present. Thus, before a poisoning campaign is begun, it will be known approximately how much bait must be placed to be lethal in the required number of feedings, and overall risk and cost factors are curbed.

Reproduction, Development, and Mortality

The Family Muridae is the most successful family in the Order Rodentia and is probably the most successful family in the entire Class Mammalia (275). As mentioned previously, murid rodents are economically, medically, and aesthetically important because many species coexist with humans. Species of greatest significance to humans, including *R. norvegicus*, *R. rattus*, *B. bengalensis*, and *M. musculus*, all destroy or contaminate foodstuffs, grain products, and other stored goods; and they all have a capacity for very rapid population increase.

Rate of Increase

Under optimal conditions of climate and abundant food, water, and harborage resources, rats and mice would have an exponential rate of increase with a maximum reproductive potential (depending on particular species) of a few hundred to many thousand descendents from one adult pair within

Table 18.3

Typical reproductive patterns of *Rattus norvegicus* (Norway rat)

Ref.	Female age at sexual maturity (days)	Prevalence of pregnancy (%)	Incidence of pregnancy (number)	Gestation period (days)	Embryos per pregnancy (number)	Litter size (number)	Production per pregnant female per year (number)	Age at weaning (days)	Heat interval (days)	Life expectancy (years)
(19)	90	—	—	22	—	8	—	—	—	0.58–1
(24)	—	—	—	24	—	3–10	—	—	—	—
(35)	75	21.4 (10.7–34.8)	4.32	22–24	8.8 (7.9–9.9)	—	38	28	—	—
(44)	85	—	—	23	—	—	—	28	—	—
(114)	63	23.7	4.7	—	8.7 (1–14)	4.7	39 (33–41)	—	—	1
(234)	—	—	4–7	22	—	8–12	20	—	—	≤1
(275)	—	19.9 (10.7–31.4)	4.01 (2.2–6.4)	22–25	8.9 (8.1–9.9)	6–9	35.7 (18.5–55.7)	—	—	—
(290)	—	—	4–6	21–23	—	6–12	20	—	4–5	—
(292)	—	—	5	—	—	8 (2–14)	—	—	—	—
Approx. range[a]	75–90	11–35	2–7	21–25	8–10	3–14	19–56	28	4–5	≤1
Approx. av.[a]	75–90	20–21	4	21–25	9	6–12	36–39	28	4–5	≤1

[a] An approximation based on the studies reviewed; due to the diverse range of methodologies used in the original sources, statistics per se were not calculated.

Table 18.4

Typical reproductive patterns of *Rattus rattus* (roof rat)

Ref.	Female age at sexual maturity (days)	Prevalence of pregnancy (%)	Incidence of pregnancy (number)	Gestation period (days)	Embryos per pregnancy (number)	Litter size (number)	Production per pregnant female per year (number)	Age at weaning (days)	Heat interval (days)	Life expectancy (years)
(19)	90	—	—	22	—	6	—	—	—	0.58–1
(35)	68	28.6 (12.9–48.8)	5.42	20–22	6.2 (3.8–7.9)	—	33.6	28	—	—
(83)	—	—	—	—	—	—	—	—	3.8	—
(112)	73	—	—	21–22	—	8	—	—	—	—
(202)	—	—	—	21–23	—	5–8	—	28–35	—	—
(234)	—	—	4–6	22	—	6–8	20	—	—	≤1
(239)	—	25	6.6	—	—	—	—	—	—	—
(275)	—	25.3 (12.9–43.3)	5.03 (2.6–8.8)	21–22	6.1 (3.8–7.5)	5–8	31.3 (12.2–56.2)	—	—	—
(292)	—	—	5	—	—	7 (2–14)	—	—	—	—
Approx. range[a]	68–90	13–49	3–9	20–23	4–8	5–14	12–56	28–35	4	≤1
Approx. av.[a]	68–90	25–29	5	20–23	6	5–8	31–34	28	4	≤1

[a] An approximation based on the studies reviewed; due to the diverse range of methodologies used in the original sources, statistics per se were not calculated.

one year. This rate is possible because of the following reproductive characteristics:

(a) early sexual maturation (usually within 3 months for both sexes);
(b) polyestrous breeding throughout the year, even under unfavorable climatic conditions if food and shelter resources are adequate;
(c) short gestation period (about 3 weeks);
(d) postparturient heat (usually within 2 days after giving birth, allowing pregnancy and lactation to occur simultaneously); and
(e) large litter size (usually no less than six).

Tables 18.3 through 18.6 list typical reproductive characteristics of selected murid rodents reported from various studies. Female age at sexual maturity (adulthood) indicates the age at

Table 18.5

Typical reproductive patterns of *Bandicota bengalensis* (lesser bandicoot rat)

Ref.	Female age at sexual maturity (days)	Prevalence of pregnancy (%)	Incidence of pregnancy (number)	Gestation period (days)	Embryos per pregnancy (number)	Litter size (number)	Production per pregnant female per year (number)	Age at weaning (days)	Heat interval (days)	Life expectancy (years)
(83)	—	—	—	—	—	—	—	—	5.3	—
(153)	—	—	—	—	2–13	—	—	—	—	—
(254)	—	—	—	—	—	—	—	—	5.2 (4–6)	—
(278)	90–100[a]	52.7 (49.4–55.8)	11.3 (10.6–12)	—	6.2 (5.3–6.4)	—	69.9 (57.2–76.7)	30	—	≤1
(278)	—	56.2	12.12	—	8	—	97	—	—	—
(278)	—	—	—	—	7.4	—	—	—	—	—
(278)	—	—	—	23	—	—	—	—	—	—
SCF[b]	60.5 (39–92)	—	—	21–23	—	6.4 (4–9)	—	<26	—	—
Approx. range[c]	39–100	49–56	11–12	21–23	2–13	4–9	57–97	<26–30	4–6	≤1
Approx. av.[c]	61	53–56	11–12	21–23	6–8	6.4	70	<26	5	≤1

[a] From one animal.
[b] Original data of S. C. Frantz.
[c] An approximation based on the studies reviewed; due to the diverse range of methodologies used in the original sources, statistics per se were not calculated.

Table 18.6

Typical reproductive patterns of *Mus musculus* (house mouse)

Ref.	Female age at sexual maturity (days)	Prevalence of pregnancy (%)	Incidence of pregnancy (number)	Gestation period (days)	Embryos per pregnancy (number)	Litter size (number)	Production per pregnant female per year (number)	Age at weaning (days)	Heat interval (days)	Life expectancy (years)
(19)	42–60	—	—	19	—	5–6	—	—	—	<1
(24)	—	—	8	19	—	3–12	46	—	—	—
(35)	42	35.3 (19.8–50.5)	7.67	19–21	5.8 (3.9–7.4)	—	44.5	25	—	—
(204)	—	—	6–10	19–21	—	5–6	≤50	21	4	1–2
(234)	—	—	8	19	—	5–6	—	—	—	<1
(252)	35–42	26.7	—	—	5.4 (1–10)	—	—	—	—	—
(275)	—	35.4 (21–50.5)	8.09 (5.1–11.4)	18–21	5.4 (3.9–6.4)	4.8	42.6 (29.1–57.2)	—	—	—
(284)	—	—	≥10	17–20	—	—	≤60	21–28	—	—
(292)	—	—	6	—	—	6 (2–13)	—	—	—	1
Approx. range[a]	35–60	20–51	5–11	17–21	1–10	2–13	29–60	21–28	4	1–2
Approx. av.[a]	35–60	27–35	6–8	18–21	5–6	4–8	43–45	21–28	4	<1

[a] An approximation based on the studies reviewed; due to the diverse range of methodologies used in the original sources, statistics per se were not calculated.

which the vaginal orifice becomes perforate and reproductive activities can begin. The prevalence of pregnancy is the percentage of adult females (necropsy sample) with macroscopically visible embryos taken over a given time (105). Incidence of pregnancy is the average number of pregnancies per adult female during the time sample, expressed here as an annual estimate. The gestation period may be slightly prolonged in females that become pregnant while still lactating. Embryos per pregnant female as well as litter size are listed in Tables 18.3 through 18.6 since some studies report embryos per se from necropsy samples whereas others report number of offspring following parturition.

Interspecies comparisons are difficult because significant differences exist even between populations of the same species. Climate, stability of food and shelter resources, and other factors have a major effect on the reproductive pattern which may also differ in the same population from year to year. Adverse conditions of very cold or hot dry climatic periods tend to reduce breeding activity and the success of rearing young unless the animals live indoors. Lesser bandicoot rats living within the protective environment of grain warehouses in Calcutta tended to have greater reproductive productivity and overall better survivability than animals living outdoors (122, 278). Studies of house mice (189, 252) and Norway rats (43, 68) have resulted in similar conclusions.

Tables 18.3 through 18.6 include a wide range of international studies; several reports are reviews of data from numerous sources (24, 35, 234, 275), whereas others represent individual studies (43, 112, 278). Approximate "range" and "average" figures are provided in an attempt to summarize information for interspecies comparisons of typical reproductive performance.

Production of Young

Of primary interest to pest managers is the production per female per year. Of the two globally distributed rat species common to the USA, R. norvegicus has the higher productivity (36–39 young/female per year). Although R. rattus has a greater prevalence and incidence of pregnancy than R. norvegicus, the small litter size results in less annual production (31–34 young/female per year). The house mouse achieves an even greater productivity (43–45 young/female per year) through a combination of higher prevalence and incidence of pregnancy and earlier

sexual maturation, even though litter size is smaller than in the two rat species.

Although less studied than the foregoing species, B. bengalensis is remarkable in that its prevalence and incidence of pregnancy (Table 18.5) are roughly twice those of the other species listed and are higher than those observed in any other murid population (275). Its estimated productivity of 70 young/female per year similarly eclipses the other species. Lesser bandicoots also appear to be weaned and to reach sexual maturity earlier than the two kinds of common rats. Under ideal environmental conditions, their reproductive characteristics could result in 15,000 descendents from one pair of bandicoots within a year. Rapid growth and development, coupled with its more aggressive nature, extensive burrowing habits (279), and climbing ability (124), have undoubtedly contributed to B. bengalensis becoming the dominant commensal rat species in some areas of South and Southeast Asia (81, 278, 302).

Mortality

Prenatal mortality, as well as mortality at later stages, is closely related to habitat and behavioral factors (see Factors Affecting Population Dynamics, below). Norway and roof rats have been estimated to have intrauterine or prenatal mortality in the range of 10–25%; for house mice it was 4–11% (275). The young of all four species represented in Tables 18.3 through 18.6 are born hairless, with eyes and ears closed, and need rather constant maternal care. By 3–4 weeks of age, they have begun to eat solid foods, make at least short exploratory excursions outside the nest, and are ready to be weaned. Studies of confined populations show postnatal (preweaning) infant mortality to be a particularly labile parameter of population dynamics. Preweaning mortality can rapidly swing from relatively low to very high rates in response to changes in habitat conditions and behavioral interactions (44, 68, 275).

Weaning

Under typical conditions, young are weaned and ultimately recruited into the adult population by the age of about 2 months (for M. musculus and B. bengalensis) or 3 months (for R. rattus and R. norvegicus). Postweaning mortality tends to be a relatively stable parameter of population dynamics but is difficult to assess in free-living populations. Basically, females tend to live longer than males, and the annual mortality rate in wild populations usually reaches

90–100% (35, 275). Under laboratory conditions, all four species may live 2 years or more, but it is doubtful if free-living commensal rodents commonly achieve such longevity.

Obviously, the reproductive patterns of *R. norvegicus, R. rattus, B. bengalensis,* and *M. musculus* have a potential for rapid population expansion. Commensal rodents can readily replace their numbers when reduced by control efforts such as poisons or traps alone. In a proper IPM program, rodent habitats are modified so as to limit food, water, and harborage resources, thereby causing stress in the population, with concomitant prenatal and postnatal mortality. The IPM effort takes advantage of natural rodent responses to changes in habitat factors.

Factors Affecting Population Dynamics

The full potential population increase is rarely achieved by free-living commensal rodents, though examples of outstanding increases or high densities have been reported (*79, 148, 156, 167, 227, 256, 261*). Mechanisms governing population growth involve three limiting factors (discussed below) interacting with three population forces (discussed below) to produce a certain resultant population size (Figure 18.35) (*24, 68, 76, 196*). The extent to which each population force influences a population to increase, decrease, or remain stable depends on limitations imposed by habitat, predation, disease, and competition.

Habitat

Nutritional, structural, and climatic factors are collectively called the habitat. The abundance and distribution of food, water, and harborage have a direct effect upon the number of rodents that can be sustained in a particular environment. These three environmental resources fundamentally define the carrying capacity, i.e., the number of rodents supported by a particular environment.

Food

Given a choice, Norway rats choose a nutritionally appropriate diet of a relatively constant caloric value (*24, 68, 257*); presumably, other commensal rodents have similar abilities. Major food resources of commensal rodents in urban areas (Figure 18.36) include improperly handled food of humans and their companion animals, garbage, and vegetation (natural, ornamental, and vegetable gardens); in rural areas, natural vegetation, seeds, and field crops become important. Food-handling and -storage facilities

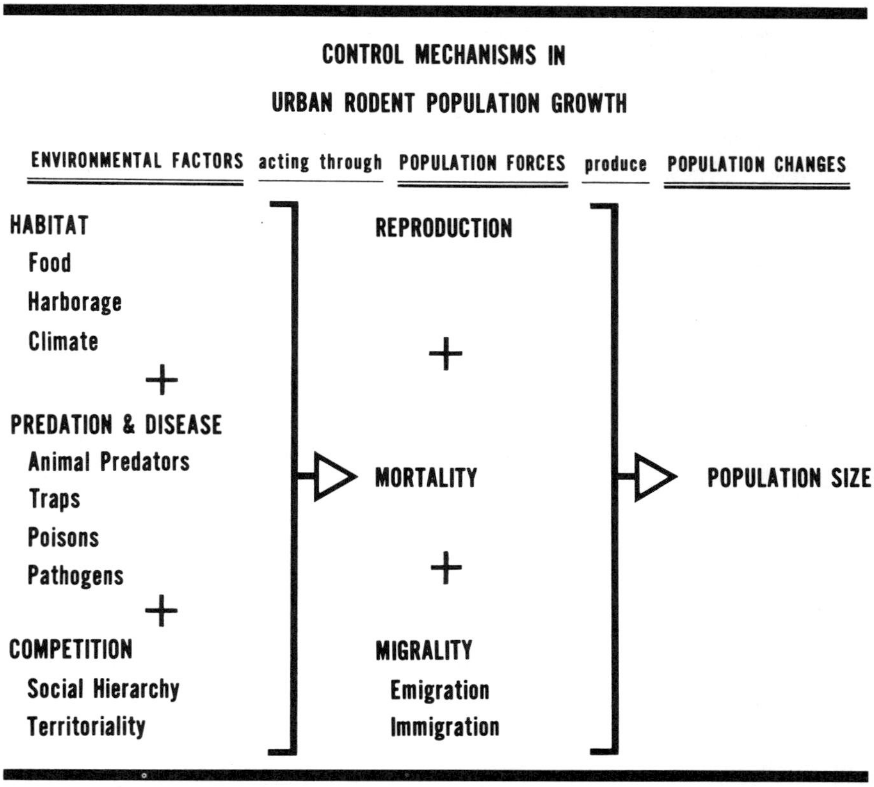

CONTROL MECHANISMS IN URBAN RODENT POPULATION GROWTH

ENVIRONMENTAL FACTORS	acting through	POPULATION FORCES	produce	POPULATION CHANGES
HABITAT Food Harborage Climate + **PREDATION & DISEASE** Animal Predators Traps Poisons Pathogens + **COMPETITION** Social Hierarchy Territoriality		**REPRODUCTION** + **MORTALITY** + **MIGRALITY** Emigration Immigration		**POPULATION SIZE**

Figure 18.35
Control mechanisms in urban rodent population growth.
SOURCE: Adapted from Frantz, S. C., and J. P. Comings. 1976. Evaluation of urban rodent infestations—an approach in Nepal. In *Proceedings of the Seventh Vertebrate Pest Conference* (University of California, Davis), C. C. Siebe, ed., pp. 279–290.

Figure 18.36
Accumulated refuse behind a large apartment complex (New York, NY); note "airmail" garbage (thrown out of windows) on fire escapes.

Figure 18.37
Spillage and rodent damage in animal feed warehouse (Kuwait). Note problem-generating storage practices: bags not organized into stacks, no inspection perimeter, spillage and rodent-damaged bags present.

(Figure 18.37) offer abundant food in open racks and bins, stacks of bagged grain, accumulated food product in improperly cleaned processing machines, and spillage.

Water

Water is often available in the form of high-moisture foods, leaking pipes, condensation collector pans of air conditioners, open sewers, and water bowls or troughs for domestic animals and pets. *R. norvegicus* and *B. bengalensis* generally need free water; *R. rattus* is more tolerant of water deprivation, and *M. musculus* can often tolerate dry conditions through utilizing metabolic water. However, all four species readily take water if it is available; hence, restricting available water coupled with the introduction of water-based or moist rodenticide baits can be useful in IPM programs.

Harborage

Harborage is usually abundant in urban and rural environments where rodents easily enter faulty or deteriorated structures, basements, attics, foundations, retaining walls, abandoned automobiles, and sewers (storm and/or sanitary). Rodents may also live in thick vegetation and burrow in gardens, lawns, river banks, crawlspaces, under buildings, under paved slabs, and in the space between walls and fences. Accumulations of household rubbish and construction materials in garages, backyards, and vacant lots commonly provide ample harborage (Figure 18.38).

Causative Conditions

Collectively, the various factors in the habitat that sustain rodent populations are the *causative conditions*. Obviously, the resources of the environment are closely intermeshed, and efforts to reduce one element may affect the others. It is axiomatic that the limiting factor(s) changes with conditions and differs between places and times, so that by the time a factor is identified as limiting, another (which may interact with still another) may have already supplanted it in importance. Therefore, for reducing rodent populations, it is most practical to modify all causative conditions to ensure that the limiting factor(s) is included (*76, 133*). Following 2 years of improvements in housing and refuse collections in Baltimore, the rat population declined about 64% (*74, 75*). In Amman, Jordan, the distribution of food, harborage, and water was found to account for no less than 79% of the variation in rat infestation between different housing types; this underscored the importance of reducing these causative conditions in any future long-term rodent control program (*208*).

Predation and Disease

Predation and disease (including malocclusions, parasites, and pathogens) can produce substantial reductions in populations of commensal rodents, but the effect is usually temporary. Predation includes the actions of humans [traps, poisons, and even use of rodents for food (*24, 95, 98, 308*)], cats, dogs, raccoons, rats themselves, birds (e.g., hawks and owls), snakes, lizards, false vampire bats, and other foes. Although predation removes individuals from a population, removal (alone) of even a large proportion of Norway rat populations in Baltimore did not cause the populations to decline to extinction; many fully recovered within 6 months (*68*).

Rodent parasites and infectious diseases can be regarded similarly; they can cause severe debility in rodents but are usually not fatal (*24, 35, 68*). However, under stressful conditions such as reduced food, water, and harborage, rodents may be more exposed to predation and have increased infection rates (resulting from physiological imbalances or wounding) which thereby augment overall population mortality.

Competition

Competition, whether between members of the same species or between different species,

Figure 18.38
Accumulated materials in a barn provide harborage for rodents (Watervliet, NY).

can be an important limiting factor. Norway rats compete with roof rats and have replaced them in many cities where both were once found (*24, 234*). Lesser bandicoot rats have similarly replaced other rat species in areas of South and Southeast Asia. Both Norway and bandicoot rats take on dominant roles in interactions with house mice; mice often restrict their activity to time periods when rats are not present. Note that when one species is controlled, the other(s) may become important unless the management strategy has adequately considered *all* species present. Where roof rats occur, they will increase as Norway rats are reduced. During the Depression, Works Progress Administration crews in southern cities of the USA reduced Norway rats on the ground level of buildings; this resulted in roof rats coming down from the upper levels.

Social Organization and Population Dynamics

The natural social behavior of Norways rats, roof rats, lesser bandicoot rats, and house mice involves hierarchy and territoriality. The organization of a social hierarchy results in some members of a colony being dominant over others in terms of, for example, access to food, harborage sites, and mates. The degree of expression of social organization appears to be a logical outgrowth of the interaction of several factors within both the ecological and the social milieux

of particular populations of rodents. Many studies of rodent social organization have been made, but relatively few have been conducted under natural circumstances so as to avoid behavioral artifacts that are often imposed by confinement.

Sociality of *R. norvegicus*

Studies of Norway rats in the wild, in outdoor enclosures, and in the laboratory are in basic agreement that a dominant (alpha) male holds top rank in the social group, moves without hesitation throughout his group, and initiates most fights (*21, 22, 44, 71, 238*). Dominant males establish territories about favorable feeding places and control burrows containing several females; lactating females also exhibit territoriality. Interlopers are driven from the general locality of a single burrow or a group of burrows. In small groups (about 20 residents) of free-living rats, interlopers are attacked, but in larger groups they are often accepted (*285*). When interlopers were released into stationary urban rat populations, social disorganization resulted, with increased mortality among residents and interlopers (*43, 68*). As a confined population increases, emigration is not possible, social rank and territory become poorly defined, and aberrant behaviors develop (*44*).

Rats utilize an array of social signals (e.g., postures, odors, sounds) that probably prevents excessive injury to any individual and ensures that a population becomes well distributed under natural circumstances (24). Free-living *R. norvegicus,* as a rule, defend only travel paths (285).

Sociality of *R. rattus*

A free-living colony of roof rats (with constant supplemental food provided) had a definite social hierarchy and maintained its integrity despite the fact that its composition was not wholly stable (112). A single dominant (alpha) male was always present and a linear male hierarchy was sometimes formed. Also present were a few mutually tolerant, top-ranking females that were subordinate to the alpha male but dominant to all other colony members. Within the colony, attacks were always directed downward on the social scale. However, serious conflicts were generally avoided because the attacked subordinates fled or adopted some form of appeasement behavior. Appeasement permits dominant rats to maintain status without the necessity of attacking subordinates to the extent of inflicting physical injury; this maintains individual health, conserves energy, and enhances overall colony survival.

The colony territory about the feeding place was defended against interlopers. Female residents took part in chasing out intruders of both sexes; resident males apparently restricted their chases to intruding males. Though territorial boundaries were fixed, colony members have been observed to explore beyond this limit under some circumstances, and thus become familiar with a larger area. Occasionally, it was possible for a persistent interloper of either sex to force its way in and ultimately become a colony member, even to the extent of displacing a dominant rat.

Sociality of *B. bengalensis*

In laboratory or pen studies, bandicoot social organization largely parallels that of Norway rats as described earlier (226). However, observations of paired interactions under field conditions indicates that *B. bengalensis* lack a well-defined, stable social organization. When these rats encountered one another, they commonly appeared to be indifferent (although perhaps very subtle social dynamics occurred) or they tested social rank only in a minor bout of physical conflict. The overabundance of grain in the study area apparently reduced the need for competition over food; harborage was also abundant, but more limited than food. Since grain was available at discrete locations (warehouses), animals were forced to gather in large aggregates (commonly 100 rats, with densitites of 2 rats/m²) when feeding.

Such grouping behavior required considerable social tolerance by the bandicoots; little energy was expended to maintain territories or dominance hierarchies. Interloper experiments in and around warehouses revealed that bandicoots were essentially nonterritorial; however, females (and cohort subadults) defended burrow entrances against some individuals, especially adult males. In addition, adult males periodically defended a small area (1–2 m²) around harborage entrances, a short-term activity apparently associated with females in heat.

Basically, the bandicoot population existed as highly fluid feeding aggregates that followed the availability of stored grain. The general lack of both territoriality and strong social ranking was believed to be due to the population's high mobility. It was precisely this mobility and social flexibility that enabled bandicoots to fully utilize the shifting food resources of the numerous grain warehouses. Continuous interactions of animals of various degrees of strangeness probably made interanimal recognition difficult and defense of a large area disadvantageous in terms of bioenergy budgeting (122, 126).

Sociality of *M. musculus*

In natural populations of house mice, females tend to nest in small groups. During pregnancy and lactation, mice become territorial, excluding adult males and strange females (63, 275, 276). Males are also territorial if population density is not excessive. Dominant males defend a system of runways and the females living therein. Outdoor territories that are limited during inclement weather (e.g., snowbound in winter) may well expand with warmer weather and increased foraging area (3).

Dominants and Subordinates

Where rank organization is highly developed, dominant rodents may restrict the access of subordinates to food resources, especially during the favored activity periods. In poisoning work, this may cause poor results due to inadequate intake by the bulk of the target population. However, if baits are widely distributed, are initially made available for at least a few days as nontoxic prebaits, and are kept replenished throughout the required number of days (depending on the action of the toxicant), most active members of the population should at least have access to the baits.

Fraternals and Strangers

As with rats (37), olfaction plays an important role in individual recognition among wild house mice (30, 270). Frequent mutual recognition, with appropriate responses regarding social rank, tends to keep physical conflict low among colony residents, but interlopers are quickly recognized and often attacked (16). Successful immigration of interlopers into an established population would probably occur more readily in a complex environment with abundant escape cover or in a less complex one where the resident population was large and the residents experienced difficulty identifying strangers (251). At high densities, a confusion effect may operate which aids in protecting submissive animals from aggressive ones (198, 244). Both effects probably operated in the warehouse bandicoot population described above (122) and may account for some differences in other studies.

Resource Instability

The real world is infinitely more complex than the best of "laboratory" enclosures. Natural habitats of commensal rodents are characteristically unstable, with changing amounts and locations of food (for example, around food-storage warehouses where products come and go at irregular intervals), changing harborage resources (as rubbish availability rises and falls), or changing mortality patterns (as pest management programs or disease outbreaks occur periodically), so that their populations often experience drastically changing resource conditions to which they must adjust their population size and/or distribution. In New York City, rats were relatively scarce even in the apparent presence of copious food and harborage resources. However, since refuse was collected daily at that time, the food resource was unstable in quantity and location (67).

Clearly, social organization (hierarchical and territorial behaviors) has a limiting function in population dynamics; that is, it limits successful breeding to individuals that are ecologically and behaviorally most successful in establishing dominance and/or territory (276). As population densities increase and space becomes more limited, a smaller proportion of individuals is able to achieve high rank or territories, and total, successful reproduction is reduced in relation to population size.

However, natural environments often enable "excess" animals to emigrate to nearby localities (even the barn next door) and establish a new colony as shown in various field and lab-oratory studies with rats and mice where growing populations have been carefully monitored. Only in studies of confined populations (with abundant food and limited harborage), where excessive growth has been allowed, has social organization totally collapsed and aberrant behaviors become commonplace. However, within unconfined populations, animals not able to establish dominance and territory are certainly more at risk to reduced fecundity (if food is limited) or social crowding mechanisms with increased infant mortality (if food is abundant).

Crowding and Density

Southwick (275) notes an important distinction between crowding and density: Populations vary considerably in density tolerance, that is, the density at which specific crowding effects appear. To varying degrees, crowded rodent populations are characterized by a high prevalence of physical conflict, reduced pregnancy rates, reduced infant survival, nest desertion, nest pollution (waste accumulation), wounding, dermatitis, and other signs of poor health (276). Although all of these characteristics are not likely to occur in any particular natural population, intensified crowding effects can be artificially induced by reducing food, water, and shelter resources (i.e., through sanitation and physical exclusion measures).

Population Factors and Patterns of Growth

Though influenced by the interacting limiting factors (Figure 18.35), the primary forces that determine the size of a rodent population at a given time are reproduction (adds individuals), mortality (subtracts individuals), and migrality (immigration adds and emigration subtracts individuals) (68).

Unsuppressed ("Normal") Populations

The natural rate of population increase in commensal rodents is rapid and characterized by a short gestation period, large litter size, rapid maturation, postparturient heat, and polyestrous breeding. When rodents are introduced to a favorable environment, reproduction will be at maximum rate. The population grows slowly at first and then at a faster rate (Figure 18.39). Growth levels out as the population approaches equilibrium with the environmental carrying capacity, primarily determined by food, water, and harborage limitations (24, 66, 68, 76, 275, 276). As these limits are approached, contact rates between animals increase radically, inter-

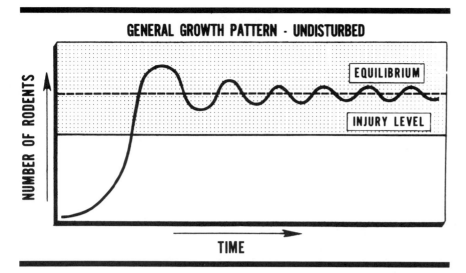

GENERAL GROWTH PATTERN - UNDISTURBED

NUMBER OF RODENTS

EQUILIBRIUM

INJURY LEVEL

TIME

Figure 18.39
General growth pattern for an undisturbed (unmanaged) rodent population.
SOURCE: Adapted from Frantz, S. C., and J. P. Comings. 1976. Evaluation of urban rodent infestations—an approach in Nepal. In *Proceedings of the Seventh Vertebrate Pest Conference* (University of California, Davis), C. C. Siebe, ed., pp. 279–290.

species competition increases, dominance hierarchies and territories are established, and mortality increases until a state of equilibrium is reached. Emigration and immigration are generally less important in determining population size than either reproduction or mortality.

In this normal (uncontrolled or undisturbed) population, the number of individuals may exceed, for brief periods of time, the carrying capacity (or equilibrium level) (Figure 18.39). This is because the natural rate of population increase is relatively faster than the regulatory action of the complex sociophysiological feedback mechanism involving direct behavioral (crowding) effects and hormonal balance (*50, 51, 68, 71, 72, 275, 276*). Then a population decreases to a point below the equilibrium level, thereby setting up an oscillation that dampens to some extent over time. The equilibrium is constantly in flux, being displaced and then returning to the original state with a constant flow of individuals through the system in a constantly changing environment. The population will continue to oscillate about the equilibrium level (carrying capacity) as long as all else in the environment remains relatively stable.

The equilibrium level for commensal rodent populations is commonly above the human injury level (Figure 18.39). That is, when the population has reached equilibrium with the environmental carrying capacity, the number of rodents is already causing economic, medical, and/or aesthetic damage and warrants IPM intervention(s). In order to maximize the potential for affecting the greatest proportion of the population, it is recommended that interventions be made during periods of low breeding (*24, 134, 234*). That is, although breeding may occur year-

round, it will be minimized (at least in outdoor populations) by adverse climatic conditions such as protracted hot or cold periods. In temperate climates, the recommended best times (in descending order of importance) for initiating IPM would be winter, summer, and fall. Similar timing would still be an important consideration for indoor control efforts since immigration from outdoor areas could otherwise be an added complication.

Temporarily Suppressed Populations

Population removal studies have shown that greatly reduced populations increase or recover more slowly than do moderately reduced populations (*68*). The critical point is that populations do tend to recover from removal interventions such as poisoning and trapping alone (Figure 18.40). Such tactics do remove individuals from the population, but do not necessarily affect the successful breeding animals, and do nothing to reduce the environmental conditions (particularly food, water, and harborage) available for remaining animals and immigrants. When such repressive measures are discontinued, the remaining population increases once again (similar to the growth pattern of newly introduced animals) to the original carrying capacity of the environment.

Considering the reproductive capabilities of rodent populations, it is obvious that removal measures alone (poisoning and trapping) are futile—the effect is temporary. In fact, such measures may serve merely to stimulate and maintain reproduction at an efficient level (*114, 169, 275*). Furthermore, as a population reestablishes itself, it will initally exceed equilibrium and, for a period, may be at a greater size (with concomitant deleterious effects) than before removal was initiated.

Permanently Suppressed Populations

When the habitat is modified so as to reduce the carrying capacity (restricting the amount of space, harborage, water, and food available), and removal measures accompany the habitat modification, the program has a much greater chance of meaningful success. Several types of interventions can usually be more effective than only one type when they are well coordinated so as to have a combined or synergistic effect. The combined effect is greater than the effect of either type used alone at different points in time.

This process should result in a lowered prevalence and incidence of pregnancy, increased age at sexual maturity, and a reduced proportion of young (*68, 71*) as the population reaches equilibrium at the reduced level of environmental resources (Figure 18.41). The objective is to lower the population equilibrium level below that which has been defined as the injury level. Because such activities negatively affect reproduction, the rodent population will remain lowered unless food, water, and harborage resources become readily available once again. Such measures are valid even if the current pest population is eliminated (an uncommon event) because the reduced (and maintained) level of resources will prevent a new population from resettling the area and growing to the previous equilibrium level.

The importance and utility of understanding rodent population dynamics has been summarized by Davis (*71*): "The general principle is to alter the environment so that the density-dependent factors will change inversely to the desired change in population. Specifically for rats, this means that in order to decrease the number of rats, the food and harborage supply should be decreased so as to increase the competition among the rats." According to Pratt et al. (*234*), "Controlling rat populations, not individual rats, is the key to a successful rodent control program in a community."

Management

Considering the foregoing, one can easily see that rodent populations have a remarkable built-in capacity for growth and survival; in the case of commensal rodents, however, people usually provide the medium in which that potential can be realized. Pest problems are not isolated, unpredictable events merely demonstrating the perversity of nature. They are a function of the design of urban industrial and rural agricultural production systems, ornamental landscapes, and architecture, as well as human behaviors and belief systems that motivate, create, and maintain those designs. In a society that all too often believes in "living better through chemistry" and has an inadequate understanding of the complexity of biological systems, it is understandable that a reliance upon simplistic rodenticidal interventions should be so appealing. The evidence of our chemical dependence is overwhelming. The annual production of biocides in the USA increased dramatically from 232 tons in 1951 to 750,000 tons in 1980 (*89*), a rate of more than 3,000%.

Beneficial poisons (such as pesticides) should be used with discretion and heightened vigilance

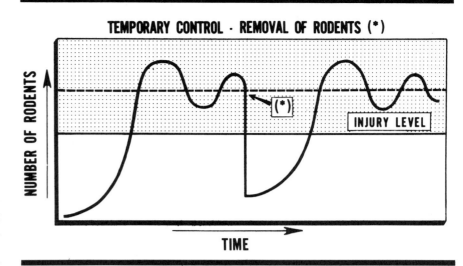

TEMPORARY CONTROL - REMOVAL OF RODENTS (*)

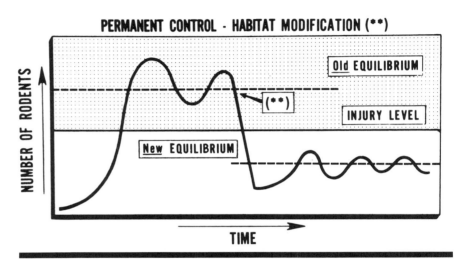

PERMANENT CONTROL - HABITAT MODIFICATION (**)

(*151*). Quantities matter. Numbers matter. Duration of time matters. Our criticism is not so much directed at technology itself as it is at the reliability of the people who manufacture, install, and operate the technology in the environment. Considering the large numbers of human, pet, and domestic animal exposures to rodenticides that occur annually in the USA and elsewhere (*40, 135, 197, 200, 269, 295*), we question the level of responsibility exercised by many who handle these products. Extrapolation from the latest published Poison Control Center data suggests that there were an estimated 17,000 + accidental (not intentional) human exposures to rodenticides in 1984 in the USA (*197*).

The numbers of cases per year of human poisonings due to all kinds of pesticides (not just rodenticides) have been estimated at 750,000 (with 1.9% mortality) (*303*) and 834,000 (with 0.36% mortality) (*194*). Levine (*194*) considers unintentional pesticide poisoning to be a major global public health problem.

Figure 18.40
Temporary control of rodent population due to removal (traps, poisons) interventions alone; asterisk indicates when removal occurred.

Figure 18.41
Permanent control of rodent population due to habitat manipulation (reduction/elimination of food, water, and harborage resources); double asterisk indicates when habitat was modified (often accompanied by removal interventions).
SOURCE: Adapted from Frantz, S. C., and J. P. Comings. 1976. Evaluation of urban rodent infestations—an approach in Nepal. In *Proceedings of the Seventh Vertebrate Pest Conference* (University of California, Davis), C. C. Siebe, ed., pp. 279–290.

Excessive pesticide use may also enhance resistance development—an ever-increasing problem (90). Rodent control failures in several European countries are already being attributed to resistance to the newer, second-generation, anticoagulant rodenticides (brodifacoum, bromadiolone, and difenacoum) (59). Similar resistance in rats has also been observed in some cities of the USA and the Middle East (A. D. Ashton and R. M. Poche, personal communications).

Scientific rigor is needed in environmental decisionmaking, of which pest management is a part. Every intervention carries some sort of risk, but by accurately monitoring the outcome of our interventions we can identify and reduce risk, and calculate to some degree the costs and benefits of the various interventions. "Control" is not synonymous with "killing." Approaches to pest management must be holistic, a combining of educational, legal, biological, physical, mechanical, and chemical interventions into a coordinated strategy to produce a sustained reduction in the problem. The Food Safety and Inspection Service, U.S. Department of Agriculture (USDA), "emphasizes proper construction, maintenance, exclusion, and sanitation as the first line of defense. Poison bait is one of the control measures permitted in the second line of defense" (M. A. Patch, Food Safety and Inspection Service, personal communication). According to Shenker (263), "In any food premises it is far more important to prevent entry of pests rather than wait until they have entered, and possibly established themselves, before taking action."

Changes must occur in the pest control industry if it is to acquire and maintain a professional status in the client community, including food companies. The industry has underpromoted its technical skills and must now "grow up fast" (59) and "be prepared to adapt" (149). There are unrestricted opportunities for those pest control firms that properly train their staff in IPM and succeed in getting their professional message across to their clients. And whether due to client demands, government restrictions, insurance premium increases, or whatever, it is evident that changes are occurring in favor of IPM. "The days of reliance on pesticides to replace sanitation are gone" (178). Cornwell (59) stated that annual fumigation of mills in the United Kingdom is being increasingly replaced by regular inspections and localized pesticide treatments. He also reported that at least one multinational pest control firm has adopted

a pest management strategy that emphasizes IPM. Thompson (287) encourages all food producers to consider IPM and to use chemical interventions "only when they are truly needed." A similar theme has been emphasized regarding pest control in health care facilities (57). And although IPM programs may be more costly than the previous chemical, quick-fix answers, Cornwell (59) states, "Our clients believe it is money well spent."

Human Behavioral Change by Educational Intervention

Organizational Factors

Brown states that "Since rodent problems are in reality 'people problems,' a well planned and executed education programme may well be the single most important factor in rodent control" (37). Major obstacles to the adoption of the IPM approach are that an examination of all circumstances that give rise to the pest problem in the first place is required, and then those basic changes must be made to the environment and in human behaviors that will result in a permanent reduction in pest habitat and food sources (222). Simply put, an ecologically literate perspective is often wanting.

It is apparent that many decisionmaking positions in the field of pest management (including private firms, in-house programs, government agencies, and other institutions) are held by individuals who have had little or no training to prepare them to handle a system that must respond in a comprehensive fashion to rodents, people, physical structures, climate, and so on. Conservatism of pest management personnel or clients may also be a problem. Individuals may resist modifying well-rehearsed routines or be suspicious that acknowledging the ineffectiveness or dangers of previous practices implies a lack of intellectual skills, power, or credibility. This attitude tends to discourage innovative approaches, particularly those that cannot be "cookbooked" and require ongoing decisionmaking.

The choice of interventions most appropriate and their subsequent efficacy requires not only a knowledge of rodent biology and behavior and of the physical environment involved, but also a knowledge of problems within the client community (e.g., food-processing plant, rural residence, urban neighborhood) where IPM is to be practiced. The human aspects of the subject sometimes assume such importance that progress seems more likely to be made by an expert in human affairs than one in rodent control (284). An elderly woman living alone in Florida was reported to have had more than a hundred rats in her home; apparently loneliness was a major factor (Figures 18.20 and 18.24). The health department had no jurisdiction unless a public health nuisance could be proved.

IPM interventions become challenging, whether the target area is a single premises or a whole city block, because the different types of actions needed are often delivered by different agencies or departments. For example, in a large office building, measures relating to structural modification might be a landlord's responsibility and would be conducted by the building superintendent's staff. These individuals would often not be the same as those conducting mechanical and chemical treatments which would be more the role of a pest control technician. Typically, none of these groups would be responsible for the day-to-day janitorial services or the handling of food and food wastes in the various offices, factors often within the tenants' jurisdiction.

Seemingly unrelated events concerned with nonrodent problems, performed by various groups unrelated to pest management, may have a direct bearing upon the successful management of pest problems at a particular time. The staff of a production kitchen may prop open an exterior door or window to increase ventilation without considering rodents or other pests. The plumber or electrician who makes a hole in a concrete wall to install a conduit would not usually be expected to consider insect or rodent access via the same hole (Figure 18.42). City planners and contractors might not be aware that major construction (e.g., building demolition, subway construction) often displaces rat populations that then show up in normally rat-free neighborhoods.

Years ago, the widescale introduction of Algerian ivy (Hedera canariensis) as an ornamental groundcover in California did not come with a warning regarding the affinity of roof rats for this plant (100). Ashton (17) noted that a roadway bump near a grain elevator resulted in small amounts of spillage from each passing grain truck. Over time, the accumulated grain enabled a significant rat population to develop; an informed and observant road repair crew might have prevented the infestation altogether.

IPM, in emphasizing systems design rather than knee-jerk reactions, aims to stabilize pest control systems and improve the predictability of control (88). Clearly, each component of an IPM program makes sense only as it fits in and works in concert with the other appropriate components, especially the feedback value of monitoring and evaluation efforts. Good communications must be established throughout a program's geographic boundaries, with a clear concept of lines of responsibility (35, 48, 79, 84, 93, 188, 193, 210, 236, 268, 311).

The advantages of in-house pest management services include improved monitoring and control over all aspects of the program, making it possible to respond rapidly to problems as they develop (9). Someone must have the authority and leverage to coordinate the various activities that occur within a program's boundaries (often including peripheral areas) into an integrated effort. This includes all interdisciplinary activities in the public and private sectors that impact on the pest population. As a corollary (especially in large communitywide programs), this coordinator must delegate adequate decision-making responsibilities to field staff. If all programmatic decisions must be made back at the main office, it will be impossible for the field staff to meet the operational needs of specific problems on a day-to-day basis.

Figure 18.42
Holes around conduits and ducts provide access routes through indoor wall for rodents and other pests (Baghdad, Iraq).

People

Populations of commensal rodents in and about human structures ultimately depend on the human occupants (and on people using adjacent premises) for their food and shelter. It is only when these individuals comprehend their role (customs, habits) in perpetuating the problem and cooperate in maintaining reduced life supports for rodents that any IPM program can have a chance for success.

The purpose of public information and training programs is to bring a sufficient amount of information to affected citizens so that they can and will intercept and control pests in their own environments (*52*). Educational interventions are designed to raise people's awareness, change attitudes, increase knowledge, and in some cases provide skills, all with the goal of modifying human behavior. Though various levels of involvement are typically necessary in pest management programs (e.g., residents, waste collectors, health code inspectors, planning boards), clients need to know that they can indeed disrupt the life cycles of rodents.

People in the infested environment are the first line of defense, a principle that holds true in the community and in food-industry plants, warehouses, and agricultural or other settings (*28, 39, 52, 97, 181, 217, 224, 237, 283*). In some cases, people are aware of causative conditions fostering rodents as well as preventive measures, but they need an organizational structure to facilitate a systematic control effort (*216*).

In developing countries, it is particularly important to extend appropriate technologies and organizational structures to women because they are often responsible for storage of food items. Not only is it socially desirable to encourage a civic sense of responsibility among urban people to keep their streets clean, but active community participation, with a major voluntary component (Figure 18.43), would greatly reduce overall costs (*55, 268*).

Figure 18.43
Volunteers clean up a vacant lot in an effort to improve community sanitation and reduce rodent problems (Troy, NY).

Media

The media needed to reach the client community in order to gain its trust and cooperation vary considerably with the information systems available and with the temporal needs (as defined by IPM monitoring and evaluation activities) of an ongoing program. As a rule, there is no single "correct" medium; combinations of media can reach a broader audience (*42, 140, 283*). In some cases, as in developing countries and in low-income communities, the selection of media may be limited, but a media "mix" is usually still possible.

Regarding the human personality as an "integrated information system," Fuglesang (*140*) classified the media according to the kinds of sensory impressions in which the information is coded:

(a) printed materials (e.g., charts, posters, pamphlets, newspapers) stimulate only vision;

(b) audio materials (e.g., radio programs, taped messages) stimulate only hearing;

(c) audiovisual materials (e.g., slidesync, film, video, puppet shows) stimulate both vision and hearing; and

(d) personal demonstration (e.g., role-playing, structured experience, apprenticeship) stimulate the full range of senses, including vision, hearing, smell, taste, touch, and even perception.

Although this classification system has limitations, it does illustrate a major point: the superiority of *face-to-face communication*.

Mass media (essentially electronic forms of communication) are often emphasized because of their conspicuous, highly quantitative effect, but they are primarily one-way channels. Information exchange via radio-television talk shows can be particularly worthwhile when audience participation can be arranged. Mass media advertisement "spots" are useful to inform audiences that specific materials are being sent to them by mail, and to direct audiences to other materials and resource people. Thus, mass media are most influential when used as part of more comprehensive communication campaigns coordinated with local media.

Local media development allows production of materials that are closely aligned with client-community concerns. Materials can be relevant to specific issues for audiences in particular cultural-geographic units. For example, locally produced booklets addressed the need for community organization to combat sanitation and rodent problems in Trinidad and Tobago (*8*) and New York (*128*); a travelling theatrical troupe

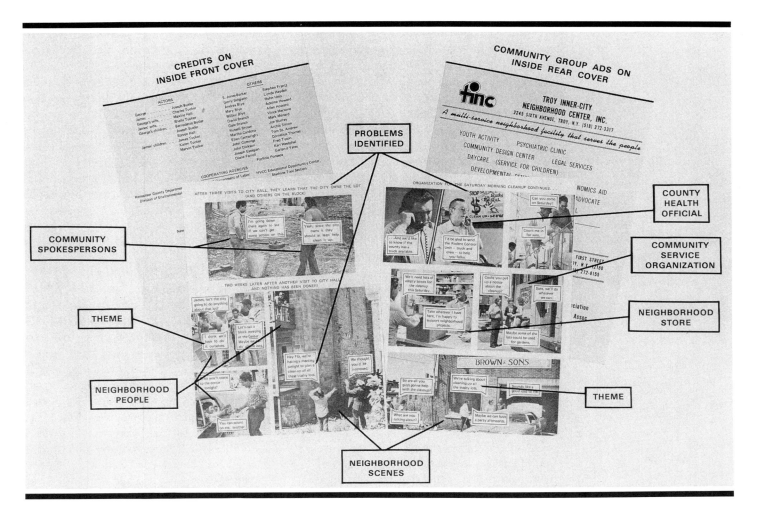

Figure 18.44
Key features of a photonovel educational tool produced by client-community/professional partnership (*128*); the product and the development process are both important (*42, 55, 248*).
SOURCE: Graphics by F. Dillon and G. McNally.

was similarly employed in Baltimore (*150*). Such materials and presentations themselves are often effective in stimulating interest and discussion at meetings, demonstrations, site visits, and so on—the media mix generates further client involvement. In fact, the method of media production is a key factor for engendering sustained educational success (*53, 224*).

Participatory Process

Local media development, in particular, lends itself to the participatory process in which a climate for mutual exchange of ideas is created between the clients (e.g., food-plant employees) and the professional intervenors (e.g., pest management staff). Mutually acceptable strategies can be developed together, with the client and professionals as partners (*183, 237*). Collaborative processes for learning and for creating learning materials strengthen a client's confidence and recognize a client's competence (*248*).

Development of photoliterature (e.g., photonovels, photoposters, etc.) with full participation of clients in all phases of production and with effective end results have been detailed elsewhere (*42, 54, 55*; Figure 18.44). The methods are relatively simple and inexpensive, stimulate collaboration between clients and professionals, and contribute to problem solving. Though largely untapped, local participatory-videography—e.g., community cable TV (*154, 155*) and even music videos (*13*)—have potential where such technology is available.

The value of the participatory process cannot be overemphasized. In a study of 36 rural development projects in Africa and Latin America for the Agency for International Development (*85*), it was concluded that the key determinants in the success of these projects were small-farmer involvement in decisionmaking and resource commitment to complement the work of outside "experts." Undoubtedly, it is important to set objectives in small but meaningful steps that are obtainable and thereby self-rewarding. The means by which objectives are achieved must be linked to the internal communication systems of the community, factory, warehouse, or other client population.

An ambitious, comprehensive outreach program/package, Community Intervention for Insect and Rodent Control (CIIRC), was recently designed to provide information and guidelines for effective, affordable vector control through community intervention (52). Sponsored by the World Health Organization and the United Nations Environment Programme, this innovative project is targeted for developing countries. However, the product's media-mix/component-materials approach is highly adaptable to end users (from high-level government officials to field staff and community residents) of diverse ethnic and geographic backgrounds and to virtually any level of development and information technology. In essence, the CIIRC package has something for everyone in an attractive, functional format; it can also provide a useful supplement for IPM programs in developed countries.

The various studies cited here emphasize that all communication ultimately ends up on the micro-level, in face-to-face, face-to-poster, face-to-loudspeaker, or face-to-screen situations. The person in the affected environment is the critical unit of behavioral change, the ultimate target of educational interventions.

Training

An integral part of the educational component of IPM is to review the methodologies of all who take part in delivering pest management services, including trainers, outreach workers, and technicians. Responsible decisionmaking can only be expected from well-informed individuals. With the goal of protecting the health and welfare of their clients, all staff members must believe in themselves and must be convinced that their knowledge and time are valuable (247). The quest for excellence gives dignity to a person and character to the parent agency or business; it gives satisfaction to clients. Provision of attractive career prospects could enhance staff motivation (268, 283).

Staff members must be trained and prepared to deal directly with their clients regarding multifarious issues [some of which may be of more acute interest than the pests per se (181, 237, 283)] crossing several lines of authority, and provide resource referrals within the community, food-industry plant, or other agency. For example, pest managers would not be maximizing their potential if they entered residential premises, placed rodenticide bait, and exited without instructing the client regarding poison hazards and precautions, the rodent holes in the basement wall that must be sealed, the kitchen wastes that must be kept in a rodentproof container, the leaking faucet that must be repaired,

the broken manhole cover that needs to be replaced, the dog food that must not be continuously exposed, and so on.

Just as clients must understand their role in preventing pest problems, the pest management staff must understand client needs and sensitivities and be willing and able to shape the program to meet these needs with creative thinking. Overall, broad-based client (constituency) support is a major key to program success.

Training and educational resource materials for rodent pest management are available from numerous organizations, some of which are listed in Appendix 18.2.

Legal/Regulatory Intervention

Legal interventions are concerned with the development, promotion, and enforcement of adequate policies, codes, and ordinances relative to managing rodent infestations and causative factors. In some ways, regulation can be considered an extension of education. That is, regulatory agencies generally encourage compliance with the law, warn offending parties of lack of compliance, and evoke legal tools as a last resort. One rat control program in southern California uses a "Premises Correction Notice" rather than a "Notice of Violation" in order to have an approach more in the spirit of cooperation between government and the public (311). Ideally, inspectors should identify "problems" before they develop and thereby prevent the need for extensive corrective measures.

Novel legal interventions reported elsewhere include a New York City law that was enacted to permit tenants to withhold rent payments where "rent-impairing" code violations, such as rat infestations, have existed for 6 months or more (4). In Belair, Maryland, a revised county ordinance would hold property owners responsible for keeping their dwellings, outbuildings, and land free of any food and harborage materials that would attract rats and insects. In Albany, New York, many rats were reported living in and about an occupied house that contained a combination of decaying animal flesh, animal feces, and garbage, much of it fermenting in several inches of water on the basement floor. The tenants were ultimately evicted and the house was demolished.

Regulations

Regulatory measures involving rodents are broadly directed at handling (methods and equipment) and storage of goods (including food), structural integrity and rodentproofing, solid-waste management, and sewerage systems. In food-industry and port facilities, such regulations are especially critical to protection of the public because food materials and products are rapidly and widely distributed, placing a large number of people at risk (*120, 192, 271*).

Compliance is encouraged through systematic, comprehensive inspections that are most effective if they are unannounced and conducted at irregular intervals and hours so that realistic appraisals of actual operating conditions are obtained. When, in community rodent management programs, indoor inspections require prior notification, the opportunity to observe unaltered, everyday conditions may be lost. However, a sharp eye often can detect that changes have just been made (e.g., a circular mark on the floor where a dog's dish had been sitting). Regardless of the client (corporate or individual), efficient legal and administrative procedures are needed to expedite the equitable handling of noncompliance and appeal cases, especially where the public's health is at risk.

An estimated 75% of all legal actions initiated by FDA against food products are based on adulteration by rodents. That is, the food contained filthy, putrid, or decomposed substances or the food had been prepared, packed, or held under insanitary conditions whereby it *may have been* contaminated by filth and rendered injurious to health. It was estimated that 90% of all food prosecution cases are based on rodent filth (*160*).

Voluntary and Involuntary Compliance

To ensure compliance, food companies should set their own standards well within the limits set by the regulatory agencies and establish some system to ensure implementation (*26, 118, 166, 281, 301*). Although many processing plants do have in-plant quality control procedures, others continue to rely on government inspectors to perform quality assurance functions that should be the responsibility of plant management (*142*). The following case history involving a bakery illustrates extreme conditions involving rodents (*160*).

In February, 1968, bakery conditions were found unsatisfactory at the first FDA inspection. A second inspection in January, 1973, revealed a highly active house mouse infestation and rodent harborages in addition to structural defects, roaches and peeling paint.

Cookies were baked and placed in trays to cool, sometimes overnight, before being decoratively iced. Mice entered the trays and deposited fecal droppings and undoubtedly urine.

The firm indiscriminately iced clean cookies as well as those with mouse droppings. Iced cookies were then left on the racks to harden during which time they were again exposed to rodents.

The firm was persuaded to voluntarily cease operations and to recall its product, after which the FDA filed an injunction through the U.S. Attorney. Two months later the injunction was lifted after the bakery had removed all food articles and was no longer engaged in illegal acts. However, clean-up and repair operations required the company to be out of production for eight months. More than two tons of recalled products were destroyed by incineration. Prosecution (another enforcement tool) was not made and the inspection reports of Nov. 1973 and Jan. 1975 showed the bakery to be "in compliance." The bakery was sold in 1976.

Regulatory Agency Problems

The above case also illustrates an important regulatory shortcoming, the long delay between initial and follow-up inspections, with concomitant exposure of many more people than necessary to rodent-contaminated food. Although high costs contribute to this shortcoming, inspections by regulatory agencies must be thorough and must occur at "bio-logical" intervals if the public's health is to be maintained. Mandatory in-plant quality control programs and civil penalties have been proposed as steps to reduce costs and curb food industry laxity (*142*). FDA laboratories must also be timely in identifying violative product samples if regulatory actions are to prevent such commodities from reaching the market (*143*).

An important point overall is that the unpleasant consequences of legal interventions are largely avoidable through proper elimination of all food, water, and harborage resources that might be used by rodents. Various local, state, national, and international guidelines help homeowners, shopkeepers, factory complexes, and countries establish and monitor standards relating to rodent infestations, e.g., *Vector Control in International Health* (*271*) and *Voluntary Industry Sanitation Guidelines for Food Distribution Centers and Warehouses* (*58*). Details regarding food-industry regulations and standards are provided elsewhere in this bulletin.

Additional regulatory shortcomings include inadequate pesticide testing and registration policies (*171, 190, 205, 229*), inadequate enforcement regarding pesticide misuse (*6, 88, 303*), and instances of corrupt practices at all levels (*231*). Size of the regulatory agency may be an important factor regarding the monitoring of corruption. Hardin (151) notes that, other things being equal, large agencies are less efficient than small; where self-interest urges individuals to evade responsibility, such acts are more feasible when the monitor is more distant. The effect of these inefficiencies and corrupt practices is to impose an external cost on the public without its consent. The real cost is charged largely to low-income residents who must live with structural defects (e.g., broken windows, crumbling ceilings, walls, and floors), neglected plumbing (e.g., unflushable toilets, stopped-up sinks, water taps that are continuously on), inadequate and inoperative electrical outlets and fixtures, and infestations of rodents and other opportunistic vermin. Improved training, education, and involvement of regulatory personnel in this very important work may be a way to help resolve this problem.

Physical Interventions

Physical interventions focus on alternative practices and human behavioral changes associated with landscape design, housekeeping, storage practices, environmental sanitation, and exclusion. The end result of physical measures is that rodents are denied access to harborage and food resources or the resources per se are eliminated.

Landscape Design

The objectives of landscaping are to create a sense of color, texture, and proper form or function with plantings that are least likely to contribute to rodent problems. Plants, shrubbery, and trees that are known to attract rodents should be minimized or avoided where possible; a professional horticulturalist would be helpful in making the proper selections. For example, expansive areas of ice plant (*Carpobrotus edulis*) are used for ground cover along California highways (Figure 18.29). This practice has been problematic because it provides food, harborage, and concealment for roof rats. Where such ground cover is necessary, breaking up the plantings in an "island" motif, with pebbled rock (<1-cm diameter) between the plantings, may reduce exploitation by rats because they would need to travel over open areas (between islands) without protective cover. Small-diameter pebbles are recommended because they cannot be used by rodents to form burrows (as might be possible with larger rocks). Combining the island plantings with other landscaping to encourage predators (e.g., raptorial birds) would undoubtedly enhance control efforts.

Inspection Strips Grass and shrubbery should be kept neatly trimmed. In fact, immediately adjacent to buildings it is desirable (particularly for the food industry) to replace vegetation with an inspection strip of pebbled rock, <1-cm diameter (166). The strip should be 60–90 cm wide and at least 10–15 cm deep. It should encompass each building on the exterior side of the foundation wall. Placing roofing paper or 3- to 5-mil polythene sheets under the pebbles discourages growth of weeds and grass. Concrete or asphalt inspection strips are suitable (though more costly) alternatives to pebbled rock, but care must be taken to ensure that rodents cannot burrow under the edges. Since rodents prefer to travel next to objects (thigmotactic behavior) such as foundation walls, the strip also provides an unobstructed flat surface for placement of IPM tools (e.g., tracking boards, bait stations, and traps). A similar inspection strip should be considered for any fence lines on a premises.

Low-growing, dense foilage can be pruned back to 15–30 cm above ground level and/or thinned in order to avoid providing desirable nesting spaces for rodents. Whenever possible, undergrowth should be trimmed to a height that will allow easy detection of any new rat holes and runs. To some degree, it may be necessary to modify our aesthetic judgment regarding manicuring of landscapes as well as landscape composition. Algerian ivy (*Hedera canariensis*) can be trimmed to a height of 30 cm without significant loss in aesthetic value (Fig. 18.45).

In other cases, Algerian ivy might be replaced with a variety of English ivy (*Hedera helix*) or with vegetation that bears foilage up off the ground and is somewhat less attractive to rodents. For rodent control purposes, it is easier to achieve aesthetically positive results in pruning English ivy than with Algerian ivy (M. A. Recht, personal communication). A gap of at least one meter should be maintained between buildings and adjacent trees and shrubs, thus preventing a direct travel connection for climbing or jumping rodents.

Rat Guards for Trees Trees such as date palms offer both food and harborage for roof rats; however, rodent utilization can be discouraged or prevented by pruning low-hanging fronds, by thinning, and by attaching a band of sheet metal around each trunk. This kind of rat guard consists of a metal band at least 45 cm wide, attached at least 1.25 m aboveground, and having a very smooth vertical seam. These features prevent rodents from breaching the barrier.

Rat guards around tree trunks must be checked annually and released as necessary to prevent girdling; occasional release also prevents moisture buildup which might lead to other pathological conditions (S. Daar, personal communication). Although laborious, the use of barriers on trees should not be an insurmountable task in most ornamental landscapes. This method has proved to be an efficient and economical way to control rat damage in coconut groves (*159*).

Housekeeping, Sanitation, Storage Practices, and Exclusion

Environmental sanitation, good housekeeping, proper storage practices, and exclusion through building design are discussed elsewhere in this volume and are, therefore, only considered briefly here. According to Penn (*228*), ''The most important factors in control of urban rats are proper storage and frequent collection of garbage and maintenance of dwellings in a rat-proof condition.'' In addition, the Food Protection and Sanitation Committee of the National Pest Control Association (*119*) has stated that ''Good housekeeping and sanitation, inside and out, are the most important factors in the control of pests in a food plant.'' The reduction of food, water, and harborage resources often occurs simultaneously when housekeeping and sanitation standards are adequate. For the food industry, numerous sanitation design standards are set by government agencies (e.g., USDA), trade associations (e.g., Baking Institute Standards Committee), and individual companies (e.g., Packaged Foods Operations Division, General Mills, Inc.).

Waste Disposal Garbage (organic wastes) and refuse (inorganic wastes) must not be allowed to accumulate in elevator and service shafts, subfloor utility trenches, and other undisturbed areas (where rodents are likely to forage or nest) during building construction or thereafter. One of us (SCF) recently discovered roof rat droppings and nesting material in a subfloor plumbing trench on the fifth floor of an unfinished office building in Baghdad; the only food resource conveniently available would have been garbage thrown away by construction workers. Similarly, during the recent refurbishing of the Statue of Liberty, it was necessary for IPM personnel of the National Park Service to direct (i.e., educate) the contractor's work crews not to deposit food wastes in various wall voids of the museum area, an area otherwise without rodent food resources.

Figure 18.45
Ground cover (*Hedera canariensis*) in this yard has been trimmed to reduce its utility to rodents (Ontario, CA); before trimming, sprinkler head (A) was nearly covered by the foliage.

If rubbish (garbage and refuse) cannot be kept out of buildings, all areas with accumulations should be cleaned at least once per week. Even more frequent cleaning is desirable because rodents attracted to such wastes may move from there to other areas throughout the structure via the many interconnecting passageways. Structural voids and other undisturbed areas—where garbage, manufacturing wastes (e.g., within processing machinery), or refuse tend to accumulate—must be eliminated, blocked off, or kept clean. Voids above dropped ceilings must also be inspected regularly and kept in compliance with these guidelines.

Rodent Exclusion Similarly, rodentproofing interventions should be directed at all doors, windows, exhaust vents, air intake units, floor drains, and eaves (*133*). Railroad trackwell doors can be a major problem in that rodents can enter the door sill on either side of the tracks when the door is closed. In trackwell construction, proofing can be accomplished by providing gaps in the track into which the specially notched door fits tightly (*166*; T. J. Imholte, personal communication). Special rubber pads and filler strip systems have also been designed to retrofit (seal) trackwell door sills (*166*); commercial products are also available, e.g., ''Pest-A-Rester'' (*182*) and ''VotRatGard'' (*298*).

It may be necessary to ratproof toilet bowl drains where rats are known to gain entry to buildings in this way (a commercial device is available for this purpose, the Ratgard Rodent

Barrier, Levenson's Inc., Omaha, NB). This can also be accomplished by feeding the pipe from the toilet bowl into a pipe section of larger diameter. For the food industry, sanitary design of processing, handling, and storage equipment is a most important consideration in preventing rodent contamination of food items in and on equipment (166, 263, 281). Designs that prevent food waste accumulation within equipment, enhance cleaning operations, and exclude rodents altogether are primary objectives.

Notification of Defects   The pest control industry has not only a responsibility to eliminate pests, but an obligation to cite identified sanitation and building deficiencies (31). In some pest management contracts, provision is made for minor holes and cracks to be repaired by pest managers; repairs of larger openings are subcontracted or are the client's responsibility. Perhaps many sanitation and pestproofing problems would be avoided if professional IPM consultants had the opportunity to collaborate more frequently with urban planners, interior designers, architects, and engineers in making land-use decisions and in the design (or rehabilitation) of structures as well as food-industry equipment. Snetsinger (272) reminds us that "it is up to the biologist to . . . fight for biologically sound planning." All too frequently, our unwillingness to face the complexity of really adequate control rules out all but temporary and costly success (39).

Tidy housekeeping practices should be encouraged indoors and outdoors. Kitchens, bathrooms, basements, and garages are of greatest significance in most residential sites; cafeterias, food preparation and processing areas, ingredient warehouses, product storage areas, loading docks, maintenance and services areas, locker rooms, and restrooms are of prime importance in commercial establishments. Where possible, stored materials should be kept no less than 46 cm off the floor (Figure 18.46) or ground to reduce the risk of such areas becoming rodent harborage. Although rodents can climb many shelving units, the added height may discourage them. When purchasing new shelving, one should consider designs that offer little harborage (e.g., open-wire grid vs solid shelves) and may be more difficult to climb (e.g., uprights of smooth metal vs rough wood).

Pallet Problems   Traditional wooden pallets hold goods about 15 cm off the floor and provide excellent rat harborage between the horizontal surfaces, especially during long-term storage (Figure 18.46). However, keeping pallets of

goods in neat rows with access aisles in between for *frequent* inspections is acceptable (119) and such goods should be rotated regularly (old goods out first). Alternatives to traditional wooden pallets are available (e.g., steel stackable pallets available from Stack Pallet Co., Inc., Gardena, CA, and custom-designed plastic pallets available from Thermodynamics Corp., Broken Arrow, OK); such designs may be less desirable for rodent harborage and should be considered

if rodent/pallet problems are significant. Racks for pallets, some of which are cantilevered, are available (e.g., from Frazier Industrial Co., Long Valley, NJ) and make it possible to keep pallets (traditional or other) above the floor and out of reach of rodents. Pallets must also be kept clean and in good repair.

As a further protective measure, all incoming materials should be carefully inspected for rodent infestation *before* being placed into storage; systematically blacklighting a fixed percentage of goods may be useful. An inspection strip at least 45 cm wide should be established along all perimeter walls and kept free of any shelving, goods, or equipment. In food-handling areas, a white band at least 15 cm wide painted along the floor next to the wall will further enhance inspections for rodent signs (234). Although this

Figure 18.46
In a study done at Rodent Control Evaluation Laboratory, New York State Health Department, Norway rats repeatedly attacked garbage in a plastic bag on a standard pallet but ignored comparable garbage offered simultaneously on an elevated (45-cm) platform (Troy, NY).

288   Stephen C. Frantz and David E. Davis

utilization of space for inspection strips may seem wasteful, it should more than pay for itself in the prevention and easier management of infestations (*299*).

With all goods off the floor and away from the walls, not only are inspections efficient, but the effectiveness of traps and baits is improved because rodents are free to run along the walls to encounter control materials placed there (*138*). In fact, bait boxes may be sought out as harborage when other harborage in the room is limited or out of reach.

Office workers keeping food materials in or on their desks should use rodentproof containers such as jars, cans, and tins with tight-fitting lids. Food wastes should be deposited only in rodentproof receptacles. All waste receptacles should be emptied at the end of each workday and washed regularly to ensure that no food scraps remain behind for rodents (or cockroaches) which become active after the people have gone. A thorough daily sweeping, vacuuming, and/or mopping will further reduce available food resources in offices, lunchrooms, and cafeterias. Similar guidelines are important in private residences and in animal quarters.

Pets and Rodents Where possible, animal food should not be exposed all day and night, as often happens when pets are kept outdoors. Rats often live near or under doghouses and feed on the dog's food at night when the dog is inactive. One feeding a day is adequate for most dogs except for the deep-chested breeds which need two feedings per day (R. F. Gruhn, personal communication). Rodents also feed on undigested particles in dog feces; hence, dog feces should be collected and disposed of daily. Bird feeders must also be designed or maintained in a way that prevents food, especially spillage, from being available to rodents.

Recently, one of us (SCF) found an unusual rodent food resource at a resort hotel in upstate New York. The hotel's dimly lit lounge/veranda was glass-enclosed and bordered by columns resting on an exterior ledge (just outside of the glass) of the veranda. At night, the columns were exteriorly illuminated from the top with white fluorescent lights. Insects attracted to the ultraviolet components of the fluorescent light flew into the glass window, were stunned on impact, and fell to the exterior ledge of the veranda. Rats living in the hollow bases of the columns (to which they had gained access through holes intended for electrical fixtures) came out at night to feed on the insects. Hotel guests who witnessed this scenario were negatively im-

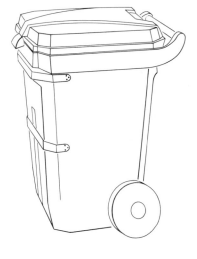

pressed, to say the least, and rapidly departed from the lounge. The hotel suffered loss of revenue as well as reputation until relatively simple corrective measures were taken.

Suggested IPM interventions included (a) replacing the fluorescent lamps with high-pressure sodium lamps (that have virtually no ultraviolet emission and are far less attractive to insects) in order to curb the rats' available food (note that a change from fluorescent to sodium lights would also require some aesthetic considerations because sodium lights would cast a yellow haze on the veranda columns); and (b) excluding rats from the columns (by repairing holes or installing electrical fixtures) after removing rats through trapping and/or poisoning, as determined by safety and aesthetic considerations. Failure to remove the rodents prior to exclusion would have resulted in rats being trapped inside the columns where they would have died and decomposed, resulting in an unacceptable odor.

Solid Waste Management

A final note concerns solid waste management inside and outside of buildings. To improve storage, collection, and disposal practices, adequate-size containers must be provided that also function to exclude rodents and other pests (Figure 18.47). Regardless of size, such containers should be thoroughly cleaned at least once per week.

Waste Collection Systems The use of fully and semi-automatic, containerized storage-disposal systems (Figure 18.48) is reported to have helped to reduce rodent problems in numerous

Figure 18.47
Garbage in plastic bags and overflowing garbage cans were easy targets for roaming dogs, resulting in spillage and increased availability to rodents (Troy, NY).

Figure 18.48
Cart for semiautomated, containerized waste-removal systems improves overall solid-waste management and reduces pest problems.

cities in the USA such as Atlanta, Georgia (F. T. Artis, personal communication), Chicago, Illinois (10), Columbus, Ohio, and Dearborn, Michigan (R. G. Benjamin, personal communication). Major manufacturers of cart systems are listed in Appendix 18.1.

Many municipalities and private haulers in the USA are adopting systems that can be designed to meet the capacity needs of clients, single-family through commercial-sized facilities (12, 14). Ease of operation and safety encourage cooperation by clients as well as waste handlers; also, the plastic trash bag as the sole container is eliminated and refuse spillage and residuals are reduced. In addition, simple improvements in existing dumpster (bulk container) design, such as a lightweight polyethlene or fiberglass lid that can be lifted easily by children and the elderly (Figure 18.49), or a self-

Figure 18.49
Providing a lightweight, easy-to-lift lid on bulk refuse container (for multifamily housing) increases the probability that wastes will be placed inside and the lid will be kept closed (San Jose, CA).

closing lid (P. Kopp, personal communication), may greatly reduce waste spillage at such receptacles. Trash compactors, especially of the large industrial-commercial variety, may be helpful for reducing the volume of paper wastes such as cardboard boxes. However, when food materials such as produce are thrown in them, the result is an unsanitary situation that is attractive to rodents and other pests (211). Careful monitoring of such compaction devices is essential to detect infestations in their early stages.

To prevent overflow, regular collection of

wastes must be scheduled to meet the capacity limitations of whatever container design is adopted for a particular client community. Although commonly overlooked, collection vehicles should be equipped with brooms and shovels (or a bulk vacuum system) to enable collectors to pick up spillage that would otherwise remain behind as a rodent food or harborage resource. Additionally, many models of vacuum sweeper-scrubbers are available to meet community and industrial needs indoors and outdoors (15). Portable, high-vacuum systems (wet/dry) can efficiently clean wall ledges of milling facilities and boot pits of conveyer systems; they can also pick up grain spills and other heavy or wet food materials or debris, e.g., from elevator pits (A. W. Holte, personal communication).

In loading or processing areas with railroad tracks, it is important to keep the trackbed free of caked grain or other food debris attractive to rodents. Legal action was taken against at least one grain-processing facility when the "resident" rodent population allegedly expanded its range from the railroad yard to include nearby residential neighborhoods. Mobile track cleaners, which scarify the trackbed and vacuum up debris for at least 1 m beyond the end of the ties, largely eliminate any rodent food resources and enhance inspection or monitoring work (11; P. Wilkerson, personal communication).

These details of exclusion and sanitation, covered elsewhere in this bulletin, combined with this general overview of rodent biology and behavioral ecology, should provide useful information for designing habitat modifications to significantly reduce rodent population numbers and depredating effects. Maximum cooperation between IPM practitioners (in-house or contracted) and other personnel (e.g., warehouse staff, kitchen crew, homeowners, government agencies) with regard to proper housekeeping, sanitation, and building techniques should result in pesticides being necessary only in limited areas on limited occasions (59). In addition, Brady (31) stated that "We believe that sanitation and building construction play a vital role in pest prevention. We further believe that it's a 90 percent housekeeping/sanitation problem and a ten percent pest control problem and *not* the other way around." However, *without continued maintenance*, rat populations will return to "environmentally-improved" areas—in 4 years for roof rats in southern California (311). Therefore, physical interventions must be properly maintained and adjusted as determined necessary by ongoing IPM monitoring.

Biological Intervention

Wodzicki (307) defined biological control as part of the natural control of animal populations, the actions of predators, pathogens or parasites on a host or prey population producing a lower general equilibrium position than would prevail in the absence of those agents. Commensal rodents are subject to attack by a wide range of biological agents, most of which, under favorable circumstances, can be significant factors in the natural regulation of rodent numbers (111, 186). The question arises whether biological agents can be effective, that is, whether they can kill (or sterilize) the host (prey) at a rate that exceeds the birth rate.

Predation

Worldwide, commensal rodents are attacked by a wide variety of native predators including Indian false vampire bats (*Megaderma lyra*), house shrews (*Suncus murinus*), various raptors, carnivorous mammals, and reptiles (24, 35, 80, 111, 122, 206, 213). Several predators have been introduced and/or specially conserved to function as biological control agents: ferrets (*Mustela putorius*), monitor lizards (*Varanus indicus*), small Indian mongooses (*Herpestes auropunctatus*), Japanese weasels (*Mustela sibirica itatsi*), barn owls (*Tyto alba*), and domestic cats and dogs (70, 102, 114, 168, 186, 294, 307).

These animals have performed as biological control agents with varying degrees of success. One problem that may occur is that as the rodent numbers diminish under predation pressure, the predator may shift to other more numerous, easier-to-catch prey. The monitor lizard, for example, readily eats crabs and smaller lizards as well as wild and domestic birds and their eggs. Similar problems were found with the mongoose which also serves as a vector of rabies. Additional difficulties may arise if the human population is prejudiced against the introduced biological agent.

Cats and Dogs It appears that cats can maintain the immediate area of farm buildings rodent-free if introduced when the rodent population is low or is reduced by other means, and if their diet is supplemented (70, 102). Jackson (168) reported that urban cats take mainly surplus rats that would have died or emigrated; therefore, cats modified the composition but not the size of the rat population (74). The general availability of urban garbage or other food resources may reduce the necessity for cats and dogs to subsist on rodents (27, 107, 168) whereas in rural areas more rodents may be eaten (64, 122, 168, 206).

Raptors and Weasels Raptors may significantly suppress rodent populations in city parks and other open areas but are probably less effective in highly urbanized areas where the carrying capacity maintains high rodent populations but provides few preferred nesting sites for most raptors.

Japanese weasels apparently kill rats even when the weasels are well fed; they tend to consume mostly murids even when other prey are available. Uchida (294) suggests that these weasels would probably not represent a rabies risk if brought from and introduced to areas that are rabies free.

Problems with Predators According to the principles of population dynamics, native or introduced predators alone could not be expected to provide a sustained rodent population reduction to humanly acceptable tolerance limits. Erlinge (111) outlined the major reasons for this failure:

(a) The reproductive potential of the predators is low compared with that of rodents.
(b) The numerical response of the predators to increasing density of rodents is delayed.
(c) The predator population can be limited by its own territorial behavior at high rodent density.
(d) The functional reactions of predators to high rodent density are undependable (i.e., surplus killing is known only in some species, viz. weasels, and among adult owls and cats providing prey to their young).
(e) The simplification and urbanization of ecosystems by humans favor rodent population stability more than the predators.

However, enough data exist to suggest that predators can play a role in rodent IPM programs by enhancing the effects of other interventions, preventing infestations from becoming well established, preventing population recovery, and preventing reinvasion in areas freed of rodents.

Enhancing Predators To more fully utilize predation in IPM, efforts must be made to increase the diversity and abundance of predators and to increase the vulnerability of rats and mice. Predator diversity can be increased through variable land use and the creation of mosaics of parks (including landscaping on the periphery of food-plant premises that favors appropriate predators). This action will also favor prey diversity, thus providing increased alternative foods for the predators and thereby enhancing predator population stability (*111, 162, 272*). Composition of the predator fauna can be important in ensuring predator pressure on different sex/age/health components of the prey population in various locations during the day and night and over a wide range of movement.

In a grain warehouse area of Calcutta, common pariah kites (*Milvus migrans*), house crows (*Corvus splendens*), and domestic dogs fed mostly on outdoor rats (*122*). Indian barn or screech owls (*Tyto alba*) and domestic cats fed on rats inside the warehouse; cats also fed on rats in the narrow passageways between warehouses. Direct observations revealed that crows appeared to select disabled (and dead) bandicoot rats, but cats (singly) and dogs (singly and cooperatively in pairs) attacked active and apparently healthy rats. The main diurnal predators were crows, with some activity by dogs and cats; kites and owls were crepuscular or nocturnal; dogs and cats were mostly nocturnal.

The number of predators can be enhanced by providing suitable harborage, nesting sites, hunting posts, and so on, as well as supplementary food when the supply of rats and mice is small. Legislation may also be necessary (e.g., the Migratory Bird Treaty Act of 1974) to protect some predators (e.g., barn owls) from people who would injure or kill them because of unwarranted fears and misunderstandings. Introductions of predators can increase predator pressure on rodent populations, but at least four prerequisite questions must be carefully considered:

(a) Does the animal actively prey on rats and mice?
(b) Can the agent adapt to the new environment (enhancements may be necessary as noted above)?
(c) Are there significant risks to nontarget species (e.g., attacks upon useful animals; zoonosis reservoir or vector)?
(d) Will the human population conserve the predator species?

The vulnerability and density of the rats and mice can be altered primarily through physical interventions as described in this chapter and elsewhere in this bulletin.

Trapping and Poisoning   Human predation in the form of trapping and poisoning undoubtedly exerts the greatest overall predator pressure on commensal rodents. These aspects have been generally integrated into various portions of this chapter and will not be detailed here since such information already forms the bulk of rodent control literature, seminars, and workshops. Of interest here are the more unusual programs reported from time to time around the globe in which the public is encouraged to hunt and kill rodents. The effect of such efforts described below is much the same as other animal predation and must be part of an organized, systematic IPM program in order to produce sustained results.

It has been reported that Egyptian authorities have, on occasion, issued to young children sticks with sharp nails in the end for skewering rats; in one village, 3,000 rats were reported killed in a 2-week period.

Hundreds of residents in the suburbs of Beijing joined together to kill rats plaguing hospitals, grain shops, food-processing factories, and poultry farms; nearly 2.3 million rats were reportedly killed in a single month.

In some cases, bounties of various sorts have been offered to encourage people to kill rodents. A woman in the Philippines won a contest by presenting the judges with 9,715 rat tails.

The problem of rats in some Chicago neighborhoods became so severe that a $1 bounty was offered for each rat killed; 550 bounties were paid in the first 2 weeks.

In an area of Indonesia, couples could be married or divorced only after handing over 10 or 23 rats, respectively; in the first 5 days of the campaign, 22,683 rats were brought to the authorities.

Finally, it is known that certain ethnic groups in various areas of the world utilize rats as protein sources, but a Brazilian health official actually made the remarkable suggestion that the "ideal way" to control a burgeoning rat population was for the public to eat rats.

Sustained reductions of commensal rodents can best be achieved through significant reductions of causal conditions (food, water, and harborage). Any other interventions that supplement the effect of the basic interventions without endangering nontarget species or the environment should also be encouraged.

However, some programs, though well intentioned, may backfire. In Bombay, the municipal rat control program included nightly forays by club-wielding workers who were given a quota of rats to kill. On at least one occasion, it was learned that a creative worker was raising rats at home to meet his nightly quota.

In some other situations where rat tails are turned in for bounty, tailless rats have been found running the streets—the local entrepreneurial response was, perhaps next month the government might give a bounty for rat ears or feet. Obviously, the more educational, people-oriented interventions were overlooked or insufficient in these latter examples. Also, "removal only" interventions (including the bounty system) will result in temporary reductions, at best, of commensal rodent populations.

Disease

Regarding the use of pathogenic bacteria to control rats and mice, there is a wealth of literature, mostly from work in Europe and the USSR (*186*). The objective has been to find one or more organisms that would cause disastrous epizootics specifically among commensal rodents. Unfortunately, rodents quickly develop immunity, sustained kills do not occur, and populations quickly recover.

Salmonellas Many bacteria that have been studied or tested for commensal rodent control are serotypes of *Salmonella* that are also dangerous to humans and domestic animals. Strains of *Salmonella* developed in the USSR and France are reported to be nonpathogenic to humans and domestic animals but are mainly for use against field rodents (not commensals) (*77, 186, 212, 296*). Although such strains are considered stable and specific, the possibility of rapid genetic changes, and the consequences of such, must be considered.

In 1967, the Joint FAO/WHO Expert Committee on Zoonoses recommended that "salmonellas should under no circumstances be used as rodenticides" (*5*). Although Barnett and Prakash (*24*) consider the pathogen method a good one, with the problem being the isolation of a suitable disease organism, at present this type of intervention offers little promise for commensal rodent control.

Chemosterilants

This approach, also known as biogenetic control, involves breaking the reproductive chain at any one of several vulnerable links to produce, at best, permanently sterile animals (*35*). Chemosterilants are chemicals that can cause permanent or temporary sterility in either sex or both sexes, or through some other physiological mechanism, reduce the number of offspring or alter the fecundity of the offspring produced (*203*). The basic principle is that sterile rodents in the population exert pressure on the fertile animals to maintain the birth rate. The sterile animals should continue to compete for environmental resources and mates and maintain dominance status within the population, but not contribute to the birth rate.

Theoretically, the birth rate would tend to fall below the death rate and the population would gradually die off. However, as the population declines, it must be assumed that immigrants would eventually colonize vacant territory unless physical interventions (e.g., exclusion) or natural physical barriers prevented such recolonization.

Essential Features Investigations have been conducted with steroidal as well as nonsteroidal compounds that are capable of producing sterility. Among the more important difficulties has been palatability, a problem that is exacerbated if the chemosterilant requires multiple feeding. Ideally, a chemosterilant should be palatable, effective in one feeding, permanent, rapidly degraded after ingestion (to prevent secondary hazards), rapidly biodegradable in the environment (to pose no long-term environmental hazard), have no adverse effect on sexual drive, possess some degree of specificity (to prevent sterility in nontarget species), and affect both sexes or at least the females.

Disadvantages In dense populations of polygamous rodents, even a few fertile males can compete effectively for fertile females against overwhelming numbers of sterile males. In dense rat populations, at least 95% of the males would need to be sterile in order to depress the pregnancy rate; at low population levels in restricted environments (e.g., sewers), the effect would be easier to achieve (*34*). Sterile males do have an impact in that any pseudopregnancies following copulation decrease the incidence of normal pregnancies. However, a given percentage of sterile females will have a far greater potential for reducing births than would sterile males. If both sexes are sterilized, the reduction effect would be compounded.

It must be remembered that a population of sterile rodents will continue to cause aesthetic damage (e.g., noise, odor), economic damage (e.g., eat food materials, gnaw stored goods), and medical damage (e.g., inflict bites, contaminate foodstuffs). Obviously, such treatments are not useful in residences, hospitals, restaurants, and food plants where the presence of rodents cannot be tolerated. However, chemosterilants may be useful where some rodents can be tolerated or where total removal of a population is unattainable or unnecessary (e.g., sewers, waterfront areas, landfills, poultry houses, livestock yards, public parks).

Supplemental Role Chemosterilants might be used on the periphery of a rat-free area (e.g., on land bordering food-plant properties) to create a buffer zone of reduced population size and to prevent reinvasion. Similarly, these compounds may be useful in conjunction with other IPM interventions to help combat population resurgence.

The future of chemosterilants is uncertain. If developed compounds are species-specific and present little or no hazard to nontarget animals, they may be safer to use in some situations than some current toxicants, e.g., where literacy is low, where toxic compounds cannot be effectively managed or regulated, where the human population objects to direct killing of rodents.

Only one chemosterilant, alpha-chlorohydrin, has been registered (restricted use only) by the U.S. EPA for use against Norway rats (*173*). This material, specific to male rodents in its sterility effect, can produce permanent sterility in sexually mature male Norway rats with a single feeding (*34, 110*). At higher doses than necessary for sterilization, alpha-chlorohydrin is lethal to Norway rats of both sexes and all age groups. This compound does not affect male libido. Problems with palatability may be mitigated somewhat by prebaiting, and its specificity enhances safe application. At this writing, alpha-chlorohydrin is temporarily unavailable in the USA, but it is sold elsewhere. Its use will probably be best suited for relatively isolated populations of Norway rats.

Mechanical Intervention

As used here, mechanical interventions are various devices that a rodent walks onto, enters, and/or releases a trigger mechanism that results in the animal being killed or captured alive. The first traps were snares, pitfalls, and deadfalls (*116, 117*). False-bottom drowning traps (large vessel partially filled with water and supplied with rodent access to the rim) were most likely a part of this early arsenal. Even earlier, sticky substances (e.g., tar, pitch)—precursors of today's rodent glue traps—were probably used to ensnare rodent pests. Some of the earliest "mechanical" rat traps were pottery traps (*94, 96*) which either strangled rats with a noose (from Mohenjo-Daro, Pakistan, 3000 B.C.E.) or enclosed rats alive by releasing a door (from Nuzi, Iraq, 2000 B.C.E.).

The invention of the iron spring allowed development of other designs. Traps from the 14th century appear very similar to modern snap traps (*116*). Today, there are myriad trap designs—spring-driven, pneumatic, electrocuting, and glue; some models are more imaginative than practical. Trap designs and their practical applications are discussed in other publications (*35, 116, 138, 234*). Although it is not the purpose of this chapter to review the world's traps, some comments and guidelines are appropriate regarding trapping techniques and professional, commercially available traps (also see Appendix 18.1).

Basic Techniques

Trapping is the preferred IPM intervention in situations where *rapid removal* of rodents is necessary and the use of rodenticides is considered undesirable (e.g., in food plants or in areas accessible by children, pets, or zoo animals), and in situations where *recovery* of the rodents is required for disease or other biological studies. However, for removal purposes, traps are rarely completely effective by themselves but may often be useful in conjunction with physical interventions. As food and harborage resources and runways are eliminated, rodents are forced to move around to find new life supports. During this exploratory process, it becomes easier to trap the animals.

Frishman's (*138*) discussions of several trapworthy circumstances are briefly summarized here.

(a) In food establishments, rodenticides should be placed in bait stations. In order to be killed, the rodent must enter the station and feed. "Once inside, why let it out?" Even with the latest anticoagulants, death for a rat or mouse will usually require several days. During much of this time, rodents will continue to feed, gnaw, and contaminate their surroundings with urine, feces, and hair. [And, occasionally, all or part of a rodent may become the "contaminant" in a box of cereal (Figure 18.2), bottle of beer, and so on.] With traps, a rodent is captured once and is thereby prevented from further contaminating actions.

(b) Rodenticides are toxic substances and can be accidentally spilled, carried away by rodents or, in some cases, blown about in the wind. Therefore, the rodenticide may no longer be where it was intended and could result in contamination of food (or it could become accessible to nontarget species).

(c) Rodent bait formulations exposed for extended lengths of time tend to lose their palatability; thus, the probability of the baits being eaten by rodents diminishes and the chance of baits becoming infested with grain insects increases. The insects may then move on to establish infestations in other foods.

(d) Rodents captured in properly utilized traps do not die in walls or other inaccessible areas, thereby alleviating associated odor and fly problems.

Other advantages of traps are that, properly used, they present little or no hazard to nontarget species; their use does not require special certifications; they are useful for capturing rodents that have become bait-shy from poisoning programs; and, since results are readily observable, they may enhance public relations. What may be considered disadvantages of trapping include the fact that more skill is required than with baiting (as commonly practiced); trapping requires labor-intensive deployment and maintenance; one must dispose of rodent bodies after capture; animals may become trap-shy; and traps are often damaged or stolen when used in publicly accessible areas.

Generally, traps are not recommended for large-scale, extended programs because of the daily manpower requirements. However, labor costs vary considerably and in much of the world trapping can be cost-effective, efficacious, and far less risky to nontarget species than rodenticides which are often difficult to regulate safely. Skillful trapping in restricted environments may provide remarkable results. Frishman (*138*) reported capturing 3,000 mice from a cold-storage facility in less than 30 days, and 120,000 rodents from a public facility in a 9-month period.

One of the first steps in choosing traps and planning a trapping program is to determine what species is to be captured. Obviously, trap size must be considered, but the behavior of the target species is also important. Mice are somewhat easier to capture than rats, but skillful application can produce good results with Norway, bandicoot, and roof rats. With rats, it is often best to expose traps unset (but baited or with bait nearby) for a few days in order to overcome a possible neophobic response. Even with mice, such exposure of traps may enhance results. After the adjustment period, traps should be kept in their same positions for setting. If trapping results decline with mice, one should move the traps and disturb the environment to stimulate the remaining mice to increase their exploratory behavior. Good results can be expected only if traps are placed liberally throughout the infested area and in specific areas likely to be frequented by rodents. Intensely trapping for a few days' duration is generally more effective than distributing traps sparsely over a wide area for weeks at a time.

Traps generally require frequent monitoring, sometimes twice daily. Traps set in the late afternoon are less likely to be tripped or disturbed by people in the area before the rodents are exposed to them. Morning visits may be required by clients who object to viewing struggling, vocalizing rodents or dead rodents in a building; it is not likely that customers in a food market would want to see trapped rodents there. The practice of frequent visits maximizes the number of available traps and allows the removal or repair of malfunctioning traps and the moving of ineffective ones. Basically, if the number of rodents caught per trap per day does not decline, then not enough traps are in use or some part of the intended area lacks traps. By keeping a simple set of daily trapping results, the pest manager can evaluate the trapping program's effectiveness.

Single-Capture/Kill Traps

Another step in choosing traps concerns what is to be the final condition of the trapped rodent—dead or alive. For capturing and killing one rodent at a time, the wood- or metal-based snap trap (break-back trap) is available worldwide. These traps are made in sizes for rats or mice (Figure 18.50).

Snap-trap effectiveness can be increased by expanding the size of the trigger (treadle) and positioning that end against the wall; this enhancement often makes it possible to trap rodents without using bait. Triggers can be expanded (size must be kept within the edges of the trap base) with cardboard, wire mesh, or other material; and they can also be purchased already adapted. If moisture is a problem with wood-based traps, the traps can be made moisture-resistant by dipping them in melted paraffin (*300*). Snap traps are commonly available, small and convenient to carry, and lend themselves to placement in strategic locations on top of the soil, on the floor (Figure 18.51), or in elevated (vertical or horizontal) positions (Figure 18.52). Be cautioned that the rat trap has a particularly strong spring, and accidental release may damage an exposed hand or foot. These traps must be kept out of the reach of children and pets.

Single-Capture/Live Traps

Traps designed to capture *one* rodent at a time and *alive* are best if it is the objective that the animal should survive (e.g., to prevent noxious odors) or if it is needed for scientific purposes. Cage-type live traps may be more effective with

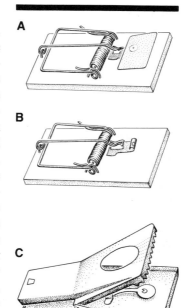
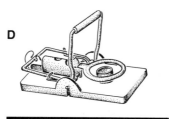

Figure 18.50
Snap trap (breakback trap) designs: (A) traditional design with standard trigger (available in rat and mouse sizes); (B) traditional design with expanded trigger (no bait necesssry); (C) Snap-E trap (available in rat and mouse sizes); (D) clothespin-type Trapper trap (available only in mouse size).

Figure 18.51
Placement of snap traps at floor level: (A) a single trap is always set with the trigger next to the wall (or toward a hole or burrow); (B) an empty box is placed in the corner next to the rodent hole to confine the runway and force rodents to cross the expanded trigger; (C) expanded trigger traps can be nailed vertically to a wall along a runway in order to catch rodents that avoid traps placed on the floor; (D, E) traps set in pairs are more likely to catch rodents than single traps; triggers can be positioned toward the wall (D) or parallel to the wall (E).

Figure 18.52
Placement of snap traps in elevated positions to catch climbing rodents: (A), traps should be placed on rubmarks along support beams. (B, C) Methods for setting traps on overhead pipes—a twisted wire passes through a drilled hole in the trap base near the trigger; a flexible wire from the other end of the trap is attached to the pipe or some other nearby object; when sprung, the trap will bounce off the pipe and hang by the flexible wire, leaving the runway clear for other rodents to encounter other traps; a hose clamp may also be used to fasten the trap to a pipe (short bolt or screw holds the base of the trap against the clamp). (D) Tree ringed with snap traps (note location of expanded trigger over runway).

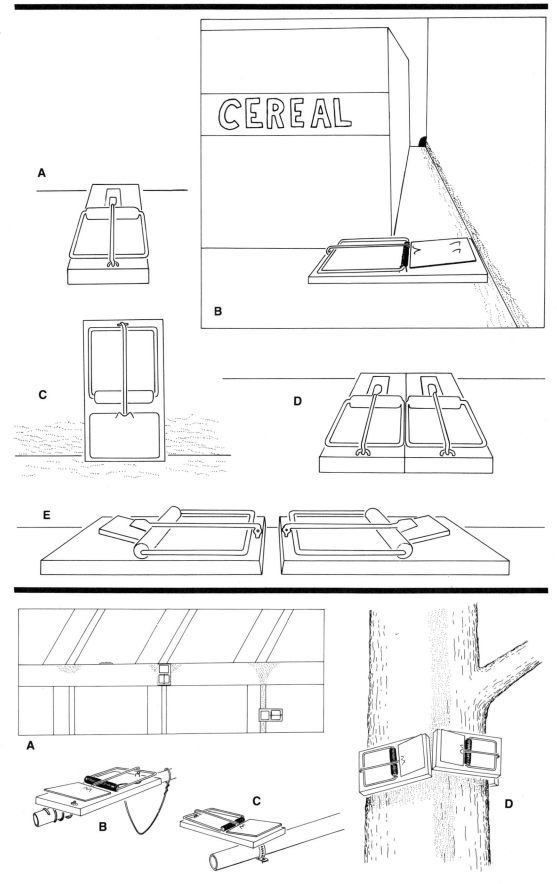

Stephen C. Frantz and David E. Davis

rats than snap traps and have none of the hazards to nontarget species. Designs with internal trigger mechanisms are particularly useful because they can be set in confined spaces (e.g., under the edge of piled construction materials) or be covered with naturally occurring materials without interfering with their operation.

The Rodent Control Evaluation Laboratory, New York State Department of Health, relied almost exclusively on the Tomahawk Nos. 102 and 201 (Figure 18.53) for trapping rats for nationwide anticoagulant rodenticide resistance surveillance (127, 129). These very durable traps have a single, spring-loaded door; rats cannot escape if the trap is upset. Properly handled, the trigger mechanism of these traps rarely needs adjustment; it will tolerate moderate exterior investigation without being set off. It cannot be tripped by wind currents. This particular design is not available for mice, but other wire-mesh, live-trap designs are available (e.g., Figure 18.54A) and may be adequate for particular tasks.

Other types of traps for live capture of rodents are constructed of galvanized steel or aluminum sheet metal; they are fabricated in rat or mouse sizes (Figure 18.54B, C). These traps also have a single, spring-loaded door; rodents cannot escape from upset traps. The internal trigger mechanism is not easily set off by exterior investigation or wind, but it requires more adjustment than the Tomahawk traps. However, these traps are useful where exposure of captured animals to the elements is problematic or where sight of the captive rodent might be of concern to sensitive observers (see Appendix 18.1).

Multiple-Capture Live Traps

Although the traps in the preceding paragraphs are called single-capture live traps, occasionally more than one animal may be captured due to chase/follow behaviors as in sexual packs (122). However, there are also a few traps designed specifically for capturing one or more rodents alive. In practice, when several rodents are captured in such traps, some captives are usually killed as a result of aggressive interactions with the other captives. Virtually all multiple-capture traps are larger and bulkier compared to the single-capture designs, but they may compensate with convenience features—they automatically reset themselves to capture numerous animals and can be monitored less frequently than single-catch models. However, to prevent deaths, foul odors, and concomitant insect infestations, these traps should be checked and

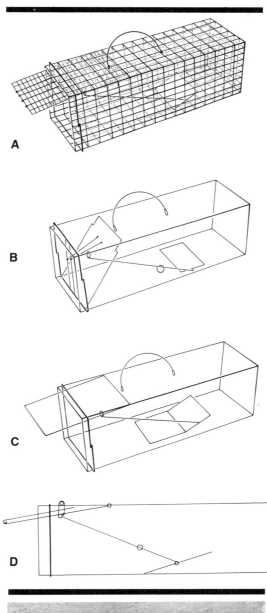

Figure 18.53
Tomahawk single-catch live trap (A) for rats is available in rigid and folding designs (the latter are easier to transport and less expensive to ship). Internal structure is shown for (B) tripped trap, (C) set trap, and (D) side view of set trap.

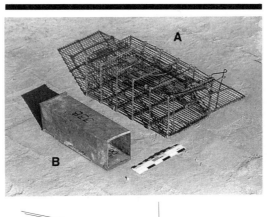

Figure 18.54
Live traps: (A) wire-mesh, single-capture trap used in south Asia to capture rats; (B) locally fabricated metal box trap (copied from the Sherman trap) used to capture house mice and house shrews in Nepal; (C) typical indoor placement of Sherman trap. (Note: Whenever possible, traps should be numbered on more than one surface to facilitate monitoring.)

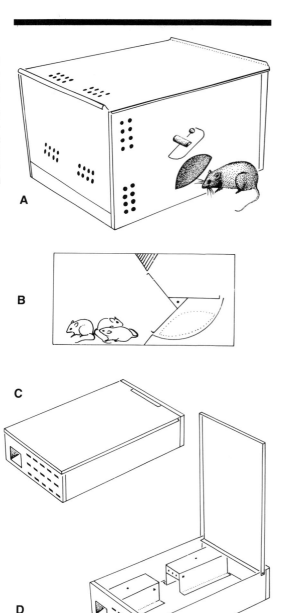

Figure 18.55
Multiple-capture metal mouse traps: (A) the Ketch-All automatic trap (both the top and the wall of holding chamber are easily removed for cleaning, and the holding chamber is also available with a transparent plastic wall to facilitate monitoring); (B) interior detail of Ketch-All; (C, D) Tin Cat trap closed (C) and open (D).

emptied every few days.

At least two galvanized sheet metal designs are available for mice, the Ketch-All and the Tin Cat (Figure 18.55; Appendix 18.1). The Ketch-All utilizes a spring-wound, paddle-wheel device that throws trapped mice into a holding compartment; an attachment is available that allows mice to "escape" into a drowning chamber. The Tin Cat utilizes a series of tunnels and counterweighted treadles that allow mice to enter, but not retreat. Both designs have been updated and improved over the past several years; they are well constructed and durable, though each design meets somewhat different needs rel-

ative to size. Both traps are equally effective under normal use (286); efficacy of both may be improved by placing the traps unset for a few days prior to trapping. An effective wire-mesh, multiple-capture mouse trap has also been reported (255).

For rats, the multiple-capture "Wonder Trap" is made of strong wire mesh in a double, one-way funnel design with a cantilevered door located in the floor of the second funnel (Figure 18.56). In Calcutta, the trap is available in a large size, roughly 45 × 25 cm; up to 30 lesser bandicoot rats have been captured overnight in a single Wonder Trap (122). This trap, developed at the Haffkine Institute (80), is commonly available in India (see Appendix 18.1).

Glueboards

Glueboards (glue traps, sticky boards) are devices consisting of a flat, rigid (or semi-rigid) base (e.g., cardboard, fiberboard, plastic) coated on one side with an extremely adhesive material that catches and holds rodents attempting to cross them, in much the same way as flypaper catches flies (136). For indoor, nondusty areas, glueboards can be particularly effective for capturing mice, but the glueboards must be checked frequently. The passage of time permits urine and feces to accumulate and this enables some mice to escape from the glueboards. Glueboards may also be useful with roof rats and small or young Norway rats, but they cannot be recommended for adult Norway rats, which are stronger and more likely to free themselves. Glueboards can also be applied in somewhat dusty environments if the boards are well covered (e.g., in a cardboard box with holes for rodent access). Similarly, outdoor applications must be protected from the effects of weather.

It is prudent to monitor glueboards at 24-hour intervals since their effectiveness may be reduced by surface contamination (dust, soil, feces, and urine) and temperature extremes. Also, laboratory studies indicate that mice do not readily die within 24 hours of capture on glueboards; hence, servicing should be frequent in order to humanely dispose of captives as well as to ensure that the rodents will not have enough time to work themselves free from the glue (139).

Glueboards should always be secured to the substrate (e.g., with double-stick tape, adhesive, tacks) to prevent captive rodents with free legs from moving the boards to less effective locations, to other areas where pets or children might become involved (36), or to where the public might become concerned (7).

Many companies have marketed glueboards over the past several years, but very little data are available on efficacy, which may vary with factors such as adhesive tackiness, viscosity, odor, and depth (*136*). The assorted designs include both rigid/nonbiodegradable and flexible/biodegradable base materials (e.g., for shaping into tubes to place on overhanging pipes); the adhesive is also available in bulk containers for fabrication of glueboards to suit particular needs (see Appendix 18.1).

It must be emphasized that trapping is generally most useful for supplementary removal of rodents in concert with other longer-term interventions such as landscape design, sanitation, and exclusion. Different types of traps can be used at different stages of an IPM program (e.g., monitoring, removal from a rodentproofed building, reinfestation prevention) and in different locations (e.g., glueboards in food-storage rooms, snap traps in nonfood storage, Ketch-All in a nonpublic corridor). Program stage, location of infestation, and trap design dictate the number of traps used and frequency of servicing them.

Electronic Intervention

The methods discussed in this section are controversial and of questionable efficacy.

Electromagnetic Devices

In an era of increasing environmental concern, people often seek a panacea to solve problems. Instruments known as electromagnetic (EM) pest control devices were one such proposed method for rodent (and insect) pest management (*108*). Manufacturers claimed that rats and mice (and a variety of other rodents and insects) were either killed or prevented from eating, drinking, or mating by the magnetic field emitted by such devices. They further claimed that EM devices would not affect pets and other beneficial species because of physiological differences due to domestication. Manufacturers also claimed that laboratory animals or caged wild animals, because of their dependence on humans for survival, were not expected to be affected by the devices' output. Although such outlandish claims might appear to have been written for people who had only recently learned to walk erect, the successful marketing of EM devices was remarkable. The first device was marketed in 1976 and within a few years there were some 30 manufacturers/distributors in the USA, with annual sales of several million dollars.

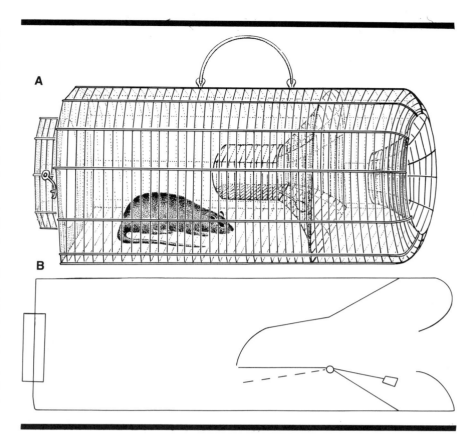

In 1977, the U.S. EPA (*108*) initiated an extensive, cooperative scientific investigation to determine if low-level EM emissions could be utilized in pest control. Results of these studies indicated clearly that there were *no* biological effects on rats, mice, and other species tested; in many cases, the devices emitted little or no electromagnetic radiation. The EM field strength at 3 meters of many devices was shown to be less than the earth's magnetic field. Several household appliances were found to generate EM fields similar to the devices. Through a variety of enforcement actions, EPA was successful by the end of 1979 in stopping the sales of and in removing from the market most EM devices. Nevertheless, court cases did carry over into the early 1980s, and EM devices continue to appear sporadically (W. W. Jacobs, personal communication).

Ultrasonic Devices

Ultrasonic devices generate sound at frequencies greater than 20 kHz. It is claimed that this level of ultrasound irritates or stresses rodents and thereby reduces the attractiveness of an area's food, water, and harborage resources. Such instruments are said to repel rodents through creation of an acoustically uncomfortable environment; in essence, ultrasonic devices are

Figure 18.56
Multiple-capture, wire-mesh Wonder Rat Trap: (A) external appearance; (B) detail of interior.

supposed to modify the habitat without physical intervention per se.

Considerable controversy surrounds the efficacy claims of these devices (161, 214). Many manufacturers/distributors in the USA have ceased operations as a result of government enforcement activities. The Canadian government has developed guidelines to restrict sales of ultrasonic devices unless efficacy claims are substantiated (185). Some problems with much of the research on ultrasonic repellers are

    (a) lack of test environments in which the normal behavior of rodents is adequately considered;

    (b) inadequate evaluation of rodent activity in rooms other than those containing an ultrasound generator;

    (c) confounding of research results by simultaneous use of other rodent control interventions;

    (d) inadequate measurement of the decibel level (ultrasound intensity, sound pressure) at the specific location where rodents would experience it (e.g., for a device mounted 2 m above a feeding area on the floor, decibel level should be measured about 5 cm above the floor, not at the device's transducer);

    (e) improper instrumentation for measuring ultrasound intensity (an industry standard is the B&K Impulse Precision Sound Level Meter Type 2209-1613 equipped with a Model 4136 microphone; see Appendix 18.1); and

    (f) lack of qualified, experienced observers.

Though data are limited, it appears that rats and mice can be moved about or have their activity areas and travel routes altered by inundating such areas with ultrasound of proper frequency and intensity (147, 265). However, ultrasound waves travel in straight paths (as does light); they do not bend around corners, nor do they pass through solid objects. The lee side of such objects, with regard to the sound source, represents a sound shadow in which a rodent will not experience the ultrasound.

From a practical standpoint, an ultrasonic device is quite limited in its effective range because the sound intensity (sound pressure) is rapidly attenuated over distance and does not reach occluded areas unless reflective surfaces are present. At best, many reflective surfaces would absorb part of the sound, and that reflected would be further diminished by total distance of travel to and from reflective surfaces. In the Canadian guidelines, a sound intensity of 90 dB is the lowest limit at which efficacy claims can be supported. In effect, this sound level requires most devices to be within a few meters of the area of expected impact.

Although it might not be impossible for ultrasonic emissions to rid a premises of rodents, it has been argued that ultrasonic devices serve to direct rodent activity to areas of sound shadows where traps or toxicants can be made available. These shadow areas are often locations (along walls, in corners, behind or beneath objects, at entrances and so on) where traps or baits would be placed even if ultrasonic devices were not utilized. Today, manufacturers generally state that ultrasound is most effective when used in conjunction with other interventions. Because these other interventions can substantially reduce or eliminate rodent infestations without "help," the extent to which simultaneous use of ultrasonic devices might add to or "coattail" upon the effects of the other methods is difficult to establish (W. W. Jacobs, personal communication). More solid research data are needed to determine if the cost of ultrasonic devices is balanced by their beneficent effects and whether or not such benefits might not be achieved more economically by means of other long-term interventions.

Chemical Intervention

An agent that incorporates a substance or mixture of substances intended to prevent, destroy, repel, or mitigate a pest is considered to be a chemical pesticide (109). A review of chemical interventions for rodent IPM will not be presented here for several reasons:

    (a) Nonlethal interventions are finding increasing favor, especially for use in situations where rodenticides are restricted because of possible food contamination or hazard to nontarget species (74).

    (b) Rodenticide application within food plants is largely contraindicated; this and other issues regarding the utility of supplementary chemical interventions in IPM programs have been addressed throughout this chapter.

    (c) Although advances in rodenticide development are important, it is unfortunate that vertebrate pest control seminars, meetings, and literature often focus on this aspect. Indeed, there seems to be an endless variety of bait formulations, the development of which is getting more and more like processed foods. Just as much of our artificially flavored, colored, scented, texturized, and stabilized foods are built to promote product acceptability (i.e., to "assist" consumers in making a choice), so has the pesticide industry cosmeticized baits. The correlation between these newly "improved" forms of bait and improved efficacy is not always clear.

(d) Information on the use of rodenticides is readily available in various publications (*35, 86, 116, 138, 173, 215, 233, 289, 292*) and in Chapter 31 of this bulletin.

(e) The purpose of this chapter is to encourage holistic rodent pest management, consisting largely of undramatic interventions that require time to manifest themselves and are of little hazard to nontarget species and the environment.

Acknowledgments

Deep appreciation is expressed to the Wadsworth Center for Laboratories and Research and to the Center for Environmental Health, New York State Department of Health, Albany, NY; to the Center for Environmental Health, Centers for Disease Control, U.S. Department of Health and Human Services, Atlanta, GA; and to the School of Hygiene and Public Health, Johns Hopkins University, Baltimore, MD. Their support made possible much of the research and associated experience described in this chapter.

Since space does not permit individual acknowledgment of the many "rodentologists" and other scientists whose various publications and communications provided the foundation for this chapter, we thank them collectively for their help. Special thanks are given to Dr. Austin M. Frishman (AMF Pest Management Services, Inc.) for suggesting this chapter, to the late Mr. Andre Ratti (Swiss Broadcasting Corp.) for encouragement to complete it, and to Dr. William W. Jacobs (Environmental Protection Agency) and Mr. Charles V. Trimarchi (Wadsworth Center) for their critical reading of the manuscript. Also, we thank the following people who provided specific information used in this chapter: F. T. Artis, A. D. Ashton, S. Daar, R. F. Gruhn, A. W. Holte, T. J. Imholte, W. W. Jacobs, P. Kopp, M. A. Patch, R. M. Poche, M. A. Recht, and P. Wilkerson. Thanks are also due to the many private citizens and pest control technicians who have aided us in our investigations internationally and brought us into contact with many practical rodent control problems.

We are also much indebted to Ms. Christine Tanner for her accuracy and patience in the laborious typing effort.

Cited References

1 Albrecht, O. W. 1980. Forecasting the solid waste stream. *Waste Age* 11(4)46–52, 76.

2 Anderson, J. W. 1954. The production of ultrasonic sounds by laboratory rats and other mammals. *Science* 119(3101)808–809.

3 Anderson, P. K. 1978. The serendipitous mouse. *Natural History* 87(4)38–43.

4 Anon. 1966. New law puts pressure on New York slum landlords. *Pest Control* 34(11)48.

5 Anon. 1967. *Joint FAO/WHO Expert Committee on Zoonoses, Third Report.* WHO Technical Report Series No. 378. World Health Organization, Geneva, Switzerland.

6 Anon. 1981. GAO says stronger enforcement needed against misuse of pesticides. *Environmental Health Newsletter* 20(21)7.

7 Anon. 1981. Using glueboards. *Pest Control* 49(12)54.

8 Anon. 1982. *NAG—Neighborhood Action Groups: A Guide to Constructive Citizen Participation.* Trinidad & Tobago Solid Waste Management Company Ltd., Port of Spain.

9 Anon. 1983. Eli Lilly uses in-house pest control approach. *Pest Control* 51(12)32–33.

10 Anon. 1985. Windy City switching to carts. *Public Works*, August, pp. 70–72.

11 Anon. 1985. Wilkerson track cleaning service. *Feed and Grain Times*, November, p. 18.

12 Anon. 1986. 1986 Cart survey. *Waste Age* 17(2)99–106 *passim*. [Note: For a list of cart manufacturers, see *Waste Age* 19(2)147 (1988).]

13 Anon. 1986. Population communication services: a new wave in education. *In Brief*, Spring, p. 7.

14 Anon. 1987. Durability is the key. *Waste Age* 18(2)93–125 *passim*.

15 Anon. 1988. 1988 Sweeping equipment guide. *Waste Age* 19(2)135–146.

16 Archer, J. 1968. The effect of strange male odor on aggressive behavior in male mice. *Journal of Mammalogy* 49(3)572–575.

17 Ashton, A. D. 1986. Performance of Vengeance (bromethalin) in urban environments. In *Proceedings of the Twelfth Vertebrate Pest Conference* (San Diego, CA, Mar. 4), T. P. Salmon, ed. University of California, Davis.

18 Attfield, R. 1983. *The Ethics of Environmental Concern.* Columbia University, New York, NY.

19 Baker, R. O. 1984. Commensal rodents—key identification characteristics and biology essential to control. In *Workshop Manual—Fundamentals of Commensal Rodent Control.* Eleventh Vertebrate Pest Conference (Sacramento, CA, Mar. 7). University of California, Davis, pp. 3–6.

20 Barnett, S. A. 1951. Damage to wheat by enclosed populations of *Rattus norvegicus*. *Journal of Hygiene* 49(1)22–25.

21 Barnett, S. A. 1958. An analysis of social behaviour in wild rats. *Proceedings of the Zoological Society of London* 130(1)107–152, 3 pl.

22 Barnett, S. A. 1963. *The Rat: A Study in Behavior.* Aldine, Chicago, IL.

23 Barnett, S. A. 1969. The feeding of rodents. In *Indian Rodent Symposium* (Calcutta, 1966), K. L. Harris, ed. Johns Hopkins University, Calcutta, India, and U.S. Agency for International Development, New Delhi, India, pp. 113–123.

24 Barnett, S. A., and I. Prakash. 1975. *Rodents of Economic Importance*. Arnold-Heinemann, New Delhi, India.

25 Batra, H. N. 1966. Feeding and general habits of common species of Indian rodents. Paper presented at Indian Rodent Symposium (Calcutta, Dec. 9, 1966) convened by Johns Hopkins University Center for Medical Research and Training, Calcutta, India, and the U.S. Agency for International Development, New Delhi, India.

26 Baur, F. J. 1977. Control points in the food industry. *Food Processing* 38(2)40–42.

27 Beck, A. M. 1972. The ecology of free-ranging dogs in Baltimore City. Dissertation, School of Hygiene and Public Health, Johns Hopkins University, Baltimore, MD.

28 Benor, D., and J. Q. Harrison. 1977. *Agricultural Extension: The Training and Visit System*. World Bank, Washington, DC.

29 Bowerman, A. M., and J. E. Brooks. 1972. What makes a palatable warfarin? *Pest Control* 40(2)22, 28, 29.

30 Bowers, J. M., and B. K. Alexander. 1968. Mice: Individual recognition by olfactory cues. *Science* 158(3805)1208–1210.

31 Brady, J. 1983. A sanitarian's view of pest control in supermarkets. *Pest Control Technology* 11(8)64–66.

32 Brooks, J. E. 1966. Roof rats in residential areas—the ecology of invasion. *California Vector Views* 13(9)69–74.

33 Brooks, J. E. 1969. *Behavior of the Norway Rat and Its Significance in Control Programs*. National Pest Control Association Technical Release 22-69, 12 pp.

34 Brooks, J. E., and A. M. Bowerman. 1971. Evaluation of U-5897 as a male chemosterilant for rat control. *Journal of Wildlife Management* 35(4)618–624.

35 Brooks, J. E., and F. P. Rowe. 1987. *Commensal Rodent Control*. Rept. No. WHO/VBC/87.949, World Health Organization, Geneva, Switzerland, 107 pp.

36 Broome, W. W., Jr. 1977. Glue traps for rodents. *Pest Control Technology* 5(10)14–16.

37 Brown, R. E., P. B. Singh, and B. Roser. 1987. Olfactory recognition of congenic strains of rats. *Annals of the New York Academy of Science* 510:202–204.

38 Brown, R. Z. 1953. Social behavior, reproduction, and population changes in the house mouse (*Mus musculus* L.). *Ecological Monographs* 23(3)217–240.

39 Brown, R. Z. 1969. Summary of lecture on rodent control in land use and town development planning (Item 7.3, AN:TCR/69/3). In *FAO/WHO Regional Training Seminar on Control of Rodents of Agricultural and Public Health Importance* (Manila, 1969). Food and Agriculture Organization, Rome, Italy.

40 Buck, W. B. 1983. A report of poisonings in the dog and cat in 1982. Presentation, 75th Annual Conference for Veterinarians, New York State College of Veterinary Medicine (Ithaca, NY, Jan. 18–20).

41 Bullard, R., and H. Shuyler. 1983. Springing the trap on post-harvest food loss. *Horizons* 2(7)26–32.

42 Cain, B. J., and J. P. Comings. 1977. *The Participatory Process: Producing Photo-literature*. Center for International Education, Amherst, MA.

43 Calhoun, J. B. 1948. Mortality and movement of brown rats in artificially supersaturated populations. *Journal of Wildlife Management* 12(2)167–172.

44 Calhoun, J. B. 1962. *The Ecology and Sociology of the Norway Rat*. U.S. Public Health Service Publ. No. 1008. Government Printing Office, Washington, DC.

45 Capretta, P. J., and L. H. Rawls. 1974. Establishment of a flavor preference in rats: Importance of nursing and weaning experience. *Journal of Comparative Physiology and Psychology* 86(4)670–673.

46 Caslick, J. W., and D. J. Decker. 1980. *Rat and Mouse Control*. Information Bulletin 163. New York State College of Agriculture and Life Sciences, Cornell University, Ithaca, NY.

47 Center for Food Safety and Applied Nutrition. 1989. *The Food Defect Action Levels*. Food and Drug Administration, Washington, DC.

48 Center for the Integration of Applied Sciences. 1981. *An Integrated Pest Management Approach to the Rats at the Berkeley Marina: Final Report*. John Muir Institute for Environmental Studies, Napa, CA.

49 Chitty, D., and H. N. Southern (eds.). 1954. *The Control of Rats and Mice*, 3 vols. Clarendon, Oxford, England.

50 Christian, J. J. 1976. Neurobehavioral endocrine regulation of small mammal populations. In *Populations of Small Mammals Under Natural Conditions*, vol. 5, D. A. Snyder, ed. Special Publication Series, Pymatuning Laboratory of Ecology, University of Pittsburgh, pp. 143–158.

51 Christian, J. J. 1980. Endocrine factors in population regulation. In *Biosocial Mechanisms of Population Regulation*, M. N. Cohen, R. S. Malpass, and H. G. Klein, eds. Yale University, New Haven, CT, pp. 55–115.

52 Cohen, H. L. 1985. Community Intervention for Insect and Rodent Control (CIIRC)—a Working Paper. WHO/CIIRC Working Group Meeting, State University of New York (Buffalo, Aug. 18–22).

53 Comings, J. 1981. *Participatory Communication in Nonformal Education.* Technical Note 17. Center for International Education, Amherst, MA.

54 Comings, J. P. 1982. *FTS Project Activity: Extension Training Materials for the Crown-Baiting Technique in the Philippines.* World Education, Boston, MA.

55 Comings, J. P., S. C. Frantz, and B. J. Cain. 1981. Community participation in the development of environmental health education materials. *Convergence* 14(2)36–44.

56 Comptroller General of the United States. 1976. Hungry Nations Need to Reduce Food Losses Caused by Storage, Spillage, and Spoilage. Report to the Congress ID-76-65.

57 Cooney, K. 1986. Once over lightly won't make it in health care. *Pest Control* 54(5)48, 50, 54, 57.

58 Cooperative Food Distributors of America, et al. 1974. *Voluntary Industry Sanitation Guidelines for Food Distribution Centers and Warehouses.* McLean, VA.

59 Cornwell, P. B. 1984. Meeting the demands of the food industry. *Pi Chi Omega News* (Suppl.), December, pp. 1–4.

60 Cowan, P. E. 1977. Neophobia and neophilia: New-object and new-place reactions of three *Rattus* species. *Journal of Comparative Physiology and Psychology* 91(1)63–71.

61 Cowan, P. E., and S. A. Barnett. 1975. The new-object and new-place reactions of *Rattus rattus* L. *Zoological Journal of the Linnean Society* 56(3)219–234.

62 Crowcroft, P. 1959. Spatial distribution of feeding activity in the wild house mouse (*Mus musculus* L.). *Annals of Applied Biology* 47(1)150–155.

63 Crowcroft, P. 1966. *Mice All Over.* Dufour, Chester, PA.

64 Davis, D. E. 1948. The survival of wild brown rats on a Maryland farm. *Ecology* 29(4)437–448.

65 Davis, D. E. 1949. A phenotypical difference in growth of wild rats. *Growth* 13:1–6.

66 Davis, D. E. 1950. The mechanics of rat populations. *Transactions of the Fifteenth North American Wildlife Conference* (San Francisco, CA, Mar. 6–9), E. M. Quee, ed. Wildlife Management Institute, Washington, DC, pp. 461–466.

67 Davis, D. E. 1950. The rat population of New York, 1949. *American Journal of Hygiene* 52(2)147–152.

68 Davis, D. E. 1953. The characteristics of rat populations. *Quarterly Review of Biology* 28(4)373–401.

69 Davis, D. E. 1956. *Manual for Analysis of Rodent Populations.* Pennsylvania State University, University Park.

70 Davis, D. E. 1957. The use of food as a buffer in a predator-prey system. *Journal of Mammalogy* 38(4)466–472.

71 Davis, D. E. 1969. Appraisal of rat populations. In *Indian Rodent Symposium* (Calcutta, 1966), K. L. Harris, ed. Johns Hopkins University, Calcutta, India, and U.S. Agency for International Development, New Delhi, India, pp. 51–57.

72 Davis, D. E. 1976. Physiological and behavorial responses to the social environment. In *Populations of Small Mammals Under Natural Conditions,* vol. 5, D. A. Snyder, ed. Special Publication Series, Pymatuning Laboratory of Ecology, University of Pittsburgh, pp. 84–91.

73 Davis, D. E., J. T. Emlen, and A. W. Stokes. 1948. Studies on home range in the brown rat. *Journal of Mammalogy* 29(33)207–225.

74 Davis, D. E., and W. T. Fales. 1949. The distribution of rats in Baltimore, Maryland. *American Journal of Hygiene* 49(3)247–254.

75 Davis, D. E., and W. T. Fales. 1950. The rat populations of Baltimore, 1949. *American Journal of Hygiene* 52(2)143–146.

76 Davis, D. E., and W. B. Jackson. 1981. Rat control. In *Advances in Applied Biology,* vol. 6, T. H. Coaker, ed. Academic Press, New York, NY, pp. 221–277.

77 Davis, D. E., and W. L. Jensen. 1952. Mortality in an induced epidemic. *Transactions. Seventeenth North American Wildlife Conference* (Miami, FL, Mar. 17–19), E. M. Quee, ed. Wildlife Management Institute, Washington, DC, pp. 151–160.

78 Davis, H., A. Casta, and G. Schatz. 1974. *Urban Rat Surveys.* Center for Disease Control, Atlanta, GA.

79 Deng, Z., and C. Wang. 1984. Rodent control in China. In *Workshop Manual—Fundamentals of Commensal Rodent Control.* Eleventh Vertebrate Pest Conference (Sacramento, CA, Mar. 7). University of California, Davis, pp. 47–53.

80 Deoras, P. J. 1966. *Rats and Their Control.* Pest Control (India) Private Ltd., Bombay, 14 pp. + 5 tables.

81 Deoras, P. J. 1966. The significance of studies on the frequency of rat populations. In *WHO Seminar on Rodents and Rodent Ectoparasites* (Geneva, Switzerland, Oct. 24–28), pp. 75–77.

82 Deoras, P. J. 1968. Cereal intake, preference, pellet exudation by *Rattus rattus* and their significance. *Zeitschrift für Angewandte Zoologie* 55(4)447–459.

83 Deoras, P. J., and G. C. Chaturvedi. 1971. Observations on duration of estrous cycle in some rats in Bombay. In *International Symposium on Bionomics and Control of Rodents* (Kanpur, India, Sept. 29–Oct. 2, 1968), S. L. Petri, Y. C. Wal, and C. P. Srivastava, eds. Science and Technology Society, Kanpur, India, pp. 57–58.

84 Department of Health, Education, and Welfare. 1974. *Proceedings. National Urban Rat Control Project Directors Meeting* (Apr. 30–May 2). Center for Disease Control, Atlanta, GA.

85 Developmental Alternatives, Inc. 1975. *Strategies for Small Farmer Development Projects.* Contract AID/CM/ta-C-73-41, vol. 1. Agency for International Development, Washington, DC.

86 Dodds, W. J., S. C. Frantz, and K. O. Story. 1986. *The Professionals Guide to Managing Poisoning by Anticoagulant Rodenticides.* Chempar Products Division, Lipha Chemicals, Inc., New York, NY.

87 Doty, R. E. 1938. The prebaited feeding station method of rat control. *Hawaiian Planters' Record* 42(2)39–76.

88 Dover, M. J. 1985. *A Better Mouse Trap: Improving Pest Management for Agriculture.* Study No. 4. World Resources Institute, Washington, DC.

89 Dover, M. J. 1985. Getting off the pesticide treadmill. *Technology Review* 88(8)52–63.

90 Dover, M. J., and B. A. Croft. 1986. Pesticide resistance and public policy. *BioScience* 36(2)78–85.

91 Drummond, D. C. 1960. The food of *Rattus norvegicus* Berk. in an area of sea wall, saltmarsh, and mudflat. *Journal of Animal Ecology* 29(2)341–347.

92 Drummond, D. C. 1971. Rodents and biodeterioration. *International Biodeterioration Bulletin* 7(2)73–79.

93 Drummond, D. 1972. Biology and control of domestic rodents. In *Vector Control in International Health*, J. V. Smith and R. Pal, eds. World Health Organization, Geneva, Switzerland, pp. 46–69.

94 Drummond, D. C. 1980. Pottery rodent traps— a provisional list. *Museum Ethnographers' Group Newsletter* 9(March)14–15.

95 Drummond, D. C. 1982. Man eats rat. *Social Biology and Human Affairs* 47(2)69–75.

96 Drummond, D. C. 1983. Mouse traps or snake houses. In *Report of the Department of Antiquities*, Nicosia, Cyprus, pp. 199–200.

97 Drummond, D. C. 1985. Developing and monitoring urban rodent control programmes. *Acta Zoologica Fennica* 173:145–148.

98 Drummond, D. C., and T. J. Crowe. 1985. Rat hunters of Burma. *Museum Ethnographers' Group Newsletter* 18(June)47–52.

99 Drummond, D. C., E. J. Taylor, and M. Bond. 1977. Urban rat control: Further experimental studies at Folkestone. *Environmental Health* 85(12)265–267.

100 Dutson, V. 1974. The association of the roof rat (*Rattus rattus*) with the Himalayan blackberry (*Rubus discolor*) and Algerian ivy (*Hedera canariensis*) in California. In *Proceedings of the Sixth Vertebrate Pest Conference* (University of California, Davis), W. V. Johnson, ed., pp. 41–48.

101 Ellerman, J. R. 1961. Rodentia (part 2). *Mammalia*, vol. 3, *The Fauna of India Including Pakistan, Burma, and Ceylon*, M. L. Roonwal, ed. Baptist Mission, Calcutta, India, pp. 815–844.

102 Elton, C. S. 1953. The use of cats in farm rat control. *British Journal of Animal Behaviour* 1(4)151–155.

103 Emlen, J. T., Jr. 1947. Baltimore's community rat control program. *American Journal of Public Health* 37(6)721–727.

104 Emlen, J. T., Jr. 1950. How far will a mouse travel to a poisoned bait? *Pest Control* 18(8)16, 18, 20.

105 Emlen, J. T., Jr., and D. E. Davis. 1948. Determination of reproductive rates in rat populations by examination of carcasses. *Physiological Zoology* 21(1)59–65.

106 Emlen, J. T., Jr., A. W. Stokes, and D. E. Davis. 1949. Methods for estimating populations of brown rats in urban habitats. *Ecology* 29(4)430–442.

107 Emlen, J. T., Jr., A. W. Stokes, and C. P. Windsor. 1949. The rate of recovery of decimated populations of brown rats in nature. *Ecology* 29(2)133–145.

108 Environmental Protection Agency. 1980. *Investigation of Efficacy and Enforcement Activities Relating to Electromagnetic Pest Control Devices.* EPA 340/02-80-001. Washington, DC.

109 Environmental Protection Agency. 1981. Criteria and Policy Notice No. 2164.2 (Mar. 10). Registration Division, Office of Pesticide Programs, Washington, DC.

110 Ericsson, R. J. 1982. Alpha-chlorohydrin (Epibloc®): a toxicant-sterilant as an alternative in rodent control. In *Proceedings of the Tenth Vertebrate Pest Conference* (University of California, Davis), R. E. Marsh, ed., pp. 6–9.

111 Erlinge, S. 1975. Predation as a control factor of small rodent populations. In *Biocontrol of Rodents*, L. Hansson and B. Nilsson, eds. Ecological Bulletin 19. Swedish Natural Science

Research Council, Stockholm, Sweden, pp. 195–199.

112 Ewer, R. F. 1971. The biology and behavior of a free-living population of black rats (*Rattus rattus*). *Animal Behaviour Monographs* 4(5)127–174.

113 Fall, M. W., and B. E. Johns. 1988. Metallic flake particle markers for determining the feeding behavior of rats at bait points. In *Vertebrate Pest Control and Management Materials* (Fifth Symposium, ASTM STP 974), S. A. Shumake and R. W. Ballard, eds. American Society for Testing and Materials, Philadelphia, PA, pp. 128–133.

114 Farhang-Azad, A., and C. H. Southwick. 1979. Population ecology of Norway rats in the Baltimore Zoo and Druid Hill Park, Baltimore, Maryland. *Annals of Zoology* 15(part 1)1–42.

115 Fertig, D. S., and V. W. Edmonds. 1969. The physiology of the house mouse. *Scientific American* 221(4)103–108, 110.

116 Fitzwater, W. D. 1979. Commensal rodents, sec. 1. *Encyclopedia of Vertebrate Pest Control*, vol. 4. National Pest Control Association, Vienna, VA.

117 Fitzwater, W. D. 1982. Bird limes and rat glues—sticky situations. *Proceedings of the Tenth Vertebrate Pest Conference* (University of California, Davis), R. E. Marsh, ed., pp. 17–23.

118 Food and Drug Administration. 1986. Food for human consumption; final rules and proposed rules. 21 CFR Part 20 et al. *Federal Register* 51(118)22458–22483.

119 Food Protection and Sanitation Committee. 1982. *Integrated Pest Management in Food Processing Plants*, parts I & II. Technical Release. National Pest Control Association, Vienna, VA.

120 Frank, J. F., and H. M. Barnhart. 1986. Food and dairy sanitation. In *Maxcy-Rosenau Public Health and Preventive Medicine*, J. M. Last, ed. Appleton-Century-Crofts, Norwalk, CT, pp. 765–807.

121 Frantz, S. C. 1972. Fluorescent pigments for studying movements and home ranges of small mammals. *Journal of Mammalogy* 53(1)218–223.

122 Frantz, S. C. 1973. Behavioral ecology of the lesser bandicoot rat, *Bandicota bengalensis* (Gray), in Calcutta. Dissertation, School of Hygiene and Public Health, Johns Hopkins University, Baltimore, MD.

123 Frantz, S. C. 1974. Evaluation of rodent infestations in Nepal—a preliminary report. *Journal of the Nepal Medical Association* 12(3–4)17–32.

124 Frantz, S. C. 1976. Note on climbing by the lesser bandicoot rat, *Bandicota bengalensis* (Gray), in Bombay. *Bulletin of Grain Technology* 14(2)130–133.

125 Frantz, S. C. 1976. The web of hunger: Rats in the granary. *Natural History* 85(2)10–21.

126 Frantz, S. C. 1977. The behavioral/ecological milieu of godown bandicoot rats—implications for environmental manipulation. *Proceedings, All India Rodent Seminar* (Ahmedabad, September 1975), Rodent Control Project, Sidhpur, Gujarat, India, pp. 95–101.

127 Frantz, S. C. 1977. *Procedures for Collecting Rats for Anticoagulant Resistance Studies—Urban Rat Control Projects*. Center for Disease Control, Atlanta, GA.

128 Frantz, S. C. 1978. A Working Neighborhood. What Does It Take? New York State Department of Health, Albany (available from the author).

129 Frantz, S. C. 1979. Procedures for sampling urban rat populations for anticoagulant resistance evaluation. In *Vertebrate Pest Control and Management Materials* (Second Symposium, ASTM STP 680), J. R. Beck, ed. American Society for Testing and Materials, Philadelphia, PA, pp. 20–28.

130 Frantz, S. C. 1980. Rodenticides and their use in an integrated pest management program. Paper presented at CDC Multi-regional Workshop for Environmental Health Program Managers and Dental Disease Prevention Programs (Cleveland, OH, June 2–4).

131 Frantz, S. C. 1984. Home range of the lesser bandicoot rat, *Bandicota bengalensis* (Gray), in Calcutta, India. *Acta Zoologica Fennica* 171:297–299.

132 Frantz, S. C. 1984. Site Visit Report—Mouse Infestation. Bureau of Management Services, New York State Health Department, Albany (available from the author).

133 Frantz, S. C. 1988. Architecture and commensal vertebrate pest management. In *Architectural Design and Indoor Microbial Pollution*, R. B. Kundsin, ed. Oxford University Press, New York, NY, Chapter 11.

134 Frantz, S. C., and J. P. Comings. 1976. Evaluation of urban rodent infestations—an approach in Nepal. In *Proceedings of the Seventh Vertebrate Pest Conference* (University of California, Davis), C. C. Siebe, ed., pp. 279–290.

135 Frantz, S. C., W. J. Dodds, and S. Kim. 1984. A Study of Accidents, Illnesses, and Deaths Resulting from the Use of Commensal Rodenticides. EPA Hearings on Rodenticide Bait Stations, Sacramento, CA, Mar. 5.

136 Frantz, S. C., and C. M. Padula. 1983. A laboratory test method for evaluating the efficacy of glueboards for trapping house mice. In *Vertebrate Pest Control and Management Materials* (Fourth Symposium, ASTM STP 817), D. E. Kaukeinen, ed. American Society of Testing and

Materials, Philadelphia, PA, pp. 209–225.

137 Frishman, A. 1982. Practical tips. *Pest Control Technology* 10(8)64.

138 Frishman, A. M. 1982. Rats and mice. In *Handbook of Pest Control*, A. Mallis, ed. Franzak & Foster, Cleveland, OH, pp. 5–77.

139 Frishman, A. M. 1983. Mouse biology and control. Presentation, Symposium on Pest Control: Challenges and Problems (May 4). Abell Waco, Ltd., and Institute of Technology, Toronto, Ontario, Canada.

140 Fuglesang, A. 1973. *Applied Communication in Developing Countries*. Dag Hammarskjöld Foundation, Uppsala, Sweden.

141 Galef, B. G., Jr., and S. W. Wigmore. 1983. Transfer of information concerning distant foods: a laboratory investigation of the "information centre" hypothesis. *Animal Behaviour* 31(3)748–758.

142 General Accounting Office. 1983. *Monitoring and Enforcing Food Safety—An Overview of Past Studies*. GAO/RCED-83-153. Washington, DC.

143 General Accounting Office. 1986. *Food and Drug Administration: Laboratory Analysis of Product Samples Needs to Be More Timely*. GAO/HRD-86-102. Washington, DC.

144 Geyer, L. A., and R. J. Barfield. 1979. Introduction to the symposium: Ultrasonic communication in rodents. *American Zoologist* 19(2)411.

145 Giraldi, A., and K. Hackett. 1983. *IPM Binder for Rats: A Management Strategy for the National Park Service*. John Muir Institute for Environmental Studies, Inc., Napa, CA.

146 Greaves, J. H. 1978. Rodents. Loss determinations by population assessment and estimation procedures. In *Postharvest Grain Loss Assessment Methods*, K. L. Harris and C. J. Linblad, eds. American Association of Cereal Chemists, St. Paul, MN, pp. 109–115.

147 Greaves, J. H., and F. P. Rowe. 1969. Responses of confined rodent populations to an ultrasound generator. *Journal of Wildlife Management* 33(2)409–417.

148 Hall, E. R. 1927. An outbreak of mice in Kern County, California. *University of California Publications in Zoology* 30:189–203.

149 Hall, R. 1986. Surprising optimism surfaces in face of changing industry. *Pest Control* 54(7)18, 19, 22, 24.

150 Hall, R. 1986. More for less. *Pest Control* 54(9)55, 58, 60.

151 Hardin, G. 1985. *Filters Against Folly*. Viking Penguin, New York, NY.

152 Harris, K. L., and C. J. Lindblad (eds.). 1978. *Postharvest Grain Loss Assessment Methods*. American Association of Cereal Chemists, St. Paul, MN.

153 Harrison, J. L., and H. C. Woodville. 1950. Variation in size and weight in five species of house-rats (Rodentia: Muridae), in Rangoon, Burma. *Records of the Indian Museum* 47(1)65–71.

154 Havens, E. G. (ed.). 1975. *Working with Film*. Access No. 14. Challenge for Change/Société Nouvelle. National Film Board of Canada, Montreal.

155 Henaut, D. T. 1971. *Community Cable TV and You*. Newsletter No. 6. Challenge for Change/Société Nouvelle. National Film Board of Canada, Montreal.

156 Hinton, M. A. C. 1931. *Rats and Mice as Enemies of Mankind*. Economic Series No. 18. British Museum (Natural History), London, England.

157 Holloway, A. H., Jr. 1947. The study of the structural-sanitary characteristics of urban residential areas in respect to their support of rodent populations. Thesis, School of Engineering, Johns Hopkins University, Baltimore, MD.

158 Hopf, H. S., G. E. J. Morley, and J. R. O. Humphries (eds.). 1976. *Rodent Damage to Growing Crops and to Farm and Village Storage in Tropical and Subtropical Regions*. Center for Overseas Pest Research, and Tropical Products Institute, London, England.

159 Hoque, M. M. 1973. Notes on rodent pests affecting coconut. *Philippines Agriculture* 56(7–8)280–289.

160 Horner, D. D. 1981. The FDA and quality assurance. *Pest Control* 49(10)40–42.

161 Howard, W. D., and R. E. Marsh. 1985. Ultrasonics and electromagnetic devices. *Acta Zoologica Fennica* 173:187–189.

162 Huffaker, C. B. 1975. Biological control in the management of pests. *Agro-Ecosystems* 2(1)15–31.

163 Hufschmidt, M. M., et al. 1983. *Environment, Natural Systems and Development: An Economic Valuation Guide*. Johns Hopkins University, Baltimore, MD.

164 Huson, L. W., and R. A. Davis. 1980. Discriminant functions to aid identification of faecal pellets of *Rattus norvegicus* and *Rattus rattus*. *Journal of Stored Products Research* 16(3–4)103–104.

165 Huson, L. W., and B. D. Rennison. 1981. Seasonal variability of Norway rat (*Rattus norvegicus*) infestation of agricultural premises. *Journal of the Zoological Society of London* 194:257–260.

166 Imholte, T. J. 1984. *Engineering for Food Safety and Sanitation*. Technical Institute of Food Safety, Crystal, MN.

167 Ishaq, M. (ed.). 1971. *Bangladesh District Gazeteers: Chittagong Hill Tracts*. Bangladesh Government, Dacca.

168 Jackson, W. B. 1951. Food habits of Baltimore, Maryland, cats in relation to rat populations. *Journal of Mammalogy* 32(4)458–461.

169 Jackson, W. B. 1969. Summary of lecture on physiology and behavior (item 3.2). In *FAO/WHO Regional Training Seminar on Control of Rodents of Agricultural and Public Health Importance* (Manila, 1969). Food and Agriculture Organization, Rome, Italy, pp. 1–5.

170 Jackson, W. B. 1979. Use of burrows for evaluating rodenticide efficacy in urban areas. In *Vertebrate Pest Control and Management Materials* (Second Symposium, ASTM STP 680), J. R. Beck, ed. American Society for Testing and Materials, Philadelphia, PA, pp. 5–10.

171 Jackson, W. B., and S. C. Frantz. 1982. The EPA efficacy data waiver debate. *Pest Control* 50(9)4, 60.

172 Jacob, M. 1985. *Suggested Guidelines for Inspection of Ships for Rodent Control.* Rept. No. WHO/VBC/85.918, World Health Organization, Geneva, Switzerland, 10 pp.

173 Jacobs, W. W. 1983. Pesticides federally registered for control of terrestrial vertebrate pests. In *Prevention and Control of Wildlife Damage,* R. M. Timm, ed. Great Plains Agricultural Council and University of Nebraska, Lincoln, pp. G1–G29.

174 Jenson, A. G. 1979. *Proofing of Buildings Against Rats, Mice and Other Pests.* Ministry of Agriculture, Fisheries and Food. Her Majesty's Stationery Office, London, England.

175 Jones, J. C. 1951. Rodent control: An essential of good sanitation. *Modern Sanitation* 3(9)24–27, 64.

176 Jones, N. B. 1978. The use of a fluorescent pigment for tracing the movements of commensal rodents. *International Biodeterioration Bulletin* 14(2)61–64.

177 Kami, H. T. 1966. Food of rodents in the Hamakua District, Hawaii. *Pacific Science* 20(3)367–373.

178 Katz, H. 1979. Integrated pest management, then and now. *Pest Control* 47(11)16, 81.

179 Kaukeinen, D. E. 1979. Field methods for census taking of commensal rodents in rodenticide evaluations. In *Vertebrate Pest Control and Management Materials* (Second Symposium, ASTM STP 680), J. R. Beck, ed. American Society for Testing and Materials, Philadelphia, PA, pp. 68–83.

180 Klimstra, W. D. 1968. Biology and behavior of the house mouse. Presentation, 35th Annual Convention of the National Pest Control Association (Salt Lake City, UT, Oct. 24).

181 Knittel, R. E. 1975. *Motivation and Attitude Change Components for Trainers of Community Health Aides.* Center for Disease Control, Atlanta, GA.

182 Knote, C. E. 1982. Physical stoppage of rodents, part III. *Pest Control* 50(11)48, 50, 52, 54.

183 Korten, D. C., and F. B. Alfonso. 1983. *Bureaucracy and the Poor: Closing the Gap.* Kumarian, West Hartford, CT.

184 Kunhardt, J. C. G. 1919. The rat problem of India. *Indian Journal of Medical Research* (Special Indian Science Congress Number), pp. 143–172.

185 Laidlaw, G. 1984. Canada has developed guidelines to restrict ultrasonics. *Pest Control* 52(3)32.

186 Laird, M. 1966. Biological control of rodents. In *WHO Seminar on Rodents and Rodent Ectoparasites* (Geneva, Switzerland, Oct. 24–28), pp. 113–120.

187 Landry, S. O., Jr. 1970. The Rodentia as omnivores. *Quarterly Review of Biology* 45(4)351–372.

188 Laughlin, P. E. 1982. The plant sanitarian. Functions in the food processing plant. *Pest Control* 50(10)39–41, 44.

189 Laurie, E. M. O. 1946. The reproduction of the house-mouse (*Mus musculus*) living in different environments. *Proceedings of the Royal Society* B133(872)248–281.

190 Lawrence, J. 1986. Inadequate regulations: The EPA limits. *ISI Press Digest* 2(Jan. 13)11.

191 LeBoulengé, E. 1985. Computer package for the analysis of capture-recapture data. *Acta Zoologica Fennica* 173:69–72.

192 Lembrez, J. 1966. Rodent control and exclusion by environmental manipulation in urban and port areas. In *WHO Seminar on Rodents and Rodent Ectoparasites* (Geneva, Switzerland, Oct. 24–28), pp. 171–175.

193 LeTendre, W. G. 1982. Miller Brewing Company—a sanitation/pest control success story. *Pest Control* 50(10)40–41.

194 Levine, R. S. 1986. *Assessment of Mortality and Morbidity Due to Unintentional Pesticide Poisonings.* Report No. WHO/VBC/86.929, World Health Organization, Geneva, Switzerland, 24 pp.

195 Levitsky, D. A. 1970. Feeding patterns of rats in response to fasts and changes in environmental conditions. *Physiology and Behavior* 5(3)291–300.

196 Lidicker, W. Z. 1976. Regulation of numbers in small animal populations—historical reflections and a synthesis. In *Populations of Small Mammals Under Natural Conditions,* vol. 5, D. A. Snyder, ed. Special Publication Series, Pymatuning Laboratory of Ecology, University of Pittsburgh, pp. 122–141.

197 Litovitz, T., and J. C. Veltri. 1985. 1984 Annual Report of the American Association of Poison Control Centers National Data Collection

System. *American Journal of Emergency Medicine* 3(5)423–450.

198 Lloyd, J. A., and J. J. Christian. 1967. Relationship of activity and aggression to density in two confined populations of house mice (*Mus musculus*). *Journal of Mammalogy* 48(2)262–269.

199 Lord, R. D. 1970. Vertebrate ecology in public health. *Public Health Reports* 85(2)105–111.

200 Lorgue, G., et al. 1986. Intoxication of domestic and wild animals by anticoagulant rodenticides—a synthesis of data from the French National Veterinary Antipoison Center. In *Proceedings of the Twelfth Vertebrate Pest Conference* (San Diego, CA, Mar. 4), T. P. Salmon, ed. University of California, Davis, pp. 82–87.

201 Majumder, S. K., M. K. Krishnakumari, and K. Muktabai. 1969. A critical appraisal of rodenticides. In *Indian Rodent Symposium* (Calcutta, 1966), K. L. Harris, ed. Johns Hopkins University, Calcutta, India, and U.S. Agency for International Development, New Delhi, India, pp. 264–280.

202 Marsh, R. E. 1983. Roof rats. In *Prevention and Control of Wildlife Damage*, R. M. Timm, ed. Great Plains Agricultural Council and University of Nebraska, Lincoln, pp. B115–B120.

203 Marsh, R. E., and W. E. Howard. 1973. Prospects of chemosterilant and genetic control of rodents. *Bulletin of the World Health Organization* 48(3)309–316.

204 Marsh, R. E., and W. E. Howard. 1981. *The House Mouse: Its Biology and Control*. Leaflet 2945. Cooperative Extension Service, University of California, Berkeley.

205 Marshall, E. 1982. EPA's high-risk carcinogen policy. *Science* 218(4576)975–978.

206 Marshall, J. T., Jr. 1962. Predation and natural selection. In *Pacific Island Rat Ecology*, T. I. Storer, ed. Museum Bulletin 225. Bernice P. Bishop Museum, Honolulu, HI, pp. 177–189.

207 Marshall, J. T., Jr. 1977. Family Muridae, rats and mice. In *Mammals of Thailand*, B. Lekagul and J. A. McNeely, eds. Association for Conservation of Wildlife, Bangkok, Thailand, pp. 395–490.

208 Meyer, A. N. 1978. *Planning Information for Amman City Rat Control Programme*. Rept. No. WHO/VBC/78.702, World Health Organization, Geneva, Switzerland, 14 pp.

209 Milan, P. P., and W. B. Jackson. 1988. Efficacy determination of rodenticide use and trapping for reduction of coconut loss in the Philippines. In *Vertebrate Pest Control and Management Materials* (Fifth Symposium, ASTM STP 974), S. A. Shumake and R. W. Ballard, eds. American Society for Testing and Materials, Philadelphia, PA.

210 Mitwally, H., and M. F. El-Sawy. 1972. The role of environmental sanitation regarding rodent control in Alexandria harbour. In *Proceedings of the First Scientific Symposium on Rodents and Their Control in Egypt* (Assiut University, Cairo), Government Printing Office, Cairo, pp. 195–213.

211 Moreland, D. 1986. Pest control in supermarkets. *Pest Control Technology* 11(8)59–62, 79.

212 Myllymäki, A. 1975. Rodent surveillance and prediction of rodent outbreaks. In *Biocontrol of Rodents*, L. Hansson and B. Nilsson, eds. Ecological Bulletin 19. Swedish Natural Science Research Council, Stockholm, Sweden, pp. 275–282.

213 Nader, J. A. 1969. Animal remains in pellets of the barn owl *Tyto alba* from the vicinity of An-Najof, Iraq. *Bulletin of the Iraq Natural History Museum* 4(1)1–7.

214 National Pest Control Association. 1978. *Rodent Control Devices*. Technical Release ESPC 041631. Dunn Loring, VA.

215 National Pest Control Association. 1979. *Rodenticide Use in Food Handling Establishments*. Technical Release ESPC 063301A. Dunn Loring, VA.

216 National Research Council. 1980. *Urban Pest Management*. National Academy Press, Washington, DC.

217 National Research Council. 1986. *Ecological Knowledge and Environmental Problem-Solving*. National Academy Press, Washington, DC.

218 Noirot, E. 1972. Ultrasound and maternal behavior in small rodents. *Developmental Psychobiology* 5(4)371–387.

219 Nowak, R. M., and J. L. Paradiso. 1983. *Walker's Mammals of the World*, 4th ed. (2 vol.). Johns Hopkins University, Baltimore, MD.

220 Nyby, J. 1983. Ultrasonic vocalizations during sex behavior of male house mice (*Mus musculus*): a description. *Behavioral and Neural Biology* 39(1)128–134.

221 Olkowski, H. 1980. What is IPM? *IPM Practitioner* 2(5)3–4.

222 Olkowski, W., H. Olkowski, and S. Daar. 1984. *Integrated Pest Management for the House Mouse*. Bio-Integral Resource Center, Berkeley, CA.

223 Palmateer, S. D. 1979. *Effect of Dyes in Efficacy of Commensal Rodenticides*. Terrestrial and Aquatic Biology Unit, Office of Pesticide Programs, Environmental Protection Agency, Beltsville, MD.

224 Palmer, P., and E. Jacobson. 1974. *Action Research: A New Style of Politics in Education*. Institute for Responsive Education, Boston, MA.

225 Parrack, D. W. 1969. A note on the loss of food to the lesser bandicoot rat, *Bandicota bengalensis*. *Current Science* 38(4)93–94.

226 Parrack, D. W., and J. Thomas. 1970. The behaviour of the lesser bandicoot rat, *Bandicota bengalensis* (Gray & Hardwicke). *Journal of the Bombay Natural History Society* 67(1)67–80.

227 Pearson, O. P. 1963. History of two local outbreaks of feral house mice. *Ecology* 44(3)540–549.

228 Penn, L. A. 1971. The problem of rodents in our modern environment. *Journal of Milk and Food Technology* 34(10)471–474.

229 Peoples, S. A., and K. T. Maddy. 1979. Poisoning of man and animals due to ingestion of the rodent poison Vacor. *Veterinary and Human Toxicology* 21(4)266–268.

230 Peterson, C., and C. Gunnerson. 1985. *Solid Waste Management: An International Perspective.* International Small-Scale Waste-to-Energy Conference (Feb. 28). Resource Recovery Report. World Bank, Washington, DC.

231 Pileggi, N. 1985. The new corruption. *New York* 18(44)42–48.

232 Pisano, R. G., and T. I. Storer. 1948. Burrows and feeding of the Norway rat. *Journal of Mammalogy* 29(4)374–383.

233 Pratt, H. D. 1983. Rodenticides: What to use where, when, and how. *Pest Control* 51(10)19, 20, 22–24.

234 Pratt, H. D., B. F. Bjornson, and K. S. Littig. 1980. *Control of Domestic Rats and Mice.* HEW Publ. No. (CDC)80-8141. Center for Disease Control, Atlanta, GA.

235 Pratt, H. D., and R. Z. Brown. 1977. *Biological Factors in Domestic Rodent Control.* HEW Publ. No. (CDC)77-8144. Center for Disease Control, Atlanta, GA.

236 Pursley, W. 1982. Small bakery accounts. *Pest Control* 50(6)78.

237 Rajagopalan, P. K., and K. N. Panicker. 1984. *Feasibility of Community Involvement in Integrated Vector Control in Villages.* Rept. No. WHO/VBC/84.903, World Health Organization, Geneva, Switzerland, 6 pp.

238 Rampaud, M. 1984. Typology of wild rat populations. *Acta Zoologica Fennica* 171:233–236.

239 Rana, B. D., R. Advani, and B. K. Soni. 1983. Reproductive biology of *Rattus rattus rufescens* in the Indian desert. *Zeitschrift für Angewandte Zoologie* 70(20)207–216.

240 Recht, M. A. 1982. The fine structure of the home range and activity pattern of free-ranging telemetered urban Norway rats *Rattus norvegicus* (Berkenhout). *Bulletin of the Society of Vector Ecologists* 7:29–35.

241 Recht, M. A., R. Geck, and G. L. Challet. 1983. The fine-structure of the home range and activity-phasing of unrestricted telemetered urban roof rats, *Rattus rattus*, in Orange County, California. *Bulletin of the Society of Vector Ecologists* 8(1)51–64.

242 Reidinger, R. F., and J. R. Mason. 1983. Exploitable characteristics of neophobia and food aversions for improvements in rodent and bird control. In *Vertebrate Pest Control and Management Materials* (Fourth Symposium, ASTM STP 817), D. E. Kaukeinen, ed. American Society of Testing and Materials, Philadelphia, PA, pp. 20–39.

243 Rennison, B. D. 1981. *Rodent Control in Kuwait.* Tolworth Laboratory, Ministry of Agriculture, Fisheries and Food, Tolworth, England.

244 Retzlaff, E. G. 1939. Studies in population physiology with the albino mouse. *Biologia Generalis* 14(2)238–265.

245 Richter, C. P., and K. H. Clisby. 1941. Phenylthiocarbamide taste thresholds of rats and human beings. *American Journal of Physiology* 134(1)157–164.

246 Riley, D. A., and M. R. Rosenweig. 1957. Echolocation in rats. *Journal of Comparative Physiology and Psychology* 50(4)323–328.

247 Rogers, R. 1986. PCT commentary: Wake up! It's time to make money. *Pest Control Technology* 14(4)44–45.

248 Roter, D. L., et al. 1981. Community-produced materials for health education. *Public Health Reports* 96(2)169–172.

249 Roubertoux, R., et al. 1984. Correlations between retrieving behavior in females and ultrasonic vocalizations in newborn mice. *Acta Zoologica Fennica* 171:101–102.

250 Rowe, F. P. 1981. Wild house mouse biology and control. *Symposia of the Zoological Society of London* 47:575–589.

251 Rowe, F. P., and R. Redfern. 1969. Aggressive behaviour in related and unrelated wild house mice (*Mus musculus* L.). *Annals of Applied Biology* 64(3)425–431.

252 Rowe, F. P., T. Swinney, and R. J. Quy. 1983. Reproduction of the house mouse (*Mus musculus*) in farm buildings. *Journal of Zoology* 199(2)259–269.

253 Sagar, P., and O. S. Bindra. 1976. Food and feeding habits of the lesser bandicoot rat. *Pest Control* 44(12)28–32.

254 Sahu, A., and B. R. Maiti. 1978. Estrous cycle of the bandicoot rat—a rodent pest. *Zoological Journal of the Linnean Society* 63(3)309–314.

255 Santini, L., B. Conti, and F. Chesi. 1984–85. Researches on alternative methods for rodent control. I. On the efficacy of ''multiple-catch'' traps against house mice infesting premises. *Frustula Entomologica* (n.s.) 7–8(20–21)669–683.

256 Saunders, G. 1986. Plagues of the house mouse in south eastern Australia. In *Proceedings of the Twelfth Vertebrate Pest Conference* (San Diego,

CA, Mar. 4), T. P. Salmon, ed. University of California, Davis, pp. 173–176.

257 Schein, M. W., and H. Orgain. 1953. A preliminary analysis of garbage as food for the Norway rat. *American Journal of Tropical Medicine and Hygiene* 2(6)1117–1130.

258 Scientific Group. 1974. Ecology and control of rodents of public health importance. *World Health Organization Technical Report Series* 553:1–42.

259 Scott, H. G., and M. R. Borom. 1977. *Rodent-Borne Disease Control Through Rodent Stoppage.* DHEW Publ. No. (CDC)77–8343. Center for Disease Control, Atlanta, GA.

260 Seal, S. C., and L. M. Bhattacharji. 1961. Epidemiological studies on plague in Calcutta. II. The role of movement of rats in the spread of plague infection. *Indian Journal of Medical Research* 49(6)1008–1018.

261 Seal, S. C., S. C. Ghosal, and P. N. Bose. 1951. Notes on the bamboo flowering, migration of rats and their unusual mortality in certain districts of North Assam. *Calcutta Medical Journal* 48(5)149–155, (6)195–199.

262 Shao Pin Yo, R. E. Marsh, and T. P. Salmon. 1986. Correlation of two census methods (food consumption and gnawing evidence) for assessing Norway rat populations. Paper presented at Fifth Symposium on Vertebrate Pest Control and Management Materials, American Society for Testing and Materials, Philadelphia, PA.

263 Shenker, A. M. 1970. Preventive pest control. *Food Manufacture* 45(6)67–69.

264 Shepherd, D. S., and J. H. Greaves. 1984. A weather-resistant tracking board. In *Workshop Manual—Fundamentals of Commensal Rodent Control.* Eleventh Vertebrate Pest Conference (Sacramento, CA, Mar. 7). University of California, Davis, pp. 112–113.

265 Shumake, S. A., et al. 1982. Variables affecting ultrasound repellency in Philippine rats. *Journal of Wildlife Management* 46(1)148–155.

266 Shuyler, H. R. 1954. The development of baits for *Rattus norvegicus,* with special reference to initial acceptability. Doctoral Dissertation Series Publ. No. 9381, University Microfilms, Ann Arbor, MI.

267 Silver, J., and W. E. Crouch. 1930. *Rat Proofing Buildings and Premises.* Farmer's Bulletin No. 1638. U.S. Department of Agriculture, Washington, DC.

268 Smith, A., and N. G. Gratz. 1984. *Urban Vector and Rodent Control Services.* Rept. No. WHO/VBC/84.4, World Health Organization, Geneva, Switzerland. 10 pp.

269 Smith, C. R. 1984. Survey of rodenticide bait-related exposures reported by the Los Angeles Medical Association Regional Poison Information Center, July 1, 1982, to June 30, 1983.

EPA Hearings on Rodenticide Bait Stations, Mar. 5, Sacramento, CA.

270 Smith, J. C. 1981. Senses and communication. *Symposia of the Zoological Society of London* 47:367–393.

271 Smith, J. V., and R. Pal (eds.). 1972. *Vector Control in International Health.* World Health Organization, Geneva, Switzerland.

272 Snetsinger, R. 1976. *Diary of a Mad Planner.* Winchester, New York, NY.

273 Snetsinger, R. 1983. *The Ratcatcher's Child—The History of the Pest Control Industry.* Franzak & Foster, Cleveland, OH.

274 Southern, H. N. (ed.). 1954. *Control of Rats and Mice,* vol. 3. *House Mice.* Oxford University, New York, NY.

275 Southwick, C. H. 1969. Reproduction, mortality and growth of murid rodent populations. In *Indian Rodent Symposium* (Calcutta, 1966), K. L. Harris, ed. Johns Hopkins University, Calcutta, India, and U.S. Agency for International Development, New Delhi, India, pp. 152–176.

276 Southwick, C. H. 1972. *Ecology and the Quality of Our Environment.* Van Nostrand Reinhold, New York, NY.

277 Spaulding, S. R., and W. B. Jackson. 1983. Field methodology for evaluating rodenticide efficacy. In *Vertebrate Pest Control and Management Materials* (Fourth Symposium, ASTM STP 817), D. E. Kaukeinen, ed. American Society of Testing and Materials, Philadelphia, PA, pp. 183–198.

278 Spillett, J. J. 1968. *The Ecology of the Lesser Bandicoot-Rat in Calcutta.* Bombay Natural History Society and Johns Hopkins University Center for Medical Research and Training, Calcutta, India.

279 Spillett, J. J. 1969. Growth of three species of Calcutta rats. In *Indian Rodent Symposium* (Calcutta, 1966), K. L. Harris, ed. Johns Hopkins University, Calcutta, India, and U.S. Agency for International Development, New Delhi, India, pp. 177–196.

280 Stern, V. M., et al. 1959. The integrated control concept. *Hilgardia* 29(2)81–101.

281 Stinson, W. S. 1977. Recent developments in the sanitary design of food processing equipment. *Food Processing* 38(2)32–36, 38.

282 Strecker, R. L., and W. B. Jackson. 1962. Habitats and habits. In *Pacific Island Rat Ecology,* T. I. Storer, ed. Museum Bulletin 225. Bernice P. Bishop Museum, Honolulu, HI, pp. 64–73.

283 Sutton, R. L., et al. 1980. *The Key to Community Outreach—Utilization of New Professionals in Community Participation Programs.* Environmental Health Services, Department of Public Health, Philadelphia, PA.

284 Taylor, K. D., D. C. Drummond, and F. P.

Rowe. 1970. *Biology and Control of Rodents*. Pest Infestation Control Laboratory, Tolworth, England.

285 Telle, H. J. 1966. Beitrag zur Kenntnis der Verhaltensweise von Ratten, vergleichend dargestellt bei *Rattus norvegicus* und *Rattus rattus*. *Zeitschrift für Angewandte Zoologie* 53(2):129–196.

286 Temme, M. 1980. House mouse behavior in multiple-capture traps. *Pest Control* 48(3)16, 18, 19.

287 Thompson, E. G. 1984. The integrated pest management approach to food protection. *Cereal Foods World* 29(2)149–151.

288 Tigner, J. R. 1972. Seasonal food habits of *Rattus rattus mindanensis* (the Philippine ricefield rat) in central Luzon. Dissertation, University of Colorado, Boulder.

289 Timm, R. 1983. Description of active ingredients. In *Prevention and Control of Wildlife Damage*, R. M. Timm, ed. Great Plains Agricultural Council and University of Nebraska, Lincoln, pp. G31–G127.

290 Timm, R. 1983. Norway rats. In *Prevention and Control of Wildlife Damage*, R. M. Timm, ed. Great Plains Agricultural Council and University of Nebraska, Lincoln, pp. B95–B114.

291 Timm, R. M., and D. D. Fisher. 1986. An economic threshold model for house mouse damage to insulation. In *Proceedings of the Twelfth Vertebrate Pest Conference* (San Diego, CA, Mar. 4), T. P. Salmon, ed. University of California, Davis, pp. 237–241.

292 Tolworth Laboratory. 1980. *Control of Rats and Mice*. Ministry of Agriculture, Fisheries and Food, Tolworth, England.

293 Tyler, P. S., and R. A. Boxall. 1984. Post harvest loss reduction programmes: A decade of activities: What consequences? *Tropical Stored Products Information* 50:4–13.

294 Uchida, T. A. 1966. *Observations on the Monitor Lizard, Varanus indicus (Daudin), as a Rat Control Agent on Faluk, Western Caroline Islands*. Rept. No. WHO/EBL/66.64, World Health Organization, Geneva, Switzerland, 34 pp.

295 Valvano, A. E., and G. C. Mitchell (eds.). 1986. *Vertebrate Damage Control Research in Agriculture*. Annual Report 1985. Denver Wildlife Research Center, Denver, CO.

296 Vashkov, V. O., and V. G. Polezhaev. 1965. *Investigations in the USSR Concerning the Biological Control of Rodents*. Rept. No. WHO/EBL/65.30, World Health Organization, Geneva, Switzerland, 4 pp.

297 Vick, F., and K. Becker. 1951. Der bautechnische Rattenschutz. *Schädlingsbekämpfung* 43(1)1–7, 43; (2)25–36.

298 Votroubek, L. C. 1983. VotRatGard. *Pest Control* 51(7)41 [advertisement].

299 Walter, V. 1982. Justifying warehouse perimeter strip. *Pest Control* 50(10)46.

300 Walter, V. 1982. The right snap trap for rats. *Pest Control* 50(12)75.

301 Walter, V. 1986. Pest control in plants under USDA. *Pest Control* 54(10)104.

302 Walton, D. W., et al. 1977. The status of *Rattus norvegicus* in Rangoon, Burma. *Japanese Journal of Sanitary Zoology* 28(4)363–366.

303 Weir, D. 1985. The global pesticide threat. *Multinational Monitor* 6(13)8, 9.

304 Westphal, K. (ed.). 1977. *Proceedings of: The 1977 Seminar on Cockroach Control* (New York, NY, Mar. 28–31). New York State Department of Health, Albany, NY.

305 Whitney, G., M. D. Stockton, and E. F. Tilson. 1971. Possible social functions of ultrasounds produced by adult mice (*Mus musculus*). *American Zoologist* 11(4)634.

306 Williams, S. (ed.). 1984. *Official Methods of Analysis*. Association of Official Analytical Chemists, Arlington, VA. [Note: See 15th edition for latest information.]

307 Wodzicki, K. 1973. Prospects for biological control of rodent populations. *Bulletin of the World Health Organization* 48(4)461–467.

308 Wodzicki, K. 1978–79. Relationships between rats and man in the Central Pacific. *Ethnomedicine* 5(3–4)433–446.

309 Wolcott, R. M., and B. W. Vincent. 1975. *The Relationship of Solid Waste Storage Practices in the Inner City to the Incidence of Rat Infestation and Fires*. EPA/530/5W/150. Environmental Protection Agency, Washington, DC.

310 Young, H., R. L. Strecker, and J. T. Emlen, Jr. 1950. Localization of activity in two indoor populations of house mice, *Mus musculus*. *Journal of Mammalogy* 31(4)403–410.

311 Zdunowski, G. 1980. Environmental manipulation in roof rat control programs. In *Proceedings of the Ninth Vertebrate Pest Conference* (University of California, Davis), J. P. Clark, ed., pp. 74–79.

Appendix 18.1. Equipment and Supplies

This section provides the names and manufacturers of various products mentioned in the text. This listing does not represent endorsement or recommendation by the authors, by the New York State Department of Health, or by the U.S. Food and Drug Administration; other products of comparable quality may be available in each case.

Bait Stations

1. High-Profile Bait Station (suitable for census-bait water jar)

Rat Cafeteria and "Jr." Rat Cafe—for mice or rats
Solvit, Inc.
7001 Raywood Road
Madison, WI 53713

2. Tamper-Resistant Bait Stations

Eaton's Plastic Tamper Proof Bait Stations—for mice or rats
J. T. Eaton and Co., Inc.
1393 East Highland Road
Twinsburg, OH 44087

Maj-ik-Box—for mice
National Institute of Pest Management
2323 Brookwood Drive
Cape Girardeau, MO 63701

Easy-Baiter/Mouseteria—for rats and mice
Sherman Technology Corporation
P.O. Box 691773
Los Angeles, CA 90069

Traps

1. Standard (& Expanded Trigger) Snap Traps—for mice or rats

Snap-E Snap Traps
Kness Manufacturing Company, Inc.
P.O. Box 70
Albia, IA 52531

Victor and Holdfast Snap Traps
Woodstream Corporation
Front and Locust Streets
Lititz, PA 17543

2. Clothespin-type—for mice

Trapper Mouse Trap
Hadley Products Company
Marietta, OH 45750

3. Cage-Type Live Traps (rigid or folding models)—for rats

Kness Manufacturing Co., Inc. (address given above)

Tomahawk Live Trap Corporation
P.O. Box 323
Tomahawk, WI 54487

4. Sheetmetal Live Trap

H. B. Sherman Traps, Inc.—for mice or rats
P.O. Box 20267
Tallahassee, FL 32316

Tomahawk Live Trap Corporation—for mice (address given above)

5. Multiple-Capture Traps—for mice

Ketch-All Automatic Mouse Trap—for mice
Kness Manufacturing Company, Inc. (address given above)

Victor, Tin Cat—for mice
Woodstream Corporation (address given above)

Wonder Rat Trap—for rats
The Deccan Rat Traps Factory
Meharun (Dist. Jalgaon)
Pin - 425001, India

6. Ready-to-Use Glueboards (*bulk glue also available)—for mice and rats

Atlantic Paste and Glue Company, Inc.
4 Fifty-Third Street
Brooklyn, NY 11232

Bell Laboratories, Inc.
33699 Kinsman Boulevard
Madison, WI 53704

J. T. Eaton and Company, Inc.* (address given above)

Sherman Technology Corporation (address given above)

Southern Mill Creek Products Company, Inc.*
P.O. Box 1096
Tampa, FL 33601

Woodstream Corporation (address given above)

Refuse Collection Carts

1. Roll-Out Carts (for semi- or fully automated refuse collection systems)

Applied Products, Inc.
P.O. Box 5338
Statesville, NC 28677

Otto Industries, Inc.
Suite 270
2300 West Glades Road
Boca Raton, FL 33431

Schaefer Systems International
Suite 200
5 Revere Drive
Northbrook, IL 60062

Zarn, Inc.
P.O. Box 1350
Reidsville, NC 27320

Electronic Equipment

1. Sound-Level Meter

Bruel & Kjaer Instruments, Inc.
185 Forest Street
Marlborough, MA 01752

Appendix 18.2 Training and Educational Materials

Europe

Ministry of Agriculture, Fisheries and Food, Tolworth Laboratory, Tolworth KT6 7NF, England.

Statens Skadedyrlaboratorium (Pest Infestation Laboratory), Skovbrynet 14, DK-2800 Lyngby, Denmark.

Overseas Development Natural Resources Institute, Storage Department, Slough SL3 7HL, England.

United Nations Food and Agriculture Organization, Plant Protective Service, Via delle Terme di Caracalla, 00100 Rome, Italy.

United Nations World Health Organization, Division of Vector Biology and Control, 1211 Geneva 27, Switzerland.

India

Central Arid Zone Research Institute, Jodhpur 342 003, India.

Central Food Technological Research Institute, Infestation Control and Pesticides, Mysore 570 013, India.

United States

Agency for International Development, Washington, DC 20523.

Bio-Integral Resource Center, P.O. Box 7414, Berkeley, CA 94707.

Centers for Disease Control, P.O. Box 2087, Fort Collins, CO 80522.

Denver Wildlife Research Center, USDA, P.O. Box 25266, Denver, CO 80225.

Food and Drug Administration, Center for Food Safety and Applied Nutrition, 200 C Street SW, Washington, DC 20204.

National Pest Control Association, 8100 Oak Street, Dunn Loring, VA 22027.

World Bank, 1818 H Street NW, Washington, DC 20433.

19

Commentary on Microbial and Invertebrate Pest Ecology

David W. Hagstrum

Chapters 5 through 15 on the ecology of microbial and invertebrate pests are of particular importance because, as suggested by Ebeling (Chapter 8), pest management is to a large extent an exercise in applied ecology. Understanding the ecology of each pest species is thus essential to effective planning, implementation, and monitoring of pest management programs. Ecology is basically the study of the distribution and abundance of plants and animals and the factors influencing their distribution and abundance. In the food industry, knowledge of factors determining seasonal changes in pest abundance allows us to anticipate *when* control will be needed, and knowledge of factors affecting pest distribution allows us to anticipate *where* control will be needed.

This chapter examines ways in which we can improve the information base for the most common pest species described in Chapters 5 through 15 and make this information base more relevant to pest management programs. For example, laboratory studies need to be designed to provide insight into the response of pests to actual conditions encountered in the food industry. To do this, we need to better characterize the food-industry environment and the distribution of pests in this environment. Comparative, quantitative reexamination of some of this information will allow us to more fully utilize available data. Ultimately, we need to undertake multidiciplinary, comprehensive studies of the food-industry ecosystem, as suggested by Mills (Chapter 5). A description of data available

for the wheat-flour sector of the food industry illustrates how a foundation is being laid for such studies.

Life History and Behavior

A comprehensive data base on the life histories and behaviors of food-industry pests is absolutely essential to the rational development of sound pest management programs, but the necessary information is often lacking, incomplete, inappropriately collected, or in need of reanalysis. Chapters 5 through 15 illustrate many of the ways in which environmental factors such as temperature, moisture, food, and natural enemies influence distribution and abundance through their effects on life-history traits (developmental times, survival, egg production, adult longevity, and diapause) and behavior (mating, oviposition, feeding, dispersal, and habitat preference). Given the large number of species inhabiting food-industry ecosystems and the diversity of environmental factors, life-history traits, and behaviors, it is not surprising that even with the many references cited in Chapters 5 through 15, all of this information is not available for any one species and many gaps in our information need to be filled.

Some data just need to be reanalyzed in a more comparative way. One example of such a comparative reanalysis is the review of energy budgets for 10 species of stored-product insects (4). Energy budgets can help us understand why some species develop faster than others and can provide a means of predicting differences in damage caused by the feeding of different species. However, to do this, we need to develop energy budgets over a range of temperatures, moistures, and diets.

A good example of comparative, quantitative rean-

Hagstrum: U.S. Grain Marketing Research Laboratory, Agricultural Research Service, U.S. Department of Agriculture, 1515 College Avenue, Manhattan, KS 66502.

alysis is one on temperature, moisture, and diet as factors affecting the developmental time of nine species of stored-grain Coleoptera (16). Many of the following generalizations about how temperature, moisture, and diet interact to determine developmental time would not have been evident without extensive quantitative reanalysis of the data. Regression equations were developed to provide a simple means to calculate developmental time in insect population models or to make quantitative comparisons of the effects of temperatures on the developmental times of different species and stages. Because temperature affects population trends through its effect on developmental time, egg production, and survival, such comparisons are extremely important in understanding and predicting differences in the population trends of different species in the same habitat.

Although enough data were available to model developmental times of the egg stage of nine species, adequate data were available to model the developmental times of the larval and pupal stages of only six species. Sufficient data were available on the effects of moisture on only four species. Because so many diets were used, diet was considered only as part of the residual variance not explained in the regression models by differences in temperature or moisture.

The tendency for moisture to affect developmental times of larvae, but not eggs or pupae, indicates that moisture may be important in altering either the rate of feeding or the efficiency with which feeding larvae can assimilate and convert their diet. Overall, the duration of the larval stage represents a larger proportion of the total developmental time than other stages. Egg, larval, and pupal stage durations at 27° averaged 15%, 66%, and 19% of total developmental times, respectively, at moistures above 12%, and 12%, 72%, and 15% of total developmental time at moistures below 12%. Similar percentages were found at other temperatures between 20° and 35°. Since the duration of a stage generally determines the number of insects in that stage when a population reaches a stable age distribution, these constant percentages for a particular stage at different temperatures indicate that the stable age distribution reached will be the same throughout this temperature range. However, because the percentages change with moisture level, the stable age distribution will vary with moisture.

At most temperatures, the order of relative influence of those factors on development was temperature > moisture > diet. However, moisture and diet influenced larval development more than temperature near the optimal temperature for development of each species. This example also points out how important the joint effects of more than one environmental factor can be in determining population dynamics.

Computer models can help us to more fully utilize available data in pest management programs. For four species, there was also sufficient information on the effects of temperature and moisture on egg production to develop computer simulation models that predict insect population trends from life history data (17, 21). Studies of *Cryptolestes ferrugineus*, *Rhyzopertha dominica*, and *Tribolium castaneum* populations in farm-stored wheat have shown that the relationship between environmental factors (temperature and moisture) and life history traits (developmental time and egg production) predicted 87%, 93%, and 96% of seasonal changes in insect density, respectively (17). Population growth rates of 12 species of *Aspergillus* and 3 species of *Penicillium*, compared over a broad range of temperatures and moistures, also were shown to differ markedly between species (1). Tolerance to low moisture was greatest close to optimal temperature.

Previous chapters in this volume give many examples of how the behavior of pest species can influence pest management programs. In this chapter, examples are given of comparative, quantitative reanalysis and the design of laboratory experiments to mimic the food-industry environment as a means of facilitating fuller utilization of behavioral data in designing pest management programs. Species interactions of stored-product Coleoptera, discussed by Arbogast in Chapter 10 and Burkholder and Faustini in Chapter 26, were reviewed earlier by Solomon (28). From a comparison of the discussions of natural enemies in these reviews, it is clear that considerable progress has been made. Ebeling in Chapter 8 also cites more significant progress in developing the use of natural enemies than is cited in the other chapters. Gordh and Hartman in Chapter 15 provided an overview of the species of hymenopterous parasites that attack insect pests of stored food.

Crowding can be a factor of major importance in population dynamics. Solomon's paper (28) also is an excellent example of comparative, quantitative reanalysis of the effects of crowding on mortality and egg production of stored-product insects. The effects of crowding increased logarithmically with population density. With a graph of eight species, he showed that mortality reached 20% at densities of 1–32 insects/g of infested material and 80% at densities of 7–130. For many of the same species, egg production was reduced 20% at densities of 0.3 to 16 insects/g and 80% reductions were observed above 3 insects/g. Overall, reproduction tended to be depressed more than mortality at a given density. He recognized that only the highest estimates of insect populations in storage were within the range in which the effects of crowding would be significant. However, he also distinguished between local densities and general densities throughout the whole bulk.

Crowding becomes important in "natural" food-industry environments when emigration or scattered food resources are limited. However, many studies on crowding have been done under unnatural conditions, which means that studies under conditions more like those encountered in the food industry are needed. Studies of the population dynamics of *T. castaneum* living on a small amount of flour in a shell vial, which restricts emigration, overemphasized the importance of cannibalism (*14*). When food is more abundant and emigration is possible, age-specific schedules of emigration are more important than cannibalism. In 60 g of flour over a 130-day period, beetles have a mean residence time of 11 days prior to emigration. Only 3.3% emigrated during the 7-day period between the eclosion of adults and their first oviposition. Thus, most females laid some eggs prior to emigration, but these eggs represented only a fraction of their fecundity and they probably spread their eggs among many locations. An additional 12% emigrated by the 8th day, and emigration was 23%/day thereafter. With densities between 1 and 15 beetles/g of flour, emigration was independent of density. Before crowding becomes important locally as a result of population growth, beetles have probably established populations at many other locations.

Another food-industry pest also has been studied under relatively natural conditions. When empty peanut storage warehouse conditions were simulated by spreading constant, low numbers of peanuts among several locations, resource distribution did not affect the number of *Ephestia cautella* eggs laid, but did reduce the number of larvae that completed development (*9*). Population growth was reduced from roughly seven- to threefold as the number of locations increased from 1 to 24. The mean number of eggs laid at a location was proportional to the number of peanuts, and larval production declined mainly as a result of females laying eggs at a smaller fraction of locations. Reduced population growth was a consequence of the population not conforming to resource distribution, with overcrowding at some locations and underutilization at others. The fact that *E. cautella* did not find all of the scattered locations with larval food in a single generation means that some of the resource was available for more gradual population growth over several generations.

These studies of *T. castaneum* and *E. cautella* illustrate how laboratory experiments can be designed to tell us more about insects in the "natural" food-industry environment and how extrapolation from these laboratory studies to this environment may require a fairly detailed characterization of that environment. The discussions of the influence of abiotic factors and crowding emphasize the importance of putting laboratory work in perspective of the actual food-industry environment.

Population Dynamics

Bin management, farm management, and marketing-channel management models for the wheat-flour industry illustrate on three different scales how we can more fully utilize available data in planning, implementing, and monitoring pest management programs. Such models would probably be useful for other pest species and sectors of the food industry. The wheat-flour sector of the food industry is used as an example to illustrate population dynamics, although ants, cockroaches, and flies are perhaps less of a concern in this sector than other sectors, because this sector has clearly taken the lead in developing the quantitative ecological data relevant to pest management.

The timing of insect population growth, the distribution of insect populations, and factors that affect population growth and distribution must be known in order to develop rational pest management programs, since they determine when and where control is needed. For farm-stored wheat, the factors affecting distribution and abundance of pests have been studied in some detail for one species. Infestation by *C. ferrugineus* of newly harvested wheat in Kansas apparently occurred after the wheat was loaded into the farm bin. Dispersal of insects into the grain mass from the surface generally resulted in a logarithmic decrease in insect abundance from the top to the bottom of the bin (*13*).

The population growth rate of *C. ferrugineus* in the top meter of wheat declined over the storage period as a result of, first, the parasite, *Cephalonomia waterstoni,* and then falling temperatures (*12*). Adult *Cryptolestes ferrugineus* showed only a slight tendency to move toward the warmer or moister parts of the bin in response to seasonal changes in environment. Temperatures were uniform and within a favorable range of 27–34° for the first 12 weeks and then declined at rates ranging from 1.3° to 2°/week. Cooling of the grain from the outside and surface inward and downward resulted in temperature gradients.

Grain moisture in the top 0.5 m declined during the first 12 weeks and increased to levels well above initial moisture. Elsewhere in the top meter, wheat retained its initial moisture level except for a tendency of grain near the wall to be drier. Lesser grain borers (*R. dominica*) were first detected after 12 weeks and were found only in the center. The infestation pattern and environmental gradients result in differences in population densities and growth rates in different parts of the grain mass.

Insect population growth rates in stored commodities have been reported for several species of insects and types of storage facilities. Hagstrum (*10*), reviewing the ecology of *E. cautella,* showed that the growth rates varied from 2- to 37-fold per month on maize, peanuts, and citrus pulp. White (*33*) reported average growth rates of 5- and 10-fold per month for *R. dominica* and *T. castaneum,* respectively, on wheat in flat storage. Hagstrum (*17*) observed similar growth rates on wheat in farm bins. With a 10-fold growth rate per month, it would take only a month for a population to again reach a particular density after 90% control and only 2 months after 99% control.

Knowledge of growth rates of these and other food-industry pests can thus help us to know when we need to again sample for pests or apply control measures. Computer simulation models are thus a fundamental tool in the development of pest management programs. Computer simulation models of insect population growth rates have been used to compare different management programs such as bin aeration (*31*) or bin aeration and insecticides (*20*). Another model simulates the effects of management programs on farm populations of stored-product insects inside and outside the grain bin (*24*). A model that simulates the effects of wheat temperature, moisture, the flow of grain, and insect control in the wheat-marketing system in the United States (USA) explained 96.5% of the month-to-month variation in *R. dominica* populations at export (*15*).

Certain aspects of pest ecology in marketing channels have been studied in some detail. Insect infestation of grain being moved from elevators to railroad cars in Canada increased steadily from May to October or November (*26*). *C. ferrugineus* was found in 38% and 53% of the lots loaded during 2 years. Grain loaded into half of the cars contained only larvae. Mites were found in grain loaded in 78% and 90% of the cars during the 2-year study. No more than 6% of the infested carloads found at the primary elevator were detected by inspectors at the terminal elevators. Less than 5% of the carloads of grain delivered directly from the farm contained *C. ferrugineus.* Loschiavo (*22*) reviewed the literature on insects in boxcars and, using probe traps in-transit, showed that 18–20% of Canadian boxcars carrying wheat were infested with *C. ferrugineus.* Several species of fungus beetles tended to be more common than *C. ferrugineus.* Wheat had moved from farms through the elevator system and reached the flour mill by railroad.

In English flour mills, populations of the two most common species, *Cryptolestes turcicus* and *Ephestia* (*Anagasta*) *kuehniella,* increased in abundance from March or April until the annual fumigation in June, July, or August (*5*). The numbers of *C. turcicus* in eight centrifugals were fairly constant during a 5-year period, with population size probably dependent upon molds that develop in damp flour residues (*6*).

Only survey information is available on Diptera (*36*) and Psocoptera (*23*) in Czechoslovakian mills. In Canada, flour in more heavily infested mills had 7.5 times as many microscopic insect fragments as the average for others and had 5.5 times more bacteria (*30*). Fungi in flour were correlated with fungi on the whole grain, and fungi in boots were correlated with insect fragments in the boots. In 11 flour mills in the USA, the counts for total bacteria ranged from 220 to 20,000/g of flour and from 15,000 to 660,000/g of wheat (*8*). Actinomycete counts ranged from 0 to 5,300/g of flour and from 0 to 300/g of wheat.

In addition to considering the flow of a commodity through the marketing channels in making management decisions, preharvest infestation and immigration from outside the food industry must also be considered. Cowpeas from 49 fields had an average density of 2.33 *Callosobruchus maculatus* adults emerging per bushel per 2 weeks of storage (*11*). From these initial infestation levels, storage populations either slowly dwindled or after 18 weeks of cool winter weather increased exponentially. The suitability of cowpea pods for *C. maculatus* oviposition varied significantly among genotypes and among three pod maturity stages (*7*).

Studies on field infestation of corn by *Sitophilus zeamais* have been reviewed (*10*). This species is also attracted to sunflowers (*34*). The moth, *Sitotroga cerealella,* also infests several grains prior to harvest (references in *11*).

The dispersiveness and widespread distribution of food-industry pests that result in field infestation of stored commodities also can result in the initial infestation of clean commodities and the reinfestation of commodities after control. To predict when and where infestations will occur, the dispersal behaviors and habitat preferences of these pest species must be understood. *S. cerealella* has been captured in pheromone traps in pastures and forest areas remote from grain fields or storages in Texas (*3*). The numbers captured decreased logarithmically with distance from an infested storage facility. From May to November, the captures near rice fields averaged 10–20 insects/trap per week, and no moths were captured during the winter. Captures inside and outside of rice stores were higher than near rice fields. In Florida, *Plodia interpunctella* and *E. cautella* were caught most frequently near peanut storage facilities (*32*). *S. cerealella* was most abundant around a wheat field. These three species were most abundant during warm months and absent in February, whereas *Ephestia kuehniella* was more abundant in January, February, and March and absent during the summer. All four species were captured in peanut warehouse yards, pastures, cornfields, feed-mill yards, and around a wheat field. In

Germany, outdoors in the city, four species of moths were trapped (35); in order of abundance, these were *P. interpunctella*, *Ephestia kuehniella*, *E. elutella*, and *E. cautella*. On the outskirts of cities and in the rural countryside, *E. cautella* was not found in traps.

In California, over 50 species of stored-product insects were trapped with food packets in nonstorage facilities such as carports, garages, equipment sheds, inactive dairy barns and poultry houses, old livestock barns, and other structures not being used to store dry food products, feed, or seed (29). In Australia, *T. castaneum* and *R. dominica* were active in the field from October to May, with an average density of 52 per million cubic meters (25). In studies of moth populations in raisins moving through marketing channels, *P. interpunctella* was found in low numbers and *Ephestia figulilella* was rarely found inside a raisin plant, although *E. figulilella* was common outside the plant (27). In all the van container loads of raisins shipped and at the receiving warehouses, only *E. cautella*, which was not found in any California raisin plant, was trapped. Studies such as these on dispersal behavior and habitat preference also are available for cockroaches (Chapter 8) and flies (Chapter 13).

Sufficient information is available to piece together an overview of pest population dynamics in the wheat-flour sector of the food industry, but this should be viewed only as a prelude to planning more comprehensive studies. The studies used in this synthesis were done over four decades during which many industrial practices have changed and over a wide geographical area with vastly different climates. Such a synthesis can, however, allow us to more fully utilize available information until more holistic studies can be conducted.

This discussion clearly shows that seasonal change in temperature is one of the most important factors determining abundance, and it emphasizes the interrelatedness of the population dynamics of different species such as insects and microbes. Given the dispersiveness of most food-industry pests, we might speculate that for most pests the time required for reinfestation will be proportional to the area considered in a management program. Brenner addresses this issue in Chapter 9 on the Asian cockroach, suggesting that control measures may be more effective in adjacent areas where property owners have no jurisdiction. Beatson-Campbell pursues this theme in Chapter 14 on ants, suggesting that spot treatments where ants are seen are ineffective and that a sufficiently large area must be treated.

Sampling and Threshold

In managing food-industry pests, detection is the first step in discovering a problem, and estimation of acutal population levels is the final step in determining

that a problem has been solved. Without an estimate of the pest density, there can be no reasonable safeguard against either overtreatment with insecticides or unacceptable levels of infestation. In the food industry, acceptable threshold infestation levels have not been defined precisely as in, for example, crop protection. Defect action levels, consumer complaints, and failed plant inspections serve as rough indicators of past pest-management failures, but pest management needs to anticipate such problems through continuous sampling, accurate estimates of pest abundance, and determination of reasonable threshold levels at which pests must be managed to prevent problems. Ebeling (Chapter 8) and Brenner (Chapter 9) mention aesthetic thresholds developed for cockroaches.

The accuracy of estimation and the probability of detection have been described for stored-product insects in relation to population density and sampling effort (18). With accuracy (reliability) expressed as a percentage of the mean insect density, each 10-fold increase in sampling effort increases the accuracy of estimation only threefold. This means that the decision to reduce by threefold the threshold level at which insect populations are managed can require a 10-fold increase in the expenditures for sampling.

With a sampling intensity of five 0.5-kg samples/1,000 bushels of wheat, mean insect densities are measured with accuracies of 0.1 ± 0.3, 1 ± 1, 10 ± 6, and 100 ± 30 insects/0.5-kg sample. At the sampling rate of 0.001% of wheat examined for insects, the calculated probabilities of detecting 0.1 and 1 insect/0.5-kg sample of wheat were 3% and 50%, respectively. With a 10-fold increase in sampling effort, the calculated probability of detecting densities of 1 insect/0.5 kg of wheat increased from 50 to 95%. With a 100-fold increase in sampling effort, the calculated probability of detecting densities of 0.1 insect/0.5 kg of wheat increased from 3 to 95%.

Quantitative comparison of sampling data from a variety of stages and species of insects, sampling devices, locations in the marketing system, times during the storage period, and types of grain suggest that these relationships between accuracy of estimation of insect density or probability of detection, population density, and sampling effort may apply to diverse populations of stored-product insects. This list has subsequently been extended to include acoustic detection of insects (19) and insect populations in empty bins (2). Such studies allow us to anticipate the sampling effort that is needed to attain the desired accuracy of estimation or probability of detection at our chosen threshold density for management decisions and provide another example of more fully utilizing available information. Sampling studies such as these would probably be beneficial to other sectors of the food industry.

Conclusion

This chapter suggests that, in the short term, pest management can be improved by more fully utilizing available information to anticipate changes in pest populations and by developing management programs based upon realistic sampling efforts and threshold density levels to indicate when management is required. However, in the long term, more comprehensive, quantitative, holistic studies of food-industry ecosystems are needed.

Cited References

1 Ayerst, G. 1969. The effects of moisture and temperature on growth and spore germination in some fungi. *Journal of Stored Products Research* 5(2)127–141.

2 Barker, P. S., and L. B. Smith. 1987. Spatial distribution of insect species in granary residues in the Prairie Provinces. *Canadian Entomologist* 119(12)1123–1130.

3 Cogburn, R. R., and K. W. Vick. 1981. Distribution of Angoumois grain moth, almond moth, and Indian meal moth in rice fields and rice storages in Texas as indicated by pheromone-baited adhesive traps. *Environmental Entomology* 10(6)1003–1007.

4 Demianyk, C. J., and R. N. Sinha. 1988. Bioenergetics of the larger grain borer, *Prostephanus truncatus* (Horn) (Coleoptera: Bostrichidae), feeding on corn. *Annals of the Entomological Society of America* 81(3)449–459.

5 Dyte, C. E. 1965. Studies on insect infestations in the machinery of three English flour mills in relation to seasonal temperature changes. *Journal of Stored Products Research* 1(2)129–144.

6 Dyte, C. E. 1966. Studies on the abundance of *Cryptolestes turcicus* (Grouv.) (Coleoptera, Cucujidae) in different machines of an English four mill. *Journal of Stored Products Research* 1(4)341–352.

7 Fitzner, M. S., et al. 1985. Genotypic diversity in the suitability of cowpea (Rosales: Leguminosae) pods and seeds for cowpea weevil (Coleoptera: Bruchidae) oviposition and development. *Journal of Economic Entomology* 78(4)806–810.

8 Graves, R. R., et al. 1967. Bacterial and actinomycete flora of Kansas-Nebraska and Pacific Northwest wheat and wheat flour. *Cereal Chemistry* 44(3)288–299.

9 Hagstrum, D. W. 1984. Growth of *Ephestia cautella* (Walker) population under conditions found in an empty peanut warehouse and response to variations in the distribution of larval food. *Environmental Entomology* 13(1)171–174.

10 Hagstrum, D. W. 1984. The population dynamics of stored-products insect pests. *Proceedings of the Third International Working Conference on Stored-Product Entomology* (Manhattan, KS, 1983), Kansas State University, Manhattan, pp. 10–19.

11 Hagstrum, D. W. 1985. Preharvest infestation of cowpeas by the cowpea weevil (Coleoptera: Bruchidae) and population trends during storage in Florida. *Journal of Economic Entomology* 78(2)358–361.

12 Hagstrum, D. W. 1987. Seasonal variation of stored wheat environment and insect populations. *Environmental Entomology* 16(1)77–83.

13 Hagstrum, D. W. 1989. Infestation by *Cryptolestes ferrugineus* (Coleoptera: Cucujidae) of newly harvested wheat stored on three Kansas farms. *Journal of Economic Entomology* 82(2)655–659.

14 Hagstrum, D. W., and E. E. Gilbert. 1976. Emigration rate and age structure dynamics of *Tribolium castaneum* populations during growth phase of a colonizing episode. *Environmental Entomology* 5(3)445–448.

15 Hagstrum, D. W., and W. G. Heid, Jr. 1988. U. S. wheat-marketing system: an insect ecosystem. *Bulletin of the Entomological Society of America* 34(1)33–36.

16 Hagstrum, D. W., and G. A. Milliken. 1988. Quantitative analysis of temperature, moisture, and diet factors affecting insect development. *Annals of the Entomological Society of America* 81(4)539–546.

17 Hagstrum, D. W., and J. E. Throne. 1989. Predictability of stored-wheat insect population trends from life history traits. *Environmental Entomology* 18(4)660–664.

18 Hagstrum, D. W., R. L. Meagher, and L. B. Smith. 1988. Sampling statistics and detection or estimation of diverse populations of stored-product insects. *Environmental Entomology* 17(2)377–380.

19 Hagstrum, D. W., J. C. Webb, and K. W. Vick. 1988. Acoustical detection and estimation of *Rhyzopertha dominica* (F.) larval populations in stored wheat. *Florida Entomologist* 71(4)441–447.

20 Longstaff, B. C. 1988. A modelling study of the effects of temperature manipulation upon the control of *Sitophilus oryzae* (Coleoptera: Curculionidae) by insecticide. *Journal of Applied Ecology* 25(1)163–175.

21 Longstaff, B. C., and W. R. Cuff. 1984. An ecosystem model of the infestation of stored wheat by *Sitophilus oryzae*: A reappraisal. *Ecological Modelling* 25(1–3)97–119.

22 Loschiavo, S. R. [1975.] The detection of insects by traps in grain-filled boxcars during transit. *Proceedings of the First International Working Conference on Stored-Product Entomology* (Savannah, GA, 1974), Stored-Product Insects Research and Development Laboratory, Savannah, GA, pp. 639–650.

23 Obr, S. 1978. Psocoptera of food-processing plants and storages, dwellings and collections of natural objects in Czechoslovakia. *Acta Entomologica Bohemoslovaca* 75(4)226–242.

24 Sinclair, E. R., and J. Alder. 1985. Development of a computer simulation model of stored product insect populations on grain farms. *Agricultural Systems* 18(2)95–113.

25 Sinclair, E. R., and R. L. Haddrell. 1985. Flight of stored products beetles over a grain farming area in southern Queensland. *Journal of the Australian Entomological Society* 24(1)9–15.

26 Smith, L. B. 1985. Insect infestation in grain loaded in railroad cars at primary elevators in southern Manitoba, Canada. *Journal of Economic Entomology* 78(3)531–534.

27 Soderstrom, E. L., et al. 1987. Detecting adult Phycitinae (Lepidoptera: Pyralidae) infestations in a raisin-marketing channel. *Journal of Economic Entomology* 80(6)1229–1232.

28 Solomon, M. E. 1953. The population dynamics of storage pests. *Transactions. IX International Congress of Entomology* 2:235–248.

29 Strong, R. G. 1970. Distribution and relative abundance of stored-product insects in California: A method of obtaining sample populations. *Journal of Economic Entomology* 63(2)591–596.

30 Thatcher, F. S., C. Coutu, and F. Stevens. 1953. The sanitation of Canadian flour mills and its relationship to the microbial content of flour. *Cereal Chemistry* 30(2)71–102.

31 Thorpe, G. R., W. R. Cuff, and B. C. Longstaff. 1982. Control of *Sitophilus oryzae* infestation of stored wheat: An ecosystem model of the use of aeration. *Ecological Modelling* 15(4)331–351.

32 Vick, K. W., J. A. Coffelt, and W. A. Weaver. 1987. Presence of four species of stored-product moths in storage and field situations in north-central Florida as determined with sex pheromone-baited traps. *Florida Entomologist* 70(4)488–492.

33 White, G. G. 1988. Field estimates of population growth rates of *Tribolium castaneum* (Herbst) and *Rhyzopertha dominica* (F.) (Coleoptera: Tenebrionidae and Bostrichidae) in bulk wheat. *Journal of Stored Products Research* 24(1)13–22.

34 Williams, R. N., and E. H. Floyd. 1971. The maize weevil in relation to sunflower. *Journal of Economic Entomology* 64(1)186–187.

35 Wohlgemuth, R., et al. 1987. The occurrence of pest moths (*Ephestia* spp. and *Plodia* sp.) outside of warehouses and food-processing factories in Germany. *Anzeiger für Schädlingskunde, Pflanzenschutz, Umweltschutz* 60(3)44–51 (in German; English summary).

36 Zuska, J., and P. Laštovka. 1969. Species composition of the dipterous fauna in various types of food-processing plants in Czechoslovakia. *Acta Entomologica Bohemoslovaca* 66(4)201–221.

20

Commentary on Vertebrate Pest Ecology

Robert M. Timm

Vertebrates offer perhaps a greater diversity of problems than do any other category of pests, not only in terms of species biology but also damage situations. Conflicts arising when birds, bats, or rodents reside in and around food storage and manufacturing establishments are more restricted, but these damage situations and their solutions often are not simple. Restrictions on toxicant use in proximity to food storage and handling further complicate control strategies.

A common theme throughout the previous chapters is the need for pest control advisors and technicians to have a broad, up-to-date, working knowledge of the pest's biology. A holistic view is needed if the person charged with solving the problem is to develop the optimum solution in terms of efficacy, cost, acceptability, and timeliness. For birds, Jackson (Chapter 16) emphasizes the need to understand their behavior and "think like a bird" in order to develop a control strategy. Constantine (Chapter 17) cautions against unreasoned fear of bats and suggests that development of the best control plans necessitates understanding their habits and behaviors. Frantz and Davis (Chapter 18) stress the necessity of an integrated approach when dealing the commensal rodents, rather than a primarily rodenticidal approach which fails to consider the factors that have led to the problem's development.

Given our excellent present knowledge of the biology, physiology, behavior, and ecology of vertebrate pests, why are efforts to control their damage often poorly planned and ineffective? A principal reason is that in the application of this knowledge, practitioners often fail to consider the human elements of the problem. Control failures are often management problems: failure to recognize and act on the sociological and psychological aspects of pest situations. We need to be more fully aware of the cultural and motivational factors that cause people to take effective action. This is particularly critical when planning vertebrate control strategies in foreign countries.

A further limitation on effective vertebrate control programs is the lack of personnel who are broadly educated in this subject. Although there is more interest in wildlife damage control among undergraduate students in the USA today than there was a decade ago, fewer than 10 colleges or universities nationwide currently offer a course in this subject. Further, the lack of graduate degree programs in this specialty continues to thwart the growth and development of the vertebrate pest control field. As a result, students may be exposed to this subject only in the content of other courses and are likely to be given information that is biased or outdated. People wishing to foster the growth of the discipline of wildlife damage control must work cooperatively to promote new undergraduate courses, as well as continuing education classes for practitioners, in this subject. Only by increasing the number of people who understand the unique and complicated nature of vertebrate pest problems will this field mature.

The future is likely to bring new challenges and new solutions to bird, bat, and rodent problems. Skills in biotechnology and the manufacture of monoclonal antibodies offer the potential for creating toxicants that will affect only target species, representing no hazard to humans or other nontargets. There remains a continuing need for research on basic problems. The literature contains surprisingly little data, for example, on the acceptance by rats and mice of bait boxes of

Timm: Hopland Field Station, University of California, Hopland, CA 95449.

323

differing materials and designs. At the same time, increased concerns about animal welfare and animal rights may result in legislation that restricts our ability to use control methodologies that appear to cause pain or suffering to animals.

The future success of vertebrate pest control probably will turn out to be a direct function of our efforts to educate ourselves, our fellow professionals, and the public. We must make a concerted effort through continuing education to give today's practitioners a more broadly based, ecological view of birds and mammals within the environment. Further, we must devote more energy toward informing the public of the need for controlling damage caused by vertebrates. People need to have a better understanding of the methods and materials used in these efforts, as well as the associated risks and benefits of conducting control versus failing to do so. It is only with the support of the citizenry that we can continue to maintain the quality and efficiency of our system of food production aided by vertebrate pest control procedures that are both economically and environmentally sound.

III
Prevention

21

Preventive Aspects of Sanitation

Darrell F. Jones

One of Webster's definitions of sanitation is "the application of measures taken to make environmental conditions favorable to health." Such measures in the early history of food protection were limited to a very few techniques. Due to their limited knowledge of food protection, our forebears protected and preserved certain foods only by cooking, drying, and salting. These efforts prevented illness, death, and, at the very least, loss of food, all caused by spoilage microorganisms, as we know now, of course, but in those early days very little was understood about microbes. The limited and rudimentary measures they could rationally use against the larger pests (insects, rodents, and birds) were carried out more because such pests represented direct competition for food supplies than for any concern for health.

This situation changed dramatically as a result of the Industrial Revolution that began in the late 1800s and early 1900s, coincident with the beginnings of commercial food processing and mass distribution. Processors began to understand that success in food preservation depends largely on avoiding contamination in raw commodities, processing and storing foods under sanitary conditions, and protecting processed foods in transit.

Today we are aware that production, transportation, and storage of food in a sanitary environment depend primarily on taking measures against elements that lead to contamination. Such

measures cannot be fully effective if they are piecemeal. We must look to root causes of contamination, no matter where they are found in the food chain, before implementing preventive programs. To accomplish this, the validity of five very basic principles of sanitation must be understood and accepted:

1. There are general forms of potential pest contamination common to all food-handling facilities. These pests are rodents, birds, insects, mites, and various kinds of microbes.
2. All foods are subject to certain specific forms of contamination or potential contamination peculiar to their origin, class, and stage of processing.
3. There are certain points in the farm-to-table food chain, within each food-processing flow, and also within structures, where the probability of contamination can be identified, measured, and modified to some degree.
4. Expertise in the various aspects of food-product protection is mandatory for the selection and use of building materials, equipment, utensils, and other accoutrements essential to sanitary operations.
5. Company dedication to preventive sanitation and follow-up is required in order for the programs to work.

Principle 1

There are general forms of potential pest contamination common to all food-handling facilities. These pests are rodents, birds, insects, mites, and various kinds of microbes.

Despite the highly advanced state of pest control technology, pests and pathogens remain with us. Although the prevalence and species of pests

Jones (Deceased): Quality Control Department, General Mills, Inc., Minneapolis, MN 55440.

may vary from region to region, pests pose the same threat to all food processors, and for two primary reasons. First, because of their ability to adapt, they have survived the most adverse conditions and have become quite cosmopolitan. Second, the extended distribution systems and dramatically increased food production and storage capacities necessary to meet our growing population's demands have greatly increased opportunities for pest infestations. Since these forms of contamination are universal, it behooves us to learn all we can about pest habits in order to protect our foods against the inevitable invasions by pests.

Pests have two things in common. First, they all require food, moisture, habitat, time, and temperature for the completion of their life cycles. The removal of any one of these requirements drastically reduces their chances for survival. One may then look to the habits of each pest in relation to each food-handling facility to determine how to make life less tenable for the pest. Each pest has, for instance, rather specific requirements for harborage. Learning what those requirements are, and reducing the available habitats, minimizes the probability of pest invasions and the establishment of pest populations. Second, all pests enter food facilities either as volunteers or as captives. The pests that walk, crawl, or fly into our buildings are obviously the volunteers, whereas those that are carried in with foods or other materials are captives.

Armed with this knowledge, we are faced with the need for at least two types of programs. One program, aimed at the volunteers, dictates a full assessment and continuous monitoring of the grounds and building exteriors to seek out and eliminate harborages and food and water supplies. Coupled with this effort is a two-pronged program of (a) reducing potential pest entryways into the buildings, and (b) utilization of traps and repellents around normal entrances that must be left open for extended periods.

The second "must" program, aimed at captives, includes (a) close inspection of all incoming foods and materials, including vehicles, with a clearly defined plan of action to be taken with positive findings; and (b) continued surveillance of building interiors for signs of pests. In the case of microscopic pests such as pathogens and microbial indicators of filth, a system of prior approvals, guarantees, and preanalytical work is mandatory.

Employees may also be a source of macroscopic as well as microscopic contaminants. Certainly there are enough documented instances of employees carrying contamination into a facility on street shoes, employees with infectious disease contaminating foods, and countless violations of FDA's Good Manufacturing Practices regarding personal hygiene to substantiate this as an important avenue of insanitation. Such concerns dictate the need for effective orientation and training programs for wage and salaried personnel to ensure awareness and enforcement of proper standards of personal hygiene and factory sanitation.

Principle 2

All foods are subject to certain specific forms of contamination or potential contamination peculiar to their origin, class, and stage of processing.

One general guide to what these specific forms of potential contamination are may be found in Chapter 6 of the FDA *Inspection Operations Manual* (1). Here, food and drink are divided into different classes. Under "Establishment Inspections" one may determine the general classes of potential contamination peculiar or specific to various kinds of foods and commodities. Levels of contamination tolerated in many domestic and imported commodities and foods may be found in the DALs (Defect Action Levels) issued by the Food and Drug Administration (FDA).

Other literature is available to the processors that outlines levels of macroscopic and microscopic contamination tolerated in domestic and imported foods. However, it is incumbent on food processors to know what specific forms of contaminants may be found in particular ingredients, processes, and products. They should be guided more by what is realistically achievable than what is allowed because, to satisfy FDA, they must also do the best they can for their customers.

Certainly within the province of Principle 2 are classes of foods with an inherent potential for contamination. Examples are *Salmonella* in egg and dairy products; aflatoxins in nuts, peanuts, cottonseed, and corn grown in certain areas; and certain species of insects in cereal grains and fruits. Other considerations should be the weather conditions during crop growth that may have increased the potential for macroscopic and microscopic contamination, and commodity storage conditions that may have permitted viable forms of contamination to flourish or that allowed new forms of contamination to invade the commodity. One last point is that one must have knowledge of what might be produced in nearby facilities that could contaminate one's own commodities or products.

Principle 3

There are certain points in the farm-to-table food chain, within each food-processing flow, and also within structures, where the probability of contamination can be identified, measured, and modified to some degree.

Our awareness of potential impurities in commodities and ingredients has led to the development of programs and techniques designed to give early warning of potential contamination so that adequate corrective action may be taken to preserve product integrity. One such concept, called Hazard Analysis Critical Control Points (HACCP), is based on the premise that there are key points within each processing system where the integrity of ingredients and products-in-process may be monitored to ensure that product integrity is being maintained. Monitoring these key points makes it possible to visually identify failures at their early stages so that required corrective action may be taken. Each key point has the means for physically separating undesirable material from the process flow, recording failures, or allowing visual examinations.

Examples of such points are:

(a) prior approvals or guarantees for sensitive ingredients to ensure, insofar as is possible, that purity is being maintained;

(b) visual examinations of ingredients, materials, and/or vehicles used to transport foods or ingredients;

(c) a sifter where the "scalpings" or "overs" are diverted to a container where they may be periodically examined;

(d) a recording thermometer where specific ranges of temperatures are required;

(e) metal detectors that reject bulk or packaged material containing metal; and

(f) magnets located at the end of the process to separate ferrous metal that may have been introduced into the process.

In Table 21.1, the typical elements of an HACCP form are shown. The action that might be taken has not been elaborated upon since there may well be several courses of action available, depending on severity or duration of the deficiency.

Two final notes on HACCP are important. First, in designing this program one must look beyond the control points already in place to where additional control points might be needed

Table 21.1
Elements of typical HACCP form

Critical Control Point	Hazard	Control	Responsibility	Documentation
1. Ingredient sifter "overs"	Foreign material	Examine "overs" every 4 hr	Ingredient handler who alerts foreman to deficiencies so that appropriate action may be taken	Document 2.A1 (example)
2. Pasteurizing recording thermometer	Improper pasteurization	Review chart every ___ minutes; take immediate action on noted deficiencies	Pasteurizing operator	Thermometer charts

or useful. Second, it may be seen that it is not a function of a critical control point to correct deficiencies. The critical point merely assists in identifying for correction those areas within a system where integrity degradation has occurred or is occurring.

Besides the processing-flow HACCP, there are also programs designed to indicate when potential contamination is probable from other sources. These programs incorporate the maintenance of buildings and equipment to exclude contamination and require appropriate actions when the programs are violated. Such programs incorporate points that are sometimes referred to as critical factors. Examples of critical factors include rodent and insect control programs, environmental sampling, employee practices and, equally important, the selection and maintenance of materials used in equipment and in building construction. Specific references can be found in FDA's Umbrella Good Manufacturing Practices and its appendices (2).

Principle 4

Expertise in the various aspects of food product protection is mandatory for the selection and use of building materials, equipment, utensils, and other accoutrements essential to sanitary operations.

Building construction and building materials, remodeling, equipment selection, layout, and installation, and the ongoing maintenance in all areas are supportive to product development and processing. All the steps should be supervised by personnel skilled in the food protection discipline to avoid mistakes that are either costly to remedy or, in some cases, impossible to permanently correct and must therefore be included in special and expensive maintenance programs.

The expertise required is not as likely to be embodied within the construction and maintenance groups, particularly in dry-food processing construction and remodeling, as it is within the product safety and sanitarian groups.

All too often equipment is constructed or installed in a manner that creates pest harborages. Nor is it uncommon to see changes made with more emphasis on getting the job done cheaply rather than properly.

Not to be ignored when discussing expertise is the role of the sanitarian. Since this aspect is covered in more detail in Chapter 49 on Integrated Pest Management, suffice it to say here that the sanitarian provides a link to all other functions. The sanitarian should be consulted on all anticipated physical changes in a food-processing system or facility.

Principle 5

Company dedication to preventive sanitation and follow-up is required in order for the programs to work.

This dedication begins with policy statements set forth at top management levels and communicated down through all management levels to the hourly employee. It is not difficult for regulatory inspectors to determine the degree of sincerity on the part of management to preventive sanitation programs. They have only to listen carefully to responses given to investigative questions and observations. Even more significant, neither is it difficult for anyone else in the plant to sense a lack of sincerity on the part of plant management. A great deal of continuous training is required if the programs are to work effectively.

It is general knowledge that today's inspectors are concerned with two broad areas of investigation during their inspections. One is the standard observation of sanitation failures noted during the inspection. Examples of such observations include the presence of pests in the facilities, with possible extensions to ingredients and/or products, cloth or metal materials fraying into product zones, wood splinters from pallets that may have punctured containers of material, personnel with open sores on hands manipulating exposed products or ingredients, and so on. The inspectors' reports reflect their satisfaction or dissatisfaction with programs or plant responses to deficient conditions by how thoroughly they detail their observations.

The other broad area of investigation requires that the inspectors, through questioning, ascertain the capability of plant personnel to anticipate and correct problems known to exist in the operation but not occurring on the day of inspection. They will be very attentive when responses are given to questions regarding problems known to occur within the industry. They will also be interested in the kinds of programs that are in effect to deal with these known problems. It behooves company management to anticipate what the regulatory inspector will be interested in and take steps to determine that preventive sanitation programs are sound and carried out in an effective manner.

Conclusion

The preventive aspects of sanitation are highly dependent upon *careful analysis, planning, training, documentation,* and *implementation,* starting with the selection and inspection of ingredients and addressing the design, construction, and maintenance of the facility and its equipment. These key preventive sanitation points need to be defined and regularly monitored. As important as systems are, it is imperative that employees and supervisors be well trained in all aspects of preventive sanitation. This can be achieved only if top management is thoroughly committed to the principles of preventive sanitation.

Cited References

1 Food and Drug Administration. 1985–90. *Inspection Operations Manual.* FDA, Rockville, MD. (Note: Updated frequently; available only from National Technical Information Service, Springfield, VA 22161.)

2 Jacobson, F. B. 1986. Current Good Manufacturing Practices. *Candy Industry* 151(12)65–90.

Selected References

Guthrie, R. K. 1988. *Food Sanitation.* AVI/Van Nostrand Reinhold, New York, NY.

Mills, R., and J. Pedersen. 1990. *A Flour Mill Sanitation Manual.* Eagan, St. Paul, MN.

Troller, J. A. 1983. *Sanitation in Food Processing.* Academic Press, New York, NY.

22
Design and Construction: Building-Out Pests

Harold George Scott

One of the most important approaches of effective management of stored-food pests is directed design and construction of food management facilities and equipment (*10, 11, 18, 19*). In conjunction with good sanitation, these techniques can give better control than any other approach (*7, 9*). Design and construction of new buildings and equipment so as to minimize or eliminate harborages and passageways for pests, as well as food accumulations, is called pest exclusion. The term is applied also to the modification of the structure of existing buildings and equipment (*17*).

Fundamentally, exclusion involves elimination of dead spaces, cracks, crevices, and other openings so that pests cannot hide in, enter, or leave a specific building, room, or piece of equipment. It also involves reduction of above-floor, upward-facing surfaces so that organic dusts conducive to insect attack cannot accumulate in hard-to-reach areas.

This approach to stored-food pest control is relatively simple and effective when applied to commensal rats, pest birds, and bats. It is more complex and somewhat less effective when applied to house mice and arthropods because of the precise sealing required to contain these tiny pests and because of the ease with which they

are reintroduced. Nevertheless, directed design and construction can limit mouse and arthropod populations and can make their detection easier and their elimination simpler. Also, as populations are "automatically" diminished by stoppage, the probability of food and by-product adulteration is reduced.

Rat Exclusion

Exclusion is only one aspect of rat control. The ultimate goal is elimination. Rarely can control be achieved by exclusion alone, and then only when the infestation is comprised entirely of rats invading a building to forage but returning outside to nest. However, even in complex situations, exclusion is a key to control since it prevents reinfestation, limits rat movement, and deprives rats of harborage (*16, 20*). Both time and ingenuity are required to exclude rats; permanent results rarely can be achieved without elimination of food and harborage (i.e., employing sanitation) (Figure 22.1).

Basic Procedures

The objectives of exclusion are to prevent rats from entering structures, to facilitate control of interior populations, and to design and construct competently and inexpensively, making certain that all work is permanent and neatly executed. Since rapid deterioration of materials requires excessive maintenance, durable materials well suited to local conditions should be selected. Slovenly construction should not be tolerated. Unsightly installations should be torn out and replaced.

Evolution of building materials complicates

Many of the English/metric "equivalents" included in this chapter are intended for use by craftsmen and maintenance personnel and are therefore not always strict arithmetic equivalents.

Scott: Program in Community Medicine, School of Medicine, Tulane University, New Orleans, LA 70112.

All drawings are by C. Feller.

Figure 22.1
Improperly stored food infested by rats.

SOURCE: U.S Department of Agriculture.

exclusion. Newer materials, seemingly equivalent to traditional items but actually of inferior quality, constantly appear. Materials long used with success are beginning to vanish from the marketplace. For example, copper insect screening tends to exclude rats but aluminum and plastic screening, through which rats easily can gnaw, is now commonplace. Development of superior products and shifts in costs and benefits of existing items also occur. As recently as the 1950s there was extensive testing of materials for rat-exclusion efficiency. Newer materials, however, rarely are tested, so exclusion personnel must be knowledgeable and selective regarding materials purchased.

Openings

For rat exclusion, all openings large enough to pass a test cylinder ½ inch (1.25 cm) in diameter should be closed if accessible to rats from the ground or by climbing, whether or not such openings are in use at the time the building is treated. However, it is often necessary to close even smaller openings and to treat places vulnerable to rats after old entryways have been blocked. For example, door treatment should include the flashing of wood sills as well as wood frames even though there are no openings large enough to permit rat passage. When deciding about treatment of potential openings, costs and effects should be balanced. For example, solid doors with good sills and frames without cracks over ⅜ inch (1 cm) can be bypassed. Frequent future inspections will reveal

any attempts by rats to gnaw through such openings.

Barriers

In order to prevent rats from reaching over, climbing, or jumping barriers such as smoothed areas of walls or rat guards, an unclimbable surface 18 inches (46 cm) wide must be provided.

Ground Floor

Rat infestation of buildings depends to a large extent upon the construction of the ground floor. It is difficult to rid food-storage places of rats harboring under wood floors. The advantage of installing cement floors should be fully explained and the responsible individuals urged to make the change. Openings to double walls are commonly located in the basement or where the joists of the ground floor rest on the foundation. These should be closed with metal, cement, or masonry. Rats may gain access to double walls through openings (chases) around pipes or wires emerging from wall spaces. All such passageways must be sealed if rat harborage in the walls is to be eliminated.

Structural Harborage

Infestation of a rat-excluded establishment is largely dependent upon in-building harborage. Any reduction in this will aid significantly in eradication. Enclosed spaces should be opened and contents rearranged so that as much harborage as possible is eliminated. Openings leading into rat harborages should be searched out and sealed up so that they cannot be opened.

Vegetation

Ornamental vegetation should be trimmed so that it does not touch the building. Vines on walls should be torn down and their roots dug up so that they will not regrow. Tree branches overhanging the building should be pruned back so that rats cannot use them to jump to and from roofs or ledges.

Foundations

When the lower floor is of material impervious to rats, treatment of foundations involves only harborage elimination. Cracks or breaks in foundation walls should be repaired with cement mortar or masonry; openings around wires and pipes should be closed with cement mortar, masonry, or metal collars. Air vents should not be sealed because they are necessary for air circulation under buildings. Galvanized hardware cloth (17-gauge) of 2 × 2 or 3 × 3 mesh is used to treat vents if the screen can be placed against a metal grill. In the absence of a grill, 18-gauge expanded metal, cemented neatly in place, can be used, especially if the vents are located in areas where they may be subject to damage.

Screening should be applied in metal or wood frames, except for vents smaller than 1 ft^2 (1,000 cm^2)—in these the screen should be bent into a basket and the edges cemented into place. Some baskets, large openings, or installations subject to abuse may require cross bars to support the hardware cloth. If wood frames are used, the hardware-cloth edges should be covered with narrow strips of galvanized metal. Framed screens can be held in place by tying them to grills with galvanized wire. Framed screening used in small vents without backing should be fastened with cement mortar. Metal frames should be made of 24- to 26-gauge galvanized metal; frame width varies from ¾ inch to 1½ inches (2–4 cm).

Curtain Walls

Sunken L-shaped walls installed around buildings can prevent rats from gaining access to spaces beneath the ground floor, shut rats off from food, and eliminate harborage. If the foundation rests on rock, curtain walls are not needed. If the soil is hardpan, a curtain wall usually is not needed even if the foundation penetrates the ground only 10–12 inches (25–30 cm). Light or sandy soils permit rats to burrow deeply. If burrows are found adjacent to buildings, they should be followed to discover if they extend under the foundation. A burrow from outside (mound of earth at exterior burrow mouth) indicates that new actions should be undertaken. A burrow from within (no mound of earth at exterior burrow mouth) should be closed and watched for fresh tunneling before additional action, such as installation of a curtain wall, is considered.

Vertical curtain walls should extend 2 feet (60 cm) below the ground, with a horizontal arm 12 inches (30 cm) wide at the bottom. Rats typically burrow down to the horizontal arm and follow it for a considerable distance without finding their way around its outer edge. However, the flange (horizontal arm) will not prevent rats from burrowing out from under the same building. Concrete is the most satisfactory material. Cement:sand mixtures vary from 1:2 to 1:4. Good construction practice usually requires that concrete slabs be 4 inches (10 cm) thick.

The space being enclosed should be ventilated unless existing vents in the foundation are adequate. In some instances it is necessary to provide wells with concrete sides and bottoms for the vents. Frames for vents should be installed when the concrete is poured. Vents should be installed at 25-foot (8-m) intervals, with 1.5 ft^2 (1,400 cm^2) for each 25 running feet plus 0.5 ft^2 (465 cm^2) for each 100 ft^2 (9.3 m^2) of building area served by the vent.

If rats subsequently burrow underneath from outside, another foot (30 cm) should be added to the width of the horizontal arm. If rats have tunneled from within, soil should be tamped down firmly into the burrow and watched for new burrowing signs.

Basements

Intensive search is often required to locate and close all passageways in basements. Openings between the basement and the ground floor should be sealed. Open drains in the floors of basements should be capped so that rats cannot gain entrance through them. To accomplish this, a perforated metal cap held in place by a hinge or a brass-screw drain cover can be installed. If drains serving areas flushed with large amounts of water are screened, they may become clogged. Thus, the hinge should operate easily so that slight water pressure will open the cover and it will fall back into place when the water ceases to flow. Openings in elevator pits should also be noted.

Basements without retaining walls, or those boarded up to prevent the earthen walls from caving in, are treated as foundations. Masonry or concrete walls and/or floors should be installed in earth basements. When basements extend only under a portion of a building, foundations should be stopped and the basement walls sealed at their junction with the floor above. Walls of complete basements can be made continuous with the exterior walls of the building on all sides. Defects in walls can be sealed by pointing up with cement mortar or masonry. All openings around pipes or similar installations should be sealed.

Horizontal Vents

To stop sidewalk ventilators, basement lighting vents, and space vents under buildings without basements, the grill should be removed, a hardware cloth screen installed, and then the grill should be replaced. If the opening is larger than required for ventilation, it can be reduced by bricking or cementing, and then screening it. If the vent is an accessway, the screen should be hinged. Sidewalk vents may open into basements or the walls may have large openings, making it difficult to install vertical screens to keep out rats. If access to the grating can be gained from the space underneath, fashion a reinforced basket of wire cloth with vertical sides 8 inches (20 cm) or more deep. Hang baskets on hooks so that any refuse that accumulates can be dumped. Horizontal vents may be over a basement that extends out under the sidewalk. Rats entering by such vents must drop several feet to the basement floor and cannot reach the grill to get out. Consequently, they rarely use these entrances. A most satisfactory means for treating sidewalk ventilators is to replace them with ½-inch (1.25-cm) gratings.

Sidewalks

Large cracks or broken places in sidewalks may allow rats to burrow into buildings; these openings should be pointed up with cement mortar. If the sidewalk does not tie into a concrete curbing, or if there is a defect in the curbing, rats may burrow under sidewalks and foundations to gain access. When evidence of such activity is discovered, curbing breaks should be repaired and the sidewalk tied into the curbing. Under extreme circumstances, it may be necessary to place a curtain wall under the sidewalk.

Meter Boxes

If a meter-box cover is broken or has holes over ½ inch (1.25 cm), rats may follow pipes or wires and gain entrance to buildings. Defective covers should be replaced. Places where rats are actually entering by following pipes or wires should be searched out and sealed.

Exterior Doors

Probably more rats gain entrance to buildings through defective doors than by any other means. However, no door should be treated unless conditions warrant. If doors are in such poor condition that they cannot be repaired at reasonable cost, they should be replaced, preferably with steel doors. Doors should be repaired so that they are sturdy and properly fitted before other rat-exclusion treatments are applied. An unused door should be handled the same as other doors because occupants may later decide to use it. Doors opening onto stairways or fire escapes should be treated as ground-floor doors.

Rats frequently enter warehouses through cracks at the bases of roll-up doors or rotted-out holes at the bases or sills of personnel doors. Mice, and sometimes rats, enter virtually unchecked and continuously through the semicircular openings formed by the "closed" corrugations of flexible roll-up doors, even when these are closed or pulled down (using a rachet under pressure) to make them "rodentproof." Sealing these corrugations is difficult since the channels on either side of the door abrade against silicone caulk or other sealants. One effective way is to mount outward sloping sheet-metal rat guards on the outboard sides of such doors. A quick, stop-gap seal for a partially damaged roll-up door flange can be made by using an old (i.e., softened) 2.5-inch (6.3-cm) fire hose lock-washered to the right-angle flange at the door base. When the door closes, its weight crushes the fire hose, creating a rodentproof seal.

Some conditions requiring flashing of doors are (a) irregularities at the bottom and along the sides for 8–10 inches (20–25 cm) from the lower edge, and (b) the presence of cracks ¼ to ½ inch (0.5–1.25 cm) wide at the bottom or sides. Wood doors at the sides and backs of buildings should be flashed, even if they are in good condition and fit well, if the establishment shows evidence of rat infestation. Flashing is held in place with galvanized nails or screws. Nails should not be more than 4 inches (10 cm) apart and should not extend all the way through the door. Brass (not aluminum) kickplates are used for front doors. Channels of 24-gauge galvanized metal, installed when flashing is not satisfactory, should cover the sides and lower edges of the door. Doors should be treated with channels whenever rats can enter under an existing door or when the condition of the door cannot be remedied by flashing (Figure 22.2).

Common conditions demanding channels are (a) doors through which rats have already gnawed openings, (b) openings over ⅜ inch (1 cm) between the sill and door, and (c) deterioration of or damage to the door. Broken or worn thresholds must be repaired. Often it is possible to build up the sills and not treat the doors. The distance between the bottom of the door and the sill should never exceed ½ inch (1.25 cm) [on new installations, ⅜ inch (1 cm)].

If a door opens over a sloping surface higher than the door sill, an 18- to 20-gauge galvanized-metal "movable dutchman" can be installed which will slide up and down according to the distance between the door and floor when opened and closed. The dutchman may be fastened at one end so that it will vary at the other, or it may slide up and down at both ends. The lower edge of the dutchman should be rounded so that it will move over the floor without catching.

Treatment of sliding doors is difficult and sometimes cannot be accomplished without extensive alteration. If it is not feasible to utilize a channel because of the irregularity of the surface over which the door slides, a movable dutchman that slides at two or more points, according to the width of the door, may be practical. Double sliding doors may not meet. A cement slot should be built at the bottom which will hold the doors in place. When doors slide into slots or follow the back of a partition, there usually is an opening wide enough to admit rats. A wood strip can be fastened to the door so that the space is filled when the door is closed. If sliding doors are constructed of grill work or iron bars that are impractical to screen, screen doors that fold back on each other can be installed.

Windows

Windows covered with quality insect screening in good condition need not be treated even though they may be accessible to rats. First-floor windows less than 30 inches (75 cm) above the ground, or windows accessible from stairways, fire escapes, roofs, rough walls, pipes, poles, wires, ladders, vines, or trees should be treated unless effectively screened against insects. Galvanized hardware cloth can be used to keep rats out of windows near the ground, those that may receive rough treatment, and those through which rats have been passing. For windows that are never opened, treatment can be limited to fastening them closed and repairing broken panes. Due caution must be taken not to obviate fire safety requirements. Hardware cloth should never be nailed directly to window frames. Screening should first be neatly framed, and then the framing fastened so the screens can be removed easily. One-inch (2.5-cm) galvanized-metal strips are usual for framing hardware cloth.

Windows barred on the outside should be screened from the inside, or the screening should be placed in the form of a basket outside the bars. To place a framed screen from inside, it may be necessary to remove the window stops and even the lower sash to insert the frame beneath the upper sash. If this is impractical, an adjustable screen can be fabricated in two sections, one of which will move in a metal groove of the other (Figure 22.3).

For windows that may swing on an axis so that they project both out and in when open, a

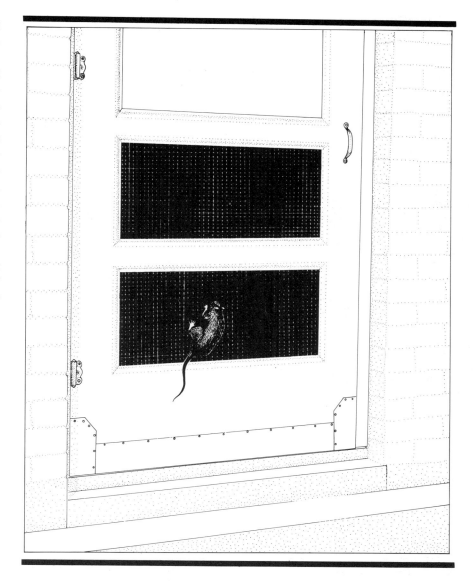

Figure 22.2
Rat-stopped door.

right-angle, 16-gauge hardware-cloth basket deep enough to permit the window to swing into it should be constructed. The free margins of the hardware cloth can be enclosed in a metal frame and the corners reinforced with right angles of galvanized metal. Skilled carpenters are needed to fabricate perfectly fitting frames. Metal channels (24–26 gauge) are suggested for the corners of wood frames. These channels, about 3 inches (8 cm) wide and fabricated of waste metal, permit onsite adjustment of the sides of frames so that they fit even though not cut exactly. This also conceals any defects at the corners and strengthens the frame.

Exterior Walls

Wall openings accessible to rats should be sealed. In fact, it is advisable to point up wall defects even though they do not permit rat passage because anyone inspecting the work is apt

Figure 22.3
Barred window rat-stopped with hardware cloth.

of the fan and creating fire and sanitation hazards. Consequently, it is best to install a ventilator protector [1½-inch (4-cm) metal strips, each of which is hinged and overlaps the one below]. When the fan operates, the protector opens. When the fan stops, the strips fall into place, closing the vents.

Roofs

Treatment of buildings is never complete without attention to roofs, especially where roof rats are present. Evidence of Norway rats and roof rats is commonly seen many stories high; rat runs have been traced from the ground to the top of 11-story buildings. If there is no likely means by which rats can gain access, no roof treatment is indicated. If, however, entryways are available (which is usually the case), eaves should be sealed and ventilators projecting above the roof surface should be screened with baskets or boxes of hardware cloth (Figure 22.5). Hoods or baskets are installed over chimney openings only when there is clear evidence that rats are entering the building by means of the chimney. Such installations must not interfere with the operation of chimneys, and occupants must be notified because accumulations of soot may interfere with the draft at some future time. Should there be a need to treat a roof drain, a hinged cover that opens easily to slight water pressure can be placed over the lower outlet.

Loading Platforms

If establishments have loading platforms with the insides boarded up so as to enclose the space underneath, the side sheeting should be removed and the area opened up. Occasionally, it may be necessary to install a curtain wall to eliminate harborage under a platform. Loading platforms are usually constructed of concrete and are often equipped with hydraulic floor lifts (levelators) that can be elevated to accommodate truck trailers of different heights when they back up to the dock. The undersides of levelators frequently are littered with trash and pallet splinters. Rats can enter the plant both at their open fronts and at the seams where they contact the dock well. Brushes should be installed between the dockwell wall and the levelator, or strips should be welded on the level of the hydraulic floor so that when the gaps at the sides are closed, the rodents cannot jump up into the warehouse. Since rats can climb rough vertical walls, the lower vertical exterior faces of concrete docks adjacent to and beneath the levelator walls should be supplied with star-bolt-anchored

to think that passageways have been overlooked. Cracks, breaks, and places where wires, pipes, and other installations pass through the wall can be blocked with cement mortar, masonry, or metal. Rough walls that may be climbed by rats do not require treatment if there are no openings admitting rats to the building after the climb. If rats may gain entrance by climbing, limited areas can be smoothed with cement mortar; it may be less expensive to seal openings than to smooth extensive rough-wall surfaces.

Ventilators in walls should be stopped only if accessible to rats. Wall vents are treated the same as foundation-wall vents, with the exception of exhaust-fan openings (Figure 22.4). Rats frequently gain access through such vents when the fan is not operating. Grease and dirt discharged by the fan tend to accumulate on hardware-cloth screens, interfering with operation metal flashings to a minimum height of 3 feet (90 cm) to serve as rat guards.

Rat Guards

Guards interposed to prevent rats from climbing or gaining entry must be wide enough to prevent rats from reaching their outer margins by jumping or climbing. Guards should be installed above heights that can be reached by passing people or vehicles to minimize probability of damage. Circular guards may be installed on wires leading to buildings. Since free-hanging guards are damaged easily, circular guards extending out 18 inches (45 cm) all around the line they guard are anchored in place by one or more arms on the side opposite that accessible to rats.

Cone-shaped circular guards can prevent rats from climbing vertical pipes, pilings, and trees. Right-angle guards are used to prevent rats from climbing pipes and wires on the sides of buildings and to stop rats from climbing surfaces. One end of the guard is fastened securely to the building; the other projects at a right angle. Perpendicular guards interpose a smooth surface over which rats cannot climb or jump. They should be 18 inches high or, if rats have access to a surface from which they can jump, 36 inches (90 cm) high. Wires and pipes up the sides of buildings can be encased in rounded galvanized-metal housings sealed at the top and bottom (Figure 22.6).

Sometimes it is as effective to install a rat guard at the bottom of a window as it is to screen the window. For example, if the window ledge is less than 30 inches (75 cm) from the ground, but it is feasible to treat the window by decreasing the foothold accessible to rats by 12 inches (30 cm), use of a 24-gauge metal strip across the bottom of the window is indicated. Window guards should be placed at an angle, the bottom of the metal being ⅜ inch (1 cm) out from and level with the windowsill while the upper edge projects 6–8 inches (15–20 cm) from the window frame. An open space between the bottom of the metal and the windowsill is essential to permit free drainage. The metal should be fastened to triangular end pieces of wood nailed firmly in place. Metal should cover the triangular blocks so that wood cannot be seen from the outside. Window guards placed below ground level may become filled with trash. They are more suitable for use on the backs of buildings than on the fronts.

Since rats can climb screening of any type, the entire surface of any screened opening must

Figure 22.4
Exhaust-fan vent rat-stopped with hinged hardware cloth.

Figure 22.5
Rat-stopped roof ventilator (diagrammatic).

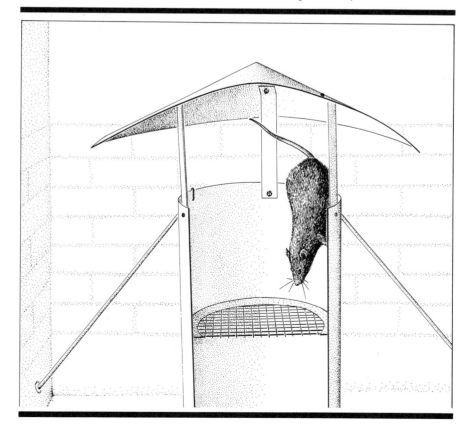

Design and Construction: Building-Out Pests   337

Figure 22.6
Rat guards installed over utility lines.

be covered to keep out rats unless some technique is used to prevent rats from climbing over the top of the screening. When treating large or high openings or those not covered at the top, an 18-inch (45-cm) strip of metal installed at the upper margin of the screening will save material. Perpendicular strips of metal, 36–48 inches (90–120 cm) wide, are used to prevent rats from gaining access to open spaces at ground level, providing that the rodents cannot tunnel under the metal. As a general rule, other types of rat exclusion are usually preferable to rat guards.

Conclusion

During construction of rat exclusion on existing buildings, many situations other than those discussed here will be encountered. Workers must use ingenuity and good judgment, remembering that nonessential work increases costs without increasing effectiveness. Techniques change and improve. Workers should make every attempt to "keep up with the field."

New Construction

The greatest opportunity for truly comprehensive and lasting rat exclusion exists in the design and construction of new buildings. Ratproofing of new construction involves both interior and exterior design. A major goal is to prevent the creation of any interior structural harborage. Some significant specifications for new construction are:

(a) Structures are built of materials impervious to rat assaults and of a design that eliminates all unnecessary enclosed spaces and nonessential openings.

(b) Openings around the edges of doors should not exceed ⅜ inch (1 cm) and all essential openings larger than ½ inch (1.25 cm), including those associated with utilities, should be rat-stopped.

(c) All openings into necessary enclosed space should rat-excluded.

(d) Inferior products, which are easily broken to expose intricate pockets and holes in which rats can harbor, should be avoided.

(e) Movable louvers that close automatically when fans are not operating and, when closed, present no opening larger than ½ inch (1.25 cm) should be used.

(f) Outside doors should be made of metal, glass, or hardwood. Sliding, swinging, and rolling doors should have no opening greater than ⅜ inch (1 cm) when closed and have double side and bottom guides to ensure a permanently tight fit.

(g) Unavoidable false ceilings and hollow walls should be sheeted in materials resistant to rat gnawing.

It is less expensive to build ratproofing into a building than to treat it after construction. Ratproofed buildings are not likely to become rat infested unless interior sanitation is poor and/or the rodents are brought in with products. Although ratproofing specifications deviate from those of fireproofing and termiteproofing, the three are complementary and construction should correlate all three.

Transportation Facilities

Ratproof construction and/or exclusion treatment of vessel watering and loading points, railcars, railroad servicing and loading points, aircraft, and air terminals are part of the overall food protection effort. Ships are designed so as to eliminate or render inaccessible every space that might afford rat harborage, avoiding enclosed spaces and arranging service lines, machinery, furniture, equipment, and fixtures so that enclosed spaces are not created. Uninspectable spaces and voids are treated for rat exclusion. To prevent passage of rats from one section of the vessel to another, the various working spaces are completely isolated from each other by means of dependable rat-exclusion techniques.

Equipment Harborages

Since equipment can be a major source of harborage and access for rats, to the usual criteria for selecting new equipment should be added the degree of ratproofing. Will the equipment involve a type of installation that will encourage passageways and harborage? Will an item essentially ratproof when properly installed shortly become a source of rat harborage or passage when side panels are removed or other modifications are made? One installer may produce a ratproof situation, whereas another, more careless, may create problems that will plague the building for years. Treatment of openings leading into the insulated space of built-in refrigerators and similar equipment is of utmost importance, as is closure of the space between the top of the refrigerator and the ceiling (ordinarily closed with hardware cloth). Damage and new problems can be detected readily with efficient inspections; rat-supporting situations that develop can be corrected with good maintenance.

Permanence of Exclusion Measures

Once a treated stucture is rat-free, it will remain so providing (a) there are no rifts in the exclusion, (b) the occupants do not leave unscreened doors and windows open, (c) rodents are not reintroduced with merchandise or by other means, and (d) the exclusion is not damaged. Rat exclusion, regardless of its thoroughness, cannot be ensured. A door may be cracked, a window broken. Workers may remove or cover exclusion, or make new openings. Additionally, rats may gnaw new openings or dig tunnels into excluded structures. This biological pressure is strongest during the first 2 weeks after treatment; the probability of reinfestation is in direct proportion to the size of the external rodent population.

The chance of rats breaking into excluded buildings can be significantly reduced by treating nearby buildings, by requiring proper refuse storage and collection, and by using rodenticides to eliminate exterior rodents while exclusion is being installed. Because constant danger of breaks exists in every structure, exclusion effectiveness depends upon repeated inspection and prompt repair. If rats do gain access to an excluded area, control should be effected before they multiply. Careful inspection during the critical first 2 weeks after treatment should locate any openings being used by rats seeking to reach former sources of food and harborage.

House Mice and Field Rodents

Although the principles of rat exclusion can be applied to house mice and field rodents, variable factors such as size and habits must be considered. For example, the adult Norway rat weighs up to 1 pound (453 g) and the adult roof rat weighs up to ¾ pound (340 g), whereas the adult house mouse weighs only about 1 ounce (28 g). Thus, all openings larger than ½ inch (1.25 cm) are closed for rats, but all larger than ¼ inch (0.6 cm) must be sealed for house mice. This may well quadruple the effort and cost. In addition, house mice are quieter, more restricted in home range, and more apt to establish themselves permanently indoors, living quietly in out-of-the-way places. Consequently, sealing of accessways becomes less important, elimination of interior harborage more important.

Rats nearly always enter buildings through tunnels or other openings, including windows and doors. House mice can and do use these entrances, but they also are introduced repeatedly with packaged goods and equipment. Therefore, building construction alone cannot keep an establishment free of house mice; there is a constant requirement for detection and elimination, tasks greatly simplified by mouseproofing.

Many stored-food installations are located in relatively undeveloped areas where field rodents abound. In such cases, invasions of these rodents can be anticipated, especially on a seasonal basis. Consideration must be given to variables such as size of entryways through which a particular species can pass, climbing ability, burrowing ability, and food choice. Control personnel should carefully search the technical literature for information on control of the specific kinds of rodents being encountered.

Birds

Some rat-exclusion techniques are applicable to pest-bird control, but important differences exist, and a number of unique approaches can be used, depending upon the species of bird involved (20). Important pest birds include pigeons, house sparrows, and starlings (12). For this discussion, pigeon exclusion is used as an example (15).

Pigeon stoppage is the modification of existing buildings and equipment, or the construction of new buildings and equipment, in such a manner that pigeons cannot enter, roost, or nest. It minimizes (a) costs of keeping structures clean, pest-free, and in good repair; (b) accumulations of pigeon excrement which might kill ornamental plants, damage structural materials, or harbor human disease organisms; (c) clogging of drain pipes by nests; (d) dangers associated with nests on fire escapes; (e) human cases of pigeon-borne disease; (f) pigeon ectoparasites such as lice, bugs, louse flies, fleas, ticks and mites, as well as nest pests such as earwigs, bugs, stored-food moths and beetles, and flies; and (g) food contamination by pigeon droppings, feathers, and nesting materials (Figure 22.7). Routine sanitation can be the major factor limiting pest-bird populations around food-storage facilities. Its concurrent use is essential if exclusion is to be effective.

Pigeon exclusion has been used widely and, when properly applied, has proven feasible and economical. Treatment should be limited to the minimum necessary to achieve desired results and should be as complete, neat, and inexpensive as possible. With thought, treatments also can be made to control other pest birds, rodents, and bats. Exclusion can give permanent control provided that (a) it is correctly applied; (b) unprotected doors, windows, and other entryways are not left open; (c) installations are not damaged; and (d) repairs and new construction maintain the exclusion.

Biological Pressure

Biological pressure is greatest the first few months after exclusion measures are taken and greater in the spring and summer than in the fall and winter. Pigeons attempting to maintain residence on a building that may have been their home for many years will roost and nest on surfaces that they ordinarily would have avoided. Usually they find these places too precarious and give them up after a short time. However, some persist and these new sites may have to be treated.

Exclusion Details

Judgment is required, balancing effect against cost. Openings not in use should be bypassed and observed for activity. If pigeons begin using them, the openings can then be treated. Pigeons trapped behind exclusion devices should be captured and removed. Treatment should improve, not detract, from the appearance of a building. Permanent work is less expensive than temporary work in the long run. Repeated inspection and prompt repair are essential. Successful exclusion requires skill, perseverance, and experience, and even highly trained personnel may make errors (which should be detected quickly

and corrected immediately).

Interiors

Interior exclusion involves blocking access to indoor roosting, feeding, and nesting places. Openings to lofts, steeples, towers, vents, and eaves should be blocked with wood, metal, glass, masonry, or hardware cloth. Whenever the purpose of the opening is ventilation, hardware cloth in the following mesh sizes should be used for exclusion: 1½-inch (4-cm) mesh for pigeons, ¾-inch (2-cm) for sparrows, ½-inch (1.25-cm) for rats, and ¼-inch (0.6-cm) for mice and bats. Expanded metal can replace hardware cloth where

especially handsome work is required. People accustomed to rodent control work tend to set pigeon exclusion devices too far in. Installations should be placed flush so as to leave no outside ledges or projections upon which pigeons can roost.

Doors and Windows

Doors through which pigeons enter should be fitted with automatic closing devices. If the door must be kept open for ventilation or for operational reasons, a replaceable flex-strip curtain of see-through plastic should be installed as a barrier. Although there is a tendency to design food plants without windows today, where windows are present they should be screened. Good

Figure 22.7
Pigeon-infested tower of building in which food was stored.
SOURCE: New York City Department of Health.

insect screening is effective in many cases. Like doors, windows should be kept closed when not in use. Even headhouse windows in grain-storage structures, which must be kept open during active grain movement, should be kept closed during inactive periods, especially over weekends when the operation is shut down.

Exteriors

Exterior exclusion involves the removal, blocking, or modification of all outdoor surfaces upon which pigeons can nest or roost. Roosting on ledges, ornaments, signs, ridges, dormers, and roof gutters can be discouraged by installing banks of sharpened, spring-tempered, stainless-steel needles. These ''porcupine wires'' are inconspicuous in most places, are permanent, and require little if any maintenance. They can be installed on flat, angular, or curved surfaces, and can be fastened to virtually any type of material. Porcupine wires consist of strips of flexible, rustproof base bars to which are welded about 120 small semicircular needles per linear foot. The strips are usually 3¾ inches (9.5 cm) high, 4½ inches (11 cm) wide, and 4 feet (122 cm) long. They come with special clips and/or fasteners designed for the particular surface upon which they are to be installed. A similar device uses looped ''wickets'' spaced parallel and interlocked to bars along their lengths.

Pigeons sometimes construct nests on porcupine wires or wickets by first building up a protective layer of sticks and/or straw. Fallen leaves may cushion these devices, enabling birds to nest on them. Roosting can be discouraged by installing hardware cloth, or less satisfactorily, fastening nets over ledges, cornices, and ornate structures. Nets can be used to stop bird damage to fruits, berries and vegetables on farms, in garden stores, and in farmers' markets and other outdoor food-managing places.

Flashing of wood, masonry, plastic, or metal installed at 45- to 50-degree angles on roosting and nesting surfaces is one of the most effective pigeon exclusion techniques. Ends of the flashing must be closed or birds will build nests under them. This technique is most effective on ledges 6 inches (15 cm) wide or less. When birds attempt to roost or nest, they tend to slip off flashings; nests under construction are easily dislodged by wind and rain. Consequently, the birds soon move elsewhere.

Control of fleas, ticks, mites, other pigeon ectoparasites, and nest pests is a necessary part of pigeon exclusion. When the birds are barred from a building, the ectoparasites left behind are without a host and may seek out and bite people. Stranded nest pests and ectoparasites often migrate into buildings.

Arthropod Pests in General

Food stored in poorly constructed buildings may become infested by insects and other arthropods flying or crawling in from nearby fields or warehouses, and poor construction contributes to rapid buildup of large populations. Thoughtfully designed and carefully constructed structures can virtually eliminate this hazard, and existing poor construction can be modified to minimize it. Meticulous, permanent sealing of cracks, crevices, and other openings in structures, equipment, and merchandise can result in marked reduction or even elimination of silverfish, cockroaches, earwigs, ants, beetles, flies, mites, and spiders. Special attention should be paid to wood-to-wood joints.

Doors and windows are, of course, important residual passages and should be made tight and self-closing. Adequate ventilation is important. High moisture content of stored food is conducive to attack by arthropods and to the formation of mold. Ventilators and doors, all properly screened, should be opened during dry weather and closed during periods of excessive humidity. During the winter it may be advantageous to permit cool air to circulate in warehouses unless this may cause damage to products susceptible to freezing. However, most grains may be subjected to extremely low temperatures without damage. Cold is useful in deterring arthropod infestations, since most arthropods develop slowly at low temperatures and many individual mites and insects are killed by freezing (*1, 2, 5, 6, 8, 13*).

Installation of arthropod exclusion should be followed by careful periodic inspection to discover where such pests are finding harborage or gaining entrance. Defects discovered should then be sealed. Maintenance must be knowledgeable and continuous. Even the best of installations can be compromised by wear and tear, damage, decomposing seals, new construction, and changing conditions of use.

Invasion by stored-product pests occurs much less rarely via structural openings than it does by (a) introduction of beetles, moths, and other pests on incoming commodities or products; (b) introductions via open warehouse doors from filthy truck trailers; or (c) dirty pallet boards. Consequently, the greatest protection from pests is a knowledgeable, regular, conscientious incoming-goods inspection by a firm's quality assurance/sanitation group.

Mold

Most foods stored in humid situations become infested with mold growths that can cause spoilage, flavor change, or otherwise render the food unsuitable for human consumption. Poor construction and management encourage rapid buildup of mold infestations. Carefully designed and constructed structures can minimize these infestations, and existing poor construction can be modified so as to prevent excessive losses. Factors involved in mold control include temperature, ventilation, moisture control and carbon dioxide management, dust and debris prevention and elimination, and minimization of other stored-food pests (3).

Total Pest Management

Exclusion of stored-food pests is a highly interrelated process. To achieve the best results, the pest manager should create integrated designs based upon all major pests being encountered, then expand these to incorporate pests of lesser significance. Such an approach results in greater effectiveness at lower cost and prevents the creation of new problems while action is being taken to solve old ones. For example, pest exclusion demands that doors and windows be kept closed when not in use. Ventilation/moisture control demands that they be kept open during periods of low humidity or cold weather. Thoughtful consideration is required in making the wisest decision in the face of such conflicts. Perhaps a self-closing, screened aperture would satisfy both objectives.

Just as exclusion should be integrated to provide for simultaneous rodent, bird, bat, insect, and mold control, so also should the exclusion effort be integrated into an overall pest management strategy: exclusion techniques and chemical, biological, and sanitation controls applied continuously from the farm through transport and manufacture to storage, distribution, and consumption (4, 7, 9, 14).

Acknowledgment

The assistance of Joseph D. Foulk, Food Safety Associates, Edmonds, WA, in the preparation of this chapter is gratefully acknowledged.

Cited References

1 Bailey, J. E. 1982. Whole grain storage. In *Storage of Cereal Grains and Their Products,* C. M. Christensen, ed. American Association of Cereal Chemists, St. Paul, MN.

2 Burrell, N. J. 1982. Refrigeration. In *Storage of Cereal Grains and Their Products,* C. M. Christensen, ed. American Association of Cereal Chemists, St. Paul, MN.

3 Christensen, C. M. (ed.). 1982. *Storage of Cereal Grains and Their Products.* American Association of Cereal Chemists, St. Paul, MN.

4 Constantine, D. G. 1982. Batproofing of buildings by installation of valvelike devices in entryways. *Journal of Wildlife Management* 46(2)507–513.

5 Foster, G. H. 1982. Drying cereal grains. In *Storage of Cereal Grains and Their Products,* C. M. Christensen, ed. American Association of Cereal Chemists, St. Paul, MN.

6 Foster, G. H., and J. Tuite. 1982. Aeration and stored grain management. In *Storage of Cereal Grains and Their Products,* C. M. Christensen, ed. American Association of Cereal Chemists, St. Paul, MN.

7 Hoyle, D. A. 1982. Built-in sanitation for food processing facilities. *Field Service Insights* 2(4)13.

8 Hyde, M. B., and N. J. Burrell. 1982. Controlled atmosphere storage. In *Storage of Cereal Grains and Their Products,* C. M. Christensen, ed. American Association of Cereal Chemists, St. Paul, MN.

9 Imholte, T. J. 1984. *Engineering for Food Safety and Sanitation.* Technical Institute of Food Safety, Crystal, MN.

10 Jenson, A. G. 1979. *Proofing of Buildings Against Rats, Mice and Other Pests.* Ministry of Agriculture, Fisheries and Food, Her Majesty's Stationery Office, London.

11 King, C. B. 1982. Facility design for bird exclusion. In *Bird Control in Food Plants—It's a Flying Shame!* F. J. Baur and W. B. Jackson, eds. American Association of Cereal Chemists, St. Paul, MN.

12 Parsons, M. A., and C. J. Stojanovich. 1965. Pest Control's pictorial key of the month: Birds. *Pest Control* 33(3)34–35.

13 Schaetzel, D. E. 1982. Bulk storage of flour. In *Storage of Cereal Grains and Their Products,* C. M. Christensen, ed. American Association of Cereal Chemists, St. Paul, MN.

14 Scott, H. G. 1961. Bats: Public health importance, identification and control. *Pest Control* 29(8)23–26.

15 Scott, H. G. 1964. Pigeon-borne disease control through sanitation and pigeon stoppage. *Pest Control* 32(9)14, 15, 19, 38.

16 Scott, H. G., and M. R. Borom. 1965. *Rodent-borne Disease Control Through Rodent Stoppage.* Communicable Disease Center, Atlanta, GA.

17 Stinson, W. S. 1978. Sanitary design principles for food processing plants. *Food Processing* 39(8)98–108.

18 Terry, N. G., et al. 1956. *The Design of Machinery and Plant in Relation to the Control of Insect Pests*. The British Food Manufacturing Industries Research Association, Leatherhead.

19 Terry, N. G., et al. 1956. *The Installation of Machinery, etc., in Relation to the Control of Insect Pests*. The British Food Manufacturing Industries Research Association, Leatherhead.

20 Timm, R. M. (ed.). 1983. *Prevention and Control of Wildlife Damage*. Wildlife Resource Committee of the Great Plains Agricultural Council and Nebraska Cooperative Extension Service, University of Nebraska, Lincoln.

Selected References

Katsuyama, A. M., and J. P. Strachan (eds.). 1980. *Principles of Food Processing Sanitation*. The Food Processors Institute, Washington, DC. (Note: New edition in preparation).

Lancaster, D. L. 1977. General sanitation criteria for the design, construction, and installation of food service equipment based on NSF standards. In *Proceedings of: The 1977 Seminar on Cockroach Control*. New York State Department of Health, Albany, NY, pp. 69–90.

23
Protecting Packages Against Insects

Henry A. Highland

The extensive use of packaging in modern food distribution systems provides a potentially effective tool in the management of insect pests. New and modified food-handling procedures, increasingly stringent sanitation standards, and increasing international trade impose a need for systems that will protect food from infestion from the time it is packed until the package is opened by the consumer. Knowledgeable selection of packaging materials can help produce packages that resist infestation. Such packages may be economically feasible when they prevent losses of food that has already incurred all costs of growing, harvesting, transporting, processing, and storing. Direct losses are not only reduced, but the company image is preserved—the consuming public holds the producer responsible no matter how or where the packaged product becomes infested.

Insect Pests

Most stored-product insects are cosmopolitan—any given species is usually found worldwide in areas with similar climatic conditions. For example, 23 species that were found in 20 warehouses in nine foreign countries have also been found in the United States (USA) (17). Warm, humid environments promote insect growth; consequently, insects are likely to be more troublesome in packages stored in warm warehouses than in cool warehouses. Obviously, insect problems are minimal in packages stored in refrigerated warehouses.

Highland: Stored-Product Insects Research and Development Laboratory, U.S. Department of Agriculture, P.O. Box 22909, Savannah, GA 31403.

The extent and incidence of package infestation increases greatly as summer progresses in the USA, and generally as one moves toward the equator. In some southeastern areas, for instance, it was once customary to remove cornmeal from retail shelves at 2- to 3-week intervals during the summer to avoid the development of noticeable populations of insects within the packages. Longer, warmer, more humid summers provide longer favorable growth periods during which insects can multiply and spread from infested to uninfested packages stored under ambient conditions. Problems with psocids increase noticeably as summer progresses. These insects can be found in or on empty paper storage containers used to hold empty primary containers such as jars or cans. The psocids often contaminate the empty jars or cans before they are filled and sealed.

Some products are more likely to become infested before consumption than others, and these products can be the foci for infestation of other products stored nearby. Dry pet food is seldom packed in insect-resistant packages. Because it is a suitable food medium for several species of insects, populations are likely to build up during storage, especially during the summer. The pet food then serves as a reservoir of insects when stored in grocery warehouses containing other infestible foods. Obviously, there is a direct relationship between length of storage, severity of infestation, and resulting cross contamination of other products if no insect control measures are utilized.

Food warehouses are, of course, not the sole site of infestations in marketing channels. Packaged foods can become infested during shipment in trucks, rail-

345

cars, and ships, or during storage at retail levels.

The problem these pests present is exacerbated by the ability of some insects (penetrators) such as *Rhyzopertha dominica* (lesser grain borer), *Tenebroides mauritanicus* (cadelle), *Lasioderma serricorne* (cigarette beetle), *Trogoderma variabile* (warehouse beetle), *Dinoderus minutus* (bamboo powderpost beetle), and *Corcyra cephalonica* (rice moth) to bore through one or more of the many flexible packaging materials in use today. Although these insects vary in their ability to penetrate, strong borers such as *R. dominica* can penetrate most flexible materials. Other species, including *Cadra cautella* (almond moth), *Trogoderma glabrum,* and *Dermestes maculatus* (hide beetle), also penetrate, but less frequently.

Other common species (invaders) that usually do not enter packages unless there is an existing opening include *Tribolium castaneum* (red flour beetle), *Oryzaephilus surinamensis* (sawtoothed grain beetle), *Cathartus quadricollis* (squarenecked grain beetle), and *Cryptolestes pusillus* (flat grain beetle). These latter species, however, can easily find the small openings in many packages as currently designed and manufactured. Adult *O. surinamensis,* for instance, can enter through openings of less than 1-mm diameter. Adult *Tribolium castaneum* can pass through openings that are 1.35 mm in diameter (7). Newly hatched larvae can, of course, enter much smaller openings. There is a direct correlation between package seal quality and the extent and swiftness of infestation (39).

These categories of penetrators and invaders, though useful, are somewhat artificial since some of the invaders become penetrators given the proper circumstances and environment. For example, fed larvae of *Cryptolestes pusillus, Cathartus quadricollis, Oryzaephilus mercator* (merchant grain beetle), and *Tribolium castaneum* did not penetrate any of eight packaging materials, but when starved, larvae of *T. castaneum* penetrated cellophane and polyethylene (6). Adults of these species penetrated some of the least resistant packages. When extremely crowded, *T. castaneum* can penetrate 10-mil polyethylene film.

In general, the more important penetrators, in order of the frequency with which they are found in packaged foods and their ability to penetrate packages, are *R. dominica, L. serricorne, Stegobium paniceum* (drugstore beetle), *Plodia interpunctella* (Indianmeal moth), *Cadra cautella, Trogoderma variabile,* and *Corcyra cephalonica. Tenebroides mauritanicus* larvae are strong penetrators but they are rarely found in processed-food handling or storage facilities. Although *R. dominica* is only infrequently found in these storage facilities, its ability to penetrate most flexible packaging materials makes it an important pest. Many

dermestid larvae (*Trogoderma* and *Dermestes* spp.) are also good penetrators.

It may be necessary at times to determine whether the insect formed ingress or egress holes. The direction of such penetrations can be determined with reasonable certainty by the characteristics of the bored hole (4). The hole on the exit side of the penetrated packaging material has a clean-cut perimeter; the diameter is usually smaller than the entrance side, and there are no frayed surfaces, scratches, or depressions around the hole on the exit side. The perimeter of the hole on the entrance side, on the other hand, is tapered, may be terraced, is usually scratched or roughened, and on plastics the perimeter is usually upturned.

Although the direction of penetration may be determined by this method, it may be difficult or impossible to use such information to determine the origin of the infestation. This difficulty arises from the ability of insects to enter even minute openings that are detectable only with great difficulty. Perforated thumb notches on paperboard cartons and partially sealed carton overwraps are illustrations of openings that are difficult to detect but are large enough to allow the entry of small larvae or adults.

Most species of stored-product insects attack packaged foods. Exceptions are the weevils [*Sitophilus granarius* (granary weevil), *Sitophilus oryzae* (rice weevil), and *Sitophilus zeamais* (maize weevil)] and the moths that attack woolens [*Tineola bisselliella* (webbing clothes moth), and *Tinea pellionella* (casemaking clothes moth)]. Some species are found in packaged foods more frequently than others. These include *Tribolium castaneum* and *Tribolium confusum* (confused flour beetle), which attack most grains and farinaceous foods, and *L. serricorne, Stegobium paniceum, Plodia interpunctella, Cadra cautella, O. surinamensis,* and *O. mercator,* which attack a wide variety of packaged foods.

Packaging Materials

Paper and cellophane are probably the least resistant to penetrating insects of all flexible packaging materials. Depending on conditions and species, some insects can penetrate kraft paper in less than a day; multi-ply construction adds little to resistance.

Almost all flexible polymer films or combinations of polymer films can be penetrated by one or more species of insects. However, there are differences between packaging films in their susceptibility to penetration by insects. Polycarbonate film is very resistant to penetration (18). However, it has not been used for packaging infestible foods because of its cost and poor moisture- and gas-barrier properties (12). In laboratory tests, polyester, cellulose diacetate, nylon, urethane, unplasticized polyvinyl chloride and some polypropylene films resisted penetration by *R. dominica* (*28*). Most other films, including cellophane, polyethylene, plasticized polyvinyl chloride, ionomer, saran, cellulose propionate, and ethylene vinyl acetate, were penetrated within 18 hours.

When small pouches were set up in beakers containing insects (*35*), both penetrators and some invaders could penetrate one or more of 18 films and laminates including polyethylene, cellophane, polyethylene/paper, saran/cellophane, polyethylene/jute, paper/foil/polyethylene, and aluminum/vinyl; however, no penetrations were found in polyester/polyethylene, polyethylene/canvas, foil/polyethylene, or paper/foil/polyethylene.

This latter foil was twice as thick (0.04 mm) as the foil in the penetrated paper/foil/polyethylene. Insects easily penetrated single thicknesses of polyethylene, cellophane, saran, cellophane laminates, and saran/pliofilm. However, laminates of saran plus polyester, with the polyester side exposed to the insects, were not penetrated (*10, 11*).

When tested as pouches against an insect population largely composed of penetrators, polypropylene/polyethylene combination films were generally the most resistant of the combination films tested (*38*). Also, coextruded polypropylene–ethylene vinyl acetate and coextruded polypropylene–polyethylene films showed more resistance to penetration than saran-coated polypropylene and polyethylene-coated cellophane films. Oriented polypropylene films were more resistant than nonoriented polypropylene films. Biaxially oriented polypropylene film has excellent insect-resistant properties when used as an overwrap on paperboard cartons but was ineffective as a pouch (*23*).

Tests have clearly demonstrated that aluminum foil can also be penetrated, although foil packages are generally more resistant to penetration than are film or paper packages. Dried soup mix pouches of 4-mil foil were penetrated within 4 days by *Tenebroides mauritanicus,* and within 2 weeks by *Trogoderma inclusum,* but not by *O. surinamensis* (*3*). Paper/foil/polyethylene pouches of cocoa powder and dried soup mix were easily penetrated in simulated warehouse storage tests (*16*).

Thicker films are more resistant to penetration than are thinner films made of the same polymer resin. Although polyethylene film is not notably resistant to penetration, unpublished packaging tests conducted at the Savannah Laboratory have shown that some resistance is exhibited by films of 5-mil or greater thickness. Thus, films vary in susceptibility to penetration depending on thickness, on the basic resin from which the film is made, on the combination of materials, on the package structure, and on the species of insects involved.

Package Structure
Shipping Containers

Although packaging materials can be carefully chosen and chemically treated to make them resistant to insect penetration, no treatment is effective if the package is not insect-tight. Bags treated with methoxychlor and constructed with tight, permanent seals remained insect-free during 24 months of exposure to heavy insect populations, but similarly treated bags with sewn, untaped closures were infested within 3 months (*22*).

Multiwall paper shipping bags are made insect-resistant by using a folded-over, stepped-end closure sealed to the opposite face of the bag with a heat-activated adhesive (*32*). Insect resistance is enhanced with permethrin or synergized pyrethrin coating on the outer ply (*28*). Another type of insect-resistant multiwall paper shipping bag, also treated with synergized pyrethrins, has synergized pyrethrins–treated paper tape heat-sealed over the stitching on both ends of the bag. These two types of bags have been used to carry millions of pounds of processed cereals to overseas destinations in the foreign aid programs of the USA.

The corrugated paper case is a major container for shipping foods produced in the USA and in other developed countries. It is difficult, however, to make these cases resistant to infestation. Well-sealed tape can be placed over all six flap junctures, or the case can be overwrapped with a heat-shrunk polyethylene film. However, neither method protects from penetrating insects for long storage periods. During storage tests lasting 2 years, permethrin-treated film was found to prevent damage to food packages in film-wrapped cases (*28*). Cases of bulk-packed raisins can be made more resistant to insects by inverting them so that the folded closures of the polyethylene liners are on the bottom, held tightly closed by the weight of the raisins (*40*).

Heat-shrunk polyethylene film overwraps, along with a polyethylene deck sheet between the load and the pallet, can be applied to pallet loads of bags or cases to provide protection from invading insects. During the heat-shrink process, the deck sheet is partially bonded to the film overwrap. Use of a black deck sheet and an infrared heat source can provide an insect-tight seal with the overwrap.

Consumer-Sized Packages

Although it is more economical to make the outer shipping container insect-resistant, it is often more desirable to insectproof consumer-sized packages. This provides protection from infestation throughout the distribution channel, including retail outlets, until the package is opened. The paperboard carton is a typical, widely used consumer package in both developed and developing countries where food is imported. These cartons can be made insect-resistant by the use of tight overwraps such as those provided by polypropylene film (23). Also, laboratory tests and reports from food-processing companies show that fully adhered paper overwraps are effective. Fully adhered paper/foil laminated overwraps also are quite resistant to insect invasion (8).

In the absence of an insect-tight overwrap, the ends of cartons must be sealed so that they are insectproof, a difficult procedure in modern high-speed operations. Beads of adhesive along the folds and edges of the end panels, or extended folded-down tabs at the sides of the end panels (the ''Van Buren ear'') provide some insect resistance. A polyethylene-coated carton having heat-sealed flaps and a polyethylene-overwrapped carton both protected raisins from infestation (37). Composite cans, made of various combinations of kraft paper, aluminum foil, polymer films, and glassine paper, and having metal ends, provide excellent protection from boring and invading insects (15).

Insecticide Treatments

Insect-repellent treatments are applied to packaging materials for food and other infestible items only in very restricted areas of usage and only with carefully controlled procedures. Thus, these procedures have a favorable cost/benefit ratio in terms of their impact on the environment when compared with the procedures used to control such insects as mosquitoes, field-crop insects, and forest insects. Nevertheless, the spatial and temporal proximity of chemicals to products destined for human consumption makes this use of pesticides potentially hazardous. These factors must be weighed against the benefits accrued by protecting the commodity from infestation when the commodity has already incurred the full costs of growing, harvesting, shipping, storing, and processing. The costs of packing, shipping, storing, and distributing the processed, packaged commodities must also be considered. It becomes obvious that to achieve the desired protection without objectionable residues, packaging treatments must be used judiciously, at minimal dosages, and with all available expertise.

Obviously, the insecticides or repellents used on packages must not migrate from the treated surface through the package wall to produce unacceptable residues in the commodity. This migration can be reduced or prevented by the use of barrier plies positioned between the treated ply and the commodity. Four plies of kraft paper prevent unacceptable levels of piperonyl butoxide in cornmeal, flour, rice, and beans stored for long periods in synergized-pyrethrin-treated bags (27). Also, saran-coated paper and greaseproof paper are effective barriers to the migra-

tion of those chemicals (24, 25). Greaseproof paper prevented movement of permethrin on paper bags during a 2-year test (28). The copolymer ethylene vinyl acetate in a coating containing synergized pyrethrins reduced the movement of piperonyl butoxide into packaged food (20).

In the United States and England, much of the early work on insect-resistant packaging was done with the so-called safe insecticides and synergists—methoxychlor, pyrethrins, allethrin, piperonyl butoxide, sulfoxide, and similar materials. More recently, other compounds have been used successfully in laboratory tests, including synthetic pyrethroids (19), an experimental antifeeding compound (33), an inert silica gel (36), and carbaryl (14). In the USA, only pyrethrins synergized with piperonyl butoxide may be used as an insect-resistant treatment. Currently, this treatment is registered for use only on multiwall paper bags or on cotton bags (2) or in the adhesive of a cellophane/polyethylene laminate for dried-fruit packages (1). Other insecticides such as malathion, DDT, dieldrin, trichlorfon, dichlorvos, lindane, and aldrin have been evaluated as treatments for food and feed packages (29–31). However, many of these attempts have had only limited success because of the failure to utilize insect-tight constructions. Also, it was shown earlier by Butterfield et al. (5) that chemicals migrated from the treated package into the contents and produced unacceptable residues.

Pouches made of laminated glassine/foil/polymer and treated with synergized pyrethrins protected cocoa powder and dried soup mix from infestation by mixed populations of common stored-product insects (21). The glassine prevented the occurrence of detectable residues in the cocoa powder or dried soup mix. In one kind of insect-resistant film, pyrethrins and piperonyl butoxide are incorporated into the polyethylene polymer used to manufacture the film (13). Polymer film treated with permethrin and heat-shrunk around bundles of small bags of cornmeal provided excellent protection from infestation (26). Only small quantities (up to 1.25 ppm) of permethrin were found in the cornmeal after 12 months of storage.

Foil-wrapped meat cubes have been protected from infestation by *Ptinus ocellus* (Australian spider beetle), *Stegobium paniceum*, *Dermestes lardarius*, and *Tribolium confusum* by packaging them in cartons treated with synergized pyrethrins (9). A mixture of 1,5a,6,9,9a,9b-hexahydro-4a(4H)-dibenzofurancarboxaldehyde (MGK Repellent 11) and N-(2-ethylhexyl)-5-norbornene-2,3-dicarboximide (MGK 264) sprayed on corrugated shipping cases prevented infestations by German cockroaches (*Blattella germanica*) (34). Since these shipping cases were not sealed tightly enough to prevent the entrance of cockroaches, short-term prevention of infestation was entirely dependent on the repellent activity of the chemicals. The development of pesticides and pesticide formulations for use in processing and packaging is inherently slow and difficult. Possible health hazards and effects on the environment must be considered. Also, manufacturers and formulators of pesticides often cannot profitably assume the task of developing materials for such uses because of the relatively small potential sales volume. However, to help conserve food, reduce costs, and maintain or raise sanitation standards, proper protective measures must be available. Measures now available include the use of packages and packaging materials that provide the most resistance to invasion and penetration by insects, along with the added protection provided by insecticide-treated, insect-resistant packages.

Cited References

1 Anonymous. 1974. Piperonyl butoxide and pyrethrins. *Federal Register* 39(210)38224–38225.

2 Anonymous. 1975. Tolerances for pesticides in food administered by the Environmental Protection Agency. *Federal Register* 40(61)14156–14164.

3 Batth, S. S. 1970. Insect penetration of aluminum-foil packages. *Journal of Economic Entomology* 63(2)653–655.

4 Brickey, P. M., Jr., J. S. Gecan, and A. Rothschild. 1973. Method for determining direction of insect boring through food packaging materials. *Journal of the Association of Official Analytical Chemists* 56(3)640–642.

5 Butterfield, D. E., E. A. Parkins, and M. M. Gale. 1949. The transfer of DDT to foodstuffs from impregnated sacking. *Journal of the Society of Chemical Industries* 68(Nov.)310–313.

6 Cline, L. D. 1978. Penetration of seven common flexible packaging materials by larvae and adults of eleven species of stored-product insects. *Journal of Economic Entomology* 71(5)726–729.

7 Cline, L. D., and H. A. Highland. 1981. Minimum size of holes allowing passage of adults of stored-product Coleoptera. *Journal of the Georgia Entomological Society* 16(4)525–531.

8 Collins, H. E. 1963. How food packaging affects insect invasion. *Pest Control* 31(10)26–29.

9 Dennis, P. O. 1962. Insect infestation. The protection of foodstuffs by synergized Pybuthrin. *Food Processing & Packaging* 31(367)131–133.

10 Gerhardt, P. D., and D. L. Lindgren. 1954. Penetration of various packaging films by common stored-product insects. *Journal of Economic Entomology* 47(2)282–287.

11 Gerhardt, P. D., and D. L. Lindgren. 1955. Penetration of additional packaging films by common stored-product insects. *Journal of Economic Entomology* 48(1)108–109.

12 Hanlon, J. F. 1971. *Handbook of Package Engineering.* McGraw-Hill, New York, NY.

13 Heselev, M. 1978. Packaging material resistant to insect infestation. British Commonwealth Patent 1,568,936. Patent Office, London, England.

14 Highland, H. A. 1967. Resistance to insect penetration of carbaryl-coated kraft bags. *Journal of Economic Entomology* 60(2)451–452.

15 Highland, H. A. 1975. *Insect Resistance of Composite Cans.* ARS-S-74. U.S. Department of Agriculture, Washington, DC.

16 Highland, H. A. 1977. Chemical treatments and construction features used for insect resistance. *Package Development and Systems* 7(3)36–38.

17 Highland, H. A. 1978. Insects infesting foreign warehouses containing packaged food. *Journal of the Georgia Entomological Society* 13(3)251–256.

18 Highland, H. A., and E. G. Jay. 1965. An insect resistant film. *Modern Packaging* 38(7)205–206, 282.

19 Highland, H. A., and P. H. Merritt. 1973. Synthetic pyrethroids as package treatments to prevent insect penetration. *Journal of Economic Entomology* 66(2)540–541.

20 Highland, H. A., R. V. Byrd, and M. Secreast. 1968. *Effect of Ethylene Vinyl Acetate on the Migration of Piperonyl Butoxide from Coatings of Synergized Pyrethrins on Kraft Paper.* ARS 51-28. U.S. Department of Agriculture, Washington, DC.

21 Highland, H. A., L. D. Cline, and R. A. Simonaitis. 1977. Insect-resistant food pouches made from laminates treated with synergized pyrethrins. *Journal of Economic Entomology* 70(4)483–485.

22 Highland, H. A., D. F. Davis, and F. O. Marzke. 1964. Insect-proofing multiwall bags. *Modern Packaging* 37(12)133, 134, 136–138, 195.

23 Highland, H. A., R. H. Guy, and H. Laudani. 1968. Polypropylene vs. insect infestation. *Modern Packaging* 41(12)113–115.

24 Highland, H. A., M. Secreast, and P. H. Merritt. 1968. Polyvinylidene-coated kraft paper as an insecticide barrier in insect-resistant packages for food. *Journal of Economic Entomology* 61(3)1459–1460.

25 Highland, H. A., M. Secreast, and P. H. Merritt. 1970. Packaging materials as barriers to piperonyl butoxide migration. *Journal of Economic Entomology* 63(1)7–10.

26 Highland, H. A., R. A. Simonaitis, and R. Boatright. 1984. Insecticide-treated film wrap to protect small packages from infestation. *Journal of Economic Entomology* 77(5)1269–1274.

27 Highland, H. A., et al. 1966. The migration of piperonyl butoxide from treated multiwall kraft bags into four commodities. *Journal of Economic Entomology* 59(3)543–545.

28 Highland, H. A., et al. 1984. Evaluation of permethrin as an insect-resistant treatment on paper bags and of tricalcium phosphate as a suppressant of stored-product insects. *Journal of Economic Entomology* 77(1)240–246.

29 Joshi, H. C., and C. L. Kaul. 1965. Studies on the

protectivity of jute bags impregnated with organic insecticides against red flour beetle and cigarette beetle. *Indian Journal of Entomology* 27(4)491–493.

30 Lal, R., P. D. Srivastava, and P. Dhar. 1960. Efficacy of insecticide impregnated jute bags against *Sitophilus oryzae* Linnaeus (Curculionidae: Coleoptera). *Indian Journal of Entomology* 22(3)204–210.

31 Langbridge, D. M. 1970. Treatment of paper to protect packaged food from insect attack. *Appita* 24(1)45–51.

32 Laudani, H., H. A. Highland, and E. G. Jay. 1966. Treated bags keep cornmeal insect-free during overseas shipment. *American Miller and Processor* 94(2)14–19, 33.

33 Loschiavo, S. R. 1970. 4'(3,3-Dimethyl-1-triazeno) acetanilide to protect packaged cereals against stored products insects. *Food Technology* 24(4)485–489.

34 Mallis, A., W. E. Esterlin, and A. C. Miller. 1961. Keeping German cockroaches out of beer cans. *Pest Control* 29(6)32–35.

35 Rao, K. M., S. A. Jacob. and M. S. Mohan. 1972. Resistance of flexible packaging materials to some important pests of stored products. *Indian Journal of Entomology* 34(2)94–101.

36 Watters, F. L. 1966. Protection of packaged food from insect infestation by the use of silica gel. *Journal of Economic Entomology* 59(1)146–149.

37 Yerington, A. P. 1971. Insect resistance research. *Good Packaging* 32(9)12–13.

38 Yerington, A. P. 1975. Insect resistance of polypropylene pouches. *Modern Packaging* 48(5)41–42.

39 Yerington, A. P. 1978. Insects and package seal quality. *Modern Packaging* 51(6)41–42.

40 Yerington, A. P. 1979. *Methods to Increase the Insect Resistance of Food Shipping Cases. USDA Advances in Agricultural Technology*. AAT-W-7. Science and Education Administration, U.S. Department of Agriculture, Oakland, CA.

Selected References

Cline, L. D., and H. A. Highland. 1976. Clinging and climbing ability of adults of several stored-product beetles on flexible packaging materials. *Journal of Economic Entomology* 69(6)709–710.

Highland, H. A. [1975.] The use of chemicals in processing and packaging of stored products to prevent infestation. *Proceedings of the First International Working Conference on Stored-Product Entomology* (Savannah, GA, 1974), pp. 254–260.

Highland, H. A. 1976. Materials, constructions and treatments for protecting packages from deterioration by insects. In *Proceedings of the Third International Biodegradation Symposium* (Kingston, RI, 1975), J. M. Sharpley and A. M. Kaplan, eds. Applied Science, London, pp. 273–278.

Highland, H. A. 1978. Insect resistance of food packages— a review. *Journal of Food Processing and Preservation* 2(2)123–130.

Wohlgemuth, R. 1979. Protection of stored foodstuffs against insect infestation by packaging. *Chemistry and Industry* 1979(10)330–334.

24
Prevention and Management of Pest Problems Associated with Transportation of Food

William H. Schoenherr and James H. Rutledge

In this chapter, each facet is approached from the viewpoint of regulatory as well as industry interests. Absolute purity and complete freedom from problems caused by pests should remain the objectives, but in reality these high standards are seldom achievable. One example is the naturally occurring pest problems that result in sanitation defects that cannot be completely removed by cleaning and processing. The ability to isolate and identify defects by means of laboratory testing procedures has progressed much more rapidly than the technological improvements to ensure the removal of defects. At present, this ability is of concern when it is related to the control of pests, since severe restrictions have been placed on the use of pesticides and pest control methodologies. It is important that the people charged with accepting or rejecting food or food-conveyance vehicles properly identify and thoroughly understand the significance of the problem. If they are unable to do so, safe and badly needed food may be diverted from normal channels of distribution. This is important today, of course,

but in years ahead it will be an absolute necessity.

The four key words in this chapter—*prevention, management, pest,* and *transportation*—should be considered in the following contexts:

> *Prevention.* The objective is to prevent a pest problem from developing during transportation of food.
> *Management.* The complete exclusion or prevention of pest problems during transportation is not always possible. Certain actions, necessary to minimize the risks, are referred to as "control measures" or "pest management."
> *Pest.* Most emphasis here is placed on the potential exposure of food and conveyances to insect pests. Mention is made of pest rodents, birds, and microorganisms, but to a much lesser degree.
> *Transportation.* Our discussion focuses on trucks and railcars since freight movement by these conveyances represents a substantial percentage of all food transported. Pest prevention and pest management procedures are often similar regardless of the type of conveyance.

The problems associated with transportation and pests are not limited to one country or to one part of the world (6, 8, 14). Imported food may already be in violation of the law when it arrives in the USA. To extend adequate control measures beyond our borders is both complicated and difficult. This does not

Schoenherr: Retired from Lauhoff Grain Company. Present address: P.O. Box 86, Inglis, FL 32649.

Rutledge: Lauhoff Grain Company, P.O. Box 571, Danville, IL 61832.

mean to imply that all imports are a problem. Some countries, in fact, have very exacting food purity requirements; many, however, do not. Therefore, food received from outside the USA should be subjected to rigid surveillance and purity requirements regardless of the point of origin or mode of transportation. Selection, rejection, cleaning, and preparation of conveyances are all a part of the prevention and management of pest problems. Within U.S. borders, the selection and preparation of conveyances for hauling food are of paramount importance.

Pests

Insect pests associated with food transport can be grouped into three categories:

(a) Primary pests are those that feed on or reproduce in the food.
(b) Secondary pests are insects that do not feed on or reproduce in food, but their presence may indicate an undesirable condition associated with the food such as wet, soured, or decomposing matter.
(c) Incidental pests are all the other insects or closely related arthropods that may enter a conveyance or otherwise come into close proximity to food.

Positive identification of the species involved is mandatory. Armed with the correct nomenclature and reference material, the manager can determine the significance of the pest. The uncanny ability of insects to show up in almost every situation means that sooner or later they will be found in most conveyances. A policy that calls for the rejection of any food or conveyance because even one insect is present will cause catastrophic problems. When the weather is warm, insects often become numerous and often enter conveyances, production plants, offices, and homes. Pest management personnel must locate and identify the pests, determine their significance and how access was gained, remove them, and set up the necessary controls to prevent future entry. If the problem is with a primary pest, then very careful inspection, sometimes supported by laboratory testing, may be advisable to determine if actual contamination of food has occurred. The high cost of food plus the projected shortages in the years ahead will eventually make this kind of quality assurance a necessity.

Rodents or contamination of rodent origin may occasionally become a cause for concern in conveyances. Rats and mice can climb; they have the ability to detect the presence of food and, if hungry, they can usually find a way to reach the food. Therefore, rodent infestation of conveyances does happen. Physical removal of the rodents does not, of course, eliminate the urine and hairs that remain behind as potential or actual contaminants. When a rodent-damaged package is noted in a conveyance, these questions should

be asked: Is the rodent actually in the conveyance? Did the damage occur prior to loading? Is there contamination by urine, hair, or feces? Inspection tools are available to help answer these questions. This brings us back to the need to establish the significance of the problem, its source, and the action to be taken.

In the case of pest birds, the species involved is not always critically significant; what is important is to determine if the conveyance or food is contaminated. Contamination can result from feathers or fecal matter coming into contact with the food. The high incidence of *Salmonella* in bird excreta gives this kind of contamination special importance. The occurrence of birds roosting, nesting, or feeding where food is produced, stored, moved, or loaded into conveyances should not be tolerated. Foot traffic, hand carts, or forklifts can carry contaminants, especially droppings, from docks into the conveyances.

Microorganisms

Since the presence of microorganisms is often difficult to determine, it is important to recognize the conditions that may serve as indicators of their presence. These indicators might include a moist condition, sour odor, staining, or change in color of the food or area being inspected. In such cases, conclusive determination may require laboratory testing. Unfortunately, testing requires time, and the decisions that need to be made often cannot be put off until the test results are in. If, for example, questionable conditions are noted about an empty conveyance that is to be used for a shipment of food, then the conveyance should either be rejected or held until samples from it can be properly analyzed. If an arriving shipment is suspect, then the sampling and analysis should be done prior to unloading. If circumstances are such that unloading is necessary and if there is a question of acceptability, then the product should be stored in such a manner that it will not contaminate other food and in such a way that it can be easily isolated should this become necessary.

Molds and bacteria are everywhere; they can always be found in empty conveyances or on shipments of food. The questions to be answered are: What is the significance, if any, of these microorganisms? What action is required?

Transportation

Railcars

Since sizable quantities of food are shipped by rail in the USA, this mode of transportation is emphasized. At least three different types of railcars are used to transport food: boxcars, closed hopper cars, and piggyback and container flatcars. Shipping containers are also moved by truck, barge, and ship. Boxcars are generally used for packaged food. Because of efforts by regulatory and private interests, considerable attention has been given to boxcar design and use (1–5, 7, 9–13, 15, 16). The practice of reserving certain boxcars for food use only should reduce the probability of contamination by chemicals and foreign materials. When closed hopper cars are lined with a smooth coating and the hatches are protected by gaskets and closing devices, bulk shipments are well protected.

It is common practice to lease railcars, especially hopper cars. This has reduced the potential for product contamination that exists when "free-run" cars are used. Bulk-loaded commodities are especially vulnerable to contamination. Residues from previous ladings may remain in hopper cars, and normal cleaning procedures may not ensure their complete removal. Regardless of the type of railcar selected for food loading, a very careful inspection is necessary and should be supported by laboratory examination of suspect materials. Containers have certain advantages over box- and hopper cars, especially in the case of exported or imported foods. Unless there is a faulty or contaminated container, the food placed into the container should reach the destination in good condition.

Trucks

Most trucks used for transporting food are of the closed-body design. The open-body design can be used for fresh produce, live animals, and grain. Common carrier vehicles present the same potential for contamination by previous lading as does the "free-run" railcar. Also, the movement of food in less than truckload lots can allow for exposure to other items carried in the same vehicle. Owned or leased trucks offer distinct advantages in the control of materials that are transported.

Closed-body tank trucks offer exceptionally good protection to the contents. Most closed-body truck trailers are not of tight enough construction to prevent the entry of microorganisms, insects, moisture, and foreign materials (e.g., road dust, exhaust fumes, and other airborne contaminants). Anyone who doubts this conclusion should try to fumigate a truck and note how quickly the gas escapes, or set up a smoke generator inside a truck and watch for the escape of smoke.

Since trucks are important and often-used vehicles for the transport of food, the same careful inspection and preparation are advisable here as were recommended for railcars (1, 2, 10). The potential for serious contamination is even greater in trucks than in railcars.

Aircraft

Considerable quantities of food are shipped to and within Alaska by air, but in the rest of the USA, the relatively high cost precludes routine air shipment of infestable foods. The major categories of food shipped by aircraft are perishable items such as seafoods, high-cost/low-bulk items, in-flight meals, food for areas that can be reached only by aircraft, and emergency food supplies airlifted into disaster areas. The last two categories have the greatest potential for infestation.

The potential for pest infestation in airlifted food is remote if the food placed on the aircraft is uninfested. This is due to rapid in-transit time, measured in hours for aircraft, in contrast to days for truck or rail, and weeks or months by ship. Because of the short time involved, the major potential pest problem is not the amount of foodstuffs eaten, damaged, or fouled in transit; it is the introduction of pests into areas where they do not occur naturally. Most important food pests are cosmopolitan, but some, for instance the khapra beetle (*Trogoderma granarium*), would cause serious problems if introduced into a new area.

There are several unique aspects to pest control on aircraft. Food on aircraft should never be fumigated. Aircraft are difficult to ventilate after fumigation; subsequent pressurization of the cabin may force unvented fumigant from dead air spaces into the cabin, endangering the crew.

Dusts or aerosols are usually used to disinsect aircraft (17–19). When properly applied, these formulations are effective against flying or crawling insects in the cabin. They provide little or no control of insects in containers or palletized freight under a tarp. The most commonly used dust is carbaryl (although there are probably other dusts labeled for aircraft use). Aircrews generally dislike dusts because they are messy. Aerosols, usually d-phenothrin, permethrin, or resmethrin, can be applied from a pressurized "bug bomb" if the aircraft has no built-in provision for disinsection. It is imperative to use special aircraft formulations so that the solvents and propellants do not etch windows, deform plastics, corrode metals, or cause other damage. Aircraft are normally not treated to control insects in freighted foods unless the insects are visible in the cabin or strict international quarantine measures are in effect.

The opportunity for cross-infestation of clean foods by infested foods aboard aircraft is minimal because of the short transit times involved. Normally, food is not stored in airports for long periods. Therefore, if food arrives at the airport free of pests, in all probability it will arrive at its destination free of pests. For this reason, port authorities should not accept visably infested foods for air shipment. Nonacceptance of infested material will keep the risk of extending the range of a pest, or of infesting clean food, to a minimum. When uninfested food is received at the airport, proper warehousing, sanitation, and pest-free-area management should all be used to minimize the risk of infestation from other sources. The pest management, sanitation, and warehousing requirements for airports do not differ significantly from those for seaports or land terminals.

Ships

Shipboard transportation of food products provides conditions that are almost ideal for insect growth and the spread of infestations from one product to another. The holds and storerooms in which products are held are virtually undisturbed for long periods, in addition to being dark and warm. Pallets of food are stored in extremely close proximity to each other to conserve valuable space. Under these circumstances, prevention and management of stored-product pest infestations are of the utmost importance. An undetected or mishandled infestation can easily result in the loss of thousands of dollars worth of food; the costs of shipping and handling up to the time of food disposition must be added to the basic value of the food.

On board the ship, prevention of insect infestation begins with proper preparation of the ship's holds, storerooms, staging areas, and conveyors. The prevention program involves three steps: (a) cleaning, (b) prophylactic treatment, and (c) inspection of products.

Before loading begins, deck gratings in all involved areas must be removed and thoroughly cleaned. Decks must be absolutely clean and free of accumulated rust, dust, and debris that might provide insect harborages. Special attention must be paid to areas frequently overlooked in routine cleaning, such as around stanchions and beneath ramp plates. Gear storage boxes or compartments frequently serve as unapproved trash receptacles, and therefore must also be cleaned.

Once all surfaces with which food might come in contact are clean, the deck should be sprayed with an approved residual insecticide such as 1% diazinon; bulkheads should also be sprayed to a height of one meter. Deck gratings may then be replaced and likewise sprayed with the insecticide.

Prior to loading products into the ship's holds, representative samples of infestable food products should be carefully inspected for signs of insect infestation. Fumigation of products, such as may have been done in transit from the producer to the pier, should not be relied upon blindly. This technique has been found to be much less than 100% effective and often merely promotes a false sense of security. Any items showing signs of infestation at the time of receipt should be rejected or, at a minimum, isolated from other infestable products and loaded into refrigerated holds to prevent growth and spread of the infestation.

Proper supervision and the use of properly trained personnel during loading will minimize breakage and spillage and will permit the most advantageous placement of pallets within holds and storerooms. Any

breakage should be spotted and removed or repaired immediately, and any resulting spillage cleaned up.

During the voyage, prevention of infestation involves a twofold program of sanitation and product inspection. Shifting cargo and unnoticed breakage can result in the exposure of the food product and its spillage onto the deck or other containers where it may attract and harbor pests. A portable wet/dry vacuum cleaner is indispensible for cleaning up such spillage; the hose can reach into narrow areas above, below, and between pallets without the necessity of shifting the entire load. Inspections of the holds and storerooms should be done weekly so that any necessary cleanup will be accomplished promptly. Although it may be difficult, the removal of spillage beneath the deck grates is essential for an effective program of insect prevention and management in ship's holds.

Infestable products should be inspected on a regular basis during shipment. Extremely light or new infestations that might easily have been overlooked during preloading inspection can balloon out of control during the course of a few weeks under the conditions prevailing in holds and storerooms.

When conducting product inspections, inspectors should first note any live or dead insects on package surfaces, on the deck, or flying about the hold. When selecting a representative sample, they should choose a package that is two or three layers from the top of the pallet (stored-product pests tend to move down and away from any light or activity). Insects infesting food products will usually be found on the underside of plastic bags, in the bottom of cartons and sacks, and under carton or box flaps. Observed frass, silk, or chewed entry/exit holes are clues to an infestation even if actual insects are not seen immediately.

Virtually any product packaged in bags or cartons is considered infestable. So also are some items that are normally packaged in jars, such as paprika or dry grated cheeses. The "top 10" infestable packaged foods are cornmeal, grits, farina, fry mix, pasta products, barley, cookie and cake mix, flour, crackers, and dry beans and peas. It should be stressed that inspection should by no means be limited to this "top 10"; such diverse items as cocoa, spices, and dried potatoes may also support insect life. Thorough inspection of any infestable item involves close attention to detail and the use of an adequate flashlight. There should be no need to actually open food containers, but cartons containing smaller packages should be opened for inspection and resealed with a suitable tape.

Products that are discovered to be infested should be immediately isolated from other infestable products and placed in refrigerated compartments or, preferably, freezers. In the event that infestation is discovered, a special effort must be made to ensure that it has not spread to other nearby products. Once all infested items are under refrigeration, the decision can be made as to disposition, based upon the level of infestation.

By following these simple guidelines, both prevention and management of food-pest infestations during shipboard transportation can be quickly, easily, and economically achieved.

Problems

Prevention

Regardless of the type of conveyance selected to ship food, the questions to ask are the same. Will the food be adequately protected from atmospheric conditions? Can it be secured to avoid pilferage? Do structural conditions exist that could damage the packages or the food? Are there any pests in the conveyances? Could the food be exposed to any conditions that could cause contamination? The final considerations are to determine the feasibility for thorough cleaning and for closing of all openings into the conveyance. The results of these considerations will permit a determination of whether the conveyance should be accepted or rejected and if the food will be adequately protected.

Management

There have been numerous studies of pest problems in railcars, some dating back to 1939 (6). Our personal observations have indicated that the problems that exist in trucks and ships are similar. A determination should be made as to the feasibility of managing the pest problems rather than rejecting the conveyance. Some conveyances can be adequately cleaned; others require a combination of cleaning and the use of pesticides. Certain pesticides are labeled for use in conveyances. Since this may change, it is advisable to examine the label to determine if there is clearance for the intended use. A lasting residual protection should not be expected from the kinds of pesticides that are authorized for use where food may be exposed. The main objective in the use of a pesticide is to kill the target pests that are present. Therefore, the sealing of all openings into the conveyance is important to prevent the entry of pests after the pesticide is no longer effective.

Unfortunately, conditions may exist that are not readily controllable. Such conditions are not necessarily limited to loading and unloading of the conveyance but allow for exposure during the transit period. This is especially the case for rail shipments where spillage of grain and other food has occurred in terminals and holding yards. When conditions of temperature and moisture are suitable, such spillage results

in heavy buildup of insects and microorganisms. Rodents and birds may also frequent such areas. The most severe potential problem is the invasion of insects into a conveyance, insects that may have been exposed to insanitary conditions. This again points to the importance of sealing all openings into conveyances. Some additional protection may be expected where the conveyance is under fumigation. In order for a fumigant to be fully effective, it must be confined, which means a well-sealed vehicle is essential.

Since many insects have the ability to fly, they may enter conveyances during the loading and unloading periods. This points up the importance of cleanliness around docks, ramps, trackage, and other areas where the conveyances may be opened or held.

Safety

There are two principal areas of personal safety concerns. One is the possibility of injury due to falling, slipping, or contacting sharp objects. The second is exposure to toxic materials. The toxic materials may be residues from earlier loads that must be removed before the conveyance can be used. Pesticide labeling states that warning signs are required for certain formulations. When warning signs are posted on a conveyance, the taking of proper precautions is mandatory.

Safety to the commodity can also have two considerations. One is the damage to the carton or package. Such damage may serve as a point of entry for pests. The second is the exposure to toxic materials. Such materials may be in the conveyance prior to your use and may represent pesticides that are intentionally used as a corrective or preventive measure, or may result from other items placed into a conveyance. The latter has concerned many shippers as well as receivers, especially where trucks or ships are used. Once again, the pesticide label clearly lists the recommendations for use. If pesticide label instructions are closely followed, then there is little chance of causing pesticide contamination of food.

Acknowledgments

We gratefully acknowledge the technical suggestions made by Stephen M. Valder and John L. McDonald on aircraft and ships, respectively.

Cited References

1 American Feed Industry Association. 1988. *Recommended Salmonella Control Guidelines for Processors of Livestock and Poultry Feeds.* AFIA, Arlington, VA.

2 American Feed Manufacturers Association. 1976. *Voluntary Transportation Guidelines Designed to Prevent the Contamination of Food, Feed, and Related Products.* AFMA, Arlington, VA.

3 Atchley, E. H. 1974. The problem of quality deterioration in the railroad freight car fleet. *Food Drug Cos-*

metic *Law Journal* 29(10)510–518.

4 Byrne, T. J. 1974. The Commission's role in car supply and its actions concerning boxcars and covered hopper cars. *Food Drug Cosmetic Law Journal* 29(10)531–536.

5 Fine, S. D. 1974. Railroad car sanitation and the FDA. *Food Drug Cosmetic Law Journal* 29(10)527–530.

6 Food Protection & Sanitation Committee. 1974. *Box Car Infestation.* Association of Operative Millers, Kansas City, MO.

7 Haderer, C. L. 1974. NAFC's view of railroad car sanitation. *Food Drug Cosmetic Law Journal* 29(10)504–509.

8 Hales, K. C. 1979. The influence of physiological and physical properties of fruit and vegetables on their overseas transport. *Annals of Applied Biology* 91(3)409–413.

9 Henderson, L. S., and H. E. Meister. 1977. *Guidelines for Pest Control in Railcars for Food Transportation.* Program Aid 1178. Agricultural Research Service, U.S. Department of Agriculture, Washington, DC.

10 Jones, D. F. 1980. Product protection in warehouses, railcars, and trucks. *Association of Food Drug Officials Quarterly Bulletin* 44(4)218–221.

11 Kirschbaum, N. E. 1974. The role of state regulatory agencies in railroad car sanitation. *Food Drug Cosmetic Law Journal* 29(10)541–544.

12 Peterson, E. L. 1974. Railroad car sanitation—a USDA perspective. *Food Drug Cosmetic Law Journal* 29(10)537–540.

13 Rutledge, J. H. 1974. Potential in-transit contamination of food products in rail cars. *Food Drug Cosmetic Law Journal* 29(10)492–498.

14 Snowdon, A. L. 1979. Disease and disorders of imported fruits and vegetables. *Annals of Applied Biology* 91(3)404–409.

15 Snyder, J. R. 1974. Rail car contamination potential—a feed manufacturing view. *Food Drug Cosmetic Law Journal* 29(10)499–503.

16 Van Slyke, W. H. 1974. Railroad program for clean cars. *Food Drug Cosmetic Law Journal* 29(10)519–526.

17 World Health Organization. 1986. Recommendations on disinsecting aircraft with permethrin. PAHO Bulletin 20(1)70–72.

18 World Health Organization. 1985. *Weekly Epidemiological Record* 60(7)45–47, 60(12)90, 60(45)345–346.

19 World Health Organization. 1987. *Weekly Epidemiological Record* 62(44)335–336.

Selected References

Halliday, W. R., N. O. Morgan, and R. L. Kirkpatrick. 1987. Evaluation of insecticides for control of stored-product pests in transport vehicles. *Journal of Entomological Science* 22(3)225–236.

Hayes, D. K. 1984. The history and present status of aircraft disinsection. In *Commerce and the Spread of Pests and Disease Vectors*, M. Laird, ed. Praeger, New York, NY, pp. 23–36.

Stauffer, J. E. 1988. *Quality Assurance of Food. Ingredients, Processing and Distribution.* Food & Nutrition, Westport, CT.

25
Commentary on Prevention

Walter Ebeling

In dealing with the preventive aspects of the ecology and management of food-industry pests, the authors of Chapters 21 through 24 addressed specifically what in a more general sense is the principal theme of this entire volume: a pest management program to reduce the use of chemicals to a bare minimum. This is not meant to disparage the importance of chemicals *when properly applied*. But whenever possible, prevention is better than control. With a few exceptions, easy chemical solutions to pest problems, unlike the pests themselves, are coming to an end. A good example is the demise of ethylene dibromide in the milling industry.

It is somewhat ironic that the huge and ambitious undertaking that Technical Bulletin 4 represents was proposed and sponsored by the Food and Drug Administration, a federal agency whose legislative mandate is to apprehend and prosecute violators of the law, not necessarily to show them how to do things right. Perhaps this is a sign of the times—a mandate for the food industry to start institutionalizing prevention and investing a generous amount of cold hard cash toward that objective. Many guidelines pointing in that direction can be found in the four chapters considered in this brief commentary.

In Chapter 21, the late D. F. Jones briefly discussed the basic sanitation principles that pertain to

(a) the type of contaminating species (rodents, birds, insects, mites, and microbes);
(b) the forms of contamination (origin, class, and place in the food-processing sequence);

(c) points in the farm-to-table food chain at which the probability of contamination can be modified;
(d) the expertise required in the various aspects of food protection; and
(e) the dedication required, on the part of food processing companies, to prevention and follow-up.

Effective orientation and training programs are essential. But their maximum impact in preventive sanitation will not be attained unless the top management of a food-industry facility is thoroughly committed to the principles embodied in the training programs.

In Chapter 22, H. G. Scott emphasized that, in addition to sanitation, proper design and construction of food management facilities and equipment are of great importance. Their purpose is to exclude pests, and this is relatively simple when dealing with commensal rats, pest birds, and bats, at least when compared with generally much smaller creatures such as mice and insects. However, building materials with a long history of successful use, such as copper screens, which tend to exclude rats, commonly vanish from the marketplace, to be replaced by aluminum and plastic screening through which rats can easily gnaw. It is encouraging to see a reliable old material—copper mesh (used to obstruct pest access points to buildings)—return to the market place (STUF-FIT, Allen Special Products, P.O. Box 605, Montgomeryville, PA 18936).

To make wise decisions on the purchase of the materials to keep out rats, IPM personnel must be knowledgeable and selective. Scott emphasized that building ratproofing into a facility is less expensive than repeatedly treating for rats after construction. The total exclusion effort should provide for an integrated

Ebeling: Department of Biology, University of California, Los Angeles, CA 90024.

357

program of rodent, bird, bat, insect, and mold control. For exclusion of insects and other arthropods, regular and repetitive inspection of incoming goods by a firm's quality assurance/sanitation group is much more important than the elimination of structural openings, though the latter should not be neglected.

In Chapter 23, H. A. Highland pointed out that some insect species (penetrators) can bore through one or more of the many flexible packaging materials. Other species (invaders) usually enter only existing openings, but can easily find them in many packages as currently designed and manufactured. A knowledgeable observer can determine whether holes formed by insects are ingress or egress holes by the characteristics of the bored holes. Highland discussed the relative merits of the many materials (mostly synthetic) that are being tested or used in packaging food products and the role of package structure in preventing penetration by insects.

Insecticides and repellents are being used to help make packages insect resistant. But their development is slow. Not only must health and environmental hazards be considered, but the potential for profitable use of acceptable formulations is limited because of a relatively small projected sales volume.

In Chapter 24, W. H. Schoenherr and J. H. Rutledge dealt with four types of conveyances—aircraft, trucks, railcars, and ships—listed here in order of increasing time spent in delivering a shipment of food products. This is also the order of increasing potential for infestation of food products in transit. The authors discuss structural features of railcars and trucks that can help to reduce infestation potential. Also, they emphasize that leasing a conveyance instead of using "free-run" or common carriers provides greater control over disinfestation and sanitation. On ships, all aspects of inspection, sanitation, and sealing of entryways are of great importance.

Some conveyances can be adequately cleaned but others require a combination of cleaning and use of pesticides labeled for application to conveyances. Those insecticides authorized for use where food may be exposed are not meant to provide residual protection but only to kill the target pests that are present.

IV
Survey and Control

26
Biological Methods of Survey and Control

27
Host-Plant Resistance to Insects in Stored Cereals and Legumes

28
Pest Resistance to Pesticides

29
Physical Methods to Manage Stored-Food Pests

30
Chemical Control of Insect Pests in Bulk-Stored Grains

31
Chemical Control of Rodent Pests in Bulk-Stored Grains

32
Chemical Methods to Control Insect Pests of Processed Foods

33
Fumigation in the Food Industry

34
Commentary on Survey and Control

26
Biological Methods of Survey and Control

Wendell E. Burkholder and Daryl L. Faustini

Stored-product insects and microorganisms are two of the major factors that directly affect the quantity and quality of grain. There is a growing concern about the need to suppress insect pests in domestic as well as foreign channels in order to upgrade grain quality. However, pesticide residues and pest resistance to insecticides are also causing concern in all market areas. This dilemma has generated a new trend toward more acceptable and more effective methods for the suppression of insect populations in storage, packaging, transportation, and distribution channels. This has led to the development of biological methods for insect survey and control in the food industry.

In both theory and practice, considerable interest has been generated in the use of pheromones as survey or monitoring tools to supplement pesticide applications. Furthermore, attention has also been directed toward the efficacy of specific biological agents and techniques as control methods: pheromones, pathogens, predators, parasitoids, insect growth regulators, chitin inhibitors, food attractants, oils, and genetic manipulation.

Burkholder: USDA ARS Stored-Product Insects Research Unit, Department of Entomology, University of Wisconsin, Madison, WI 53706.

Faustini: Research Center, Philip Morris USA, P.O. Box 26583, Richmond, VA 23261.

This contribution from the College of Agriculture and Life Sciences, University of Wisconsin, Madison, was supported by a cooperative agreement between the University of Wisconsin and the Agricultural Research Service, U.S. Department of Agriculture.

Survey Methods

One of the most important aspects of a pest management program is the establishment of a survey or scouting plan designed to find insect populations. Direct sampling of grain or packaged products will give absolute levels of insect populations and damage. Biological monitoring with natural products such as pheromones is an effective sampling procedure to determine population trends (*42*). Generally, the stored-grain environment is devoid of such environmental factors as wind, rain, and extreme temperature fluctuations, thus affording a more stable milieu for pheromone monitoring systems. Direct sampling and biological sampling used in concert provide a sound basis for pest management decisions.

An ideal survey program can offer the food-industry manager a relatively precise method for determining the need for control. Factors to consider are the presence of insects, a number index to establish a treatment threshold, location of the insect infestation, species present, and the insect's developmental stage. Some insects are merely incidental invaders that cause only temporary concern. Therefore, numbers and distribution of pest species within a warehouse are particularly important.

In storage and processing environments, the index for a treatment threshold is often the presence of one insect. In many instances when the commodity is sampled, the adult threshold level is low and larval forms are difficult to find. This is particularly true if the insects are internal grain feeders or are harbored within food packages or finished goods.

Pheromone traps are excellent devices for discovering low population levels of adult insects. These traps can often be used in conjunction with other sampling and control methods such as ultraviolet light, sticky-board traps, physical probing, or direct sampling of the product.

The use of pheromone traps for monitoring insects in storage facilities has been receiving considerable attention in recent years. Pheromone identification for a number of stored-product pests, especially Lepidoptera and Coleoptera, is continually expanding (Table 26.1). Lepidopteran pheromones have been available since (Z,E)-9,12-tetradecadien-1-ol acetate (ZETA) was isolated and identified as a component of the sex pheromone of the almond moth (*Cadra cautella*) and the Indianmeal moth (*Plodia interpunctella*) (9, 11, 56), the Mediterranean flour moth (*Anagasta kuehniella*) (10, 56), the tobacco moth (*Ephestia elutella*) (9), and the raisin moth (*Cadra figulilella*) (8). ZETA-baited traps have been shown to elicit attraction and other behavioral responses in males of each of these species, as has been shown in experimental trapping studies (11, 90, 92, 93).

Another component isolated from females of *P. interpunctella* is an alcohol, (Z,E)-9,12-tetradecadien-1-ol (ZETOH) (100). Traps containing a 60:40 mixture (acetate:alcohol) attracted more *P. interpunctella* than the acetate component alone. The alcohol component, also isolated from the abdominal glands of females of *C. cautella* (54, 89), *E. elutella*, and *A. kuehniella* (54), appears to inhibit the attraction behavior of *C. cautella* males (90, 101). Female *C. cautella* also produce another acetate component, (Z)-9-tetradecen-1-ol acetate (ZTA) (7, 98), which appears to enhance the attractiveness of ZETA (42, 90, 91).

When used together in the same trap, the sex pheromones of the Angoumois grain moth (*Sitotroga cerealella*) (114; Table 26.1) and of the Indianmeal moth (ZETA) were effective for the simultaneous capture of males of more than one species (115). These studies show that at least these two sex attractants can be dispensed from the same trap to monitor more than one species of stored-product moths in a warehouse.

Several dermestid species (Coleoptera) have been monitored in Wisconsin (2, 3) with pheromone traps constructed from four 9-cm² pieces of single-backed corrugated paper, stacked four layers high (17) and treated with 85 mg malathion and either 0.25 mg (E,Z)-3,5-tetradecadienoic acid (megatomoic acid), a sex attractant of the black carpet beetle [*Attagenus unicolor*

(=*megatoma*)] (16, 99), or (Z)-14-methyl-8-hexadecen-1-ol, a sex attractant component of several species of *Trogoderma* (94). The traps treated with pheromones caught significantly more male insects than the control traps in a cargo terminal, a grain elevator, and a milling company. These traps also caught *T. variabile*, a species not previously recorded in that warehouse. The seasonal emergence of *A. unicolor* over a 2-year period was also recorded through the use of the trap program.

More recently, trapping studies have been conducted in other states with an improved formulation of *Trogoderma* pheromone (2, 3, 29). The Animal and Plant Health Inspection Service (APHIS) has been utilizing the *Trogoderma* pheromone in a program to detect khapra beetles (*T. granarium*) in port facilities throughout the United States (USA). In the western USA where *Trogoderma* populations are generally more prevalent, trapping studies have been conducted using a modified cardboard trap at food warehouses. Traps were placed at strategic locations to verify the source of infestation. The preliminary results indicate that floor traps are more effective than those placed 1–2 m above the floor.

The cigarette beetle (*Lasioderma serricorne*) is the major pest of stored and processed tobacco. Studies have shown that pheromone traps are effective monitors for pinpointing cigarette beetle infestations in warehouses as well as processing areas (30, 31). Traps offer several advantages over conventional methods:

(a) They can be placed in processing equipment and areas more likely to harbor cigarette beetles.

(b) They can compete with tobacco odors generated during cigarette manufacturing.

(c) They do not require additional expenses to operate (i.e., electricity).

(d) Examination of the trap contents is less laborious since only the target insect is trapped (33).

Males of the lesser grain borer (*Rhyzopertha dominica*) produce a two-component aggregation pheromone (51, 119) (Table 26.1). Field studies indicate a potential for this pheromone in monitoring lesser grain borer populations. Trap catches of these beetles around storage bins were inversely proportional to the height of the trap (26).

An aggregation pheromone (4,8-dimethyldecanal) (105) has been isolated from two other species of stored-product pests, the confused flour beetle (*Tribolium confusum*) and the red

Table 26.1

Major pheromone components of some stored-product insects

Taxa	Sex	Pheromones	References
Anobiidae			
Lasioderma serricorne	female	4,6-Dimethyl-7-hydroxy-nonan-3-one;	(21, 22, 24)
		2,6-Diethyl-3,5-dimethyl-3,4-dihydro-2H-pyran;	(62, 63)
		2,3-cis-2,3-Dihydro-3,5-dimethyl-2-ethyl-6-(1-methyl-2-oxobutyl)-4H-pyran-4-one;	(23)
		2,3-cis-2,3-Dihydro-3,5-dimethyl-2-ethyl-6-(1-methyl-2-hydroxybutyl)-4H-pyran-4-one	(23)
Stegobium paniceum	female	2,3-Dihydro-2,3,5-trimethyl-6(1-methyl-2-oxobutyl)-4H-pyran-4-one	(58, 59)
Bostrichidae			
Rhyzopertha dominica	male	1-Methylbutyl (E)-2-methyl-2-pentenoate[a];	
		1-methylbutyl (E)-2,4-dimethyl-2-pentenoate[a]	(51, 119)
Bruchidae			
Acanthoscelides obtectus	male	(E)-(−)-Methyl-2,4,5-tetradecatrienoate	(44, 45)
Callosobruchus chinensis	female	Multicomponent	(43, 111, 112)
Cucujidae			
Cryptolestes ferrugineus	male	(E,E)-4,8-Dimethyl-4,8-decadien-10-olide[a];	
		(3Z,11S)-3-dodecen-11-olide[a]	(6, 121)
Cryptolestes pusillus	male	(Z)-3-Dodecenolide[a];	
		(Z)-5-tetradecon-13-olide[a]	(76)
Cryptolestes turcicus	male	(Z,Z)-5,8-Tetradecadien-13-olide[a]	(76)
Oryzaephilus mercator	male	(Z,Z)-3,6-Dodecadien-11-olide[a];	
		(Z)-3-dodecen-11-olide[a]	(84)
Oryzaephilus surinamensis	male	(Z,Z)-3,6-Dodecadien-11-olide[a];	
		(Z,Z)-3,6-dodecadienolide[a];	
		(Z,Z)-5,8-Tetradecadien-13-olide[a]	(84)
Curculionidae			
Sitophilus oryzae	male	(R*,S*)-5-Hydroxy-4-methyl-3-heptanone[a]	(82, 96)
Sitophilus zeamais	male	(R*,S*)-5-Hydroxy-4-methyl-3-heptanone[a]	(82, 96)
Dermestidae			
Anthrenus flavipes	female	(Z)-3-Decenoic acid	(20, 36)
Anthrenus verbasci	female	(Z)-5- and (E)-5-Undecenoic acid	(55)
Attagenus brunneus (= elongatulus)	female	(Z,Z)-3,5-Tetradecadienoic acid	(4, 5, 37)
Attagenus unicolor (= megatoma)	female	(E,Z)-3,5-Tetradecadienoic acid	(19, 99)
Trogoderma glabrum	female	(E)-14-Methyl-8-hexadecen-1-ol;	
		(E)-14-methyl-8-hexadecenal	(19, 29, 122)
Trogoderma granarium	female	92:8 (Z:E)-14-Methyl-8-hexadecenal	(29, 61)
Trogoderma inclusum	female	(Z)-14-Methyl-8-hexadecen-1-ol;	
		(Z)-14-methyl-8-hexadecenal	(19, 29, 94)
Trogoderma variabile	female	(Z)-14-Methyl-8-hexadecenal	(29)
Gelechiidae			
Sitotroga cerealella	female	(Z,E)-7,11-Hexadecadien-1-ol acetate	(50, 114)
Pyralidae			
Amyelois transitella	female	(Z,Z)-11,13-Hexadecadienal	(25)
Anagasta kuehniella	female	(Z,E)-9,12-Tetradecadien-1-ol acetate	(10)
	female	(Z,E)-9,12-Tetradecadien-1-ol	(54, 57)
Cadra cautella	female	(Z,E)-9,12-Tetradecadien-1-ol acetate	(11, 56)
	female	(Z,E)-9,12-Tetradecadien-1-ol	(54, 89, 90, 116)
	female	(Z)-9-Tetradecen-1-ol acetate	(7, 9)
Cadra figulilella	female	(Z,E)-9,12-Tetradecadien-1-ol acetate	(8)
Ephestia elutella	female	(Z,E)-9,12-Tetradecadien-1-ol acetate	(9)
	female	(Z,E)-9,12-Tetradecadien-1-ol	(54)
Plodia interpunctella	female	(Z,E)-9,12-Tetradecadien-1-ol acetate	(11, 56)
	female	(Z,E)-9,12-Tetradecadien-1-ol	(101, 116)
Tenebrionidae			
Tribolium castaneum	male	4,8-Dimethyldecanal[a]	(105, 106, 107)
Tribolium confusum	male	4,8-Dimethyldecanal[a]	(105, 106, 107)

[a]Aggregation pheromone; all others are sex pheromones.

flour beetle (*T. castaneum*). Larvae of *T. castaneum* are attracted both by contact and by vapor to synthetic 4,8-dimethyldecanal (*78*). The medium may be conditioned by the aggregation pheromone and by quinones, depending upon the number and sex of the insects and the duration of occupation. These conditions dictate whether or not the medium will be attractive to *T. castaneum* larvae (*77*). Quinones regulate population density by counteracting the effect of the aggregation pheromone. Under stressful conditions of overcrowding and lack of sufficient food resources, the pheromone is inhibited (*32*).

Males of the granary weevil (*Sitophilus granarius*) (*33*), the maize weevil (*S. zeamais*) (*117*), and the rice weevil (*S. oryzae*) (*82, 96*) produce aggregation pheromones (Table 26.1). The aggregation pheromone of male rice weevils attracts males as well as females (*81*).

Recent studies have shown that several stored-product insects utilize macrolide molecules as aggregation pheromones. Males of the rusty grain beetle (*Cryptolestes ferrugineus*) produce such a pheromone (*114, 121*). Two macrolides isolated from volatiles and frass of flat grain beetles (*Cryptolestes pusillus*) are both aggregation pheromones (*76*). The sawtoothed grain beetle (*Oryzaephilus surinamensis*) and the merchant grain beetle (*O. mercator*) also produce macrolide aggregation pheromones (*84*).

Pheromones have been identified in several of the seed beetles (Bruchidae): a male-produced sex pheromone in the bean weevil (*Acanthoscelides obtectus*) (*44, 45*), and a female sex attractant in the southern cowpea weevil (adzuki bean weevil) (*Callosobruchus chinensis*) (*43*) and the cowpea weevil (*C. maculatus*) (*95*). More complete knowledge about the pheromones of these insects will lead to the development of more effective pest management strategies for the future.

Control Methods

Mating Disruption with Pheromones

The use of pheromones for control depends on the efficacy of traps in capturing the attracted populations or in luring the insects to an area where other suppression methods can affect them. Mating disruption with pheromones may also produce some suppression, but this technique has the greatest impact when used in combination with other methods.

Disruption techniques have been utilized to reduce mating of *Plodia interpunctella* in enclosed environments in the presence of a synthetic pheromone (*100*). The reduction was not as effective at high population densities (10 pairs/m^2) as at low densities (0.2 pairs/m^2) on ceiling and wall surfaces. Similar experiments have been conducted using the pheromone of *Attagenus unicolor* (*16*). Mating disruption with (Z,Z)-11,13-hexadecadienal, the synthetic sex pheromone of the navel orangeworm (*Amyelois transitella*), also appears to be a promising technique of population suppression (*25*).

Pheromones with Pathogens

Another innovative strategy involves the use of pheromones as lures in devices that contain insect pathogens, a technique that is especially successful against insects in stored products. The habitats of these pests provide ideal conditions for inducing epizootics, especially when the pathogens have desiccation-resistant spores. Even when the food resource is limited to a confined niche, highly concentrated populations of insects may still occur under conditions that have excellent potential for the creation of epizootics. The pathogen would not kill the insects immediately, but an infected, spore-ladened insect returning to its natural habitat would infect others of its kind. This is an especially promising method for long-term control of stored-product insect pests when pathogen exposure can be enhanced by using an effective pheromone-baited device.

The combination of pheromones with pathogens or other biotic agents for suppression of dermestid beetles (*18, 19*) has been tested against populations of *Trogoderma glabrum* (*98*). A sex pheromone (14-methyl-8-hexadecenal) from the beetles was combined with the protozoan pathogen *Mattesia trogodermae* (Neogregarinida: Ophryocystidae). Since failure at any point to bring the insects into contact with the pathogens would contravene the desired effect of this strategy, the following conditions were established:

(a) Adult males used as the test population had to emerge synchronously, prior to the females;
(b) The adult male population had to be situated downwind from the sites where the pheromone and pathogenic spores were available;
(c) Adult males that came to the lure had to be redistributed among emerging females; and
(d) Dead adults had to be available as food for the larvae.

In this model system (*98*), subsequent generations of *T. glabrum* were substantially suppressed by a single introduction of *M. trogodermae* spores into high-density populations (32 P$_1$ adult males/m^2). The treated populations increased only fourfold in the first posttreatment generation and fell below pretreatment levels in the second generation. Meanwhile, the control population increased 24-fold in the first generation and 100-fold in the second generation. The low-density treated populations (2 P$_1$ adults/m^2) and the low-density controls both increased 12-fold during the first generation.

These results demonstrated that, with the dense population, (a) a 48-hour exposure to the pheromone plus spores was sufficient to achieve distribution of effective doses within a radius of 1.25 meters around the sites of the treatment, and (b) spore transfer was enhanced because attracted males tried to copulate with the pheromone source. However, spore transfer to the subsequent generations was mainly the result of the larvae ingesting either dead, contaminated adults or larval food that adults had contaminated by contact. Although the maximum distance that the spores were transferred was limited by the size of the experimental arenas (2 × 2 meters), it is likely that in a natural environment the pathogen would be dispersed substantially farther since flight by contaminated adults is possible.

Burges and Hurst (*15*) noted sudden and spectacular mortality of storage moths in a maize storage facility in Kenya when the insects were exposed to *Bacillus thuringiensis*. However, mortality of similar moths exposed in a similar way in laboratory jars was only progressive. As a result, they suggested that cadavers of larvae are the most potent source of infective materials because they contained many more spores and crystals of *B. thuringiensis* than did adults, frass, or eggs. Spores from dead larvae are capable of rapidly killing larvae that feed on them. When spores and crystals of *B. thuringiensis* were spread over the surface of grain in the jars rather than applying the material to one point source on the surface, significantly more spores and crystals were required to start an epizootic.

Epizootics may arise because infected larvae migrate into the storage area from adjacent stored grain or from residues of food from local farms or in transport vehicles, terminal stores, or bags contaminated with frass and insect bodies. In fact, the study showed that healthy larvae sometimes feed on larval cadavers even when other food is present. Thus, the most susceptible larvae would succumb first and provide inoculum

to infect the more resistant larvae. Such natural infections with *B. thuringiensis* rarely curb moth damage to food; the main effect is to limit moth reproduction (*15*). Predictable control and adequate protection can probably be obtained only by admixing a lethal dose (e.g., 2 × 10^9 spores/200 g of food) that will kill most young larvae (*14*).

Since the severe mortality of moth larvae in the Kenya study (*15*) was enhanced by high concentrations of spores and crystals in the larval cadavers, it might be possible to mimic nature by means of a simulated larval cadaver (SLC). The introduction of a few SLCs might generate an epizootic; the initial rapid kill of feeding larvae would provide real cadavers, thus accelerating the suppression of the insect population. The sporozoans *Glugea gasti* and *Mattesia grandis* have been used in a similar manner against the boll weevil (*Anthonomus grandis*) (*73*). Also, a feeding stimulant has been used to increase the effectiveness of a nuclear polyhedrosis virus of *Heliothis* species (*79*).

B. thuringiensis has been used under experimental conditions to prevent infestations of *Plodia interpunctella* and *Cadra cautella* in stored corn and wheat (*70*). Treatment of the surface layer to a depth of 100 mm was more effective than treatments to depths of 33 or 67 mm and as effective as treatments of the entire grain mass. The formulation was less effective in controlling *Sitotroga cerealella*; doses that gave complete control of *P. interpunctella* and *C. cautella* reduced emergence of adult *S. cerealella* by only about one-third. The viability of *B. thuringiensis* and granulosis virus was only slightly reduced one year after treatment of wheat in a farm grain bin. With proper timing of application, either pathogen might be able to protect the grain from *P. interpunctella* for one year; residual activity of the pathogens would probably extend protection even longer (*52*). Some laboratory-reared strains of *P. interpunctella* have developed resistance to *B. thuringiensis* (*71*).

Crystals and spores of a new strain of *B. thuringiensis* (var. *tenebrionis*), isolated from yellow mealworms (*Tenebrio molitor*), when applied to larvae of several kinds of beetle pests, produce a dosage-dependent pathogenic reaction (*53*).

Granulosis virus from *P. interpunctella* also protected dried nuts (*46, 47*) and grains (*69*) from infestation by this moth. An aqueous suspension of the granulosis virus controlled Indianmeal moths in stored in-shell almonds for

134 days (48). Feeding damage to the treated nuts was substantially reduced, and the percentage of rejected nuts was decreased by as much as 88%. The virus suspension used in the treatment lost no activity when stored at 80° for 18 months.

A noncrystalliferous, aerobic, spore-forming bacterium, *Bacillus cereus*, has been isolated from the cigarette beetle (*L. serricorne*) (34). This bacterium is pathogenic to larvae exposed to the spores (113). Studies are under way to combine this pathogen with a sex pheromone in an attempt to control small infestations.

The use of pheromones and attractants in baits with pathogens appears promising. Baits could consist of pieces of corrugated paper, paper straws, or a natural material such as wheat straw that are coated with the pathogen and the attractant. The ideal system would be a laminated structure made of safe, biodegradable materials. Adjuvants or stickers of the type currently used in the pesticide industry might be useful in binding the pathogen to the attractant. Distribution of the baits in the walls or cracks of empty bins, under conveyors, or in other areas where residual populations exist would enhance population suppression.

Another method of effectively distributing a pathogen among stored-product insects would be to provide a pheromone-baited or light-baited device with an open reservoir containing a pathogen such as *B. thuringiensis*. The insects attracted by the pheromone or light would become dusted with the pathogen and would distribute it within the insect population and habitat. Light has been used successfully as an attractant to induce night-flying insects to disseminate viral material in California cotton fields (38). The virus-dusted insects aggregated and dispersed in response to a timer that turned the light on and off in 15-minute cycles.

Predators

The predatory warehouse pirate bug, *Xylocoris flavipes* (Hemiptera, Anthocoridae), and several other anthocorid bugs of the subfamily Lyctocorinae frequently occur as predators in storage ecosystems (1). These insects are promising agents for suppression of Coleoptera and Lepidoptera in stored products since they prey on most stages of many of these species (49, 60, 86). Also, these predators have a high capacity to increase their numbers and to consume all available prey, or to reduce their numbers by cannibalistic behavior, thus reducing their own populations when prey is scarce. The difficulty is that *X. flavipes*, though effective against many unprotected insects, is incapable of penetrating hard materials such as seeds. Therefore, it is ineffective against weevils that infest grains and pulses. Nevertheless, its role in an integrated suppression program for certain commodities warrants further study.

Parasitoids

Two common parasitoids, *Habrobracon hebetor* (Hymenoptera, Braconidae) and *Venturia canescens* (Hymenoptera, Ichneumonidae), are being evaluated for suppression of pest insects in stored products (87). In a study of the suppression and regulation of *Cadra cautella* populations by the action of *V. canescens*, the adults attacked host larvae on the food surface, but did not appreciably penetrate the food surface (108). *H. hebetor* can attack host larvae in a deep layer of food, but its population declines because of the unfavorable environment. Takahashi (108) has suggested that procedures should be developed to maintain the population level of the parasitoid. We believe that parasitoids could be attracted by either a food source (nectar) or by their own pheromones.

Hagstrum and Smittle (40) suggested that insect growth regulators or systems for increasing the incidence of diapause might be a way to retain hosts in a susceptible stage for parasitism. They also suggested that parasitized (paralyzed) host larve could be released at the same time as the parasitoids (41). Thus, prolongation of suitability of the parasitoid-paralyzed host would provide the parasitoid with considerable time to relocate hosts that crawl away after envenomation, but before paralysis is complete.

The furniture carpet beetle (*Anthrenus flavipes*) possesses a supra-anal organ that serves as a defense mechanism against the wasp, *Laelius pedatus* (Hymenoptera, Bethylidae), a parasitoid associated with dermestid larvae (66). However, several other species of *Anthrenus* lack this structure and thus can serve as hosts. The life history and behavior of *L. pedatus* associated with *Anthrenus verbasci* have been described by Mertins (74). Another bethylid, *Cephalonomia tarsalis*, a parasitoid of the sawtoothed grain beetle (*O. surinamensis*), has been investigated as a possible means of suppressing the host insect (85).

There are several promising pteromalid (Hymenoptera) parasitoids of grain and pulse weevils. *Anisopteromalus calandrae* has long been known to be of considerable economic importance in the control of grain weevils (28) and several other stored-product pests. *Lariophagus distinguendus* and *Choetospila elegans* are other cosmopolitan parasitoids of grain weevils.

Perhaps the most promising use of beneficial insects as agents in the suppression of pest populations would be with commodities such as seed corn, peanuts, and other unprocessed materials that may already contain predators and parasitoids.

Insect Growth Regulators and Chitin Inhibitors

Chemicals that have juvenile hormone activity (IGRs) have been studied in a closed environment and have been relatively successful against several stored-product moths and beetles (*39, 65, 72, 102, 103, 120*). For example, methoprene and hydroprene, applied at a rate of 20 parts per millon (ppm) to diets, prevented pupation of *Tribolium castaneum* and substantially reduced pupation of *T. confusum* (*65*). Methoprene at 1 ppm or higher in rolled oats or cornmeal prevented emergence of adult grain beetles (*O. mercator* and *O. surinamensis*). Adult granary weevils (*Sitophilus granarius*) that had fed on wheat treated with hydroprene at 10 and 20 ppm produced very few F_1 adults. Treatments with both compounds at 1 and 5 ppm reduced the population of F_1 adult weevils (*S. granarius* and *S. oryzae*) but not enough to provide useful control (*65*).

Eggs of the cigarette beetle are sensitive to IGRs (*118*). Methoprene can be useful in controlling *Lasioderma serricorne* on packaged commodities (*68*). It prevents the larval forms of the cigarette beetle as well as the tobacco moth from developing into normal pupae or adults. When applied directly to tobacco at concentrations of 10 ppm, just prior to storage, the IGR prevented adult emergence for a period of 4 years (*67*).

Food Attractants

Several investigations have been directed toward isolating and identifying chemical compounds in foods that are attractive to several stored-product insects. Several volatile components of wheat germ oil responsible for initiating aggregation activity by *Trogoderma glabrum* have been identified (*80*). Eight fractions that stimulate attraction and aggregation in adult *O. surinamensis* have been isolated from rolled oats (*35*). Certain beetle and food volatiles produced various olfactory effects dependent upon the age of the beetles and their population density (*83*). Extracts of brewers'

yeast, flour, and wheat germ have been tested for components that elicit aggregation and feeding responses in *Tribolium confusum* (*64*). Wheat germ triglycerides elicit aggregation behavior in *T. confusum* (*109, 110*) and certain fatty acids are attractive to *T. castaneum* and *Trogoderma granarium* (*27*).

Oils

A component of lemon oil is toxic to the cowpea weevil (*Callosobruchus maculatus*) (*104*). Certain vegetable oils have protected stored beans from bruchid attack (*97*); other oils applied at 5 mL/kg have protected cowpeas from *C. maculatus* (*75*). Some vegetable oils (e.g., cottonseed, soybean, maize, and peanut) are toxic and repellent to *Sitophilus granarius* in wheat (*88*).

Genetic Manipulation

There is some potential for genetic control of stored-product insect populations (*12*). One experiment demonstrated that a population of *Cadra cautella* could be controlled by the release of cytoplasmically incompatible males (*13*).

Conclusion

A more rational basis for survey and control of stored-product insects in the food industry would be to integrate biological techniques into control programs. These techniques could be used for monitoring pest populations and as a supplement to pesticide applications. As pheromones of stored-product insects become commercially available, their use can be integrated with other natural biological agents for insect control. The end result would be an environmentally sound system for monitoring and control of stored-product insects. The system promises to be economically more desirable in that pesticide treatments may be limited to areas of special concern and routine treatments may be delayed until biological monitoring establishes that chemical treatment is necessary.

Cited References

1 Arbogast, R. T. 1979. The biology and impact of the predatory bug *Xylocoris flavipes* (Reuter). *Proceedings of the Second International Working Conference on Stored-Product Entomology* (Ibadan, Nigeria, 1978), Agency for International Development, Washington, DC, pp. 91–105.

2 Barak, A. V. 1989. Development of a new trap to detect and monitor khapra beetle (Coleoptera: Dermestidae). *Journal of Economic Entomology* 82(5)1470–1477.

3 Barak, A. V., and W. E. Burkholder. 1976. Trapping studies with dermestid sex pheromones. *Environmental Entomology* 5(1)111–114.

4 Barak, A. V., and W. E. Burkholder. 1977. Behavior and pheromone studies with *Attagenus elongatulus* Casey (Coleoptera: Dermestidae). *Journal of Chemical Ecology* 3(2)219–237.

5 Barak, A. V., and W. E. Burkholder. 1977. Studies on the biology of *Attagenus elongatulus* Casey (Coleoptera: Dermestidae) and the effects of larval crowding on pupation and life cycle. *Journal of Stored Products Research* 13(4)169–175.

6 Borden, J. H., et al. 1979. Aggregation pheromone in the rusty grain beetle, *Cryptolestes ferrugineus* (Coleoptera: Cucujidae). *Canadian Entomologist* 111(6)681–688.

7 Brady, U. E. 1973. Isolation, identification and stimulatory activity of a second component of the sex pheromone system (complex) of the female almond moth, *Cadra cautella* (Walker). *Life Sciences* 13(3)227–235.

8 Brady, U. E., and R. C. Daley. 1972. Identification of a sex pheromone from the female raisin moth, *Cadra figulilella*. *Annals of the Entomological Society of America* 65(6)1356–1358.

9 Brady, U. E., and D. A. Nordlund. 1971. *cis*-9,*trans*-12 Tetradecadien-1-yl acetate in the female tobacco moth *Ephestia elutella* (Hübner) and evidence for an additional component of the sex pheromone. *Life Sciences* 10, pt. 2(14)797–801.

10 Brady, U. E., D. A. Nordlund, and R. C. Daley. 1971. The sex stimulant of the Mediterranean flour moth *Anagasta kuehniella*. *Journal of the Georgia Entomological Society* 6(4)215–217.

11 Brady, U. E., et al. 1971. Sex stimulant and attractant in the Indian meal moth and in the almond moth. *Science* 171(3973)802.

12 Brower, J. H. [1975.] Potential for genetic control of stored-product insect populations. *Proceedings of the First International Working Conference on Stored-Product Entomology* (Savannah, GA, 1974), pp. 167–180.

13 Brower, J. H. 1980. Reduction of almond moth populations in simulated storages by the release of genetically incompatible males. *Journal of Economic Entomology* 73(3)415–418.

14 Burges, H. D. 1964. Insect pathogens and microbial control of insects in stored products. I. Test with *Bacillus thuringiensis* Berliner against moths. *Entomophaga. Mémoire Hors Série* 2:323–327.

15 Burges, H. D., and J. A. Hurst. 1977. Ecology of *Bacillus thuringiensis* in storage moths. *Journal of Invertebrate Pathology* 30(2)131–139.

16 Burkholder, W. E. 1973. Black carpet beetle: Reduction of mating by megatomoic acid, the sex pheromone. *Journal of Economic Entomology* 66(6)1327.

17 Burkholder, W. E. 1974. Programs utilizing pheromones in survey or control: Stored product pests. In *Pheromones*, M. C. Birch, ed. American Elsevier, New York, NY, pp. 449–452.

18 Burkholder, W. E., and G. M. Boush. 1974. Pheromones in stored product insect trapping and pathogen dissemination. *EPPO Bulletin* 4(4)455–461.

19 Burkholder, W. E., and R. J. Dicke. 1966. Evidence of sex pheromones in females of several species of Dermestidae. *Journal of Economic Entomology* 59(3)540–543.

20 Burkholder, W. E., et al. 1974. Sex pheromone of the furniture carpet beetle, *Anthrenus flavipes* (Coleoptera: Dermestidae). *Canadian Entomologist* 106(8)835–839.

21 Chuman, T., K. Kato, and M. Noguchi. 1979. Synthesis of (±)-serricornin, 4,6-dimethyl-7-hydroxy-nonan-3-one, a sex pheromone of cigarette beetle (*Lasioderma serricorne* F.). *Agricultural and Biological Chemistry* 43(9)2005.

22 Chuman, T., et al. 1979. 4,6-Dimethyl-7-hydroxy-nonan-3-one, a sex pheromone of the cigarette beetle (*Lasioderma serricorne* F.). *Tetrahedron Letters* 25:2361–2364.

23 Chuman, T., et al. 1983. Serricorone and serricorole, new sex pheromone components of cigarette beetle. *Agricultural and Biological Chemistry* 47(6)1413–1415.

24 Coffelt, J. A., and W. E. Burkholder. 1972. Reproductive biology of the cigarette beetle, *Lasioderma serricorne*. 1. Quantitative laboratory bioassays of the female sex pheromone from females of different ages. *Annals of the Entomological Society of America* 65(2)447–450.

25 Coffelt, J. A., et al. 1979. Isolation, identification, and synthesis of a female sex pheromone of the navel orangeworm, *Amyelois transitella* (Lepidoptera: Pyralidae). *Journal of Chemical Ecology* 5(6)955–966.

26 Cogburn, R. R., W. E. Burkholder, and H. J. Williams. 1984. Field tests with the aggregation pheromone of the lesser grain borer (Coleoptera: Bostrichidae). *Environmental Entomology* 13(1)162–166.

27 Cohen, E., V. Stanic, and A. Shulov. 1974. Olfactory and gustatory responses of *Trogoderma granarium, Dermestes maculatus* and *Tribolium castaneum* to various straight-chain fatty acids. *Zeitschrift für Angewandte Entomologie* 76(3)303–311.

28 Cotton, R. T. 1923. *Aplastomorpha vandinei* Tucker, an important parasite of *Sitophilus oryza* (L.). *Journal of Agricultural Research* 23(7)549–556.

29 Cross, J. H., et al. 1976. Porapak-Q collection

of pheromone components and isolation of (*Z*)- and (*E*)-14-methyl-8-hexadecenal, sex pheromone components, from females of four species of *Trogoderma* (Coleoptera: Dermestidae). *Journal of Chemical Ecology* 2(4)457–468.

30 Faustini, D. L. 1985. Stored insect control: A biotechnical approach to cigarette beetle monitoring. *Tobacco Reporter* 112(5)44, 46–47.

31 Faustini, D. L., et al. 1990. Combination-type trapping for monitoring stored-product insects— A review. *Journal of the Kansas Entomological Society* 63(4)539–547.

32 Faustini, D. L., and W. E. Burkholder. 1987. Quinone-aggregation pheromone interaction in the red flour beetle. *Animal Behaviour* 35(2)601–603.

33 Faustini, D. L., et al. 1982. Aggregation pheromone of the male granary weevil, *Sitophilus granarius* (L.). *Journal of Chemical Ecology* 8(4)679–687.

34 Fletcher, L. W., and J. S. Long. 1971. A bacterial disease of cigarette beetle larvae. *Journal of Economic Entomology* 64(6)1559.

35 Freedman, B., et al. 1982. Olfactory and aggregation responses of *Oryzaephilus surinamensis* (L.) to extracts from oats. *Journal of Stored Products Research* 18(2)75–82.

36 Fukui, H., et al. 1974. Identification of the sex pheromone of the furniture carpet beetle, *Anthrenus flavipes* LeConte. *Tetrahedron Letters* 40:3563–3566.

37 Fukui, H., et al. 1977. Isolation and identification of a major sex-attracting component of *Attagenus elongatulus* (Casey) (Coleoptera: Dermestidae). *Journal of Chemical Ecology* 3(5)539–548.

38 Gard, I. E., and L. A. Falcon. 1978. Autodissemination of entomopathogens: Virus. In *Microbial Control of Insect Pests: Future Strategies in Pest Management Systems*, G. A. Allen, C. M. Ignoffo and R. P. Jaques, eds. University of Florida, Gainesville, FL, pp. 46–54.

39 Gonen, M., and A. Schwartz. 1979. A controlling effect of a juvenile hormone analogue on *Ephestia cautella* (Wlk.) by nondirect application. *Proceedings of the Second International Working Conference on Stored-Product Entomology* (Ibadan, Nigeria, 1978), Agency for International Development, Washington, DC, pp. 106–112.

40 Hagstrum, D. W., and B. J. Smittle. 1977. Host-finding ability of *Bracon hebetor* and its influence upon adult parasite survival and fecundity. *Journal of Environmental Entomology* 6(3)437–439.

41 Hagstrum, D. W., and B. J. Smittle. 1978. Host utilization by *Bracon hebetor*. *Journal of Environmental Entomology* 7(4)596–600.

42 Haines, C. P. 1976. The use of synthetic sex pheromones for pest management in stored-product situations. *Pesticide Science* 7(6)647–649.

43 Honda, H., and I. Yamamoto. 1976. Evidence for and chemical nature of a sex pheromone present in azuki bean weevil, *Callosobruchus chinensis* L. *Proceedings of a Symposium on Insect Pheromones and Their Applications* (Nagaoka and Tokyo, Japan), Agriculture, Forestry and Fisheries Research Council, Ministry of Agriculture and Forestry, Tokyo, p. 164.

44 Hope, J. A., D. F. Horler, and D. G. Rowlands. 1967. A possible pheromone of the bruchid, *Acanthoscelides obtectus* (Say). *Journal of Stored Products Research* 3(4)387–388.

45 Horler, D. F. 1970. (−) Methyl *n*-tetradeca-*trans*-2,4,5-trienoate, an allenic ester produced by the male dried bean beetle, *Acanthoscelides obtectus* (Say). *Journal of the Chemical Society* C(6)859–862.

46 Hunter, D. K., S. J. Collier, and D. F. Hoffmann. 1973. Effectiveness of a granulosis virus of the Indian meal moth as a protectant for stored inshell nuts: Preliminary observations. *Journal of Invertebrate Pathology* 22(3)481.

47 Hunter, D. K., S. J. Collier, and D. F. Hoffmann. 1975. Compatibility of malathion and the granulosis virus of the Indian meal moth. *Journal of Invertebrate Pathology* 25(3)389–390.

48 Hunter, D. K., S. J. Collier, and D. F. Hoffmann. 1977. Granulosis virus of the Indian meal moth as a protectant for stored inshell almonds. *Journal of Economic Entomology* 70(4)493–494.

49 Jay, E., R. Davis, and S. Brown. 1968. Studies on the predacious habits of *Xylocoris flavipes* (Reuter) (Hemiptera: Anthocoridae). *Journal of the Georgia Entomological Society* 3(3)126–130.

50 Keys, R. E., and R. B. Mills. 1968. Demonstration and extraction of a sex attractant from female Angoumois grain moths. *Journal of Economic Entomology* 61(1)46–49.

51 Khorramshahi, A., and W. E. Burkholder. 1981. Behavior of the lesser grain borer *Rhyzopertha dominica* (Coleoptera: Bostrichidae). Male-produced aggregation pheromone attracts both sexes. *Journal of Chemical Ecology* 7(1)33–38.

52 Kinsinger, R. A., and W. H. McGaughey. 1976. Stability of *Bacillus thuringiensis* and a granulosis virus of *Plodia interpunctella* on stored wheat. *Journal of Economic Entomology* 69(2)149–154.

53 Krieg, A., et al. 1983. *Bacillus thuringiensis* var. *tenebrionis*: ein neuer, gegenüber Larven von Coleopteren wirksamer Pathotyp. *Zeitschrift für Angewandte Entomologie* 96(5)500–508 (English summary).

54 Kuwahara, Y., and J. E. Casida. 1973. Quantitative analysis of the sex pheromone of several

phycitid moths by electron-capture gas chromatography. *Agricultural and Biological Chemistry* 37(3)681–684.

55 Kuwahara, Y., and S. Nakamura. 1985. (Z)-5- and (E)-5-Undecenoic acid: Identification of the sex pheromone of the varied carpet beetle, *Anthrenus verbasci* L. (Coleoptera: Dermestidae). *Applied Entomology and Zoology* 20(3)354–356.

56 Kuwahara, Y., et al. 1971. Sex pheromone of the almond moth and the Indian meal moth: *cis*-9,*trans*-12-Tetradecadienyl acetate. *Science* 171(3973)801–802.

57 Kuwahara, Y., et al. 1971. The sex pheromone of the Mediterranean flour moth. *Agricultural and Biological Chemistry* 35(3)447–448.

58 Kuwahara, Y., et al. 1975. Studies on the isolation and bioassay of the sex pheromone of the drugstore beetle, *Stegobium paniceum* (Coleoptera: Anobiidae). *Journal of Chemical Ecology* 1(4)413–422.

59 Kuwahara, Y., et al. 1978. Chemical studies on the Anobiidae: Sex pheromone of the drugstore beetle, *Stegobium paniceum* (L.) (Coleoptera). *Tetrahedron* 34(12)1769–1774.

60 LeCato, G. L., J. M. Collins, and R. T. Arbogast. 1977. Reduction of residual populations of stored-product insects by *Xylocoris flavipes* (Hemiptera: Anthocoridae). *Journal of the Kansas Entomological Society* 50(1)84–88.

61 Levinson, H. Z., and A. R. Bar Ilan. 1967. Function and properties of an assembling scent in the khapra beetle *Trogoderma granarium*. *Rivista di Parassitologia* 28(1)27–42.

62 Levinson, H. Z., et al. 1981. The pheromone activity of anhydroserricornin and serricornin for male cigarette beetles (*Lasioderma serricorne* (F.). *Naturwissenschaften* 68(3)148–149.

63 Levinson, H. Z., et al. 1982. Suppressed pheromone responses of male tobacco beetles to anhydroserricornin in presence of serricornin. *Naturwissenschaften* 69(9)454–455.

64 Loschiavo, S. R. 1965. The chemosensory influence of some extracts of brewers' yeast and cereal products on the feeding behavior of the confused flour beetle, *Tribolium confusum* (Coleoptera: Tenebrionidae). *Annals of the Entomological Society of America* 58(4)576–588.

65 Loschiavo, S. R. 1976. Effects of the synthetic insect growth regulators methoprene and hydroprene on survival, development or reproduction of six species of stored-products insects. *Journal of Economic Entomology* 69(3)395–399.

66 Ma, M., W. E. Burkholder, and S. D. Carlson. 1978. Supra-anal organ: A defensive mechanism of the furniture carpet beetle, *Anthrenus flavipes* (Coleoptera: Dermestidae). *Annals of the Entomological Society of America* 71(5)718–723.

67 Manzelli, M. A. 1982. Management of stored-tobacco pests, the cigarette beetle (Coleoptera: Anobiidae) and tobacco moth (Lepidoptera: Pyralidae), with methoprene. *Journal of Economic Entomology* 75(4)721–723.

68 Marzke, F. O., J. A. Coffelt, and D. L. Silhacek. 1977. Impairment of reproduction of the cigarette beetle, *Lasioderma serricorne* (Coleoptera: Anobiidae) with the insect growth regulator, methoprene. *Entomologia Experimentalis et Applicata* 22(3)294–300.

69 McGaughey, W. H. 1975. A granulosis virus for Indian meal moth control in stored wheat and corn. *Journal of Economic Entomology* 68(3)346–348.

70 McGaughey, W. H. 1976. *Bacillus thuringiensis* for controlling three species of moths in stored grain. *Canadian Entomologist* 108(1)105–112.

71 McGaughey, W. H. 1985. Insect resistance to the biological insecticide *Bacillus thuringiensis*. *Science* 229(4709)193–195.

72 McGregor, H. E., and K. J. Kramer. 1975. Activity of insect growth regulators, hydroprene and methoprene, on wheat and corn against several stored-grain insects. *Journal of Economic Entomology* 68(5)668–670.

73 McLaughlin, R. E., et al. 1969. Development of the bait principle for boll weevil control. IV. Field tests with bait containing a feeding stimulant and the sporozoans *Glugea gasti* and *Mattesia grandis*. *Journal of Invertebrate Pathology* 13(3)429–441.

74 Mertins, J. W. 1980. Life history and behavior of *Laelius pedatus*, a gregarious bethylid ectoparasitoid on *Anthrenus verbasci*. *Annals of the Entomological Society of America* 73(6)686–693.

75 Messina, F. J., and J. A. A. Renwick. 1983. Effectiveness of oils in protecting stored cowpeas from the cowpea weevil (Coleoptera: Bruchidae). *Journal of Economic Entomology* 76(3)634–636.

76 Millar, J. G., A. C. Oehlschlager, and J. W. Wong. 1983. Synthesis of two macrolide aggregation pheromones from the flat grain beetle, *Cryptolestes pusillus* (Schönherr). *Journal of Organic Chemistry* 48(23)4404–4407.

77 Mondal, K. A. M. S. H. 1985. Response of *T. castaneum* larvae to aggregation pheromone and quinones produced by adult conspecifics. *International Pest Control* 27(3)64–66.

78 Mondal, K. A. M. S. H., and G. R. Port. 1984. Response of *Tribolium castaneum* larvae to synthetic aggregation pheromone. *Entomologia Experimentalis et Applicata* 36(1)43–46.

79 Montoya, E. L., C. M. Ignoffo, and R. L. McGarr. 1966. A feeding stimulant to increase effectiveness of, and a field test with, nuclear-polyhedrosis virus of *Heliothis*. *Journal of Invertebrate Pathology* 8(3)320–324.

80 Nara, J. M., R. C. Lindsay, and W. E. Burkholder. 1981. Analysis of volatile compounds in wheat germ oil responsible for an aggregation response in *Trogoderma glabrum* larvae. *Journal of Agricultural and Food Chemistry* 29(1)68–72.

81 Phillips, J. K., and W. E. Burkholder. 1981. Evidence for a male-produced aggregation pheromone in the rice weevil. *Journal of Economic Entomology* 74(5)539–542.

82 Phillips, J. K., et al. 1985. (R*,S*)-5-Hydroxy-4-methyl-3-heptanone. Male-produced aggregation pheromone of *Sitophilus oryzae* (L.) and *S. zeamais* Motsch. *Journal of Chemical Ecology* 11(9)1263–1274.

83 Pierce, A. M., J. H. Borden, and A. C. Oehlschlager. 1983. Effects of age and population density on response to beetle and food volatiles by *Oryzaephilus surinamensis* and *O. mercator* (Coleoptera: Cucujidae). *Environmental Entomology* 12(5)1367–1374.

84 Pierce, A. M., et al. 1984. Aggregation pheromones in the genus *Oryzaephilus* (Coleoptera: Cucujidae). *Proceedings of the Third International Working Conference on Stored-Product Entomology* (Manhattan, KS, 1983), Kansas State University, Manhattan, pp. 107–120.

85 Powell, D. 1938. The biology of *Cephalonomia tarsalis* (Ash.), a vespoid wasp (Bethylidae: Hymenoptera) parasitic on the sawtoothed grain beetle. *Annals of the Entomological Society of America* 31(1)44–49.

86 Press, J. W., B. R. Flaherty, and R. T. Arbogast. 1975. Control of the red flour beetle, *Tribolium castaneum*, in a warehouse by a predaceous bug, *Xylocoris flavipes*. *Journal of the Georgia Entomological Society* 10(1)76–78.

87 Press, J. W., B. R. Flaherty, and R. T. Arbogast. 1977. Interactions among *Nemeritis canescens* (Hymenoptera: Ichneumonidae), *Bracon hebetor* (Hymenoptera: Braconidae), and *Ephestia cautella* (Lepidoptera: Pyralidae). *Journal of the Kansas Entomological Society* 50(2)259–262.

88 Qi, Y.-T., and W. E. Burkholder. 1981. Protection of stored wheat from the granary weevil by vegetable oils. *Journal of Economic Entomology* 74(5)502–505.

89 Read, J. S., and P. S. Beevor. 1976. Analytical studies on the sex pheromone complex of *Ephestia cautella* (Walker) (Lepidoptera: Phycitidae). *Journal of Stored Products Research* 12(1)55–57.

90 Read, J. S., and C. P. Haines. 1976. The functions of the female sex pheromones of *Ephestia cautella* (Walker) (Lepidoptera, Phycitidae). *Journal of Stored Products Research* 12(1)49–53.

91 Read, J. S., and C. P. Haines. 1979. Secondary pheromone components and synergism in stored-products Phycitinae. *Journal of Chemical Ecology* 5(2)251–257.

92 Reichmuth, C., et al. 1976. Untersuchungen über den Einsatz von pheromonbeköderten Klebefallen zur Bekämpfung von Motten im Vorratsschutz. *Zeitschrift für Angewandte Entomologie* 82(1)95–102 (English summary).

93 Reichmuth, C., et al. 1978. Die Fängigkeit pheromonbeköderter Klebefallen für Speichermotten (*Ephestia elutella* Hbn.) in unterschiedlich dicht befallenen Getreidelägern. *Zeitschrift für Angewandte Entomologie* 86(2)205–212 (English summary).

94 Rodin, J. W., et al. 1969. Sex attractant of female dermestid beetle *Trogoderma inclusum* Le Conte. *Science* 165(3896)904–906.

95 Rup, P. J., and S. P. Sharma. 1978. Behavioural response of males and females of *Callosobruchus maculatus* (F.) to the sex pheromones. *Indian Journal of Ecology* 5(1)72–76.

96 Schmuff, N. R., et al. 1984. The chemical identification of the rice weevil and maize weevil aggregation pheromone. *Tetrahedron Letters* 25(15)1533–1534.

97 Schoonhoven, A. V. 1978. Use of vegetable oils to protect stored beans from bruchid attack. *Journal of Economic Entomology* 71(2)254–256.

98 Shapas, T. J., W. E. Burkholder, and G. M. Boush. 1977. Population suppression of *Trogoderma glabrum* by using pheromone luring for protozoan pathogen dissemination. *Journal of Economic Entomology* 70(4)469–474.

99 Silverstein, R. M., et al. 1967. Sex attractant of the black carpet beetle. *Science* 157(3784)85–87.

100 Sower, L. L., W. K. Turner, and J. C. Fish. 1975. Population-density-dependent mating frequency among *Plodia interpunctella* (Lepidoptera: Phycitidae) in the presence of synthetic sex pheromone with behavioral observations. *Journal of Chemical Ecology* 1(3)335–342.

101 Sower, L. L., K. W. Vick, and J. H. Tumlinson. 1974. (Z,E)-9,12-Tetradecadien-1-ol: A chemical released by female *Plodia interpunctella* that inhibits the sex pheromone response of male *Cadra cautella*. *Environmental Entomology* 3(1)120–122.

102 Staal, G. B. 1977. Insect control with insect growth regulators based on insect hormones. In *Natural Products and the Protection of Plants*, G. B. Marini-Bettolo, ed. Elsevier, New York, NY, pp. 353–377.

103 Strong, R. G., and J. Diekman. 1973. Comparative effectiveness of fifteen insect growth regulators against several pests of stored products. *Journal of Economic Entomology* 66(5)1167–1173.

104 Su, H. C. F. 1976. Toxicity of a chemical component of lemon oil to cowpea weevils. *Journal of the Georgia Entomological Society* 11(4)297–301.

105 Suzuki, T. 1980. 4,8-Dimethyldecanal: The aggregation pheromone of the flour beetles, *Tribolium castaneum* and *T. confusum* (Coleoptera: Tenebrionidae). *Agricultural and Biological Chemistry* 44(10)2519–2520.

106 Suzuki, T. 1981. Identification of the aggregation pheromone of flour beetles *Tribolium castaneum* and *T. confusum* (Coleoptera: Tenebrionidae). *Agricultural and Biological Chemistry* 45(6)1357–1363.

107 Suzuki, T., and K. Mori. 1983. (4*R*,8*R*)-(−)-4,8-Dimethyldecanal: The natural aggregation pheromone of the red flour beetle, *Tribolium castaneum* (Coleoptera: Tenebrionidae). *Applied Entomology and Zoology* 18(1)134–136.

108 Takahashi, F. 1973. An experimental study on the suppression and regulation of the population of *Cadra cautella* (Walker) (Lepidoptera: Pyralidae) by the action of a parasitic wasp, *Nemeritis canescens* (Gravenhorst) (Hymenoptera: Ichneumonidae). *Memoirs of the College of Agriculture, Kyoto University* 104:1–12.

109 Tamaki, Y., S. R. Loschiavo, and A. J. McGinnis. 1971. Effect of synthesized triglycerides on aggregation behaviour of the confused flour beetle, *Tribolium confusum*. *Journal of Insect Physiology* 17(7)1239–1244.

110 Tamaki, Y., S. R. Loschiavo, and A. J. McGinnis. 1971. Triglycerides in wheat germ as chemical stimuli eliciting aggregation of the confused flour beetle, *Tribolium confusum* (Coleoptera: Tenebrionidae). *Journal of Agricultural and Food Chemistry* 19(2)285–288.

111 Tanaka, K., et al. 1981. Copulation release pheromone, erectin, from the azuki bean weevil (*Callosobruchus chinensis* L.). *Journal of Pesticide Science* 6(1)75–82.

112 Tanaka, K., et al. 1982. Synthesis of erectin, a copulation release pheromone of the azuki bean weevil, *Callosobruchus chinensis* L. *Journal of Pesticide Science* 7(4)535–537.

113 Thompson, J. V., and L. W. Fletcher. 1972. A pathogenic strain of *Bacillus cereus* isolated from the cigarette beetle, *Lasioderma serricorne*. *Journal of Invertebrate Pathology* 20(3)341–350.

114 Vick, K. W., et al. 1974. (*Z-E*)-7,11-Hexadecadien-1-ol acetate: The sex pheromone of the Angoumois grain moth, *Sitotroga cerealella*. *Experientia* 30(1)17–18.

115 Vick, K. W., et al. 1979. Investigation of sex pheromone traps for simultaneous detection of Indianmeal moths and Angoumois grain moths. *Journal of Economic Entomology* 72(2)245–249.

116 Vick, K. W., et al. 1981. Recent developments in the use of pheromones to monitor *Plodia interpunctella* and *Ephestia cautella*. In *Management of Insect Pests with Semiochemicals*, E. R. Mitchell, ed. Plenum Press, New York, NY, pp. 19–40.

117 Walgenbach, C. A., et al. 1983. Male-produced aggregation pheromone of the maize weevil, *Sitophilus zeamais*, and interspecific attraction between three *Sitophilus* species. *Journal of Chemical Ecology* 9(7)831–841.

118 Walker, W. F., and W. S. Bowers. 1970. Synthetic juvenile hormones as potential coleopteran ovicides. *Journal of Economic Entomology* 63(4)1231–1233.

119 Williams, H. J., et al. 1981. Dominicalure 1 and 2: Components of aggregation pheromone from male lesser grain borer *Rhyzopertha dominica* (F.) (Coleoptera: Bostrichidae). *Journal of Chemical Ecology* 7(4)759–780.

120 Williams, P., and T. G. Amos. 1974. Some effects of synthetic juvenile insect hormones and hormone analogues on *Tribolium castaneum* (Herbst). *Australian Journal of Zoology* 22(2)147–153.

121 Wong, J. W., et al. 1983. Isolation and identification of two macrolide pheromones from the frass of *Cryptolestes ferrugineus* (Coleoptera: Cucujidae). *Journal of Chemical Ecology* 9(4)451–474.

122 Yarger, R. G., R. M. Silverstein, and W. E. Burkholder. 1975. Sex pheromone of the female dermestid beetle *Trogoderma glabrum* (Herbst). *Journal of Chemical Ecology* 1(3)323–334.

Selected Reference

Anonymous. 1991. Parasitic and predaceous insects used to control insect pests; proposed exemption from a tolerance. *Federal Register* 56(2)234–235.

27

Host-Plant Resistance to Insects in Stored Cereals and Legumes

Philip Dobie

The insect species that attack stored products are highly specialized and successful exploiters of the storage environment. Many species are capable of totally destroying stored crops, but, fortunately, farming and storage systems usually ensure that losses are not catastrophic. Small-scale farmers have traditionally incorporated the use of fairly resistant varieties in their farming systems as a means of limiting storage losses, their local varieties presumably having been selected for resistance over many generations of exposure to the pest complex.

In recent years, producers and marketers of food have sometimes experienced very high losses due to the introduction of improved varieties of crops that have less intrinsic resistance to storage pests than traditional varieties. Unfortunately, scant attention has been paid by most plant breeders to the problem of postharvest resistance, mainly because breeding for preharvest qualities that result in high yield has always been a priority. Plant breeders sometimes lack knowledge of postharvest constraints upon crop utilization, although it has frequently been observed that susceptible varieties are unpopular with growers, especially in countries where funds or materials for pest control are often unavailable.

Even in the developed countries, it is inefficient to

Dobie: Overseas Development Natural Resources Institute, Central Avenue, Chatham Maritime, Chatham, Kent ME4 4TB, United Kingdom.

depend entirely upon the use of pesticides to control pests; moreover, restrictive legislation is making pesticide use increasingly difficult. It is thus imperative that postharvest biologists and plant breeders find ways of working together to improve levels of crop resistance to postharvest attack. This chapter attempts to draw together information available on postharvest resistance to insect attack and to place it on a firm biological base.

Pest Population Increase

Stored-product pests damage produce seriously as a result of massive numbers of insects attacking the commodity. In most cases, initial infestations are caused by relatively few insects that rapidly reproduce and increase in numbers to produce damaging populations. Rapid population increase results from (a) high birth rate (i.e., many eggs laid over a short period of time), (b) rapid development, and (c) low death rate (i.e., few insects dying before reaching sexual maturity and producing offspring).

A resistant variety should (a) cause a reduction in the rate of egg-laying, (b) cause high mortality of developing insects, and/or (c) extend the developmental period of the pest. Possible ways in which this could be brought about are (a) reduction of the birth rate, (b) prolongation of the developmental period, and (c) increase of the death rate.

Reduction of the birth rate can be achieved by (a) selection of varieties with mechanical barriers that prevent access of insects to the seed or fruit upon which they feed, thus reducing the number of insect eggs laid upon the crop; (b) selection of varieties that repel the insects or that are unattractive to them; (c) selection of varieties that are for some reason unsuitable for oviposition (e.g., too hard for species that chew holes in which to lay eggs or too rough for species that stick their eggs onto a surface).

The pest's developmental period can be extended by selecting crop varieties upon which insects develop slowly. These could be hard-textured varieties, partially toxic varieties, or nutritionally inadequate varieties.

Increase of the death rate can be accomplished by (a) selection of varieties into which the larvae hatching from eggs are unable to penetrate, making it impossible for them to feed; and (b) selection of varieties that are nutritionally inadequate for, or toxic to, the feeding insects.

Potential Sources of Resistance to Insects

Food Quality

Species of stored-product insects differ widely in their ability to infest various commodities. This ability may be strongly affected by the condition and quality of the infested product. Factors relating to the quality of the grain that are liable to affect susceptibility are damage, impurities, moisture content, and developmental defects in kernels.

Damage

For convenience, stored-product insects may be classified as primary or secondary pests, depending upon their ability to infest sound, undamaged commodities. Some species such as certain Bruchidae and the grain weevils (*Sitophilus* spp.) are specialized for attacking whole, undamaged, dry seeds. These primary pests are not greatly affected by damage to individual kernels of grain. In contrast, the secondary pests benefit from damage to a commodity that renders it more susceptible to their attack. Generally, such pests are associated with comminuted commodities, but also readily attack whole seeds if they are damaged. Thus, careless harvesting, drying, and handling may damage grain and make it susceptible to attack by a variety of insects that would otherwise cause little damage.

Certain species cannot be categorized as either primary or secondary pests since they are capable of attacking whole seeds but only under certain conditions. For example, *Rhyzopertha dominica* (lesser grain borer), frequently regarded as a major primary pest of cereals, is a rather inefficient pest of whole, undamaged grains. Survival of *R. dominica* is higher on damaged wheat than on undamaged wheat because of the greater number of first-instar larvae that penetrate the damaged grain and become established in the endosperm (*11*). In undamaged grain, the larvae enter the kernels at the embryo end where the outer layer is most frequently lacerated or loose. Clearly, the condition of the grain is of critical importance to the amount of damage that the pest may cause.

Impurities

Insects that prefer damaged commodities benefit from the presence of organic impurities in an otherwise sound product. The impurities probably assist in the maintenance of a pest population that may eventually cause damage to the grain as a result of repeated attempts at penetration (*5*). In some cases, certain species may feed upon the frass and detritus caused by the activity of another species, or in the case of *R. dominica* infesting good-quality grain, some of the larvae may be able to survive in the frass produced by individuals of the same species. Adult female *R. dominica* often lay their eggs directly in frass, which in the case of this species contains a high proportion of undigested grain (*15*).

Moisture Content

Stored-product insects are able to live in fairly dry environments where no free water is available and the moisture content of the food is typically less than 16% by weight. However, their ability to infest a product is steadily reduced as the moisture content is lowered. Grain with less than 10% moisture is essentially immune to significant damage by all of the major pest species except *Trogoderma granarium* (khapra beetle). The influence of available moisture on survival and development is usually a direct effect upon the metabolism of the insect, but high moisture content may make otherwise inaccessible food available. For example, a high moisture content in wheat causes swelling of the embryo which facilitates penetration by *Cryptolestes* species that selectively attack the germ.

Developmental Defects in the Grain

Grain harvested early or from plants that have been exposed to adverse conditions in the field (e.g., drought, flood, frost) is subject to various defects. Usually, there is a lack of storage materials laid down in the endosperm of the cotyledons, resulting in undersized and wrinkled grains. Such grains are commonly damaged and are easily infested by secondary pests. They may be nutritionally inferior to sound grains (there appears to be no published information on this with relevance to insect infestation), but their presence in appreciable quantities in grain assists in maintaining a population of secondary pests that may eventually cause significant damage.

Physical Resistance

Protective Husks and Seed Coats

An effective natural barrier that prevents insect attack provides a stored commodity with a valuable form of resistance. Whole seeds all have outer layers that are to some extent effective barriers against secondary pests that cannot penetrate undamaged seeds. Primary pests, however, are well adapted to penetrate these layers. Certain varieties of some crops have natural barriers that can reduce infestation levels. The barriers may be of two types: (a) those that prevent all forms of access by the insect to the seed and (b) those that prevent a feeding stage of the insect from penetrating the seed.

Barriers of the first type are found in the pods of legumes and the sheathing leaves of maize ears. Some of the bruchid beetles that attack peas and beans in store are relatively unimportant pests of pulses stored in the pod because the adults cannot easily penetrate the pods; however, these same bruchids can become important pests when the pulses are shelled. In maize, the value of a complete, well-fitting set of sheathing leaves for reducing prehusking infestation by *Sitophilus* species has been recognized for many years (*23, 33, 44*).

Examples of natural barriers to the penetration of seeds by primary pests are rare, presumably because co-evolution of the pests with the crops has resulted in the development of species that are well adapted to penetrate seed coats. In some cases, however, features of seed coats have been associated with difficulties experienced by insects in penetrating seeds. For example, the orientation of cells in the testas of cowpeas (*Vigna unguiculata*) affects the ability of hatching larvae of *Callosobruchus maculatus* (cowpea weevil) to bore into the seeds. This orientation of cells varies among several cowpea varieties (*53*). There is little evidence that primary cereal pests are directly affected by hard testas.

Grain Hardness

The ability of stored-product insects to feed upon and develop on produce is often limited by the internal hardness of the commodity. Unfortunately, the effects of grain hardness are frequently difficult to distinguish from those caused by nutrition-related characteristics that may be associated with hardness. Cereals especially vary in hardness; the endosperm of the grains can be quite variable. Hard maize varieties have been shown to be relatively resistant to infestation by *Sitophilus* species (*82*).

Suitability of Seed Surfaces for Oviposition

Sitophilus species prepare their own egg-laying sites by chewing small oviposition chambers in cereal grains and, therefore, are not affected by the surface texture of the grain as long as they can obtain sufficient purchase with their legs to support themselves during the hard work of chewing. Other pests, however, have to seek oviposition sites; thus, the surface characteristics of the seeds may be important. *Sitotroga cerealella* (Angoumois grain moth) lays eggs in any suitable crack or crevice in the grain surface and also especially beneath the glumes that sometimes adhere to maize kernels after harvest. *Callosobruchus* species glue their eggs to the surface of pulse seeds; it has been shown that rough-coated seeds are less suitable for oviposition than smooth seeds (*53*).

Nutrition-Related Resistance

Stored grains may have high resistance to insect pests because of the absence of vital nutrients or the presence of compounds that adversely affect insect development. The nutrition requirements of stored-product insects and humans are broadly similar; crops lacking in insect nutrients are probably also unsuitable foods for humans. Of greater interest is the possibility that some varieties of crops contain compounds that are toxic to insects but harmless to humans after the grain is cooked. In maize, both sugar content (*82*) and amylose content (*67*) influence susceptibility to *Sitophilus* species.

In legumes, it has been postulated that enzyme inhibitors protect the seeds against insect attack. Recently, it has been shown that a variety of cowpeas known to be resistant to attack by *C. maculatus* (*81*) contains high levels of protease inhibitor, and that when incorporated into diets for *C. maculatus* larvae, the inhibitor is toxic (*30*).

A considerable amount of work has been done on the importance of other secondary compounds found in various legume seeds in the protection of the seeds against insect attack (*39, 54*). Much of this information is not relevant to agricultural crops because most of the highly toxic secondary compounds were found in wild legumes and are not present in crop legumes. However, it has been suggested that the phytohemagglutinins found in *Phaseolus vulgaris* seeds may be responsible for the inability of *C. maculatus* to infest them, whereas the common pest of *P. vulgaris* seeds, *Acanthoscelides obtectus* (bean weevil), has developed metabolic pathways that permit it to metabolize the toxin (*29, 40*).

Techniques for Measuring Resistance

The resistance of a crop may be measured either to satisfy scientific curiosity or to obtain information for a crop-breeding program to improve resistance. The purpose for which resistance measurements are made will influence the measurement technique selected for use. For the purpose of crop improvement programs, measures of susceptibility may be derived from investigations involving only one generation of progeny, even though the effect of the natural selection of insects adapted to attack particular varieties cannot be investigated.

Free-Choice and No-Choice Methods

Many workers have developed free-choice methods (*46, 90*) whereby a sample of insects is allowed to choose between two or more varieties of a crop being tested and to infest those chosen. The measure of intensity of infestation of each sample (usually the number of successfully developing F_1 adults) is, therefore, some form of composite of the results of the preference of the adults for particular samples and the rate of increase of the population upon those samples. Under field conditions, a choice is seldom possible. A large warehouse may contain more than one variety of crop, but many species developing within a bulk commodity will not emerge from the bulk unless the insect density is very high. Thus they are unlikely to exercise choice.

In most situations, a silo, bin, warehouse, or farm store will contain only one variety of crop. Free-choice techniques are, therefore, of little practical significance unless the results of a free-choice test can be shown to be related to the results obtained in a no-choice test where the insects are caged upon a single variety. This has been shown to be true under some circumstances (*46*); in such cases, free-choice techniques can be very rapid. However, there are practical problems in interpreting the results because the presence of a susceptible sample may affect the result obtained from adjacent resistant samples. Free-choice techniques may be of considerable interest to a research worker testing insect behavior, but if the purpose of testing is to try to predict the performance of varieties in storage, then a no-choice technique (*20, 22, 32, 66, 78, 86*) is likely to be more appropriate.

Positive and Negative Screening

The entomologist investigating resistance as part of a breeding program is likely to be faced with very large numbers of lines of breeding materials to be tested. It will, therefore, be necessary to develop one or more screening techniques to identify resistant lines. A logical approach would be first to screen all of the lines using a "rough" or "negative" screening technique, the purpose of which is to eliminate the most susceptible lines. The negative screen would be followed by a "detailed" or "positive" screening of the remaining lines to positively identify those that are resistant. The negative screen should discard only the most susceptible lines. In order to avoid the risk of accidentally discarding resistant lines, the screening technique should be designed to be very conservative in discarding samples. The positive screen, on the contrary, should positively identify resistant lines and should, therefore, be conservative in identifying resistance.

Insect Life Cycle and Its Influence

The components of the insect life cycle that contribute to the buildup of an insect population are (a) rate of oviposition, (b) egg-to-adult developmental period, and (c) rate of mortality.

Rate of Oviposition

If an insect species tends to lay more eggs on some varieties of a crop than on others, then it may be of interest to estimate the egg-laying potential of the species on different varieties. Ideally, this estimate will indicate the likely rate of addition of individuals to the population per unit time when a population is infesting the crop. The optimum age of adults when oviposition should be measured must be determined from biological experiments. Differences in egg-laying potential on different varieties may be of interest if the differences can be related to a characteristic of the crop under investigation. In other cases, it may be sufficient not to measure oviposition separately but to measure adult progeny produced per parent, which results from the combined effects of oviposition rate and mortality rate of the offspring during development.

Egg-to-Adult Developmental Period

Insects tend to develop more slowly on resistant varieties that on susceptible varieties of a crop. The developmental period is most conveniently estimated by counting the daily emergence of mature adults developing from a group of eggs laid at a known time. It is then possible to derive a measure of the average developmental period. Tedious daily counting can often be eliminated by counting less frequently and fitting the data to a mathematical model previously shown to describe the emergence pattern of the species being used (*24*).

Mortality

If a known number of eggs has been laid on or placed on a trial substrate, then a measure of mortality during the developmental period can be derived. If the parent insects are allowed to lay eggs in an uncontrolled manner, then the information on mortality will be combined with the information on oviposition to give a measure of effective productivity.

Review of Information on Specific Crops

Cereals

The most important cosmopolitan insect pests of whole cereals are *Sitophilus* species, *R. dominica*, and *Sitotroga cerealella*. These are all highly specialized and very effective grain pests, and no varieties very highly resistant to their attacks are known. Nevertheless, cultivars of all cereals vary greatly in their susceptibility to attack, and those of lower susceptibility are likely to suffer much less damage as a result of insect infestation than others.

In addition to the major primary pests, a few others may be of local importance. For example, *Prostephanus truncatus* (larger grain borer), a bostrichid beetle closely related to *R. dominica*, may be a major pest of cereals in the southern USA and Central America and has recently become established in Tanzania (*38*), Togo (*43*), and several other African nations. *Pagiocerus frontalis*, a scolytid beetle, is a frequent pest of maize stored at high altitudes in the Andean regions of South America. Some other insects, such as *Ephestia elutella* (tobacco moth), may be important pests of seed grains because they commonly feed selectively on the grain embryo and thus destroy the germination capabilities of the seeds.

Maize (Corn)

Sitophilus species and *Sitotroga cerealella* may infest maturing maize in the field. The infestation may begin at moisture contents as high as 60% (*33*). At this stage, the completeness of the sheathing leaves is of greatest importance in preventing infestation. Although this has been recognized for many years (*44*), many varieties and hybrids that have been produced recently by plant breeders are unsatisfactory in this regard. This is especially regrettable since the necessary features of maize ears that confer resistance have been identified and described (*23, 42*) and it has been shown that ears that are resistant to *Sitophilus* attack are also resistant to infestation by preharvest pests such as *Heliothis* species (*88, 94*). Moreover, maize ears with deficient husk cover are liable to suffer severe rotting of the tips if exposed to late rains before harvest. Investigators have frequently reported that in most parts of the world, local indigenous varieties of maize have husk covers superior to those of newly introduced cultivars (*68, 80*). This is clearly an area where plant breeders could make considerable progress.

After harvesting and shelling, maize grain of different cultivars varies in susceptibility to pests, and this variability may be quite considerable. Many workers have studied varieties of maize that were of local interest, and a few attempts have been made to study comprehensive collections of maize in order to discover lines with high levels of resistance. Several workers have also attempted to determine the factors that govern resistance in maize. A search for resistance in a world collection of maize varieties housed at Chapingo, Mexico, revealed that varieties from the Chandelle race (a coastal tropical flint race), and from an unknown race were resistant to *Sitophilus zeamais* (maize weevil); it was concluded that lowland tropical maize varieties may be a good source of resistance (*90*). This makes intuitive good sense to a biologist because it is in these regions that infestation pressure is most severe, and this has probably resulted in both artificial and natural selection for resistance. Many indigenous maize varieties from localized geographical areas are more resistant to infestation than newly developed varieties (*22, 28, 68*).

Little progress has been made in determining which factors are responsible for grain resistance. It is known that hard kernels are more resistant to infestation than soft kernels, although it is not clear whether this is due simply to the difficulty experienced by the pest in chewing into the grain, or whether there are further nutrition characteristics associated with hardness (*22, 82*). Sugar content of kernels has been shown to be negatively correlated with resistance to *S. zeamais*, whereas fat content apparently has no effect (*23, 82*). There is some evidence to indicate that the total protein content of the kernel is positively correlated with resistance (*23*). High-amylose maize has been shown to be resistant to *Sitophilus* species and to *Sitotroga cerealella* (*56–59, 67*). It is difficult to separate the effects of high amylose content from those of grain hardness because high-amylose kernels are invariably hard. However, the inclusion of amylose in artificial diets based on comminuted constituents depressed the survival of *S. cerealella* (*59*).

In recent years, considerable advances have been made in improving maize cultivars carrying the double recessive "opaque" (o_2) gene. This gene results in kernels with excellent protein quality, having high levels of both tryptophan and lysine. Unfortunately, opaque kernels are characteristically soft and starchy. Preliminary observations indicated that opaque maize was invariably susceptible to infestation by insects. More recent investigations have shown that this is not always true (*77*) and that grain hardness can be improved by selection without affecting protein quality, so that the susceptibility of opaque maize can be as low as that of normal maize (*23*).

There is some doubt as to whether the outer skin (the combined pericarp and testa) of the maize kernel provides a barrier to insect penetration. Damaged maize kernels are more susceptible to *Sitophilus zeamais*; sound coats probably retard oviposition (*77*). However, it has been shown more recently that the same numbers of eggs were laid in varieties of differing susceptibility, and that although females of *S. zeamais* lay more eggs in damaged kernels, they do not necessarily lay the additional eggs in the damaged parts of each kernel (*22, 23*). It is possible that damaging the kernels permits adult insects to feed more easily, thus increasing their rate of oviposition. Removing the pericarp increases oviposition. Oviposition may also be affected by a chemical stimulant in the maize grain (*35*). An ethanol extract from a susceptible variety of maize stimulated the oviposition of *Sitophilus oryzae* (rice weevil) when the extract was incorporated in starch pellets (*36*).

Pigmentation of maize appears to be related only occasionally to resistance; mottled maize from Bolivia was resistant to *Sitotroga cerealella* (*41*). There is no consistent difference between white and yellow maize.

Studies of the inheritance of resistance to *Sitophilus* attack have shown that resistance is inherited in a complex manner. When inbred lines were crossed, their reciprocal, F_1 progeny sometimes differed in resistance and sometimes were similar; resistance was inherited from both parents (*92*). Resistance is influenced by the interaction of several factors controlled by more than one gene locus (*34*).

Wheat

Various studies of wheat resistance have been made, but there have been few attempts to determine the factors that govern resistance. During one study of a world collection of wheat, 53 varieties resistant to *Sitophilus oryzae* and 47 resistant to *R. dominica* were identified (*62*). It was also shown that the hardness of the grain, as measured by its Pelshenke value (a measure of gluten strength), was correlated with resistance to *S. oryzae* but not to *R. dominica*. Hardness, however, cannot be the only factor affecting resistance; one of the varieties most resistant to *S. oryzae*—Kalyansona (HD 1593)—is extremely soft (*10*). In other trials it was shown that the relative resistance of six wheat varieties to *S. oryzae* remained unchanged after 3 years of storage at 26.5° and 60% relative humidity (*12*).

Rice

Rice may be stored unhusked (rough rice or paddy), dehusked (brown rice), or fully milled (polished). Paddy is relatively resistant to insect attack, and the factor of greatest importance in determining that resistance is the tightness of the glumes surrounding the grain (*14–16*). There is a relationship between glume defects and oviposition by *Sitotroga cerealella* (*19*), *Sitophilus oryzae*, and *Sitophilus zeamais* (*73*). The presence of hairs on the husk (*50, 91*) and the thickness and silica content of the husk (*63*) have been related to resistance to *Sitophilus* species. *Sitotroga*

cerealella penetrates paddy husks through the abscission scar (the point where the grain was previously attached to the plant); varietal differences in susceptibility to attack through that route have been demonstrated (*18*). Polished rice lacks the protection of the glumes and tends to be more susceptible to insect attack; nevertheless varieties differ in degree of susceptibility.

Several pest species—*Plodia interpunctella* (Indianmeal moth), *R. dominica*, *S. oryzae*, *Cadra cautella* (almond moth), *Lasioderma serricorne* (cigarette beetle), *Tribolium castaneum* (red flour beetle), and *Oryzaephilus surinamensis* (sawtoothed grain beetle)—have been tested on several rice varieties (*48*). The Dawn variety was resistant against all species except *O. surinamensis* and *L. serricorne*; the Labelle variety was resistant to *R. dominica* (*47*).

The factors governing resistance in milled rice have not been studied extensively. Amylose content is not related to resistance to *Sitotroga cerealella* (*1*); high protein levels are probably related to low resistance (*17*). Raw rice is more susceptible to *Sitophilus oryzae* than parboiled rice (*88*).

Barley

Very little detailed information on resistance in barley has been published, but there have been reports on the relative resistance of several varieties to infestation by *R. dominica*, *S. oryzae*, *T. castaneum*, *T. confusum*, *O. surinamensis*, and *Cryptolestes turcicus* (*84, 85*); on barley varieties somewhat resistant to *Tribolium confusum* (*45*) and *Trogoderma granarium* (*83*); on the resistance of seven varieties of barley to attack by *Sitophilus granarius* (granary weevil) (*9*); and on the resistance of varieties with intact husks to attack by *S. oryzae* (*83*).

Sorghum

Attempts have been made to breed varieties of sorghum resistant to infestation by a *Sitophilus* species (*26*). After first identifying varieties of sorghum that were resistant to infestation, it was shown that resistance was associated with a hard outer layer of corneous endosperm surrounding a soft inner starchy layer. Subsequently, it was found possible to select resistant lines from advanced generations derived from the F_1 of a susceptible strain crossed with a resistant strain and then back-crossed onto the susceptible parent (*27*). Subsequent work confirmed the importance of hard endosperm (*20, 71, 72, 74*).

A study of a world sorghum collection showed that varieties that, after harvest, retained glumes that tightly covered the seeds were resistant (*69*), that the resistance of all varieties of sorghum does not respond to changes in humidity in the same manner, and that resistance was not associated with a low moisture content at equilibrium (70). Abrading the outer layer of the grain renders it more susceptible to attack by *Sitophilus oryzae*. Grain that has been previously exposed to *S. oryzae* is more susceptible to subsequent attack (93).

Triticale

Triticale is a new cereal developed by crossing wheat (*Triticum*) and rye (*Secale*). Although not yet of importance as a commercial crop, it is showing good yield and drought-resistance potential and is likely to be exploited widely in the future. In a comparative study of resistance to *Sitophilus zeamais*, *S. oryzae*, and *S. granarius* of five triticale varieties and representative bread wheats, durum wheat, barley, and maize, it was found that all the triticale varieties were more susceptible than all other crops tested (25). Much of this susceptibility was probably associated with the soft, badly formed endosperm that triticale varieties often have; my own tests on varieties with harder endosperms indicated that they were much more resistant.

Legumes

The dried edible seeds of legumes are frequently attacked by beetles of the family Bruchidae. There are several genera of stored-product bruchids (87) associated with a range of host plants. Most species are associated with several particular host species, and other host plants are relatively resistant to them. Thus, *Callosobruchus maculatus* and *C. chinensis* are associated with cowpeas (*Vigna unguiculata*), adzuki beans (*Vigna angularis*), chick-peas (*Cicer arietinum*), lentils (*Lens culinaris*), and green gram (*Vigna radiata*) but normally do not attack common field beans (*Phaseolus vulgaris*). Cowpeas are also attacked by *Callosobruchus analis*, *C. phaseoli*, and *C. rhodesianus*. Black gram seeds (*Vigna mungo*) are attacked by *C. maculatus*, *C. chinensis*, and *C. rhodesianus*. Dolichos beans (*Lablab purpureus*) are the most common hosts of *C. phaseoli*. *C. subinnotatus* is restricted to attacking Bambara groundnuts (*Vigna subterranea*). The other two common storage species, *Acanthoscelides obtectus* and *Zabrotes subfasciatus* (Mexican bean weevil), are both major pests of common field beans and lima or butter beans (*Phaseolus lunatus*), although both are capable of attacking Bambara groundnuts. *Z. subfasciatus* can attack adzuki beans and has been recorded on cowpeas.

Certain species of legumes are resistant to certain species of Bruchidae but it is of more practical interest to search for varieties that show resistance to known pests. Surprisingly little work has been done in this direction although advances have been made in determining the nutrition factors that affect insect development in certain legumes.

Cowpeas

The major pests of cowpeas are *Callosobruchus* species. Infestation can begin in the field. However, the pest normally goes through only one generation in the field. When dry, cowpeas in the pod (especially those with thick pods) are very resistant to infestation (2). High mortality of eggs occurs after oviposition on green or mature pods. Oviposition preferences may be affected by the surface texture of oviposition sites (50). After shelling, or accidental shattering of the pods, the peas become highly susceptible to bruchid attack, but certain varietal differences in susceptibility have been observed. This resistance has been attributed to low rates of oviposition and to poor adult survival (64). *Callosobruchus* species lay more eggs on smooth-coated seeds than on rough ones (13, 53), and the arrangement of cells in the testa of smooth cowpeas facilitates penetration by first-instar larvae (52).

"Secondary" chemical compounds in the seeds of legumes may be of importance in combating insect attack. Several compounds have been found in non-host legumes that are toxic to *Callosobruchus* species. For example, phytohemagglutinins from black beans (*P. vulgaris*) are toxic to *C. maculatus* (40). Until recently, little progress had been made in identifying components in cowpeas (or any other host crop) that are detrimental to *Callosobruchus* development. Protease inhibitors, common in legume seeds, may be insect defense compounds, but investigations using commercial trypsin inhibitors from soybeans (*Glycine max*) have shown them to be ineffective (40).

It has been suggested that bruchids lack gut proteases because they feed on legume seeds that are themselves rich in proteases (3). However, larvae of *C. maculatus* have been shown to have proteolytic activity (6) and a variety of cowpeas from northern Nigeria, shown to be highly resistant to attack by *C. maculatus* (81), has about three times the usual level of trypsin inhibitor. The inhibitor extracted from this variety is highly toxic to *C. maculatus* at the levels found in the resistant variety (30). Trypsin inhibitors, therefore, could be an important source of resistance in cowpeas. However, concentrations of sulfur-bearing amino acids interact with the trypsin inhibitor (29). Inheritance of resistance has been shown in experiments to be controlled by one or two major genes with modifiers (65, 66, 86). The color of cowpea seeds apparently has no effect upon susceptibility (8). *Zabrotes subfasciatus*, which normally does not attack cowpeas, has recently become a widespread pest of this crop in Uganda (21). This demonstrates that some species adapt to plants other than their normal hosts.

Common Beans

Common beans are most frequently attacked by *A.*

obtectus and *Z. subfasciatus*. Most work on resistance has been carried out with *A. obtectus*. *A. obtectus* lays eggs loosely in the food commodity; after hatching, the first-instar larvae bore into the beans. Damaged beans are more likely to be attacked than sound beans (*75*). Beans contain an extractable component that stimulates oviposition (*60*). The composition of this component is unknown, as is whether varieties exist that lack this ovipositional stimulant.

Several studies of resistance of different varieties have been made, and differences in susceptibility have been reported (*7*). In a study of more than 4,000 bean varieties, only low levels of resistance were found (*76*), but later studies demonstrated excellent resistance in wild bean varieties originating from Mexico (*77*). It has been suggested that *A. obtectus* can feed on beans because it is unaffected by a soluble heteropolysaccharide found in beans that is toxic to *C. chinensis* (*4*). High levels of this heteropolysaccharide partly inhibited *A. obtectus* development. In fact, the resistance of one variety to *A. obtectus* is due to the presence of a complex heteropolysaccharide (*55*). In another study it was found that a glycoprotein from red beans inhibited amylase activity in several insects (*61*). The wild varieties of Mexican beans appear to be good sources of resistance to the Bruchidae. A novel protein in some varieties has been associated with resistance to *Z. subfasciatus* (*31*).

The value of searching for antimetabolites in legume seeds has recently been demonstrated. The gene causing synthesis of the trypsin inhibitor that makes cowpeas resistant to *Callosobruchus maculatus* has been transferred to tobacco plants (*Nicotiana tabacum*). As a result, the foliage of the tobacco plants was made resistant to *Heliothis virescens* (*37, 49*). Clearly, there is great potential for improving the resistance of crops through biotechnology.

Conclusion

Considerable research has been done on many aspects of host-plant resistance in stored cereals and pulses. Unfortunately, much of the work has been performed in entomology laboratories and not in collaboration with plant breeders who could use the results obtained to improve the storage qualities of crops. To make practical progress in postharvest varietal improvement, it will be necessary to

(a) screen large numbers of wild varieties and commercial varieties (cultivars) of crops to identify those that are resistant;

(b) study the resistant varieties to discover the cause of resistance (such studies may involve entomologists, botanists, biochemists, and others);

(c) study the inheritability of resistance and incorporate resistant lines into breeding programs to develop new resistant lines; and

(d) study the effect of the introduction of resistant varieties upon pest populations and upon consumers.

Clearly, such investigations cannot be performed by individual workers; they will require the establishment and support of interdisciplinary teams.

Cited References

1 Abraham, C. C., et al. 1972. Relative susceptibility of different varieties of paddy to infestation by the Angoumois grain moth *Sitotroga cerealella* Olivier (Gelechiidae: Lepidoptera), as influenced by the amylose content of the endosperm. *Bulletin of Grain Technology* 10(4)263–266.

2 Akingbohungbe, A. E. 1976. A note on the relative susceptibility of unshelled cowpeas to the cowpea weevil (*Callosobruchus maculatus* Fabricius) (Coleoptera: Bruchidae). *Tropical Grain Legume Bulletin* 5(July)11–13.

3 Applebaum, S. W. 1964. Physiological aspects of host specificity in the Bruchidae. I. General consideration of developmental compatability. *Journal of Insect Physiology* 10(5)783–788.

4 Applebaum, S. W., and M. Guez. 1972. Comparative resistance of *Phaseolus vulgaris* beans to *Callosobruchus chinensis* and *Acanthoscelides obtectus* (Coleoptera: Bruchidae); the differential digestion of soluble heteropolysaccharide. *Entomologia Experimentalis et Applicata* 15(2)203–207.

5 Ashby, K. R. 1961. The population dynamics of *Cryptolestes ferrugineus* (Stephens) (Col., Cucujidae) in flour and on Manitoba wheat. *Bulletin of Entomological Research* 52(2)363–379.

6 Baker, A. M. R. 1978. Protease inhibitors and the biological basis of insect resistance in *Vigna unguiculata*. Dissertation, University of Durham, England.

7 Bastidas, R. C., H. F. Sanchez, and V. G. Bravo. 1973. Resistencia de cinco variedades de frijol almacenado al ataque del gorgojo mayor (*Acanthoscelides obtectus* Say). *Fitotécnica Latinoamericana* 9(1)36–39.

8 Bastos, J. A. M. 1969. Influencia da cor de feijao de corda, *Vigna sinensis* Endl. no ataque do gorgulho *Callosobruchus analis*. *Turrialba* 19(2)296–297.

9 Bell, A. C., and J. P. Moore. 1984. Varietal susceptibility of winter barley attack by the grain weevil *Sitophilus granarius*. *Tests of Agrochemicals and Cultivars* 5(*Annals of Applied Biology* 104, Suppl.)110–111.

10 Bhatia, S. K. 1976. Resistance to insects in stored grain. *Tropical Stored Products Information* 31:21–35.

11 Birch, L. C. 1945. The mortality of the immature stages of *Calandra oryzae* L. (small strain) and *Rhizopertha dominica* Fab. in wheat of different moisture content. *Australian Journal of Experimental Biology and Medical Science* 23(2)141–145.

12 Boles, H. P., and R. L. Ernst. 1976. Susceptibility of six wheat cultivars to oviposition by rice weevils reared on wheat, corn or sorghum. *Journal of Economic Entomology* 69(4)548–550.

13 Booker, R. H. 1965. Pests of cowpea and their control in Northern Nigeria. *Bulletin of Entomological Research* 55(4)663–672.

14 Breese, M. H. 1960. The infestibility of stored paddy by *Sitophilus sasakii* (Tak.) and *Rhyzopertha dominica* (F.). *Bulletin of Entomological Research* 51(3)599–630.

15 Breese, M. H. 1963. Studies on the oviposition of *Rhyzopertha dominica* (F.) in rice and paddy. *Bulletin of Entomological Research* 53(4)621–637.

16 Breese, M. H. 1964. The infestibility of paddy and rice. *Tropical Stored Products Information* 8:289–299.

17 Chatterji, S. M., R. C. Dani, and S. Govindaswami. 1977. Evaluation of rice varieties for resistance to *Sitotroga cerealella* Oliv. (Lepidoptera: Gelechiidae). *Journal of Entomological Research* 1(1)74–77.

18 Cogburn, R. R., C. N. Bollich, and S. Meola. 1983. Factors that affect the relative resistance of rough rice to Angoumois grain moths and lesser grain borers. *Environmental Entomology* 12(3)936–942.

19 Cohen, L. M., and M. P. Russell. 1970. Some effects of rice varieties on the biology of the Angoumois grain moth, *Sitotroga cerealella. Annals of the Entomological Society of America* 63(4)930–931.

20 Davey, P. M. 1965. The susceptibility of sorghum to attack by the weevil *Sitophilus oryzae* (L.). *Bulletin of Entomological Research* 56(2)287–297.

21 Davies, J. C. 1972. A note on the occurrence of *Zabrotes subfasciatus* Boh. (Coleoptera, Bruchidae), on legumes in Uganda. *East African Agriculture and Forestry Journal* 37(4)294–299.

22 Dobie, P. 1974. The laboratory assessment of the inherent susceptibility of maize varieties to post-harvest infestation by *Sitophilus zeamais* Motsch. (Coleoptera: Curculionidae). *Journal of Stored Products Research* 10(3)183–197.

23 Dobie, P. 1977. The contribution of the Tropical Stored Products Centre to the study of insect resistance in stored maize. *Tropical Stored Products Information* 34:7–22.

24 Dobie, P. 1978. A simple curve describing the development pattern of some beetles breeding on stored products. *Journal of Stored Products Research* 14(1)41–44.

25 Dobie, P., and A. M. Kilminster. 1978. The susceptibility of triticale to post-harvest infestation by *Sitophilus zeamais* Motschulsky, *Sitophilus oryzae* (L.) and *Sitophilus granarius* (L.). *Journal of Stored Products Research* 14(2–3)87–93.

26 Dogget, H. 1957. The breeding of sorghum in East Africa. I. Weevil resistance in sorghum grains. *Empire Journal of Experimental Agriculture* 25(97)1–9.

27 Dogget, H. 1958. The breeding of sorghum in East Africa. II. The breeding of weevil-resistant varieties. *Empire Journal of Experimental Agriculture* 26(101)37–46.

28 Fortier, G., et al. 1983. Local and improved corns (*Zea mays*) in small farm agriculture in Belize, C. A.; their taxonomy, productivity and resistance to *Sitophilus zeamais. Phytoprotection* 63(2)68–78.

29 Gatehouse, A. M. R., and D. Boulter. 1983. Assessment of the antimetabolic effects of trypsin inhibitors from cowpea (*Vigna unguiculata*) and other legumes on development of the bruchid beetle *Callosobruchus maculatus. Journal of the Science of Food and Agriculture* 34(4)345–350.

30 Gatehouse, A. M. R., et al. 1979. Biochemical basis of insect resistance in *Vigna unguiculata. Journal of the Science of Food and Agriculture* 30(10)948–958.

31 Gatehouse, A. M. R., et al. 1987. Role of carbohydrates in insect resistance in *Phaseolus vulgaris. Journal of Insect Physiology* 33(11)843–850.

32 Giga, D. P., and R. H. Smith. 1981. Varietal resistance and intraspecific competition in the cowpea weevils *Callosobruchus maculatus* and *C. chinensis* (Coleoptera: Bruchidae). *Journal of Applied Ecology* 18(3)755–761.

33 Giles, P. H., and F. Ashman. 1971. A study of pre-harvest infestation on maize by *Sitophilus zeamais* Motsch. (Coleoptera, Curculionidae) in the Kenya highlands. *Journal of Stored Products Research* 7(2)69–83.

34 Gomez, L. A. 1983. Mechanisms and inheritance of resistance of selected maize genotypes, *Zea mays* L., to the rice weevil, *Sitophilus oryzae* (L.). *Dissertation Abstracts International* B43(12)(pt. I)3837B.

35 Gomez, L. A., et al. 1983. Relationship between some characteristics of the corn kernel pericarp and resistance to the rice weevil (Coleoptera: Curculionidae). *Journal of Economic Entomology* 76(4)797–800.

36 Gomez, L. A., et al. 1983. Chemosensory responses of the rice weevil (Coleoptera: Curculionidae) to a susceptible and resistant corn genotype. *Journal of Economic Entomology* 76(5)1044–1048.

37 Hilder, V. A., et al. 1987. A novel mechanism of insect resistance engineered into tobacco. *Nature* 330(6144)160–163.

38 Hodges, R. J., et al. 1983. An outbreak of *Prostephanus truncatus* (Horn) (Coleoptera: Bostrichidae) in East Africa. *Protection Ecology* 5(2)183–194.

39 Janzen, D. H., H. B. Juster, and E. A. Bell. 1977. Toxicity of secondary compounds to the seed-eating larvae of the bruchid beetle *Callosbruchus maculatus. Phytochemistry* 16(2)223–227.

40 Janzen, D. H., H. B. Juster, and I. E. Liener. 1976. Insecticidal action of phytohemagglutinin in black beans on a bruchid beetle. *Science* 192(4241)795–796.

41 Kempton, J. H. 1917. Protective coloration in seeds of Bolivian maize. *Journal of Heredity* 8(5)200–202.

42 Kirk, V. M., and A. Manwiller. 1964. Rating dent corn for resistance to rice weevils. *Journal of Economic Entomology* 57(6)850–852.

43 Krall, S. 1984. A new threat to farm-level maize storage in West Africa: *Prostephanus truncatus* (Horn) (Coleoptera: Bostrichidae). *Tropical Stored Products Information* 50:26–31.

44 Kyle, C. H. 1918. *Shuck Protection for Ear Corn.* Bulletin 708. U.S. Department of Agriculture, Washington, DC.

45 Loschiavo, S. R., A. J. McGinnis, and D. R. Metcalf. 1969. Nutritive value of barley varieties assessed with

the confused flour beetle. *Nature* 224(5216)288.

46 McCain, F. S., W. G. Eden, and D. N. Singh. 1964. A technique for selecting for rice weevil resistance in corn in the laboratory. *Crop Science* 4(1)109–110.

47 McGaughey, W. H. 1973. Resistance to the lesser grain borer in "Dawn" and "Labelle" varieties of rice. *Journal of Economic Entomology* 66(4)1005.

48 McGaughey, W. H. 1974. Insect development in milled rice: effects of variety, degree of milling, parboiling and broken kernels. *Journal of Stored Products Research* 10(2)81–86.

49 Meeusen, R. L., and G. Warren. 1989. Insect control with genetically engineered crops. *Annual Review of Entomology* 34:373–381.

50 Messina, F. J. 1984. Influence of cowpea pod maturity on the oviposition choices and larval survival of a bruchid beetle *Callosobruchus maculatus*. *Entomologia Experimentalis et Applicata* 35(3)241–248.

51 Nwana, I. E., and D. T. Akibo-Betts. 1982. The resistance of some rice varieties to damage by *Sitophilus zeamais* Motschulsky during storage. *Tropical Stored Products Information* 43:10–15.

52 Nwanze, K. F., and E. Horber. 1976. Seed coats of cowpeas affect oviposition and larval development of *Callosobruchus maculatus*. *Environmental Entomology* 5(2)213–218.

53 Nwanze, K. F., E. Horber, and C. W. Pitts. 1975. Evidence for ovipositional preference of *Callosobruchus maculatus* for cowpea varieties. *Environmental Entomology* 4(3)409–412.

54 Osborn, T. C., et al. 1988. Insecticidal activity and lectin homology of arcelin seed protein. *Science* 240(4849)208–210.

55 Osborn, T. C., M. Burow, and F. A. Bliss. 1988. Purification and characterization of arcelin seed protein from common bean. *Plant Physiology* 86(2)399–405.

56 Peters, L. L., M. L. Fairchild, and M. S. Zuber. 1972. Effect of corn endosperm containing different levels of amylose on Angoumois grain moth biology. 1. Life cycle, certain physiological responses and infestation rates. *Journal of Economic Entomology* 65(2)576–581.

57 Peters, L. L., M. L. Fairchild, and M. S. Zuber. 1972. Effect of corn endosperm containing different levels of amylose on Angoumois grain moth biology. 2. Physical and chemical properties of experimental corn. *Journal of Economic Entomology* 65(2)581–584.

58 Peters, L. L., M. L. Fairchild, and M. S. Zuber. 1972. Effect of corn endosperm containing different levels of amylose on Angoumois grain moth biology. 3. Interrelationship of amylose levels and moisture content of diets. *Journal of Economic Entomology* 65(4)1168–1169.

59 Peters, L. L., M. S. Zuber, and V. Fergason. 1960. Preliminary evidence of resistance of high-amylose corn to the Angoumois grain moth. *Journal of Economic Entomology* 53(4)573–574.

60 Pouzat, J. 1976. Oviposition behaviour of the bean bruchid in the presence of food plant extracts. Demonstration of interactions between taste and touch. *Comptes Rendus*

Hebdomadaires des Séances de l'Académie des Sciences D282(22)1971–1974.

61 Powers, J. R., and J. Culbertson. 1981. Inter-action of insect amylase with a bean (*Phaseolus vulgaris*) glycoprotein. *Cereal Foods World* 26(9)485.

62 Pradhan, S., et al. 1984. *Resistance to Two Major Stored Grain Pests in World Collection of Wheat*. Research Bulletin (new series) 1. Division of Entomology, Indian Agricultural Research Institute, New Delhi.

63 Prakash, A. 1984. Varietal resistance of stored rice grains to *Sitophilus oryzae* Linn. (Curculionidae: Coleoptera). *Bulletin of Grain Technology* 20(1)8–12.

64 Ramalho, F. S., and R. C. de Sena. 1975. Resistencia relativa de differentes variedades de *Vigna* ao *Callosobruchus maculatus* (Fabr.) (Bruchidae: Coleoptera) no Estado de Pernambuco. *Ciência e Cultura* (São Paulo) 27(11)1244.

65 Redden, R. J. 1983. The inheritance of seed resistance to *Callosobruchus maculatus* F. in cowpea (*Vigna unguiculata* L. Walp.). Analyses of percentage emergence and emergence periods of bruchids in F_4 seed generations of two reciprocal crosses. *Australian Journal of Agriculture* 34(6)697–705.

66 Redden, R. J., P. Dobie, and A. M. R. Gatehouse. 1983. The inheritance of seed resistance to *Callosobruchus maculatus* F. in cowpea (*Vigna unguiculata* L. Walp.). I. Analyses of parental, F_1, F_2, F_3 and backcross seed generations. *Australian Journal of Agriculture* 34(6)681–685.

67 Rhine, J. R., and R. Staples. 1968. Effect of high-amylose field corn on larval growth and survival of five species of stored-grain insects. *Journal of Economic Entomology* 61(1)280–282.

68 Rodriguez, R. R. 1976. Determinación de daño causado por plagas de almacén a variedades de maíz en Yucatán. *Agricultura Técnica en México* 3(12)442–446.

69 Rogers, R. R., and R. B. Mills. 1974. Evaluation of a world sorghum collection for resistance to the maize weevil *Sitophilus zeamais* Motsch. (Coleoptera: Curculionidae). *Journal of the Kansas Entomological Society* 47(1)36–41.

70 Rogers, R. R., and R. B. Mills. 1974. Reactions of sorghum varieties to maize weevil infestations under three relative humidities. *Journal of Economic Entomology* 67(5)692.

71 Russell, M. P. 1962. Effects of sorghum varieties on the lesser rice weevil, *Sitophilus oryzae* (L.). I. Oviposition, immature mortality and size of adults. *Annals of the Entomological Society of America* 55(6)678–685.

72 Russell, M. P. 1966. Effects of four sorghum varieties on the longevity of the lesser rice weevil, *Sitophilus oryzae* (L.). *Journal of Stored Products Research* 2(1)75–79.

73 Russell, M. P. 1968. Influence of rice variety on oviposition and development of the rice weevil, *Sitophilus oryzae*, and the maize weevil, *S. zeamais*. *Annals of the Entomological Society of America* 61(5)1335–1336.

74 Russell, M. P., and M. M. Rink. 1965. Some effects

of sorghum varieties on the development of a rice weevil *Sitophilus zeamais* (Coleoptera: Curculionidae). *Annals of the Entomological Society of America* 58(5)763.

75 Sandner, H., and M. Pankanin. 1975. Influence of artificial damage to bean seeds upon the fecundity of the common bean weevil *Acanthoscelides obtectus* Say (Coleoptera, Bruchidae). *Polskie Pismo Entomologiczne* 45(3–4)615–623.

76 Schoonhoven, A. V., and C. Cardona. 1982. Low levels of resistance to the Mexican bean weevil in dry beans. *Journal of Economic Entomology* 75(4)567–569.

77 Schoonhoven, A. V., C. Cardona, and J. Valor. 1983. Resistance to the bean weevil and the Mexican bean weevil (Coleoptera: Bruchidae) in noncultivated common bean accessions. *Journal of Economic Entomology* 76(6)1225–1259.

78 Schoonhoven, A. V., et al. 1972. Resistance in corn kernels to the maize weevil *S. zeamais* Motsch. *Proceedings of the North Central Branch of the Entomological Society of America* 27:108–110.

79 Schoonhoven, A. V., C. E. Wassom, and E. Horber. 1972. Development of maize weevil on kernels of opaque-2 and floury-2, nearly isogenic corn inbred lines. *Crop Science* 12(6)862–863.

80 Schulten, G. M. 1976. Insects in stored maize ears. *Abstracts on Tropical Agriculture* 2(6)9–17.

81 Singh, S. R. 1978. Resistance to pests of cowpeas in Nigeria. In *Pests of Grain Legumes*, S. R. Singh, H. E. van Emden, and T. A. Taylor, eds. Academic Press, New York, NY.

82 Singh, D. N., and F. S. McCain. 1963. Relationship of some nutritional properties of the corn kernel to weevil infestation. *Crop Science* 3(3)259–261.

83 Singh, V., B. N. Mathur, and S. K. Sharma. 1972. A note on the relative susceptibility of some barley (*Hordeum vulgare*) varieties to *Trogoderma granarium* Everts. *Madras Agricultural Journal* 59(1)42–43.

84 Sinha, R. N. 1969. Reproduction of stored grain insects on varieties of wheat, oats and barley. *Annals of the Entomological Society of America* 62(5)1011–1015.

85 Sinha, R. N. 1971. Multiplication of some stored products insects on varieties of wheat, oats and barley. *Journal of Economic Entomology* 64(1)98–102.

86 Smith, C. M. 1989. *Plant Resistance to Insects: A Fundamental Approach*. Wiley, New York, NY.

87 Southgate, B. J. 1978. The importance of Bruchidae as pests of grain legumes, their distribution and control. In *Pests of Grain Legumes*, S. R. Singh, H. E. van Emden, and T. A. Taylor, eds. Academic Press, New York, NY.

88 Starks, K. J., et al. 1966. Damage to corn by pink scavenger caterpillar and its relationship to corn earworm and rice weevil damage. *Journal of Economic Entomology* 59(4)931–934.

89 Sudhakar, T. R., and N. D. Pandey. 1983. Relative resistance of raw and parboiled rice varieties to the rice weevil, *Sitophilus oryzae* (L.). *Indian Journal of Entomology* 43(3)279–282.

90 Vanderschaaf, P., D. A. Wilbur, and R. H. Painter. 1969. Resistance of corn to laboratory infestation of the larger rice weevil *Sitophilus zeamais*. *Journal of Economic Entomology* 62(2)352–355.

91 Virmani, S. S., P. K. B. Menon, and A. S. Gobeh. 1980. Varietal resistance to rice weevil. *International Rice Research Newsletter* 5(4)8.

92 Widstrom, N. W., et al. 1983. Dent corn inbred sources of resistance to the maize weevil (Coleoptera: Curculionidae). *Journal of Economic Entomology* 76(1)31–33.

93 Williams, J. O., and R. B. Mills. 1980. Influence of mechanical damage and repeated infestation of sorghum on its resistance to *Sitophilus oryzae* (L.) (Coleoptera: Curculionidae). *Journal of Stored Products Research* 16(2)51–53.

94 Wiseman, B. R., W. W. McMillian, and N. W. Widstrom. 1970. Husk and kernel resistance among maize hybrids to an insect complex. *Journal of Economic Entomology* 63(4)1260–1262.

28
Pest Resistance to Pesticides

Bruce R. Champ and Edward Highley

Resistance to pesticides is widespread among food-industry pests. Most of the major pest species of stored-food products have been reported resistant, in some degree, to most pesticides in common use. The species concerned are primarily pests of grain or grain products but also include pests that attack products of animal origin. They encompass invertebrate pests, such as insects and mites, and vertebrates, such as rodents.

As regards the main insect and mite pests—the group of greatest significance to the food industry—the problem has grown rapidly from three species in 1960 to eight species in 1970, and currently involves more than 20 species. Pesticide resistance in cockroaches has been a significant constraint in the control of these universal food-industry pests for many years. Even though resistance of commensal rodents to anticoagulant agents was first reported more than 20 years ago, resistance currently seems to represent no serious threat to long-term control, thanks to the availability of effective alternatives.

Given the genetic nature of the phenomenon, it is inevitable that resistance to pesticides will increase among food-industry pests and will involve not only existing pesticides but also any new materials brought forward to deal with pest problems. The best that can be hoped for is to retard the rate of development and the spread of resistance among pest populations. Resistance must be managed and, indeed, resistance

management should be considered an integral part of the whole pest management process.

It is critical that the development and spread of resistance be contained. There are few alternatives to currently used residual pesticides, and the emergence of resistance to phosphine—perhaps the most useful food-industry fumigant—spells a potentially disastrous situation, particularly for the preservation of food stocks in the developing world.

The Nature of Resistance

Resistance is a genetic phenomenon. It arises from selection among individuals in a pest population that are more tolerant of a toxicant. The result is a general elevation of the tolerance of the population to levels that allow survival of individuals after treatment with doses of pesticide calculated to kill normally susceptible individuals of the species.

An increase in tolerance may be offset to some degree by increasing dosages of the pesticide. However, in time, a portion of the population will tolerate the toxicant at concentrations greater than those permitted by economics and allowable residue limits in commodities. When this occurs, control measures based on the particular material are useless.

The fact that resistance results *only* from selection permits the following deductions of practical importance in its management.

(a) The rate of selection is dependent on the number of individuals available for selection. It therefore follows that any reduction in the number available (e.g., through hygiene and good warehouse practices) re-

Champ and Highley: Postharvest Research Program, Australian Centre for International Agricultural Research, GPO Box 1571, Canberra ACT 2601, Australia.

duces genetic variability and the probability of resistance developing.

(b) When control measures achieve complete kills, there is no selection. Therefore, the dose applied should be sufficient to kill the most tolerant individuals in any normal population and the full dose should be available to every individual.

Fumigation, when properly performed, gives kills approximating 100% and, for this reason, has been effective in delaying resistance development. This is one of the reasons for recommending fumigation and why it should always be used following failure of residual pesticides. Therefore, the development of resistance to commonly used fumigants, already alluded to, has to be viewed very seriously.

The application of pesticides at dosages exceeding estimated minimum effective dosages may retard the rate of development of resistance in a population. This is a desirable approach when the pest species concerned characteristically shows high levels of variability in response to the pesticide in use or is prone to increased tolerance from nonspecific causes.

Resistance Mechanisms and Modeling

Although the mechanisms of resistance are by no means clear, certain well-established features both highlight the complexity of the problem and enable planning of effective countermeasures.

(a) Pesticides fall into well-defined chemical classes, and the development in a population of resistance to a particular material may confer on the population high-level resistance to related pesticides. Low-level resistance to unrelated classes of pesticides may also develop concurrently.

(b) Pests may be exposed to a number of different types of pesticides and this, together with cross-infestation by strains carrying other types of resistances developed elsewhere, may result in a complex of unrelated resistances being present in the same strain.

(c) Strains of pests in which resistance to a particular class of pesticides has developed will develop high-level resistance to unrelated pesticides with greater facility than will strains not previously exposed to pesticides.

(d) Resistance in field populations *may* diminish when the pests are no longer exposed to the pesticides.

These observations lead to the following practical conclusion: When resistance has been observed, the pesticide in use should be replaced by an unrelated material. When a changeover is undertaken, further resistances may occur and the availability of reserve types of pesticides for emergencies must be considered. A knowledge of the resistance mechanism involved is useful in choosing possible alternative materials. In considering countermeasures to resis-

tance, it is essential to know at least the characteristic cross-resistance patterns for the material involved.

Population geneticists have been modeling resistance to pesticides and strategies to overcome it for some years. The results, summarized by Collins (*17*), are interesting and occasionally appear to be, superficially at least, at variance with current conventional wisdom. The models are based on standard population genetics theory relating to the spread of a gene for resistance in a population and the relative fitnesses of susceptible (SS), resistant (RR), and heterozygous (RS) individuals. The models deal with discrete populations and take no account of dispersal, a key factor in the spread of resistance among pest populations during the movement of foodstuffs in trade.

Simple genetic models indicate that the *only* factor accelerating the spread of resistance within a particular pest population is high population growth rate. This aligns well with the observation that resistance arises *only* from selection—the larger the population, the greater the selection pressure.

Factors deduced from modeling that delay the spread of resistance are

(a) low pesticide dosages;
(b) the presence of refuges for pests;
(c) the creation of effective recessiveness of the resistance gene by the use of high dosages of pesticide;
(d) the immigration or release of susceptible individuals; and
(e) alternating the use of different pesticides.

The suggestion that low pesticide dosages and the presence of refuges (i.e., a situation where large sections of the population are not treated) retard the spread of resistance may present conceptual difficulties. However, once again, the development of resistance rests only on selection, and low dosages and nontreatment presumably reduce selection pressure. Unfortunately, they may also lead to inadequate levels of control. They do, however, suggest alternative strategies that have high potential for practical application. A low kill rate of susceptible individuals resulting from the application of low pesticide dosages can be made compatible with adequate control by reducing the population growth rate by nonpesticidal means. One such means is by reducing temperature (*20*), but there are many others, all falling within the ambit of management and reinforcing the notion that management of resistance must be an integral part of the whole food-pest management system.

Distinguishing Between Resistance and Other Phenomena

With residual pesticides, the first indication of resistance in practical usage of materials is a progressive reduction in the time for which residual materials remain effective. A chemical protectant applied to stored grain, for example, may—when first introduced—give a long period of protection. Subsequently, however, this period is reduced, even though residue analyses may reveal *no change* in the normal decay pattern of the pesticide on the commodity.

Early warning indications of resistance such as this often go unnoticed, particularly when management is lax, and the problem is not recognized until there are more obvious signs of resistance, such as large-scale buildup of pest populations in treated commodities. By this time, the operational life of the pesticide concerned is finished.

Resistance can be imputed when living pests are found after a properly conducted fumigation in which the product of the Concentration $\times$ Time for the fumigant was satisfactory.

Regular monitoring of the tolerance status of the major pest species of particular commodities is a valuable means of early detection of resistance and can provide unequivocal evidence of the contribution of resistance to control failures.

Breakdown of control resulting from any of the following factors does not constitute resistance:

(a) use of pesticide formulations that have deteriorated before use;
(b) incomplete coverage because of inadequate equipment or treatment;
(c) use of unstable preparations;
(d) disregard of compatibility recommendations when mixtures are used;
(e) temperature-dependent variations in the susceptibility of pests (dosages effective at normal temperatures may be inadequate during extremely hot or cold weather);
(f) presence of very large numbers of pests, indicating that enough survivors exist to cause considerable economic damage; and
(g) loss of pesticides at abnormally high rates because of extreme conditions of temperature, moisture, or exposure to light.

Although not indicative of resistance, the incidence of any or all of these factors may contribute to the development of resistance by increasing selection pressure.

Methods of Monitoring for Resistance

As already noted, failure to control pests can occur for many reasons, and confirmation of the presence or absence of pesticide resistance is essential for implementation of effective countermeasures. All expressions of resistance involve changes in tolerance to the pesticides and, however slight, may be detected and measured provided that appropriate methods are used to compare either response levels at similar doses, or doses producing similar responses.

The most satisfactory basis for monitoring resistance is to expose pests to single doses of pesticide that would be expected to kill normally susceptible pests. The 99.9% level is used as the discriminating dosage. Pest survival at these dosages is indicative of resistance and a need for more detailed confirmatory tests. The discriminating dosages in the primary tests are established from the dosage-mortality relationships of known susceptible strains of the pest, taking into account normal strain variability.

Various methods have been proposed for determining resistance in the major food pests. The Food and Agriculture Organization (FAO) of the United Nations has promoted the introduction of a range of standardized test methods for detection and measurement of resistance. Details of these and other test methods can be found in the literature (e.g., *1–6, 9–11, 13, 14, 16, 18, 19, 22–25*).

Although these tests generally are performed in the laboratory so as to allow reproducible responses under standard and defined conditions, it is desirable that they also simulate field tests of chemicals. However, this cannot be done at the expense of sensitivity and reproducibility and, since pesticides are often used in the field in very different formulations and ways, simulating these conditions and seeking common grounds for comparison are difficult. Essentially, monitoring for resistance involves determining changes in physiological response to pesticides, and the ideal test is one in which the only variable factor is the dosage of pesticide.

In general, methods for detecting resistance must

(a) be efficient and yield adequate measures of the level of resistance;
(b) give rapid results so as to allow contermeasures to be implemented quickly if resistance is discovered;
(c) give consistent results and allow large numbers of tests to be performed quickly;
(d) be able to be performed by relatively inexperienced personnel after minimal instruction, and require simple and readily available equipment;
(e) be able to deal with all sizes of samples, including those where only one or two individual pests are available;
(f) preserve the samples for future confirmatory, comparative, and cross-resistance testing.

The procedures to be followed in a resistance test method are as follows:

(1) Selecting an appropriate life-history stage to obtain responses representative of the tolerance status of the species concerned (adult insects usually satisfy the requirements most closely in terms of uniformity of response);
(2) selecting an appropriate type of test to measure responses;

(3) establishing base response data from known susceptible strains;

(4) using discriminating doses to screen samples for resistance either on the field-collected specimens or on their progeny and, if the results are not considered unequivocal, using supplementary discriminating tests on the progeny of the few survivors from the primary test (susceptible reference strains should be included in all tests as a check on procedures);

(5) completely defining resistance by comparing graded dosage-mortality responses of susceptible and resistant strains;

(6) determining cross-resistance patterns;

(7) establishing correlations between resistance levels of different life-history stages; and

(8) correlating laboratory resistance data with responses achieved in simulated field use of the pesticides concerned.

Overview of the Occurrence of Resistance

Residual Pesticides

As noted at the beginning of this chapter, most of the major pest species in stored-food products have been reported resistant, in some degree, to most pesticides in common use. The resistances involve about 80 pesticides and related compounds, including fumigants, the so-called biological insecticides, and all classes of residual pesticides. Resistance is seen as a worldwide problem affecting and threatening, in some measure, the efficacy of all pest control programs involving pesticides.

The resistance problem with storage pests was first highlighted internationally at the FAO Symposium on Resistance of Agricultural Pests to Pesticides, held in Rome in September 1969. During the course of that symposium, attention was drawn to the deteriorating situation in pest control in stored-food products throughout the world, particularly in developing countries. The FAO Working Party of Experts on Resistance of Pests to Pesticides tackled the problem in a practical way by sponsoring, during 1972–73, an FAO Global Survey of Pesticide Susceptibility of Stored Grain Pests. The survey defined the general nature of the resistance problem and identified resistances that either were localized in occurrence or whose significance had not been appreciated.

The FAO survey (15), based on data from 1,700 samples obtained from 90 countries, was a valuable supplement to local surveys and research on resistance that had been in progress in a number of laboratories throughout the world. The potential for spread of resistance was emphasized by data provided on movement of stored products in international trade. Resistances were recorded in infestations that had undoubtedly been imported into particular countries with the various commodities from which the samples were

taken, and infestations sampled from ships and from cargoes being unloaded were nearly always found to be resistant. It was quite clear that the resistant strains were being widely dispersed in infested commodities in world trade.

A detailed summary of the known resistances of stored-product insects to residual pesticides and related compounds is given in tabular form by Champ (12). The tables record widespread resistance to DDT, lindane, the cyclodienes, the pyrethrins and synthetic pyrethroids, and malathion and other organophosphorus insecticides. An interesting new record of resistance since then has been that of stored-product moths to the biological insecticide *Bacillus thuringiensis* (21).

Fumigants

Champ (12) also reported field occurrences of resistance to methyl bromide, phosphine, and hydrogen cyanide, the main materials used for fumigating food storages and stored products. Low-level resistance to carbon dioxide—the main "controlled atmosphere" alternative to conventional fumigants—has already been demonstrated in laboratory studies.

Fumigation is a powerful and generally reliable technique used primarily as a control for established infestations of pests in commodities or in storage or handling premises. It is the preferred method for controlling infestations of storage pests because of the high probability of successful treatment in a properly performed exercise. Also, as mentioned previously, fumigation helps to delay resistance development and combats resistance to residual insecticides. Indeed, the current spread of resistance to residual pesticides is forcing greater reliance on fumigants in order to establish and maintain acceptable levels of pest control.

For all these reasons, the high-level resistance to phosphine, perhaps the most useful of the very limited range of materials available, must be viewed with concern. It seems likely that high-level resistance to phosphine, first reported from Bangladesh, is the result of poor fumigation practices. Unsatisfactory sealing of storages, leading to rapid loss of gas and consequent inadequate exposure of the pests, has been proposed as the reason for the development of resistance. Some strains have already been widely dispersed through trade to other parts of the world. Just as poor management has led to phosphine resistance and all the resultant problems, it is only through good management and careful, effective use of this fumigant that the problems can be overcome. Various agencies (8) are therefore promoting codes of practice that will ensure properly performed fumigations.

Cockroaches

Cockroaches are universal pests of foodstuffs. Infestations are immediately obvious, and pest numbers are directly related to the level of hygiene. Because the presence of cockroaches is totally unacceptable in most circumstances, control programs are mandatory. Although it is often clear that management attention to housekeeping and sanitation would overcome the problem, less onerous solutions such as the use of pesticides are frequently chosen. In practice, very large quantities of pesticides have been used with scant attention to hygiene to reduce pest numbers to less noticeable levels.

It is not surprising then that pesticide resistance has been a significant constraint in cockroach control through the years. This is especially true of the major domestic species, *Blattella germanica*, for which significant resistances to members of all the major classes of pesticides have been recorded. Resistance in other cockroach species has been recorded on a limited scale only.

Anticoagulant Resistance in Rodents

Resistance of commensal rodents to anticoagulants was first reported more than 20 years ago. Since then, considerable attention has been focused on the problem, and many basic studies have been undertaken. A WHO Standardized Test for resistance has been developed and has been in use for some years (7). Currently, few countries are monitoring regularly for rodent resistance, and the situation appears generally stable. Effective new rodenticides are available, more critical attention is being given to control programs, and the rate of spread of resistance appears to be less than was first expected.

Summary and Conclusions

Resistance to pesticides is a real problem in the management of food-industry pests, particularly those infesting stored cereal grains and related basic foodstuffs. Careful monitoring and appropriate management approaches can minimize the problem, but management is often indifferent to proper care in the use of pesticides to control infestations.

Management failure may reflect a lack of knowledge of the implications of poor food-storage practices. In the developing world, enormous efforts are being made by many national and international agencies to devise appropriate methods of pest control and to train managers and operators in their use. Almost without exception, these methods are promoted as components of pest management strategies which also stress hygiene and other good storage practices. The aim is to reduce general background levels of pest occurrence, thereby minimizing opportunities for selection for resistance.

In general, pest control programs using chemicals should aim for complete kills in order to minimize the chance that resistance will develop. If less than complete kills are commercially or economically acceptable, nonchemical methods should be sought. It should be realized by all concerned that protection of foodstuffs against pests in storage can be based only on well-tried and proven principles of sound storage practice if the impact of pesticide resistance is to be minimized and the long-term efficacy of controls is to be achieved. The principles have been clearly understood for a long time, and it is mandatory that all controls be based on them. It should also be realized that pest resistance to pesticides is an international problem, given the free movement of commodities—and pests—in international trade.

Cited References

1 Anon. 1967. *Methods for the Detection and Measurement of Resistance in Agricultural Pests.* First Session, FAO Working Party of Experts on Resistance of Pests to Pesticides. Meeting Report PL/1965/18:1–106. Food and Agriculture Organization, Rome.

2 Anon. 1968. *International Collaborative Program for the Development of Standardised Tests for Resistance in Important Pests of Agriculture.* Item 5. Third Session, FAO Working Party of Experts on Resistance of Pests to Pesticides. Meeting Report PL/1967/M/8:9–11. Food and Agriculture Organization, Rome.

3 Anon. 1969. Recommended methods for the detection and measurement of resistance of agricultural pests to pesticides. 1. General principles. *FAO Plant Protection Bulletin* 17(4)76–82.

4 Anon. 1970. Recommended methods for the detection and measurement of resistance of agricultural pests to pesticides. Tentative method for adults of the red flour beetle, *Tribolium castaneum* (Herbst). FAO Method No. 6. *FAO Plant Protection Bulletin* 18(5)107–113.

5 Anon. 1974. Recommended methods for the detection and measurement of resistance of agricultural pests to pesticides. Tentative method for adults for some major beetle pests of stored cereals with malathion or lindane. FAO Method No. 15. *FAO Plant Protection Bulletin* 22(5–6)127–137.

6 Anon. 1975. Recommended methods for the detection and measurement of resistance of agricultural pests to pesticides. Tentative method for adults of some major beetle pests of stored cereals, with methyl bromide and phosphine. FAO Method No. 16. *FAO Plant Protection Bulletin* 23(1)12–15.

7 Anon. 1982. *Instructions for Determining the Susceptibility or Resistance of Rodents to Anticoagulant Rodenticides.* Rept. No. WHO/VBC/82.843, World Health Organization, Geneva, Switzerland, pp. 1–9.

8 ASEAN Food Handling Bureau/Australian Centre for International Agricultural Research. 1989. *Suggested Recommendations for the Fumigation of Grain in the*

ASEAN Region. Part 1. *Principles and General Practice*. AFHB (Level 3, G14 & G15, Damansara Town Centre), 50490 Kuala Lumpur, Malaysia.

9 Busvine, J. R. 1980. Recommended methods for measurement of pest resistance to pesticides. Method for mite pests of stored products (e.g., *Acarus siro* L.). FAO Method No. 19. *FAO Plant Production and Protection Paper* 21:111–113.

10 Busvine, J. R. 1980. Recommended methods for measurement of pest resistance to pesticides. Method for lepidopterous larval pests of stored products and tentative method for detecting resistance in adults of stored-product lepidopterous pests. FAO Method No. 22. *FAO Plant Production and Protection Paper* 21:123–127.

11 Champ, B. R. 1968. A test method for detecting insecticide resistance in *Sitophilus oryzae* (L.) (Coleoptera, Curculionidae). *Journal of Stored Products Research* 4(2)175–178.

12 Champ, B. R. 1986. Occurrence of resistance to pesticides in grain storage pests. In *Pesticides and Humid Tropical Grain Storage Systems*, B. R. Champ and E. Highley, eds. ACIAR Proceedings No. 14, Australian Centre for International Agricultural Research, Canberra, pp. 229–255.

13 Champ, B. R., and M. J. Campbell-Brown. 1970. Insecticide resistance in Australian *Tribolium castaneum* (Herbst)—I. A test method for detecting insecticide resistance. *Journal of Stored Products Research* 6(1)53–70.

14 Champ, B. R., and M. J. Campbell-Brown. 1970. Insecticide resistance in Australian *Tribolium castaneum* (Herbst) (Coleoptera, Tenebrionidae)—II. Malathion resistance in eastern Australia. *Journal of Stored Products Research* 6(2)111–131.

15 Champ, B. R., and D. E. Dyte. 1976. *Report of the FAO Global Survey of Pesticide Susceptibility of Stored Grain Pests*. FAO Plant Production and Protection Series No. 5. Food and Agriculture Organization, Rome.

16 Cogan, P. M. 1982. A method for the rapid detection of malathion resistance in *Plodia interpunctella* (Hübner) (Lepidoptera: Pyralidae) with further records of resistance. *Journal of Stored Products Research* 18(3)121–124.

17 Collins, P. J. 1986. Genetic analysis of fenitrothion resistance in the sawtoothed grain beetle, *Oryzaephilus surinamensis* (Coleoptera: Cucujidae). *Journal of Economic Entomology* 79(5)1196–1199.

18 Coveney, R. D., and P. A. Corban. 1970. Simple field test for organophosphorus resistance. *Tropical Stored Products Centre Report* 1965–1966:24.

19 Kumar, V., and F. O. Morrison. 1964. Macdonald College test kits for testing the susceptibility of stored products insect pests to residual insecticides. *Canadian Entomologist* 96(1–2)122.

20 Longstaff, B. C. 1988. Temperature manipulation and the management of insecticide resistance in stored grain pests: A simulation study for the rice weevil, *Sitophilus oryzae*. *Ecological Modelling* 43(2)303–313.

21 McGaughey, W. H., and R. W. Beeman. 1988. Resistance to *Bacillus thuringiensis* in colonies of Indianmeal moth and almond moth (Lepidoptera: Pyralidae). *Journal of Economic Entomology* 81(1)28–33.

22 Rajak, R. L., M. Ghate, and K. Krishnamurthy. 1973. Bioassay technique for resistance to malathion of stored product insects. *International Pest Control* 15(6)11–13, 16.

23 Redfern, R., and J. E. Gill. 1978. The development and use of a test to identify resistance to the anticoagulant difenacoum in the Norway rat (*Rattus norvegicus*). *Journal of Hygiene* 81(3)427–431.

24 Tyler, P. S., and N. Evans. 1981. A tentative method for detecting resistance to gamma-HCH in three bruchid beetles. *Journal of Stored Products Research* 17(3)131–135.

25 White, N. D. G., and S. R. Loschiavo. 1985. Testing for malathion resistance in field-collected populations of *Cryptolestes ferrugineus* (Stephens) and factors affecting reliability of the tests. *Journal of Economic Entomology* 78(3)511–515.

Selected References

Aamir, M. M. I. 1984. Développement de la résistance au lindane et au fénitrothion en conditions contrôlées chez *Tribolium confusum* Duv. (Coleoptera—Tenebrionidae). *Bulletin de l'Académie des Sciences Agricoles et Forestières* (Bucharest) 13:133–141.

Ahuja, D. B. 1985. Cross-resistance characteristics of a laboratory selected pirimiphos-methyl resistant strain of *Tribolium castaneum* (Herbst) to some insecticides. *Journal of Entomological Research* 9(2)174–178.

Ahuja, D. B. 1987. Cross resistance pattern of a pirimiphos-methyl resistant strain of *Tribolium confusum* Duv. to some insecticides. *Entomon* 12(2)131–135.

Amos, T. G., P. Williams, and R. L. Semple. 1977. Susceptibility of malathion-resistant strains of *Tribolium castaneum* and *T. confusum* to the insect growth regulators methoprene and hydroprene. *Entomologia Experimentalis et Applicata* 22(3)289–293.

Anon. 1959. Variation in insect resistance. *Pest Infestation Research* 1958:25–26.

Anon. 1970. *Pest Resistance to Pesticides in Agriculture. Importance, Recognition and Countermeasures*. Rept. No. AGP:CP/26. Food and Agriculture Organization, Rome.

Anon. 1974. *Fumigants—Mode of Action, Use and Residue Analysis*. Research Branch Report 1973:186. Research Institute, Agriculture Canada, London, Ontario.

Anon. 1978. Cross-resistance. *Pest Infestation Control Laboratory Report* 1974–76:118.

Anon. 1978. Monitoring for resistance in the United Kingdom. *Pest Infestation Control Laboratory Report* 1974–76:115–116.

Anon. 1978. Resistance to acaricides. *Pest Infestation Control Laboratory Report* 1974–76:54–55.

Anon. 1978. Survey of fumigant resistance in the United Kingdom. *Pest Infestation Control Laboratory Report* 1974–76:85.

Anon. 1979. Anticoagulant resistance studies. Warfarin Research Laboratory Quarterly Report (September). Environmental Studies Center, Bowling Green State University, Bowling Green, OH.

Anon. 1981. Acaricide resistance in storage mites. *Storage Pests* 1979:56.

Anon. 1981. Cross resistance. *Pest Infestation Control Laboratory Report* 1977–79:75–76.

Anon. 1981. Detection of resistance to fumigants. *Storage Pests* 1979:43.

Anon. 1981. Fumigant resistance in *Oryzaephilus surinamensis*. *Storage Pests* 1980:74.

Anon. 1981. Malathion resistance in imported *Oryzaephilus surinamensis*. *Storage Pests* 1979:53–54.

Anon. 1981. Measuring resistance. *Storage Pests* 1979:55.

Anon. 1981. Monitoring for resistance in the United Kingdom. *Pest Infestation Control Laboratory Report* 1977–79:74–75.

Anon. 1981. Resistance in moths. *Storage Pests* 1980:79.

Anon. 1981. Resistance in other beetles. *Storage Pests* 1979:54–55; 1980:76–77.

Anon. 1981. Resistance in storage mites. *Storage Pests* 1980:45–46.

Anon. 1981. Resistance tests for liquid fumigants. *Pest Infestation Control Laboratory Report* 1977–79:52–53.

Anon. 1981. Resistance to acaricides. *Pest Infestation Control Laboratory Report* 1977–79:32.

Armstrong, J. W., and E. L. Soderstrom. 1975. Malathion resistance in some populations of the Indian meal moth infesting dried fruits and tree nuts in California. *Journal of Economic Entomology* 68(4)505–507.

Arthur, F., J. L. Zettler, and W. R. Halliday. 1988. Insecticide resistance among populations of almond moth and Indianmeal moth (Lepidoptera: Pyralidae) in stored peanuts. *Journal of Economic Entomology* 81(5)1283–1287.

Attia, F. I. 1976. Insecticide resistance in *Cadra cautella* in New South Wales, Australia. *Journal of Economic Entomology* 69(6)773–774.

Attia, F. I. 1977. Insecticide resistance in *Plodia interpunctella* (Hübner) (Lepidoptera: Pyralidae) in New South Wales, Australia. *Journal of the Australian Entomological Society* 16(2)149–152.

Attia, F. I. 1981. Insecticide resistance in pyralid moths of grain and stored products. *General and Applied Entomology* 13(1)3–8.

Attia, F. I. 1984. Insecticide and fumigant resistance in insects of grain and stored products in Australia. *Proceedings of the Third International Working Conference on Stored-Product Entomology* (Manhattan, KS, 1983), Kansas State University, Manhattan, KS, pp. 196–208.

Attia, F. I. 1984. Multiple and cross-resistance characteristics in phosphine-resistant strains of *Rhyzopertha dominica* and *Tribolium castaneum*. In *Controlled Atmosphere and Fumigation in Grain Storages*, B. E. Ripp et al., eds. Elsevier, Amsterdam, pp. 49–53.

Attia, F. I., and T. Frecker. 1984. Cross-resistance spectrum and synergism studies in organophosphorus-resistant strains *Oryzaephilus surinamensis* (L.) (Coleoptera: Cucujidae) in Australia. *Journal of Economic Entomology* 77(6)1367–1370.

Attia, F. I., and H. G. Greening. 1981. Survey of resistance to phosphine in coleopterous pests of grain and stored products in New South Wales. *General and Applied Entomology* 13(1)93–97.

Attia, F. I., E. Shipp, and G. J. Shanahan. 1979. Survey of insecticide resistance in *Plodia interpunctella* (Hübner), *Ephestia cautella* (Walker) and *E. kuehniella* Zeller (Lepidoptera: Pyralidae) in New South Wales. *Journal of the Australian Entomological Society* 18(1)67–70.

Bansode, P. C. 1974. Studies on the development of resistance to malathion in *Sitophilus oryzae* (L.). *Entomology Newsletter* (Indian Agricultural Research Institute) 4:8.

Bansode, P. C., and S. K. Bhatia. 1976. Selection for resistance to malathion in the rice weevil, *Sitophilus oryzae* (L.). *Bulletin of Grain Technology* 14(2)118–123.

Barwal, R. N., and R. L. Kalra. 1988. Nature of lindane resistance in the laboratory and field stains of *Tribolium castaneum* Herbst (Coleoptera: Tenebrionidae). *Insect Science and Its Application* 9(1)47–53.

Bedingfield, W. D. 1952. Insecticide resistant cockroaches. *Pest Control* 20(4)6.

Beeman, R. W., and S. M. Nanis. 1986. Malathion resistance alleles and their fitness in the red flour beetle (Coleoptera: Tenebrionidae). *Journal of Economic Entomology* 79(3)580–587.

Beeman, R. W., W. E. Speirs, and B. A. Schmidt. 1982. Malathion resistance in Indianmeal moths (Lepidoptera: Pyralidae) infesting stored corn and wheat in the North-Central United States. *Journal of Economic Entomology* 75(6)950–954.

Bell, C. H., B. D. Hole, and P. H. Evans. 1977. The occurrence of resistance to phosphine in adult and egg stages of strains of *Rhyzopertha dominica* (F.) (Coleoptera: Bostrichidae). *Journal of Stored Products Research* 13(1)91–94.

Bell, C. H., et al. 1984. An investigation of the tolerance of stages of khapra beetle (*Trogoderma granarium*, Everts) to phosphine. *Proceedings of the Third Internationl Working Conference on Stored-Product Entomology* (Manhattan, KS, 1983), Kansas State University, Manhattan, KS, pp. 329–340.

Bengston, M. 1986. Grain protectants. In *Pesticides and Humid Tropical Grain Storage Systems*, B. R. Champ and E. Highley, eds. ACIAR Proceedings No. 14, Australian Centre for International Agricultural Research, Canberra, pp. 277–289.

Bhatia, S. K., and S. Pradhan. 1968. Studies of resistance to insecticides in *Tribolium castaneum* (Herbst). I. Selection of a strain resistant to p,p'DDT and its biological characteristics. *Indian Journal of Entomology* 30(1)13–32.

Bhatia, S. K., and S. Pradhan. 1970. Studies on resistance to insecticides in *Tribolium castaneum* (Herbst). II. Cross-resistance characteristics of the p,p'DDT-resistant strain.

Indian Journal of Entomology 32(1)32–38.

Bhatia, S. K., and S. Pradhan. 1972. Studies of resistance to insecticides in *Tribolium castaneum* (Herbst). V. Cross-resistance characteristics of a lindane-resistant strain. *Journal of Stored Products Research* 8(2)89–93.

Bhatia, S. K., T. D. Yadav, and P. B. Mookherjee. 1971. Malathion resistance in *Tribolium castaneum* in India. *Journal of Stored Products Research* 7(3)227–230.

Binns, T. J. 1983. Evaluation of insecticides against a malathion-resistant and a susceptible strain of the rice weevil, *Sitophilus oryzae. Tests of Agrochemicals and Cultivars. Annals of Applied Biology* 102(suppl. 4)30–31.

Binns, T. J. 1986. The comparative toxicity of malathion, with and without the addition of triphenyl phosphate, to adults and larvae of a susceptible and two resistant strains of *Tribolium castaneum* (Herbst) (Coleoptera: Tenebrionidae). *Journal of Stored Products Research* 22(2)97–101.

Binns, T. J., and G. W. Pemberton. 1981. Lindane-resistant strains of the leather beetle *Dermestes maculatus* in the United Kingdom. *Annals of Applied Biology* 99(1)25–28.

Binns, T. J., and P. S. Tyler. 1978. Lindane resistance in *Dermestes maculatus* Deg. (Coleoptera: Dermestidae). *Journal of Stored Products Research* 14(1)19–23.

Blackith, R. E., and B. S. Gorringe. 1953. Response of pests to fumigation. I. Toxicity of mercury vapour to the eggs of *Calandra granaria* (L.). *Bulletin of Entomological Research* 44(2)217–224.

Blackman, D. G., and J. I. Peckover. 1976. Incidence of resistance in the United Kingdom. Other beetles. The long headed flour beetle. *Pest Infestation Control Laboratory Report* 1971–73:84.

Bond, E. J. 1973. Increased tolerance to ethylene dibromide in a field population of *Tribolium castaneum* (Herbst). *Journal of Stored Products Research* 9(1)61–63.

Bond, E. J. 1984. Resistance of stored product insects to fumigants. *Proceedings of the Third International Working Conference on Stored-Product Entomology* (Manhattan, KS, 1983), Kansas State University, Manhattan, KS, pp. 303–307.

Bond, E. J., and C. T. Buckland. 1979. Development of resistance to carbon dioxide in the granary weevil. *Journal of Economic Entomology* 72(5)770–771.

Bond, E. J., and E. Upitis. 1972. Persistence of tolerance to methyl bromide in *Sitophilus granarius* after cessation of selection. *Journal of Stored Products Research* 8(3)221–222.

Borah, B., and B. S. Chahal. 1979. Development of resistance in *Trogoderma granarium* Everts to phosphine in the Punjab. *FAO Plant Protection Bulletin* 27(3)77–80.

Brun, L. O., and F. I. Attia. 1983. Resistance to lindane, malathion and fenitrothion in coleopterous pests of stored products in New Caledonia. *Proceedings of the Hawaiian Entomological Society* 24(2–3)211–215.

Burden, G. S., C. S. Lofgren, and J. H. Eastin. 1959. Malathion-resistance in a laboratory strain of the German cockroach. *Pest Control* 27(2)38.

Caliboso, F. M., P. D. Sayaboc, and M. R. Amoranto. 1986. Pest problems and the use of pesticides in grain storage in the Philippines. In *Pesticides and Humid Tropical Grain Storage Systems*, B. R. Champ and E. Highley, eds. ACIAR Proceedings No. 14, Australian Centre for International Agricultural Research, Canberra, pp. 17–29.

Carter, S. W. 1975. Laboratory evaluation of three novel insecticides inhibiting cuticle formation against some susceptible and resistant stored products beetles. *Journal of Stored Products Research* 11(3–4)187–193.

Carter, S. W., P. R. Chadwick, and J. C. Wickham. 1975. Comparative observations on the activity of pyrethroids against some susceptible and resistant stored products beetles. *Journal of Stored Products Research* 11(3–4)135–142.

Champ, B. R. 1967. The inheritance of DDT resistance in *Sitophilus oryzae* (L.) (Coleoptera: Curculionidae) in Queensland. *Journal of Stored Products Research* 3(4)321–334.

Champ, B. R. 1979. Pesticide resistance and its current significance in control of pests of stored products. *Proceedings of the Second International Working Conference on Stored-Product Entomology* (Ibadan, Nigeria, 1978), Agency for International Development, Washington, DC, pp. 159–181.

Champ, B. R., and J. N. Cribb. 1965. Lindane resistance in *Sitophilus oryzae* (L.) and *Sitophilus zeamais* Motsch. (Coleoptera, Curculionidae) in Queensland. *Journal of Stored Products Research* 1(1)9–24.

Champ, B. R., and R. G. Winks. 1982. Infestation and degradation—the grain drain. *Proceedings of the First International Grain Trade Transportation and Handling Conference* (London, England), pp. 153–161.

Cichy, D. 1971. The role of some ecological factors in the development of pesticide resistance in *Sitophilus oryzae* L. and *Tribolium castaneum* Herbst. *Ekologia Polska* 19(36)563–616.

Clarke, T. H., and D. G. Cochran. 1959. Cross-resistance in insecticide-resistant strains of the German cockroach, *Blattella germanica* (L.). *Bulletin of the World Health Organization* 20(5)823–833.

Cline, L. D., et al. 1984. Continuous exposure to sublethal doses of synergized pyrethrins: Effects on resistance and repellency in *Tribolium confusum* (Coleoptera: Tenebrionidae). *Journal of Economic Entomology* 77(5)1189–1193.

Cochran, D. G. 1961. Further studies on cross-resistance in the German cockroach. *Bulletin of the World Health Organization* 24(4–5)557–561.

Cochran, D. G., and M. H. Ross. 1964. Resistance to Telodrin in the German cockroach, *Blattella germanica. Journal of Economic Entomology* 57(4)485.

Collins, P. J. 1985. Resistance to grain protectants in field populations of the sawtoothed grain beetle in Southern Queensland. *Australian Journal of Experimental Agriculture* 25(3)683–686.

Collins, P. J., and D. Wilson. 1986. Insecticide resistance in the major coleopterous pests of stored grain in southern

Queensland. *Queensland Journal of Agricultural and Animal Sciences* 43(2)107–114.

Collins, P. J., and D. Wilson. 1987. Efficacy of current and potential grain protectant insecticides against a fenitrothion-resistant strain of the sawtoothed grain beetle, *Oryzaephilus surinamensis* L. *Pesticide Science* 20(2)93–104.

Dal Monte, G. 1969. World survey for resistance. Item 5/10/5. Report of the 4th Session, FAO Working Party of Experts on Resistance of Pests to Pesticides. Meeting Report PL/1968/M/10:31. Food and Agriculture Organization, Rome.

De Lima, C. P. F. 1987. Insect pests and postharvest problems in the tropics. *Insect Science and Its Application* 8(4–6)673–676.

Du Chanois, F. R. 1956. A preliminary report on cockroach control at naval installations. *Pest Control* 24(12)9–12, 36.

Dyte, C. E. 1974. Problems arising from insecticide resistance in storage pests. *EPPO Bulletin* 4(3)275–289.

Dyte, C. E. 1979. The importation of insecticide-resistant strains of stored-product pests. *Annals of Applied Biology* 91(3)414–417.

Dyte, C. E., and D. G. Blackman. 1967. Selection of a DDT-resistant strain of *Tribolium castaneum* (Herbst) (Coleoptera, Tenebrionidae). *Journal of Stored Products Research* 2(3)211–228.

Dyte, C. E., and D. G. Blackman. 1972. Laboratory evaluation of organophosphorus insecticides against susceptible and malathion-resistant strains of *Tribolium castaneum* (Herbst) (Coleoptera, Tenebrionidae). *Journal of Stored Products Research* 8(2)103–109.

Dyte, C. E., and R. Forster. 1970. Lindane resistance in the groundnut borer. *Pest Infestation Research* 1969:43.

Dyte, C. E., and R. Forster. 1970. Lindane resistance in the maize weevil. *Pest Infestation Research* 1969:43.

Dyte, C. E., and R. Forster. 1973. Studies on insecticide resistance in *Oryzaephilus mercator* (Fauv.) (Coleoptera, Silvanidae). *Journal of Stored Products Research* 9(3)159–164.

Dyte, C. E., R. Forster, and S. Aggarwal. 1976. The rust-red flour beetle. Resistance to juvenile hormone mimics. *Pest Infestation Control Laboratory Report* 1971–73:79–80.

Dyte, C. E., and D. Halliday. 1983. Problems of development of resistance to phosphine by insect pests of stored grains. *EPPO Bulletin* 15(1)51–57.

Dyte, C. E., K. A. Mills, and N. R. Price. 1983. Recent work on fumigant-resistant insect strains (paper 19). *Proceedings of the Sixth British Pest Control Conference* (Cambridge, England).

Dyte, C. E., D. G. Rowlands, and J. P. Edwards. 1976. Incidence of resistance in the United Kingdom. The rust-red flour beetle. Resistance to piperonyl butoxide. *Pest Infestation Control Laboratory Report* 1971–73:82–83.

Dyte, C. E., et al. 1973. Non-specific resistance in the rust-red flour beetle. *Pest Infestation Control* 1968–70:118–119.

Dyte, C. E., et al. 1976. Incidence of resistance in the United Kingdom. The saw-toothed grain beetle. Cross-resistance studies. *Pest Infestation Control Laboratory Report* 1971–73:74.

Dyte, D. E., and D. R. Wilkin. 1965. Increased resistance of stored-products insects to insecticides. The saw-toothed grain beetle and carbaryl. *Pest Infestation Research* 1964:34.

Edwards, J. P., and J. E. Short. 1984. Evaluation of three compounds with insect juvenile hormone activity as grain protectants against insecticide-susceptible and resistant strains of *Sitophilus* species (Coleoptera: Curculionidae). *Journal of Stored Products Research* 20(1)11–15.

Evans, N. J. 1985. The effectiveness of various insecticides on some resistant insect pests of stored products from Uganda. *Journal of Stored Products Research* 21(2)105–109.

Fernando, H. E. 1967. Summary of cases of resistance to pesticides in agricultural pests. Item 7/16/21. Report of the 1st Session, FAO Working Party of Experts on Resistance of Pests to Pesticides. Meeting Report PL/1965/18:78. Food and Agriculture Organization, Rome.

Fisk, F. W., and J. A. Isert. 1953. Comparative toxicities of certain organic insecticides to resistant and non-resistant strains of the German cockroach, *Blattella germanica* (L.). *Journal of Economic Entomology* 46(6)1059–1062.

Godavaribai, S., K. Krishnamurthy, and S. K. Majumder. 1962. Bacterial spores with malathion for controlling *Ephestia cautella*. *Pest Technology* 4(7)155–158.

Gorham, J. R. 1962. Insecticide resistance-susceptibility tests with German cockroaches (*Blattella germanica* L.) in Puerto Rico. *Journal of Agriculture of the University of Puerto Rico* 46(3)219–225.

Gough, H. C. 1939. Factors affecting the resistance of the flour beetle, *Tribolium confusum* Duv., to hydrogen cyanide. *Annals of Applied Biology* 26(3)533–571.

Gouhar, K. A., et al. 1980–81[1984]. Development of resistance to carbaryl, malathion and lindane in a strain of *Callosobruchus maculatus* (Fab.) (Coleoptera: Bruchidae). *Bulletin of the Entomological Society of Egypt (Economic Series)* 12:5–10.

Grayson, J. M. 1961. Resistance to diazinon in the German cockroach. *Bulletin of the World Health Organization* 24(4–5)563–565.

Grayson, J. M. 1965. Resistance to three organophosphorus insecticides in strains of the German cockroach from Texas. *Journal of Economic Entomology* 58(5)956–958.

Greaves, J. H., R. Redfern, and R. E. King. 1974. Some properties of calciferol as a rodenticide. *Journal of Hygiene* 73(3)341–351.

Halliday, W. R., F. H. Arthur, and J. L. Zettler. 1988. Resistance status of red flour beetle (Coleoptera: Tenebrionidae) infesting stored peanuts in the Southeastern United States. *Journal of Economic Entomology* 81(1)74–77.

Hashimoto, Y., and J. Fudami. 1964. Resistance to insecticides in the almond moth, *Ephestia cautella* Walker. I. Development of methyl-parathion resistance. *Japanese Journal of Applied Entomology and Zoology* 8(1)62–67.

Heather, N. W. 1986. Sex-linked resistance to pyrethroids in *Sitophilus oryzae* (L.) (Coleoptera: Curculionidae). *Journal of Stored Products Research* 22(1)15–20.

Heather, N. W., and D. Gauldie. 1982. Selection of rice weevil for resistance to candidate grain protectants. *Annual Report of the Queensland Department of Primary Industries* 1981–82:140.

Heather, N. W., and D. Wilson. 1983. Resistance to fenitrothion in *Oryzaephilus surinamensis* (L.) (Coleoptera: Silvanidae) in Queensland. *Journal of the Australian Entomological Society* 22(3)210.

Heuvel, M. J. van den, and D. G. Cochran. 1965. Cross resistance to organophosphorus compounds in malathion- and diazinon-resistant strains of *Blattella germanica*. *Journal of Economic Entomology* 58(5)872–874.

Holborn, J. M. 1957. The susceptibility to insecticides of laboratory cultures of an insect species. *Journal of the Science of Food and Agriculture* 8(3)182–188.

Hoppe, T. 1981. Testing of methoprene in resistant strains of *Tribolium castaneum* (Herbst) (Col., Tenebrionidae). *Zeitschrift für Angewandte Entomologie* 91(3)241–251.

Horton, P. M. 1984. Evaluation of South Carolina field strains of certain stored-product Coleoptera for malathion resistance and pirimiphos-methyl susceptibility. *Journal of Agricultural Entomology* 1(1)1–5.

Howe, R. W. 1962. The entomological problems of assessing the success of a fumigation of stored produce. *Proceedings of the 11th International Congress of Entomology* (Vienna, Austria, 1960) 2:288–290.

Jackson, W. B., and A. D. Ashton. 1979. Present distribution of anticoagulant resistance in the United States. Paper presented at 8th Steenbock Symposium, Madison, WI.

Joubert, P. C., and P. R. de Beer. 1968. The toxicity of contact insecticides to seed infesting insects (No. 6). Tests with bromophos on maize. *Technical Communication* (Department of Agriculture, South Africa) 84:1–9.

Kay, I. R., and P. J. Collins. 1987. The problem of resistance to insecticides in tropical insect pests. *Insect Science and Its Application* 8(4–6)715–721.

Keller, J. C., P. H. Clark, and C. S. Lofgren. 1956. Results of USDA-sponsored research tests on cockroach control. *Pest Control* 24(9)12–14.

Keller, J. C., P. H. Clark, and C. S. Lofgren. 1956. Susceptibility of insecticide-resistant cockroaches to pyrethrins. *Pest Control* 24(11)14–15, 30.

Kem, T. R. 1975. Studies on the development of resistance to phosphine in *Tribolium castaneum* (Herbst). *Entomology Newsletter* (Indian Agricultural Research Institute) 5:6–7.

Kühne, H. 1967. Co-Resistenz und Resistenz-Vererbung eines gegen Dieldrin resistenten Laboratoriumstammes der Kleidermotte [*Tineola bisselliella* (Hum.), Lep.]. *Zoologische Beiträge* 13(2–3)397–407.

Kühne, H., and G. Becker. 1965. Züchtung giftresistenter Kleidermotten [*Tineola bisselliella* (Hum.), Lep.]. *Zeitschrift für Angewandte Entomologie* 56(1)61–89.

Kumar, V., and F. O. Morrison. 1963. The susceptibility levels of certain stored product pest populations to chemicals used for their control. *Phytoprotection* 44(2)101–105.

Kumar, V., and F. O. Morrison. 1965. Recording the susceptibility levels of current stored product pest populations to current insecticides. *Proceedings of the 12th International Congress of Entomology* (London, England, 1964), pp. 656–657.

Kumar, V., and F. O. Morrison. 1967. Carbamate and phosphate resistance in adult granary weevils. *Journal of Economic Entomology* 60(5)1430–1434.

La Hue, D. W. 1969. Control of malathion-resistant Indian meal moths *Plodia interpunctella* (Hübner) with dichlorvos resin strips. *Proceedings of the North Central Branch of the Entomological Society of America* 24:117–119.

Lakocy, A. 1970. Actual problems of the resistance of agricultural pests to pesticides in Poland. *Biuletyn Instytutu Ochrony Róslin* 47:89–103.

Le Patourel, G. N. J., and M. A. Salama. 1986. Mechanism of gamma-HCH resistance in a strain of granary weevil (*Sitophilus granarius* L.). *Pesticide Science* 17(5)503–510.

Lindgren, D. L., and L. E. Vincent. 1965. The susceptibility of laboratory-reared and field-collected cultures of *Tribolium confusum* and *T. castaneum* to ethylene dibromide, hydrocyanic acid, and methyl bromide. *Journal of Economic Entomology* 58(3)551–555.

Lloyd, C. J. 1969. Studies on the cross-tolerance to DDT-related compounds of a pyrethrin-resistant strain of *Sitophilus granarius* (L.) (Coleoptera: Curculionidae). *Journal of Stored Products Research* 5(4)337–356.

Lloyd, C. J. 1973. The toxicity of pyrethrins and five synthetic pyrethroids, to *Tribolium castaneum* (Herbst), and susceptible and pyrethrin-resistant *Sitophilus granarius* (L.). *Journal of Stored Products Research* 9(1)77–92.

Lloyd, C. J., and E. A. Parkin. 1963. Further studies on a pyrethrin-resistant strain of the granary weevil, *Sitophilus granarius* (L.). *Journal of the Science of Food and Agriculture* 14(9)655–663.

Lloyd, C. J., D. G. Rowlands, and G. Ruczkowski. 1976. Incidence of resistance in the United Kingdom. The rust-red flour beetle. Resistance to synthetic pyrethroids. *Pest Infestation Control Laboratory Report* 1971–73:81–82.

Lloyd, C. J., and G. E. Ruczkowski. 1980. The cross-resistance to pyrethrins and eight synthetic pyrethroids, of an organophosphorus-resistant strain of the rust-red flour beetle *Tribolium castaneum* (Herbst). *Pesticide Science* 11(3)331–340.

Maeda, O. 1958. Development of DDT resistance in the flour beetle, *Tribolium confusum* Duv. *Botyu-Kagaku* 23(2)66–74.

Mallis, A., W. C. Easterlin, and R. J. Astor. 1966. Resistance in a large cockroach *Periplaneta brunnea* Burmeister in Florida. *Pest Control* 34(6)22.

Mathlein, R. 1952. Undersöknigar över uppkomst av DDT-resistens hos kornivel, *Calandra granaria* L. *Statens Växtskyddsanstalt Meddelande* 62:1–20 (English summary).

McDonald, I. C., and D. G. Cochran. 1968. Carbamate

cross resistance in a carbaryl-resistant strain of the German cockroach. *Journal of Economic Entomology* 61(3)670–673.

McGaughey, W. H. 1985. Insect resistance to the biological insecticide *Bacillus thuringiensis. Science* 229(4709)193–195.

Mello, E. J. R. 1972. Constatacão de resistencia ao DDT e lindane em *Sitophilus oryzae* (L.) em milho armazenado, na localidade de Capinipolis, Minas Gerais. *Reports of the Biological Institute,* São Paulo, Brazil.

Mills, K. A. 1983. Resistance to the fumigant hydrogen phosphide in some stored-product species associated with repeated inadequate treatments. *Mitteilungen der Deutschen Gesellschaft für Allgemeine und Angewandte Entomologie* 4(1–3)98–101.

Monro, H. A. U., E. Upitis, and E. J. Bond. 1972. Resistance of a laboratory strain of *Sitophilus granarius* (L.)(Coleoptera, Cuculionidae) to phosphine. *Journal of Stored Products Research* 8(3)199–207.

Morallo-Rejesus, B. 1974. Survey of Philippine populations of rice weevil complex [*Sitophilus oryzae* (L.) and *Sitophilus zeamais* Motsch.] for resistance to insecticides. In *Advances in Food and Agriculture Research and Development,* part I, pp. 295–299.

Muggleton, J. 1986. Selection for malathion resistance in *Oryzaephilus surinamensis* (L.)(Coleoptera: Silvanidae): Fitness values of resistant and susceptible phenotypes and their inclusion in a general model describing the spread of resistance. *Bulletin of Entomological Research* 76(3)469–480.

Muggleton, J. 1987. Insecticide resistance in stored product beetles and its consequences for their control. In *Stored Products Pest Control,* T. J. Lawson, ed. British Crop Protection Council Monograph 37, pp. 177–186.

Nakakita, H., and J. Kuroda. 1986. Differences in phosphine uptake between susceptible and resistant strains of insects. *Journal of Pesticide Science* 11(1)21–26.

Nakakita, H., and R. G. Winks. 1981. Phosphine resistance in immature stages of a laboratory selected strain of *Tribolium castaneum* (Herbst) (Coleoptera: Tenebrionidae). *Journal of Stored Products Research* 17(2)43–52.

Navarro, S., et al. 1986. Malathion resistance of stored-product insects in Israel. *Phytoparasitica* 14(4)273–280.

Navarro, S., R. Dias, and E. Donahaye. 1985. Induced tolerance of *Sitophilus oryzae* adults to carbon dioxide. *Journal of Stored Products Research* 21(4)207–213.

Odeneal, J. F. 1961. Insect resistance survey—II. *Soap and Chemical Specialties* 37(May)103, 105, 107.

Osman, N., and B. Morallo-Rejesus. 1985. Insecticide resistance in *Sitophilus zeamais* Mots. and *Rhizopertha dominica* (F.) in Indonesia. *Pertanika* 8(1)1–7.

Parkin, E. A. 1965. The onset of insecticide resistance among field populations of stored product insects. *Journal of Stored Products Research* 1(1)3–8.

Parkin, E. A., and P. E. Bright. 1965. Increased resistance of stored-product insects to insecticides. The groundnut borer and dieldrin. *Pest Infestation Research* 1967:34.

Pasalu, I. C., and S. K. Bhatia. 1974. Specific nature of

malathion resistance in *Tribolium castaneum* (Herbst) in India. *Bulletin of Grain Technology* 12(3)229–231.

Pinniger, D. B. [1975.] The behaviour of insects in the presence of insecticides: The effect of fenitrothion and malathion on resistant and susceptible strains of *Tribolium castaneum* Herbst. *Proceedings of the First International Working Conference on Stored-Product Entomology* (Savannah, GA, 1974), pp. 301–308.

Price, L. A., and K. A. Mills. 1988. The toxicity of phosphine to the immature stages of resistant and susceptible strains of some common stored product beetles, and implications for their control. *Journal of Stored Products Research* 24(1)51–59.

Price, N. R. 1984. Active exclusion of phosphine as a mechanism of resistance in *Rhyzopertha dominica* (F.) (Coleoptera: Bostrychidae). *Journal of Stored Products Research* 20(3)163–168.

Price, N. R. 1986. Action and inaction of fumigants. In *Pesticides and Humid Tropical Grain Storage Systems,* B. R. Champ and E. Highley, eds. ACIAR Proceedings No. 14, Australian Centre for International Agricultural Research, Canberra, pp. 203–210.

Prickett, A. J. 1980. The cross-resistance spectrum of *Sitophilus granarius* (L.) (Coleoptera: Curculionidae) heterozygous for pyrethrin resistance. *Journal of Stored Products Research* 16(1)19–25.

Rajak, R. J., and P. S. Hewlett. 1971. Effects of some synergists on the insecticidal potency of phosphine. *Journal of Stored Products Research* 7(1)15–19.

Ramsey, P. R., and T. K. Farley. 1978. Mating and fecundity in malathion-resistant and susceptible strains of the Indian meal moth, *Plodia interpunctella. Annals of the Entomological Society of America* 71(4)513–516.

Rassmann, W. 1988. Insecticide resistance in stored products pests. *Gesunde Pflanzen* 40(1)39–42 (in German, English summary).

Reierson, D. A., et al. 1988. Insecticide resistance affects cockroach control. *California Agriculture* 42(5)18–20.

Rennison, B. D., and A. C. Dubock. 1978. Field trials of WBA 8119 (PP 581, brodifacoum) against warfarin-resistant infestations of *Rattus norvegicus. Journal of Hygiene* 80(1)77–82.

Ricci, M. 1948. Sull'azione del DDT su *Blatta orientalis. Rivista di Parrasitologia* 9(3)143–167 (English summary).

Roslavtseva, S. A. 1988. World distribution of arthropod populations resistant to insecto-acaricides. *Agrokhimiya* 2:121–136 (in Russian).

Salama, M. A. El-M. 1987. Resistance to gamma-HCH lindane in strains of the grain weevil, *Sitophilus granarius* (L.). Thesis, Imperial College of Science and Technology, London. *Index to Theses Accepted for Higher Degrees in the Universities of Great Britain and Ireland* 36(1)291.

Saleem, M. A. 1986. Toxicity of fenvalerate against a malathion resistant and a susceptible strain of *Oryzaephilus surinamensis* (L.). *Biologia* [Lahore] 32(3)119–125.

Saleem, M. A., and R. M. Wilkins. 1984. Precocene-I: An

anti-juvenile hormone, a potential 4th generation insecticide against a malathion-resistant strain of *Oryzaephilus surinamensis* (L.). *Pakistan Journal of Zoology* 16(2)195–201.

Saleem, M. A., and R. M. Wilkins. 1984. Studies on the cross-resistance to NRDC 143 of a malathion-resistant strain of *Oryzaephilus surinamensis* (L.) (Coleoptera: Cucujidae). *Pakistan Journal of Zoology* 16(2)203–213.

Santos, F. L., and C. Y. Lim-Sylianco. 1985. DDT resistance in three local populations of rice weevils. *Philippine Journal of Science* 114(3–4)151–157.

Saxena, J. D., and S. K. Bhatia. 1980. Laboratory selection of the red flour beetle, *Tribolium castaneum* (Herbst) for resistance to phosphine. *Entomon* 5(4)301–306.

Saxena, J. D., and S. K. Bhatia. 1985. Phosphine susceptibility of lindane, pp'DDT, and malathion resistant strains of *Tribolium castaneum* (Herbst). *Indian Journal of Entomology* 47(3)356–357.

Schmid, W. 1987. Permethrin resistance in the carpet beetle *Anthrenus flavipes* Casey (Col., Dermestidae). *Anzeiger für Schädlingskunde Pflanzenschutz Umweltschutz* 60(1)9–15.

Sevintuna, C., and A. J. Musgrave. 1961. Observations on males and females of *Sitophilus granarius* (L.), the granary weevil, GG strain, exposed for six generations to allethrin and piperonyl butoxide. *Canadian Entomologist* 93(7)545–552.

Shukla, R. M., and A. S. Srivastava. 1982. Studies on the resistance to insecticides in *Cadra cautella* Walker. III. Selection of a strain resistant to lindane. *Bulletin of Grain Technology* 20(3)110–114.

Shukla, R. M., and A. S. Srivastava. 1982. Studies on the resistance to insecticides in *Cadra cautella* Walker. IV. Cross-resistance characteristics of the lindane-resistant strain. *Bulletin of Grain Technology* 20(3)160–163.

Shukla, R. M., and A. S. Srivastava. 1984. Studies on the resistance to insecticides in *Cadra cautella* Walker. I. Selection of a strain resistant to malathion. *Indian Journal of Entomology* 46(2)117–121.

Shukla, R. M., and A. S. Srivastava. 1984. Resistance to insecticides in *Cadra cautella* Walker. II. Cross-resistance characteristics of the malathion-resistant strain. *Indian Journal of Entomology* 46(2)201–204.

Speirs, R. D., L. M. Redlinger, and R. Jones. 1971. DDT-resistant red flour beetles from a Georgia peanut sheller. *Journal of Economic Entomology* 64(5)1328–1329.

Speirs, R. D., and J. L. Zettler. 1969. Toxicity of three organophosphorus compounds and pyrethrins to malathion-resistant *Tribolium castaneum* (Herbst) (Coleoptera, Tenebrionidae). *Journal of Stored Products Research* 4(4)279–283.

Stables, L. M. 1980. The effectiveness of some recently developed pesticides against stored-product mites. *Journal of Stored Products Research* 16(3–4)143–146.

Stables, L. M. 1984. Effect of pesticides on three species of *Tyrophagus*, and detection of resistance to pirimiphos-methyl in *T. palmarum* and *T. putrescentiae*. In *Acarology VI*, vol. 2, D. A. Griffiths and C. E. Bowman,

eds. Horwood/Halstead/Wiley, Chichester, England, pp. 1026–1033.

Stables, L. M., and D. R. Wilkin. 1981. Resistance to pirimiphos-methyl in cheese mites. *Proceedings of the 1981 British Crop Protection Conference (Pests and Diseases)* 2:617–624.

Subramanyam, B., P. K. Harein, and L. K. Cutkomp. 1989. Organophosphate resistance in adults of red flour beetle (Coleoptera: Tenebrionidae) and sawtoothed grain beetle (Coleoptera: Cucujidae) infesting barley stored on farms in Minnesota. *Journal of Economic Entomology* 82(4)989–995.

Summer, W. A., II, P. K. Harein, and B. Subramanyam. 1988. Malathion resistance in larvae of some southern Minnesota populations of the Indianmeal moth, *Plodia interpunctella* (Lepidoptera: Pyralidae), infesting bulk-stored shelled corn. *Great Lakes Entomologist* 21(3)133–137.

Taylor, R. W., and D. Halliday. 1986. The geographical spread of resistance to phosphine by coleopterous pests of stored products. *Proceedings, 1986 British Crop Protection Conference*, pp. 607–613.

Thind, B. B., and J. P. Edwards. 1986. Laboratory evaluation of the juvenile hormone analogue fenoxycarb against some insecticide-susceptible and resistant stored products beetles. *Journal of Stored Products Research* 22(4)235–241.

Tyler, P. S., and T. J. Binns. 1976. Incidence of resistance in the United Kingdom. Other beetles. The hide beetle. *Pest Infestation Control Laboratory Report* 1971–73:84–85.

Tyler, P. S., R. W. Taylor, and D. P. Rees. 1983. Insect resistance to phosphine fumigation in food warehouses in Bangladesh. *International Pest Control* 25(1)10–13, 21.

Vincent, L. E., and D. L. Lindgren. 1967. Susceptibility of laboratory and field-collected cultures of the confused flour beetle and red flour beetle to malathion and pyrethrins. *Journal of Economic Entomology* 60(6)1763–1764.

Wallbank, B. E. 1984. Fenitrothion resistance in *Oryzaephilus surinamensis* (L.), sawtoothed grain beetle, in New South Wales. *Proceedings of the 4th Australian Applied Entomological Conference* (Adelaide), pp. 507–511.

Wang, S. C., and T. Y. Ku. 1982. Status of maize weevil resistance to insecticides in Taiwan. *Plant Protection Bulletin* (Taiwan) 24:59–68.

Watanabe, T., and K. Kurokawa. 1986. The tolerance to methyl bromide of stored product insects, 4: Almond moth, *Ephestia cautella* (Walker). *Ministry of Agriculture, Forestry and Fisheries, Plant Protection Service (Yokohama) Research Bulletin* 22:79–86.

Webb, J. E., Jr. 1961. Resistance of some species of cockroaches to organic insecticides in Germany and France, 1956–59. *Journal of Economic Entomology* 54(4)805–806.

White, N. D. G., and R. J. Bell. 1988. Inheritance of malathion resistance in a strain of *Tribolium castaneum* (Coleoptera: Tenebrionidae) and effects of resistance genotypes on fecundity and larval survival in malathion-treated wheat. *Journal of Economic Entomology* 81(1)381–386.

White, N. D. G., and F. L. Watters. 1984. Incidence of malathion resistance in *Tribolium castaneum* and *Cryptolestes ferrugineus* populations collected in Canada. *Proceedings of the Third International Working Conference on Stored-Product Entomology* (Manhattan, KS, 1983), Kansas State University, Manhattan, KS, pp. 290–302.

Wilkin, D. R. 1966. Insects infesting farm stores. Tolerance to insecticides of farm strains of *O. surinamensis. Pest Infestation Research* 1965:24.

Wilkin, D. R. 1973. Resistance to lindane in *Acarus siro* from an English cheese store. *Journal of Stored Products Research* 9(2)101–104.

Wilkin, D. R., and J. A. Hope. 1973. Evaluation of pesticides against stored product mites. *Journal of Stored Products Research* 8(4)323–327.

Wilkin, D. R., et al. 1976. Incidence of resistance in the United Kingdom. Mites. *Glycyphagus destructor. Pest Infestation Control Laboratory Report* 1971–73:86.

Winks, R. G. 1969. Resistance to the fumigant phosphine in a strain of *Tribolium castaneum* (Herbst). *Insect Toxicology Information Service* 12:178.

Winks, R. G. 1986. The significance of response time in the detection and measurement of fumigant resistance in insects with special reference to phosphine. *Pesticide Science* 17(2)165–174.

Winks, R. G., and C. J. Waterford. 1986. The relationship between concentration and time in the toxicity of phosphine to adults of a resistant strain of *Tribolium castaneum* (Herbst). *Journal of Stored Products Research* 22(2)85–92.

Yana, A. 1967. Note sur l'efficacite d'un insecticide de la serie des organophosphores: Le Sumithion (fénitrothion) sur quelques souches Tunisiennes d'insectes des graines. *Documents Techniques* (Institut National de la Recherche Agronomique de Tunisie) 25:1–15.

Zettler, L. 1982. Insecticide resistance in selected stored-product insects infesting peanuts in Southeastern United States. *Journal of Economic Entomology* 75(2)359–362.

Zettler, J. L., et al. 1973. *Plodia interpunctella* and *Cadra cautella* resistance in strains to malathion and synergized pyrethrins. *Journal of Economic Entomology* 66(5)1049–1050.

29

Physical Methods to Manage Stored-Food Pests

F. L. Watters

Stored grain and processed foods are subject to attack by pests that feed on various products stored under a wide range of conditions. Insects, mites, and rodents destroy and contaminate foods intended for human or animal consumption. These pests live in granaries, trucks, grain elevators, railway cars, terminal elevators, cereal processing plants, food warehouses, bakeries, and homes. Most have become adapted to living in one or more links of the food storage → processing → distribution chain and they compete with humans for the food they need.

Chemical pesticides are relied on extensively to control pests in the food industry. However, the high costs of developing new chemicals to replace those withdrawn from use because of pest resistance (132) and the imposition of regulations to prevent environmental damage or human health risks have led to increased interest in nonchemical methods of pest control.

Physical methods are becoming more widely recognized as alternatives or adjuncts to chemical pesticides for the control of stored-food pests. In several countries, the use of low temperatures, aeration, and hermetic storage are well-

Watters: Research Station, Agriculture Canada, 195 Dafoe Road, Winnipeg R3T 2M9, Canada. Present address: 59 Caton Place, Victoria BC V9B 1L1, Canada.

Contribution No. 971.

established components of certain storage-pest control programs (38, 54, 83, 136). Research is progressing in the application of other physical methods such as ionizing and nonionizing radiation, atmospheric gases, and physical barriers for the protection of stored foods. Several comprehensive reviews have been published which deal with general or specific aspects of physical control (18, 110, 135, 140). The present review concentrates on recent research conducted on physical methods of pest control related to the solution of problems in the storage, processing, and distribution of cereal grains in the postharvest food system. Although fruits, vegetables, and fish form part of the postharvest system, these products are not included in this review.

Temperature

Stored-product insects and mites develop and reproduce within a temperature range usually characteristic of each species. Optimum and minimum temperatures of a number of the major stored-product insect species are known (71), as are the physical limits and optima for the development of microorganisms, mites, and insects that commonly occur in stored grain (118), and the effects of temperature and humidity on the growth of grain mite populations (51).

Insects and mites may survive beyond their reproductive and developmental temperature limits. When these limits are exceeded, the thermal death point will depend on the duration of exposure at a given temperature. The ability of insects to acclimate to high (64) or low (121) temperatures influences their response to temperature extremes. Insects die more rapidly at high than at low temperatures. The upper thermal death points of several major stored-product insects have been determined (62).

Effects of Temperature Extremes

The susceptibility of insects and mites to low and high temperatures is an important means of pest control. Heat has been used for insect control in flour mills in the United States (USA), especially in the warmer regions before the advent of modern insecticides (50). Good control of insects was achieved by maintaining all parts of a mill at 50–55° for 10–12 hours. The Mediterranean flour moth (*Anagasta kuehniella*) and the tobacco moth (*Ephestia elutella*), both important flour mill pests, become infertile at 30° (22).

High Temperatures

Insects may be killed by short exposures (>3 minutes) of infested foods to high temperatures (>60°) produced by infrared, radiofrequency, or microwave radiation devices, and fluidized-bed techniques (55, 77, 104, 137). However, all parts of the infested product must be maintained uniformly at specified temperatures for the required period to kill all stages of insects and mites.

The heating of grain and flour to control infestations is not without risk of damage to the products. Germination of soft winter wheat can be reduced by as much as 8% following irradiation for 10 seconds in a microwave oven at a frequency of 2.45 GHz (80). The milling and baking quality of wheat remains unimpaired after heating grain to 45° for 90 minutes to control flour beetles (26). However, bread baked from flour milled from wheat heated to 55°, 65°, and 80° had reduced loaf volumes (67). Flour made from soft red winter wheat that was stored for 1 or 2 months at 49° to control rice weevils (*Sitophilus oryzae*) had a higher ash content than controls, and there was also some discoloration and a decrease in diameter of cookies baked from the flour (77).

Low Temperatures

Low temperatures are used actively or passively in temperate regions of the world to control insects in stored grain and grain products. In the grain-growing regions of the northern United States and Canada, grain stored in farm granaries of about 40-ton capacity or less cools naturally shortly after the fall harvest to below the reproductive thresholds (15–20°) of most stored-product insects. In January and February, temperatures of small bulks of grain (<15 tons) decrease to about −20° (119).

Certain insect species may acclimatize rapidly to low temperatures, thus enabling them to survive. Adults of the rusty grain beetle (*Cryptolestes ferrugineus*), an important stored-grain pest in Canada, survived 9 days' exposure to −12° after they were acclimatized at 15°, whereas no insects survived when exposed directly to −12° following their removal from cultures at 27° (122).

During winter on the Canadian prairies, warm, infested grain can be cooled and the infestation controlled by mixing the warm grain with cold grain from the periphery of a grain bin. Exposure of warm grain to low ambient temperatures for the time taken to transfer the grain to another bin is not alone sufficient to control insects (133).

Forced-Air Cooling Grain can be cooled more rapidly in temperate regions by using forced air from fans connected to perforated ducts of different shapes and sizes. Refrigeration units can be used in regions where environmental temperatures are not sufficiently low to cool grain below the reproductive and developmental thresholds of storage pests. Burrell (39) described such equipment and stated that the low temperatures obtained reduced the likelihood of grain damage even though the pests themselves were not killed.

Freeze-Outs Probably the most widely practiced method using low temperatures in Canada is in the milling industry where freeze-outs are effective in controlling insect pests in flour mills. Meteorological forecasts are used by millers to determine periods of subzero temperatures that will control insects in milling equipment.

The success of low temperatures for pest control in flour mills is evidenced by the fact that it is usually not necessary to use pest control chemicals for 2–5 months after a freeze-out.

Figure 29.1 shows temperatures attained during a freeze-out at Winnipeg in 1962. The main reservoirs for reinfestation within a mill are heated areas such as boiler rooms which are not cooled sufficiently to control insects. Eggs of the confused flour beetle (*Tribolium confusum*) are susceptible to low temperatures and can be killed by exposing them at 10 or 15° for 15 days (*134*). Shorter exposures of 7, 8, or 9 hours at −10° killed 95% of the eggs of the sawtoothed grain beetle (*Oryzaephilus surinamensis*), the red flour beetle (*Tribolium castaneum*), and the almond moth (*Cadra cautella*), respectively (*97*).

Aeration

Aeration of stored grain involves the forced movement of air through the intergranular spaces. This is usually accomplished by a fan fitted to a duct that leads into a perforated floor or one or more perforated ducts of various shapes and sizes. The airflow rate may range from 3 to 30 m³/hr per metric ton (*41*) and is dependent on the dimensions and speed of the fan, the power of the motor, and the resistance of the grain to the flow of air. Various aeration systems are available commercially. They may be used to improve the storability of grain by incorporation of refrigeration systems that chill the grain (*56, 101*), cool it by using air at ambient temperatures (*52, 66, 73, 108*), or dry it with a batch dryer (*59, 87*).

Drying and Cooling Grain

Insect infestations can be controlled or arrested by drying and cooling grain (*6, 13 23, 37, 102*). Aeration systems also can be used in airtight bins to improve the distribution of fumigants through bulk grain (*24, 25, 58, 82, 91*); the fumigant vapors are then discharged from the grain after the pests have been controlled. However, because of corrosion and flammability hazards, it is dangerous to distribute phosphine gas in this manner (*25, 91*).

The value of aeration in preserving the quality of grain stored in metal or concrete silos is becoming more widely recognized, especially when grain is harvested and stored at moisture contents higher than those recommended for safe storage. In temperate regions, aeration with ambient air cools grain below temperatures at which many storage pests develop and reproduce. In nonaerated grain bulks, insect and mite infestations often occur in zones where moisture accumulates as a result of translocation; the use of grain aerators prevents the formation of moist zones where storage pests aggregate and cause grain to heat. Thermal relays and humidistatic controls are usually needed to stop aeration fans during warm or humid weather.

Moisture Loss Through Aeration

Although aeration by itself is not usually rec-

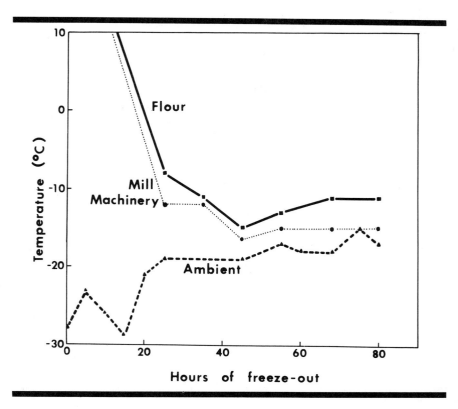

Figure 29.1
Record of temperatures in infested flour and mill machinery during a freeze-out at Winnipeg, Manitoba.

ommended for drying grain, the forced circulation of cool, dry air throughout a grain bulk does result in loss of moisture. Recent experiments in Manitoba showed that aeration of 45 tons of stored rapeseed with a ¾-HP (0.56-kW) motor reduced the moisture content from 11.67 to 8.22% from September to November 1978, compared to a reduction of 11.83 to 9.57% in a similar quantity of nonaerated rapeseed (R. N. Sinha, personal communication).

Precautions with Oilseeds

It is advisable to incorporate adequate temperature and humidity controls to regulate the operation of the fan motor because movement of excessive amounts of warm, moist air through stored oilseeds may cause a total loss of grain due to heating. Fava beans (*Vicia faba*) aerated in a steel bin left unattended heated to 260° in Manitoba (*88*). Moysey (*93*) has described some of the precautions required in the drying of oilseeds to prevent damage to seed quality. Foster (*59*) has discussed the theory of heated-air drying of stored grain and described different types of air dryers and how they might be used to achieve maximum efficiency with minimum damage to grain caused by cracking and breaking of kernels.

Modified Atmospheres

Manipulation of the proportions of gases that

form the intergranular environment of stored grain and processed foods has not been fully exploited for the control of pests. Recent work in Australia (12, 20) and Israel (6, 102) has demonstrated the practicality of this technique for controlling insects in stored grain. Different methods have been used to produce the combinations of O_2, CO_2, and N_2 that control insect pests. The atmosphere of shelled or in-shell peanuts stored in small test chambers has been purged using N_2 or CO_2 from pressurized steel cylinders (111). In a field trial, two gas-fired, exothermic generators were used to produce a modified atmosphere, low in O_2, which displaced air in the intergranular spaces in 20,000 bushels (544 tons) of wheat stored in a concrete silo 18 ft (5.5 m) in diameter and 130 ft (39.7 m) tall (123). The modified atmosphere contained <0.1% O_2, 8.5–11.5% CO_2, and the remainder mostly N_2.

Purging Intergranular Gases

Intergranular gases can also be purged with CO_2 supplied in tanks (under pressure in steel cylinders) (74, 75) or as dry ice. CO_2 applied as dry ice was used to disinfest 19 tons of bagged wheat shipped from Australia to West Germany in a freight container (20). Concentrations of up to 50% CO_2 were maintained in the containers for 10 days prior to the sea voyage. Navarro et al. (102) used a propane gas generator fitted with a catalytic converter combustion chamber to modify the atmospheres in steel bins containing 1,106 to 1,252 tons of wheat. O_2 content was lowered to 0.2–1.2%, and CO_2 levels were raised to 9.8–13.3%.

Airtight Structures

Flexible airtight structures made of butyl-rubber (a copolymer of isobutylene and isoprene) and EPDM (ethylene propylene diene monomer) were effective in controlling infestations of five species of stored-grain insects in 900 tons of wheat, provided the fabric remained undamaged throughout the storage period (99). Infestations at the apex of grain bulks enclosed in airtight bags are mainly due to perforations in the fabric and accumulations of moisture. They can be controlled readily by applying appropriate fumigants.

Insecticidal Effects

In a review of the use of modified atmospheres for insect control, Banks (19) reported that the insecticidal effects of low O_2 atmospheres could be enhanced by increasing the temperature. For example, complete insect mortality was obtained in a few days at 35°, whereas several weeks were required when the temperature was held at 15°. Low O_2 concentrations (2–5%) at 30° kill more adult *Tribolium castaneum* than the same O_2 levels at 26° (42). Toxic atmospheres containing <2% O_2 may be obtained by purging the grain with nitrogen or exhaust gases produced by burning hydrocarbons in air. Raisins stored up to 12 months at 24 ± 1° and 50% relative humidity (rh) in an atmosphere composed of <1% O_2, 9–9.5% CO_2, 86–89% N_2 were equal or superior in flavor to samples stored at normal atmospheres under similar conditions (65). No adverse effects of the use of modified atmospheres have been reported in the literature that has been reviewed.

Physical Forces

Stored grain and grain products are subjected to various physical forces during the course of their transportation, processing, and marketing. Pneumatic conveyors, grain augers, elevators, and centrifugal impact machines subject insect pests in grain and food particles to forces that impair development or cause death. The effectiveness of physical force in controlling infestations depends partly on physical factors such as particle velocity, distance of travel, and force of impact. Other factors such as temperature, the degree of protection afforded immature stages of insects within kernels, and the size and relative fragility of insects at different developmental stages also influence effectiveness of physical forces. Nevertheless, the transfer or turning of infested grain by elevating and dropping it into an empty bin is often used to achieve temporary control of infestations.

Pest Control Associated with Grain Handling

Several workers have made field observations to determine the efficacy of practical grain-handling practices to control insects. When a pneumatic system was used to transport 122 tons of infested grain to a grain irradiator, most adults of the rice weevil (*Sitophilus oryzae*), the lesser grain borer (*Rhyzopertha dominica*), and *Cryptolestes* species were killed; emergence of F_1 generation beetles from incubated samples was greatly reduced in comparison to that of the controls (48). Various mites (*Acarus siro*, *A. farris*, *Glycyphagus destructor*, and *Tyrophagus longior*) were controlled by moving infested barley through a conveying system (141). The method provided no long-term protection, however, because 4 to 19% of the mites survived the treatment.

Effects of Turning Grain

Turning of infested, farm-stored wheat controlled infestations of *Cryptolestes ferrugineus* in winter when the average grain temperature was reduced from 25° to <14° (*95*). In studies on the use of a farm auger to control heavy infestations (133 adults/150 g) of *C. ferrugineus*, movement of wheat through a 4-m-long auger at a rate of 6.5 tons/hr failed to control adults or larvae within the kernels (*139*). In experiments in a terminal elevator to study the effects of pneumatic delivery systems for conveying grain samples to inspection offices, mortalities of 73% for *C. ferrugineus*, 65% for *Oryzaephilus surinamensis* and *Tribolium castaneum*, and 22% for *Sitophilus granarius* (granary weevil) were reported (*36*). The grain speed was 7.2–11 m/sec through a 39- to 99-ft (11.9- to 30.2-m) pipe 5 cm in diameter. There was no correlation between mortality and tube length or number of bends in the tube.

Effects of Physical Forces

Laboratory experiments have been conducted to study the effects of percussion, stress, and physical disturbance on insects (*15, 16, 85*). Percussive forces resulting from insects (*S. granarius*) impinging on solid surfaces at a speed of 45.72 m/sec killed adults but permitted survival of 1–2% of insects within the wheat kernels; about 20% of the kernels were damaged by the treatment (*15*). Repetitive treatments of internally infested grain impinging on surfaces after being subjected to speeds of 12.8 m/sec resulted in lower mortalities of insects, depending on stage of development, without noticeable damage to the wheat (*16*). Adults of *C. ferrugineus* were more likely to be killed than those of *S. oryzae* or *S. granarius* when infested wheat was dropped 14.1 m through a tube at a speed of 16.6 m/sec (*85*). Repetitive dropping of infested wheat increased insect mortality. Emergence of *C. ferrugineus* from wheat dropped seven times down the tube was about half that of controls.

Entoleters

Impact machines such as entoleters have been used by the cereal-processing industry for many years to eliminate insects from wheat and flour. Infested material is fed into the center of a rotor revolving at high speed. The material is spun outward by centrifugal force into the path of stationary or moving studs inside the walls of the machine. The speed of the machine can be adjusted to avoid damage to sound kernels; however, kernels that have been structurally damaged by insects would be broken on impact. Broken kernels are separated from sound kernels by aspirators and sifters. Cotton (*49*) reported 99% control of free-living insects in wheat fed into an entoleter operating at 1,750 rpm. A bulk-flour entoleter operating at 3,500 rpm may be installed directly above the packer bins to kill all stages of insects that might otherwise be present in flour. In the pasta industry, treatment of semolina with an entoleter at 2,900 rpm completely controlled adults and eggs of *T. castaneum*, *O. surinamensis*, *S. granarius*, and *Stegobium paniceum* (drugstore beetle) (*70*). Insect fragments were only partially removed by sieving.

Physical Barriers

Stored grain and processed foods must be protected from infestation for as long as possible during storage and distribution. The enclosure of food commodities in a suitable protective container is the first step toward preservation of food quality. Natural protection from infestation is afforded to certain varieties of corn that have the cob completely enclosed by a husk. In developing countries, corn is often stored with the husk intact (*54*). Likewise, storage of pecans in undamaged shells provides complete protection from insects during storage (*76*). Virtually no protection is afforded to cereal grains packed in burlap or jute sacks. Woven polypropylene bags tested in an experimental shipment of 1,000 tons of rice provided better protection against rodents and, probably, insects than burlap bags and took up less storage space (*7*).

Protective Packaging

Much research has been done on the effectiveness of various packaging materials as barriers to insect infestation of food products (see Chapter 23 for additional details). Sealed foil prevented infestation of cartons containing processed white or brown rice (*143*). Addition of a cellophane overwrap to standard packages reduced infestation by only 1%. Hot-melt-glue-sealed packages had a lower infestation level than cold-glue-sealed or heat-sealed ones. Workers in this field generally agree that the primary need in developing insectproof packages is for better sealing of closures rather than introduction of major changes in package construction.

Factors Affecting Penetration

Aluminum foil can be penetrated by certain species of insects (21). Confectionary products packaged in paper containers can be protected from insects by an outer wrapping of polypropylene or aluminum/film laminates (117). Larvae of a flour moth can penetrate polyethylene-coated (0.032-mm) paper sacks (84). Tightly sealed polypropylene film overwraps significantly increase resistance of paperboard cartons (69). However, polypropylene film did not protect against infestation as well when used as a loose wrap or as a pouch.

Composite cans made of various combinations of paper, foil, and polymer films provided excellent protection from infestation during 29 months of exposure to boring and nonboring insects (68).

Availability of Food

The availability of food outside a package may affect the extent to which insects attempt to penetrate packaging materials. In tests of seven flexible packaging materials, it was found that larvae of seven species of stored-product insects were more likely to penetrate food packages when no other food was available (45). Packages made from 16.5-micron aluminum foil, polyester, and polypropylene were most resistant to penetration. Larvae of the cadelle (*Tenebroides mauritanicus*) and the warehouse beetle (*Trogoderma variabile*), however, penetrated all materials tested.

Slick Packaging

Packages constructed of flexible materials that are difficult to climb may inhibit the ability of insects to infest packaged foods (44). Polypropylene and polyethylene were the most difficult materials for insects to climb. However, accumulation of dust on packages may enable insects to crawl up otherwise slippery vertical surfaces. Laboratory tests on the penetration of packaging materials by insects do not take into account the behavior of insects under warehouse conditions where freedom of movement may influence infestation of foods in different types of packages (14).

Underground Storage

Underground pits have been used as food containers in the Mediterranean region and Middle East since Greek and Roman times (61). The sides of the pit can be constructed of seaweed, grass matting, or baked mud, depending on the materials available in the region. These linings form barriers to insect and rodent attack and may assist in forming a hermetic seal that further increases the protection. Polyethylene, concrete, or metal can also be used as liners (10).

Temporary Storage Structures

Aboveground grain storages may be constructed of butyl-rubber/EPDM fabricated as a liner inside a supporting steel weldmesh wall (100). The liner forms a barrier to insects and provides hermetic conditions to reduce O_2 content and control infestations in the grain. The weldmesh also affords some protection from rodents. A portable grain bin fabricated from polyethylene was used to store 27 tons of wheat successfully for 2 years on a farm in Manitoba (94). Only mice caused minor damage—at the bottom.

The widespread use of cargo containers to transport grain and processed foods makes it possible to package products in less robust cartons (121). Cargo containers provide a barrier to insect and rodent infestations of enclosed commodities during transit. Containers can also be sealed tightly to facilitate application of fumigants to disinfest cereal products.

Electrical Apparatus

Various types of electrical apparatus have been designed to modify the behavior of food pests. Some devices such as electrocuters have been successfully used to control flies in confined spaces such as dairy barns (9) and food-industry plants (8). One survey found that 53% of large independent bakeries in Great Britain relied on electrified grids for pest control (90). The grids are operated at high voltages (4,500 V) and low currents (0.01 amp) and are usually used in conjunction with a black or blue light source to attract insects to the grid (5). Electrocuters have provided satisfactory control of flying insects in food preparation, cooking, and packing areas of fish plants (60).

Other electrical devices have been developed to emit low-level electromagnetic waves to control insects and repel rodents. However, one such device developed to control the pine vole was reported to be ineffective (40). In addition, the U.S. Environmental Protection Agency tested several devices and reported that they did not perform according to the manufacturers' claims (11).

Ionizing Radiation

Two types of ionizing radiation have been used to control insects in stored foods: Gamma radiation produced by the nuclear disintegration of radioactive isotopes such as cobalt-60, and high-speed electrons emitted from a heated cathode and accelerated in an electric field. The penetrating power of gamma radiation is much greater than that of accelerated electrons which can penetrate only about 1.7 cm when one side of the material is irradiated. Both types of radiation cause the ejection of orbital electrons from atoms of material being irradiated. The loss of electrons from some atoms and their acquisition by others result in the formation of positive and negative ions. High-energy radiation may also rupture chemical bonds and form reactive free radicals that may interact with organic molecules and cause chemical changes.

Ionizing radiation has been evaluated in many countries to determine the lethal and sterility effects on various species and stages of stored-product insects. Much research has also been done on application methods and their effects on irradiated products. Recent reviews and assessments (17, 18, 125, 126, 130, 131, 142) have listed the advantages and limitations of ionizing radiation for insect control in stored foods.

Effects of Dose and Sensitivity

The survival period of irradiated insects is related to the applied dose and the radiosensitivity of the species. Unlike chemical treatments which kill insects within hours or days after exposure, depending on the nature and toxicity of the chemical, irradiated insects may survive for several weeks after exposure at doses recommended for commercial treatment of stored foods. Most adults of *Sitophilus granarius* and *S. zeamais* (maize weevil) die within 4 weeks after irradiation at 30 krad (32). Adults treated at 10 krad fail to reproduce.

Larvae of *Trogoderma inclusum* and *T. variabile* fail to develop to the adult stage when irradiated at 10 krad but many survive for more than 14 weeks (30). Though a few adults develop from pupae irradiated at 100 krad, all pupae and adults are sterilized by 30 krad. Eggs and larvae of *T. destructor* fail to form adults after irradiation at 5 krad (28); pupae are more resistant—some have formed adults after irradiation at 5–30 krad.

Induced Sterility

Radiation-induced sterility of stored-product insects has prompted research on the release of sterile adults for control of pests. Brower (27) listed the advantages and limitations of techniques for the genetic control of stored-product insects. Two major disadvantages are that the sterilized insects continue to feed on stored products after they have been released, and that pest populations are controlled slowly (31).

Brower (29) found that irradiation of eggs of the Indianmeal moth (*Plodia interpunctella*) for mass production of either sterile or partially sterile adults was not feasible. However, irradiation of 24-hr-old adults of *P. interpunctella* at 50 krad did not decrease sexual competitiveness of adults (4); overwhelming unirradiated populations in the laboratory with irradiated males and females at ratios of 15:1 resulted in a high percentage of infertility in the population.

Age Effects

The age at which adults are irradiated for sterility may determine whether insects can recover their fertility. Newly emerged males of the almond moth (*Cadra cautella*) irradiated at 40 krad recovered fertility, but adults irradiated on the third day after emergence remained sterile (63). Studies with flour beetles (*Tribolium castaneum* and *T. confusum*) indicated that because of the large numbers of sterile insects needed, the sterile male technique does not seem promising for the control of stored-product insects (57).

Radiation Combinations

Gamma radiation has been combined with nonionizing radiation or insecticides to increase the effectiveness of insect control. Gamma radiation supplemented by infrared or microwave radiation reduced by 96–99% the emergence of adults of *R. dominica* from irradiated eggs, larvae, and pupae (78). The resulting mortalities were 20% higher than those predicted for gamma irradiation alone. Similarly, when eggs, larvae, and pupae of the Angoumois grain moth (*Sitotroga cerealella*) were irradiated with both gamma radiation at 10 krad and microwave radiation at 2,450 MHz, good control was obtained, irrespective of the sequence in which the treatments were applied (127).

Radiation Plus Insecticides

Gamma radiation (at 10 krad or higher) combined with insecticide or fumigant applications produced higher mortalities than either treatment alone (46, 47). The main advantage of combining the treatments appears to be more rapid death. The opposite result occurred at lower doses (5 krad); the radiation apparently suppressed the toxicity of malathion to *T. castaneum* (47).

Gamma Radiation from Cobalt-60

When infested wheat and flour were treated with a commercial-scale cobalt-60 irradiator at 27.4–41.2 krad, live insects (probably postirradiation invaders) were observed in 13.6-ton lots of grain 6 months later (48). When infested white flour packed in metal cans or in treated, multiwall paper bags was irradiated at ca 40 krad, no living insects were found in the irradiated flour after 3–14 months (128). In further tests, irradiated white or yellow cornmeal infested with *P. interpunctella*, *T. castaneum*, and *R. dominica* was irradiated at 15.8–62.5 krad (129). Though some insects survived 35.5 krad after one month, all insects were dead 3 months after irradiation at 15.8 krad.

Accelerated Electrons

Accelerated electrons are another form of ionizing radiation that have been used to control insect pests in stored foods. Though not as penetrating as gamma radiation, the dose rate can be adjusted to enable the rapid treatment of large quantities of bulk commodities at rates of 200–400 tons/hr. Comparison of cobalt-60 gamma radiation and accelerated electrons for the disinfestation of corn showed that at 15 or 25 krad the two types of radiation were equally effective in preventing the development and emergence of *Sitophilus* species (1). However, a dose of 25 krad was necessary to obtain complete control of *T. castaneum* (138). Studies on the effect of the intensity of radiation, expressed as dose rate, on *T. castaneum* and *Prostephanus truncatus* (larger grain borer), using cobalt-60 and accelerated electrons, have shown that at 25 krad there is no significant difference in the lethal effect of various dose rates (2, 3). High radiation intensities, but still lower than the doses approved for the commercial treatment of wheat and flour (125), caused lower hatches of the eggs of *T. castaneum* (33, 98).

Practical Concerns About Radiation

Vas (130, 131, 142) summarized the technological status of irradiation as a food preservation measure and emphasized the need to harmonize national legislation in various countries to faciliate wider acceptance and usage of radiation in the food industry. Deitch (53) considered that radiation treatment of wheat, wheat flour, and potatoes in the USA was unlikely to become commercial in the near future because well-equipped and ample storage facilities are already on hand and can be utilized more cheaply. Bailey (17) suggested that it may be more appropriate to irradiate infested foods in food-importing countries that have approved the irradiation process rather than in food-producing countries where international clearances of irradiated products must be taken into considertion.

The costs and reliability of electron accelerators have been discussed by Morganstern (92). Typical costs were 13 cents/ton of grain for a 6,000-hr/year operation; 15 cents/ton for a 4,000-hr operation. Flour could be treated for 2.75 cents/50-lb (22.7-kg) bag. Irradiated grain or flour must be stored in insect-free warehouses or containers to prevent reinfestation.

Nonionizing Radiation

Infrared, radiofrequency, and light radiation are the main forms of nonionizing radiant energy that have been investigated for the control of stored-product insects. Infrared and radiofrequency devices are used to heat infested foods to temperatures that will kill all stages of insects. Light in the visible and invisible parts of the electromagnetic spectrum is used to control the activity of insect pests, usually by attracting them to insect traps.

Infrared Heaters

Laboratory tests with infrared heaters have shown that adults of 12 species of insects in soft winter wheat were controlled at $65° ± 1.5°$ (81). The heat was transferred to the insects by radiation, convection, and conduction through air and wheat. In further experiments on a larger scale, Kirkpatrick and Cagle (79) used infrared heaters to treat 13 tons of hard red winter wheat infested with four species of stored-product insects. The average grain temperature of 55° after treatment was insufficient to kill immature stages.

Exposure to Electric Fields

A more effective way of heating stored foods to control insects is a high-frequency electric field that generates heat uniformly within the product. The amount of heat produced depends on the physical and electrical characteristics of the food components and on the frequency and power output of the generator. Though much research has been done on this form of insect control, it has not been widely used in the food industry mainly because of the high capital cost of installations and the fact that effective pest control programs are already in place. Most of the present programs, however, rely heavily on the use of insecticides and fumigants. High-frequency heating provides an alternative to the use of chemicals for insect control in the food industry, and the research that has already been done is basic for the development of new technology using this form of radiation.

Electrical Properties of Insects

Information is now available on the electrical characteristics of insects and various kinds of grain and processed foods (103–106). Because the value of the dielectric properties of insects at certain frequencies may be higher than those of cereals and cereal products, insects may absorb more electrical energy and thus be heated at a higher rate than the surrounding food medium. Dielectric properties of insects and foods are dependent on the frequency of the generator, however, and determination of their values at different frequencies will make it possible to predict the frequency range at which selective heating of insects is most likely to occur. A frequency of 39 MHz completely controlled adult *Sitophilus oryzae* 8 days after wheat had been heated to 40° (106). Microwave frequencies of 2.45 to 12 GHz (1 GHz = 10^9 cycles/sec) were much less effective than 39 MHz for controlling adult *S. oryzae*, probably because of the lower values of dielectric properties at microwave frequencies (105, 107) and the uneven distribution of microwave energy in containers of wheat or flour (137).

Radiofrequency Radiation

The biological effects of radiofrequency radiation on living organisms are due mainly to heat damage to tissues and fluids (124). Studies with the yellow mealworm (*Tenebrio molitor*) showed that irradiation of eggs in a radiofrequency field at 39 MHz impaired embryonic development (113). Further studies reported adverse effects on reproductive tissues and fecundity of *T. molitor* (114–116). Carpenter and Livstone (43) proposed evidence of nonthermal effects of microwave radiation (10 GHz) on pupae of *T. molitor* which resulted in abnormal development. However, these results may also be explained by the selective heating of different tissues within insects, a phenomenon that apparently occurred when flesh fly larvae (*Sarcophaga crassipalpis*) were exposed to microwave radiation at 2.375 GHz (144).

Microwave Radiation

Insects may also respond to the intensity gradient in a nonuniform electromagnetic field produced by microwave radiation (2.375 GHz) by avoiding regions that are more intensely irradiated (109). This may explain the survival of mobile stages of insects exposed in a high-frequency electromagnetic field. Practical trials were conducted with a 0.896-GHz generator that produced a microwave beam directed at a conveyer belt moving through a heated tunnel 8-m long (72). Higher insect mortalities were obtained in wheat than in cocoa beans. Addition of hot air to the tunnel in which the grain was irradiated decreased the rate of heat loss from the commodities and increased insect mortality.

Similar results were obtained when irradiated wheat samples were insulated to reduce the rate of heat loss after irradiation (137). There was no change in fat or moisture content in wheat and cocoa beans after irradiation, but the peroxide level of the cocoa beans was lowered (72). Cost of treatment per ton was calculated at U.S. $0.70 (1980). This compares favorably with the costs of fumigation in England. Smith (120) quoted amortization and operating costs of microwave heating equipment at U.S. $0.0033 to $0.02 per pound (0.454 kg) of food. In addition, he reported enhanced quality of irradiated foods such as pasta and desiccated vegetables.

A microwave radiation device has been developed to dry stored grain to safe storage limits (67). The heat produced within the grain was sufficiently high to control certain stored-product insects (*Tribolium confusum*, *Sitophilus granarius*, and *Cryptolestes ferrugineus*). The loaf volume of bread baked from flour milled from irradiated wheat was greatly reduced when the wheat temperature reached 55° or higher. Further work showed that wheat could be dried at 45° in an irradiator powered by a microwave generator at 2.45 GHz or a high-frequency generator at 13 MHz. The costs of the two systems were similar and the milling qualities of irradiated wheats were not impaired (26). The rate of germination for wheat irradiated at 2.45 GHz was reduced by 8% after exposure for 10 sec in open containers (80). The low capacity of this grain drier probably accounts for its lack of acceptance by the grain industry.

Visible and Ultraviolet Light

Visible light and ultraviolet radiation have been used to influence insect behavior and control infestations. Physiological and behavioral responses of stored-product insects are altered by exposure to visible-light wavelengths of 350–770 nm (87). It may be possible to utilize this type of response behavior for the development of more effective insect control techniques. Ultraviolet light has been used to attract flying insects to traps or electrified grids; it has also been used to disinfest insect-rearing facilities that had become infested with the straw itch mite (*Pyemotes tritici*), a parasite of stored-product insects in cultures (*35*).

Sound

Sound has been used experimentally to control the activities of insects and rodents, but the results have not proved sufficiently promising to date for wide application in the food industry. It has been suggested that sound might be used to attract insects to traps, to repel them from food sources, or to jam their natural communication systems (*110*). The audible range for the normal human ear is 20 Hz to 20 kHz. Certain insects and rodents have hearing ranges that extend beyond those of humans. Sounds in those extreme ranges can be detected only by special instruments (*112*).

Various ultrasound devices have been developed to expel rodents from buildings where food is stored. Tests of the effects of two different types of ultrasonic generators (frequency ranges, 41–48 kHz) on the feeding activity of Norway rats in an infested building showed that neither type of generator would be effective in expelling the rats from warehouses and preventing them from feeding at distances of 1.5–31.3 m from the sound source (83).

The effects of low-frequency sound from 10 to 24 Hz on the emergence of a moth (*Plodia interpunctella*) have been studied by Mullen (*96*). He concluded that sound reduced adult production by 21–26%. He suggested that accelerated activity of larvae exposed to low-frequency sound might enhance the effectiveness of chemical control.

Conclusion

Most, if not all, pest control programs applied at various links in the postharvest food system involve some use of physical methods. In many cases, physical methods can be combined with the judicious use of chemicals to enhance the effectiveness of both. Improvements in the design of postharvest storage, transportation, and processing facilities or machinery should include elimination of unnecessary recesses and spaces where accumulations of cereals and processed foods provide refuge and sustenance for insect and rodent pests (see Chapter 22 for additional details). Improved equipment design must be accompanied by good management practices that emphasize effective sanitation and hygiene to ensure that work areas are free from dust and food residues that encourage infestations.

Some of the physical methods reviewed in this chapter refer to equipment and devices that will require further research before they can be implemented. Capital-intensive installations required for electromagnetic radiation devices are considered as possible alternatives for use in pest control programs that presently rely heavily on chemicals. Although radiation devices are effective physical means of controlling insect pests, they do not confer any residual protection to products that have been treated. Therefore, other physical methods such as insect barriers or insectproof containers and effective sanitary practices must be used to protect irradiated products from reinfestation. In some cases, it will probably also be necessary to supplement physical control measures with insecticidal treatments. But as we acquire more information on physical controls and on the behavior and ecology of food pests, we shall be able to reduce our dependence on chemicals and thus limit the impact of their detrimental effects such as resistance and unwanted residues.

Cited References

1 Adem, E., et al. 1978. Comparison of ^{60}Co gamma radiation and accelerated electrons for suppressing emergence of *Sitophilus* spp. in stored maize. *Journal of Stored Products Research* 14(3)135–142.

2 Adem, E., R. M. Uribe, and F. L. Watters. 1979. Dose rate effects on survival of two insect species which commonly infest stored corn. *Radiation Physics and Chemistry* 14:663–670.

3 Adem, E., R. M. Uribe, and F. L. Watters. 1979. Responses of *Prostephanus truncatus* (Coleoptera: Bostrichidae) and *Tribolium castaneum* (Coleoptera: Tenebrionidae) to gamma radiation from ^{60}Co. *Canadian Entomologist* 111(10)1111–1114.

4 Ahmed, M. Y. Y., E. W. Tilton, and J. H. Brower. 1976. Competitiveness of irradiated adults of the Indian meal moth. *Journal of Economic Entomology* 69(3)349–352.

5 Alosi, C. 1968. Nuova attrezzatura per la lòtta agli insètti alati nelle centrale del latte. *Industrie Alimentari* 7(44)87–88.

6 Annis, P. S. and J. V. S. Graver. 1986. Use of carbon dioxide and sealed storage to control insects in bagged grain and similar commodities. In *Pesticides and Humid Tropical Grain Storage Systems*, B. R. Champ and E. Highley, eds.

Australian Centre for International Agricultural Research, Canberra.

7 Anon. 1970. New packaging material used in test shipment of 1,000 tons of rice to Indonesia. *Rice Journal* 73(4)20–22.

8 Anon. 1973. Electrocuters solve flying insect problem. *Food Processing* 34(9)58.

9 Anon. 1974. Insects shocked to death. *American Dairy Review* 36(4)57.

10 Anon. 1976. Protecting stored grain: heat, cold, or suffocation. *Rural Research* 91(June)18–20.

11 Anon. 1979. EPA stops sale of electromagnetic repellers. *Pest Control* 47(5)54–55.

12 Anon. 1979. Grain pests hit by nitrogen and carbon dioxide. *Rural Research* 102(March)20–22.

13 Armitage, D. M., and N. J. Burrell. 1978. The use of aeration spears for cooling infested grain. *Journal of Stored Products Research* 14(4)223–226.

14 Ashman, F. 1975. Pest control and storage problems in feed commodities in tropical and subtropical countries. In *Proceedings of the Tropical Products Institute Conference on Animal Feeds of Tropical and Subtropical Origin* (London, England), Tropical Products Institute, London, pp. 262–272.

15 Bailey, S. W. 1962. The effects of percussion on insect pests of grain. *Journal of Economic Entomology* 55(3)301–304.

16 Bailey, S. W. 1969. The effects of physical stress in the grain weevil *Sitophilus granarius*. *Journal of Stored Products Research* 5(4)311–324.

17 Bailey, S. W. 1979. The irradiation of grain: An Australian viewpoint. In *Symposium on the Protection of Grain Against Insect Damage During Storage* (Moscow, 1978), Division of Entomology, CSIRO, Canberra, Australia, pp. 136–138.

18 Banks, H. J. 1976. Physical control of insects—recent developments. *Journal of the Australian Entomological Society* 15(1)89–100.

19 Banks, H. J. 1979. Recent advances in the use of modified atmospheres for stored product pest control. In *Proceedings of the Second International Working Conference on Stored-Product Entomology* (Ibadan, Nigeria, 1978), pp. 198–217.

20 Banks, H. J., and A. K. Sharp. 1979. *Trial Use of CO_2 to Control Insects in Exported Containerized Wheat*. CSIRO (Australia) Division of Entomology, Report No. 15, 11 pp.

21 Batth, S. S. 1970. Insect penetration of aluminum-foil packages. *Journal of Economic Entomology* 63(2)653–655.

22 Bell, C. H. 1975. Effects of temperature and humidity on development of four pyralid moth pests of stored products. *Journal of Stored Products Research* 11(3–4)167–175.

23 Bhatnagar, A. P., and A. S. Bakshi. 1975. Aeration studies on the storage of wheat grains in a 50-tonne outdoor metal bin. *Punjab Agricultural University Journal of Research* 12(2)189–199.

24 Bond, E. J. 1973. Chemical control of stored grain insects. In *Grain Storage—Part of a System*, R. N. Sinha and W. E. Muir, eds. AVI, Westport, CT.

25 Bond, E. J. 1984. *Manual of Fumigation for Insect Control*. FAO Plant Production and Protection Paper 54. Food and Agriculture Organization, Rome.

26 Boulanger, R. J., W. M. Boerner, and M. A. K. Hamid. 1969. Comparison of microwave and dielectric heating systems for the control of moisture content and insect infestations of grain. *Journal of Microwave Power* 4(3)194–208.

27 Brower, J. H. [1975.] Potential for genetic control of stored-product insect populations. In *Proceedings of the First International Working Conference on Stored-Product Entomology* (Savannah, GA, 1974), pp. 167–180.

28 Brower, J. H. 1975. Radiosensitivity of *Tribolium destructor* Uyttenboogaart (Coleoptera: Tenebrionidae). *Journal of Stored Products Research* 11(3–4)223–227.

29 Brower, J. H. 1975. Sterility of adult Indian meal moths and their progeny reared from gamma irradiated eggs. *Environmental Entomology* 4(5)701–704.

30 Brower, J. H., and E. W. Tilton. 1972. Gamma-radiation effects on *Trogoderma inclusum* and *T. variabile*. *Journal of Economic Entomology* 65(1)250–254.

31 Brower, J. H., and E. W. Tilton. 1973. Weight loss of wheat infested with gamma-radiated *Sitophilus oryzae* (L.) and *Rhyzopertha dominica* (F.). *Journal of Stored Products Research* 9(1)37–41.

32 Brown, G. A., J. H. Brower, and E. W. Tilton. 1972. Gamma radiation effects on *Sitophilus zeamais* and *S. granarius*. *Journal of Economic Entomology* 65(1)203–205.

33 Brown, G. A., and R. Davis. 1973. Sensitivity of red flour beetle eggs to gamma radiation as influenced by treatment age and dose rate. *Journal of the Georgia Entomological Society* 8(3)153–157.

34 Browne, S. 1989. Food irradiation. *Journal of Environmental Health* 51(5)269–270.

35 Bruce, W. A., and P. T. M. Lum. 1979. The effects of UV radiation on stored-product insects. In *Proceedings of the Second International Working Conference on Stored-Product*

Entomology (Ibadan, Nigeria, 1978), pp. 271–277.

36 Bryan, J. M., and J. Elvidge. 1977. Mortality of adult grain beetles in sample delivery systems used in terminal grain elevators. *Canadian Entomologist* 109(2)209–213.

37 Burrell, N. J. 1967. Grain cooling studies—II: Effect of aeration on infested grain bulks. *Journal of Stored Products Research* 3(2)145–154.

38 Burrell, N. J. 1974. Aeration. In *Storage of Cereal Grains and Their Products,* C. M. Christensen, ed. American Association of Cereal Chemists, St. Paul, MN.

39 Burrell, N. J. 1974. Chilling. In *Storage of Cereal Grains and Their Products,* C. M. Christensen, ed. American Association of Cereal Chemists, St. Paul, MN.

40 Byers, R. E. 1979. Field evaluation of a commercial magnetic device for pine vole control. *Pest Control* 47(1)22–23.

41 Calderon, M. 1972. Aeration of grain—benefits and limitations. *EPPO Bulletin* 6:83–94.

42 Calderon, M., and S. Navarro. 1979. Increased toxicity of low oxygen atmospheres supplemented with carbon dioxide on *Tribolium castaneum* adults. *Entomologia Experimentalis et Applicata* 25(1)39–44.

43 Carpenter, R. L., and E. M. Livstone. 1971. Evidence for nonthermal effects of microwave radiation: Abnormal development of irradiated insect pupae. *IEEE Transactions on Microwave Theory and Techniques MTT* 19(2)173–178.

44 Cline, L. D. 1978. Clinging and climbing ability of larvae of eleven species of stored-product insects on nine flexible packaging materials and glass. *Journal of Economic Entomology* 71(4)689–691.

45 Cline, L. D. 1978. Penetration of seven common flexible packaging materials by larvae and adults of eleven species of stored-product insects. *Journal of Economic Entomology* 71(5)726–729.

46 Cogburn, R. R., and H. B. Gillenwater. 1972. Interaction of gamma radiation and fumigation on confused flour beetles. *Journal of Economic Entomology* 65(1)245–248.

47 Cogburn, R. R., and R. D. Speirs. 1972. Toxicity of malathion to gamma-irradiated and non-irradiated adult red flour beetles. *Journal of Economic Entomology* 65(1)185–188.

48 Cogburn, R. R., E. W. Tilton, and J. H. Brower. 1972. Bulk-grain gamma irradiation for control of insects infesting wheat. *Journal of Economic Entomology* 65(3)818–821.

49 Cotton, R. T. 1963. *Pests of Stored Grain and Grain Products.* Burgess, Minneapolis, MN.

50 Cotton, R. T. 1964. High temperatures—how they affect insect pests. *Northwestern Miller* 270(Mar. 16)30–32.

51 Cunnington, A. M. 1976. The effect of physical conditions on the development and increase of some important storage mites. *Annals of Applied Biology* 82(1)175–178.

52 Cuperus, G. W. et al. 1986. Insect populations in aerated and unaerated stored wheat in Oklahoma. *Journal of the Kansas Entomological Society* 59(4)620–627

53 Deitch, J. 1975. United States potentials and problems for insect disinfestation of wheat and sprout inhibition of potatoes by irradiation. In *Requirements for the Irradiation of Food on a Commercial Scale. Proceedings of a Panel on the Commercialization of Irradiated Food Items Accepted for Human Consumption* (Vienna, Austria, Mar. 18–20, 1974), pp. 13–29.

54 De Lima, C. P. F. 1979. A review of the use of physical storage procedures in East Africa: Aspects for improvement and extension. In *Proceedings of the Second International Working Conference on Stored-Product Entomology* (Ibadan, Nigeria, 1978), pp. 237–241.

55 Dermott, T., and D. E. Evans. 1978. An evaluation of fluidized-bed heating as a means of disinfesting wheat. *Journal of Stored Products Research* 14(1)1–12.

56 Donahaye, E., S. Navarro, and M. Calderon. 1974. Studies on aeration with refrigerated air. III. Chilling of wheat with a modified chilling unit. *Journal of Stored Products Research* 10(1)1–8.

57 Erdman, H. E. 1974. Productivity modifications to flour beetles (Coleoptera: Tenebrionidae) by gamma radiation (^{60}Co) of one or both sexes and by the addition of radiated males or females to population. *Researches on Population Ecology* 16:52–58.

58 Food and Agriculture Organization. 1985. *Manual of Pest Control for Food Security Reserve Grain Stocks.* Plant Production and Protection Paper 63. FAO, Rome.

59 Foster, G. H. 1973. Heated-air grain drying. In *Grain Storage—Part of a System,* R. N. Sinha and W. E. Muir, eds. AVI, Westport, CT.

60 Franz, R. 1973. Electronic units solve insect control problem. *Food Processing* 34(2)30.

61 Gilman, G. A., and R. A. Boxall. 1974. The storage of food grains in traditional underground pits. *Tropical Stored Products Information* 28:19–38.

62 Girish, G. K. 1965. Effect of temperature on the development of stored grain insect pests. *Bulletin of Grain Technology* 3(4)142–154.

63 Gonen, M. 1975. Effects of gamma radiation on *Ephestia cautella* (Wlk.) (Lepidoptera: Phycitidae)—IV. Sensitivity of maturing sperm in the adult to a sterilizing dose. *Journal of Stored Products Research* 11(2)97–101.

410   F. L. Watters

64 Gonen, M. 1977. Susceptibility of *Sitophilus granarius* and *S. oryzae* (Coleoptera, Curculionidae) to high temperature after exposure to supra-optimal temperature. *Entomologia Experimentalis et Applicata* 21(3)243–248.

65 Guadagni, D. G., C. L. Storey, and E. L. Soderstrom. 1978. Effect of controlled atmosphere on flavor stability of raisins. *Journal of Food Science* 43(6)1726–1728.

66 Hall, D. W. 1970. *Handling and Storage of Food Grains in Tropical and Subtropical Areas*. FAO Agricultural Development Paper No. 90. Food and Agriculture Organization, Rome.

67 Hamid, M. A. K., and R. J. Boulanger. 1969. A new method for the control of moisture and insect infestations of grain by microwave power. *Journal of Microwave Power* 4(1)11–18.

68 Highland, H. A. 1975. *Insect Resistance of Composite Cans*. ARS-S-75. U.S. Department of Agriculture, Washington, DC.

69 Highland, H. A., R. H. Gug, and H. Laudani. 1968. Polypropylene vs insect infestation. *Modern Packaging* 41(12)113–115.

70 Hoppe, T. 1978. Einsatz von Prallmaschinen in der Teigwarenindustrie zur alternativen Bekämpfung von vorratsschädlichen Insekten. *Lebensmittel Wissenschaft Technologie* 11(3)169–172.

71 Howe, R. W. 1965. A summary of estimates of optimal and minimal conditions for population increase of some stored products insects. *Journal of Stored Products Research* 1(1)177–184.

72 Hurlock, E. T., B. E. Llewelling, and L. M. Stables. 1979. Microwaves can kill insect pests. *Food Manufacture* 54(Aug.)37, 39.

73 Hyde, M. B., and N. J. Burrell. 1973. Some recent aspects of grain storage technology. In *Grain Storage—Part of a System*, R. N. Sinha and W. E. Muir, eds. AVI, Westport, CT.

74 Jay, E. G. 1980. Methods of applying carbon dioxide for insect control in stored grain. *Advances in Agricultural Technology, Southern Series (USDA)*, S-13:1–7.

75 Jay, E. G., and G. C. Pearman, Jr. 1973. Carbon dioxide for control of an insect infestation in corn (maize). *Journal of Stored Products Research* 9(1)25–29.

76 Joubert, A. J., and P. C. Joubert. 1969. Protecting pecan and macadamia nuts. *Farming in South Africa* 45:41–48.

77 Kirkpatrick, R. L. [1975.] The use of infrared and microwave radiation for control of stored-product insects. In *Proceedings of the First International Working Conference on Stored-Product Entomology* (Savannah, GA, 1974), pp. 431–437.

78 Kirkpatrick, R. L., J. H. Brower, and E. W. Tilton. 1973. Gamma, infra-red and microwave radiation combinations for control of *Rhyzopertha dominica* in wheat. *Journal of Stored Products Research* 9(1)19–23.

79 Kirkpatrick, R. L., and A. Cagle. 1978. Controlling insects in bulk wheat with infrared radiation. *Journal of the Kansas Entomological Society* 51(3)386–393.

80 Kirkpatrick, R. L., and J. R. Roberts, Jr. 1971. Insect control in wheat by use of microwave energy. *Journal of Economic Entomology* 64(4)950–951.

81 Kirkpatrick, R. L., and E. W. Tilton. 1972. Infrared radiation to control adult stored-product Coleoptera. *Journal of the Georgia Entomological Society* 7(1)73–75.

82 Kluijver, H. 1980. Pest control in unmalted grains in the tropics. *MBAA Technical Quarterly* 17(3)149–155.

83 LaVoie, G. K., and J. F. Glahn. 1977. Ultrasound as a deterrent to *Rattus norvegicus*. *Journal of Stored Products Research* 13(1)23–28.

84 Liepe, H.-U. 1969. Schädlingsbefall in Cerealienpackungen. *Verpackungs-Rundschau* 20(9)1287–1288 (English summary).

85 Loschiavo, S. R. 1978. Effect of disturbance of wheat on four species of stored-product insects. *Journal of Economic Entomology* 71(6)888–893.

86 Lum, P. T. M. [1975.] Effect of visible light on the behavior of stored-product insects. In *Proceedings of the First International Working Conference on Stored-Product Entomology* (Savannah, GA, 1974), pp. 375–382.

87 McKenzie, B. A., et al. 1972. *Dryeration—Better Corn Quality with High Speed Drying*. Cooperative Extension Service AE-72. Purdue University, West Lafayette, IN.

88 Mills, J. T. 1980. Bin fires: a case history. *Country Guide* (West ed.) 99(8)27–28.

89 Mills, R. B. 1979. Potential and limitations of the use of low temperatures to prevent insect damage in stored grain. In *Proceedings of the Second International Working Conference on Stored-Product Entomology* (Ibadan, Nigeria, 1978), pp. 244–259.

90 Milward, A. F., A. J. B. Rudge, and J. K. Taylor. 1973. The use of pesticides in bakeries. *Baking Industries Journal* 6(Aug.)8–10.

91 Monro, H. A. U. 1969. *Manual of Fumigation for Insect Control*. FAO Agricultural Studies No. 79. Food and Agriculture Organization, Rome.

92 Morganstern, K. H. 1978. Economics of electron accelerators in the preservation of food by irradiation. In *Proceedings of an International Symposium on Food Preservation by Irradiation Organized by IAEC, FAO, and WHO* (Wageningen, The Netherlands, 1977), vol. 2, pp. 267–283.

93 Moysey, E. B. 1973. Storage and drying of oilseeds. In *Grain Storage—Part of a System*, R. N. Sinha and W. E. Muir, eds. AVI, Westport, CT.

94 Muir, W. E., et al. 1980. Emergency farm structures for storing grain—a multidisciplinary evaluation. *Transactions of the American Society of Agricultural Engineering* 23:208–213, 217.

95 Muir, W. E., G. Yaciuk, and R. N. Sinha. 1977. Effects on temperature and insect and mite populations of turning and transferring farm-stored wheat. *Canadian Agricultural Engineering* 19(1)25–28.

96 Mullen, M. A. 1973. Low frequency sound affecting the development of the Indian meal moth. *Journal of the Georgia Entomological Society* 8(4)320–321.

97 Mullen, M. A., and R. T. Arbogast. 1979. Time-temperature-mortality relationships for various stored-product insect eggs and chilling times for selected commodities. *Journal of Economic Entomology* 72(4)476–478.

98 Nair, K. K., and Subramanyam. 1963. Effects of variable dose-rate on radiation damage in the rust-red flour beetle, *Tribolium castaneum* Herbst. In *Radiation and Radioisotopes Applied to Insects of Agricultural Importance*. International Atomic Energy Commission, Vienna, Austria.

99 Navarro, S. 1977. Distribution and abundance of insects in butyl-rubber/EPDM silos containing wheat. *Proceedings of the XV International Congress of Entomology* (Washington, DC, 1976), pp. 680–687.

100 Navarro, S., and E. Donahaye. 1976. Conservation of wheat grain in butyl-rubber/EPDM containers during three storage seasons. *Tropical Stored Products Information* 32:13–23.

101 Navarro, S., E. Donahaye, and M. Calderon. 1973. Studies on aeration with refrigerated air—II. Chilling of soybeans undergoing spontaneous heating. *Journal of Stored Products Research* 9(4)261–268.

102 Navarro, S., M. Gonen, and A. Schwartz. 1979. Large-scale trials on the use of controlled atmospheres for the control of stored grain insects. In *Proceedings of the Second International Working Conference on Stored-Product Entomology* (Ibadan, Nigeria, 1978), pp. 260–270.

103 Nelson, S. O. 1972. Possibilities for controlling stored-grain insects with RF energy. *Journal of Microwave Power* 7(3)231–239.

104 Nelson, S. O. 1973. Insect-control studies with microwaves and other radiofrequency energy. *Bulletin of the Entomological Society of America* 19(3)157–163.

105 Nelson, S. O. 1976. Microwave dielectric properties of insects and grain kernels. *Journal of Microwave Power* 11(4)299–303.

106 Nelson, S. O. 1977. Dielectric properties of wheat and possibilities for control of stored-grain insects by dielectric heating. *Proceedings of the Tenth National Conference on Wheat Utilization Research* (Tucson, AZ).

107 Nelson, S. O., and L. E. Stetson. 1974. Possibilities for controlling insects with microwaves and lower frequency RF energy. *IEEE Transactions on Microwave Theory and Techniques MTT* 22(12)1303–1305.

108 Noyes, R. T., B. L. Clary, and G. W. Cuperus. [1987.] Maintaining Quality of Stored Grain by Aeration. OSU Extension Facts No. 1100. Oklahoma State University, Stillwater.

109 Ondráček, J., et al. 1976. Importance of antennae for orientation of insects in a non-uniform microwave electric field. *Nature* 260(5551)522–523.

110 Osmun, J. V. 1972. Physical methods of pest control. *Journal of Environmental Quality* 1(1)40–45.

111 Press, A. F., Jr., and P. K. Harein. 1967. Mortality of *Tribolium castaneum* (Herbst) (Coleoptera, Tenebrionidae) in simulated peanut storages purged with carbon dioxide and nitrogen. *Journal of Stored Products Research* 3(2)91–96.

112 Pye, J. D. 1979. Why ultrasound? *Endeavour* 3(n.s.)(2)57–62.

113 Rai, P. S., et al. 1972. Lethal effects of radiofrequency energy on eggs of *Tenebrio molitor* (Coleoptera: Tenebrionidae). *Annals of the Entomological Society of America* 65(4)807–810.

114 Rai, P. S., et al. 1974. Cytopathological effects of radiofrequency electric fields on reproductive tissue of adult *Tenebrio molitor* (Coleoptera: Tenebrionidae). *Annals of the Entomological Society of America* 67(4)687–690.

115 Rai, P. S., et al. 1975. Effects of radiofrequency electrical treatment on fecundity of *Tenebrio molitor* L. (Coleoptera: Tenebrionidae). *Annals of the Entomological Society of America* 68(3)542–544.

116 Rai, P. S., et al. 1977. Spermatozoan activity and insemination in *Tenebrio molitor* following radiofrequency electrical treatment (Coleoptera: Tenebrionidae). *Annals of the Entomological Society of America* 70(2)282–284.

117 Sacharow, S. 1970. Frucht- und Nusskonfekt verlangen ausgereifte Verpackungen. *Neue Verpackung* 20(4)442–443.

118 Sinha, R. N. 1973. Interrelations of physical, chemical and biological variables in the deterioration of stored grains. In *Grain Storage—Part of a System*, R. N. Sinha and W. E. Muir, eds. AVI, Westport, CT.

119 Sinha, R. N., et al. 1979. Storability of farm-stored hulless oats in Manitoba. *Canadian Journal of Plant Science* 59(4)949–957.

120 Smith, F. J. 1977. How processing with microwave heat affects food qualities. *Food Product Development* 11(2)60, 62, 78.

121 Smith, K. G. 1974. Containerization of cargo and its effects on the control of stored product insect pests in international trade. *Tropical Stored Products Information* 28:31–36.

122 Smith, L. B. [1975.] The role of low temperatures to control stored food pests. In *Proceedings of the First International Working Conference on Stored-Product Entomology* (Savannah, GA, 1974), pp. 418–430.

123 Storey, C. L. 1973. Exothermic inert-atmosphere generators for control of insects in stored wheat. *Journal of Economic Entomology* 66(2)511–514.

124 Thomas, A. M. 1952. *Pest Control by High-Frequency Electric Fields—Critical Resume.* Technical Report W/T23. British Electrical and Allied Industrial Research Association, Leatherhead, England.

125 Tilton, E. W. [1975.] Achievements and limitations of ionizing radiation for stored-product insect control. In *Proceedings of the First International Working Conference on Stored-Product Entomology* (Savannah, GA, 1974), pp. 354–361.

126 Tilton, E. W. 1979. Current status of irradiation for use in insect control. In *Proceedings of the Second International Working Conference on Stored-Product Entomology* (Ibadan, Nigeria, 1978), pp. 218–221.

127 Tilton, E. W., et al. 1972. Combination of gamma and microwave radiation for control of the Angoumois grain moth in wheat. *Journal of Economic Entomology* 65(2)531–533.

128 Tilton, E. W., J. H. Brower, and R. R. Cogburn. 1974. Insect control in wheat flour with gamma irradiation. *International Journal of Applied Radiation and Isotopes* 25(7)301–305.

129 Tilton, E. W., J. H. Brower, and R. R. Cogburn. 1978. Irradiation disinfestation of cornmeal. *Journal of Economic Entomology* 71(4)701–703.

130 Urbain, W. M. 1986. *Food Irradiation,* Academic Press, Orlando, FL.

131 Vas, K. 1977. Food irradiation—technical and legal aspects. *Food and Nutrition* 3(3)28.

132 Watson, D. L. 1979. An industry viewpoint on professionalism and ethics in entomology. *Bulletin of the Entomological Society of America* 25(3)235–236.

133 Watters, F. L. 1963. The cooling of heating grain by transfer during cold weather. *Journal of Economic Entomology* 56(2)215–219.

134 Watters, F. L. 1966. The effects of short exposures to sub-threshold temperatures on subsequent hatching and development of eggs of *Tribolium confusum* Duval (Coleoptera, Tenebrionidae). *Journal of Stored Products Research* 2(2)81–90.

135 Watters, F. L. 1972. Control of storage insects by physical means. *Tropical Stored Products Information* 23:13–28.

136 Watters, F. L. 1976. *Insects and Mites in Farm-Stored Grain in the Prairie Provinces.* Canada Department of Agriculture Publ. No. 1595, 25 pp.

137 Watters, F. L. 1976. Microwave radiation for control of *Tribolium confusum* in wheat and flour. *Journal of Stored Products Research* 12(1)19–25.

138 Watters, F. L., E. Adem, and R. Uribe. 1979. Potential of accelerated electrons for insect control in stored grain. In *Proceedings of the Second International Working Conference on Stored-Product Entomology* (Ibadan, Nigeria, 1978), pp. 278–286.

139 Watters, F. L., and M. Bickis. 1978. Comparison of mechanical handling and mechanical handling supplemented with malathion admixture to control rusty grain beetle infestations in stored wheat. *Journal of Economic Entomology* 71(4)667–669.

140 Whitney, W. K. [1975.] A general survey of physical means for control of storage pests. In *Proceedings of the First International Working Conference on Stored-Product Entomology* (Savannah, GA, 1974), pp. 385–411.

141 Wilkin, D. R. 1975. The effects of mechanical handling and the admixture of acaricides on mites in farm-stored barley. *Journal of Stored Products Research* 11(2)87–95.

142 World Health Organization. 1988. *Food Irradiation, a Technique for Preserving and Improving the Safety of Food.* WHO, Geneva, Switzerland.

143 Yerington, A. P. 1973. The effects of moisture content and construction on the insect resistance of rice packages. *Rice Journal* 76(8)22–24.

144 Žďárek, J., J. Ondráček, and J. Ďatlov. 1976. Differential sensitivity to microwaves in the fleshfly larva tissues. *Entomologia Experimentalis et Applicata* 20(3)270–274.

30
Chemical Control of Insect Pests in Bulk-Stored Grains

Phillip K. Harein

The chemical protection of stored grains has been reviewed by Snelson (6) and by scientists at the Overseas Development Natural Resources Institute (2). Newer chemical protectants are being tested against insect pests of stored grains and legumes (1, 3). To help protect people, pets, livestock, wildlife, and the environment from harm, the use of pesticides is regulated by several laws, such as the Federal Insecticide, Fungicide and Rodenticide Act (FIFRA) and the Federal Food, Drug and Cosmetic Act (FD&C Act). The instructions given below will assist the user in complying with the regulatory requirements for rate and method of application and any resulting residues.

Treatment Before Storage

Any grain that is to be stored for more than 6 months can become seriously infested. Thus, the key to good storage is anticipating and preventing problems through good bin management.

New grain should never be placed on old grain unless the old grain is largely free from insect infestation. Market grain should be stored away from the farmstead, but if this is not possible, it should be stored as far away as possible from feed rooms and feed bins, both likely sources of new infestations.

Before the grain is treated with protectants, the storage structure should be virtually free of insect-infested grain. Leftover grain should be removed from

Harein: Department of Entomology, University of Minnesota, St. Paul, MN 55108.
Modified from Raney et al. (5).

the bins, and floors and walls should be swept and vacuumed. This cleanup is most effective if done in early spring or immediately after the bin has been emptied.

All grain-handling equipment, including augers, combines, trucks, and wagons, also should be cleaned and grain residues removed before harvest. Places where seeds, livestock feed, and pet foods are stored can be serious sources of infestations. Grain and feed accumulations that are frequently overlooked include empty feed sacks, dusts created by the feed grinders, seed litter from the haymowers, feed left in animal self-feeders, and grain-based rodenticides.

Bin Wall, Ceiling, and Floor Treatments

As soon as the bin is cleaned, it can be treated with protective insecticides. It is best to treat during the warmer months when insects are active, but if treatments have been applied more than 3 months earlier, an additional treatment should be applied 2–3 weeks before new grain is placed in the bin. The treatment will kill insects emerging from their hiding places (cracks, crevices, under floors, and in aeration systems) as well as those crawling or flying in from the outside.

Malathion (premium grade) used for this purpose in metal bins will remain effective for 4–8 weeks at 27°. In wooden bins, malathion may remain effective for up to 6 months. Chlorpyrifos-methyl is registered for treatment of bins used to store barley, oats, rice, sorghum, wheat, and any other grains on which chlorpyrifos-methyl has been approved for use. Pirimiphos-methyl is not registered for use as a prestorage bin treatment.

The spray should be applied to as many surfaces as possible, especially joints, seams, cracks, ledges, and corners. Walls and floors should be sprayed to the point of runoff. Cracks and crevices should be subjected to a coarse spray at a pressure of under 30 pounds per square inch (psi).

The area beneath the bin, the bin supports, and a 6-foot border around the outside foundation should be sprayed. Outside surfaces, especially cracks and ledges near the door and fans, should be treated as well as pertinent areas in cleaned harvesting equipment, elevators, augers, trucks, or wagons.

Insecticides, formulations, concentrations, and rates of application approved for this and other uses are subject to change. Label directions should always be followed to comply with current recommendations.

Empty Metal Grain Bins

The increased use of metal bins with perforated floors for grain drying and aeration has contributed to a serious insect problem in farm-stored grain. Grain dockage (broken kernels, grain dust, and chaff) sifts through the floor perforations and collects in the subfloor plenum creating a favorable environment for insect development. If possible, the perforated floors should be removed to clean the plenum area and spray it with an insecticide. When the floor or the screen over aeration ducts cannot be removed, the area may be treated with a fumigant.

Treatment During Storage

Insect infestation is prevented or reduced by treating the product *as it is moved into storage* with one of three approved insecticides (see labels for application rates, permitted target commodities, and other essential information):

(a) Chlorpyrifos-methyl (Reldan) [emulsifiable concentrate (EC) or dust] is registered for application to wheat, oats, rice, sorghum, and barley (*not* corn). It works well even on high-moisture grains and is effective against insects resistant to malathion.

(b) Malathion (EC or dust) is labeled for use on corn and other small grains. Water is required as a diluent for malathion EC. Water added to the grain at the maximum recommended rate of 5 gallons per 1,000 bushels will increase the moisture content of the grain by less than 0.1%.

(c) Pirimiphos-methyl (Actellic) (EC) is registered for application to corn and sorghum.

Effectiveness

The effectiveness of treating bulk grain depends on at least five factors.

(a) *Proper mixing.* Application so thorough and complete that the protectant reaches nearly every kernel is not necessary. This became apparent when the drip-on application procedure for liquid protectants was found to be adequate on the basis of both insect kill and residue analysis. One disadvantage of emulsifiable formulations is that most of them must be agitated to avoid settling. Consequently, gravity-flow or "drip-on" applicators and pressure-type sprayers must be shaken periodically to ensure that the formulation is mixed evenly. Power sprayers do not have this problem because the formulation is agitated continuously.

(b) *A fresh spray mixture.* Only enough insecticide for one day's use should be mixed; excess insecticide mixture should *not* be used for the next day's treatment. The concentrated insecticide, mixed spray, or insecticide dust should be kept cool and not stored in direct sunlight. Fresh dust formulations should be used, and carry-over from one year to the next should be avoided.

(c) *Point of application.* Protectants can be added to the grain just before it reaches final storage by applying them at either end of the elevating equipment. Less residual deposits are lost with grain dust, chaff, and broken kernels if the protectant is applied at the end of the auger just before the grain drops into the bin. Also, grain that is treated and then transferred long distances through numerous grain-handling systems (e.g., pneumatic systems, belt augers, conveyors, spouts, legs) before storage will have less insecticide residue when the grain is finally dropped into the bin. Any insecticide left in the handling systems will help reduce insects in these areas.

(d) *Application pressure.* If other than a gravity-flow system is used, the spray pressure should be as low as possible, preferably 10–20 psi. With low spray pressures, larger spray droplets are produced. The larger droplets fall on the grain and are less likely to drift off into the air.

(e) *Moisture and temperature of the grain.* Most protectant failures occur because of excessive grain moisture and/or temperatures. Grain should not be treated if its moisture content (mc) is above 13% (16% for corn) and its temperature is above 32.5°. If warm grain is treated, it should be cooled by aeration as soon as possible. The operation of an aeration system will not remove the protectant from the grain. A 10-ppm (parts per million) treatment with malathion on grain at 10% mc and 15.7° can degrade to 6.3 ppm in one year. The same treatment on 14% mc grain at 27° degrades to 0.2 ppm in one year. As a general rule, the higher the moisture content and the higher the temperature, the faster the rate of malathion degradation. A 3-ppm residue will prevent the development of most stored-grain insects (except the Indianmeal moth, *Plodia interpunctella*), but grain treated at 14% and 16% mc will be protected for only short periods, depending on the temperature.

Application Procedures

Liquid Formulations

Any low-pressure sprayer that can be calibrated to deliver a known volume of liquid is suitable for application of protectants. This includes compression-type sprayers and gasoline-engine-driven power sprayers. Garden-type compression sprayers can be used for treating small lots of grain, whereas power sprayers and metering-type sprayers are generally more useful when large bulks of grains are to be treated.

It is important to have the correct orifice size in the sprayer nozzle, because both orifice size and pressure regulate the rate of insecticide flow. Every manufacturer of spray nozzles has nozzle charts giving the capacity in gallons per minute and the spray angle for each size of orifice. Nozzle selection can also determine the spray particle size.

A simple gravity or drip-on applicator (with no moving parts) can be purchased or constructed. An application system can be built by fitting two brass valves and polyethylene tubing in sequence to an opening in the bottom of a plastic jug. These fittings are readily obtainable at a plumbing supply store. The upper shutoff cock on the jug serves as the on/off valve, and the lower needle valve regulates the amount of insecticide flowing through the plastic tubing. The needle valve is first calibrated to the desired flow, based on the rate of grain delivery into storage. It then can be kept at the same setting, without the need for fine adjustment each time the flow is turned on.

The gravity-feed applicator is used to treat grain as it is unloaded from a truck by means of an auger. The tubing is taped horizontally along the auger tube at the pick-up end, with the end of tubing extending ¼ inch beyond the end of the auger tube so that insecticide flows directly onto the grain. The plastic container can be suspended from the top of the grain bin or auger and must be agitated periodically to keep the formulation mixed. See Raney et al. (5) for detailed directions on the construction of a drip-on applicator.

Installation of Nozzles Sometimes, two or more nozzles are needed to obtain the proper application rate or to get adequate coverage on a wide belt. When spraying grain on a moving belt, the nozzle should be kept 6–8 inches (15.2–20.3 cm) above the belt, and the spray should be angled against the flow of grain. The nozzle should be set so that the spray pattern covers the entire width of the stream of grain but does not touch the belt. It is extremely important to reduce and direct air movement above the belt around the spray nozzle to keep the spray from drifting off into the air instead of onto the grain. This can be done by placing a baffle across the belt several inches above the grain. This baffle deflects the air and allows the spray to fall directly on the grain.

Calibration of Sprayers When the turning rate of the grain passing on the belt, auger, or conveyor is known (i.e., the amount of grain passing a point in a given time), a nozzle should be chosen from the manufacturer's chart which will deliver the gallons per hour needed in the range of 10–20 psi. For example, if the turning rate of the grain is 5,000 bu/hr and the amount of insecticide to be applied is 2 gallons per 1,000 bushels, then a nozzle with a capacity of 10 gal./hr at 15 psi is required.

The nozzle should deliver 10 gal./per hr at as low a pressure as possible. It should be installed on the sprayer and, using plain water, the sprayer should be operated at the pressure recommended. All water delivered in 10 minutes should be caught, weighed or measured, and the amount multiplied by 6 to determine the amount delivered per hour. The output of the nozzle will vary slightly from the rated output. If the amount collected was more than a gallon, the pressure should be decreased slightly and the test rerun. If the amount collected was less than a gallon, the pressure should be increased slightly and the test rerun. This procedure should be repeated until the sprayer delivers the correct amount. Calibration of the pressurized garden-type sprayer can be done as above, or the entire unit can be weighed on a platform or hanging scale to determine the amount delivered. One gallon of water weighs about 8.3 pounds.

Dust Formulations

For small lots of grain, a special dust applicator can be used. Several different models, available from various manufacturers, utilize the 1% or 2% formulations of malathion (consult the label for specific application rates). These dust formulations are sometimes applied prior to binning by spreading them evenly over the grain surface while the grain is in a truck. Then it is mixed in with a shovel and further mixed as it falls into the auger hopper. A large dust applicator using the 6% formulation is more suitable for treating large quantities of grain as in a grain elevator. See Ramzan et al. (4) for a description of an experimental field trial of a 5% dust formulation of malathion. The results of many other studies germane to this chapter and to this Bulletin in general have been summarized by Stein (7, 8).

Safety

Insecticides are poisonous. They should be used only when needed and should be handled with extreme care. Before selecting or using pesticides, current labels should be consulted and directions and precautions followed carefully.

When handling or mixing any insecticide concentrate, contact with the skin, eyes, nose, and mouth should be avoided. In case of spills, skin should be washed and contaminated clothing changed immediately. Eyes should be flushed with plenty of water for 15 minutes and medical attention sought.

To date, the insecticides mentioned have not been designated as "restricted" under FIFRA. Malathion, Reldan, and Actellic can be applied safely without special protective clothing or devices when they are diluted and used as directed on their respective labels. However, appropriate respiratory protection is recommended when these insecticides are applied in the enclosed head spaces of grain storage bins. Grain bins are dangerous places in which to work; see Raney et al. (5) for a discussion of important safety precautions.

Surface Treatment

Immediately after the bin is filled and the grain leveled, a surface treatment (top dressing) of malathion should be applied as a grain protectant. A surface treatment can also be applied when the grain is going to be stored through a warm season or after a general fumigation to help prevent insect reinfestation. The treatment will help control insects that enter the grain through roof openings.

Surface treatments are effective, but the following limitations and conditions apply:

(a) They will not control insects already in the storage bin; thus, the grain must not be infested prior to surface treatment.

(b) The storage structure must be insect-tight below the treated 2 inches of grain.

(c) The surface treatment should not be disturbed since it provides the protective barrier against insect infestations.

In addition, malathion surface treatment probably will not control or prevent an infestation of the Indianmeal moth or the almond moth (*Cadra cautella*) because they are resistant to malathion. Reldan and Actellic are effective against insects resistant to malathion, but neither has label approval by EPA for surface treatments. To get the best results from a top dressing of malathion, the grain bulk should be fumigated first. Then, the required amount of diluted insecticide (based on the surface area to be protected and the application rates given on the label) should be mixed and this quantity divided into two equal portions. One portion should be applied to the grain surface and raked in to a depth of 4 inches. The second portion should be applied to the grain surface and special care taken thereafter to make sure that the surface remains undisturbed.

The same equipment used to spray the bin wall can be used to apply the surface spray. Surface treatments should be applied only one time during a storage year.

Indianmeal Moth Control

Indianmeal moths infest exposed areas of the grain mass and grain residues, such as the grain surface, aeration ducts, and materials beneath false floors. The worst infestations are found on the grain surface. The larvae produce masses of webbing that constrict air flow and make fumigation or surface treatments less effective. Damage by these insects to farm-stored grain is low compared to that caused by beetles or weevils. Compared to other grain pests, Indianmeal moths usually cause serious economic damage only when they infest seed grains.

To aid in control of the adult moths that emerge or invade the overspace, a dichlorvos (DDVP) resin strip (Pest STRIP, BIO-STRIP, Reno, NV, and Loveland Industries, Greely, CO) should be suspended above the grain surface. These strips slowly release vapor that kills the adult moths. They are most effective when used at the rate of one strip for every 1,000 ft³ of air space in situations where the space in which the strips are hung has minimal or no air exchange or movement.

Cited References

1 Arthur, F. H., and H. B. Gillenwater. 1990. Evaluation of esfenvalerate aerosol for control of stored product insect pests. *Journal of Entomological Science* 25(2)262–267.

2 Food and Agriculture Organization. 1985. *Manual of Pest Control for Food Security Reserve Grain Stocks.* FAO Plant Production and Protection Paper 63. FAO, Rome, Italy.

3 Giga, D. P., and P. Zvoutete. 1990. The evaluation of different insecticides for the protection of maize against some stored product pests. *International Pest Control* 32(1)10–13.

4 Ramzan, M., B. S. Chalal, and B. K. Judge. 1989. Field evaluation of synthetic pyrethroids for the protection of stored wheat seed against storage pests. *International Pest Control* 31(4)87–89.

5 Raney, H. G., et al. 1987. *Management of On-Farm Stored Grain.* Cooperative Extension Service, College of Agriculture, University of Kentucky, Lexington, KY.

6 Snelson, J. T. 1987. *Grain Protectants.* Australian Centre for International Agricultural Research, Canberra.

7 Stein, W. 1987. New results about stored products protection (animal pests). I. *Journal of Plant Diseases and Protection* 94(6)649–668 (in German).

8 Stein, W. 1988. New results about stored products protection (animal pests). II. *Journal of Plant Diseases and Protection* 95(6)651–669 (in German).

31
Chemical Control of Rodent Pests in Bulk-Stored Grains

Robert M. Timm

The concept of holistic rodent management is fully developed in Chapter 18. The focus in this chapter is on the chemical tools often used in rodent control and on some practical matters associated with their use [see also Raney (1) and Timm (2) for additional details].

Each grain storage area has its own unique geography and design, but certain features are common to all. Generally, there are large areas for parking or open space adequate for vehicles to load or unload grain. Railroad sidings or tracks can be found at larger grain elevators. Storage sheds, maintenance areas, office buildings, and other structures make preplanning essential for effective control.

Planning a Program

To facilitate systematic planning for rodent control, a map or diagram of the area should be drawn. It can be useful in determining prime areas of possible entry by rodents, such as drainage ditches or runoff pathways. In the process of making the diagram, the periphery of structures and adjacent areas should be inspected to see if rodent signs (e.g., tracks, trails, droppings, fresh gnawing) are present. Areas of sig-

nificant rodent activity can be indicated on the map, and control activities should be concentrated in these areas.

Sanitation

Clean, orderly facilities will not support large numbers of rats or mice. Attention to two key elements that rodents require for survival—*food* and *shelter*—is important in limiting the habitat's carrying capacity for these species. Sanitation involves good housekeeping, including proper storage and handling of food materials. Warehouses, granaries, grain mills, silos, port facilities, and similar structures may provide excellent habitats for commensal rodents. Grains should be stored in rodentproof structures whenever possible.

Shelter is an invitation for a rodent problem (especially when the shelter is near a food source). Although it may be difficult when structures are not rodentproof to prevent rodents from feeding on grains, it *is* possible to achieve good rodent control by taking away their shelter. When mice and rats are unable to hide or nest, they cannot remain in any location for very long. Regular removal of debris and control of weeds around structures will reduce the amount of shelter available to rodents.

In some instances, a strip of heavy gravel placed adjacent to building foundations or other structures will reduce rodent burrowing, especially that of Norway rats, at these locations. Gravel should be at least 1 inch (2.5 cm) in

Timm: Nebraska Cooperative Extension Service, Department of Forestry, Fisheries, and Wildlife, University of Nebraska, Lincoln, NE 68583. Present address: Hopland Field Station, University of California, 4070 University Road, Hopland, CA 95449.

419

Table 31.1
Nonanticoagulant rodenticides and some of their useful characteristics for controlling rats and mice

Common name	Chemical name	Active ingredient used in food bait (%)	Mode of action	Time to death	Bait acceptance	Bait shyness	Human hazard	Swine hazard	Rodents controlled: House mice	Norway rats	Roof rats
Bromethalin	n-Methyl-2,4-dinitro-n-(2,4,6-tribromophenyl)-6-trifluoromethyl benzenamine	0.01	Central nervous system depression and paralysis	2–4 days	good	none reported	moderate	unknown	yes	yes	yes
Cholecalciferol	9,10-Seco-5Z,7E,10(19)-cholestatrien-3β-ol	0.075	Mobilizes calcium, resulting in death from hypercalcemia	3–4 days	fair–good	none reported	low to moderate	unknown	yes	yes	yes
Red squill	Scilliroside glycoside[a]	10.0	Heart paralysis	24 hr	poor–fair	moderate to high	low	low	no	yes	no
Strychnine	Strychnine	0.25–1.0	Tetanic convulsions leading to respiratory failure	0.25–3 hr	fair	moderate to high	moderate to high	moderate	yes	yes	no
Zinc phosphide	Zinc phosphide	1.0–2.0	Phosphine gas enters circulatory system; heart paralysis, gastrointestinal and liver damage	0.5–20 hr	fair	moderate to high	moderate	moderate	yes	yes	yes

NOTE: Rodenticides such as ANTU, arsenic trioxide, and phosphorus are registered and available in some states; they are rarely used today because of their limited availability and low efficacy in most situations.
[a]Principal active ingredient.

diameter and laid in a band at least 2 feet (0.6 m) wide and 6 inches (15 cm) deep. In any event, keeping the periphery of buildings and other structures clear of weeds and debris (including stacked lumber, firewood, and other stored materials) will discourage rodent activity and allow easier detection of rodent signs. Cleaning up also puts rodents under stress. This makes it more likely that they will accept lethal quantities of baits (rodenticides) when these are used as part of a control program.

Water sources should be eliminated wherever possible. Rats require water daily for their survival, unless they are feeding on very moist foods. Although house mice can survive without water, they will drink it when they can. Drainage ditches and runoff areas must be constructed so that water flows quickly away from structures and so that no standing water accumulates. Where no sources of water are present, rats cannot thrive. Further, reduction of water sources enhances the effectiveness of liquid baits.

Types of Rodenticides

Both anticoagulant and nonanticoagulant rodenticides are available for rodent control. Although a variety of ready-to-use baits are available, knowledgeable rodent control personnel often prefer to mix their own baits using rodenticide concentrates. In most situations, ready-to-use commercial baits are preferred because they have proven efficacy and do not require that the applicator handle the concentrated toxicant, a more hazardous material.

Nonanticoagulant Rodenticides

Some nonanticoagulant rodenticides (Table 31.1) knock down a rodent population more quickly because they are effective with a single feeding and are relatively rapid in action. All will control anticoagulant-resistant rodent populations. They may be preferred where rodents are abundant or where it is difficult to get them to accept a bait for several days in succession (as may be necessary with some anticoagulants) because of competing food items. Use of some formulations is restricted by the U.S. Environmental Protection Agency (Restricted-Use Pesticides).

With most nonanticoagulant rodenticides, prebaiting with unpoisoned bait for several days before the rodenticide is offered will increase bait acceptance, thereby increasing control success. Because ''bait shyness'' or ''poison shyness'' may develop following a sublethal ingestion of some single-dose rodenticides such as zinc phosphide or strychnine, it is best not to use these more than twice per year at a given location, and preferably only once. Because non-anticoagulant rodenticides are generally more rapid in action and because first aid and antidotes are often less effective, some of these materials may be potentially more hazardous to humans, pets, or livestock if accidentally ingested.

Bromethalin is a slow-acting toxicant in comparison to most other nonanticoagulant compounds. For this reason, bait shyness is less of a problem than with other faster-acting compounds. Bromethalin can cause both chronic and acute effects. Acute effects, which occur following ingestion of a single dose equal to two or more times the LD_{50} or by generous bait consumption, include tremors, one or two episodes of clonic convulsions, prostration, and death (usually within 18 hours). Chronic effects occur following a single dose equal to the LD_{50} or after consumption of multiple smaller doses. They include lethargy, hind leg weakness, loss of muscle tone, and paralysis. These effects appear to be reversible if ingestion of the toxicant is discontinued. Formulated bromethalin baits can be hazardous to nontarget species, so care must be taken in their placement and use.

Cholecalciferol (vitamin D_3) also acts as either a chronic or acute toxicant. At the concentration used in the currently marketed baits [0.075% active ingredient (a.i.)], rodents need to feed over several days. Death from hypercalcemia occurs 3 or 4 days after ingestion of a lethal dose. Bait shyness is not reported to be a problem.

Red squill is a relatively safe and selective toxicant for use against Norway rats. The poison itself acts as an emetic, which provides some degree of protection to certain nontarget species that might accidentally consume the bait. Since rats are unable to vomit, they cannot rid themselves of the toxicant once it is ingested. In the past, one problem has been variation in the quality of the material, which is derived from a plant (*Urginea maritima*). Red squill must be stored in a sealed container, because moisture will cause it to lose potency. Efforts to develop highly refined scilliroside—the active ingredient in red squill—are under way and may result in a more effective rodenticide in future years.

Strychnine, although not recommended for rat control, can be used effectively against house mice, which apparently are not so likely to reject strychnine baits. It is usually formulated on can-arygrass seed (*Phalaris canariensis*) or other dry bait materials. Since strychnine is toxic to most species of mammals and birds, it must be used with extreme care. People unfamiliar with its use should not attempt house mouse control without professional assistance.

Zinc phosphide is a dark grey powder, insoluble in water, that has been used extensively in the control of rodents. It is available in concentrates and in ready-to-use dry baits. Its strong garlic-like odor appears to be attractive to rodents that are not bait-shy. Oils and fats make excellent binders for zinc phosphide and increase absorption of the toxicant when ingested. Although it is somewhat emetic, thus providing a safety factor to some nontarget species, care must be used to avoid accidental poisoning of pets, livestock, and wildlife.

Anticoagulant Rodenticides

Anticoagulant rodenticides (Table 31.2) cause death as a result of internal bleeding, which occurs as the animal's blood loses its clotting

Table 31.2

Anticoagulant rodenticides for controlling commensal rodents

Common or trade name	Chemical name	Active ingredient used in food bait (%)
Hydroxycoumarins		
Brodifacoum[a]	3-{3-[4'-Bromo(1,1'-biphenyl)-4-yl]-1,2,3,4-tetrahydro-1-naphthalenyl}-4-hydroxy-2H-1-benzopyran-2-one	0.005
Bromadiolone[a]	3-{3-[4'-Bromo(1,1'-biphenyl)-4-yl]-3-hydroxy-1-phenylpropyl}-4-hydroxy-2H-1-benzopyran-2-one	0.005
Coumafuryl (Fumarin)	3-(α-Acetonylfurfuryl)-4-hydroxycoumarin	0.025
Prolin	3-(α-Acetonylbenzyl)-4-hydroxycoumarin + sulfaquinoxaline (0.025%)	0.025
Warfarin	3-(α-Acetonylbenzyl)-4-hydroxycoumarin	0.025
Indandiones		
Chlorophacinone	2-{(p-Chlorophenyl) phyenylacetyl}-1,3-indandione	0.005
Diphacinone	2-Diphenylacetyl-1,3-indandione	0.005
Pindone (Pival)	2-Pivalyl-1,3-indandione	0.025
Valone (PMP)	2-Isovaleryl-1,3-indandione	0.055

NOTE: All anticoagulants have the same mode of action and all have a delayed time to death, although some are slightly faster-acting than others. All anticoagulants are effective against house mice, Norway rats, and roof rats. They all are considered to have good bait acceptance, to produce no bait shyness, and to be low hazards to humans.
[a]Effective against warfarin-resistant rats and mice.

ability and capillaries are destroyed. Prior to death, the animal exhibits increasing weakness due to blood loss. Animals killed by anticoagulants may show extreme lack of color of the skin, muscles, and viscera. Hemorrhage may occur in any part of the body.

Several anticoagulant compounds have been registered for the control of commensal rodents. The active ingredients are used at very low levels, and bait shyness does not occur, primarily because of their slow action. With the exceptions noted below, multiple feedings over a period of several days are usually required to cause death. Relatively low, chronic doses are fatal, whereas the same amount of toxicant ingested at a single feeding may produce no significant effects. Feeding does not have to be on consecutive days, but if the toxicant is consumed daily, death may occur as early as the third or fourth day.

Anticoagulants have been the most preferred materials for controlling commensal rodents since World War II. They are easy to use, come in a variety of formulations, and if used properly are safe around livestock, pets, and humans. Anticoagulants have the same effect on nearly all warm-blooded animals, but sensitivity to specific toxicants varies among species. If misused, anticoagulants can be lethal to nontarget animals such as dogs, pigs, and cats. Additionally, residues of anticoagulants that are present in the bodies of dying or dead rodents can cause toxic effects in predators and scavengers. In general, however, the potential secondary hazard from anticoagulants is low. Treatment of accidental anticoagulant toxicity may include administration of vitamin K or blood transfusion.

When multiple-dose anticoagulant rodenticides are used, bait must be available continuously until all rodents stop feeding. This usually takes at least 2 weeks, and 100% control is possible. However, this is rarely the case with nonanticoagulant rodenticides, and hence, anticoagulants are often used as a follow-up to other types of control.

Bromadiolone and brodifacoum are two somewhat new anticoagulants. They are exceptional in their ability to cause death after a single feeding, although as with other anticoagulants, death usually does not occur for at least 4 days. Because they are effective in a single dose, they may be somewhat more hazardous to certain susceptible nontarget species than are some other anticoagulants. A technique of pulsed baiting with these compounds has a number of advantages over standard baiting methods. Numerous small bait placements are made at approximately 1-week intervals. This reduces the total amount of bait used, making it less likely that a single rodent will ingest far in excess of a lethal dose. Further, it reduces the potential hazard to nontarget species.

Chlorophacinone and diphacinone, both indandiones, are similar in potency and are more toxic than the earlier-developed anticoagulants (e.g., warfarin and coumafuryl). Thus, they are formulated at lower concentrations.

Coumafuryl, a hydroxycoumarin, is approximately equal to warfarin in effectiveness. Pindone is less potent than chlorophacinone or diphacinone and is regarded as slightly less effective than warfarin against Norway rats. It has some properties that resist the growth of mold and insects in prepared baits. Valone is less palatable to rodents than the other compounds, but it is inexpensive to produce. For this reason, it has been used primarily in tracking powders.

Warfarin was the first marketed anticoagulant and is, therefore, the best known and most widely used of its generation of compounds. Its effectiveness is sometimes diminished by the presence of impurities that reduce bait acceptance. This has been resolved with the development of microencapsulated warfarin.

The development of Prolin was an attempt to improve the effectiveness of warfarin by adding an equal concentration of an antibacterial agent to inhibit vitamin K–producing microbes that live in the gut of rodents. Its superiority over warfarin has been difficult to prove outside the laboratory.

Resistance to Anticoagulants

Within any population of rats or mice, some individuals are less sensitive to anticoagulants than others. Where anticoagulants have been used over long periods at a particular location, the potential exists for resistance to the lethal effects of baits to develop in a population. Such populations of Norway rats have been identified at many locations throughout the United States (USA) since 1971.

With some exceptions, resistance to any single anticoagulant means that the rodent will be resistant to the other compounds. Two exceptions are brodifacoum and bromadiolone, to which no resistance has been reported to date in the USA. These compounds are lethal to rodents that are resistant to any or all of the other anticoagulants. In addition, all of the nonanticoagulant rodenticides are effective in controlling anticoagulant-resistant rodents.

Reasons for Failure of Anticoagulant Baiting

Anticoagulant resistance is only one (and perhaps the least likely) reason that anticoagulant baits may fail to control rodents. Failure with baits that are highly accepted may be attributed to one or more of the following reasons:

(a) The time period that bait is exposed is too short.
(b) Insufficient bait is made available, and none remains from one baiting to the next (insufficient replenishment of bait).
(c) Too few bait stations are used; some are too far apart. In some situations, stations may have to be within 25–50 feet (7.6–15.2 m) of one another. For house mice, bait stations should be placed no more than 10 feet (3 m) apart, and preferably 6–8 feet (1.8–2.4 m).
(d) The control program does not cover a large enough area, permitting rodents to move in from untreated adjacent areas.
(e) Some rodents are genetically resistant to the anticoagulant. Although this is unlikely, resistance should be suspected if about the same amount of bait is taken daily for a number of weeks.

Failure to achieve control with anticoagulant baits that are poorly accepted may be due of one or more of the following reasons:

(a) The bait used is a poor choice or is formulated improperly; other foods are more attractive to the rodents.
(b) The bait stations are not placed properly. Other foods are more convenient to the rodents.
(c) Other foods for rodents are abundant.
(d) The bait has become moldy, rancid, insect-infested, or contaminated with other material that reduces acceptance.

Old bait should be discarded periodically and replaced with fresh bait.

Occasionally, rodents may accept bait well and an initial population reduction will be achieved. Then, at some point, bait acceptance stops, although some rodents are still present. In such instances, it is likely that the remaining animals never accepted the bait either because of its formulation or location of placement. The best strategy is to switch to a different bait formulation, make bait placements at different locations, and/or use other control methods such as traps and burrow fumigants.

Baits

Selection and Formulations

Contrary to popular belief, rats and mice prefer fresh, high-quality foods and will reject spoiled or inferior food items when given a choice. Therefore, rodent baits should be made from high-quality food materials.

Corn, oats, wheat, and barley are the grains most preferred by Norway rats. House mice often prefer foods high in fat, protein, or sugar, as well as seeds and grain. Canarygrass seed, in particular, is a good bait for house mice. Preference will vary between populations and among individual rodents. Bait materials similar to what rodents are accustomed to eating are often a good choice, particularly if the normal food of the rodents is limited or can be made less available to them.

Experienced rodent control personnel often prefer to mix their own baits, a common mixture being ground cereal with 5% powdered sugar and 3–10% vegetable oil. To this is added the proper amount of a toxicant concentrate. Certain anticoagulants and single-dose poisons can be purchased in concentrate form for use in formulating baits.

Under some conditions, baits made with fruits, vegetables, meat, or fish may be highly accepted. However, use of such bait materials may increase the risk of poisoning cats, dogs, livestock, and other nontarget species. To determine bait preference of rodents, a bait-choice test should be conducted by placing about 4 oz (115 g) of each of several nontoxic candidate baits about 1 foot (30 cm) apart in several locations where rodents are present. Check baits the next day to find out which foods were most readily accepted. Rats are suspicious of new objects and novel foods, and therefore, they may not accept a new bait until the third or fourth day. House mice are nibblers; they usually do not hesitate to sample various types of food they encounter.

The ready-to-use baits most available to the public contain anticoagulant rodenticides. Several types are available. Grain-based baits in a loose meal or pelleted form are available in bulk or are packaged in small, 4- to 16-oz (112- to 454-g) plastic, cellophane, or paper "place packs." These packets keep bait fresh and make it easy to place baits into burrows, walls, or other locations. Rodents gnaw into these bags to feed on the attractive bait. Pelleted baits can more easily be carried by rodents to another location. Such hoarding of food by rats is not uncommon, and it may result in some of the bait being moved to places where it is undetected or difficult to recover and may, if accessible, be a hazard to nontarget species.

Anticoagulant baits also have been formulated into paraffin or wax blocks. These are particularly useful in sewers or where moisture may cause loose grain baits to spoil. Rodents accept paraffin block baits less readily than loose or pelleted grain baits.

Sodium salts of anticoagulants are available as concentrates to be mixed with water, making a liquid bait. Since rats require water daily, they can be drawn to water stations in some situations. Water baits are particularly useful in grain storage structures, warehouses, and other such locations where water is scarce. Rodents are more easily able to detect anticoagulants in water baits than in food baits; therefore, up to 5% sugar is sometimes added to liquid baits to increase acceptance of the bait solution. Since water is consumed by most animals, water baits should be used in ways that prevent nontarget animals from drinking them.

Bait Stations

Bait stations (bait boxes) may increase both the effectiveness and the safety of rodenticides. They came into general use after the development of the first anticoagulants which required that a continuous supply of bait be made available to rodents. Bait stations are useful because they

(a) protect bait from moisture and dust;
(b) provide a protected place for rodents to feed, allowing them to feel more secure;
(c) keep other animals (e.g., pets, livestock, desirable wildlife) and children away from bait which may be hazardous to them;
(d) allow placement of the bait in some locations where it would otherwise be difficult because of weather or potential hazards to nontarget animals;
(e) help prevent the accidental spilling of bait;
(f) permit inspection of the bait to determine whether rodents are feeding on it.

Types Bait stations can contain solid baits (food baits), liquid baits, or both. Manufactured bait boxes made of plastic, cardboard, or metal are sold to pest control companies and to the public in sizes for rats or mice. Some farm supply and agricultural chemical supply stores have them in stock or can order them.

Bait boxes can be built from scrap materials, and homemade stations can be designed to fit particular needs. Sturdy materials should be used so they cannot be knocked out of place easily or damaged. Where children, pets, or livestock are present, the stations should be constructed so that the bait is accessible only to rodents. Locks, seals, or concealed latches are often used to make bait boxes more nearly tamperproof. In some situations, stations should be secured in place. Where the rodenticide label so specifies, use tamperproof bait boxes. Clearly label all bait boxes or stations with "POISON" or "RODENT BAIT—DO NOT TOUCH," or some similar warning.

Design Bait stations should be large enough to allow several rodents to feed at once. They can be as simple as a flat board nailed at an angle to the bottom of a wall or a length of pipe into which bait can be placed. More elaborate stations are completely enclosed and can hold liquid as well as solid baits. A hinged lid with a childproof latch can be used for convenience in inspecting permanent stations.

Bait stations should have at least two openings approximately 2.5 inches (6 cm) in diameter for rats, and 1 inch (2.5 cm) for mice. Making these two holes on opposite sides of the station is preferable, because the rodent can see an alternate escape route as it enters the station.

Maintenance As stated above, anticoagulant baits must be fresh and of high quality. Rats, in particular, will reject spoiled or stale foods. Enough fresh bait should be provided for rodents to eat all they want. When the bait boxes are first placed, they should be checked daily and fresh bait added as needed. After a short time, rodent numbers and feeding will decline, and the boxes will need to be checked only every 2 weeks or once a month. If the bait becomes moldy, musty, soiled, or insect-infested, the box should be emptied, cleaned, and then refilled with fresh bait. Spoiled or uneaten bait should be disposed of in accordance with the label. All label directions for the product being used should be followed.

Placement Proper placement of bait stations is just as important as using the appropriate bait. Rodents will not visit bait stations, regardless of their contents, if they are not conveniently located in areas where rodents are active.

Where possible, bait should be placed between the rodents' source of shelter and their food supply: near rodent burrows, against walls, or along travel routes used by the rats. Since rats are often suspicious of new or unfamiliar objects, it may take several days for them to enter and feed in bait stations. House mice usually investigate and utilize bait stations quickly.

On farmsteads, bait station placement depends on building design and use. Bait boxes may be placed in attics or along the floor or alleys where rodents are active. Rodent tracks, visible on dusty surfaces, and rodent droppings clearly indicate rodent activity. Bait stations should never be placed where livestock, pets, or other animals can knock them over. Spilled bait may be a potential hazard, particularly to smaller animals.

Where buildings are not rodentproof, permanent bait stations can be placed inside buildings, along the outside of building foundations, or around the perimeter of the area. When maintained regularly with fresh anticoagulant bait, these bait stations will help keep rodent numbers at a low level. Rodents moving in from nearby areas will be controlled before they can reproduce and cause serious damage.

Tracking Powders

Toxic dusts or powders have been successfully used for many years to control rats and mice. When rodents walk over a patch of toxic powder, they pick some of it up on their feet and fur and later ingest it while grooming. Tracking powders are useful in controlling rodents where food is plentiful and good bait acceptance is difficult to achieve. Rodents are more likely to ingest a lethal amount of poorly accepted toxicant applied by this method than if it is mixed into a bait material. There is little likelihood of toxicant shyness developing when using tracking powders.

Because the amount of material a rodent may ingest while grooming is small, the concentration of active ingredient in tracking powders is considerably higher than in food baits that utilize the same toxicant. Therefore, these materials can be more hazardous than food baits. For the most part, tracking powders are used by professional pest control operators and others trained in rodent control. Tracking powders containing either single-dose poisons or anticoagulants are commercially available, although some are Restricted-Use Pesticides.

Tracking powders should be placed in rodent burrows, along runways, in wall voids, behind boards along walls, or on the floor of bait stations. Placement can be aided by using various types of sifters, shakers, or blowers. Dampness may cause the powder to cake and lessen its effectiveness. Care must be taken to place tracking powders only where they cannot contaminate food or animal feed, or where nontarget animals cannot come into contact with them. They should not be placed where rodents can track the material onto food intended for use by humans or domestic animals. Because of potential hazards to children and pets, tracking powders are not generally recommended for use in and around homes. Where possible, they should

be removed after the rodent control program is completed.

Fumigants

Fumigants (poisonous gases) are most commonly used to control rats in their burrows at outdoor locations. Various compounds such as methyl bromide, chloropicrin, calcium cyanide, aluminum phosphide, and gas cartridges are registered for this purpose. The incendiary gas cartridge burns, producing carbon monoxide and other gases that suffocate rodents in their burrows. Anhydrous ammonia is not recommended for use as a burrow fumigant because it is not registered for this purpose.

Because most fumigants are highly toxic to humans and other animals, they should be used only by people familiar with the necessary precautions. They should not used in any situation that might expose the occupants of a building to the fumes. Only licensed structural pest control operators should use fumigants in buildings or other structures.

To fumigate rat burrows, the burrow opening should be closed with soil or sod immediately after introduction of the fumigant. Rat burrows often have multiple entrances, and all openings must be sealed in order for fumigants to be effective. Fumigants are less effective in soils that are very porous or dry.

Safety Precautions

Certain general safety precautions should be followed in addition to those appearing on the labels of products. All rodenticides should be considered dangerous enough to cause death in nontarget animals, and baits should be placed where only rodents can get them. There are no known rodenticides that do not present some degree of hazard to animals other than rodents. The anticoagulants and some nonanticoagulant rodenticides may present some hazard to predators or scavengers that feed on the carcasses of poisoned rodents. Therefore, care must be taken to keep baits out of the reach of domestic animals or nontarget wildlife. Rodent carcasses should be handled with rubber gloves or long tongs, and buried or incinerated when found. As an added safety precaution, dogs or cats should be confined or well fed while baiting operations are in progress.

All bait containers and stations should be labeled clearly with appropriate warnings, and unused bait should be kept in its original container. Bait and concentrates should be kept in a locked cabinet out of the reach of children or animals, and appropriate warnings posted on the outside of cabinet doors. If baits are stored with other chemicals, they should be packaged in airtight containers to prevent absorption of foreign chemical odors that could reduce bait effectiveness. Label directions on all rodenticides should be followed carefully. Except when permanent bait stations are used, all uneaten bait should be removed and destroyed at the end of the poisoning program.

Maintaining Control

Once control is achieved, some managers fail to pay attention to rodent control for a couple of months. Unfortunately, this habit leads to the undoing of all the work it took to control the rodents initially. A few rodents are likely to survive even the most thorough control effort. Also, rodents from nearby fields or structures may invade storage facilities at any time. These rodents will multiply quickly if not kept in check with an ongoing control program. Therefore, it is important to establish permanent bait stations in or around structures. Fresh anticoagulant bait in these stations will control invading rodents before breeding populations become established.

Rodent control should be a regular and continual part of a grain storage operation. An effort should be made to put aside an hour or two each month, after control has been achieved, to check and refill bait stations and inspect facilities for fresh rodent signs.

Acknowledgments

I thank Rex E. Marsh and Robert M. Corrigan for assistance in developing and reviewing the information contained in this chapter. Credit is also due to J. E. Brooks, H. D. Pratt, and W. E. Howard, whose earlier publications have served as a basis for some of the information reported here.

Cited References

1 Raney, H. G. 1987. *Management of On-Farm Stored Grain*. Cooperative Extension Service, College of Agriculture, University of Kentucky, Lexington.

2 Timm, R. M. (ed.). 1983. *Prevention and Control of Wildlife Damage*. Great Plains Agricultural Council and Cooperative Extension Service, University of Nebraska, Lincoln.

32
Chemical Methods to Control Insect Pests of Processed Foods

Fred J. Baur

The three main chemical methods for insect control in and around food-handling establishments are fumigation, space treatment, and application of residual insecticides. Fumigation, which utilizes insecticides in gaseous form and requires a sealed facility, is covered in Chapter 33. Space treatments involve aerosolized insecticides which, when dispersed into the air, make direct contact with and immediately affect the targeted insects. Fumigation and space treatments tend to be used as corrective measures to limit the spread of pests or to eliminate an infestation. Residuals are insecticides in liquid or solid form applied to surfaces. Although they also have a distinct value in correcting a problem, their use has a much stronger preventive thrust.

This chapter addresses issues of the need for chemical control, whether the use of such control is reasonable, and which chemicals are to be used. Chemical control is put into perspective with other control techniques as a part of a totally integrated insect control program. This integrated view is necessary for every food operation, but since each operation has the potential for different problems, decisions on the selection of control measures must be made on a case-by-case basis.

Baur: 1545 Larry Avenue, Cincinnati, OH 45224. Formerly with The Procter and Gamble Company, Cincinnati, OH.

Need for Chemical Control

Humans are probably the greatest threat to the continued existence of human life on this globe. Insects run at least a strong second. They are prime competitors for our food, clothing, and shelter, and the diseases they carry kill us by the millions (fortunately, the most dangerous vectors usually are not found in food establishments). Species are innumerable, numbers are incalculable, habits and behavior are variable, and some insects are virtually microscopic in size. They are such a significant challenge and a threat to us and our food supply that we are compelled to control them by every reasonable measure available to us.

Reasonableness of Chemical Control

Chemical control is reasonable primarily because of its cost-effectiveness. Other reasons include (a) the prompt response—insect population numbers are reduced and can be held to acceptable levels; (b) the effectiveness for a broad range of species; (c) the increase in available information on how to use the chemicals safely and effectively; (d) the surveillance of their use by regulatory agencies such as the U.S. Environmental Protection Agency (EPA), the Food and Drug Administration (FDA), and the U.S. Department of Agriculture (USDA); and (e) the growing professionalism within the pest control industry which is vital because not all food establishments can provide the needed internal expertise and knowledge.

Use of insecticides is an essential part of most, if not all, insect control programs. However, insecticides are toxic materials and therefore must be handled in a manner that is safe to the applicator, other people, the foods and operations that have necessitated their use, and the environment. An integrated pest management (IPM) system is the necessary approach.

Integrated Pest Management

IPM is a systems approach that considers all reasonable methods to avoid pest problems and combines the control or suppression procedures that best suit the particular need at a particular time. As situations change, the combination of methods can be changed, and this is often the case.

Generally speaking, greater care is needed in selecting and implementing control measures under the following circumstances: the larger the facility is, the more complex it is, the poorer the facility is from physical and maintenance standpoints, the more attractive the materials are, the more conducive the environment is for pest activity, and any combinations thereof. Consumer and regulatory expectations must also be considered since these vary with the commodity, the facility, and the governmental entities involved.

IPM originated about 35 years ago in California when it was discovered that insects harmful to crops were becoming resistant to insecticides (7). Also, formerly innocuous insects were becoming serious pests because some of their natural enemies were being eliminated.

Attention really focused on IPM when, in 1972, the Council on Environmental Quality (CEQ), recognizing that chemical pesticides had become the predominant method of controlling pests and that such use was adversely influencing the environment including untargeted organisms, published a report, entitled "IPM," on alternative methods. This report stimulated increased national and international interest in the systems approach as an economically efficient, environmentally preferable approach to pest control, particularly in agriculture.

In 1976, CEQ began a more comprehensive review and widened the scope to include urban pest control. In 1977, President Carter instructed the Council to review IPM and recommend actions. The resulting report (3) stated that chemical pesticides are and will continue to be of considerable importance in food pro-

duction and defined IPM as "the selection, integration, and implementation of pest control based on predicted economic, ecological, and sociological consequences. IPM seeks maximum use of naturally-occurring pest controls, including weather, disease agents, predators, and parasites. In addition, IPM utilizes various biological, physical, and chemical control and habitat modification techniques."

Prevention

The intent of IPM in food establishments is not to eradicate pests but to control pests in a more effective manner. Foremost in any control program is prevention. The expression, "an ounce of prevention is worth a pound of cure," has special pertinence to the food industry. Most preventive measures are common to food-industry pests. Undergirding all preventive measures is sanitation and underpinning sanitation is cleanup. Insects require food to survive. The quantities of food they require are miniscule when compared to our own food needs; therefore, cleanup must to be almost fastidious. Other important preventive steps, include:

(a) Using clean design in the contruction and any subsequent alterations of a facility. Effective cleanup requires that the facility or operation (grounds, building, equipment) be readily cleanable and maintained that way. Insects seek shelter, food, and a place to propagate. Therefore, it is essential to eliminate the cracks, crevices, and voids where they can live and breed. Some insects are quite small and even the ubiquitous cockroach can quickly enter a surprisingly small crack such as often occurs at the floor/wall junction.

(b) Keeping insects out by carefully inspecting incoming goods including equipment, pallets, and packaging supplies; maintaining a tight building including screening of windows, keeping doors closed, providing air curtains; making employees aware of the fact that under certain conditions they can inadvertently carry insects into a food establishment; avoiding attractants such as outside lighting close to buildings or odors emanating from the food-handling facility.

(c) Making landscaping attractive but not inviting to or providing harborage for insects.

(d) Controlling trash handling.

(e) Avoiding long-term dead storage of food-stuffs, storage of uncleaned equipment, and so on.

(f) Promptly disposing of damaged packages of food materials.

(g) Using insecticides judiciously, particularly the

residuals. This means reading the label and following directions!

(h) Installing, maintaining, and using, whenever possible, process-type equipment such as entoletors or sifters, which kills or removes insects.

(i) Studying prevention as described in this book and in other references on good manufacturing practices. Of particular use is a treatise on insect control published by the American Association of Cereal Chemists (2).

Alternatives to Chemical Control

The advantages and disadvantages of some of the alternatives to chemical control are listed in Table 32.1. Three chemical methods are included for comparison. Not surprisingly, alternatives to chemicals vary in their state of development and potential. They do, however, offer the means to minimize the use of chemicals and, perhaps, in some instances, to eliminate their use entirely.

Considerations in the Use of Chemical Methods

Background Information

The active ingredients in pesticides are registered by EPA for "general use," "restricted use," or both. However, a specific product or formulation may be registered for general use or restricted use but not for both. EPA assigns the use category after considering possible effects of pesticides on the environment; if it produces adverse effects, even when used properly, the product is classified as restricted. The purpose of the restricted-use classification is that certain products can be purchased and applied only by certified applicators or by individuals under direct supervision of a certified applicator. Pesticide products classified as restricted are so identified on the label of the container in which they are marketed.

Insecticides are generally described as nonresidual or residual. Nonresidual (general-use or contact) insecticides are applied as space or spray treatments in which the insecticides are dispersed into the air to kill flying insects and exposed crawling insects on contact. Residual insecticides are applied directly to surfaces as general, spot, or crack and crevice (C&C) treatments to obtain effects lasting at least several hours or sometimes several days or longer.

General treatment is the application of insecticides to broad expanses such as walls, floors, and ceilings or as an outdoor treatment.

Table 32.1

Value comparison of various approaches to insect control

Approach	Advantages	Disadvantages
Biological control (parasites, pathogens, predators)	EPA registers microbial biocontrol agents but not arthropod predators or parasitic nematodes and arthropods; use of arthropod predators and parasites may not be permitted in FDA-regulated facilities (but see ref. 9)	Additional insects undesirable (more suitable for raw commodities); lacks scope, versatility (more study needed)
Cold	Residue-free food	Change must be abrupt; long time requirement
Fumigation	Effective kill of all stages	Expensive; possible "off" quality of food; requires shutdown time; safety risk including possible residues; long time requirement
Heat	Effective kill; residue-free food; sealing facility less critical	Requires shutdown time; long time requirement
Irradiation	Effective kill; residue-free food	Expensive (large initial capital outlay); lacks scope, versatility
Modified atmospheres	Cost-competitive with fumigation; effective kill (eggs); residue-free food	Lacks general applicability; long time requirement
Protective packaging	Necessary for proper distribution of commodity	More study needed (film wraps, insecticides); most consumer packaging can be penetrated by insects
Residuals	Broad-spectrum, good kill; good flexibility; inexpensive, fast, long-lasting; low risk to applicator	Safety, including possible residues
Space treatment	Can be automatic; faster than fumigation; less training required; uses general, not restricted, pesticides	Food and food-contact surfaces need protection/cleaning; limited kill (only exposed insects; not all species, not all stages)
Traps (bait, light, sticky, pheromone)	Diagnostic and monitoring	Incomplete kill; maintenance, monitoring required

SOURCE: Baur, F. J. (ed.). 1984. *Insect Management for Food Storage and Processing.* American Association of Cereal Chemists, St. Paul, MN.

Spot treatment is application to noncontiguous level areas not to exceed 2 ft² per spot. Food or food-handling equipment must not come into contact with these treated areas, and as a general rule, the treatment surfaces should not be touched by workers in a food establishment. Spot treatments may be applied to floors, walls, and at the bases or undersides of equipment.

C&C treatment is the application of small amounts of insecticides into cracks and crevices

Table 32.2

Registration status of insecticides cleared for use in food areas of food-handling establishments

	Application type				Authority	
Compound	C&C[a]	Spot	General/space	Comments	PR-73-4[b]	FAR[c]
Acephate (Orthene)	1% spray	1% CLP[d]	—	Reg. No. 239-2464; tolerance = 0.02 ppm	—	193.10
Allethrin (includes bioallethrin)	—	—	Nonresidual contact and space spray	Plant not in operation; Reg. No. 4816-554	—	—
Bendiocarb (Ficam)	Spray or dust		Spray or dust	—	—	193.152
Borax	Yes	—	—	—	+	—
Boric acid	Yes	—	—	—	+	—
Chlorpyrifos (Dursban)	2%	0.5% CLP spray, 2% paint	—	Also, 10% adhesive strips, 36/1,000 ft²	+	193.85
Diatomaceous earth	Dust	Dust	—	Food must be removed or covered during treatment	—	193.135
Diazinon	2% spray or dust, 1% Mcap[e]	2% spray or dust, 1% Mcap	—	Spot treatment limited to 20% of floor surface	+	193.142
Dichlorvos (DDVP)	Yes	—	Nonresidual contact and space treatment	For space treatment, plant must not be in operation	+	—
Fenthion (Baytex)	Yes	—	—		+	—
Fenvalerate (Pydrin)	Yes	Yes	1 gal 0.25%/1,000 ft², general treatment	Food must be covered or removed during treatment	—	193.97
Malathion	Yes	—	—	—	+	—
MGK 264[f]	Yes	Yes	Yes	Food must be covered or removed during treatment	+	193.320
Piperonyl butoxide[f]	Yes	Yes	Yes	Food must be covered or removed during treatment	+	193.360
Propetamphos (Safrotin)	Yes	Yes	—	Reg. No. 11273-22, 45; tolerance = 0.1 ppm	—	193.375
Propoxur	Yes	—	—	Reg. No. 3125-146, 6754-33	+	—
Pyrethrins	Yes	Yes	Yes	Food must be removed or covered; tolerance = 1 ppm	+	193.340
Resmethrin[g]	Yes	Yes	Yes	Food must be removed or covered; tolerance = 3 ppm	—	193.464
Silica gel	Yes	Yes	—	—	+	—
Trichlorfon	Yes	—	—	—	+	—

NOTE: Excludes baits and fumigants.
[a]Crack and crevice.
[b]Pesticide Registration Notice 73-4. Pluses indicate authority under the listed PR; minuses indicate another authority.
[c]Food Additive Regulation (21 CFR Part 193).
[d]Coarse low-pressure spray.
[e]Mcap = Microencapsulated flowable formulation.
[f]Synergist.
[g]Allethrin, bioallethrin, fenvalerate, and resmethrin are pyrethroids. Cyfluthrin (Baythroid, Tempo), another pyrethroid, is now labeled for use in edible-product areas of certain USDA-inspected establishments (see label for specific use information).

(such as floor/wall or equipment support junctions) in which insects hide or through which they may enter a facility. Such openings may lead to voids such as hollow walls, equipment legs and bases, motor housings, and switch boxes. These sites often contain food for insects and provide places where insects proliferate. Only C&C and/or spot treatments with residuals are permitted in food-handling establishments.

A food-handling establishment is an area or place other than a private residence in which food is held, processed, prepared, and/or served. Food-handling establishments may be classified as (a) plants, bakeries; (b) processing facilities for upgrading raw agricultural products such as grains, spices, meat, poultry; or (c) food service facilities such as restaurants and institutional kitchens.

Food, as defined by the Federal Food, Drug and Cosmetic Act (FD&C Act), consists of (a) articles used for food or drink for humans and animals, (b) chewing gum, and (c) materials used for components of (a) and (b). Food areas of a food establishment include places for receiving, storing, serving, preparing, and packing food as well as enclosed processing equipment. Generally, residual insecticides can be applied when the food establishment is in operation unless such use is specifically prohibited by the label. Appropriate care must be exercised to prevent contamination of food or food-contact surfaces. Before applying an insecticide in a food-handling establishment, the applicator must read, understand, and follow all instructions on the label of the pesticide product.

Choice of Chemicals

Table 32.2 summarizes the registration status of insecticides by type of application in food areas of food-handling establishments. For space treatments, the selection is limited to allethrins, fenvalerate, resmethrin, and pyrethrins. Dichlorvos has long been approved and used for space treatments, but its use is under review by EPA. These materials must make direct contact with the insects, and hence are often referred to as contact sprays. Dichlorvos is unique in that besides killing insects on direct contact, it is highly volatile and therefore becomes a gaseous insecticide. In this form its action is similar to that of a fumigant but it lacks the commodity pepenetrability required of an effective fumigant.

Table 32.3 lists the EPA-approved chemicals registered and labeled for use as residuals in food establishments. Note that if the processing

plant is under USDA jurisdiction, the chemical must be on the USDA-approved list and must be cleared for use by the USDA inspector in charge prior to application.

Table 32.3 provides some insight into choosing the appropriate material for the various spe-

Table 32.3

EPA-approved chemicals registered and labeled for use as residual sprays in the food-handling and food-processing industry

	Acephate	Bendiocarb	Borax	Boric acid	Chlorpyrifos	Dichlorvos	Diatomaceous earth	Diazinon	Fenthion	Fenvalerate	Hydramethylnon[a]	Malathion[b]	Propetamphos	Propoxur	Pyrethrins[c]	Resmethrin	Silica gel	Trichlorfon
Angoumois grain moth									X			X			X	X		
Ants	X	X	X	X	X	X	X	X		X		X	X	X	X	X	X	
Black carpet beetle		X								X		X				X		
Cadelle					X				X	X					X		X	
Carpet beetles		X			X	X				X		X		X	X			
Centipedes		X			X			X	X					X	X	X	X	
Cereal moths		X													X			
Cigarette beetle					X	X		X	X				X	X	X			
Cockroaches	X	X	X	X	X	X	X	X	X	X	X	X	X	X	X	X	X	X
Confused flour beetle	X	X			X	X		X	X			X			X	X		
Crickets	X	X			X	X	X	X	X			X	X	X	X	X		
Dark mealworm					X											X		X
Dermestids	X	X							X						X			
Driedfruit beetle									X				X		X			
Drugstore beetle		X			X			X		X		X		X	X			X
Drywood termites		X												X	X			X
Earwigs	X	X			X	X		X		X				X	X	X	X	
Firebrat/silverfish	X	X	X	X	X	X	X	X	X	X		X	X	X	X	X	X	X
Fleas		X			X	X	X		X			X	X	X	X			X
Flies					X	X	X	X	X			X			X			X
Flour beetles	X				X	X		X				X		X	X			
Flying moths						X				X					X	X		
Fruit flies (*Drosophila*)										X		X			X	X		
Grain beetles		X			X										X	X		
Grain mites															X	X	X	
Grain weevils					X				X			X		X	X	X		
Granary weevil												X			X			X
House fly		X						X	X			X			X	X		
Indianmeal moth	X				X	X		X	X			X			X	X		
Lesser grain borer		X				X						X			X	X		
Mediterranean flour moth					X	X			X			X			X	X		
Millipedes		X			X			X		X				X	X	X	X	
Mosquitoes		X			X	X		X	X			X		X	X	X		
Red flour beetle	X				X	X		X		X		X	X		X	X	X	
Rice weevil	X				X	X		X		X		X			X			
Sawtoothed grain beetle		X			X	X		X	X			X		X	X	X	X	
Sowbugs/pillbugs	X	X			X			X		X		X		X	X	X	X	X
Spiders		X			X	X		X		X		X		X	X	X	X	
Tobacco moth		X				X									X			

NOTE: Pesticides approved as of Aug. 10, 1987.
[a]Cannot be used in food areas of USDA-inspected facilities.
[b]Pest resistance is becoming increasingly more serious, especially in such commodities as oilseeds, nuts, and grains.
[c]Synergized.

cies (as listed on the label). The following considerations should also help in the selection process:

(a) *Whether it is registered by EPA* (Tables 32.2, 32.3). This is a dominant factor not only from the legal standpoint but also from the standpoint of the label information that EPA requires from the manufacturer, information which describes the proper use of the product. Applicable state and/or local laws, which deal with specifics and therefore require careful checking and study, should also be considered.

(b) *Whether its use is permitted in a food establishment.* This and other use restrictions are found on the label and in information provided by the manufacturer.

(c) *Whether tolerances exist for the candidate insecticide for any food commodities present in the facility.* Every effort must be made to avoid possible contamination (see section below on Tolerances and Action Levels).

(d) *Whether its use is permitted for the target insect and/or whether spraying is allowed at the sites involved.* Again, the label provides the answers.

(e) *Whether the target insect has developed resistance to or a tolerance for the insecticide* (6; Table 32.4). Resistance of insects to insecticides continues to grow. Contributing factors include misuse of insecticides, cross-resistance, and failure to try other similarly effective insecticides in any given problem situation. Whatever the factors, the potential exists for the development of pest resistance to all insecticides. This alone clearly indicates that several management tools should be used and that insecticide use should be kept to a minimum.

(f) *Whether the selected insecticide will be compatible with the facility and its operations.* Possible concerns include whether people can continue to work during application or whether the plant must be shut down, the danger of damage to equipment (wiring, surfaces), the potential for odor contamination of food products, and the risk of fire. Sometimes other formulations or physical forms of the same insecticide can alleviate these concerns.

Table 32.4

Insect pests that exhibit physiological resistance to certain pesticides

Insect pests	DDT	Cyclodienes	Organophosphates	Carbamates	Fumigants[a]	Other
Almond moth	x	x	x		PH_3	Pyrethrins
Angoumois grain moth		x	x			
Ants		x				
Cockroaches						
German	x	x	x	x		Pyrethrins
Oriental	x	x				
Dermestids						
Hide beetle		x	x			
Khapra beetle			x			
Fleas						
Dog	x	x				
Cat	x	x				
Human	x	x				
Flour beetles						
Confused		x			MB, PH_3	Pyrethrins
Red	x	x	x	x	MB, PH_3	Bioresmethrin
Grain beetles						
Merchant		x	x			
Sawtoothed			x	x		
House fly	x	x	x	x		Pyrethroids
Indianmeal moth	x	x	x		PH_3	Pyrethrins
Lesser grain borer		x	x		PH_3	
Mosquitoes[b]	x	x	x	x		Pyrethroids
Weevils						
Granary	x	x	x	x	MB, PH_3	Pyrethroids
Maize	x	x	x	x		
Rice	x	x	x	x	CPN, PH_3	

[a]PH_3, phosphine; CPN, chloropicrin; MB, methyl bromide.
[b]Virtually all species of mosquitoes tested are resistant to DDT and the cyclodienes; most are resistant to the organophosphates, and some are resistant to carbamates.
SOURCE: Georghiou, G. P., and T. Saito (eds.). 1983. *Pest Resistance to Pesticides.* Plenum Press, New York, NY.

(g) *Whether the effectiveness (toxicity, stability) of the insecticide will be adversely affected by environmental conditions in the plant.* Concerns include type, frequency, and scope of cleanup; ambient temperatures and relative humidities; nature of processing (dry vs. wet); nature of interior surfaces such as floors and walls; and high levels of ultraviolet light. Table 32.5 provides further information on these aspects.

(h) *Whether the control needed is corrective, preventive, or both.* The use of residuals is both corrective and preventive, space treatments tend to be more corrective than preventive, and fumigants are only corrective.

A proper selection is not as complicated as it may seem. Comparable needs require comparable control. Needs tend to be repetitive within a given facility. Hence, having handled a problem once, the next need is often more easily handled. Chlorpyrifos, diazinon, and propoxur are probably the most commonly used and generally effective residuals.

The varying nature of the food establishment, the kind of pest problem that needs to be solved, and the comparative strengths of residual sprays versus space sprays influence the choice of treatment. Food establishments subject to infestations by the more common stored-food insects (''pantry pests'') would usually be sprayed with residuals. Pantry pests, especially beetles, are hard to reach and to control by space treatments. Many species in this group do not fly and are therefore less susceptible to space treatments. Yet there are occasions when space treatments seem more appropriate. Usually this happens when a previously unknown infestation is found during an inspection. Space treatments can be used to prevent migration of insects from an infested area to an uninfested area. Some warehouse workers favor space treatments because the results are immediately noticeable and may be more cost-effective than residual applications. For example, automatic foggers in peanut warehouses can effectively control certain kinds of food-infesting moths.

Space treatment is often the application mode of choice in restaurants and institutional kitchens where the greatest insect risks are from ''environmental species'' such as cockroaches and flies. Whereas cockroach control is often achieved only by means of residuals, fly control frequently requires a space spray (electrocutors may also be useful). The optimum goal in the control of flies is to eliminate breeding sites, but this is not always feasible. Another reason for the use of space treatment is that the insecticides in some of the space sprays are in the general-use category and therefore do not require the high levels of knowledge and training needed for the use of residuals. Institutional kitchens, especially where government surplus raw materials are used, have an elevated risk of infestation by stored-food insects. Large bakeries are similar to food-processing plants; smaller ones are more comparable to restaurants/kitchens.

Target Species

The various pest species and the chemicals approved for use in their control are listed in Table 32.3. A similar list (Table 32.6) has been compiled by the National Pest Control Association (NPCA). The information in Table 32.3 comes from the manufacturers of insecticides, and is primarily from laboratory tests, whereas the information in Table 32.6 comes mainly from field experience (there is, of course, some overlap of data between the two tables). Also, Table 32.6 is mainly concerned with the environmental species rather than with food-infesting species.

Table 32.5

Approximate stability of organic residuals

Insecticide	Adverse chemical exposures in facility environments	Applied toxicity duration[a] (days)		
		NPCA[b]	General literature	Manufacturers' information
Acephate	alkalinity	>15	5–10, 42–56	7–49
Bendiocarb	alkaline surfaces (concrete), water, pH sensitive	>15	>4–5	7–35, 30–60
Carbaryl		1–15		
Chlorpyrifos	water	>15	2, 10–14, 21	7, (56–70)
Diazinon	acidity, sunlight, water	>15	1, 14	7, (42, 60–89)
DDVP	acidity, alkalinity, water	1–15	1–3	
Fenthion	alkalinity	>15		
Malathion	alkalinity	1–15	1–3	7
Propetamphos		1–15	14	7
Propoxur	alkaline surfaces	>15	2	7–35
Resmethrin	air, sunlight	1–15		
Trichlorfon	alkalinity	>15		

NOTE: Inorganics—boric acid, diatomaceous earth, silica gel—are essentially stable.
[a]Information is for sprays; dusts or baits last longer; encapsulation prolongs effective toxicity (examples in parentheses).
[b]National Pest Control Association data.

Timing and Frequency

The frequency of application of insecticides is clearly dependent on need; that, in turn, is a function of many interrelated factors, including

(a) how well the other management techniques are keeping insects at an acceptably low level;

(b) what one's own particular risks are from the standpoints of the type and condition of the facility, the food materials involved in the operation, the permitted action levels/tolerances, the consumers and their concerns, and the risks one is willing to take;

(c) what the life cycles of the species needing control are, keeping in mind the environmental conditions (e.g., temperature, water, food);

Table 32.6

Selected insecticides recommended by NPCA for use on arthropod pests, 1986

Insecticide	Effect[a]	Ants	Booklice	Carpet beetles	Clover mites	Cockroaches	Crickets	Earwigs	Flies	Grain/cereal beetles	Millipedes	Silverfish	Spiders	Springtails
Baygon bait	C					1	2							
Baygon residual spray	C	1		2	2	1	1	1	2	1	1	1	1	
Boric acid dust	C	2				2						2		
Cypermethrin	C,D	1				1	1		1			1	1	
DDVP bait	C									1				
DDVP contact	B	2		2		2	2	2	2	2	2	2	2	
DDVP resin strip	D								1	2				
DDVP space	B	2		2		2	2		1	2		1	2	
Diazinon dust	C	1		2		2				2		2	2	
Diazinon encapsulated	C,D	1				1		2				1	1	
Diazinon granules	C,D	2			2		2	2			2			2
Diazinon residual spray	C,D	1		1	1	1	1	1	1	1	1	1	1	1
Dursban dust	C	1				2						2		
Dursban encapsulated	C	1		1	1	1	1	1		1		1	1	
Dursban granules	C,D	2			2		2	2			2			
Dursban residual spray	C	1		1	1	1	2	1	1	1	1	1	1	
Fenvalerate	C,D	1				1	1	1	1	1		1	1	
Ficam D dust	C	1				1	2	2			2	1		
Ficam Plus	C,D	1		2		1	1	1		2	1	1	1	
Ficam W	C,D	1		2		1	1	1		2	1	1	1	
Hydroprene	C					1								
Killmaster II	C,D	1				1	2			2		1	1	
Malathion dust	C	2			2	2	2	2		2		2	2	2
Malathion residual spray	B	2		2	2	2	2	2	2	1	2	2	2	2
MaxForce	C,D					1								
Methoprene	C,D	1												
Orthene	C					1	1	2		2		2		
Permethrin	D								1					
Pyrethrin contact	A	2		2	1	1	2	2	1	1	2	2	1	
Pyrethrin dust	A	2				2				2		2	2	
Pyrethrin space	A	2	1	2	2	1	2	2	1	2	2	1	1	
Resmethrin contact	B	2		2		1	2	2		1		1	1	
Resmethrin space	B			2		1			1	2			1	
Safrotin	C,D	1				1	1	2		2		2	1	
Safrotin dust	C	1		1		1	1	1				1	1	
Sevin dust	C,D	2		1				2			2			
Sevin granules	C,D	2					2	2			2			
Sevin residual spray	B	2						2			2			
Silica gel	C	2										2		

NOTE: 1 = Material of choice of proven value to pest control operators; 2 = alternative material, useful in some situations. Updates available from National Pest Control Association, 8100 Oak Street, Dunn Loring, VA 22027 [telephone (703) 573–8330].

[a]A = short residual (<1 day), for quick knockdown; B = moderate residual (1–15 days), "general" spray; C = long residual (>15 days), usually limited to crack & crevice, spot treatments, or baits; D = long residual (>15 days), "general" treatments.

(d) what evidence has been noted (e.g., numbers, species, locations), all of which is gained from the ongoing, essential inspection program;

(e) what the nature of the past control record has been on a continuing basis;

(f) what the stabilities (Table 32.5) of the chosen insecticides are as they are influenced by the conditions in the food facility (these conditions vary greatly and are, of course, beyond the control of the suppliers of the pesticide products).

No matter which treatment system (space, residual, or both) is selected, at least one treatment per month should be made (8), except during the winter in ambient-temperature warehouses located in a temperate climate or in a cold-storage facility, which require less frequent applications. Some food facility managers apply space treatments daily, but this practice would be difficult to justify in many instances. Two reasons to speed up an approximately monthly schedule would be sufficient insect evidence and probable deterioration of the residual applications already in place.

As a general rule, insecticides should be applied at dusk or dawn since those are the times that insect activity tends to be the greatest. Since these are also the times when it is easiest to restrict human activity in most facilities, the probability of exposing people to insecticides is diminished.

Personnel

Residual insecticides should be applied only by a certified applicator or by a person under the direct supervision of a certified applicator. If the residual is labeled for restricted use only, then only a certified applicator can make the treatment. Certification usually requires passing a test, usually state-administered, on basic application techniques, pesticide safety, label comprehension, and precautionary measures. Most states require certified applicators to attend pertinent seminars or presentations on a regular basis in order to qualify for renewal of certification and to remain current on new developments in pesticide application. General-use insecticides do not require a certified applicator.

For food plants, food warehouses, larger bakeries, and so on, it is important that the applicator be certified. The educational experience involved in attaining certification is beneficial regardless of the classification of the insecticide chosen for use. As a general rule, this position is filled in most larger facilities by one of their own employees, frequently in conjunction with a licensed pest control operator (PCO). Smaller facilities often retain a PCO on contract to apply pesticides. However, since the ultimate responsibility for insect control resides with the management of the facility, management personnel should acquire a working knowledge of proper control techniques.

Space Treatment

EPA has defined space treatment as "the dispersal of insecticides into the air by foggers, misters, aerosol devices or vapor dispensers for control of flying insects and exposed crawling insects." Organic (petroleum) solvents are used to make a solution of allethrins, dichlorvos, pyrethrins, pyrethroids, and/or resmethrin.

Factors to be considered in selection of the equipment include desired size of the droplets, volume of the space to be treated, time available for treatment, degradation characteristics of the insecticide, and operational costs.

There are three means used to generate aerosols (also called mists or fogs) (2): thermal, mechanical, and vapor.

Thermal Foggers Thermally generated fogs tend to be composed of droplets of 1 micron or less. Such small droplets are easily carried by air and disperse quickly and widely. They do not readily stick to surfaces, nor do they enter cracks and crevices. Therefore, their main effect is on flying insects. Because of the heat involved, there may be some degradation of the pesticidal chemical during thermal generation. Although this method is considered the least expensive approach, its lower level of effectiveness usually does not justify its use against stored-product pests.

Mechanical Generators This equipment uses compressed air or air turbulence to create an aerosol. Careful adjustment of the generator is required to prevent nozzle clogging or the production of droplets so large that a "wet" fog results. This is undesirable because rapid settling of heavier droplets makes the treatment less effective and leaves wet residues on surfaces which is potentially hazardous to employees and food products. Compressed-air units are more suitable for dispensing dichlorvos formulations.

The production of air turbulence is the physical basis for the increasingly popular ultra-low-volume (ULV) or ultra-low-dosage (ULD) units. If properly operated (5), these units generate droplets in the range of 1–30 microns (mostly 10–20 microns). Droplets of this size are ideal; they disperse throughout the treated space, impinging upon and killing flying insects and exposed crawling insects. More time is required to perform ULV/ULD applications, and initial equipment costs are greater than with thermal foggers, but less insecticide is needed. Cost for ULV/ULD may be higher than for other application systems but this is more than balanced by increased effectiveness.

Aerosol Dispenser (Bug Bomb) A third type of aerosol generator makes use of a vapor, often chlorofluorohydrocarbons, under pressure as the propellant. Total-release units, once activated, dispense their entire contents in a short period of time. Automatic units dispense measured quantities of 10- to 20-micron droplets at timed intervals. These automatic units work best when the temperature is around 24°. Higher temperatures cause too rapid a delivery of the insecticide. Lower temperatures tend to increase droplet size and to deliver more insecticide than is required. Compared to the thermal and ULV/ULD systems, costs of operating the automatic dispensers are higher, especially in larger facilities, but these costs may be offset by savings on application time and applicator personnel.

As a general rule, aerosols should be applied only when a food facility is unoccupied, thus avoiding unnecessary exposure to people. Ventilation (admixture of outside air) should be kept to a minimum during space treatments, but fans can be used to facilitate the uniform dispersal of droplets.

Residual Treatment

Space treatments are often useful and effective, but they are not cure-alls. It is unlikely that space treatments alone could eliminate a pest population. To accomplish this requires the use of a variety of management tools, especially residual insecticides. A basic point to remember about residuals is that in food-handling establishments they may be applied only as C&C and spot treatments (2). Residual insecticides are usually applied in one of three physical forms: liquid suspensions or emulsions, dusts, and baits. Regardless of which form is used (all three forms are often used in a given facility), the secret to success with residuals is proper placement.

Liquid Suspensions or Emulsions Water is often (but by no means always) the diluent or carrier used to make emulsions (from emulsifiable concentrates) or suspensions (from wettable powders). The resulting liquid insecticide is usually applied by means of a compressed-air sprayer (4), often the 1-gallon model that can be carried over the shoulder and can be

equipped with a variety of nozzles especially suited for applying residuals in food-handling establishments. The use of a brush for spot treatments with liquid residuals permits the operator to place the insecticide more precisely than is usually possible with a compressed-air sprayer. The brush should be cleaned after use and should be stored pending future reuse in an appropriate and properly labeled container.

The effectiveness of liquid residuals may be increased by using them in combination with synergized pyrethrins. The flushing action of the pyrethrins enhances the probability that the pest insects will cross treated surfaces (this flushing action can also be used advantageously as an inspection technique).

Encapsulated residuals (Table 32.5) tend to be more stable than nonencapsulated residuals and are generally less repellent to insects. Repellency, a cause for increasing concern especially with the use of liquid residuals, can be somewhat alleviated if the applicator gives special attention to careful formulation and thorough application.

Dusts Hand-held dusters are versatile and convenient for C&C application. Older models, because of their small capacity, require frequent refilling. Newer models have a larger capacity and a continuous supply of compressed air to float the dust. Dusts are especially useful for the treatment of wall voids, but they may also be used on porous surfaces where liquid residuals are too readily absorbed. Exposure to water usually renders dusts ineffective. Heating a dust before use may enhance its dispersibility. Many kinds of inorganic and organic dusts have been approved for use in food-handling establishments (2).

Baits Baits and dusts are both dry formulations but the particle size of baits is much larger, permitting the pests to pick up and carry away the particles of bait. This may result in product contamination. Baits usually consist of an attractive food mixed with a pesticide. Baits have been used for the chemical control of insects in food-handling establishments only at very modest levels, but for special situations such as voids, baits at least should be considered as one potentially effective element in a pest management program.

Labels and Labeling

Labels contain a lot of important information of various kinds (1, 2). They must be read and understood as a part of the pesticide selection process, and they must be followed for reasons of safety, effectiveness, and compliance.

Labeling

The meanings of the words "label" and "labeling" are frequently confused and misunderstood. "Labeling" is all the information that a pesticide manufacturer provides regarding its product, including the label on the product package and all associated printed information pertaining to a particular product.

Label

The "label" is the information printed on or attached to the container of pesticide. To the manufacturer, the label is a license to sell the product. To the regulatory agencies, it is a means by which control is maintained over the distribution, sale, storage, use, and disposal of the product. To the buyer or user, it is the main source of facts that govern the correct and legal use of the product and any special safety measures needed. Every pesticide label must contain specific information relating to the correct use of the product. There are 14 categories of specific information that are required by law to be printed on a pesticide label.

(a) *Brand name or trade name.* The brand name is the name used by the manufacturer in its promotions and advertisements and is usually the most readily identifiable name for the product. It usually appears on the label in large print but is not necessarily repeated in the active ingredient section.

(b) *Type of formulation.* Different types of pesticide formulations (e.g., liquids, emulsifiable concentrates, wettable powders, dusts) require different methods of handling. The label tells what type of formulation the product is.

(c) *Common name.* A common name is used to more easily identify pesticides that have complex chemical names. A chemical made by more than one company is, of course, sold under several brand names, but the same common name or chemical name is used on all packages containing that chemical (see Table 32.7).

(d) *Type of pesticide.* This item identifies the active ingredient—insecticide, herbicide, fungicide, a combination, or whatever.

(e) *Ingredient statement.* This statement defines the chemical makeup of the product. The amount of active ingredient is given as a percentage by weight or as pounds per gallon of concentrate and can be listed by either the chemical name or the common name. Inert ingredients need not be named, but the label must show what percentage of the contents they constitute.

Table 32.7

Common names, brand names, and chemical categories of residual organic insecticides

Common name[a]	Registered brand or trade name[b]	Chemical type
Acephate	Orthene	Organophosphate
Bendiocarb	Ficam, Tatoo	Carbamate
Chlorpyrifos	Dursban, Killmaster, Lorsban	Organophosphate
Diazinon	Diazide, Spectracide, Knox Out	Organophosphate
DDVP	Pest STRIP	Organophosphate
Fenthion	Baytex	Organophosphate
Fenvalerate	Pydrin	Pyrethroid
Malathion	Cythion	Organophosphate
Propetamphos	Safrotin	Carbamate
Propoxur	Baygon	Carbamate
Resmethrin	Chrysron	Pyrethroid
Trichlorfon	Dipterex, Proxol	Organophosphate

NOTE: This list is not all-inclusive. Many other insecticides are on the market, but most are for agricultural uses.
[a]EPA-accepted.
[b]Listed on the label.

(f) *Net contents.* The net contents may be expressed in gallons, pints, pounds, or other units of measure.

(g) *Name and address of manufacturer.* The manufacturer or distributor of the pesticide must put the name and address of the company on the label so that the user will readily know who made or sold the product.

(h) *Registration and establishment numbers.* The registration number shows that the product has been registered with EPA. The establishment number tells which plant made the chemical. This number does not have to be on the label, but it will be printed somewhere on each container.

(i) *Signal words and symbols.* The signal words and symbols are designed to provide the user with information about the human toxicity or hazard of the active insecticide present in the formulation. These words are set by law; it is incumbent upon each manufacturer to use the correct signal words or symbols on every label (see Table 32.8). All products must bear the statement, "Keep out of reach of children." The symbol of skull and crossbones is used on all highly toxic materials along with the signal words "DANGER" and "POISON."

(j) *Precautionary statement.* This statement relates the ways in which the pesticide may be poisonous to humans and animals. It also relates any special steps required to avoid poisoning (e.g., protective equipment needed). If the pesticide is highly toxic, proper treatment for poisoning is given.

(k) *Statement of practical treatment.* This infor-

Table 32.8

Meaning of signal words on pesticide labels

Signal word	Toxicity	Approximate amount that will kill the average person
DANGER	Highly toxic	A taste to a teaspoonful
WARNING	Moderately toxic	A teaspoon to a tablespoon
CAUTION	Low toxicity or comparatively free from danger	An ounce to more than a pint

mation tells what emergency first-aid measures are needed if swallowing, inhaling, or getting the pesticide on the skin or in the eyes would be harmful. It also tells what type of exposure requires medical attention.

(l) *Statement of classification.* The label must state whether the pesticide is for general use or restricted use. If the pesticide causes minimal or no probable harm to the applicator or to the environment when it is used exactly as directed, it is labeled as a general-use pesticide with the wording "General Classification." If the pesticide could cause some human injury or environmental damage even when used as directed, it is classified as restricted use with the label reading, "Restricted use pesticides for retail sale to and application only by certified applicators or persons under their direct supervision." This statement must be at the top of the front panel of the label.

(m) *Directions for use.* This section of the label tells the correct way to apply the pesticide. It lists the specific pests that the product is registered to control, the crop, the animal, or other items that the product can be used on, whether it is for general or restricted use, in what form the product should be applied, and how much, when, and where the product should be applied. Other parts of this section include the misuse statement (a reminder that it is a violation of federal law to use the product in a manner inconsistent with the label); the reentry statement (if required, this indicates how much time must pass before a pesticide-treated area is safe for reentry by a person without protective equipment or clothing); the category of applicator (if required for the pesticide); and, if needed, a statement limiting use to certain categories of commercial applicators.

(n) *Storage and disposal instructions.* Specific instructions are given for the storage and disposal of the pesticide as formulated in a given labeled container (See section on Safety for more information about storage).

Tolerances and Action Levels

A tolerance is the amount of residue of a pesticide legally permitted in food, animal feed, or raw agricultural commodities. It represents the maximum residue allowable in the commodity as a result of pesticide application to the agricultural commodity. The setting of tolerances is the responsibility of EPA, under the authority of the FD&C Act. Section 408 deals with Pesticide Petitions (PP) or tolerances for raw agricultural commodities. In the absence of a separate food-additive tolerance, the tolerance level for an agricultural commodity must not be exceeded, as a result of processing, in any finished product. Frequently, registrants for an insecticide also submit a Food Additive Petition (FAP) to gain acceptance of tolerances for finished products (section 409). Table 32.9 lists examples of approved finished-product tolerances for EPA-approved insecticides as listed in the April 1987 issue of Title 21 the *Code of Federal Regulations* (CFR), Part 193.

A food can become contaminated with an insecticide as a result of insecticide use in and around a food establishment. Section 406 of the FD&C Act authorized FDA to establish tolerances for residues of pesticides as added poisonous or deleterious substances that are required in the production of food or that cannot otherwise be avoided by Good Manufacturing Practices.

In the absence of a tolerance under section 406, and provided that a tolerance is not in effect under sections 408 or 409, FDA may establish an action level for unavoidable pesticide residues in food. The level at which an FDA action level is established is based on EPA's recommendation. Action levels are established and revised according to criteria specified in 21 CFR Parts 109 and 509 and are revoked when a regulation establishing a tolerance for the same substance and use becomes effective. Action levels and tolerances represent limits at or above which FDA may take legal action to remove adulterated products from the market.

Where no established action level or tolerance exists, FDA may take legal action against the product at the minimal detectable level of the contaminants. Action levels and tolerances are established on the premise that certain poisonous or deleterious substances are unavoidable; they do not represent permissible levels of contamination where such contamination is avoidable. The blending of a food (or feed) containing a substance in excess of an action level or tolerance with another food (or feed) is not permitted, and the final product from blending is unlawful regardless of the level of the contaminant. The only insecticide approved by EPA for use in food establishments that now has an action level established by EPA (as of January 1985) is fenthion; the level is 0.3 ppm for ground red peppers.

Section 402(a)(2) of the FD&C Act describes the conditions under which a food containing a pesticide residue shall be deemed adulterated. Such food is considered by FDA to be actionable when (a) the pesticide residue level exceeds an established tolerance or an established action level; or (b) there is evidence clearly demonstrating that a pesticide residue is present due to misuse, regardless of whether there exists a tolerance or action level. Strict adherence to the label directions will prevent both misuse of the insecticide and excessive applications that may result in residuals that exceed tolerances. The best continuing source of information for agricultural commodities is *The Pesticide Chemical News Guide* published by Food Chemical News. Updates are released as needed. The task of establishing tolerances passed from FDA to EPA in 1970. Although EPA now establishes the tolerances for pesticide residues in raw agricultural commodities and in finished food or feed, FDA has retained the responsibility for their enforcement, except for meat, poultry, and eggs, which are enforced by USDA.

Safety

The primary hazards from the use of insecticides arise from misuse or not following the directions and precautions as they are given on the labels. The details of these directions and precautions depend not only on the degree and nature of the toxicity of the insecticide, but also on the stability of the toxicant.

All insecticides must be considered toxic to humans and animals (2). The application should be completely preplanned to protect the food materials, employees, and the facility, and it should be conducted in a manner that does not endanger the applicator(s). Insecticides should be stored so that they cannot be misused, or containers and unused materials disposed of in ways that do not endanger the environment and/or the food chain.

Some general principles of pesticide storage and disposal are:

(a) Place pesticides in a designated area reserved for pesticides.
(b) Control that area by lock and key and restrict distribution of the key.

Table 32.9

Finished-product tolerances for some insecticides

Insecticide	Tolerance (ppm)	Notes on uses around food
Acephate	0.02	In or on all foods
Diatomaceous earth	unlimited	Food must be removed or covered; can be used in processing or storage areas
DDVP	0.5	On packaged or bagged nonperishable processed foods, dried figs
Fenitrothion	0.5	Limited to spot treatments and crack & crevice treatments
Fenthion	1.0	In or on foods
MGK 264	10.0	Cannot be used directly on stored grains
Piperonyl butoxide	10.0	On dried foods of 4% or less fat
Propetamphos	0.1	On foods
Pyrethrins	1.0	On dried foods of 4% or less fat
Resmethrin	3.0	In or on foods

SOURCE: 40 CFR Part 193 (1987).

(c) Do not locate the pesticide storage area near food or feed products or operations.
(d) Store herbicides away from insecticides and fungicides.
(e) Keep the storage area cool (but above 0°) and as dry as possible.
(f) Provide good ventilation.
(g) Protect from rain, moisture, and excessive heat (such as direct sunlight or storage near heaters).
(h) Completely empty containers if at all possible. If contents are not completely used, reclose original containers as tightly as possible.
(i) Do not reuse containers, but dispose of them in an approved landfill. Triple rinsing of the container is required, and disposal of the rinsings must be carefully controlled whether through use as a diluent in a subsequent preparation or by discarding.
(j) Never dispose of used containers in areas where drainage is poor or leaching is likely to occur.
(k) Open dumping of pesticides or pesticide containers is prohibited.
(l) Have a fire extinguisher and gas mask handy.
(m) Keep a complete inventory of all pesticides in an office separate from the storage area.
(n) Keep inventories moving on a first-in, first-out basis, and avoid storage periods of more than 2 years.
(o) Storage and disposal of fumigants require special precautions (consult Chapter 33 and read the labeling associated with each fumigant product).

These principles are already well known, but they cannot be overemphasized.

Traps

Next to preventive measures, the most important aspect of a complete insect control program is early detection of possible problems, hence the necessity for an ongoing, appropriate inspection program. To assist in this, attractant traps should be used (see Chapter 26). All such traps include an attractant and a means of capturing and retaining the insects. A selection can be made from light (insect electrocutor) traps, adhesive (sticky) traps, food (bait), and/or pheromone traps. The insect electrocutor light trap monitors flying insects, both food-infesting and "environmental" species. This kind of trap, properly utilized (2), can also achieve a degree of control of certain pest populations. Adhesive traps and food traps catch flying and nonflying food-infesting insects as well as "environmental insects." Pheromone traps tend to be species- or group-specific for the major moth and beetle pests of stored food (2).

Mites

Mites present some risk to food establishments (see Chapter 6). Some, such as the cheese, driedfruit, brown flour, and grain mites, infest and damage food. Others such as clover mites may invade a food-handling facility. The main preventive measure against mites is dryness; the relative humidity inside a food facility should be kept at <70% and the moisture content of infestable foods should be <12%. Mite control by chemical means can be achieved by fumigation with methyl bromide (not phosphine), contact sprays (pyrethrins), or residuals (most organic phosphates). For applications outdoors or to the external surface of a building to control clover mites, dicofol (Kelthane) or chlorobenzilate should be used. Check the label for special reference to mites.

New Materials

Although at least one promising new synthetic pyrethroid, praeallethrin (MGK Company, Minneapolis, MN), is being tested at the Stored-Product Insect Research and Development Laboratory, Savannah, GA (F. H. Arthur, personal communication, 1991), the prospects of new insecticides being approved for food-establishment use are not bright. Economics is the key factor. The costs for developing and registering a new chemical pesticide run around $20 million to $30 million plus an additional $10 million once the new material is on hand. Furthermore, the commercial payoff is in the marketing of materials for agricultural rather than for food-establishment use.

Summary

Residual insecticides and space treatments are two chemical approaches to insect control. The appropriate use of residual insecticides is encouraged since residuals have both preventive and corrective properties. In view of the known disadvantages of chemicals, such as dangers to humans, products, and the environment, and the growing resistance to them by insects, all approved techniques of insect control should be considered to solve any given pest problem.

Good sanitation is basic, and the basis of sanitation is cleanup. Other approaches that merit consideration include biological control, the use of heat, and protective packaging. The importance of reading and following pesticide labels cannot be overemphasized. Many pesticides cannot be used around food; in others, the maximum residue allowable in a commodity—the tolerance level—has been set by EPA.

Insofar as the approval of possible new insecticides is concerned, the future is bleak. High development costs and regulatory requirements are the main factors preventing the entry of new pesticides into the food-industry market.

Cited References

1 Anonymous. 1984. Read the Label. *Pest Control Technology* 12(11)48–49.

2 Baur, F. J. (ed.). 1984. *Insect Management for Food Storage and Processing.* American Association of Cereal Chemists, St. Paul, MN.

3 Bottrell, D. R. 1979. *Integrated Pest Management.* Council on Environmental Quality, Washington, DC.

4 Brehm, W., and L. Krzak. 1983. How to properly care for your B&G sprayer. *Pest Control Technology* 11(7)68, 74, 89.

5 Frishman, A. 1982. Practical tips. *Pest Control Technology* 10(7)60.

6 Georghiou, G. P., and T. Saito (eds.). 1983. *Pest Resistance to Pesticides.* Plenum Press, New York, NY.

7 Koehler, C. S. 1985. Safe ways to control backyard pests. *American Forests* 91(6)13–16.

8 Russell, B. 1991. Government news: Residues and residuals are complicated issues. *Pest Control* 59(2)8.

9 U.S. Environmental Protection Agency. 1991. Parasitic and predaceous insects used to control insect pests; proposed exemption from a tolerance. *Federal Register* 56(2)234–235.

33
Fumigation in the Food Industry

Vernon E. Walter

Even the cleanest food plant can have insect or other pest problems occasionally. In some cases, these insects or other pests will be living in hidden, protected areas such as wall voids, false ceilings, or inside equipment where normal sprays and fogs cannot reach them. Even though these infestations may be hard to see and certainly are hard to eliminate, they must be controlled. The only form of pesticide that can penetrate into hollow block or brick walls or other protected areas is a fumigant.

A fumigant is a chemical that, at a required temperature and pressure, can exist in the gaseous state in sufficient concentration to be lethal to a given pest organism. The key part of this definition is that the fumigant is a gas. This means that it exists as single small molecules that can move between fibers of wood, between aggregate particles of concrete blocks, or through small openings in equipment. It also means that certain laws of physics apply to its movement and that these movements can be predicted if we understand the rules.

Advantages and Limitations of Fumigants

The chief advantage of a fumigant is that it can penetrate to almost any area and be lethal to a wide spectrum of pests. This permits control

of pests in their harborage instead of where they roam. Better control and quicker control can thus be obtained, and there will be less need for other pesticides to be used in the area.

There are, however, many limitations to the use of fumigants. All are poisonous to humans. Some fumigants can corrode certain metals, and some are flammable or even explosive under certain conditions. Some fumigants can leave illegal residues in a product or may create objectionable odors in some products.

Availability of Fumigants

Recently, in the United States (USA), the use of several fumigants has been banned. In most situations, the choice is between methyl bromide and hydrogen phosphide (phosphine). In special cases, however, ethylene oxide, propylene oxide, hydrogen cyanide, and chloropicrin can be used. In addition, vikane (sulfuryl fluoride) can be used if there is no chance that foods will be contaminated.

In at least one other country, the United Kingdom, there has been some study of the use of methyl chloroform (10) as a fumigant to replace some of the former uses of ethylene dibromide. Although methyl chloroform is widely used in the USA as an industrial solvent and degreaser, it also has been used as a fumigant in a few minor ways such as treatment of ant hills. It is also currently the inert diluent in some 5% dichlorvos (DDVP) formulations. Recent studies have suggested that methyl chloroform may be

Walter: Abash Insect Control Service, 110 North Sixteenth Street, McAllen, TX 78501.

441

a carcinogen; it is doubtful, therefore, that any U.S. manufacturer would spend money to try to register this material as a fumigant.

Reference manuals on this subject usually describe the properties of an "ideal fumigant" (6), but such a substance does not exist and probably never will exist. All have advantages and disadvantages. Since manufacturers seem to have little interest in creating any new fumigants, knowledge of how fumigant gases work is necessary so that wise choices can be made among the few remaining ones.

Conditions That Affect Fumigation

Fumigants as well as insects behave differently under different conditions (2, 8). Unless these responses to various conditions are understood, the fumigator will not be able to realize why one job is successful and another fails.

Temperature

Temperature is the most important environmental factor influencing the action of fumigants on insects. New fumigant labels require that the temperature be above 4.5° for any fumigation, but even product temperatures in the range of 10–20° may result in failures or other problems unless the effects of temperature are understood.

Sorption

Sorption is inversely proportional to temperature; therefore, a fumigant gas sorbs into a cold product more than it does into a warm product. So much gas may be sorbed into the upper or more peripheral layers of a product that insufficient gas reaches the more isolated areas.

Insects and related arthropods are cold-blooded; their respiratory and other activities depend directly on temperature. Below about 10°, most insects respire very little, and their metabolic functions are slowed down enough that they may not be killed by a normally fatal concentration of fumigant gas around them.

Molecular Motion

As the product and surrounding air warm up, insects can be killed with lower concentrations of gas. This is because the insects are breathing faster and because gas molecules move more rapidly in warm air than in cold air. This causes the gas to diffuse rapidly and evenly throughout the fumigated area.

In cold air the molecules not only move more slowly, but they also might not diffuse evenly to all areas where the insects are hiding. Even if only a few insects survive the fumigant treatment, the population can soon increase to dangerous levels.

Low-Temperature Problems

Some fumigant gases cannot be properly evolved in cold air. Methyl bromide must be above 3.6° to reach its boiling point. If the cylinder is colder than this, the natural vapor pressure of the gas will be less than one-half of what it would be at 25.2°, and the smaller stream of chemical coming out might remain as a liquid until it touches a warmer surface or warmer air. This could cause damage since liquid methyl bromide is a strong solvent. Methyl bromide can be warmed with a heat exchanger placed between the cylinder and the discharge point. Best results occur if the product to be treated is also warmed well ahead of the fumigation so that the infesting insects are breathing at a normal rate.

Requirements for Phosphine

Phosphine gas is released from aluminum phosphide formulations by the actions of both moisture and heat in the air. If these two conditions are absent, gas release will be slowed or prevented. In particular, ambient temperature and product temperature must be considered in the case of phosphine. There must be enough heat (and moisture) in the air to release the required dosage of gas within the scheduled time. Sometimes control of the insects does not occur because cool product temperatures affect insect respiration and gas penetration, even though the monitored levels showed Time × Concentration readings in the lethal range.

In addition to the steps necessary to evolve the gas, the fumigator must alter the time of exposure and/or the concentration of the gas. With methyl bromide, either choice is possible. With phosphine, little benefit is gained from increased dosage and concentration, but the kill can be enhanced with a longer exposure time. Pesticide labels take these facts into consideration in suggesting different dosages for different temperatures.

Movement of Gases

A fumigant is only effective if it contacts and enters the body of the insect or other pest. From relatively few points of origin, fumigants must spread out to reach all areas in order to be effective.

Graham's Law

The movement of gas molecules is governed by several laws. Graham's law states that the speed of movement of molecules in a vacuum is inversely proportional to the square root of its density and that densities are proportional to their molecular weights (mw). Only a few fumigations are done in vacuum chambers but the principles are useful for normal treatments. Fumigants listed in Exhibit 33.1 as having higher molecular weights will move slower than ones with lower molecular weights. Thus, phosphine (mw 34) will diffuse faster on normal air currents than methyl bromide (mw 95).

Fumigant Equilibrium

Another law states that gases move from regions of higher concentration to regions of lower concentration until there is an equilibrium. The fumigator will try to obtain equilibrium as quickly as possible through the choice of fumigants and through the use of fans.

There is air movement in virtually any structure at all times. This can be measured with simple smoke tubes or sophisticated instruments. If the normal air currents are not sufficient for the fumigant chosen, fans can be used. The number of fans used will vary with the fumigant manufacturer's recommendation and the experience of the fumigator.

Fumigators often place plastic sampling hoses in various parts of fumigated structures and then monitor the gas concentration at various times. These samples tell them when equilibrium has been reached and will later tell them the rate of loss of the gas.

Phosphine does not require extra fans since it is only 1.2 times the weight of air. Methyl bromide normally requires fans for at least the first 30 minutes to get the air and gas mixed.

Dosage and Concentration

Dosage is the total quantity of gas introduced into a region. Concentration is the quantity of gas that exists in a given space at a given time [concentration = quantity (usually mass) per unit volume].

The dosage is calculated to try to obtain the correct concentration for the time that the space will be exposed to the fumigant gas. Dosage varies with respect to different temperatures, differences in size of structures, and anticipated loss of gas. The manufacturer's published guidelines on dosage indicate a range of dosages to allow the fumigator to use his or her judgment on any particular job.

Leakage and Sorption

Concentration is affected by gas loss due to leakage and sorption. All buildings or other enclosed spaces leak gas to some extent. This can be affected by outside wind as well as the type of building construction and the degree of sealing of openings. Concentration is also affected by loss from sorption. The spongelike action of sorption can be a serious problem in grocery warehouses storing large amounts of charcoal. It is also a problem in treating some grain that has a lot of fine dust particles mixed with it. Sorption can occur in several ways. The gas can be combined chemically with the product, it can be held within grain or product, or it can be held just on the surface of a material. Obviously, the target insects will be unaffected by sorbed gases.

Concentration × Time

Fumigators know that kill depends on reaching a certain Time (in hours) × Concentration (in ounces/1,000 ft^3) factor. With one gas, this factor might be 100. In this case, the fumigator could hold 50 ounces per 1,000 ft^3 of this gas for 2 hours or hold 2 ounces per 1,000 ft^3 for 50 hours. The actual choice would probably reflect a higher concentration in the first few hours and a lower concentration as sorption and leakage occurred.

Applications of methyl bromide and some other gases can be varied by changing either the concentration or the time. Methyl bromide fumigations can also be monitored, and extra gas can be introduced through preplaced hoses. This allows a great deal of flexibility in the fumigation.

Phosphine does not seem to work better if the concentration is doubled above the recommended amount and the time is cut in half. Phosphine fumigations can be prolonged, though, and a good kill can be obtained with relatively low concentrations of the gas.

How Fumigants Kill

A small amount of fumigant may actually penetrate the external skeleton of the insect, but the primary lethal effect comes from the gas that enters the body by way of the tracheal system. This means that any condition that causes an insect to "breathe" rapidly also causes it to take in more fumigant gas; conversely, any condition that reduces the respiration rate also reduces the amount of gas entering the tracheal system.

Insects respire rapidly in warm weather and slowly in cold weather. They respire rapidly in the active larval and adult stages and slowly in the egg, pupal, or diapause/resting stages. Some evidence shows that carbon dioxide (CO_2) can increase the respiration rate of insects, and it is sometimes added to another fumigant to try to achieve a better kill. After the fumigant gets inside the insect, it interferes with enzyme actions or affects nerve tissue (11). In humans and insects, phosphine attaches to heme iron in the presence of oxygen and inhibits electron transport (5). Garry and colleagues (5) suspect that phosphine may be genotoxic in humans, but confirmation of this assertion will require much more experimental work.

Fumigations are normally performed only when the insects are warm and active—this is when their feeding and other activities are obvious to quality control personnel. Nevertheless, some treatments must be performed for quarantine or other reasons at temperatures down to 4.5°. Since insects under such conditions respire slowly, and the fumigant gas will be moving more slowly, more gas must be used to achieve the same kill. High kill rates cannot be obtained with some insects during their diapause/resting stages because of their low rate of respiration.

Phosphine fumigations that take a week or longer achieve some of their excellent kills because the insects metamorphose into other more susceptible or more active stages. Insects freshly emerged from the egg are usually easier to kill than they are in the egg stage.

Sealing Structures for Fumigation

Fumigation has sometimes been described as the "art of sealing." Leakage is obviously the chief reason for low concentration readings when proper dosage is used. Fumigant gases are designed to penetrate most materials, including some building materials, to reach hidden insect infestations. They quickly go through cinder block walls or wood and plaster walls. Even brick and concrete-block walls can permit excessive loss of fumigant gases.

Any structure can be modified to hold fumigant gases but some, such as residential structures, must be covered with gas-retaining tarps. These tarps can either be easily-torn polyethylene sheets or more durable and more expensive vinyl-coated nylon.

Each type of fumigation presents its own problems; the techniques of sealing are discussed with each type of fumigation.

Measuring Fumigant Concentrations

Fumigators must be able to measure the amount of gas left in a structure after aeration to know that it is safe to allow people back into the structure or even to allow them work next to a structure that is being fumigated. Fumigators also must be able to determine the concentrations at various times in order to protect themselves. Some gases present serious risk of harm to the operator through skin absorption alone if the concentration is still high. Knowing the concentration of the fumigant gas at various times during the fumigation enables the fumigator to find and correct excessive leaks. It also enables the fumigator to alter the time or to add more gas to obtain the correct factor of Time × Concentration.

Colorimetric Systems

Several kinds of instruments are available to measure various gases in the air; more are being developed. Colorimetric (detector) tubes draw 0.1-liter volumes of air through freshly opened glass tubes containing appropriate reactive chemicals. When a specified number of 0.1-liter volumes of air has been drawn through the tubes, the concentration of the gas is read in terms of a change in color of the chemicals in the tubes against calibration colors on the sides of the tubes. The reactive chemicals in the tubes vary for the gas and/or the concentration being tested.

Instruments have been designed to use these tubes for either single readings with hand-operated pumps or for longer exposures with electrical pumps. Colorimetric tubes cost $2.00 to 3.00 each; hand-operated pumps cost about $150 each. This is a reasonable cost for measurements that are usually within 10–25% of the readings given by the most sophisticated equipment.

Halide Detectors

Halide leak detectors are used by technicians to spot leaks in refrigeration systems. Their wide use and simple construction make them the least expensive tool per measurement for thoses gases for which they are designed. Air and gases are pulled through a hose into a gas flame. The flame heats a special copper ring upon which the gases react. Any gas that contains bromine, iodine, chlorine, or fluorine causes a blue or green halo around the flame. The concentration of the gas can be estimated according to the color of the flame.

Halide detectors are useful in locating leaks in a structure that is being fumigated, but they cannot give an accurate reading of the gas concentration during a fumigation and certainly cannot read down to 5 ppm to show when safe reentry is possible without protection. Since halide detectors use an open flame, they cannot be used with flammable or explosive gases. They cannot detect phosphine. Their only use at the present is with methyl bromide.

Thermal Conductivity

Thermal conductivity instruments (e.g., Fumiscope) are used to monitor the concentration of gases such as methyl bromide. They are fairly expensive and must be recalibrated regularly. However, they are not sensitive enough to read down to the 5-ppm level.

Infrared Devices

Infrared instruments are among the most accurate devices available to monitor concentrations of gas. They are also the most expensive and the most delicate of the instruments. They utilize the principle that many gases have characteristic infrared absorption properties sufficiently intense to be measured by infrared spectrophotometers. Problems can occur with this and other instruments when several different gases are present in the atmosphere or if there is too much water vapor in the air. Special filters are needed in these cases.

Exhibit 33.1 suggests some instruments to measure each gas. As many as three different instruments might be needed on a methyl bromide fumigation to spot leaks, check concentration, and then to "clear" the building for safe reentry.

New electronic equipment has been developed to measure low concentrations of phosphine. Although quite expensive, these units are useful additions to equipment. Similar units for methyl bromide are available in Japan and also seem to be very accurate.

Criteria for Choosing Fumigants

Time

Time is often an overriding consideration. When there are less than 48 hours available for exposure, phosphine would not be the choice with the exception of spot fumigation. Normally, with phosphine, at least 72 hours are needed to kill all life stages of insects.

Sulfur

Certain items with a free sulfur radical, such as some rug padding, some cinder blocks, certain types of photographic chemicals, and other listed materials, should not be fumigated with methyl bromide because a bad odor may be produced. If there is doubt about the materials, a test fumigation at even higher concentrations should be tried first.

Metals

Metals such as copper, copper alloys, gold, and silver can be corroded by phosphine (3). This is particularly true when the metal is exposed to high concentrations of phosphine coupled with high temperatures and high humidities. Corrosion can cause electrical contact points to fail to operate. This problem occurs rarely but can be serious and can even continue after aeration in some cases. Telephones and other delicate electrical equipment should be removed. Covering these items with plastic bags will not prevent corrosion and, indeed, may enhance the effect.

Commodity

Potatoes and some other vegetables or fruits can be affected by aluminum phosphide fumigants. This could result in potatoes being discolored black or purple with a subsequent loss of market value.

Germination

Protection of seed germination capacity is important in some cases. Barley, for example, must germinate to permit the production of malt. Methyl bromide can reduce the germination rate of some seeds at some moisture levels. Phosphine does not diminish the germination capacity of any known seeds, and thus, it is the fumigant of choice where germination is important.

Residues

Multiple methyl bromide fumigations (and sometimes even a single one) can result in residues of inorganic bromides (9) that exceed the Raw Agricultural Commodity Tolerance or the Food Additive Tolerance. Most grains, other than oats, can be safely fumigated several times with methyl bromide, but there could have been previous fumigations of the product or of the original grain that the fumigator would not know about. If there is concern about possible excess residues, phosphine would be the fumigant of choice. The low permitted levels of phosphine are virtually never reached since the gas is not readily retained.

Packaging

Insects in products sealed in plastic bags or other hard-to-penetrate packages might survive

Exhibit 33.1

Currently available fumigants

	Vikane (sulfuryl fluoride)	Methyl Bromide	Hydrogen Cyanide[a]	Ethylene Oxide[a]	Aluminum Phosphide	Magnesium Phosphide	Dichlorvos DDVP[b]
Major Uses	Dwellings, buildings, construction materials, vehicles (not aircraft)	Space; grain, static vehicles, foods	Space; grain, commodities	Commodities, spices, surgical equipment, certain vehicles	Grain, vehicles, buildings	Spot; space; vehicles	Space; contact spray
Gas Evolved From	Compressed gas cylinders	Compressed gas cylinders	Compressed gas cylinders	Compressed gas cylinders	Tablets, pellets, sachets, PrePacs	Tablets, Magtoxin Plates, Fumi-Cel, Fumi-Strip	Liquid & strips release vapor; aerosols also used
Speed of Kill	Very quick	Very quick	Extremely quick	Quick	Slow (2–3 days)	Slow	Quick
Penetration	Excellent	Very good	Fair	Fair	Excellent	Excellent	Poor
Ease of Aeration	Very good (but monitor)	Good	Fair	Good	Very good	Very good	Very good
Mixed with Other Gases	Chloropicrin used as warning agent	Chloropicrin may be used as warning if foods and drugs are not exposed	No	Was often mixed with carbon dioxide to make carboxide[a]	Usually not	Usually not	May be sprayed on exterior surface of machinery after spot fumigation
Sorption	Not a problem	May be a problem	Yes	High in presence of water	Slight (but do not sell products for 48 hours)	Not a problem	Low
Molecular Weight of Gas	102	94.94	27.03	44.05	34.04	34.04	221
Specific Gravity of Gas (air = 1)	1.3 at 77°F (25°C)	3.27 at 0°C	.9 (HCN is the only fumigant lighter than air)	1.521	1.214	1.214	7.6
Solubility in Water	Very low; 0.075% at 25°C	Very low; 1.5g/100g at 20°C	Infinite at all temperatures	Infinite at 0°C	Very slightly soluble cold H_2O	Very slightly soluble in cold H_2O	Slight (hydrolyzes slowly in H_2O)
Latent Heat of Vaporization	79.5 BTU/pound at −67°F (−55.2°C)	61.52 cal/g	210 cal/g	139 cal/g	102.6 cal/g	102.6 cal/g	—
Odor	None	None at normal concentration	Almond-like	Mustard-like, irritating	Carbide, garlic or ammonia	Carbide or garlic	Slightly acidic
Boiling Point	−55.2°C (−67°F) at 760 mm Hg	3.6°C (38°F)	26°C (78.8°F)	10.7°C (51.25°F)	−87.4°C (−125.32°F)	−87.4°C (−125.32°F)	120°C at 14 mm/Hg
Skin Absorption of Gas	Not a problem (but liquid can freeze skin on contact)	Yes (slight)	Yes (rapid)	Yes	Negligible	Negligible	Yes
Chronic Poison (all are acute)	No	Yes	No	No	No	No	Possible
Threshold Limit Value	5 ppm	5 ppm now suggested	10 ppm	1 ppm	.3 ppm	.3 ppm	350 ppm in 80% Trichlorethylene solvent
Short Term Exposure Limit	10 ppm	Not listed on label	0 ppm	0 ppm	Not listed on label	Not listed on label	None
Skin Blistering	None from gas	Yes	No	Yes	No	No	No
Flammability	Nonflammable under ordinary conditions but can form corrosive acid	Nonflammable under ordinary conditions but can form corrosive acid	Yes (when 6–41% by volume in air)	Yes (when 3–80% by volume in air)	Self combustible over 17,900 ppm or 1.79% in air (rare)	Self combustible over 17,900 ppm or 1.79% in air (rare)	No (but solvent may be flammable)
Effect on Germination	Not registered for seed treatment; do not use on seeds	Variable with kind of seed and moisture content	Safe for most seeds at low moisture content	Harmful to many kinds of seeds	None known	None known	None known (but usually not used on seeds)
Reacts with	Unreactive with most materials; does react with strong bases	As gas, with sulfur compounds; liquid reacts with aluminum and magnesium	Levulose in dried fruits under certain conditions	Certain vitamins, salt, rubber	Copper, gold, silver, brass, 3M copy paper	Copper, gold, silver, brass, 3M copy paper	May be corrosive on black iron and mild steel; possible effects on other metals
Detection	Monitor with Fumiscope; clearance with Interscan VK50 Miran 101, Vikane Detector with Kitagawa tubes	Monitor with Fumiscope; leaks with halide leak detector; clear with detector tubes	Detector tubes; acetate or methyl orange papers	Detector tubes	Detector tubes; GasTech phosphine detector	Detector tubes; GasTech phosphine detector	Detector tubes

(continued)

Exhibit 33.2 (continued)
Currently available fumigants

	Vikane (sulfuryl fluoride)	Methyl Bromide	Hydrogen Cyanide[a]	Ethylene Oxide[a]	Aluminum Phosphide	Magnesium Phosphide	Dichlorvos DDVP[b]
Respiratory Protection	SCBA only	SCBA now used	Gas mask; full-face SCBA now recommended	SCBA recommended	0–0.3 ppm—none required 0.3–15 ppm—full face mask with MSA GMHS-SSW cannister >15 ppm or unknown—SCBA only	0–0.3 ppm—none required 0.3–15 ppm—full face mask with MSA GMHS-SSW cannister >15 ppm or unknown—SCBA only	SCBA only
Remarks	Not to be used on food or drugs; these must be removed before fumigation; excellent control of drywood termites	Good space fumigant for food-industry plants; good with grains when recirculated; kills eggs of powderpost beetles	No longer used in USA; very dangerous to applicator; has caused explosions; even empty cylinder explosive	Manufacturer no longer recommends use as fumigant	Our most penetrating fumigant; poor on mites; best grain fumigant for insect control	Forms hydrogen phosphide gas as does aluminum phosphide; Mg phosphide comes off faster; Mag-toxin may be used as spot fumigant in some states	Often used as space treatment; effective on exposed insects; often ineffective on spiders

[a]No longer available in the USA.
[b]DDVP included in this chart because it has some properties similar to fumigants and because of its recent increased usage.

a methyl bromide treatment. Phosphine, however, would probably give good kills in such a situation, and is the fumigant of choice where penetration is very important.

Methyl Bromide

Methyl bromide (Exhibit 33.2) has a long history of use on a variety of pests in many situations, and it has been used very successfully in most. Methyl bromide is available in 1-lb or 1.5-lb cans, in 50-, 100-, and 200-lb cylinders, and in bulk "pigs." The dosage is usually 1–3 lb/ 1,000 ft³ but this depends on many different factors. The label provides guidelines on dosage.

Whenever possible, methyl bromide gas should be injected into a structure through plastic tubes; used in this way, the can or cylinder can be opened outside for added safety. Part of a cylinder can be discharged by placing the cylinder on scales and then setting the beam back for the required amount. The discharge is continued until the new weight is indicated.

The total dosage of all compressed gases is available as soon as the gases are dispersed; in contrast, phosphine is generated over 2 or more days. Methyl bromide, like other compressed gases, exhibits the latent heat-of-evaporation phenomenon and may remain a liquid for a period of time after release. Because the liquid form is a strong solvent, damage may occur if precautions are not taken.

Methyl bromide is sometimes passed through a copper coil immersed in hot water to ensure complete vaporization. It also can be discharged into a pan or over a plastic sheet to provide a safe evaporating surface. Slow release may cause the valve to freeze.

Methyl bromide should never be released through aluminum or magnesium pipes because an explosive compound could be generated when liquid methyl bromide contacts these metals in the absence of oxygen. Only copper, steel, or plastic (polyethylene) pipe should be used.

Methyl bromide fumigations can be accelerated by increasing the dosage. As long as the same factor of Time × Concentration is met, the fumigation should be successful. Thus, 1 lb of methyl bromide/1,000 ft³ in a 24-hour period could be changed to 2 lb in 12 hours, and the result should be the same. Some fumigators hesitate to vary the dosage much more than twofold for fear of possible additional residue problems.

Methyl bromide may cause serious burns if it contacts the skin. Gloves, rings, or bandages should *not* be worn during fumigation or aeration. If the methyl bromide should contact any clothing, the clothing should be removed immediately and not worn again until it has been aerated and washed. Since shoes can retain methyl bromide for a very long time, they should be discarded if contaminated.

Methyl bromide can affect the liver and kidneys (7). Effects of the consumption of alcoholic beverages during the same period that exposure to methyl bromide occurs would be multiplicative on these vital organs. Because alcohol could also affect vital decisionmaking or otherwise interfere with the ability of the fumigator, abstinence for 24 hours before a fumigation, during the fumigation, and for 24 hours after the last potential exposure is recommended.

Hydrogen Phosphide Gas (Phosphine)

Aluminum Phosphide

Phosphine is the most penetrating gas but it cannot be compressed safely and sold in cylinders. Manufacturers have come up with several methods of generating the gas on location from various compounds. One manufacturer compresses aluminum phosphide, paraffin, and ammonium carbonate into tablets or pellets that release phosphine when exposed to warm moist air at the rate of 1 gram for each tablet and 0.2 gram for each pellet.

Another manufacturer makes tablets and pellets from a mixture of aluminum phosphide, ammonium carbamate, and urea. Again, the amount of phosphine released is the same. The two original manufacturers of phosphine-producing materials have been joined by a number of others manufacturing similar products.

The pellets and tablets are available in resealable flasks or they can be obtained in bags or precounted plastic strips to facilitate railcar fumigation.

Magnesium Phosphide

One manufacturer has marketed products prepared from magnesium phosphide. Phosphine gas is still the product generated, again by the action of warm moist air, but the gas can be evolved somewhat more quickly and possibly at lower temperatures. In addition to tablets and pellets, plates and strips are available with this product. The pellets are also available packaged in a paper strip that can be used in spot fumigations (discussed later).

Limitations

One limitation of phosphine is that both adequate air temperature and sufficient air moisture must be present to generate the gas. In cold dry winter conditions, weeks may be needed instead of days to create the phosphine gas. Leakage might exceed this slow release in practical situations.

Another limitation is the long time required for a kill. At least 2 days at temperatures above 26.9° are required, and 3 or more days are more commonly used to ensure a kill.

Under conditions of high humidity, high temperature, and high gas concentration, phosphine can corrode copper, gold, and silver (3); this can seriously damage electrical equipment. This phenomenon does not occur often but should be considered with every treatment. Temperature and humidity readings should be taken before each application where products or equipment could be damaged. Even if the readings are low, it is better to remove portable equipment.

The slow release of phosphine is considered to be a safety factor. It is important to measure gas concentrations from the outset and not just assume that there are no dangerous levels in the first hour or 2. Proper respiratory equipment should be either worn on the job or be immediately available if needed.

Disposal

The claylike dust remaining after the gas has evolved can be left in grain if normal cleaning later on would remove it and other dusts. If aluminum phosphide is used in food warehouses or transportation vehicles, the dust should be recovered and disposed of according to the manufacturer's suggestion. One such method is to slowly stir the dust into a barrel of water and detergent outside the building. This releases any remaining phosphine. Workers doing this task should be protected with proper respiratory equipment.

After use, spent packaged materials in bags or plastic and paper strips should be stored in the perforated drums or wire baskets provided by the manufacturer. They should be kept in these containers until there is no chance of any gas remaining.

The final residue can be taken to a landfill in most states, but it is better to contact the landfill management first to explain what the material is and what steps have been taken to deactivate it. If conditions permit, spent magnesium phosphide plates or strips are often buried in a trench dug at the job site. This avoids any transportation problems. Once covered with soil, the residual dust will release any small amount of remaining gas into the soil; this should cause no problem if the disposal site is sufficiently distant from occupied areas.

Ethylene Oxide

Ethylene oxide fumigant is more than just an insecticide; it is also a bactericide used in hospitals to sterilize surgical instruments. In addition, it is used by food processors to fumigate imported spices that might be contaminated or infested.

Ethylene oxide is explosive and must be handled with care; it is often mixed with carbon

Exhibit 33.2

COMMODITIES UNSUITED FOR METHYL BROMIDE FUMIGATION

Methyl bromide is a fumigant which has proven its value in protecting a wide variety of stored commodities from rodent or insect damage. But methyl bromide must be used with discretion. There are materials which should not be exposed to methyl bromide, or should be exposed only under very carefully controlled conditions.

The most common reason for avoiding fumigation of a material with methyl bromide is the off-odor resulting from a reaction of the fumigant with certain sulfur compounds. These odors usually persist indefinitely and in most cases there is no practical way to remove them.

Some materials have sorptive qualities or a solvent effect which will reduce the methyl bromide concentration in the fumigated area to the point of ineffectiveness. Other reasons for caution with methyl bromide exposure involve phytotoxicity (toxicity to growing plants), destruction of seed viability, the possibility of illegal residues, and rapid deterioration after fumigation of commodities such as fresh fruits and vegetables.

The following is a list of materials which should not be exposed to methyl bromide:

1. *Foodstuffs*
   a. Iodized salt stabilized with sodium hyposulfite.
   b. Full fat soya flour.
   c. Certain baking sodas, cattle licks (i.e., salt blocks), or other foodstuffs containing reactive sulfur compounds.
   d. Fresh fruits and vegetables.†

Note: Never exceed the recommended dosage or exposure period for food or feedstuff commodities. Prior to repeated fumigation, have the food commodity analyzed for inorganic bromide residues.

2. *Seeds, Bulbs, and Plants*
   a. Seeds and bulbs to be used for planting.†
   b. Nursery stock and other living plants.†

3. *Pets*
   (All pets, including fish and birds.)

4. *Rubber Goods*
   a. Sponge rubber.
   b. Foam rubber, as in rug padding, pillows, cushions, mattresses, and some car seats.
   c. Rubber stamps and other similar forms of reclaimed rubber.

5. *Furs*

6. *Horsehair*

7. *Feathers*
   (Especially in feather pillows.)

8. *Leather Goods*
   (Particularly white kid or other leather goods tanned with sulfur processes.)

9. *Woolens*
   (Extreme caution should be used in the fumigation of Angora woolens. Some adverse effects have been noted on woolen socks, sweaters, shawls, and yarn.)

10. *Viscose Rayon*
    (Those rayons processed or manufactured by a process in which carbon bisulfide is used.)

11. *Vinyl*

12. *Paper*
    a. Silver polishing papers.
    b. Certain writing and other papers cured by sulfide processes.
    c. Photographic prints and blueprints stored in quantity.
    d. "Carbonless" carbon paper.
    e. Blueprint papers.

†For specific information on procedures to prevent commodity injury, contact
The Industrial Fumigant Company or the U.S.D.A.

13. *Cellophane‡*

14. *Photographic Chemicals*
    ("Darkroom" chemicals, but not cameras or film.)

15. *Rug Padding*
    (Foam rubber, felt, etc.)

16. *Cinder Blocks*

17. *Mixed Concrete*
    (Occasionally picks up odors.)

18. *Mixtures of Mortar and/or Soil Used for Chinking Log Cabins*

19. *Charcoal*
    (Methyl bromide is readily absorbed by charcoal. This may not only contaminate such materials, but may reduce the concentration of the gas in the fumigated area to the point of ineffectiveness.)

‡*In the event of uncertainty about the possible presence of reactive sulfur compounds, conduct a trial fumigation of a small quantity of the material in question.*

Readers who have had adverse experiences with other materials or products which they believe should be added to the list of "Commodities Unsuited for Fumigation with Methyl Bromide" are urged to write to Industrial Fumigant Company, P.O. Box 1200, 601 East 159th Street, Olathe, Kansas 66061.

RESTRICTED USE
PESTICIDE

For retail sale to and use only by Certified Applicators or persons under their direct supervision and only for those uses covered by the Certified Applicator's certification.

NOTICE: Seller warrants that the product conforms to its chemical description and is reasonably fit for the recommended purposes stated on the label when used in accordance with directions under normal conditions of use, but neither this warrany nor any other warranty of MERCHANTABILITY or FITNESS for a PARTICULAR PURPOSE, express or implied, extends the use of this product contrary to label instructions, or under abnormal conditions, or under conditions not reasonably foreseeable to Seller, and Buyer assumes the risk of any such use.

dioxide at the rate of 9 parts CO_2 to 1 part ethylene oxide. For some sterilization uses, it must be used in its pure form. This requires trained workers and adequate safeguards. Worker exposure to ethylene oxide is now highly restricted because of reports that this chemical may be a carcinogen. The only possible substitute for ethylene oxide is propylene oxide, but this chemical is endowed with a similar set of disadvantages.

Ethylene oxide is normally used in the food industry only in fumigation chambers. It is shipped as a compressed gas in cylinders. The gas is fed into the fumigation chamber by means of tubing. The cylinder must be grounded to avoid sparks from static electricity.

Ethylene oxide–carbon dioxide mixtures are used to fumigate vehicles such as railcars. An entire cylinder is used per car in this type of work, and the cylinder must be fastened down so that it cannot tip over during discharge.

Carbon Dioxide

Carbon dioxide has been used in fumigation mixtures for many years (14). It causes insects to "breathe" faster and thus "inhale" more of the fumigant. It also helps suppress the risk of fire or explosion in some fumigant mixtures.

Carbon dioxide is now gaining favor as a fumigant in its own right. When it constitutes about two-thirds of the total weight of the air, it is a slow but effective fumigant. It has been used for a number of years in Australia in long-term pit storage of grain. One food company in the USA has done considerable testing of carbon dioxide as a fumigant for in-transit fumigation. Both the dry-ice form and the compressed gas have been tested. It has definite advantages in that there is no residue problem and the railcar is much safer to open on arrival.

Carbon dioxide has limitations similar to all other fumigants. Its action is very slow; 4–10 days are required to obtain kill of all stages of insects. Massive amounts of the gas are needed to kill insects in a large mill. An entire railroad tanker load may be needed to treat a large plant. The gas is first introduced to purge the existing atmosphere, and then carbon dioxide must be continually added to maintain the concentration.

Other inert gases such as nitrogen have been tested but seem to be less effective than carbon dioxide. There has been some use of equipment to burn up the oxygen in the air and leave an atmosphere of carbon dioxide and other inert gases. This is done on grain rather than on processed foods, and in some cases may be a low-cost method of obtaining control.

In-Transit Fumigation of Products or Ingredients

Products or ingredients transported by railcar or truck may arrive at their destination infested with a small population of insect pests. It is possible, although not common with processed foods, that the product was infested when it was loaded into the vehicle. The vehicle can be infested before being loaded, and surveys show that this does indeed happen. Previous shipments may have been infested, leaving residual populations in trucks or cars. Railcars are parked more hours than they are in motion during most shipments. These rail sidings may have infested spillage or may otherwise be infested. The insects at the siding may fly or crawl through cracks in the doors or enter through holes, tears, or vents in the walls or floors of the vehicle. When it is done right, in-transit fumigation has been a very effective technique to combat railcar infestations (12, 13). The procedure also has been used successfully in ships (4).

Label Restrictions

At the time of this writing, in-transit fumigation is legal in railcars, "piggyback" truck trailers on flat cars, in barges, and in ships. It is not legal to fumigate truck trailers while they are in transit. Because the food industry uses many more trucks than railcars, there have been requests to reconsider this law and permit in-transit truck-trailer fumigation.

In the USA, phosphine is the only gas labeled for in-transit fumigation of processed foods. The slow release of phosphine gas over 3 or more days fits well with the long period of travel that rail shipments normally take. The continued release of the gas helps maintain the concentration, replacing what may be lost due to leakage (if the leaks are not too great) or sorption.

Sealing, Temperature, and Humidity

In-transit fumigations can be very effective if the sealing is done properly and if the temperature and humidity are adequate to dissipate the gas from the tablets, pellets, bags, or other manufacturer-supplied units. The temperature of the product should be maintained above 4.5°. In reality, temperatures above 15.7° are probably needed for short trips. Eggs, pupae, and other dormant stages are easier to kill if both the product and the ambient air are warm.

The humidity in the air is just as important

as the temperature since the breakdown of aluminum phosphide requires moisture. In cold, dry areas, the breakdown and dissipation of the gas could take weeks instead of days, and the concentration might not be high enough to kill insects that would be semidormant at these temperatures.

Most failures to control insects during in-transit fumigation are caused by failure to maintain the gas concentration for a sufficient time. This is usually due to holes in the car that were not discovered or to ineffective sealing of the doors or hatches. Drain holes in the four corners of some railcars can be missed. When a forklift is used to open the door of a railcar, the prongs may slip and gouge a hole in the sidewall. All of these can cause leakage of gas.

A far too common method of sealing doors on railcars is to use masking tape between the door and the door frame. It is easy to miss an opening at the bottom of the door since there are many different types of doors. Even if this area is sealed properly, the tape will often break the first time that the railcar is "humped." This jarring pulls the tape loose or breaks it at the seam. Even small leaks can cause the concentration to remain too low to be effective.

A far better sealing method is to glue 4-mil polyethylene sheeting to one door frame before the car is loaded and then glue another precut poly sheet to the other door frame after loading and treatment. The door can then be carefully closed, and it can bounce freely on its overhead track without affecting the seal of the plastic to the door frame.

Before shippers undertake in-transit fumigation, they should have a written agreement with the receiver stating that the receiver does indeed want the fumigation done and that the receiver will furnish a properly qualified person to open the car and measure the gas level to determine if it is low enough to permit entry and/or unloading, and that the residual material will be disposed of according to the manufacturer's instructions. The shipper should always placard the car to warn anyone that the gas is inside. Opening instructions are often placed in a glued-on envelope next to the placard. The fumigator may provide extra safety insurance by using special door seals or placing a second placard on the inner seal.

The manufacturers have developed convenient packages that make this type of treatment easy and relatively safe. Since hopper cars are usually much tighter than boxcars, they have a better record of control, but even old boxcars can be successfully fumigated if properly sealed.

Bin or Silo Fumigation

Any storage container for cereal grains and cereal products can become infested; bins and silos are no exceptions. It is difficult to force a fumigant gas through a large mass of flour or similar finely ground material. Food processors often try to control infestations by fumigating the silos or bins when they are empty, and the gas can be effectively dispersed throughout the entire volume.

Some breweries routinely fumigate the headspace at the top of silos because the Indianmeal moth (*Plodia interpunctella*) and some other insects that might occur in the ingredients would be primarily in the upper area. There is no thought here that penetration by the gas might reach the bottom of the silo. Indeed, the dosage is based on treating the upper area only.

If bins are treated inside a building, there may be a problem of exhausting the gas at the end of fumigation. The solution to this problem must be planned before the fumigation begins. The size of the fumigated bin should be considered in relation to the surrounding area and the distance to the outside to decide whether or not a dangerous concentration might exist when the remaining gas is released. It may be necessary to install temporary ducts and fans.

Tarp Fumigation

The food processor may discover that just one small lot is infested and needs to be fumigated. This lot can be placed over a nonporous surface such as concrete, covered with polyethylene or other gas-retaining cover, and fumigated (*1*).

Failures may occur if the fumigation is done over loose, dry soil, gravel, or even asphalt, since some of the gas may seep through these surfaces and be lost. The gas can also escape through holes or tears in tarps, through the bottom seal of the tarp, or through canvas or other tarp fabrics that were not designed to contain a gas.

Successful jobs can be done on concrete if the concrete does not have any open joints in the area being used. The material should be placed on pallets to permit air flow, covered with a 2-mil or 4-mil polyethylene sheet or a clean vinyl-coated nylon tarp, and the edges sealed to the floor with tape or with weights such as sand snakes (sand-filled cloth tubes). Methyl bromide can be fed into the prepared area with ¼-inch polyethylene tubing. Phosphine-producing materials can be placed under the tarp in trays or other containers.

If a large area is being treated with methyl bromide, a fan should be put under the tarp at the release point and run for at least 30 minutes after the gas release. This ensures dispersion. Fans are not needed with phosphine since it is only 1.2 times the weight of air; it disperses easier than methyl bromide (about 3.5 times the weight of air).

If fumigations must be done over porous sub-surfaces, a tarp can be placed on the ground, the material to be fumigated carefully placed on the tarp and then covered with another tarp. The tarps are sealed by overlapping a foot of each tarp and rolling them into a tight continuous seam that is held together with spring steel clamps at 1-foot intervals.

Methyl bromide tarp treatments should last from 2 hours for sensitive fresh fruits and vegetables (but see Exhibit 33.1) to 24 hours for most cereal products. Phosphine fumigations normally run for at least 3 days to ensure kill of all stages of the insects.

General Fumigation

General fumigation refers to the fumigation of an entire building or a large section thereof. This is done when an infestation is believed to be scattered widely over the area and is hidden too deep to permit other means of control to be effective. General fumigations are sometimes done routinely on the first available holiday in the spring to ensure complete kill of virtually all pests before the warm summer temperatures permit rapid pest reproduction.

Selection of Fumigant

Phosphine is the fumigant of choice if at least 3 days are available and if good penetration is needed. Phosphine might be contraindicated if sensitive electrical equipment cannot be removed.

Methyl bromide would be chosen if only 24 hours were available for fumigant exposure, and if none of the products would be harmed by exposure to this gas. The list of products that may be harmed should always be checked (Exhibit 33.2).

Importance of Sealing

Most brick or concrete buildings can be sealed adequately by covering doors, windows, and vents with 4-mil polyethylene. Metal buildings usually leak too much gas at the top of the wall, at the base, and between the metal panels to permit effective fumigation if only sealing procedures can be used.

Sealing is the hardest and most important part of any fumigation. Certain techniques have proven helpful. Some vents on the roof can be sealed with plastic bags taped to the base. To avoid punctures and tears, the vent first can be covered with a cloth bag or another plastic bag. Steel single-piece doors can be sealed with 2- or 3-inch-wide tape around the edges. Most commonly used fumigant gases can penetrate tape to some extent, but the loss will be small on a large job. Wooden doors or roll-up doors cannot hold gas; such openings should be covered with polyethylene glued to the door frame. Small pieces of plastic can be held up with wide masking tape, but professional fumigators often use a special aerosol to make the surface tacky before they put the tape on. For large sections of plastic, glue such as Bondmaster is painted on the door frame and then the plastic is pressed onto the sticky surface.

Escape of Phosphine

It should be remembered that fumigants such as phosphine can penetrate through virtually any sealing material. The objective is merely to try to reduce the loss as much as possible. Indeed, there will be loss through brick walls, through concrete block walls, and through many other types of construction materials.

The success of a fumigation is dependent on holding a sufficient concentration of gas for a sufficient time. The gas concentration can be maintained only if leakage is held to a minimum. The better fumigators often double-tarp the doorways (the gas reading between the two tarps is often almost as high as that inside the fumigated area). Obviously, gas loss can be reduced by using two tarps.

Special Considerations with Methyl Bromide

Whenever possible, methyl bromide should be introduced from the outside of the building to be fumigated. Quarter-inch plastic hoses can be laid to lead the gas to key areas. The gas is usually discharged in front of a fan. Sometimes the cylinders are placed throughout the plant and then opened as the crew works from the back of the building to the exit point. This method requires careful planning and rehearsal to be sure that the applicator team does not cross back through a treated area and that no applicator will be exposed to an area where the gas concentration is above the STEL (Short-Term Exposure Limit—the concentration of gas that can be tolerated no more than 15 minutes). Applicator personnel should wear self-contained breathing to the site of the general fumigation, it may be necessary to cancel the fumigation job. Gas can travel from a fumigated building to another nearby that the team will not be delayed by a sticking valve.

Since the gas quickly cools upon release from the cylinder, it could fall to the floor in liquid form. This liquid form is a strong solvent and can damage floor tiles or other materials and, therefore, precautions are needed. Some fumigators use tubing and special nozzles to atomize the gas near the ceiling level. *Caution:* Aluminum or magnesium pipes should *not* be used since methyl bromide liquid passing through them can form explosive material. Some fumigators spread out plastic sheets for about 20 feet in front of the cylinder to catch any liquid that falls and allow it to become a gas. Heat exchangers are used outside buildings; indoors they would be too slow and would possibly subject the operator to extra exposure.

How to Use Phosphine

Phosphine tablets or pellets are placed on large sheets of paper that have been distributed evenly throughout the plant. This permits the fumigator to easily recover the dust left after the fumigant has evolved and remove it for disposal.

Magnesium phosphide plates are merely leaned up against pallets or walls so that both sides of the plates are exposed. Strips are unfolded enough so that no two sides are touching (the zigzag pattern remaining will keep them standing). At the end of the fumigation, these can be collected and carried to a previously dug trench where they can be buried.

The remaining dust from the tablets or pellets will have to be neutralized with water and detergent according to the manufacturer's instructions. Accidents have occurred when poorly trained workers have placed the remaining material in tight plastic bags or thrown it into a dumpster and shut the door. It is a rare fumigation in which readings even approach 1,000 ppm. Fire or explosions cannot occur until the concentration reaches 17,900 ppm. This is likely to occur only in a very small area. The plastic bag or the dumpster provides the small area where this concentration can be reached even though there was only a small amount of gas left in the material.

Since large amounts of gas will be used in all general fumigations, the fumigator must consider where leaking gas will go and where the released gas at the end of the fumigation will go. If there is an occupied building very close

apparatus, not gas masks. It is common practice to quickly open and close each cylinder on the final walk-through before the release to be sure building through shared heating ducts, sewers, or similar means. These should always be located and sealed.

Chamber Fumigation

Some food processors receive ingredients or returned merchandise that has a high probability of some infestation. This often occurs with imported spices and with some imported nutmeats. When these conditions occur, some processors install fumigation chambers so that all or selected products can be fumigated.

Most fumigation chambers operate at normal atmospheric pressure, but some are designed as vacuum chambers for better penetration or quicker kill. Most fumigation chambers are separate buildings or possibly separate, isolated rooms. They can be made of a number of different materials that can retain the gas, but solid concrete is the best. The chamber must be fitted with a large, properly gasketed door, fans to distribute the gas, and a chimney to exhaust the gas.

Vacuum chambers must be specially constructed of steel plates that can stand the pressure. Gas under negative pressure penetrates quickly; it can be exhausted with a series of aerations and then drawing vacuum again.

Chamber fumigation is used to kill microorganisms as well as insects. Ethylene oxide is an excellent sterilant either by itself or mixed with carbon dioxide (to make it less flammable).

Spot Fumigation of Food-Processing Equipment

Many types of food-processing equipment used in the cereal and baking industries are hard to clean completely; they often have small residues of static foodstuffs in remote areas. These hard-to-clean and hard-to-inspect areas can provide food and breeding sites for stored-product insects.

The food industry has often used spot fumigation of this type of equipment to be sure that no live insects remain inside the equipment to contaminate the products. For many years, liquid fumigants containing mixtures of ethylene dibromide (EDB) and methyl bromide or ethylene dichloride (EDC) and carbon tetrachloride were used. Use of EDB was banned in 1984, and most of the production of EDC and carbon tetrachloride ended in 1985.

These products apparently did leave residues in foods that were not detected in early tests that only went down to 1 part per million. In today's sophisticated testing down to 1 part per billion or less, we indeed can find some residue. This technique, coupled with animal tests that indicate that at least EDB is a probable carcinogen, doomed the continued use of these products that had been dependable parts of sanitation programs for over 40 years.

The liquid fumigants gave a good kill in only 24 hours and stayed close to where they were placed because of their heavy molecular weight. No other fumigants have quite the same characteristics, so there is no exact replacement.

Some manufacturers have been able to redesign equipment so that it can be cleaned better, and this is the long-term answer for most. However, in some cases, equipment conversion cannot be economically done at this time.

Fortunately, a new spot fumigant is available (only in some states at the time of this writing). Magtoxin is a magnesium phosphide tablet that gives off phosphine gas over a period of about 36 hours. Since phosphine gas is 1.2 times the weight of air, it moves away from the point of application (if it can), but the tablets continue to generate gas enough to maintain a killing concentration at the needed spot.

Proper sealing of the equipment to confine the gas to the desired location has always been necessary, but is far more important with the lightweight phosphine gas. The gas can dissipate to other areas of the equipment, to connected equipment, or to areas outside the equipment.

The gas is generated quickly if temperature and humidity are as high as they normally are in food-processing plants. This means that careful training of the fumigator is necessary so that he or she will know what and how to seal and how to move quickly during application to minimize his or her exposure to the toxic gas. Obviously, proper respiratory protection is needed anytime the gas concentration is above the STEL.

The treated equipment is only an "island" of treatment; therefore, reinfestation can occur from areas outside the equipment or from infested product moving through the equipment. Effective control programs usually combine spot fumigation with space treatment in the rooms outside the equipment and possibly applications of residual pesticides in probable hiding places of stored-product insects.

Special Situations

Cheese Rooms

Cheese infested with mites may need to be fumigated. When methyl bromide is used under normal cheese storage conditions, it should be heated with a heat exchanger to ensure complete volatilization, but this will not change the temperature of the room or the product, both of which should be as warm as quality control will permit. The cold will cause slower dispersion, which must be countered with more fans. It will cause the mites to "breathe" less, which must be countered with a higher Time $\times$ Concentration factor. The cold will also cause more sorption and slower desorption. This must be countered with more fans during aeration and a planned longer aeration.

Some cheese storage rooms are built in a series with entrance or exit only through the adjoining cold room. If the infested product is in one of the center rooms, it would be dangerous to try to fumigate that room and then exhaust the gas through several cold rooms. Dangerous amounts of gas could be sorbed into the other coolers and injure someone at a later date. The products should be shifted so that fumigations occur only in the outside room, unless all other rooms through which the gas will have to pass are held at temperatures above $21.3°$.

Caves

Caves are used for food storage in a number of areas and they can become infested. They are easy to fumigate but difficult to aerate. Fumigations should be avoided whenever other methods are likely to be effective. If a fumigation must be done, it will be necessary to install large plastic tubes about 3 feet in diameter, hook them up to fans, and place these at various points throughout the cave so that fresh air can be pumped to all areas as fumigant-containing air is led to the outside.

Warehouses with Cold-Storage Areas

Many food warehouses contain a section for cold storage or frozen storage. When these cold storage areas are equipped with normal insulating panels, the fumigant gas will probably seep into the cold area and then be very difficult to remove. Warehouses of this type should not be fumigated if any other control method is likely to be effective.

If the warehouse must be fumigated, all freezer doors should be sealed on both sides with 4-mil polyethylene, and some arrangement should be made to ventilate the cold area to the outside every 4–6 hours during and after the fumigation until product is no longer "gassing off." If this cannot be done, the freezer area will remain dangerous to all who enter it for a week or more.

The gas apparently penetrates the thick insulating panels at the joints or perhaps throughout the entire panel. Cold-storage areas could probably be built with solid concrete walls around all sides and then insulated on at least one side; this would probably stop the gas flow.

The circulation of outside air into coolers could have a detrimental effect on the product; this must be monitored constantly with temperature probes.

Although small areas such as switchboard panels have been protected from gas contact with a positive pressure system, it would take a very large quantity of chilled air continuously supplied to protect rooms the size of most cold-storage rooms. The best answer is to remove the cold products to a temporary location and then warm the cold storage areas so that they will aerate quickly, permitting early and safe reuse.

Safety Rules for Fumigators

1. Read and understand the label and all labeling before starting any new job. The label may have changed and a moment to reread it is worth the time for review, if nothing else.

2. Gain practical training before doing your first fumigation or any new type of fumigation.

3. Review all products in the structure, all structural members, and all equipment for anything that can be damaged by the fumigant.

4. Consider all legal fumigants in view of contents and time available. Choose the appropriate one with emphasis on safety.

5. Plan the release of the gas to avoid human exposure as much as possible.

6. Know symptoms and emergency treatments for exposure to the fumigant to be used.

7. Have proper respiratory protection on hand and have all personnel properly trained in its use.

8. Take the label with appropriate instructions to physicians or to the nearest hospital before starting a fumigation.

9. Notify police and fire departments before starting a fumigation.

10. Rehearse a typical accident ahead of time so all parties will react automatically during a stressful actual incident.

11. Be sure you have adequate fumigation insurance before doing any treatment.

12. Plan for the aeration before releasing gas. Check for adjoining buildings or buildings connected by common vents.

13. Check for weather conditions anticipated during fumigation period and aeration. Know

effects of all types of weather on both the fumigation and the aeration.

14. Take temperature of the commodities in representative areas. Check these temperatures against the label for prohibitions or changes necessary.

15. Take relative humidity readings particularly if using hydrogen phosphide. Know when the relative humidity would be too low to release enough gas within the time available. Know when the relative humidity is so high that copper and other heavy metals could be damaged.

16. Be sure that you have at least one other trained person with you when doing *any* fumigation. Do not consume any alcohol before or during the job or for 24 hours thereafter.

17. Properly seal the structure by sealing all openings or by covering with gasproof tarps sealed together.

18. Have applicator personnel equipped with appropriate protective clothing, but know what gloves and other clothing are contraindicated by each fumigant.

19. Know all applicable laws including any local ordinances.

20. Be sure that crew can work without undue fear of pesticides or the wearing of protective respiratory equipment. There should be respect for the fumigant but no false bravado that could endanger all.

21. Use lifelines if entering tanks or other similar areas.

22. Rehearse actual release if first time for this crew.

23. Always work in pairs in sight of each other. However, the safety person should remain in a position of relative safety, and if one person is in a railcar, the other should be outside.

24. Carefully release the gas with the least possible human exposure. Release compressed gas from the outside if possible. Put out slow-release gas such as phosphine quickly enough to ensure that exposure will be below the STEL or wear protective equipment.

25. Seal exit door and secure all doors from accidental entry by use of supplementary locks, guards, warning gases, or a combination of these.

26. Post proper warning signs on all possible entry points.

27. Check structure for leaks with appropriate leak detection devices.

28. If any fumigant has spilled on clothing, remove clothing and wash. Get medical attention if necessary.

29. If any member of the crew shows symp-

toms, don't take chances—get medical attention.

30. Monitor gas levels so that job is terminated only when the proper Time × Concentration has been met.

31. Start aeration with regard to amount of gas left and where it *might* go. Consider its movement to other buildings or its staying in the building in the event of a temperature inversion with no wind.

32. Clear structure by gas measurement only. Never trust a set number of hours for clearance. There are too many variables.

33. Make final clearance tests with representatives of labor and management to avoid undue fears from either group.

34. Check products for excessive residue if there is any reason to suspect problems.

35. Remove signs when all gas is below the STEL.

36. Notify the appropriate authorities when all is clear again.

37. Get needed rest after filling out final written reports but do not drink alcohol for another 24 hours.

Cited References

1 Anon. 1981. Tarpaulin fumigations. *Pest Control* 49(11)82.

2 Bond, E. J. 1984. *Manual of Fumigation for Insect Control.* Plant Production and Protection Paper 54. Food and Agriculture Organization, Rome.

3 Bond, E. J., T. Dumas, and S. Hobbs. 1984. Corrosion of metals by the fumigant phosphine. *Journal of Stored Products Research* 20(2)57–63.

4 Davis, R., and R. H. Barrett. 1986. In-transit shipboard fumigation of grain: Research to regulation. *Cereal Foods World* 31(3)227–229.

5 Garry, V. F., et al. 1989. Human genotoxicity: Pesticide applicators and phosphine. *Science* 246(4927)251–255.

6 Harein, P. K. 1982. Chemical control alternatives for stored-grain insects. In *Storage of Cereal Grains and Their Products,* C. M. Christensen, ed. American Association of Cereal Chemists, St. Paul, MN.

7 Hayes, W. J., Jr. 1982. *Pesticides Studied in Man.* Williams & Wilkins, Baltimore, MD.

8 Heuser, S. G. [1975.] Factors influencing dosage and choice of toxicant in stored-product fumigation. *Proceedings of the First International Working Conference on Stored-Product Entomology* (Savannah, GA 1974), pp. 246–253.

9 Heuser, S. G. 1975. The occurrence and significance of bromide residues in foodstuffs in relation to fumigation practice. *Tropical Stored Products Information* 29:15–20.

10 Hole, B. D., C. H. Bell, and C. R. Bowley. 1985. The toxicity of methyl chloroform to stored product insects. *Journal of Stored Products Research* 21(2)95–100.

11 Price, N. R. 1985. The mode of action of fumigants. *Journal of Stored Products Research* 21(4)157–164.

12 Taylor, R. W. D. 1989. Phosphine—a major grain fumigant at risk. *International Pest Control* 31(1)10–14.

13 U.S. Department of Defense, Armed Forces Pest Management Board. 1987. *Hydrogen Phosphide Fumigation of Subsistance with Aluminum Phosphide.* Technical Information Memorandum 11. Washington, DC.

14 Wainman, H. E., B. Chakrabarti, and P. R. Warre. 1983. The use of methyl bromide/carbon dioxide mixture as a space fumigant. *International Pest Control* 26(6)174–175, 180.

34
Commentary on Survey and Control

Robert Davis

In reading this section on Survey and Control, one comes away with two basic thoughts. First, we can and need to do a better job of pest management; second, research is moving forward at a very fast pace on alternatives to our use of conventional chemical pesticides and on techniques for the safer use and application of approved chemical pesticides. The chapters on biological methods, host resistance, pesticide resistance, physical control, the three chapters on chemical control of insects in grain and processed foods, the one on rodent control, and the chapter on fumigation are informative and adequately cover the areas of consideration.

The use of pheromones is the primary focus of the chapter on biological methods by Burkholder and Faustini. The application of pheromones as a survey tool is perhaps the use that first comes to mind for these biorational materials. However, there are other biological methods that we also need to begin to integrate into our postharvest pest management programs. The use of pathogens, insect predators, insect parasites, insect growth regulators, and toxicants from natural plant products are all being investigated and are beginning to offer interesting potential. Some of these biological interventions are compatible with our conventional chemical pesticides and with each other. There is, of course, no question that much additional research is required before we will realize the full potential of these emerging pest management tools.

Davis (Retired): Stored-Product Insects Research and Development Laboratory, USDA Agricultural Research Service, P.O. Box 22909, Savannah, GA 31403.

However, we need to begin to look for ways to integrate these biological survey and control techniques into our pest management programs.

Mr. Dobie, in Chapter 27 on host-plant resistance, presents us with an excellent review of potential causes of resistance that may occur or develop in the postharvest stages of grains and legumes. He also discusses current techniques for measuring resistance and provides an excellent review of the available information on resistance in the major cereals and legumes. He concludes his thesis with the statement that unfortunately most research on host-plant resistance has been academic and has been conducted in entomological laboratories without the collaboration of plant breeders. Mr. Dobie feels that four things must be undertaken before practical progress in host resistance of postharvest commodities can be effected, namely, (a) screening of large numbers of wild varieties and commercial cultivars to locate sources of resistance; (b) study of resistant varieties to locate the cause of resistance; (c) study of the inheritability of resistance; and (d) study of the effects of the introduction of resistant varieties on pest populations and consumers. I find no fault with these objectives. However, I feel that much can also be learned by studying the physiology of the insect pest. Knowledge of how the insect utilizes resistant and nonresistant varieties should also add immensely to our ability to safeguard our harvests.

Australian scientists Champ and Highley are acknowledged experts in the area of insecticide resistance and its management in postharvest pests. In Chapter 28, they have put together an excellent summary of the world situation and the perils that will face mankind if we don't soon make a quantum jump forward in the seesaw battle between man's newest insecticide and the postharvest pest's resistance to it.

The authors state that the solution to this problem will involve continuous monitoring of the status of insect resistance in field populations and that control attempts should always be directed at securing 100% kills. Here again, it is believed that an integration of conventional chemicals with nonchemical alternatives will also contribute to a delaying action. It will be essential to attack this problem at the most basic level, the gene, before we can ever hope to find lasting relief.

In Chapter 29, Dr. Watters presents a very complete picture of our current knowledge about physical methods of managing stored-product pests. Interestingly, some of our oldest pest management approaches— hermetic storage, drying, physical barriers and smoke— are all physical methodologies and are still used today with only modification of techniques. Many of these physical control interventions need to be expanded in our present-day pest management programs. The use of both high and low temperatures and even freezing temperatures, where ambient conditions can be utilized, is very effective. The use of aeration to achieve both drying and cooling has been successful in some temperate and subtropical situations. The two physical approaches that would seem to offer the widest acceptance and would be most effective are the use of modified atmospheres (carbon dioxide, nitrogen, or combustion gases) and irradiation. Both are currently approved for use on most raw and processed commodities throughout most of the world. Other physical approaches are less well understood and still require some quantification. Dr. Watters concludes his theme with the statement that many of the presently available physical methods can be combined with the judicious use of pesticide chemicals to enhance the effectiveness of both.

Dr. Harein, in Chapter 30 on chemical control of insects in bulk grain, approaches his subject in a straightforward manner with the theme that sanitation comes first, followed by properly applied chemical pesticides. This approach cannot be faulted, and if it were universally adhered to we would have fewer losses and better-quality grain and grain products with fewer instances and lower levels of resistance. He stresses the applicable FDA and EPA regulations and laws and the importance of adhering to the label.

Chapter 31 on chemical control of rodents, by Dr. Timm, is also a straightforward narrative on the importance of sanitation and the use of anticoagulant and nonanticoagulant baits. The author also discusses the selection of baits and formulations, bait station use, tracking powders, fumigants, and the problems of maintaining continuous control. This narrative provides an excellent summary for the pest management specialist who needs to learn or review the basics of rodent control.

Chapter 32 by Mr. Baur, on chemical methods to control insects in processed foods, is well presented and carries the reader through a short course on two of the three main chemical methods for insect control: space treatments and residuals. The sections on labels/labeling and tolerances/action levels are quite comprehensive. The author encourages the appropriate use of residuals as being both preventive and corrective in their action. Still, with all the disadvantages of pesticide chemicals, the reader is encouraged to consider all approved pest management interventions for each given pest problem.

Again, as in the previous chapter, Mr. Walter provides in Chapter 33 an excellent short course on the application and use of fumigants in the food industry. Both Baur and Walter have drawn heavily on their years of experience in association with various agribusinesses in the USA.

In today's food industry, hydrogen phosphide gas (phosphine) is by far the fumigant of choice in most instances. Methyl bromide runs a distant second, and there are really no other choices. A fumigant usually offers the only corrective procedure that can be used in a known time frame and which can guarantee an acceptable kill to the customer or a governmental quarantine official. Fumigants are toxic gases and are therefore restricted-use pesticides. To purchase and apply them, the fumigator must be certified. Because of the hazard of using fumigants, Mr. Walter ends his theme with a list of safety rules for fumigation which the applicator would be wise to review before applying a fumigant.

I have found that even without the inclusion of the remaining sections in this Technical Bulletin, Chapters 23 thru 30, with their extensive bibliographies, are capable of standing by themselves and would be a valuable reference book for anyone interested in pest management in and around raw or processed foods.

V

Health Considerations

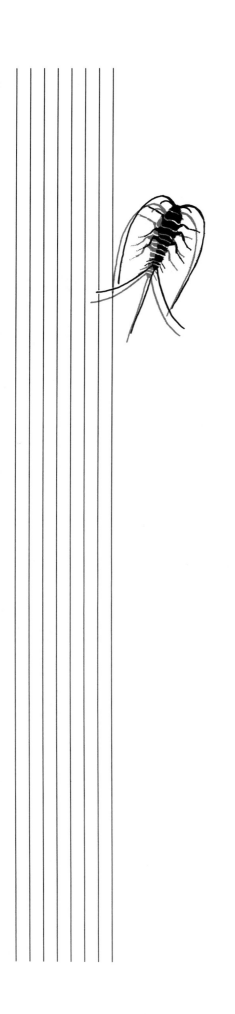

35
Nutrition Changes Caused by Pests in Food

Harold George Scott

Foods shift in nutritional value as a result of pest infestation. Occasionally, these shifts result in nutrition gain, but in nearly every case, infestation leads to nutrition loss. The nature and extent of these changes have not been studied adequately. Summarized below are the salient facts about six factors often commonly involved: chemical change, water balance, heating, dilution, consumption, and acceptability.

Annual losses in the United States of America (USA) caused by insects and other arthropods infesting postharvest foods and food products have been estimated at $2 billion to $6 billion in 1980 dollars; that caused by rodents and other vertebrates, at $2.5 billion. These figures, developed chiefly from shifts in weight and/or volume, do not adequately portray the nutrition degradation, which may be three times as great; nor do they take into account damage to crops in the field, pest-initiated fires, costs of pest control, or losses attributable to human and domestic-animal diseases (17, 18) associated with stored-food pests.

Studies show food losses to be as great as 25% of the annual harvested product in technologically advanced, temperate-zone nations such as the USA. Such losses reach 70% in some tropical countries, such as Niger, that lack high-quality storage, transportation, and processing systems. Usual losses of total product often amount to 35–50% in tropical countries (44) and 9–13% in the USA (22, 30; see also Chapter 2). The control of these losses is the most rapid and least

costly way to substantially increase food supplies for the earth's undernourished millions (1, 14, 20, 28).

Grain storage specialists have used such criteria as alterations in weight (26), volume, uric acid content (29, 38, 43), milling suitability, and baking quality to estimate losses. Although each of these criteria has some validity, none accurately reflects what is probably the most important consideration—nutrition depletion. Some investigators have recently turned to an approach that involves the computation of accurate energy budgets (40, 42). By measuring exact amounts of nutrients consumed by each pest during its life history, the actual and potential calories used by that species can be determined. When caloric consumption and degree and length of infestation are correlated with chemical change, water balance, heating, dilution, and acceptability, reliable estimates of nutrition loss emerge.

Chemical Change

Foods undergo inherent chemical change with time (33). Some such as fresh fruits may decompose rapidly whereas others such as grains usually change slowly. Apples, for example, deteriorate within a few months even under ideal conditions. In one study, pest-free wheat, stored cold and dry for 30 years, showed some increase in free fatty acids and some diminution in baking quality, but there was essentially no change in the levels of thiamine and protein (32).

Pests infesting foods during extended periods of storage produce heat and chemicals that accelerate the inherent aging processes. Fat breakdown, associated with rancidity and off-flavor, is promoted by insect

Scott: Program in Community Medicine, School of Medicine, Tulane University, New Orleans, LA 70112.

attack, especially when the pests break off small particles of the food, introduce microorganisms, or raise the temperature or moisture levels. Uninfested wheat stored at the recommended temperature and humidity loses 7% of its thiamine in 6 months; pest-infested wheat loses 65% during the same period (6).

Although some chemical changes are innocuous and a few might even be beneficial, the vast majority represent major defects in and/or dangers to human and animal nutrition (35, 39), with mycotoxicoses (34), allergies, intestinal disorders, and other acute health problems in addition to the more insidious effects of nutrition loss. Pig feed infested with mites (*Acarus siro*) lost about one-fifth of its nutritive value over a period of 9 weeks. In a related feeding experiment, three measurable factors were significantly reduced in pigs that received the mite-infested diet: daily live-weight gain, feed/gain ratio, and nitrogen retention (5).

Water Balance

Infestation so alters the physical and chemical makeup of food that the water content shifts upward. In most situations, this creates extensive dilution of the product and thereby causes a notable reduction in the nutritional value per unit of weight. This dilution is associated primarily with the fact that most pest degradations cause food to become more hygroscopic. Also, the metabolic water of the food, released by biochemical reactions within the pests, is added to food via pest excreta and carcasses. Pests may also damage packaging, thus allowing moisture to enter.

Although some volumetric increase of the product may result from this new moisture, greater increase occurs in mass, causing some one-third of the weight of stored food to become water. At a desirable 12% moisture level, the wheat in a "big-John" freight car, loaded to capacity (90,000 kg), contains some 11,000 kg of water. At a common but undesirable 22% moisture level, the wheat in the same fully loaded freight car holds nearly 20,000 kg of water (23). Where food is sold by weight, as is increasingly the case, added water becomes a major dilution of economic as well as nutrient value (23).

Maximum moisture for safe storage of grains is about 13.5% (7). Above this level, mold damage and stored-food mite populations develop rapidly, resulting in lowered baking, malting, and feeding quality, reduced viability in seed grain, and increased susceptibility to injury from chemicals used to treat seed (25). Germination levels of 84% in wheat stored at 13.5% moisture drop to 7% at 16.5% moisture. In Nigeria, maize with 22% moisture suffers 40–48% damage, whereas that kept at 12% shows only 9–16% damage.

Water is seldom uniformly distributed in foods, some parts of each storage unit being much wetter and therefore much more affected than others.

Heating

When stored-food pests become concentrated in grain, the heat generated by their metabolic processes causes hot spots which can completely rob kernels of their power to germinate, thus making them useless as seed. Besides accelerating pest infestations, heating results in more rapid chemical deterioration of the grain, with consequent degradation of nutritional value. The presence of dockage (broken grain, dust, weed seeds, and other foreign matter) increases the probability and degree of heating. When cool weather comes, the water vapor that rises from the hot spots condenses on cooler grain surfaces where it promotes molding, sprouting, and further heating (3, 24, 41).

If grain temperatures at time of loading are variable, so that warmer and cooler layers are sandwiched in storage, the temperatures will gradually rise throughout the bulk to the highest level (3). The elevated temperatures cause translocation of water so that moisture content in the top layers increases, often to values above the safe level of 13.5% (7).

Temperature is one of the most significant limiting factors affecting stored-food arthropods and fungi. Growth and development of most of these arthropods slow and essentially stop when descending temperatures reach about 17° (7, 27). With increasing temperatures above that level, population growth is rapid until about 35°, when it slows down. Metabolic heating in grain perpetuates and accelerates development of pest populations even when ambient temperatures are low.

Dilution

Foods infested with arthropod pests (e.g., beetles, moths, mites) undergo continuous dilution as a result of the presence of the living arthropods themselves, their excreta, webbing, exoskeletons cast off during molting and metamorphosis, and the dead bodies of previous generations (11, 12). Rats, mice, bats, and other mammalian pests accomplish dilution via urine, feces, saliva, hairs, carcasses, and substances tracked in on feet and fur (21). Pigeons and other pest birds accomplish dilution by means of their feces, regurgitated pellets, feathers, carcasses, and tracked substances. Fungi and other contaminant microorganisms dilute foods with their excretions and their dead bodies.

After long periods of infestation, the diluents generated by pests may equal or even exceed the amount of original food remaining. Although some of these diluents have limited nutritional value for humans and domestic animals, that value usually represents only a minute fraction of the nutritional value of the original food. In The Gambia, "protected" and "insecticide-treated" shelled peanuts developed a population of 90,000 to 100,000 living insects per sack during 16 weeks of storage. Feces, exoskeletons, dead insects, and frass added to the weight of the insects themselves constituted a major portion of the "peanuts" when they were shipped (19).

Pests are active at every step in stored-food management. For example, as grain moves from farm to table, rodent-dropping and/or hair contamination per sample shifts from about 6% at harvest to 23% in the warehouse, 12% after sift cleaning, 3% after air cleaning, 16% after milling, and 16–30% after home storage. Conservative estimates indicate that 32% of American farms have rats and 55% have mice; 76% of grain warehouses have rats and 80% have mice. Each rat deposits 25,000 droppings per year; each mouse, 17,500.

Massive nutrition dilution also occurs when people accidentally or deliberately mix nonfood material with food, and because of the large amount of spillage and wastage that occurs during storage, transport, and processing. In the Far East, wastage and spillage during rice processing and distribution amount to an incredible 20–25% of the harvested crop (4).

Consumption by Pests

Pests, especially rodents, directly consume vast quantities of stored foods, thereby depriving people and livestock of these nutritional resources. A pair of rats eats 25 kg of grain per year and contaminates an additional 250 kg in the process. Also, insect-damaged grain germinates so poorly that, if used for seed, it must be applied in much greater quantities to give a good stand, resulting in the loss of the nutritional value of the excess kernels.

In untreated bins of Kansas wheat, insects can destroy the wheat at the rate of at least 1% per month by weight from July to November. In an experimental wheat sample with 30% of the kernels infested with rice weevils (Sitophilus oryzae), loss in weight at 4 weeks was about 4%; at 7 weeks, about 10%. In Nicaragua, 15% of the weight of the national crop of corn is reportedly consumed by stored-food insects. Weight increases from feces, pest carcasses, and water absorption are not considered in these figures. Thus, the actual loss in nutritional value is much greater than the percentages indicate (16, 45).

As great as the loss from direct consumption of food by storage pests is, this effect is magnified by the selective nature of their feeding (10). Generally speaking, pests do not consume all of a product at a uniform rate (31). Indeed, most choose carefully the most nutritious portions, leaving the least beneficial fraction largely untouched and therefore often giving the observer an immediate impression that the food is

unaffected, nutritionally as good as ever. For example, weevils (Curculionidae) develop in the endosperm of wheat, doing little damage to the germ or seed coat (9). Spider beetles (Ptinidae) eat the coat and little of the underlying endosperm (2). Tobacco moths (Ephestia elutella) attack the germ of one kernel, then move on to the germ of another without eating the seed coat or exposed endosperm (37).

Nigeria annually produces approximately 1 million metric tons of cowpeas. Soon after harvest, 4% of the cowpeas show emergence holes of bruchid beetles. This rate of damage rises steadily until, after several months, 60–70% of the seeds have emergence holes and 20% of the nutritional value of the peas has been consumed (8). In Niger, 93% of cowpeas stored in jute sacks without special precautions were attacked by bruchid beetles in 2 months. Subsequent attacks by other beetles (Trogoderma spp.) completely emptied the peas so that only the outer husks were left. In Nigeria, weight loss of sorghum is estimated at 7–7.5% during a 9-month storage period, with 20% of the grains damaged and largely depleted of nutrients (15).

Each rice weevil larva during development metabolizes about 14 mg of a kernel of wheat into carbon dioxide, water, heat, and excreta. During the process, the weevil consumes about two-thirds of the endosperm, an especially nutritious portion of the kernel. In contrast, the lesser grain borer (Rhyzopertha dominica) consumes almost the entire kernel, leaving only the bran coat and dust.

Each granary weevil (Sitophilus granarius) consumes 62.6% of the caloric content of a wheat kernel, about five-sixths of which goes into metabolism. In dense populations, adult weevils scour out partly used grains until all that remains is seed coat, a few traces of endosperm, and a fine powder of fecal pellets (36).

During the course of development, each larva of the tobacco moth consumes the germ of some 348 kernels of wheat. Since each 100 liters of untreated wheat in Kansas during September may contain 2,000 to 8,500 adult insects, constant pest feeding obviously causes major changes in the quality of wheat stored without protection.

Even though a particular lot of wheat may have lost less than 10% of its weight or even may have gained weight due to water absorption, and even though the lot may superficially appear unaffected, it may easily have lost 60% of its caloric value, with the majority of its most nutritious elements having been consumed. This is one of the most poorly understood factors in stored-food management, yet clearly one of the more significant.

Acceptability

In addition to the revulsion associated with actually

seeing pests, pest fragments, or pest filth in food, infestations usually cause caking and produce noticeable changes in color, texture, taste, and odor. These changes often make food unacceptable to purchasers and result in greatly lowered economic value. In Senegal, stocks of peanuts become unacceptable in the market place after hosting three generations of the groundnut bruchid (*Caryedon serratus*) over a period of 15 weeks (*13*).

When nonacceptability develops prior to retail sale, the value of the food may sometimes be partially reclaimed by using it as livestock feed or for alcohol-fuel production. However, when rejection occurs after retail sale, it nearly always results in product destruction with consequent total loss of nutritional and economic value.

Summary

Loss incurred in stored food as a result of pest infestation is immense, indeed monumental, representing 9–25% of the annual harvested product in technologically advanced, temperate-zone nations such as the USA, and 35–70% in tropical countries, such as Niger, where storage, transportation, and processing systems are underdeveloped.

Nutrition degradation of stored food by food pests is one of the greatest ongoing assaults on our civilization. This problem, which should be receiving high-priority attention by every nation, rich or poor, large or small, is either widely ignored or given only token recognition.

Cited References

1 Anon. 1974. The world food crisis. *Time*, Nov. 11, pp. 64–84.

2 Appert, J. 1953. Les insectes de l'arachide au Senegal. *Bulletin du Centre de Recherch Agronomique* (Bambey, Senegal) 7:67–75.

3 Armstrong, M. T., and R. W. Howe. 1963. The sawtoothed grain beetle (*Oryzaephilus surinamensis*) in home-grown grain. *Journal of Agricultural Engineering Research* 8(3)256–261.

4 Bond, E. J. [1975.] Future needs and developments for control of stored product insects. In *Proceedings of the First International Working Conference on Stored-Product Entomology* (Savannah, GA, 1974), pp. 317–322.

5 Braude, A. G., et al. 1980. Effect of flour mite infestation (*Acarus siro* L.) on nutritive value of pig diets. *Veterinary Record* 106(2)35–36.

6 Bronswijk, J. E. M. H. van, and R. N. Sinha. 1971. Interrelations among physical, biological, and chemical variates in stored-grain ecosystems: a descriptive and multivariate study. *Annals of the Entomological Society of America* 64(4)789–803.

7 Burrell, N. J. 1974. Aeration. In *Storage of Cereal Grains and Their Products*, C. M. Christensen, ed. American Association of Cereal Chemists, St. Paul, MN.

8 Caswell, G. H. [1975.] The development and extension of nonchemical control techniques for stored cowpeas in Nigeria. In *Proceedings of the First International Working Conference on Stored-Product Entomology* (Savannah, GA, 1974), pp. 63–67.

9 Coombs, C. W., and G. E. Woodroffe. 1963. The influence of food condition on the longevity of *Sitophilus granarius* (L.) (Col., Curculionidae). *Entomologist's Monthly Magazine* 49(1191–1193)145–146.

10 Cotton, R. T., and D. A. Wilbur. 1982. Insects. In *Storage of Cereal Grains and Their Products*, C. M. Christensen, ed. American Association of Cereal Chemists, St. Paul, MN.

11 Daniel, V. A., et al. 1977. Effect of insect infestation on the chemical composition and protein efficiency ratio of the proteins of kaffir corn and green gram. *Indian Journal of Nutrition and Dietetics* 14(2)38–42.

12 Daniel, V. A., et al. 1977. Effect of insect infestation on the chemical composition and the protein efficiency ratio of the proteins of Bengal gram and red gram. *Indian Journal of Nutrition and Dietetics* 14(3)70–73.

13 Deuse, J. P. L., and J. G. Pointel. [1975.] Assessment of research at farm level storage in Francophone Africa. In *Proceedings of the First International Working Conference on Stored-Product Entomology* (Savannah, GA, 1974), pp. 85–91.

14 Ebeling, W. 1975. *Urban Entomology*. Agriculture & Natural Resources, University of California, Oakland, CA.

15 Giles, P. H. 1964. The insect infestation of sorghum stored in granaries in northern Nigeria. *Bulletin of Entomological Research* 55(3)573–588.

16 Giles, P. H., and O. Leon V. [1975.] Infestation problems in farm-stored maize in Nicaragua. In *Proceedings of the First International Working Conference on Stored-Product Entomology* (Savannah, GA, 1974), pp. 68–76.

17 Gorham, J. R. 1975. Filth in foods: Implications for health. *Journal of Milk and Food Technology* 38(7)409–418.

18 Gorham, J. R. 1979. The significance for human health of insects in food. *Annual Review of Entomology* 24:209–224.

19 Green, A. A. 1959–60. The control of insects infesting groundnuts after harvest in The Gambia. I–III. *Tropical Science* 1(3)200–205; 2(1–2)44–54; 2(3)130–133.

20 Hall, D. W. 1969. Food storage in developing countries. *Tropical Science* 11(4)298–318.

21 Harris, K. L., and F. J. [Baur]. 1982. Rodents. In *Storage of Cereal Grains and Their Products*, C. M. Christensen, ed. American Association of Cereal Chemists, St. Paul, MN.

22 Howe, R. W. 1965. Losses caused by insects and mites in stored foods and feedingstuffs. *Nutrition Abstracts & Reviews* 35(2)285–303.

23 Hunt, W. H. 1974. Moisture—its significance, behavior, and measurement. In *Storage of Cereal Grains and Their Products*, C. M. Christensen, ed. American Association of Cereal Chemists, St. Paul, MN.

24 Hyde, M. B., and C. G. Daubney. 1960. A study of grain storage in fossae in Malta. *Tropical Science* 2(3)115–129.

25 Machacek, J. E., et al. 1961. Effect of a high water content in stored wheat, oat, and barley seed on its germinability, susceptibility to invasion by moulds, and response to chemical treatment. *Canadian Journal of Plant Science* 41(2)288–303.

26 Murthy, N. K., and R. Kokilavani. 1980. Biodeterioration of stored, insect infested jowar (*Sorghum vulgare*) and ragi (*Eleusine coracana*). *Indian Journal of Nutrition and Dietetics* 17(6)201–204.

27 Navarro, S. [1975.] Aeration of grain as a non-chemical method for the control of insects in the grain bulk. In *Proceedings of the First International Working Conference on Stored-Product Entomology* (Savannah, GA, 1974), pp. 341–353.

28 Parkin, E. A. 1956. Stored product entomology. *Annual Review of Entomology* 1:223–240.

29 Parvathappa, H. C., et al. 1970. Physical and biochemical changes in sorghum (*Sorghum vulgare*). *International Biodeterioration Bulletin* 6(3)95–99.

30 Pimentel, D. 1976. World food crisis: Energy and pests. *Bulletin of the Entomological Society of America* 22(1)20–26.

31 Pingale, S. W., M. N. Rao, and M. Swaminathan. 1954. Effect of insect infestation on stored grain. I. Studies on soft wheat. *Journal of the Science of Food and Agriculture* 5(1)51–54.

32 Pixton, S. W., M. B. Hyde, and G. Ayerst. 1964. Long-term storage of wheat. *Journal of the Science of Food and Agriculture* 15(3)152–161.

33 Pomeranz, Y. 1982. Biochemical, functional, and nutritive changes during storage. In *Storage of Cereal Grains and Their Products,* C. M. Christensen, ed. American Association of Cereal Chemists, St. Paul, MN.

34 Pushpamma, P., and M. U. Reddy. 1979. Physico chemical changes in rice and jowar stored in different agro-climate regions of Andhra Pradesh. *Bulletin of Grain Technology* 17(2)97–108.

35 Rajan, P., et al. 1975. Effect of insect infestation on the chemical composition and nutritive value of maize and cowpea. *Indian Journal of Nutrition and Dietetics* 12(10)325–332.

36 Richards, O. W. 1947. Observations on grain-weevils, *Calandra* (Col., Curculionidae). I. General biology and oviposition. *Proceedings of the Zoological Society of London* 117(1)1–43.

37 Richards, O. W., and N. Waloff. 1946. The study of a population of *Ephestia elutella* Hübner (Lep., Phycitidae) living on bulk grain. *Transactions of the Royal Entomological Society of London* 97(11)253–298.

38 Sharma, S. S., V. K. Thapar, and G. S. Simwat. 1979. Biochemical losses in stored wheat due to infestation of some stored grain insect-pests. *Bulletin of Grain Technology* 17(2)144–147.

39 Shehnaz, A., and F. Theophilus. 1975. Effect of insect infestation on the chemical composition and nutritive value of Bengal gram (*Cicer arietinum*) and field bean (*Dolichos lablab*). *Journal of Food Science and Technology* 12(6)299–302.

40 Sinha, R. N. 1973. Ecology of storage. *Annales de Technologie Agricole* 22(3)351–369.

41 Sinha, R. N. 1975. Effect of dockage in the infestation of wheat by some stored-product insects. *Journal of Economic Entomology* 68(5)699–703.

42 Sinha, R. N., and A. Campbell. 1975. Energy loss in stored grain by pest infestation. *Canada Agriculture* 20(2)15–17.

43 Swaminathan, M. 1977. Effect of insect infestation on weight loss, hygienic condition, acceptability and nutritive value of foodgrains. *Indian Journal of Nutrition and Dietetics* 14(7)205–216.

44 Taylor, T. A. [1975.] Motivation and method in ensuring protection of tropical produce from grower to his market. In *Proceedings of the First International Working Conference on Stored-Product Entomology* (Savannah, GA, 1974), pp. 11–14.

45 White, G. D. 1953. Weight loss in stored wheat caused by insect feeding. *Journal of Economic Entomology* 46(4)609–610.

36

Food Pests as Disease Agents

Robert A. Wirtz

Arthropod food pests are usually viewed as posing little hazard to human health when acting as disease agents. Apparently, the consumption of small herbivorous arthropods is not a serious hazard to human health. We routinely consume these organisms and their products with no apparent ill effects (8, 13, 21, 25, 52). There are recognized instances, however, where consumption of or exposure to food-pest arthropods or their products poses a significant potential health hazard to humans (2, 17, 44, 60).

Currently, two major problems exist in determining whether or not health hazards can be associated with the consumption of arthropods and arthropod products. The first of these is the diversity of both the arthropods and the materials they infest. The second is the paucity of reliable research data available in this area.

Food-pest arthropods range from the ubiquitous and seemingly innocuous thrips and mites (the consumption of which is impossible to avoid when eating many fresh and prepared fruits and vegetables) to the moths and beetles that attack grain products and other stored foods. Arthropod products range from cast skins and cuticular parts, common to all arthropods, to specialized defensive setae and defensive chemicals produced by some beetles and mites that infest stored foods.

Current attitudes on the health hazards of arthropod consumption are influenced more by indirect evi-

dence, supposition, and cultural and aesthetic considerations than by clearly designed and well-executed experiments. Much of the early work, especially studies involving the feeding of insects to humans, was poorly designed. Many animal feeding studies reported in the recent literature, although better designed and executed, have resulted in inconclusive or conflicting results. In these studies it has also been difficult to determine whether harmful effects were due to the arthropods, to microorgansims present in the rearing media, or to indirect effects such as nutritional or caloric losses from the infested commodity (10, 26).

The Food and Drug Administration (FDA), responsible for protecting the public's health, has established defect action levels (DALs) for natural and unavoidable defects in foods, which allow for the consumption of arthropods and arthropod parts on the "basis of no hazard to health" (11). Using the assumption that "too much" filth causes disease and that any quantity less than "too much" is harmless, FDA has established DALs for approximately 100 foods (11). This assumption that consuming arthropods and their products is harmless is based more on experience than experiment. Enforcement of established DALs has not required specific proof that "too much" filth bears a cause-and-effect relationship to disease. Although the relationship is usually assumed and has generally been upheld in the courts (22), the intent of Congress in making the law was to prevent adulteration *per se*. However, if any risk to human health were discovered to be definitely associated with a given exposure, the DAL would be revised in favor

Wirtz: Department of Entomology, Walter Reed Army Institute of Research, Washington, DC 20307.

of consumer health (22). In fact, the DALs currently in force are under constant review and are periodically adjusted as technology improves. This has led some individuals to suggest that the FDA *ad hoc* policy of reducing tolerances in food for small insects that offer no public health threat contributes to increased insecticide use which may pose a health hazard, reduces environmental quality, and increases food costs (47).

The identification of problem areas is often compounded by the methods used for the analysis of arthropod-contaminated food products. Currently approved methods usually emphasize the number of whole arthropods or fragments identified in the sample (11, 12). For example, cornmeal is rejected only if it contains more than an average of one whole insect or 50 insect fragments per 50 grams (11). Although toxic products of arthropod origin are likely to represent a greater potential health hazard than fragments, there are no guidelines or methods for using these as rejection criteria. This may result in the disposal of infested food products that are safe for human consumption or, conversely, in the consumption of hazardous food supplies.

An exception to this lack of policy exists in the U.S. Armed Forces. They continue to use the number of insects in an infested commodity as the rejection criterion, but they also recognize the potential health hazards associated with the consumption of dermestid hastisetae and of tenebrionid flour beetle secretions. The Armed Forces consider the presence of one *Trogoderma* larva in the product or an infestation of three adult *Tribolium* flour beetles per pound within the product as justification for condemnation (36).

Documented or suspected health hazards currently exist in the following areas: consumption of living arthropods and fragments, consumption of excretory or secretory products, and consumption of (or exposure to) allergenic substances of arthropod origin.

Consumption of Arthropods and Arthropod Fragments

This area of concern includes the consumption of living and dead insects and mites as well as exuviae and portions of the exoskeleton such as spines and setae. In many case studies, it is difficult to determine whether the symptoms reported were in response to consuming the living arthropod or consuming portions of the exoskeleton. There are cases, however, where symptoms are attributed to specific cuticular structures (6, 38, 41). Consumption of infested foods may result in symptoms caused by toxic chemical products or in allergic reactions; these areas are discussed below as separate topics.

The accidental consumption of living arthropods is referred to as enteric (gastrointestinal) arthropodiasis.

Depending on the type of insect or mite consumed, different terms are often used—i.e., myiasis for the consumption of fly larvae, canthariasis for beetles, scoleciasis for moths, and acariasis for mites (23). The type of arthropodiasis is usually related to the food consumed, with most cases of accidental enteric myiasis resulting from the consumption of fruits and vegetables containing fly eggs and larvae. Instances of canthariasis and scoleciasis are usually associated with the consumption of infested grain products and other stored or dried foods. The consumption of uncooked foods and the increasing use of convenience foods requiring little or no cooking before eating may result in an increase in the incidence of enteric arthropodiasis. The reported incidence of arthropodiasis is much higher among children than adults (44); most cases reported in the United States (USA) are the result of the parent noticing larvae in a child's stool or diaper (44).

The majority of instances involving consumption of arthropod-contaminated food do not result in clinical illness. However, in some cases, the symptoms can be quite severe, depending on the species and number of arthropods ingested, as well as their location in the digestive tract. Severe cases have resulted in fever, nausea, vertigo, abdominal pain, and bloody diarrhea (14, 44). The incidence of enteric arthropodiasis is low in the USA; it is much more prevalent in countries where pest infestation rates are higher and sanitation standards are less stringent (44).

Enteric myiasis is the most common diagnosis in reported cases of intestinal upset from accidental insect consumption. Approximately 50 species of fly larvae have been identified in feces or vomit from patients with enteric myiasis. These larvae are primarily from the families Drosophilidae, Calliphoridae, Muscidae, and Sarcophagidae (23, 44). These insects deposit eggs or larvae on fresh fruits and vegetables, cheese, cold meats, and other foods that are consumed without further cooking. The drosophilid flies, often associated with fresh fruits and vegetables, are probably ingested more frequently than any other kind of fly. However, no deleterious effects from the consumption of these flies have been documented (21). In experimental studies, gastrointestinal disturbances were reported by 50 of the 60 human volunteers after swallowing three species of fly larvae in gelatin capsules (27).

The reported incidence of enteric canthariasis is less frequent than that of myiasis (14, 44). Foods commonly infested with beetles are processed or prepared in a manner that makes the ingestion of living insects less likely. The consumption of these insects is a reflection of local infestation rates, and most incidences of canthariasis in the USA are due to the ingestion of larvae or adults of the families Dermestidae, Tenebrionidae, Curculionidae, and Cucujidae (*10, 23*). The consumption of cuticular fragments from these insects presents no documented health hazard, although the number of such fragments is often used by FDA to establish DALs.

Large pieces of cuticle and spines can produce intestinal blockage or irritation but these are seldom encountered in foods. At least one study has correlated the consumption of large quantities of insect-infested maize with digestive upset in humans in Africa (*26*). No such symptoms were demonstrated in experiments in which human volunteers consumed flour beetles, rice weevils, and their dejecta (*20, 37*).

In a series of laboratory studies, rats fed diets of beetle-infested grain developed changes in liver tissue and showed slower growth rates than did control animals (*50, 62*). However, the infested grain showed marked chemical changes resulting from infestation; these were accompanied by lowered nutritive and caloric values. The consumption of foods infested with dermestid beetles poses unique health hazards because of the specialized setae (hastisetae) present on the larvae of these insects. These setae have been associated with intestinal trauma in two cases of enteric canthariasis as well as eye irritation, dermatitis, and allergic reactions (*14, 38, 41*). A single dermestid larva can release over 3,000 hastisetae into the material it is infesting (*38*).

There are also few reported cases of scoleciasis; however, since several species of Lepidoptera are major pests of dried or stored foods, the eggs and larvae of these insects are probably ingested more often than the number of cases indicates. The Indianmeal moth (*Plodia interpunctella*) frequently infests candy bars in the USA. These larvae are evidently consumed with no serious effects (*20*) (note, however, further comments about this insect in the section below on allergy).

The ability of mites to infest food products, even when such items seem to be well protected, is often underestimated (*7, 28, 56, 60*). The majority of the mites infesting stored and processed foods belong to the families Acaridae (= Tyroglyphidae), Carpoglyphidae, and Glycyphagidae (*28, 56, 60*). A given species will usually be found on a specific type of product. These infestations may also be accompanied by one or more predator species, often belonging to the family Cheyletidae (*60*). In addition to the more commonly infested materials such as flour products and processed cereals, these mites infest dried fruit, cheese, spices, and smoked and dried meat and fish products (*28, 60*).

There are numerous reports of intestinal acariasis in the literature (*60*), the majority of which cite the presence of mites or mite eggs in stool samples, often in association with diarrhea. The primary symptoms

attributed to enteric acariasis are vomiting and diarrhea. Authors are divided as to the significance of the presence of mites in human feces (*24, 34, 60*). One group concludes that these arthropods have no pathological significance; others believe that there is a direct correlation of symptoms to the number and persistence of intake as well as to individual sensitivity and mite species.

There is one food in which mites or their products definitely appear to be responsible for intestinal upsets. The distinctive flavor of Milbenkäse or "mite cheese" supposedly is imparted by mites reared for that purpose. These mites must be wiped off the cheese before it is eaten. To avoid intestinal upsets, only small quantities should be consumed initially. However, some individuals become sensitive to this product (*60*).

The results of animal studies on mite-infested foods are also inconsistent (*18*). When conducting such studies, the investigator must ensure that the symptoms are due to the mite population and not to microorganisms that are also present. In a recent study, four species of mites were isolated from the material they were reared in and fed to seven species of laboratory and farm animals for 6–21 days at levels of 50,000 to 1 million mites per day. The test animals showed no symptoms of disease at the end of the experiment, and no organic lesions were found at necropsy. Therefore, it appears that the mites used in these studies had no pathologic effects, even when consumed in large numbers. Subsequent bacteriological examination of mite-contaminated feeds led these investigators to conclude that previous reports of disease from infested feeds were brought about not by the mites but by the microorganisms present in these materials (*18*). A second series of studies, in which pigs, chickens, and ducks were fed large numbers of grain mites (*Acarus siro*), gave similar results (*58, 59*).

The psychosomatic reactions to arthropodiasis are probably responsible for some of the physical symptoms exhibited by humans (*19, 44*). However, serious health hazards associated with this type of entomophobia are probably rare. In a recent review on entomophobia in an urban area, no mention was made of cases for which treatment was required because of arthropods in foods that had been consumed (*42*).

Consumption of Arthropod Excretory and Secretory Products

Members of the arthropod phylum are noted for the diversity of chemical compounds they produce and release into their environments (*15*). When these arthropods infest materials to be used for human consumption, potential health hazards can exist. Although the chemical products of some food-infesting arthropods have been identified, few of these pests have been examined in this respect.

In some instances of arthropod infestation, the source of the infestation can be determined by its odor. Flour beetles (*Tribolium* spp.) impart a sharp, penetrating odor and discolor the infested material (*45*). Bread made from flour infested with sawtoothed grain beetles (*Oryzaephilus surinamensis*) and yellow mealworms (*Tenebrio molitor*) has a "chemo-phenolic" taste and odor (*55*). Infestations by the lesser grain borer (*Rhyzopertha dominica*) impart a honey-like odor to infested wheat, whereas the Angoumois grain moth (*Sitotroga cerealella*) and the common grain mite (*A. siro*) impart a disagreeable odor to badly infested products and render them unpalatable (*49*). Other acarid mites give a distinctive mint-like odor to the infested commodities (*14*). Some species of predatory mites that commonly attack grain and stored-food pest arthropods produce toxins that may cause skin lesions, asthma, nausea, and other symptoms (*28*). The characteristic odor of flour beetles (*Tribolium* spp.) is attributed to the benzoquinones produced by these insects (*33*), but the odoriferous materials released by the other arthropods mentioned above have not been identified.

Insects have often been characterized as uric acid excretors. However, the diversity of excretory products is greater than generally realized, and some of these products may adversely affect health if consumed. In the absence of quantitative data, it is unjustified to assume that uric acid is the sole or even the predominant nitrogenous metabolite of an arthropod. This is especially true of insects that harbor gut microbes and insects feeding on high-protein diets (*9*).

Nitrogenous excretory products from tryptophan metabolism have been identified in American (*Periplaneta americana*) and Madeira (*Leucophaea maderae*) cockroaches (*5, 39, 40*). Several of these metabolites have been identified as carcinogenic or mutagenic in mice and are suspected as the responsible agents of tumor-like lesions found in the guts of anal-blocked cockroaches. At least one of these metabolites is attributed to the gut-inhabiting microorganisms of the American cockroach (*40*).

Uric acid has been identified or presumed to be present in the excreta of numerous insect pests of food. The uric acid content of an infested commodity increases with the infestation level for many stored-product insects. Several authors (*4, 16, 53, 61*) have reported methods for using uric acid concentration as a reliable indicator of insect contamination. Although consuming small quantities of uric acid is not generally viewed as a health hazard, it has been associated with an undesirable off-flavor and bitter taste of infested flour (*26, 63*).

In a series of taste tests involving infested whole wheat flour from local shops in India, a level of 15 mg of uric acid per 100 grams of flour was adopted as the limit of acceptability (*26*). Such heavily infested products would probably be disposed of in developed nations, but in poorer countries, citizens are forced to sift out the insects and larger fragments and consume the "cleaned" product containing the excretory wastes (*26*).

Quinone and quinone derivatives are given off by some species of cockroaches (*15*), and benzoquinones have been identified as secretory products of at least 17 species of tenebrionid beetles (*31*), many of which are common pests of flour and other stored products. Adult flour beetles (*Tribolium* spp.) may contain from 30 μg to over 500 μg benzoquinone per insect depending on species, age, and sex of the insect (*66*). There is also evidence to indicate that the benzoquinone contamination per beetle could be much higher than these levels since the materials are released by the beetles under stress and the storage gland is then replenished (*30, 51*). Repeated stresses could, therefore, greatly increase the quinone levels in the infested commodity. These benzoquinones impart a disagreeable taste, odor, and appearance to infested flour and also impair its baking qualities (*55*).

The benzoquinones are highly reactive compounds that show bactericidal, enzyme-inhibiting, fungistatic, mutagenic, and carcinogenic activities (*29, 32, 35*). For laboratory animals, the acute oral lethal dosage (LD_{50}) of the hydroquinone–quinone system covers the range of 50–300 mg per kg of body weight (*43, 57*). The benzoquinones are difficult to detect by color or odor at low concentrations. Present analytical methods for the discovery of insect infestations are based upon actual counts of insects, insect fragments, or damaged grain kernels. Usually, no distinction is made between quinone-secreting insects and other stored-product insect pests (*12*).

The reported toxic and carcinogenic effects of the benzoquinones and the possible levels of these compounds in stored foods indicate a potential hazard. This is especially so in the countries where high infestation rates of quinone-releasing flour beetles are the norm.

Allergic Reactions to Arthropod Food Pests

Since the consumption of contaminated food is unavoidable, our immune systems are constantly being challenged by material of arthropod origin. The incidence of allergic reactions is affected by the frequency and degree of arthropod infestations, which varies with species, product, environmental factors, and geographic locations; the chemical composition of the sensitizing substances; and the routes of exposure, which are usually ingestion, inhalation, or direct skin contact (17, 54).

Consumption of arthropod-contaminated foods without allergenic symptoms is not an assurance that these materials are not activating the immune system. A significant proportion of the population responds subclinically to these exposures. Skin tests using extracts of insect pests of food resulted in a positive test rate of 25% for "nonallergic" individuals, and other skin-test experiments resulted in 6–24% positive reactions to extracts from six common species of food-infesting insects (2). The Indianmeal moth gave the highest percentage of positive responses in the study by Bernton and Brown (2).

Cockroach extracts have also given positive results in skin tests (65). Cockroaches are highly suspected as a source of ingestant allergens (2, 3), and the close association of these insects with human foods is well established. A relationship between the severity of infestation and duration of exposure has been reported for individuals sensitive to cockroach allergens (17).

Food-pest arthropods can also function as inhalant allergens (46); allergic symptoms (rhinitis, asthma) are often observed in food inspectors and in individuals rearing arthropods (14, 38, 41, 54). In a recent study by the Entomological Society of America on insect allergies, 28.7% of the institutions contacted reported that at least one individual had developed an inhalant allergy while working with arthropods (64). Mill dust allergy is an occupational hazard associated with arthropod fragments in the dust (48).

Dermatitis is also a well-documented allergic response to food-pest arthropods (1, 14, 60). As with inhalant allergies, most cases of arthropod-induced dermatitis are observed in individuals whose occupation brings them into contact with food pests (20, 48). The best-documented cases of contact dermatitis, often involving numerous individuals, have been among handlers of mite-infested foods. Vanilla itch, copra itch, grocers itch, and grain itch are the common terms applied to mite-induced skin inflammations in the several industries (14, 60). Moths, beetles, and cockroaches have also been implicated in cases of contact dermatitis (17).

Summary

The consumption of food-infesting arthropods and their products does not usually pose a significant health hazard to the consumer. Each of us is a confirmation of this statement since we routinely consume numerous herbivorous arthropods in the fresh and prepared foods we eat. There are, however, some cases of documented or suspected health problems resulting from the consumption of or allergic reaction to living arthropods, their fragments, or chemical products.

Although the majority of cases of arthropod consumption do not result in clinical illness, symptoms of fever, nausea, abdominal pain, and bloody diarrhea have been reported in the more severe cases. The consumption of arthropod excretory and secretory products with suspected or demonstrated toxic or carcinogenic properties poses potential health hazards. Allergic reactions to food-infesting arthropods are often viewed only as occupational health hazards. However, a significant proportion of the population exhibits detectable sensitivity to allergic materials of arthropod origin.

There is a definite need for continued research in the field of health hazards associated with arthropod-infested foods, especially in the area of chemical products and their effects on the consumer. After reliable information is available in these areas, it is essential that it be utilized to establish realistic rejection and acceptance criteria for use by the food industry and governmental regulatory agencies.

Acknowledgments

The author thanks Mrs. L. Applewhite, Technical Publications Editor, and Dr. G. H. G. Eisenberg, Chief, Division of Cutaneous Hazards, Letterman Army Institute of Research, for their comments on the content and format of this chapter.

Cited References

1 Alexander, J. O. 1972. Mites and skin diseases. *Clinical Medicine* 79(4)14–19.

2 Bernton, H. S., and H. Brown. 1967. Insects as potential sources of ingestant allergens. *Annals of Allergy* 25(7)381–387.

3 Bernton, H. S., and H. Brown. 1970. Insect allergy: The allergenicity of the excrement of the cockroach. *Annals of Allergy* 28(11)543–547.

4 Bhattacharya, A. K., and G. P. Waldbauer. 1970. Use of the faecal uric acid method in measuring the utilization of food by *Tribolium confusum*. *Journal of Insect Physiology* 16(10)1983–1990.

5 Bignell, D. E., and D. E. Mullins. 1977. A preliminary investigation of the effects of diets on lesion formation in the hindgut of adult female American cockroaches. *Canadian Journal of Zoology* 55(7)1100–1109.

6 Bodenheimer, F. S. 1951. *Insects as Human Food*. Dr. W. Junk, The Hague.

7 Boese, J. 1977. Mites associated with stored products: Selected references published from 1960 through 1975. *FDA By-Lines* 5:229–261, 1:25–30 (1978).

8 Bristowe, W. S. 1932. Insects and other invertebrates for human consumption in Siam. *Transactions of the Royal Entomological Society of London* 80(2)387–404.

9 Bursell, E. 1967. The excretion of nitrogen in insects. *Advances in Insect Physiology* 4:33–67.

10 Busvine, J. R. 1980. *Insects and Hygiene*. Methuen, London, England.

11 Center for Food Safety and Applied Nutrition. 1989.

The Food Defect Action Levels. Food and Drug Administration, Washington, DC.

12 Cotton, R. T. 1959. Effects and detection of insect and rodent infestation of cereals. In *The Chemistry and Technology of Cereals as Food and Feed.* AVI, Westport, CT.

13 DeFoliart, G. R. 1975. Insects as a source of protein. *Bulletin of the Entomological Society of America* 21(3)161–163.

14 Ebeling, W. 1975. *Urban Entomology.* Agriculture & Natural Resources, University of California, Oakland.

15 Eisner, T., and J. Meinwald. 1966. Defensive secretions of arthropods. *Science* 153(3742)1341–1350.

16 Farn, G., and D. Smith. 1963. Rate of excretion of uric acid by the rust-red flour beetle. *Journal of the Association of Official Agricultural Chemists* 46(3)517–521.

17 Frazier, C. A. 1969. *Insect Allergy.* Green, St. Louis, MO.

18 Gesztessy, T., and L. Nemeseri. 1970. Veterinary-hygienic aspects of mite-contaminated feed. I. Experiments on chickens, guinea-pigs, rabbits, and sheep. II. Experiments on albino mice, swine and cattle. *Acta Veterinaria Scientiarum Hungaricae* 20(1)29–33, (4)401–403.

19 Gorham, J. R. 1975. Filth in foods: Implications for health. *Journal of Milk and Food Technology* 38(7)409–418.

20 Gorham, J. R. 1976. A rational look at insects as food. *FDA By-Lines* 5:231–241.

21 Gorham, J. R. 1978. Insects as food. *Bulletin of the Society of Vector Ecologists* 3(1976)11–16.

22 Gorham, J. R. 1979. The significance for human health of insects in food. *Annual Review of Entomology* 24:209–224.

23 Harwood, R. F., and M. T. James. 1979. *Entomology in Human and Animal Health.* Macmillan, New York, NY.

24 Hinman, H. E., and R. G. Kampmeier. 1934. Intestinal acariasis due to *Tyroglyphus longior. American Journal of Tropical Medicine and Hygiene* 14(4)355–362.

25 Hoffmann, W. E. 1947. Insects as human food. *Proceedings of the Entomological Society of Washington* 49(9)233–237.

26 Howe, R. W. 1965. Losses caused by insects and mites in stored foods and feedingstuffs. *Nutrition Abstracts & Reviews* 35(2)285–303.

27 Kenney, M. 1945. Experimental intestinal myiasis in man. *Proceedings of the Society for Experimental Biology and Medicine* 60(2)235–237.

28 Krantz, G. W. 1978. *A Manual of Acarology.* O.S.U. Book Stores, Corvallis, OR.

29 Ladisch, R. K. 1965. Quinone toxins and allied synthetics in carcinogenesis. *Proceedings of the Pennsylvania Academy of Science* 38(2)144–149.

30 Ladisch, R. K. 1966. Assay of tetritoxin in flour beetles *Tribolium* and method of detecting the toxin on cereal foods. *Proceedings of the Pennsylvania Academy of Science* 39(2)36–41.

31 Ladisch, R. K., S. K. Ladisch, and P. M. Howe. 1967. Quinoid secretions in grain and flour beetles. *Proceedings of the Pennsylvania Academy of Science* 41:213–219.

32 Ladisch, R. K., M. S. A. Suter, and G. F. Froio. 1968. Sweat gland carcinoma produced in mice by insect quinones. *Proceedings of the Pennsylvania Academy of Science* 42:87–91.

33 Loconti, J. D., and L. M. Roth. 1953. Composition of the odorous secretion of *Tribolium castaneum. Annals of the Entomological Society of America* 46(2)281–289.

34 Manson-Bahr, P., and W. J. Muggleton. 1945. Significance of mites and their eggs in human feces. *Lancet* 1(6334)81–82.

35 Mayer, R. L. 1950. Compounds of quinone structure as allergens and cancerogenic agents. *Experientia* 6(7)241–250.

36 Military Standard. 1984. *Guidelines for Detection, Evaluation, and Prevention of Pest Infestation of Subsistence.* MIL-STD-904A. Department of Defense, Washington, DC.

37 Mills, H. B., and J. H. Pepper. 1939. The effect on humans of the ingestion of the confused flour beetle. *Journal of Economic Entomology* 32(6)874–875.

38 Mills, R. B., and G. J. Partida. 1976. Attachment mechanisms of *Trogoderma* hastisetae that make possible their defensive function. *Annals of the Entomological Society of America* 69(1)20–33.

39 Mullins, D. E., and D. G. Cochran. 1973. Nitrogenous excretory materials from the American cockroach. *Journal of Insect Physiology* 19(5)1007–1018.

40 Mullins, D. E., and D. G. Cochran. 1983. Tryptophan metabolite excretion by the American cockroach. *Comparative Biochemistry and Physiology* 44B(2)549–555.

41 Okumura, G. 1967. A report of canthariasis and allergy caused by *Trogoderma* (Coleoptera: Dermestidae). *California Vector Views* 14(3)19–22.

42 Olkowski, H., and W. Olkowski. 1976. Entomophobia in the urban ecosystem, some observations and suggestions. *Bulletin of the Entomological Society of America* 22(3)313–317.

43 Omaye, S. T., R. A. Wirtz, and J. T. Fruin. 1979. Acute oral toxicity of substituted *p*-benzoquinones and 1-pentadecene in the male rat. Letterman Army Institute of Research Technical Note No. 12, pp. 1–7.

44 Palmer, E. D. 1970. Entomology of the gastrointestinal tract: A brief review. *Military Medicine* 135(3)165–176.

45 Payne, N. M. 1925. Some effects of *Tribolium* on flour. *Journal of Economic Entomology* 18(5)737–744.

46 Perlman, F. 1958. Insects and inhalant allergens. *Journal of Allergy* 29(4)302–328.

47 Pimentel, D., et al. 1977. Pesticides, insects in foods, and cosmetic standards. *BioScience* 27(3)178–185.

48 Pratt, H. D., K. S. Littig, and H. G. Scott. 1975. *Household and Stored-Food Insects of Public Health Importance and Their Control.* DHEW Publ. No. (CDC)75-8122. Center for Disease Control, Atlanta, GA.

49 *Principal Storage Pests.* 1983. Degesch America, Weyers Cave, VA.

50 Rajan, P., et al. 1976. Effect of feeding diet containing insect infested maize and cowpea on the growth and histological changes in livers of albino rats. *Nutrition Reports International* 13(4)347–354.

51 Roth, L. M., and R. B. Howland. 1941. Studies on the gaseous secretion of *Tribolium confusum* (Duval). I. Abnormalities produced in *Tribolium confusum* (Duval) by exposure to a secretion given off by the adults. *Annals of the Entomological Society of America* 34(1)151–175.

52 Ruddle, K. 1973. The human use of insects: Examples from the Yukpa. *Biotropica* 5(2)94–101.

53 Sen, N. P. 1968. Uric acid as an index of insect infestation in flour. *Journal of the Association of Official Analytical Chemists* 51(4)785–791.

54 Shulman, S. 1967. Allergic responses to insects. *Annual Review of Entomology* 12:323–346.

55 Smith, L. W., Jr., et al. 1971. Baking and taste properties of bread made from hard wheat flour infested with species of *Tribolium, Tenebrio, Trogoderma* and *Oryzaephilus. Journal of Stored Products Research* 6(4)307–316.

56 Solomon, M. E. 1943. *Tyroglyphid Mites in Stored Products. 1. A Survey of Published Information.* Department of Scientific and Industrial Research, London, England; see also 1944 Supplement, pp. 1–7.

57 Suter, M. S. A., and R. K. Ladisch. 1963. Toxicity of insect quinones to mice. *Proceedings of the Pennsylvania Academy of Science* 37:137–141.

58 Swabowicz, A., et al. 1958. Toxicity of *Tyroglyphus farinae* to animals. IV. Experiments on chickens and ducks. *Medycyna Weterynaryjna* 14(9)556–558 (in Polish, English summary).

59 Swabowicz, A., K. Miedzobrodzki, and W. Schmidt. 1958. The toxicity of *Tyroglyphus farinae* for animals. II. Experiments on pigs. *Medycyna Weterynaryjna* 14(6)344–346 (in Polish, English summary).

60 TerBush, L. E. 1972. The medical significance of mites of stored food. *FDA By-Lines* 2:57–70.

61 Venkat Rao, S., et al. 1959. An improved method for the determination of uric acid in insect infested foodstuffs. *Annals of Biochemistry and Experimental Medicine* 19(8)187–188.

62 Venkat Rao, S., et al. 1960. The effect of feeding diets containing insect-infested jowar on the growth and composition of blood and liver of albino rats. *Annals of Biochemistry and Experimental Medicine* 20(5)135–142.

63 Venkatrao, S., et al. 1960. Effect of insect infestation on a stored field bean (*Dolichos lablab*) and black gram (*Phaseolus mungo*). *Food Science* 9(Mar.)79–82.

64 Wirtz, R. A. 1980. Occupational allergies to arthropods—documentation and prevention. *Bulletin of the Entomological Society of America* 26(3)356–360.

65 Wirtz, R. A. 1984. Allergic and toxic reactions to nonstinging arthropods. *Annual Review of Entomology* 29:47–69.

66 Wirtz, R. A., S. L. Taylor, and H. G. Semey. 1978. Concentrations of substituted *p*-benzoquinones and 1-pentadecene in the flour beetles *Tribolium madens* (Charp.) and *Tribolium brevicornis* (LeC.) (Coleoptera, Tenebrionidae). *Comparative Biochemistry and Physiology* 61C(2)287–290.

37
Food Pests as Disease Vectors

J. Richard Gorham

Much of the background information pertinent to this subject has been presented in the preceding chapter and in other publications (28, 29) to which the reader is referred. The focus of this chapter is on the roles of food pests as potential vectors.

Commercial sterilization, when properly accomplished, kills any human pathogens that might be present in foods. Therefore, only post-processing situations are treated here. High-risk situations occur when pathogens get into food after the last biocidal step in food manufacture. What may happen is that uncovered food or food packaged in some kind of packaging material that can be breached by pests or their metabolic products is exposed to pests, a violation of section 403(a)(3) of the Federal Food, Drug, and Cosmetic Act (FD&C Act). Even if the packaging were not breached by either the pests or their metabolic products, the co-existence of the food and the food pests in the same place would constitute a violation of section 403(a)(4) of the FD&C Act.

The "Highways" and "Byways" of Transmission

The general epidemiological picture surrounding most cases of food- or waterborne disease involves a direct fecal-oral route. To break this simple chain of events is a major goal of and a continuing challenge to food and water sanitation.

Animals that feed on sources of human pathogens may mechanically carry, either externally or internally, those pathogens to food or food-contact surfaces. Since we can readily observe the regularity with which such pests visit sources of pathogens as well as food or food-contact surfaces, we might assume that vectored transmission of foodborne disease is a

Gorham: Food and Drug Administration, Washington, DC 20204.

common occurrence. Such is not the case. The vast majority of reported cases of foodborne illness can be traced to the direct fecal-oral route. But, taking a worldwide view, the vast majority of cases of foodborne illness are never seen by a physician or any other health professional, much less investigated in any scientific way. So it may be that vectors could be more important than reports from highly developed countries would indicate. Further, it appears that the explanation of a direct fecal-oral route may be accepted too readily, and that the potential role of vectors is too rarely considered seriously.

It is true that the known incidence of natural infection of commensal rats, mice, and cockroaches with *Salmonella* species runs about 1.2% (59, 73), and this would lend credence to the conclusion that vectors may be relatively unimportant in the spread of foodborne pathogens. Yet the numerous studies (1, 3, 7, 10–12, 16, 17, 20, 21, 24–27, 32, 33, 35, 37, 39, 42, 48, 49, 52, 68–70, 72) in which pathogens have been routinely recovered from the external surfaces of many pests suggests at the very least that these pests pose a continual threat to the safety of food and that their presence around food is incompatible with even modest standards of sanitation.

Three Sets of Potential Vectors

There are three general groups of potential vectors associated with food, but these groups are not consistently distinct from each other; there is considerable overlap among them.

(a) *Inadvertent vectors*. These are the birds and bats. Their potentially pathogen-laden feces may become airborne, foodborne, or waterborne, or they may be "accidentally" deposited directly on food or food-contact surfaces (6, 10, 55, 61).

(b) *Opportunistic vectors*. These are cockroaches, flies, ants, mice, and rats. Their mobility and their continuous searching for food and water may bring them into contact with both food and food-contact surfaces and with such potential sources of pathogens as floors, drains, sewers, feces, wounds, and discarded surgical dressings.

(c) *Obligatory vectors.* These are the various beetles, moths, and mites that commonly infest stored foods.

The terms inadvertent (accidental), opportunistic, and obligatory are used to describe the pest's relationships to human food, not its capabilities as a vector. Obligatory vectors, for example, are those pests dependent on stored foods—they derive all their food, water, and shelter from them. Opportunistic vectors can survive without human food but will utilize it whenever they get the chance. Inadvertent vectors may or may not eat human food; they may simply use storage structures for places to rest, nest, or roost.

Dissemination of Pathogens

It is often assumed that the pantry pests (referred to earlier as obligatory vectors), compared to some other potential vectors (e.g., flies, ants, cockroaches, mice), are less of a threat to food safety from a microbiological point of view because their sphere of existence is circumscribed by the food itself—food that is very likely to be pathogen-free after manufacture. The effect of these pantry pests, then, is further assumed to result in merely a diminished aesthetic value of the food rather than a risk to food safety. These assumptions would be safe to make except for two considerations:

(a) The pantry pests have to come to the food from some harborage site external to the food. That harborage site and the route of migration between it and the food may or may not be pathogen-free.

(b) Other potential vectors besides the obligatory ones readily visit food already infested by the pantry pests. The opportunistic or accidental vectors may bring pathogens to points within the limited ranges of the pantry pests where they may be picked up and disseminated within the food by these pests (3, 18, 19, 36, 49, 60).

Examples from Veterinary Pathology

Two examples are given here to demonstrate that food pests—insects in both cases—can vector pathogens in ways germain to the subject of this review.

Porcine Bronchopneumonia

Pigs in a veterinary hospital isolation unit contracted pneumonia after being exposed to ants (*Monomorium pharaonis*) that had carried the causative agent (*Bordetella bronchiseptica*) into the isolation unit (7).

Avian leucosis

Under experimental conditions, chickens came down with avian leucosis after eating beetles (*Alphitobius diaperinus*) that had acquired the virus by feeding on dead or moribund chickens or on the feces of chickens infected with the virus (23).

Examples from Tropical Medicine

Chagas Disease

People infected by the protozoan (*Trypanosoma cruzi*) that causes Chagas disease often live in thatched huts. Triatomid bugs feed at night on these people, then retreat to the thatched roof to rest and digest the blood meal. Their pathogen-laden feces rain down indiscriminately on food and water below. The pathogens can remain alive and infective in the fecal pellets up to 10 days. Any susceptible person who eats unheated food runs a risk of contracting Chagas disease (51).

Bolivian Hemorrhagic Fever

Wild grass mice (*Calomys callosus*) of Bolivia have inapparent infections but shed the viruses continually in their urine. People may become infected by breathing the airborne viruses or by consuming unheated food or water contaminated during the nightly foraging by the mice (63). Lymphocytic choriomeningitis, another disease of mice (*Mus musculus*), may be transmitted in the same manner (9, 31, 74).

Histoplasmosis and Ornithosis

It is possible to make at least a tenuous connection between food pests and the transmission of histoplasmosis and ornithosis (61). However, there is insufficient evidence for infection *per os*; the respiratory route is the major pathway to infection.

Toxoplasmosis

Domestic cats should not be admitted to food storage facilities, especially not in situations where cats can defecate directly on food materials. Infected cats shed the oocysts of *Toxoplasma gondii* in their feces (40). Sporulated oocysts are infective *per os*.

Leptospirosis

The circumstances surrounding the transmission of leptospirosis in the slaughterhouse, meat-packing, and poultry-processing industries are well known. In the food-service industry and in the home, human beings may come into contact with leptospires shed in the urine of rats and mice (9). Numerous studies have shown that leptospirosis is common in *Rattus norvegicus*, and somewhat less common in *Mus musculus* and *Rattus rattus* (42, 68). The percentage of naturally infected rats ranges up to 52% (8). Longevity of shed leptospires on foods varies from 30 minutes to 13 days (50). They survive best on cool, moist, slightly alkaline substrates, or in cool liquids. The chances of acquiring leptospirosis under such conditions would appear to be very small, but some cases can be explained in no other way.

Aflatoxicosis

Aspergillus fumigatus and *A. niger*, both aflatoxin producers, occur as natural infections in certain cockroaches. Mites, cockroaches, and some other insects mechanically disseminate mold spores in stored grains. Thus, these pests contribute indirectly to the potential production of aflatoxins in stored foods (*3, 49, 56*).

Amebiasis

It is often recommended that flies should be controlled as an auxillary measure in the suppression of outbreaks of amebiasis. Although both flies and cockroaches sometimes have natural infections of *Entamoeba histolytica,* and both flies and cockroaches have feeding habits conducive to the transfer of amoebic cysts from fecal material to food, there is little objective evidence for vectoral transmission of amebiasis. The circumstantial evidence is, however, persuasive (*32, 58*).

Enteric Diseases

The fecal-oral route is the predominant pathway for the transmission of both enteric viral diseases and enteric bacterial diseases. This mode of transmission may be direct—feces to hands to mouth—or sometimes food and/or water are intermediate in the cycle of events. It is conceivable that vectors may sometimes add another dimension to transmission. Viral and bacterial pathogens can be almost routinely isolated from natural populations of the insects, mites, rodents, and birds that are often found in or around our food. Experimental studies of vectoral capacity have repeatedly demonstrated that many food pests are competent to carry pathogens from sources of contamination to food or food-contact surfaces (*2, 4, 13, 15, 18, 31, 34, 38, 41, 45, 54, 58, 59, 63–65, 68, 74, 75*).

Viral Diseases

The epidemiology of foodborne bacterial diseases has been studied in detail, but until recently little was known about viral pathogens in food. Now a long list of foodborne viruses has been compiled (*44, 57*) and some information is being developed about how foods become contaminated with viruses and about the significance of these viruses in human disease. Food pests probably play no more than a minor role in the epidemiology of foodborne viral diseases. However, several kinds of food-associated pathogenic viruses have been isolated from food pests (*1, 31, 63, 74*).

Rotaviral Enteritis

Rotaviral enteritis, caused by a reovirus, occurs often in localized epidemics in infants, young children, and elderly adults. In 1982 and 1983 in China, there were two epidemics of acute diarrhea affecting more than 12,000 adults (*66, 67*). A rotavirus-like agent, distinct from the virus causing rotaviral enteritis, was implicated as the causative organism. The virus was probably spread by the fecal-oral route involving a waterborne phase and person-to-person phase. Serologic studies of animals within the epidemic zones showed that 36% of the pigs and 47% of the rats tested had been exposed to this new rotavirus.

Another rotaviral agent, also antigenically distinct from the virus causing rotaviral enteritis, was isolated from three adults and three children, all suffering from gastroenteritis, during 1983 and 1984 in Baltimore, Maryland. This same agent causes diarrheal disease in infant rats (*22*). A similar, perhaps identical, agent was isolated from spontaneous cases of diarrhea in infant rats (*71*). Serologic studies of 30 wild rats (*Rattus norvegicus*) in Baltimore showed that 30% had been exposed to a rotavirus-like agent (J. Eiden, Johns Hopkins University, 1985, personal communication).

Studies are under way to further characterize these rotavirus-like agents, to determine their modes of transmission among animals and humans, and to elucidate the significance of rats associated directly with people, with food or water consumed by people, or with food-contact surfaces. Although the information gathered thus far does not clearly implicate rats in the epidemiology of human disease caused by rotavirus-like agents, it does strongly suggest that rat-human and rat-human-food contacts should be prevented to the greatest possible extent.

Other recent sudies of viruses in *Rattus norvegicus* and various other rodents lend support to the idea that contacts between rodents and people should be avoided whenever possible. Korean haemorrhagic fever (KHF) virus and certain other viruses closely related to it are apparently shed in the feces, urine, and saliva of rodents. Rodents eating or walking about may deposit these viruses on exposed food or food-contact surfaces.

Hantaan virus, a close relative of KHF virus, has been found in *R. norvegicus* in many places in the Western Hemisphere but has not yet been identified as the cause of human disease in this region. It is certain, however, on the basis of serological evidence, that humans have been exposed to *Hantavirus* in Baltimore, Maryland (*14*). This situation and its potential implications for the food industry will have to be followed carefully in the future.

Bacterial Diseases: Listeriosis

Although listeriosis does not fall precisely under

the category of enteric disease, the infection may at least sometimes be acquired *per os*. During the course of investigations surrounding outbreaks of listeriosis associated with cheese in California, Food and Drug Administration scientists recovered *Listeria monocytogenes* from ants (*Iridomyrmex humilis*) present in manufacturing facilities producing the cheese (A. R. Olsen, FDA, 1989, personal communication). The significance of these findings in relation to the epidemiology of listeriosis is unknown, but they suggest that ants at the very least may have some role in the spread of bacteria within food-manufacturing establishments.

Salmonellosis

Salmonellosis would probably continue to exist as a disease of humans and animals even in the absence of rats, mice, cockroaches, flies, ants, and other food-associated pests. Person-to-person passage via the fecal-oral route is probably the major mode of transmission. Vectors, however, have been so often implicated in either major or minor roles in the transmission of this disease that on this basis alone their presence in any phase of food handling should be considered a threat to public health.

Rats and Mice

A large-scale survey of rodent pellets showed that viable *Salmonella* bacteria were present in only 1.2% of the pellets. The bacteria could remain viable in these pellets from naturally infected rats for at least 148 days at room temperature. The infection is readily passed from rat to rat. In one test, three out of four rats were infected with just 15 bacteria per rat (*73*). Rats were implicated in an epidemic of salmonellosis in 1935. There were 208 cases with three deaths. All cases were traced to a bakery where *Salmonella* bacteria were isolated from the bakery goods and from rats trapped in the bakery (*62*).

Insects

Cockroaches have been implicated on the basis of circumstantial evidence in several epidemics of salmonellosis (*31, 46, 47*). Ants (*7*), cockroaches (*5*), and flies (*32*) are well known for their ability to mechanically carry *Salmonella* bacteria on the surfaces of their bodies. Flies and cockroaches also harbor populations of *Salmonella* in their alimentary canals (*32, 52, 59*). These bacteria are deposited wherever these insects leave feces and vomitus. The bacteria remain viable for varying periods of time, up to 199 days in the feces (*53*) and up to 60 days in the dead bodies of cockroaches (*43*). The potential danger of this situation has been summarized by Klowden and Greenberg (*43*):

The post-mortem persistence of *Salmonella* in the carcasses of cockroaches infected before their death is important in view of the minimal levels of "filth" composed of insect parts and rodent hairs which food processors are allowed . . . to include in their foods.

Cockroach remains are undoubtedly major contributions to this tolerable filth. Unpasteurized foods, such as spices and packaged cereals, might be "seasoned" with a portion of infected cockroach carcass, and become a route of *Salmonella* dissemination. The incorporation of contaminated cockroach debris into the environment of the home, restaurant, and food processing plant is another avenue which has not been explored.

Conclusion

Although relatively little is known about the health significance of insect, rodent, and bird filth in food, the information we do have indicates that food pests must be excluded from all situations in the food industry. More laboratory research is needed to elucidate the pathogenic significance of filth in food. Cases of filth-associated human illness should be investigated and reported in the technical literature. Regulatory and quality control personnel in the food industry should be alert for opportunities to fill in the gaps in our knowledge of foodborne filth and its relation to human disease.

Cited References

1 Asahina, S., et al. 1963. Detection of polioviruses from flies and cockroaches captured during the 1961 epidemics in Kumamoto Prefecture. *Japanese Journal of Sanitary Zoology* 14(1)28–31 (in Japanese, English summary).

2 Ash, N., and B. Greenberg. 1980. Vector potential of the German cockroach (Dictyoptera: Blattellidae) in dissemination of *Salmonella enteritidis* serotype *typhimurium*. *Journal of Medical Entomology* 17(5)417–423.

3 Aucamp, J. L. 1969. The role of mite vectors in the development of aflatoxin in groundnuts. *Journal of Stored Products Research* 5(3)245–249.

4 Barber, M. A. 1914. Cockroaches and ants as carriers of the vibrios of Asiatic cholera. *Philippine Journal of Science* B9(1)1–4.

5 Bartlett, P. G. 1967. Roach control—key to *Salmonella* control. *Soap & Chemical Specialties* 43(9)62–64, 66.

6 Baur, F. J., and W. B. Jackson (eds.). 1982. *Bird Control in Food Plants—It's a Flying Shame!* American Association of Cereal Chemists, St. Paul, MN.

7 Beatson, S. H. 1972. Pharaoh's ants as pathogen vectors in hospitals. *Lancet* 1(7747)425–427.

8 Bell, I. 1973. Rodent control is necessary for food protection. *Journal of Environmental Health* 36(1)75–77.

9 Benenson, A. S. (ed.). 1990. *Control of Communicable Diseases in Man*. American Public Health Association, Washington, DC.

10 Berg, R. W., and A. W. Anderson. 1972. Salmonellae and *Edwardsiella tarda* in gull feces: A source of con-

480   J. Richard Gorham

tamination in fish processing plants. *Applied Microbiology* 24(3)501–503.

11 Brühl, W., and M. E. A. Fuchs. 1973. Schaben als Vektoren humanpathogener und toxinbildender Pilze. *Mykosen* 16(6)215–217 (English summary).

12 Burgess, N. R. H., S. N. McDermott, and J. Whiting. 1973. Aerobic bacteria occurring in the hindgut of the cockroach, *Blatta orientalis*. *Journal of Hygiene* 71(1)1–7.

13 Burgess, N. R. H., S. N. McDermott, and J. Whiting. 1973. Laboratory transmission of Enterobacteriaceae by the oriental cockroach, *Blatta orientalis*. *Journal of Hygiene* 71(9)9–14.

14 Childs, J. E., et al. 1988. Evidence of human infection with a rat-associated *Hantavirus* in Baltimore, Maryland. *American Journal of Epidemiology* 127(4)875–878.

15 Clot, J., and C. Vago. 1970. Recherches sur le passage de bactéries pathogènes pour les invertébrés et les vertébrés (homme, animal), a travers le tube digestif de dictyoptères disséminateurs (expériences sur *Blabera fusca*). *Annales de Recherches Vétérinaires* 1(1)31–40 (English summary).

16 Cornwell, P. B., and M. F. Mendes. 1981. Disease organisms carried by oriental cockroaches in relation to acceptable standards of hygiene. *International Pest Control* 23(3)72–74.

17 Crawford, R. E., L. A. McDermott, and A. J. Musgrave. 1960. Microbial isolations from the granary weevil *Sitophilus granarius* (L.) (Coleoptera: Curculionidae). *Canadian Entomologist* 92(8)577–581.

18 Crumrine, M. H., V. D. Foltz, and J. O. Harris. 1971. Transmission of *Salmonella montevideo* in wheat by stored-product insects. *Applied Microbiology* 22(4)578–580.

19 De Las Casas, E. 1977. Stored product insects and microorganisms in grain ecosystems. *University of Minnesota Agricultural Experiment Station Technical Bulletin* 310:31–34.

20 Doom, R., and M. E. Curtice. 1966. Identification of the characteristic flora of *Supella supellectilum*. *Proceedings of the South Dakota Academy of Science* 45:255–259.

21 Eichler, W. 1973. Synanthrope Aspekte sur Ökologie der Pharaoameise. *Deutsche Entomologische Zeitschrift* 20(4–5)425–432.

22 Eiden, J., S. Vonderfecht, and R. H. Yolken. 1985. Evidence that a novel rotavirus-like agent of rats can cause gastroenteritis in man. *Lancet* 2(8445)8–11.

23 Eidson, C. S., et al. 1966. Induction of leukosis tumors with the beetle *Alphitobius diaperinus*. *American Journal of Veterinary Research* 27(119)1053–1057.

24 Frishman, A. M., and I. E. Alcamo. 1977. Domestic cockroaches and human bacterial disease. *Pest Control* 45(6)16, 18, 20, 46.

25 Fuchs, M. E. A. 1976. Zur Verbreitung humanpathogener und toxinbildender Pilze durch Schaben. *Zeit-schrift für Angewandte Entomologie* 82(1)89–93 (English summary).

26 Fuchs, M. E. A., H. Messmer, and J. D. Haase. 1976. Zur taktilen und exkretorischen Verbreitung von Pilzen und Bakterien durch Schaben. *Bundesgesundheitsblatt* 19(21)329–331 (English summary).

27 Garg, R. K. 1977. The flora of the alimentary canal of the hide-beetle, *Dermestes frischii* Kugelann (Dermestidae: Coleoptera). *Journal of Natural History* 11(1)97–99.

28 Gorham, J. R. 1975. Filth in foods: Implications for health. *Journal of Milk and Food Technology* 38(7)409–418.

29 Gorham, J. R. 1979. The significance for human health of insects in food. *Annual Review of Entomology* 24:209–224.

30 Graffar, M., and S. Mertens. 1950. Le rôle des blattes dans la transmission des salmonelloses. *Annales de l'Institut Pasteur* 79(5)654–660.

31 Gratz, N. G. 1988. Rodents and human disease: A global appreciation. In *Rodent Pest Management*, I. Prakash, ed. CRC, Boca Raton, FL.

32 Greenberg, B. 1973. *Flies and Disease*, vol 2. *Biology and Disease Transmission*. Princeton University, Princeton, NJ.

33 Greenberg, B., and M. Sanati. 1970. Enteropathogenic types of *Escherichia coli* from primates and cockroaches in a zoo. *Journal of Medical Entomology* 7(6)744.

34 Griffitts, S. D. 1942. Ants as probable agents in the spread of *Shigella* infections. *Science* 96(2490)271–272.

35 Harein, P. K., and E. De Las Casas. 1968. Bacteria from granary weevils collected from laboratory colonies and field infestations. *Journal of Economic Entomology* 61(6)1719–1720.

36 Husted, S. R., et al. 1969. Transmission of *Salmonella montevideo* from contaminated to clean wheat by the rice weevil. *Journal of Economic Entomology* 62(6)1489–1491.

37 Ipinza-Regla, J., G. Figueroa, and I. Moreno. 1984. *Iridomyrmex humilis* (Formicidae) y su papel como posible vector de contaminación microbiana en industrias de alimentos. *Folia Entomológica Mexicana* 62:111–124 (English summary).

38 Ipinza-Regla, J., G. Figueroa, and J. Osorio. 1981. *Iridomyrmex humilis* "hormiga Argentina" como vector de infecciones intrahospitalarias. I. Estudio bacteriológico. *Folia Entomológica Mexicana* 50:81–96 (English summary).

39 Ishiyama Cervantes, V. 1967. Investigación de *Salmonella* en cucarachas de diferentes zonas de Lima. *Biota* 7(53)1–16.

40 Jacobs, L., and J. K. Frenkel. 1982. Toxoplasmosis. In *CRC Handbook Series in Zoonoses, Section C: Parasitic Zoonoses*, vol. 1, part 1, L. Jacobs, ed. CRC, Boca Raton, FL.

41 Jones, E. R., and H. D. Wright. 1936. *B. aertrycke* food poisoning due to contamination of food with excreta of mice. *Lancet* 1(230)22–23.

42 Kavtaradze, K. N., A. D. Berstein, and G. Y. Kvaratskheliya. 1957. Sources of leptospirosis in the Georgian ASSR. *Journal of Microbiology, Epidemiology and Immunobiology* 28(9–10)1276–1279.

43 Klowden, M. J., and B. Greenberg. 1977. Effects of antibiotics on the survival of *Salmonella* in the American cockroach. *Journal of Hygiene* 79(3)339–345.

44 Larkin, E. O. 1981. Food contaminants—viruses. *Journal of Food Protection* 44(4)320–325.

45 Lawson, F. A., and J. E. Johnson. 1970. Coxsackie A12 in *Periplaneta americana*—preliminary report (Blattaria: Blattidae). *Journal of the Kansas Entomological Society* 43(4)435–440.

46 Mackerras, I. M., and M. J. Mackerras. 1949. An epidemic of infantile gastro-enteritis in Queensland caused by *Salmonella bovis-morbificans* (Basenau). *Journal of Hygiene* 47(2)166–181.

47 Mason, D. 1985. *Hospitals Can Damage Your Health.* British Pest Control Association, London, England.

48 Mielke, U., and W. Bohm. 1977. Gesundheitsschädlinge als Vektoren von Bakterien und Pilzen. *Zeitschrift für Gesamte Hygiene* 23(6)367–370.

49 Misra, C. P., C. M. Christensen, and A. C. Hodson. 1961. The Angoumois grain moth, *Sitotroga cerealella*, and storage fungi. *Journal of Economic Entomology* 54(5)1032–1033.

50 Mitrofanova-Perfil'eva, E. B. 1957. The significance of food products in the epidemiology of the leptospiroses. *Journal of Microbiology, Epidemiology and Immunobiology* 28(9–10)1272–1276.

51 Nuñez, J. C. G. 1973. Triatomids in relation to Chagas disease in Latin America. A paper represented on 27 November at the annual meeting of the Entomological Society of America, Dallas, TX.

52 Okafor, J. I. 1981. Bacterial and fungal pathogens from the intestinal tract of cockroaches. *Journal of Communicable Diseases* 13(2)128–131.

53 Olson, T. A., and M. E. Rueger. 1950. Experimental transmission of *Salmonella oranienburg* through cockroaches. *Public Health Reports* 65(16)531–540.

54 Ostrolenk, M., and H. Welch. 1942. The common house fly (*Musca domestica*) as a source of pollution in food establishments. *Food Research* 7(3)192–200.

55 Palfreman, M. K. 1975. Birds and health. *Proceedings of the Fourth British Pest Control Conference* (St. Heiler, Jersey), Paper No. 21, pp. 1–3.

56 Ragunathan, A. N., D. Srinath, and S. K. Majumder. 1974. Storage fungi associated with rice weevil (*Sitophilus oryzae* L.). *Journal of Food Science and Technology* 11(1)19–22.

57 Rao, V. C. 1976. Virus transmission through foods. *Journal of Food Science and Technology* 13(6)1–7.

58 Roberts, E. W. 1947. The part played by the feces and vomit drop in the transmission of *Entamoeba histolytica* by *Musca domestica*. *Annals of Tropical Medicine and Parasitology* 41(1)129–142.

59 Rueger, M. E., and T. A. Olson. 1969. Cockroaches (Blattaria) as vectors of food poisoning and food infection organisms. *Journal of Medical Entomology* 6(2)185–189.

60 Schuster, D. J., R. B. Mills, and M. H. Crumrine. 1972. Dissemination of *Salmonella montevideo* through wheat by the rice weevil. *Environmental Entomology* 1(1)111–115.

61 Scott, H. G. 1961. Pigeons—public health importance and control. *Pest Control* 29(9)9–20, 60–61.

62 Staff, E. J., and M. L. Grover. 1936. An outbreak of *Salmonella* food infection caused by filled bakery products. *Food Research* 1(5)465–479.

63 Steele, J. H. 1981. Bolivian hemorrhagic fever (BHF). In *CRC Handbook Series in Zoonoses, Section B: Viral Zoonoses*, vol. 2, G. W. Beran, ed. CRC, Boca Raton, FL.

64 Stek, M., Jr. 1982. Cockroaches and enteric pathogens. *Transactions of the Royal Society of Tropical Medicine and Hygiene* 76(4)566–567.

65 Stek, M., R. V. Peterson, and R. L. Alexander. 1979. Retention of bacteria in the alimentary tract of the cockroach, *Blattella germanica*. *Journal of Environmental Health* 41(4)212–213.

66 Tao, H., et al. 1984. Waterborne outbreak of rotavirus diarrhoea in adults in China caused by a novel rotarvirus. *Lancet* 1(8387)1139–1142.

67 Tao, H., et al. 1985. Seroepidemiology of adult rotavirus. *Lancet* 2(8450)325–326.

68 Tsai, C.-C., W. D. Kundin, and J. W. Fresh. 1971. The zoonotic importance of urban rats as a potential reservoir for human leptospirosis. *Journal of the Formosan Medical Association* 70(1)1–4.

69 Ulewicz, K. 1976. Über die epidemiologische Bedeutung der Schabe *Blattella germanica* (L.) bei Übertragung von Infektionen auf den Menschen. *Schädlingsbekämpfer* 28(9)129–134 (English summary).

70 Ulewicz, K., A. Kunert, and P. Michniewski. 1976. Bacterial flora in roaches *Phyllodromia germanica* L. caught on ships. *Wiadomosci Parazytologiczne* 13(6)745–749 (in Polish, English summary).

71 Vonderfecht, S. L., et al. 1984. Infectious diarrhea of infant rats produced by a rotavirus-like agent. *Journal of Virology* 52(1)94–98.

72 Weeks, R. 1970. The relationship of bird habitats to disease producing fungi. *Proceedings. Fifth Bird Control Seminar* (Bowling Green, OH), pp. 128–133.

73 Welch, H., M. Ostrolenk, and M. T. Bartram. 1941. Role of rats in the spread of food poisoning bacteria of the *Salmonella* group. *American Journal of Public Health* 31(4)332–340.

74 Winkler, W. G., and V. J. Lewis. 1981. Lymphocytic choriomeningitis. In *CRC Handbook Series in Zoonoses, Section B: Viral Zoonoses*, vol. 2, G. W. Beran, ed. CRC, Boca Raton, FL.

75 Zebe, H., R. Sanwald, and E. Ritz. 1972. Insect vectors in serum hepatitis. *Lancet* 1(7760)1117–1118.

38

Commentary on Health Considerations: Overt and Insidious Implications of Food Pests and Health

Richard J. Brenner

In the course of mankind's cultural evolution, agriculture has grown from small scattered plots, typical of subsistence farming, to well-organized and extensive systems of crop production and storage. Irrigation and climate-controlled storage systems, orchestrated by modern technology, have allowed us to grow and harvest genetically engineered plants in areas previously inhospitable to most plants and animals. Thus, in many instances, availability of food and water is no longer a major constraint on survival of pestiferous arthropods and vertebrates that also find these commodities palatable. The advent of rapid transport has allowed these pests to expand their geographic ranges beyond their native distributions to areas outside the influence of natural predators and parasites, further enhancing the probability of successful colonization and exploitation of our food resources.

The previous chapters in this section clearly demonstrate that our food supply is at risk of becoming infested at virtually every stage from preharvest to consumption. Physical and chem-

Brenner: Medical and Veterinary Entomology Research Laboratory, USDA Agricultural Research Service, P.O. Box 14565, Gainesville, FL 32604.

ical processing (heating and sterilization) eliminates some pests and reduces risks from pathogenic microorganisms (see Chapter 37) but other health threats are more insidious and not so easily eliminated or managed (47). Thus, maintaining bountiful supplies of healthful products requires strict vigilance during production, harvesting, and storage of crops, warehousing, shipping, and storage by consumers. Figure 38.1 presents a comprehensive overview of the health implications of food pests. It summarizes much that has been presented in the preceding three chapters, and also contains some additional references. Although not an exhaustive review of literature, it is intended to provide a rapid entry point into the published data.

Pests as Modifiers of Commodities

In many cases, the presence and activities of the pests result in changes in temperature, humidity, and chemical composition that may affect viability of seed, baking quality, or palatability (see Chapter 35). In other instances, key components of grains may be consumed, thereby reducing the nutritional value. Finally, the entire grain may be consumed, in which case the effect of pests is felt even more keenly by

483

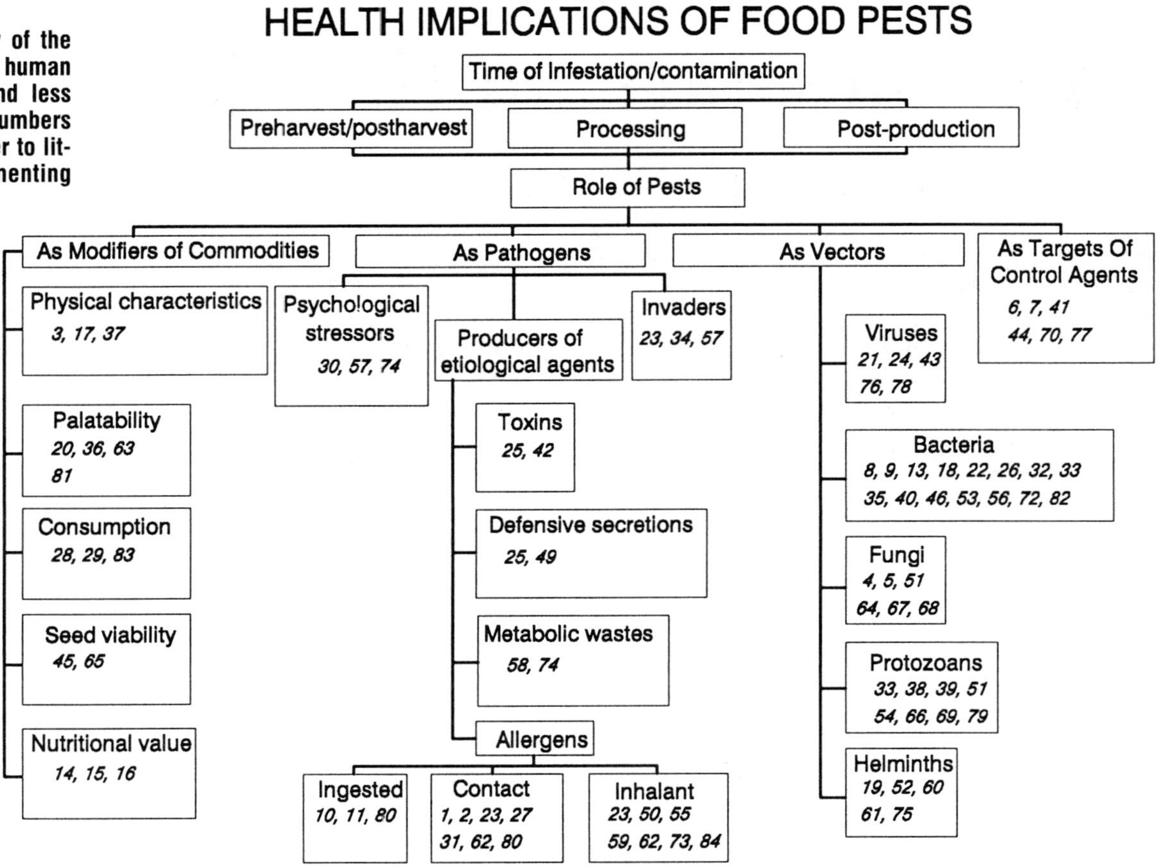

Figure 38.1
Comprehensive overview of the
effects of food pests on human
health includes overt and less
apparent implications. Numbers
below each category refer to literature citations documenting
each effect.

HEALTH IMPLICATIONS OF FOOD PESTS

Figure 38.1 Comprehensive overview of the effects of food pests on human health includes overt and less apparent implications. Numbers below each category refer to literature citations documenting each effect.

the human community that must survive on reduced yields and compromised nutrition.

Pests as Vectors

Food pests also function as vectors of a variety of pathogenic microorganisms, or appear
to have potential in the role of maintaining and
transmitting pathogens. Many references have
been included in this category that describe natural isolations of pathogens from field populations of cockroaches (15). Whether cockroaches
and other arthropods are significant vectors has
been debated for years. Unlike hematophagous
arthropods, whose very act of feeding is often
responsible for *direct* transmission, omnivorous
insects and mites can serve only to transmit microbes *indirectly* by contaminating foods or food-
contact surfaces. Because there are so many
sources of *Salmonella* organisms, for example,
it would be difficult to unequivocally demonstrate substantial vectorial capabilities of cockroaches (15). Yet, the disgust and stigma brought
on by the presence of cockroaches and rodents
may also exact a psychological toll that could
lead to clinical manifestations (71).

In light of these contentions, it is far easier
to demonstrate the significance for human health
of these pests in their varied roles as producers
of etiological agents, including toxins, defensive secretions, defensive hairs, metabolic wastes
and allergens (more on this category later). Direct ingestion of arthropods has been recorded
(most commonly with fruits and vegetables that
receive little or no postharvest processing), reportedly resulting in a variety of clinical symptoms and pathologies including diarrhea, nausea,
and intestinal lesions (see Chapter 37).

Pests as Targets of Control Agents

Misuse or overuse of pesticides in an effort
to control food pests also adversely affects human health. Pyrethrins and synthetic pyrethroids, commonly used to control a variety of
household insects, can cause transient respiratory irritations.

During a 5-year period at two poison control
centers near Washington, DC, the Environmental Protection Agency (EPA) recorded over 780

cases of incidental ingestion of boric acid, a material commonly used to control cockroaches. Children under the age of 6 comprised 80% of the cases. Although no long-term health effects resulted, many of the patients were taken to emergency rooms where they received various treatments including stomach pumping and hemodialysis (44). Clearly, the emotional anguish and the value of medical treatment constitute a significant cost to human health.

Similarly, ineffective electronic pest control devices are not only costly but may have harmful effects on humans. The public spent over $50 million in 1983 for ultrasound devices that allegedly kill cockroaches (77), despite numerous studies that clearly demonstrate their ineffectiveness (6, 7, 41); in some cases the emissions of ultrasound exceeded established safety levels, perhaps further jeopardizing human health (70).

Arthropod Allergens

Recent trends and research in allergic diseases suggest that this may be of greater significance to human health than has been previously recognized. Allergy to cockroaches and other arthropods has been documented only within the past three decades. Bernton and Brown (10, 11) and Bernton et al. (12) demonstrated that cockroaches deposit measurable quantities of allergen during the simple acts of crossing a dinner plate or chewing on a biscuit, and that these allergens are viable even after exposure to 100° (boiling water) for over an hour. Allergies, or hypersensitizations, are known to result from dermal contact with or inhalation of allergenic materials; it has been generally accepted that ingestion of allergens also may result in hypersensitization (27).

The likelihood of developing allergies to arthropods is related to many factors, including genetic predisposition (e.g., atopy and asthma), the source of allergens (i.e., species of arthropod), and the duration of exposure to them (27). The incidence of allergic disease in general is on the rise. Massicot and Cohen (48) summarized data from the late 1970s and early 1980s indicating that

(a) 17% of the population of the United States (USA) is affected;
(b) that more boys are affected by asthma than girls (under the age of 5);
(c) women suffer more commonly than men;
(d) incidence of atopic dermatitis is 6.9 people per 1,000;
(e) insect allergy affects 4 people per 1,000; and

(f) severe reactions to insects occur most commonly in people at least 30 years old.

These trends have also been observed internationally (48).

Asthmatics are at greatest risk. Recent data indicate that the most common allergy of asthmatics is to house dust mites (*Dermatophagoides* spp.), followed by allergies to cockroaches (39). In an article published in July 1989, the Asthma and Allergy Foundation of America provided a grim summary of the trends: Deaths from asthma are on the rise, and those at greatest risk are the poverty-stricken in urban environments. Among several theories presented by the authors is the idea that "a broadening of environmental triggers of asthma and more severe forms of the disease, not clearly understood by medical science, may be occurring" (2).

Even more disturbing, and perhaps more pertinent to the subject of this book, is the theory of "pan allergies," where developing allergies to one or a few arthropods results in allergic response to a great number of insects and other arthropods such as crustaceans (5). Of particular significance, Bernton and Brown (10) found that 30% of people with allergies and 25% of "nonallergic" people showed positive skin tests to various food pests, especially *Plodia interpunctella*, the Indianmeal moth, one of the more common invaders of foods sold in vending machines. This suggests that even a substantial portion of the "nonallergic" population may have subclinical allergies and be at risk for developing clinical symptoms.

Conclusions

Given that the population of the USA is aging, that hypersensitization is, in part, related to chronic exposure to allergenic components of arthropods, and that the theory of pan allergies is valid, this overview reveals profound implications that extend well beyond whether pests infesting food or contaminating food-preparation surfaces may cause acute gastroenteritis. In short, our goal should be to prevent the first allergy by reducing risks of chronic exposure. Perhaps our greatest risk, indeed, is associated with the food supply. Diligent efforts and adequate safeguards are essential if food is to play the role of enhancer of our immunological systems rather than the role of suppressor.

Cited References

1 Alexander, J. O. 1927. Mites and skin diseases. *Clinical Medicine* 79(4)14–19.

2 Anonymous. 1989. Researchers developing data

to help explain rising asthma deaths. *The Asthma and Allergy Advance*. The Asthma and Allergy Foundation of America, Washington, DC, June–July, pp. 1, 2, 6, 7.

3 Armstrong, M. T., and R. W. Howe. 1963. The saw-toothed grain beetle (*Oryzaephilus surinamensis*) in home-grown grain. *Journal of Agricultural Engineering Research* 8(3)256–261.

4 Aucamp, J. L. 1969. The role of mite vectors in the development of aflatoxin in goundnuts. *Journal of Stored Products Research* 5(3)245–249.

5 Baldo, B. A., and R. C. Panzani. 1988. Detection of IgE antibodies to a wide range of insect species in subjects with suspected inhalant allergies to insects. *International Archives of Allergy and Applied Immunology* 85(3)278–287.

6 Ballard, J. B., and R. E. Gold. 1982. Ultrasonics: No effect on cockroach behavior. *Pest Control* 50(6)24, 26.

7 Ballard, J. B., and R. E. Gold. 1983. The response of male German cockroaches to sonic and ultrasonic sound. *Journal of the Kansas Entomological Society* 56(1)93–96.

8 Bartlett, P. G. 1967. Roach control—key to *Salmonella* control. *Soap & Chemical Specialties* 43(9)62–64, 66.

9 Beatson, S. H. 1972. Pharaoh's ants as pathogen vectors in hospitals. *Lancet* 1(7747)425–427.

10 Bernton, H. S., and H. Brown. 1967. Insects as potential sources of ingestant allergens. *Annals of Allergy* 25(7)381–387.

11 Bernton, H. S., and H. Brown. 1970. Insect allergy: The allergenicity of the excrement of the cockroach. *Annals of Allergy* 28(11)543–547.

12 Bernton, H. S., T. T. McMahon, and H. Brown. 1972. Cockroach asthma. *British Journal of the Diseases of the Chest* 66(1)61–66.

13 Bitter, R. S., and O. B. Williams. 1949. Enteric organisms from the American cockroach. *Journal of Infectious Diseases* 85(1)87–90.

14 Braude, A. G., et al. 1980. Effect of flour mite infestation (*Acarus siro* L.) on nutritive value of pig diets. *Veterinary Record* 106(2)35–36.

15 Brenner, R. J., P. G. Koehler, and R. S. Patterson. 1987. Health implications of cockroach infestations. *Infections in Medicine* 4(8)349–355, 358, 359, 393.

16 Bronswijk, J. E. M. H. van, and R. N. Sinha. 1971. Interrelations among physical, biological and chemical variates in stored-grain ecosystems: A descriptive and multivariate study. *Annals of the Entomological Society of America* 64(4)789–803.

17 Burrell, N. J. 1974. Aeration. In *Storage of Cereal Grains and Their Products*, C. M. Christensen, ed. American Association of Cereal Chemists, St. Paul, MN.

18 Cao, G. 1906. Nuove osservazioni sul passaggio dei microrganismi a travèrso l'intestino di alcuni insètti. *Annali d'Igiene Sperimentale* 16(n.s.)339–368.

19 Chandler, A. C. 1926. Some factors affecting the propagation of hookworm infections in the Asansol Mining Settlement with special reference to the part played by cockroaches in mines. *Indian Medical Gazette* 61:209–212.

20 Deuse, J. P. L., and J. G. Pointel. [1975.] Assessment of research at farm level storage in Francophone Africa. In *Proceedings of the First International Working Conference on Stored-Product Entomology* (Savannah, GA, 1974), pp. 85–91.

21 Dow, R. P. 1955. A note on domestic cockroaches in south Texas. *Journal of Economic Entomology* 48(1)106–107.

22 Eads, R. B., et al. 1954. Studies on cockroaches in a municipal sewerage system. *American Journal of Tropical Medicine and Hygiene* 3(6)1092–1098.

23 Ebeling, W. 1975. *Urban Entomology*. Agriculture & Natural Resources, University of California, Oakland.

24 Eidson, C. S., et al. 1966. Induction of leukosis tumors with the beetle *Alphitobius diaperinus*. *American Journal of Veterinary Research* 27(119)1053–1057.

25 Eisner, T., and J. Meinwald. 1966. Defensive secretions of arthropods. *Science* 153(3742)1341–1350.

26 El-Kholy, S., and M. A. Gohar. 1945. Cockroaches as possible carriers of disease. *Journal of the Royal Egyptian Medical Association* 28(3):82–87.

27 Frazier, C. A. 1969. *Insect Allergy*. Green, St. Louis, MO.

28 Giles, P. H. 1964. The insect infestation of sorghum stored in granaries in northern Nigeria. *Bulletin of Entomological Research* 55(3)573–588.

29 Giles, P. H., and O. Leon V. [1975.] Infestation problems in farmed-stored maize in Nicaragua. In *Proceedings of the First International Working Conference on Stored-Product Entomology* (Savannah, GA, 1974), pp. 68–76.

30 Gorham, J. R. 1975. Filth in foods: Implications for health. *Journal of Milk and Food Technology* 38(7)409–418.

31 Gorham, J. R. 1976. A rational look at insects as food. *FDA By-Lines* 5:231–241.

32 Graffar, M., and S. Mertens. 1950. Le rôle des blattes dans la transmission des salmonelloses. *Annales de l'Institut Pasteur* 79(5)654–660.

33 Greenberg, B. 1973. *Flies and Disease*, vol 2. *Biology and Disease Transmission*. Princeton University, Princeton, NJ.

486   Richard J. Brenner

34 Harwood, R. F., and M. T. James. 1979. *Entomology in Human and Animal Health*. Macmillan, New York, NY.

35 Hatcher, E. 1939. The consortes of certain North Carolina blattids. *Journal of the Elisha Mitchell Scientific Society* 55(2):329–334.

36 Howe, R. W. 1965. Losses caused by insects and mites in stored foods and feedingstuffs. *Nutrition Abstracts & Reviews* 35(2)285–303.

37 Hunt, W. H. 1974. Moisture—its significance, behavior, and measurement. In *Storage of Cereal Grains and Their Products*, C. M. Christensen, ed. American Association of Cereal Chemists, St. Paul, MN.

38 Jacobs, L., and J. K. Frenkel. 1982. Toxoplasmosis. In *CRC Handbook Series in Zoonoses, Section C: Parasitic Zoonoses*, vol. 1, part 1, L. Jacobs, ed. CRC, Boca Raton, FL.

39 Kang, B., and C. Morgan. 1980. Incidence of allergic skin reactivities of asthmatics to inhalant allergens. *Clinical Research* 28(2)426A (abstract).

40 Klowden, M. J., and B. Greenberg. 1977. Effects of antibiotics on the survival of *Salmonella* in the American cockroach. *Journal of Hygiene* 79(3)339–345.

41 Koehler, P. G., R. S. Patterson, and J. C. Webb. 1986. Efficacy of ultrasound for German cockroach (Orthoptera: Blattellidae) and oriental rat flea (Siphonaptera: Pulicidae) control. *Journal of Economic Entomology* 79(4)1027–1031.

42 Krantz, G. W. 1978. *A Manual of Acarology*. O.S.U. Book Stores, Corvallis, OR.

43 Leibovitz, A. 1951. The cockroach, *Periplaneta americana*, as a vector of pathogenic organisms. I. The acid-fast organisms; a preliminary report. *Boletín de la Oficina Sanitaria Panamericana* 30(1)30–41.

44 Litovitz, T. L., et al. 1988. Clinical manifestations of toxicity in a series of 784 boric acid ingestions. *American Journal of Emergency Medicine* 6(3)209–213.

45 Machacek, J. E., et al. 1961. Effect of high water content in stored wheat, oat, and barley seed on its germinability, susceptibility to invasion by moulds, and response to chemical treatment. *Canadian Journal of Plant Science* 41(2)288–303.

46 Mackerras, I. M., and M. J. Mackerras. 1949. An epidemic of infantile gastro-enteritis in Queensland caused by *Salmonella bovis-morbificans* (Basenau). *Journal of Hygiene* 47(2)166–181.

47 Mason, D. 1985. *Hospitals Can Damage Your Health*. British Pest Control Association, London, England.

48 Massicot, J. G., and S. G. Cohen. 1986. Epidemiologic and socioeconomic aspects of allergic diseases. *Journal of Allergy and Clinical Immunology* 78(5)954–958.

49 Military Standard. 1984. *Guidelines for Detection, Evaluation, and Prevention of Pest Infestation of Subsistence*. MIL-STD-904A. Department of Defense, Washington, DC.

50 Mills, R. B., and G. J. Partida. 1976. Attachment mechanisms of *Trogoderma* hastisetae that make possible their defensive function. *Annals of the Entomological Society of America* 69(1)20–33.

51 Misra, C. P., C. M. Christensen, and A. C. Hodson. 1961. The Angoumois grain moth, *Sitotroga cerealella*, and storage fungi. *Journal of Economic Entomology* 54(5)1032–1033.

52 Morischita, K., and K. Tsuchimochi. 1926. Experimental observations on the dissemination of diseases by cockroaches in Formosa. *Journal of the Medical Association of Formosa* 255:566–599 (in Japanese, English summary).

53 Nicewicz, N., W. Nicewicz, and R. Kowalik. 1946. Bacteriological analysis of microorganisms from the alimentary tracts of bed bugs, house flies, and cockroaches. *Annales Universitatis Mariae Curie-Sklodowska (Lublin)* C1:35–38 (in Polish, English summary).

54 Nuñez, J. C. G. 1973. Triatomids in relation to Chagas disease in Latin America. A paper presented on 27 November at the annual meeting of the Entomological Society of America, Dallas, TX.

55 Okumura, G. 1967. A report of canthariasis and allergy caused by *Trogoderma* (Coleoptera: Dermestidae). *California Vector Views* 14(3)19–22.

56 Olson, T. A., and M. D. Rueger. 1950. Experimental transmission of *Salmonella oranienburg* through cockroaches. *Public Health Reports* 65(16)531–540.

57 Palmer, E. D. 1970. Entomology of the gastrointestinal tract: A brief review. *Military Medicine* 135(3)165–176.

58 Payne, N. M. 1925. Some effects of *Tribolium* on flour. *Journal of Economic Entomology* 18(5)737–744.

59 Perlman, F. 1958. Insects and inhalant allergens. *Journal of Allergy* 29(4)302–328.

60 Porter, A. 1929. Some remarks on the hookworm problem in South Africa. *South African Journal of Science* 26(Dec.)396–401.

61 Porter, A. 1930. Cockroaches as vectors of hookworms on gold mines of the Witwatersrand. *Journal of the Medical Association of South Africa* 4(Jan.11)18–20.

62 Pratt, H. D., K. S. Littig, and H. G. Scott. 1975. *Household and Stored-Food Insects of Public Health Importance and Their Control*. DHEW Publ. No. (CDC)75-8122. Center for Disease Control, Atlanta, GA.

63 *Principal Storage Pests*. 1983. Degesch America, Weyers Cave, VA.

64 Ragunathan, A. N., D. Srinath, and S. K. Majumder. 1974. Storage fungi associated with rice weevil (*Sitophilus oryzae* L.). *Journal of Food Science and Technology* 11(1)19–22.

65 Richards, O. W., and N. Waloff. 1946. The study of a population of *Ephestia elutella* Hübner (Lep., Phycitidae) living on bulk grain. *Transactions of the Royal Entomological Society of London* 97(11)253–298.

66 Roberts, E. W. 1947. The part played by the feces and vomit drop in the transmission of *Entamoeba histolytica* by *Musca domestica*. *Annals of Tropical Medicine and Parasitology* 41(1)129–142.

67 Roth, L. M., and E. R. Willis. 1957. The medical and veterinary importance of cockroaches. *Smithsonian Miscellaneous Collections* 134:1–147.

68 Sartory, A., and A. Clerc. 1908. Flore intestinale de quelques orthoptères. *Comptes Rendus Hebdomaires des Séances et Mémoires, Société de Biologie* (Paris) 64:544–545.

69 Schneider, R. F., and G. W. Shields. 1947. Investigation on the transmission of *E. histolytica* by cockroaches. *Medical Bulletin* [Standard Oil Co., New Jersey] 7:119–121.

70 Schreck, C. E., J. C. Webb, and G. S. Burden. 1984. Ultrasonic devices: Evaluation of repellency to cockroaches and mosquitoes and measurement of sound output. *Environmental Science and Health* A19:521–531.

71 Schrut, A. H., and W. G. Waldron. 1963. Psychiatric and entomological aspects of delusory parasitosis. *Journal of the American Medical Association* 186(4)429–430.

72 Shrewsbury, J. F. D., and G. J. Barson. 1948. A study of the intestinal flora of *Blattella americana*. *Journal of Pathology and Bacteriology* 60(3)506–509.

73 Shulman, S. 1967. Allergic responses to insects. *Annual Review of Entomology* 12:323–346.

74 Smith, L. W., Jr., et al. 1971. Baking and taste properties of bread made from hard wheat flour infested with species of *Tribolium, Tenebrio, Trogoderma* and *Oryzaephilus*. *Journal of Stored Products Research* 6(4)307–316.

75 Sondak, V. A. 1935. Cockroaches as carriers and hosts of parasitic worms in Leningrad and its environs. In *Parazity, Perenoschiki i Iadoviye Zhivotnye*, E. N. Pavlovskogo, ed. pp. 316–327 (in Russian, English summary).

76 Steele, J. H. 1981. Bolivian hemorrhagic fever (BHF). In *CRC Handbook Series in Zoonoses, Section B: Viral Zoonoses*, vol. 2, G. W. Beran, ed. CRC, Boca Raton, FL.

77 Stubbs, J. 1983. Pest control devices from an EPA point of view. Paper presented on November 30 at the annual meeting of Entomological Society of America, Detroit, MI.

78 Syverton, J. T., et al. 1952. The cockroach as a natural extrahuman source of poliomyelitis virus. *Federation Proceedings* 11(1, pt. 1)483.

79 Tejera, E. 1926. Las cucarachas como agentes de deseminación de gérmenes patógenos. *Revista de la Sociedad Argentina de Biología* 2:243–256.

80 TerBush, L. E. 1972. The medical significance of mites of stored food. *FDA By-Lines* 2:57–70.

81 Venkatrao, S., et al. 1960. Effect of insect infestation on stored field bean (*Dolichos lablab*) and black gram (*Phaseolus mungo*). *Food Science* 9(Mar.)79–82.

82 Wedberg, S. E., C. D. Brandt, and C. F. Helmboldt. 1949. The passage of microorganisms through the digestive tract of *Blaberus craniifer* mounted under controlled conditions. *Journal of Bacteriology* 58(5)573–578.

83 White, G. D. 1953. Weight loss in stored wheat caused by insect feeding. *Journal of Economic Entomology* 46(4)609–610.

84 Wirtz, R. A. 1980. Occupational allergies to arthropods—documentation and prevention. *Bulletin of the Entomological Society of America* 26(3)356–360.

VI
Regulation and Inspection

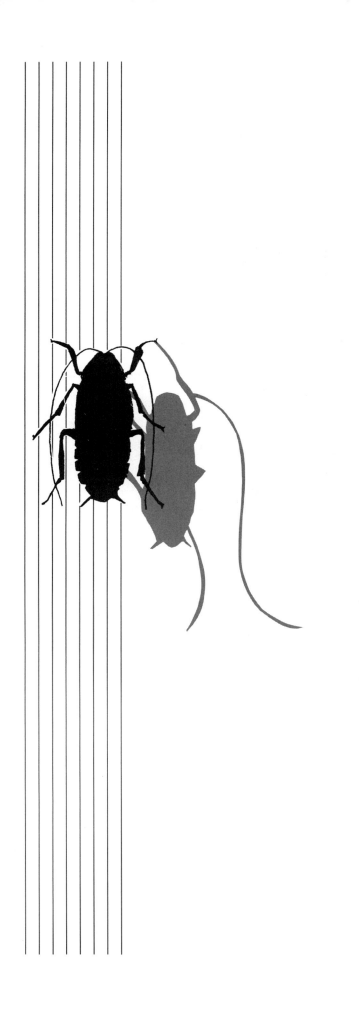

39

The Food and Drug Administration and the Regulation of Food Sanitation

Paris M. Brickey, Jr.

The Food and Drugs Act of 1906 was probably one of the most far reaching enactments by Congress to ever affect the American consumer as well as the American food processor. The Act made adulteration of food a violation of federal law. Around the turn of the century, home canning and preservation of foods began to shift over to commercial processing. When this occurred, the consumer began to lose personal control of the sanitary environment of food processing. When the Federal Food, Drug, and Cosmetic Act (FD&C Act) (3) was passed in 1938, food sanitation became the responsibility of commercial operators, and the presence of food pests such as birds, rodents, insects, and other arthropods in a food or food establishment constituted a violation of the Act.

Although this book is entitled *Ecology and Management of Food-Industry Pests*, "pests" are included in this chapter in the much broader category of "filth." As District Judge Harris stated in *United States v. Roma Macaroni Factory* (6), the "word 'filthy,' as used in the Act, should be construed to have its usual and ordinary meaning, and should not be confined to any scientific or medical definition." District Judge Chesnut, first quoting *Webster's International Dictionary* and then *Corpus Juris*, defined "filthy" in his ruling on *United States v. 667 Cases of Canned Herring Roe* (6) as follows:

"Filthy means defiled with filth, whether material or moral; nasty; disgustingly dirty; polluting; foul;

impure; obscene. Secondary meaning: disgraceful; disgusting; low." And then I turn to *Corpus Juris* as a handy definition from a legal viewpoint of "filthy" and find it is defined as "Containing or involved in filth; contemptible; defiled by sinful practices; foul; dirty; low; mean; morally foul; noisome; polluted; scurvy; that which is nasty, dirty, vulgar, indecent, offensive to the moral sense; morally depraving and debasing."

History of Filth

In 1907, Harvey W. Wiley (1), in his annual report to the Secretary of Agriculture, stated that sanitation inspections were being initiated on local dairy farms. In 1908, he referred to a new investigation, "The Bacteriology of the Fly and Its Danger." Investigations of insect filth in figs, olives, and cocoa beans, mold in ketchup, and microbial contamination in water, milk, and fish were reported on in succeeding years. Sanitation inspections were also initiated in the grain, fish, and canning industries. Wiley emphasized in each of his annual reports the fact that sanitation problems in the food industry were of extreme importance.

Filth Sections of the FD&C Act

The Act defines "prohibited acts" as follows: "The following acts and the causing thereof are hereby prohibited in the introduction or delivery for introduction into interstate commerce of any food, drug, device, or cosmetic that is adulterated or misbranded" [section 301(a)]; "the adulteration, mutilation, destruction, obliteration, or removal of the whole or any part of the labeling of, or doing of any other act with

Brickey: Food and Drug Administration, Washington, DC 20204.

respect to, a food, drug, device, or cosmetic'' [section 301(k)].

Section 402 elaborates on the meaning of adulteration. A food shall be deemed to be adulterated ''if it consists in whole or in part of any filthy, putrid, or decomposed substance, or if it is otherwise unfit for food'' [section 402(a)(3)]; and ''if it has been prepared, packed, or held under insanitary conditions whereby it may have become contaminated with filth, or whereby it may have been rendered injurious to health'' [section 402(a)(4)]. These two sections spell out what constitutes a violation of the Act with regard to adulteration by filth.

Sections 402(a)(3) and (4) are the two basic parts of the FD&C Act under which domestic regulatory actions are taken against adulterated foods. There is, however, another very important section of the Act that permits the FDA to regulate imported foods. Section 801(a) provides for the delivery of imported food samples by the Secretary of the Treasury to the Secretary of Health and Human Services for the purpose of determining if an article has been manufactured, processed, packed, or held under insanitary conditions. The collection and analytical authorities of that section have been delegated to the Food and Drug Administration (FDA) by the Treasury Department to determine compliance or noncompliance with the Act.

Judge Harris ruled in *Roma Macaroni Factory* (7) that

> The Act does not provide that persons shall avoid putting filthy food substances into interstate commerce, or preparing such substances under unsanitary conditions, if it is possible; it provides that it shall not be done at all. A party who cannot prepare food products under sanitary conditions must cease putting such products into interstate commerce.

U.S. States Circuit Court of Appeals Judges Soper and Dobie and District Judge Chesnut in *338 Cartons, More or Less, of Butter, 1209 Cartons, More or Less, of Butter, and 318 Cartons, More or Less, of Butter, Bowser Sales & Trading Corporation, Claimant v. United States''* (6) ruled that the ''courts have recognized that adulteration of foodstuffs may be so slight as to come under maximum *de minimis non curat lex*, but no case has been found wherein food found to be filthy under the Act has not been condemned.''

The two cases cited above clearly show that filthy foods will not be tolerated by the courts, thereby implying that regulatory actions will be taken when such food is encountered.

Regulatory Actions

Regulatory actions are usually taken by FDA as a last resort to remove an adulterated product from commercial channels and/or to bring a food establishment into compliance with the Act. Although these actions are usually quite successful, they can be extremely costly to the taxpayer. For this reason the FDA has developed many educational programs to help importers and domestic food processors comply with the Act.

Consumer Affairs Officers throughout FDA provide guidance to industry, consumer groups, and others in the form of workshops, seminars, and so on. Officials in FDA districts are always available for consultation with importers and food processors to discuss sanitation/adulteration problems. The FDA's Center for Food Safety and Applied Nutrition has developed an industry information/liaison program.

The Industry Activities Section is responsible for developing industry information materials in response to specific compliance problems. These informative and motivational materials elaborate on FDA policies, procedures, guidelines, and regulations. The Industry Activities Section also develops educational workshops, seminars, and other informative approaches, many of which are in cooperation with industry trade associations. As discussed in Chapter 40, FDA investigators point out to industry any insanitary practices/conditions observed during the inspection of an establishment.

Many guidelines have been formulated by FDA to assist the food industry. The first Trade Notice was issued in 1916 regarding mold in tomato products. Another one in 1923 dealt with mold and insect filth in cocoa beans. These Notices continued to be issued until about 1963, at which time the FDA Administrative Guidelines came into effect. The guidelines were classified as ''confidential'' by high-level management at FDA until 1972 when Commissioner Charles Edwards decided that the Guidelines should be made available to the public in the form of ''The Food Defect Action Levels'' (DALs) (2).

The preface of that publication states that the

> food defect action levels contained in this list are set on the basis of no hazard to health. Any products that might be harmful to consumers are acted against on the basis of their hazard to health, whether or not they exceed the action levels. In addition, poor manufacturing practices will result in regulatory action, whether the product is above or below the defect action level. The defect action levels are set because it is not now possible, and never has been possible, to grow in open fields, harvest, and process crops that are totally free of natural defects.

Usually more than one violative inspection is required before regulatory action is initiated. Once a sample analysis is complete, the FDA District Office sends a ''Report of Analysis'' (FD 1551) to the firm. A ''Notice of Adverse Findings'' letter may be issued giving the food processor an opportunity to meet with FDA district officials to discuss any corrective actions to be taken to comply with the Act. Follow-up inspection of a violative firm may be conducted. When all else fails, appropriate regulatory action may be the only option left to FDA.

As stated previously, the taking of regulatory actions can be costly and time-consuming for both industry and government. Once evidence has been collected through inspection and/or analysis, the responsible FDA district determines if regulatory action is the preferred choice of action. If so determined, the district then makes the recommendation. Evidence is examined by compliance and scientific experts and then the district forwards the case to the Office of the General Counsel of the U.S. Department of Health and Human Services. It is reviewed there and then forwarded to the Department of Justice for further action. Basically there are five types of enforcement actions that FDA may take, depending upon the violation.

Seizure

The great majority of court actions are seizures, the purpose of which is to take harmful, filthy, spoiled, or misbranded products off the market. If the goods are in very bad condition, the court may order them destroyed. If it seems likely that they can be brought into compliance with the law, the court may allow them released for sorting, cleaning, relabeling, repackaging, and so on. Some types of filth, such as soluble filth, may render the food unreconditionable. In "*338 Cartons, more or less, of Butter* (6), cited above, the Court of Appeals ruled that the

> evidence showed that the process proposed by the claimant for renovating the butter (which contained maggots, fly eggs, among other things) would remove only the insoluble insect parts, and that the portion of the contamination which had become soluble could not be separated. There is not a scientific method by which the insect fat could be detected in the finished product, but such fat would probably be in the finished product.

Since FDA requests seizure only on the basis of scientific evidence, only occasionally does the owner of the seized goods protest the government's charges of violation. When this happens, the case goes to trial in federal court. Many times a consent decree is entered into with both sides agreeing to acceptable actions to bring a product or processing plant into compliance. Should such a consent decree be unsuccessful, forfeiture of the foods or their conversion into animal feed may result.

Prosecution

A second type of federal court action is the criminal prosecution of a firm and the individuals alleged to be responsible for violation of the FD&C Act. Maximum penalties can be very severe. The Criminal Fine Enforcement Act of 1984 (Pub. L. 98-596) provided new fines for federal law violations perpetrated after January 1, 1985. Although this act is an amendment to Title 18 of the *U.S. Code* (not the FD&C Act), the Justice Department has advised FDA that it applies to fines under the FD&C Act.

Effective November 1, 1987, the following fines from Pub. L. 98-596 became applicable to the FD&C Act for each offense:

(a) A fine of up to $100,000 for a misdemeanor by a corporation or $25,000 by an individual that did not result in death.
(b) A fine of up to $250,000 for a misdemeanor perpetrated by an individual that results in death, or a felony.
(c) A fine of up to $500,000 for a misdemeanor perpetrated by a corporation that results in death, or a felony.

The maximum imprisonment for a misdemeanor under the Act remains 1 year for each offense or 3 years after a previous conviction or for any conviction with intent to defraud or mislead.

Injunction

The third type of court action is an injunction. This action restrains firms or individuals from trafficking in violative goods. Like seizure proceedings, the injunction is a civil action. However, either civil or criminal contempt action may be brought in court when a violation of the restraining order is charged. Injunction petitions are occasionally made to the court when it is found that such large quantities of filthy or decomposed foods are in an establishment that it would be a burden on FDA to keep them under surveillance so that seizure could be accomplished after delivery for interstate shipment.

Recall

A fourth type of enforcement action is a recall. Recalls are an effective means of removing from the market a product that is in violation of the law. Recalls may be on a voluntary basis by the manufacturer or they may be made at the request of FDA. They are classified as follows:

> *Class I.* A situation in which there is reasonable probability that the use of, or exposure to, a violative product will cause serious adverse health consequences or death.
> *Class II.* A situation in which use of, or exposure to, a violative product may cause temporary or medically reversible adverse health consequences or where the probability of serious adverse health consequences is remote.

Class III. A situation in which use of, or exposure to, a violative product is not likely to cause adverse health consequences.

Detentions

The fifth type of action is unique in that it applies only to imports. As a rule, it is based on the results of sample analysis, wharf examinations, and so on—not evidence from inspections. This is because FDA has neither the authority nor the manpower to inspect foreign food manufacturers. Many imports are placed on what FDA calls a "block list." When this happens, a food product upon arrival in the USA is automatically detained without analysis. According to the FDA *Regulatory Procedures Manual (4)*, the criteria for "blocklisting" a product are as follows:

> *Automatic detention* is the administrative act of detaining an entry of a specified article without physical examination solely on the basis of information regarding its *past* violative history and/or other information indicating that the product may be violative. Automatic detention actions are implemented through the issuance of Import Alerts.
>
> Automatic detention was first used in 1974 when it was called "blocklisting." It is an effective action used against *severe or chronic violations or violators.* It is also effective for those importers that expect FDA to serve as a quality control laboratory. Often, it is only after a Notice of Detention has been issued that importers attempt to assure that their products are in compliance with the law. Automatic detention action effectively takes FDA out of the quality control laboratory business and properly places the responsibility for determining quality and compliance with the law on an importer or broker.
>
> Automatic detention should be recommended whenever there is information suggesting that a significant number of shipments of a particular product or products offered for import may be violative. The recommendation may be based on such information as the violative history of a product, manufacturer, shipper, grower, or geographic area/country. It also may be based on other information such as knowledge of polluted waters or knowledge of poor manufacturing or handling practices within a manufacturing facility or geographic area/country. An automatic detention may be invoked without any previous detentions if it is adequately supported.

Criteria for removal from automatic detention are:

> *For specific product(s) from an individual manufacturer or shipper.* Documentation that establishes that the most recent five consecutive shipments entered were without a violation. The documentation should consist of examination records (analytical worksheets) from private laboratories, foreign government facilities, and/or FDA.
>
> *For specific product(s) from a country of a specific geographic area.* Documentation that estab-

lishes that the most recent 12 consecutive shipments entered without a violation. These 12 shipments should include a representative number of manufacturers/shippers from the geographic area or country offering that product(s) for entry.

Other Detentions

If a food product is not in an "automatic detention," detention can still occur when analysis shows the product to be in violation. When this occurs, the product may be reconditioned to bring it into compliance. To attempt reconditioning, an importer must submit a "proposal for reconditioning" (form FD 766) to the detaining FDA district, fully outlining the reconditioning procedure. If FDA believes that the procedure will work and that the product can be brought into compliance, it will allow the attempt to be made. However, if the proposed reconditioning procedure is not likely to work or the product is contaminated with filth that cannot be removed, the product must be converted to animal food, destroyed, or reexported to a country willing to accept the product.

A good example of reexporting, where the defendent claimed that his adulterated product was exempt from federal jurisdiction because he intended to export to Chile, is defined in *United States v. an Article of Food Consisting of 6,796 Boxes More or Less, Labeled in Part: "River Enriched Rice . . ."* (7). The courts found that rice destined for export was not exempt from condemnation. A finding in favor of exemption would have required the claimant to prove that the food was not adulterated according to the meaning of that term in Chilean law. Reconditioning is not a right but rather a privilege. Therefore, as long as FDA is not arbitrary and capricious, there is no other alternative except destruction or reexport should FDA deny approval of the FD 766.

One can easily see from the above references to the law and to the various regulatory actions that FDA plays a very important role in ensuring the cleanliness of products used by the American consumer, that is, every one of us.

Conclusion

The passage of the FD&C Act had a far reaching effect on food sanitation as evidenced by the actions of Congress to ensure that the American consumer receives clean and wholesome food (5). The sections of the Act and the various enforcement actions available to FDA were carefully spelled out in the law. There is no other country in the world that protects its consumers as carefully as does the United States of America.

Cited References

1 Anonymous. 1951. *Federal Food, Drug and Cosmetic*

Law Administrative Reports, 1907–1949. Commerce Clearing House, Chicago, IL, pp. 3–259.

2 Center for Food Safety and Applied Nutrition. 1989. *The Food Defect Action Levels*. Food and Drug Administration, Washington, DC.

3 Food and Drug Administration. 1989. *Federal Food, Drug, and Cosmetic Act, as Amended, and Related Laws*. HHS Publ. No. (FDA)89-1051. U.S. Department of Health and Human Services, Rockville, MD.

4 Food and Drug Administration. 1988. *Regulatory Procedures Manual*. Part 9, Import Procedures. Chapter 9-25, Automatic Detention [TN 88-5 (4/11/88)]. Government Printing Office, Washington, DC.

5 Janssen, W. F. 1988. Golden anniversary of the FD&C Act: Consumers "never had it so good." *Journal of the Association of Food and Drug Officials* 52(3)59–60.

6 Kleinfeld, V. A., and C. W. Dunn. [1949]. *Federal Food, Drug, and Cosmetic Act Judicial and Administrative Record 1938–1949*. Commerce Clearing House, New York, NY, pp. 30–33, 222–224, 348–350.

7 Kleinfeld, V. A., A. H. Kaplan, and S. A. Weitzman. 1979. *Federal Food, Drug, and Cosmetic Act Judicial Record 1975–1977*. Food and Drug Law Institute, Washington, DC, pp. 25–28, 43–45.

40
Inspection Activities of the Food and Drug Administration

Donald R. Martin

The Food and Drug Administration (FDA or Agency) conducts many types of inspections of food processors. These vary from the simple sanitation inspections to the more complex Hazard Analysis Critical Control Points (HACCP) approach for inspections of low-acid canned-food processors, acidified-food processors, convenience-food manufacturers, and dried-yeast processors.

Inspections are conducted to see if firms are in compliance with all the requirements of the Federal Food, Drug, and Cosmetic Act (FD&C Act), the Fair Packaging and Labeling Act (FPLA), and the other acts (2) enforced by the Agency and the regulations promulgated under these acts in Title 21 of the *Code of Federal Regulations* (21 CFR). The inspections cover the broad aspects of the law and the specific points of the Title 21 regulations.

When inspecting low-acid canned-food (LACF) producers, FDA determines whether the firms are in total compliance with 21 CFR Part 108 (Emergency Permit Control), 21 CFR Part 110 (Current Good Manufacturing Practice in Manufacturing, Processing, Packing, or Holding Human Foods), 21 CFR Part 113 (Thermally Processed Low-Acid Foods Packaged in Hermetically Sealed Containers), and any other applicable regulations such as 21 CFR Part 130 (Food Standards General) and 21 CFR Part 155 (Canned

Martin (Retired): Division of Field Investigations, Office of Regional Operations, Food and Drug Administration, Rockville, MD 20857.

Vegetables). If FDA inspects a producer of acidified foods, 21 CFR Parts 108, 110, and 114 (Acidified Foods Regulations) would apply.

Inspections are becoming more and more complex and time-consuming both for the FDA investigator and for the industry inspected. This is due to the complexity of the regulations (see Chapter 39) and the advanced technical nature of the industry being inspected. There are many avenues of contamination; those mentioned in this chapter represent just a few that are looked for by FDA investigators.

Preparing for the Inspection

Most FDA inspections begin in the District Office when the investigator is given the assignment to inspect a firm. At that time, the investigator obtains a copy of the file FDA maintains on each firm previously inspected. The file is reviewed to determine what products the firm manufactures, the hours of operation, the inspection and regulatory history, and any other necessary information. The investigator then reviews any pertinent references, which may cover the firm's manufacturing operation, products, or procedures, to fine tune knowledge of the processes. Next, FDA Compliance Programs and the *Inspection Operations Manual* (1) are reviewed to make sure all necessary information is obtained. Finally, any specialized inspection equipment or sampling supplies are gathered for the inspection. When this process is completed, the investigator is ready to proceed to the firm.

When an investigator arrives at the plant, credentials are shown and a written Notice of Inspection

497

(FDA-482) is presented to top-level management at the facility to be inspected. Other necessary forms are issued to management at this time or at the appropriate time during the inspection. The investigator will then normally hold a brief discussion concerning the reason for the inspection, products manufactured, operational procedures, responsibility of various persons, and other pertinent information. Next, a brief tour of the plant may be conducted so that the investigator can become familiar with the physical layout and the product flow. After this is completed, the actual inspection of the plant can begin.

Inspection of Raw-Material Storage

Many investigators start the inspection in the raw-material receiving area to determine what type of quality control samples, tests, specifications, and records are kept on incoming raw materials. For example, raw materials such as nonfat dried milk, dried yeast, or other products should meet suitable microbiological standards.

Other raw-material ingredients are also checked to determine if they are in proper condition, e.g., fresh produce not decomposed, seafood organoleptically examined for decomposition.

After these records and sampling procedures have been reviewed, the investigator will conduct a physical inspection of the raw material storage area. This physical inspection typically covers many facets, among which are the following:

(a) Observation of the perimeter of storage areas for any openings leading directly to the outside (unstopped pipe chases, holes in the walls, broken or open unscreened windows, openings under doors) that could furnish access for insects, rodents, and birds to the product storage area. This also includes a check of the area outside the building for evidence of debris, rodent harborages, rodent burrows, or other signs of animal activity.

(b) Inspection of bagged and boxed stored materials for evidence of rodent or insect activity. This would include rodent pellets, nests, gnawed bags or cartons, live or dead insects, evidence of insect tunneling.

(c) Observation of storage practices for dangerous chemicals (e.g., pesticides, cleaning compounds), noting whether or not they are stored in areas away from food products so that no direct or indirect contamination of the foods or other products can occur.

(d) Review of color additives to determine if these additives are allowed for use in foods. If older additives are noted, status is determined and, if banned, voluntary correction may be accomplished.

(e) Noting whether or not raw materials requiring refrigeration or freezing are stored at the proper temperatures.

Production-Line Inspection

The investigator then inspects the manufacturing or production area. This encompasses the many facets of the manufacturing operation including, but not limited to, the following points:

(a) The condition and types of various ingredients to determine what colors and food additives are used in the food product being manufactured and whether they are used at the proper levels.

(b) Cleanliness of the equipment, its construction for ease of cleaning, and suitability for the job being done.

(c) Employee practices to determine if hand-sanitizing solutions and equipment sanitizers are available and used as needed.

(d) Formulation of the product to see if it is made and labeled properly.

(e) Proper processing of the product, e.g., cooked at proper temperature and for the proper length of time.

(f) Proper production cycle in the context of undue time lags in the process, i.e., where products are held for times and at temperatures that could facilitate bacterial growth, or handling practices that could lead to the introduction of bacteria and/or filth to the product.

(g) Satisfactory packaging operation regarding proper labeling, net contents, can seamer operation, or that critical factors such as fill of container or viscosity of the product are met.

(h) Cleanliness of the overall area; the presence of rodents, insects, or other sources of contamination such as the misuse of pesticides.

Coverage of the manufacturing area (and the entire plant, for that matter) is all-encompassing. Inspection is made for the presence of vermin such as rodents, insects, household pets, birds, and of employee practices to determine whether filth and/or bacteria are added to the product by practices such as handling insanitary objects and then in-process foods without first washing and sanitizing the hands, or if in-process products are dropped on the floor and then placed back into the production flow without proper cleaning or destruction. The investigator looks for nested raw-material or in-process food containers where the bottom of a container touches the floor and then the bottom touches the food without the container first being sanitized. A check is made for the reuse of cartons, drums and boxes, and in-process or finished-product storage containers (e.g., chicken or egg crates being used to store intermediate products), or the use of pesticide, drug, or cleaning-supply fiber drums to store raw materials without proper cleaning and liners.

Packaging and Storage Inspections

After checking the raw materials and production

areas, the investigator may proceed to the finished-product packaging and storage areas. In this area inspection covers many things including, but not limited to, the points listed below:

(a) Suitability of equipment used for packaging—proper function, ease of cleaning, and whether it is actually sanitized between products so that there is no chance of cross-contamination.
(b) Proper labeling of the product—ingredients listed in descending order as required, dual net weight statement, nutrition labeling.
(c) Quality control of net contents and/or volume.
(d) Overall cleanliness of the area, including the possible presence of vermin.
(e) Storage of the finished product—freezers working properly, storage area clean, free from vermin.

Quality Assurance Inspection

After the packaging and finished-product storage areas, the firm's quality assurance section is covered and finished-product assay records are reviewed for many items including the following:

(a) Assays for enrichment, if applicable, to determine if final product complies with any nutrition labeling claims or enrichment standards for enriched products.
(b) Assays for presence of bacteria such as *Salmonella, Staphylococcus, Escherichia coli,* if the product is susceptible to bacterial contamination.
(c) Fill of containers and/or net-weight records.
(d) Assays for compliance with the Standards if a standardized product has been manufactured.
(e) Lot records for canned goods properly recorded by day or by production unit to determine any potential problems with possible underprocessed lots, can-seam problems, or other deviations.

Wrap-Up of Inspection

When the physical inspection of the plant is completed, and prior to leaving the plant, the investigator performs several operations including:

(a) Issuance of the Inspectional Observations form (FDA-483), a listing of any objectionable conditions noted during the inspection of the plant.
(b) Issuance of the Receipt for Samples form (FDA-484) for any samples collected during the inspection if these samples are to be analyzed for filth or any other contamination that could lead to a charge of adulteration of the product being manufactured by the firm.
(c) Discussion with the management of the firm of any adverse conditions noted during the inspection. The discussion includes any insanitary or objectionable conditions noted during the inspection and warnings of any apparent violations of laws or regulations. These discussions are not technical in nature but do include the law, Good Manufacturing Practices, and minor label comments such as absence of mandatory labeling (technical label comments are not covered; comments of this type must be made by the Agency).

After the issuance of the mandatory forms and the discussion of the inspection with the firm's top management, the investigator concludes the inspection of the plant.

Cited References

1 Food and Drug Administration. 1985. *Inspection Operations Manual.* FDA, Washington, DC (updated frequently; available only from National Technical Information Service, Springfield, VA).

2 Food and Drug Administration. 1989. *Federal Food, Drug, and Cosmetic Act, as Amended, and Related Laws.* HHS Publ. No. (FDA)89-1051. Department of Health and Human Services, Rockville, MD.

41
Regulatory and Residue Issues in the U.S. Environmental Protection Agency

Joseph G. Cummings and Randolph B. Perfetti

This chapter deals with controls exerted by the U.S. Environmental Protection Agency (EPA) over the use of pesticides in the food industry. To be consistent with the theme of this compendium, it would seem appropriate to exclude purely agricultural uses of pesticides, although in the broadest sense that would be part of the food industry. Treatment is therefore limited to the legal aspects of pesticide applications to produce after it leaves the farm gate. Even this distinction is not clear-cut because existing federal laws on food safety and pesticide registration were not designed to differentiate between applications to growing crops or to commodities in storage and processing. The law does make a distinction between residues on "raw agricultural commodities" and on "processed foods," which is explained below. However, those definitions cannot be used as a basis for differentiation between "food industry" versus agricultural uses. Many commodities treated during storage or at some stage in food processing are raw agricultural commodities within the meaning of the law (e.g., stored grains or perishable produce on loading docks of canneries are raw commodities).

The universe of sites within the food industry to which pesticide treatments may be made is almost endless. In general, however, they can be divided into three broad categories: (a) sites for the open preparation of food for consumption such as restaurant kitchens, bakeries, and galleys; (b) food processing sites such as bottling plants, flour mills, and canneries; and (c) food distribution and storage sites such as grain elevators, vehicles of transport, and grocery stores.

The majority of chemical pesticides used in the food industry are probably insecticides but also included are sprout inhibitors, bactericides, fungicides, and rodenticides. The scope of formulations used includes, but is not limited to, fumigants, aerosols, fogs, sprays, powders, controlled-release tapes, automatic mechanical dispensers, and baits.

Regardless of the locale or mode of application, the same federal laws are applicable. However, given the multiplicity of factors that affect the tendency of the foods to incur residues, the risk assessment for an industry use is sometimes more difficult than in the case of straightforward field sprays to growing crops, and different criteria for safety must be applied depending on the type of application and pesticide as well as the category of food-handling establishment involved.

Early perceptions of health risks from pesticide residues were concentrated on agricultural uses. The 1950 Spray Residue Hearings in the Congress and the subsequent "Miller Bill" or Pesticide Amendment of 1954 to the Federal Food, Drug, and Cosmetic Act (FD&C Act) dealt entirely with residues on raw agricultural commodities. It was not until the late 1960s that some attention was focused on residues contributed to foods from a variety of pest control measures performed beyond the point of regulation for raw

Cummings (formerly with EPA): 407 Willington Drive, Silver Spring, MD 20904.

Perfetti: OPP, HED, RCB (TS-769), Environmental Protection Agency, Room 810, Crystal Mall II, 1921 Jeff Davis Highway, Washington, DC 20460.

agricultural commodities. The principal concern at that time was with insecticide treatments in "food handling establishments," a term slightly more circumscribed than "food industry." Such uses were for the most part housekeeping controls for crawling and flying insects but also included the more specialized uses. With few exceptions, control was exercised under the Federal Insecticide, Fungicide, and Rodenticide Act (FIFRA), with reliance placed on product label warnings and use precautions to preclude health risks. For example, a product recommended as a space spray in bakeries might contain a warning to cover food and to wash down food-contact surfaces after the application, or more simply, to "avoid contamination of food."

It was generally recognized that, given the numerous avenues by which a chemical could be transferred to foods from such applications, the label warnings alone were of dubious value. An obvious need was that each of the uses registered at that time (and any future proposals) be subjected to the more structured risk assessments already required of agricultural uses through tolerance petitions under the FD&C Act.

Accordingly, meetings were held with the National Pest Control Association, food industry representatives, and the concerned government agencies. In 1973, EPA issued a policy statement on use of residual insecticides in food-handling establishments (38 Fed. Reg. 21,685, Aug. 10, 1973). The notice also defined food-handling establishments, nonfood areas, food areas, various types of treatments, and listed certain insecticides that were registered for those purposes. It required that residue data be gathered and Food Additive Petitions be submitted by registrants before October 1974. Subsequently, the Federal Working Group on Pest Management (an interdepartmental group) appointed a Research Panel to develop research protocols for residue investigations. A suggested general protocol was submitted by that group to the EPA Administrator in December 1973; it served as a guide for petitions until the Residue Chemistry guidelines, which also address this subject, were finalized (3).

Applicable Laws

There are three federal laws that bear directly on the use of pest control chemicals in the food industry: FIFRA, the FD&C Act, and the Meat and Poultry Inspection Act. EPA has primary responsibility for administration of FIFRA. This function was transferred from the U.S. Department of Agriculture (USDA) in the departmental reorganization of 1970. Under the same Executive Order, certain responsibilities under the FD&C Act relating to establishment of pesticide tolerances (section 408) and food additive tolerances (section 409) were transferred from the Food and Drug Administration (FDA) to EPA. Under a formal Inter-

departmental Agreement, the USDA maintains compliance with pesticide tolerances established by EPA for meat, poultry, and meat products (42 Fed. Reg. 56,629, Oct. 27, 1977). The USDA also publishes a manual that prescribes the conditions under which pesticides may be used in premises under supervision of the Meat and Poultry Inspection Act.

These three federal laws more or less directly control pesticide usage. Certain other laws administered by EPA could indirectly impinge on pesticide usage in the food industry. For example, EPA establishes National Primary Drinking Water Regulations under authority of the Safe Drinking Water Act of 1974, as amended in 1978 (Pub. L. 96-502), which includes standards for the maximum contaminant levels permitted for pesticides. (Further modifications of the Act are being considered by Congress.) Those standards are relevant to the extent that waters containing those levels of pesticides could be used in food processing. Discharges of toxic pollutants into public waters are controlled under the Federal Clean Water Act as amended in 1977, and reenacted in 1987 (Pub. L. 95-217). Certain uses in the food industry, particularly those involving continuous processes, would require discharge permits from the EPA Office of Water Enforcement and Permits.

Pesticide Tolerances and Registration

Broadly speaking, FIFRA controls the use of pesticides through a registration process, and the FD&C Act controls the level of pesticide contaminants that may legally be present in foods and feeds. The latter is accomplished through the establishment of tolerance levels by EPA and a surveillance system maintained by FDA and USDA.

Under the food safety system prevailing in the United States (USA) since adoption of the Pesticide Amendment of the FD&C Act in 1954, any proposal for registration under FIFRA for a new use involving food or feed must be accompanied by a proposal for a safe residue tolerance under the FD&C Act. If the proposed use is on a raw agricultural commodity, the petition for a tolerance is submitted under the provisions of section 408, FD&C Act, whether the use site is on a farm or in the food-processing industry. If the pesticide treatment is to be made directly to a processed food, the petition is submitted under section 409 (Food Additive Amendment, FD&C Act). If the treatment is to a raw commodity but residues survive processing and concentrate in any fraction of a processed food (e.g., flour), tolerances are required under both sections 408 and 409.

Raw agricultural commodities are defined as fresh fruits in their unpeeled natural form, whether or not washed, colored, or otherwise treated; vegetables in their raw or natural state, whether or not they have been stripped of outer leaves, waxed, or prepared for use in fresh salads; grains; nuts; eggs; raw milk; meat; and other similar agricultural produce. Excluded from the definition of a raw commodity are foods that have been processed, fabricated, or manufactured by cooking, freezing, dehydrating, or milling. The exclusion from the definition of raw commodities in the regulations [40 CFR 180.1(e)] places the latter foods under section 409 as processed foods.

The Office of Pesticide Programs (OPP) in EPA holds the responsibilities for both registration under FIFRA and the establishment of pesticide tolerances under the FD&C Act. The data requirements for registration are so voluminous as to be well beyond the scope of this chapter other than to note salient points. The Agency has published in the *Federal Register* and through the National Technical Information Service (NTIS) a series of proposed guidelines for data requirements for registration ranging from product chemistry to hazard evaluation for humans and domestic animals [these guidelines have now been finalized (*1, 2*; see also 49 Fed. Reg. 42,856, Oct. 24, 1984).

The registration process has been further complicated by several FIFRA amendments since 1972 which require among other things that all registered products be reregistered under generally tighter data requirements. The Agency announced that under recent legislative authority, the registration process would be simplified by adoption of a generic registration standard system. Using this system, the Agency is in the process of reregistering all registered pesticides.

Tolerance proposals are presently reviewed in the Hazard Evaluation Division of OPP. The tolerance can be no higher than necessary to cover the residues likely to result from the proposed use. This is the basis for the requirement for controlled residue studies to demonstrate the residues that will be incurred in foods and feeds. As noted previously, early attention was concentrated on residues from agricultural usage. As a result, a considerable body of knowledge was developed in government agencies and by industry petitioners on how to conduct residue field trials on growing crops and food animals.

Residue studies predictive of food industry uses of pesticides are sometimes more difficult to design. This is particularly true of treatments in food-handling areas where it is difficult to reproduce in a few test sites the multiplicity of physical features in the sites where the product will actually be used. Further complication is introduced by the human factor. How conscientiously will the actual applicator observe precautions to "direct spray to cracks and crevices" and "avoid splatter to food contact surfaces"?

Another variable is the large number of foods that may be potential recipients of the residue, each with a different propensity to absorb. In recognition of these difficulties, the Residue Chemistry Branch, Hazard Evaluation Division, reviews and comments on specific protocols for residue studies before a study is undertaken by a registrant.

Other food industry uses such as the metered addition of protectants to stored grains are relatively straightforward and present no unusual problems in designing residue studies. However, in common with all of the studies, it is necessary to conduct investigations on the nature of the residues. This means looking for altered products other than the parent compound which may occur as a result of the use of a particular pesticide.

The findings of the review chemists on the identity and level of the terminal residues provide the exposure estimate used in the toxicological evaluation (also currently performed in the Hazard Evaluation Division). The toxicological evaluation proceeds to a risk assessment for the general population of the USA. A further refinement of the toxicological risk assessment evaluation to any subgroups that may be at special risk is partially in place at this time.

A comprehensive discussion of the factors involved in the toxicological evaluation of a proposed residue tolerance is not appropriate here. There is still controversy in the scientific community concerning the extrapolation of effects on test animals to humans, particularly when it is not known if there is a threshold effect level. Nevertheless, the present general toxicity data requirements may be summarized as follows:

(a) acute oral, dermal, and inhalation studies;
(b) irritation and sensitization studies;
(c) chronic and oncogenic studies;
(d) teratogenicity and reproduction studies;
(e) mutagencity studies;
(f) neurotoxicity evaluations;
(g) subchronic studies;
(h) special testing.

For more details on toxicology data requirements, the reader is referred to the published guidelines (*2*; see also 49 Fed. Reg. 42,856).

One final but important requirement for a pesticide tolerance is that there be an analytical method suitable to enforce the proposed tolerance. The burden is on the petitioner to provide the method, which is validated in the EPA laboratories and subsequently published in the FDA *Pesticide Analytical Manual*, Volume II (*4*). As a practical matter, multiresidue methods are routinely used in the FDA and USDA regulatory programs. The specific method developed by the petitioner is usually used for confirmatory purposes and for those special cases where a particular residue cannot be determined by multiresidue methods. The method to be used for enforcement should not require equipment or reagents not readily available and should permit competent analysts to apply it successfully without undue modification.

Temporary Tolerances

A temporary tolerance for pesticide residues on raw agricultural commodities may be granted under 40 CFR 180.31, for a period designed to allow the marketing of a commodity produced while a pesticide is being tested under an experimental-use permit issued under section 5 of FIFRA. Although this provision of the law was primarily designed for agricultural testing, it is also applicable to any food-industry use where the treated commodity is a raw commodity as defined in section 408 of the FD&C Act.

There is no parallel provision in section 409 for tolerances in processed foods for experimental uses. However, an administrative accommodation has been made by which EPA will issue food additive regulations with the understanding that they will be revoked after a period of time specified in the experimental permit. The data requirements for temporary tolerances are less stringent than for permanent tolerances, but there must be sufficient information to permit a conclusion that the public health will be protected during the specified period. As with permanent tolerances, there must be an analytical method to enforce temporary tolerances.

Exemption from the Requirements of a Tolerance

Under section 408(c), FD&C Act, an exemption from the requirement of a tolerance may be granted when the total quantity of the pesticide in all raw commodities in which it is used will involve no hazard to public health. This provision of the law is usually reserved for certain compounds that are so innocuous by nature that risks would be negligible regardless of the amounts present.

Summary

Uses of pesticidal chemicals in the food industry are subject to the same federal laws as are agricultural uses. The interlocking framework of the three laws that regulate pesticide usage have been discussed with emphasis on the responsibilities assigned to the U.S. Environmental Protection Agency. Prospective sponsors for uses of pesticides in the industry are referred to the Registration Division, Office of Pesticide Programs, EPA, for specifics on the procedures for obtaining registration under FIFRA and tolerances under the FD&C Act.

Cited References

1 Environmental Protection Agency. 1982. *Pesticide Assessment Guidelines Subdivision D, Product Chemistry.* EPA-540/9-82-018 (available from National Technical Information Service, Springfield, VA, No. PB83-153890).

2 Environmental Protection Agency. 1982. *Pesticide Assessment Guidelines Subdivision F, Hazard Evaluation, Human and Domestic Animals.* EPA-540/9-82-025 (available from National Technical Information Service, Springfield, VA, No. PB83-159916).

3 Environmental Protection Agency. 1982. *Pesticide Assessment Guidelines Subdivision O, Residue Chemistry.* EPA-540/9-82-023 (available from National Technical Information Service, Springfield, VA, No. PB83-153981).

4 Food and Drug Administration. 1989–90. *Pesticide Analytical Manual*, 2 vols. Washington, DC (available from National Technical Information Service, Springfield, VA, No. VB-119; vol. 1 is 1990, vol. 2 is 1989).

42

Regulatory and Inspection Functions in the U.S. Department of Agriculture

Mary Kenney, David Orr, and Michael J. Shannon

Animal and Plant Health Inspection Service

Many dangerous plant pests are "hitchhikers" that can be artifically spread long distances. Plant pests are often disseminated widely around the world because of inadequate controls applied in the country of origin. Since controls are lacking in many countries, destination countries must depend on their own laws and regulations to protect against pests and to retard pest spread within their borders.

The agency of the U.S. Department of Agriculture (USDA) that is responsible for keeping foreign pests from entering the United States (USA) is the Animal and Plant Health Inspection Service (APHIS) through its Plant Protection and Quarantine (PPQ) program. Various statutes and acts give the Secretary of Agriculture the authority to:

(a) take emergency action against plant pests that are new or not widespread and against weeds designated as noxious;

(b) establish restrictive and prohibitory regulations against imports that may introduce exotic plant pests and designated noxious weeds;

(c) establish regulations for cooperative federal/state programs to prevent the spread of introduced plant pests and designated noxious weeds;

(d) carry out cooperative federal/state or international suppression, control, and eradication measures against designated plant pests; and

(e) provide expert certification for domestic plants and plant products upon request.

PPQ Program

All food products entering the USA are subject to being detained by the U.S. Customs Service until they are inspected by PPQ to determine the presence of exotic plant pests and weeds and to ensure that conditions specified by regulation are met prior to entry. (To determine whether imported plant materials may be subject to specific permits or other entry requirements, contact the PPQ offices located at all major ports of entry or write to Permit Unit, PPQ, APHIS, USDA, Room 638, Federal Building, Hyattsville, MD 20782.)

Although the greatest threats of exotic pest introduction exist with imported plants, fruits, and vegetables, this discussion focuses on risks posed by imported stored products. Such products imported from foreign areas have been found contaminated with undesirable plants and soil, seeds of noxious exotic weeds, and insect pests

Kenney: Agricultural Marketing Service, U.S. Department of Agriculture, Washington, DC 20205. Present address: General Accounting Office, Room 4073, 441 G Street, NW, Washington, DC 20548.

Orr: Federal Grain Inspection Service, U.S. Department of Agriculture, Washington, DC 20205.

Shannon: Animal and Plant Health Inspection Service, U.S. Department of Agriculture, Washington, DC 20205.

that do not occur in the USA. When certain exotic pests are discovered, they must be eliminated by a specific treatment process before materials are released to importers. Such treatments generally occur at the port of entry. Even though exotic pests may occur at levels below those of concern from a product quality standpoint, actions may still be required to eliminate the risk of pest introduction.

Stored-Product Insect Pests

Stored-product insects are among the most difficult pests to exclude. These insects have become widely distributed by dissemination in host materials suitable for shelter, feeding, or reproduction. Most of the important cosmopolitan stored-product insect pests are already widely distributed in the USA and are not subject to quarantine exclusion measures when found in incoming products. The one major pest in this group that does not occur in this country is the khapra beetle, *Trogoderma granarium* (see Chapter 10 for additional information on this insect).

Khapra Beetle

Khapra beetles were found in grain storages and other food-industry sites in the southwestern USA and Mexico in 1954. An extensive cooperative campaign eradicated the pest by 1966, and no further infestations in the grain industry have been reported since that time. In 1980, an infestation of a spice warehouse was discovered in New Jersey. Subsequent surveys disclosed 25 infested premises in six states. All infestations were successfully eradicated. Such actions are taken under authority of the Plant Quarantine Act (7 USC 151–165 and 167), and the Federal Plant Pest Act (7 USC 150aa–jj).

The khapra beetle is one of the world's worst pests of stored grain and grain products. Under hot, dry conditions, this dermestid rapidly builds up large populations, and complete destruction of the host material can occur in a short period of time. Feeding and contamination result in loss of weight and lowering of quality of grain or stored products. Countries where the pest does not occur often impose severe restrictions on imported products originating in infested countries. A khapra beetle infestation in the USA would pose a serious threat to our grain industry. Khapra beetles occur generally throughout the Middle East, Southeast Asia, and Africa. They are transported into this country on a variety of materials, many of which have been shown in laboratory studies to be less-than-ideal food

Table 42.1

Interceptions of khapra beetles from cargoes (FY 1971 to 1981)

Intercepting ports	Number of interceptions	Percentage of all interceptions
New York	871	65
San Francisco	121	9
Charleston	99	7.4
Baltimore	55	4.1
New Orleans	34	2.5
Savannah	34	2.5
Houston	31	2.3
Philadelphia	26	1.9
Others[a]		5.3

Countries of origin	Number of interceptions	Percentage of all interceptions
India	586	44
Sudan	317	24
Pakistan	198	15
Iran	63	5
Others[b]		22

Hosts[c] identified with five or more interceptions	Number of interceptions
Gum	132
Sheepskins	121
Artware	100
Capsicum	81
Cotton piece goods	47
Curcurbits	41
Cumin	40
Carpets	34
Foodstuffs	27
Automobiles	24
Celery seed	17
Cassia	16
Brassware	15
Hibiscus	10
Peanuts	8
Rice	8
Pistachios	8
Wooden screens	7
Coffee	6
Niger-seed	5
Goatskins	5
Containers	5

[a] Twenty-five ports with 11 or fewer interceptions.
[b] Thirty-two countries with 24 or fewer interceptions.
[c] Total identified hosts = 801, or 60% of all hosts. Identified hosts with less than 5 interceptions = 44. Total unidentified hosts = 535.

sources. A recent 10-year review provides an overview of pathways for introduction of this pest (Table 42.1). Interceptions totaled 1,336 during FY 1971–81; they averaged 101.7 per year.

Contamination of the products noted in Table 42.1 can be attributed to inadequate sanitation in the countries of origin, use of pest-contaminated, used burlap bagging, and/or shipment of commodities in infested carriers. If the insects are not discovered and eliminated at ports of entry, then infestations of an importer's premises can result. When this has happened in the past, emergency actions have been taken to eliminate the infestations.

PPQ Work Units responsible for inspecting commodities at ports of entry examine all imported products suspected as carriers for this pest. As a backup, PPQ Work Units throughout the country locate and inspect firms receiving potentially infested materials. These inspections are often conducted by or in cooperation with state plant pest regulatory agencies. Inspections involve examination of stored commodities and of debris accumulated in walls and on floors, and placement of attractant traps.

When any infestation is confirmed, eradicative measures are applied under the authority of the Federal Plant Pests Act (7 USC 150dd) to prevent spread of the beetles. Because this particular pest is so easily spread by commerce, materials that have previously left the premises are backtracked to determine if further infestations have occurred.

Prevention and Control of Infestations Because of their habit of hiding in cracks and crevices inaccessible to pesticide contact, khapra beetles are more difficult to eliminate than other stored-product pests. Beetles in these locations are often likely to be in a hibernating or resting state, the most difficult stage to kill. Therefore, the primary eradication method for this pest is fumigation. Gastight tarpaulins are placed over the infested structure and fumigant concentrations are maintained at prescribed levels for specific periods. The time required for such treatments varies with the temperature and the commodities present, but ranges from 12 to 36 hours. Generally, structures and their contents are released from quarantine after treatment and after complete aeration of the fumigated enclosure. Fumigations are performed by commercial firms according to specified procedures and under the direct supervision of PPQ personnel.

Importers receiving high-risk materials from khapra beetle-infested countries can prevent infestation by this pest in their premises by communicating concerns about sanitation in packing, transportation, and storage sites in contracts with foreign suppliers; avoiding the use of burlap bagging, especially used burlap; and promptly disposing of used packaging materials from imported shipments.

Intensive port-of-entry inspections, backed up by timely detection and elimination of infestations, have kept this pest from establishing a permanent foothold in the USA.

Federal Grain Inspection Service

The Federal Grain Inspection Service (FGIS) inspects agricultural commodities under the Agricultural Marketing Act of 1946 (7 CFR Part 68). FGIS is responsible for inspecting those commodities assigned to it by the department. The commodities include flour, cornmeal, soybean meal, grits, bulgur, pasta products, vegetable oils and shortening, bread, bakery products and mixes, breakfast cereals, beans, peas, lentils, rice, and other edible products (the complete list is available from FGIS).

FGIS inspects agricultural commodities for compliance with contract requirements. Most contracts require products to be processed in a sanitary manner under sanitary conditions or according to Food and Drug Administration (FDA) requirements. FGIS performs plant sanitation inspections under this authority and at the request of applicant firms. It uses for this service its *Sanitation Inspection Handbook for Beans, Peas, Lentils, and Processed Commodities.* The bases for this handbook are 21 CFR Part 110 (formerly Part 128), *Current Good Manufacturing Practice in Manufacturing, Processing, Packing, or Holding of Human Foods;* Military Standard 668, *Sanitary Standards for Food Plants;* and Military Standard 1105, *Sanitary Standards for Bakeries.* The FGIS handbook was reviewed by FDA before it was placed in effect. FDA and FGIS cooperate in plant sanitation inspections through FGIS Instruction 906-2, "Implementation of the FGIS-FDA Memorandum of Understanding" (MOU).

Initial sanitation inspections are performed when plants have not had a previous contract for which FGIS had inspection responsibility and when plants have not been inspected and found sanitary by USDA, FDA, or state or local authorities within the previous 3 months. Subsequent inspections are performed (a) in 2 or 6 months, depending on the last inspection results; (b) when plant operations have been shut down for 2 weeks or longer; (c) when plants have been renovated or significantly altered since the last inspection; or (d) as frequently as may be warranted by unusual circumstances (e.g., fire, flood).

Sanitation inspections are unannounced and are usually performed in the presence of a responsible plant representative. When performing these inspections, FGIS inspectors complete a Sanitation Inspection Report (SIR) in duplicate. Exhibits 42.1–42.3 are the SIRs for processed products; beans, peas, and lentils; and overpackers, respectively. (Overpackers are specialized firms that assemble small packages of food into larger packages for shipment, often to overseas destinations). Inspectors give the applicant for inspection a copy of the completed form immediately after the inspection is completed.

Defects listed on SIRs have been assigned points or categorized as "critical." The range of points is based on the importance of the item to the maintenance of good sanitation in the plant. For example, the defect item "presence of evidence of rodents in plant" is assigned a point range of 0 to 5, whereas defect item "waste not properly stored or contained" is assigned a point range of 0 to 3. In performing inspections, inspectors score the defect item on the basis of their estimate of its importance. The most serious defects are listed as "critical." These are scored on a "yes" or "no" basis. Critical items scored "yes" require immediate action by plant management. Plant operations are considered insanitary if even a single critical defect is found or if the plant receives a 76 or higher (processed products), 46 or higher (beans, peas, and lentils), and 21 or higher (overpacker) SIR score.

FGIS Handbook Items

According to the FGIS handbook,

(a) Insects and rodents are capable of transmitting a number of diseases to man through contamination of food. Their presence in a plant creates a potential public health hazard. The only way to guard against this is by effective vermin control.

(b) Elimination and destruction of insects and rodents in and around plants are vital to good sanitation. Two principles should be followed: Prevent their breeding and prevent their entrance into the plant.

(c) If pests gain entrance to plants, this is an indication that the preventive measures have not been entirely successful, and the management and inspectors should determine where they have failed and act to prevent a recurrence.

Recommended Sanitation Practices

FGIS inspectors look at or for the following, as applicable, for pest control when performing sanitation inspections.

Outside Grounds

There should not be any disorderly, haphazard accumulation of useless materials, such as plant refuse or discarded equipment, which could provide a refuge and breeding place for pests. Suitable containers or facilities must be provided for the collection of such materials. Grounds should be policed periodically and vegetation should be kept under control. There should be adequate drainage to prevent insect breeding.

Plant Construction

Floors are ideal places in which food may lodge, decompose, and support the growth of insects. Therefore, floors should be maintained to eliminate all open cracks and crevices. Depressions or low areas that accumulate moisture should be eliminated.

Walls and ceilings must be maintained free of scaling and loose paint because this can be a source of product contamination and can also shelter insects.

Walls, floors, and ceilings that have been penetrated by rodents should be repaired or replaced with rodentproof material such as concrete or brick. Tunnels may be blocked with 17-gauge hardware cloth, glass, metal, or other rodentproof material.

Stone or brick walls should have the joints pointed up flush and smooth, and all cracks, crevices, openings around pipes, and other similar gaps should be sealed tight.

Floor drain grates or strainers should be in good repair and should remain in place so as to prevent the entrance of rats through drainage lines.

All windows, doorways, and other openings that could admit vermin should be equipped with effective pest screens or barriers. Insect-excluding air curtains should be installed over exterior shipping or receiving doorways of food-handling areas.

Pest Control

Rodents, birds, insects, and other animals should not be present in plants. Furthermore, there should not be any evidence of these pests. There must be an effective pest control program. Any rodenticides and insecticides must be used in compliance with FDA and Environmental Protection Agency (EPA) regulations and applied in a safe, acceptable manner. Poison baits, if used, must be contained in anchored, tamper-resistant stations.

Storage Areas

Materials and equipment stored on outside

Exhibit 42.1
Sanitation Inspection Report
(Processed Products)

SANITATION INSPECTION REPORT
(PROCESSED PRODUCTS)

SANITARY INSPECTION OF (Name and Address of Plant)	INSPECTOR
	DATE INSPECTED
PLANT OWNED BY (Company or Individual)	INSPECTOR ACCOMPANIED BY (Name and Title)

SANITARY DEFECTS	Assigned Defect Points	Inspr's Defect Points	SANITARY DEFECTS	Assigned Defect Points	Inspr's Defect Points
I. PREMISES			**V. VENTILATION**		
A. Outside premises not well drained.	3		B. Evidence or presence of excessive condensate on walls, ceilings, equipment or pipelines.	4	
B. Outside premises not free of weeds, clutter, unused equipment or other materials.	3		C. Areas where steam or hot air vapors are predominant and not effectively exhaust-ventilated.	3	
C. Waste not properly stored or contained.	3		D. Cooling and air-conditioning units not equipped with filtering devices; filters not changed or cleaned when necessary.	3	
D. Outside premises not free of harborage or breeding places for insects, rodents, birds, or other animals.	4		E. Windows opening to the outside not screened with 14-mesh or finer screen (during presence of flies or other flying insects).	3	
E. Approaches to docks not clean or treated to minimize dust.	3		F. Exhaust fans not screened or equipped with self-closing louvers.	3	
II. RAW MATERIALS			**VI. WATER SUPPLY**		
A. Raw materials used in the product not handled in a sanitary manner that would prevent adulteration of finished product.	Critical		A. Lack of satisfactory evidence that the water supply is potable.	Critical	
III. CONSTRUCTION OF BUILDING			B. Presence of a cross-connection between water lines and sewage system (not allowed).	Critical	
A. Plant construction does not exclude probable entrance of insects, rodents, birds, or other animals.	5		**VII. DISPOSAL OF WASTES**		
B. Walls, partitions, ceilings, and exposed structural supports, in poor repair.	3		A. Garbage, trash, and dry waste materials, inside or outside the building, not satisfactorily contained.	4	
C. Floors not smooth and readily cleanable.	3		B. Waste is not disposed of frequently enough to prevent unsightliness or undesirable odors.	4	
D. Floors that frequently get wet not constructed to prevent pooling or are not kept dry.	3		**VIII. TOILET, HAND-WASHING, AND DRESSING ROOM FACILITIES**		
E. Exposed, unprotected overhead sewer lines present in product area.	Critical		A. Toilet facilities not provided for each sex.	3	
F. Wooden floors, where present, not coated with a suitable floor seal that effectively minimizes cracks and crevices.	5		B. Toilets and urinals not water-flushed.	3	
G. Walls not sealed at juncture with floor.	3		C. Toilet room(s) not equipped with self-closing doors.	3	
H. Walls and ceiling not free of flaked paint, nor in good repair.	3		D. Toilet rooms open directly into a processing or food storage room where food is exposed.	Critical	
I. Windows not intact.	3		E. Cold and hot, warm, or tapped water not provided at wash basins.	3	
J. Entranceways to production rooms not equipped with self-closing doors nor provided with other effective means to prevent entrance of vermin.	3		F. Hand cleaner and towels not provided at or near basin.	3	
K. Screen doors, where present, not self-closing.	3		G. Adequate hand-washing signs not provided.	3	
L. Utility rooms or areas, including boiler rooms and maintenance shop, not isolated enough or separated from production rooms.	4		H. Adequate dry waste receptacles not provided.		
IV. LIGHTING			I. Toilet facilities not working.	5	
A. Production rooms not adequately lighted (20-foot candles).	3		J. Toilet and locker room not satisfactorily ventilated.	4	
B. Places, where detailed inspection of the product or ingredients is required, not adequately lighted (50-foot candles).	4		K. Toilet and dressing rooms not kept in a clean orderly condition.	5	
C. Places, where detailed inspection of the processes or operations is required, not adequately lighted (50-foot candles).	3		**IX. CONSTRUCTION AND REPAIR OF EQUIPMENT**		
D. Storage areas not adequately lighted (20-foot candles).	3		A. Equipment product contact surfaces not constructed of smooth, nontoxic, corrosion-resistant, odorless materials; wood, where used, not fine, close grained, high-density type.	5	
E. Toilets and dressing rooms not adequately lighted (20-foot candles).	3		B. Equipment product zone surfaces not free of cracks, crevices, pits, or other imperfections (such as knots, in wood).	5	
F. Light bulbs, fixtures, skylights, or other glass suspended over food in any stage of preparation not of a safety type or otherwise protected where essential to prevent contamination of the product during production.	5		C. Adequate space not allowed around equipment for cleaning and inspection purposes.	4	
V. VENTILATION			D. Equipment not designed to protect the product zone against contamination and extraneous materials.	5	
A. Plant not reasonably free of undesirable odors, hot air vapors or dust.	3		E. Equipment not free of parts or areas inaccessible for cleaning and sanitary inspection.	4	
			F. Electrical system not constructed or positioned to prevent insect harborage.	3	
			G. Equipment not kept in good repair.	2 for each item not in good repair.	

FORM FGIS-952 (1-84)   (Replaces Form IN-244 (8-77) which may be used until exhausted)

SANITARY DEFECTS	Assigned Defect Points	Inspr's Defect Points	SANITARY DEFECTS	Assigned Defect Points	Inspr's Defect Points
X. CLEANING			**XIII. STORING AND STORAGE FACILITIES**		
A. Equipment not cleaned at frequencies necessary to prevent contamination of the product.	Critical		A. Supplies and products not adequately protected from contamination nor from becoming spoiled.	5	
B. A dust-free method of cleaning not used wherever possible.	3		B. Storage areas not well ventilated and free from objectionable odors.	3	
XI. CONTROL OF INSECTS AND ANIMALS			C. Storage areas not clean and dry.	3	
A. Presence or evidence of any rodents in plant.	5		D. Materials not stored in an orderly manner in suitable, covered or closed containers.	3	
B. Presence of any birds or any other animals including domestic types in plant.	5		E. Supplies and product not protected against unfavorable temperatures and humidity.	3	
C. Presence of live insects in processing or product storage areas (two or more).	Critical		F. Adequate skids or pallets not used where needed.	3	
D. Presence or dead insects in processing equipment (two or more).	Critical		**XIV. PERSONNEL**		
E. Presence of live or dead insects in other plant areas (three or more).	5		A. Employees not wearing garments suitable for work being performed.	4	
F. Rodenticides, insecticides, and other control or eradication material including applicatory equipment not used in a safe acceptable manner.	Critical		B. Personnel in contact with unwrapped product or ingredients not using proper headwear, including protection from facial hair.	4	
G. Evidence that effective pest control not exercised where needed.	3		C. Fingernail polish, costume jewelry, and wrist watches worn by plant personnel working on processing line.	3	
H. Type of insecticides and rodenticides not in compliance with Federal Food and Drug Administration regulations.	4		D. Gloves, if worn, not kept in a sanitary condition.	3	
I. Poisoned baits, if used, not adequately secured.	5		E. Storage of employees. personal effects in production room.	3	
XII. COOLING AND REFRIGERATION FACILITIES			F. Employees not washing hands after contamination.	Critical	
A. Perishable supplies and products not adequately protected from contamination or from becoming spoiled.	5		G. Failure of employees to be hygienically clean: fingernails not kept clean and trimmed.	4	
B. Storage areas not free of visible mold and objectionable odors.	3		H. Employees affected with or a carrier of a communicable or infectious disease not excluded from product areas.	Critical	
C. Storage areas not reasonably clean.	3		I. Plant employees having an infectious wound, sore, or lesion on hands, arms, or other exposed parts of the body, not excluded from contacting ingredients, products, or product zones.	5	
D. Where required, refrigeration facilities not properly cooling and keeping perishable supplies and products at temperatures not exceeding 45° F.	5		J. Plant personnel not instructed in acceptable hygienic practices and proper sanitary rules of food handling.	5	
E. Adequate skids or pallets not used when needed.	3		K. Personnel not prohibited from expectorating, eating, or smoking in product areas.	5	
F. Items not stored in an orderly, easily accessible manner in suitable, covered or closed containers.	3				
	TOTAL				

PLANT OPERATIONS ARE CONSIDERED TO BE INSANITARY IF ONE OR MORE CRITICAL DEFECTS ARE FOUND OR IF THE PLANT RECEIVES A SIR SCORE OF 76 OR HIGHER.

REMARKS

SIGNATURE OF INSPECTOR	DATE

Exhibit 42.2
Sanitation Inspection Report (Beans, Peas, and Lentils)

FORM FGIS-952-1 (1-84) (Replaces Form IN-245 (7-81) which may be used until exhausted)

SANITATION INSPECTION REPORT
(BEANS, PEAS, AND LENTILS)

SANITARY INSPECTION OF (Name and Address of Plant)

INSPECTOR

DATE INSPECTED

PLANT OWNED BY (Company or Individual)

INSPECTOR ACCOMPANIED BY (Name and Title)

SANITARY DEFECTS	Assigned Defect Points	Inspr's Defect Points	SANITARY DEFECTS	Assigned Defect Points	Inspr's Defect Points
I. PREMISES			**VI. WATER SUPPLY**		
A. Outside premises not well drained.	3		A. Lack of satisfactory evidence that the water supply is potable.	Critical	
B. Outside premises not free of weeds, clutter, or unused equipment or other materials.	3		B. Presence of a cross-connection between water lines and sewage system (not allowed.)	Critical	
C. Waste not properly stored or contained.	3		**VII. DISPOSAL OF WASTES**		
D. Outside premises not free of harborages or breeding places for insects, rodents, birds, or other animals.	4		A. Garbage, trash, and dry waste material, inside or outside the building, not satisfactorily contained.	4	
II. RAW MATERIALS			B. Waste is not disposed of frequently enough to prevent unsightliness or undesirable odors.	4	
A. Beans, peas, and lentils not handled in a sanitary manner that would prevent adulteration.	Critical		**VIII. TOILET, HAND-WASHING, AND DRESSING ROOM FACILITIES**		
III. CONSTRUCTION OF BUILDING			A. Toilet facilities not provided for each sex.	3	
A. Plant construction does not exclude probable entrance of insects, rodents, birds, or other animals.	3		B. Toilets and urinals not water-flushed.	3	
B. Walls, partitions, ceilings, and exposed structural supports, in poor repair.	3		C. Toilet room(s) not equipped with self-closing doors.	3	
C. Floors not smooth and readily cleanable or impervious under normal usage.	3		D. Toilet room(s) open directly into a processing or food storage room where food is exposed.	Critical	
D. Floors that frequently get wet not constructed to prevent pooling or are not kept dry.	3		E. Cold and hot, warm or tepid water not provided at wash basins.	3	
E. Exposed, unprotected overhead sewer lines present in product area.	Critical		F. Hand cleaner and towels not provided at or near each basin.	3	
F. Walls not sealed at juncture with floor.	3		G. Adequate hand-washing signs not posted.	3	
G. Walls and ceilings not free of flaked paint, nor in good repair.	3		H. Adequate dry waste receptacles not provided.	3	
H. Windows not intact.	3		I. Toilet facilities not working.	5	
I. Entranceways to production rooms not equipped with self-closing doors nor provided with other effective means to prevent entrance of vermin.	3		J. Toilet and locker room not satisfactorily ventilated.	4	
J. Screen doors, where present, not self-closing.	3		K. Toilet and dressing rooms not kept in a clean orderly condition.	5	
IV. LIGHTING			**IX. CONSTRUCTION AND REPAIR OF EQUIPMENT**		
A. Production rooms not adequately lighted (20-foot candles).	3		A. Adequate space not allowed around equipment for cleaning and inspection purposes.	4	
B. Places, where detailed inspection of the product or ingredients is required, not adequately lighted (50-foot candles).	4		B. Equipment not designed to protect the product zone against contamination or extraneous materials.	5	
C. Places, where detailed inspection of the processes or operations is required, not adequately lighted (50-foot candles).	4		C. Equipment not free of parts or areas inaccessible for cleaning and sanitary inspection.	4	
D. Storage areas not adequately lighted (20-foot candles).	3		**X. CLEANING AND SANITIZING TREATMENT**		
E. Toilets and dressing rooms not adequately lighted (20-foot candles).	3		A. Equipment not cleaned at frequencies necessary to prevent contamination of the product.	Critical	
F. Protective shields for lights not present in areas where essential to protect the product during production.	5		**XI. CONTROL OF INSECTS AND ANIMALS**		
V. VENTILATION			A. Presence or evidence of any rodents in plant.	5	
A. Plant not reasonably free of undesirable odors.	3		B. Presence of any birds or any other animals including domestic types in the plant.	5	
B. Evidence or presence of excessive condensate on walls, ceilings, equipment, or pipeline.	4		C. Presence of live insects in processing or product storage areas (two or more).	Critical	
C. Windows opening to the outside not screened with 14-mesh or finer screen (during presence of flies or other flying insects)	3		D. Presence of dead insects in processing equipment (two or more).	Critical	
D. Exhaust fans not screened or equipped with self-closing louvers.	3		E. Presence of live or dead insects in more than minimal number in other areas (three or more).	5	
			F. Rodenticides, insecticides, and other control or eradication material including applicatory equipment not used in a safe acceptable manner.	Critical	

SANITARY DEFECTS	Assigned Defect Points	Inspr's Defect Points	SANITARY DEFECTS	Assigned Defect Points	Inspr's Defect Points
XI. CONTROL OR INSECT AND ANIMALS (Cont.)			**XII. PERSONNEL (Cont.)**		
G. Evidence that effective pest control, not exercised where needed	3		F. Failure of employees to be hygienically clean; fingernails not kept clean and trimmed.	4	
XII. PERSONNEL			G. Employees affected with or a carrier of a communicable or infectious disease not excluded from product areas.	Critical	
A. Employees not wearing garments suitable for work being performed.	4		H. Plant employees having an infectious wound, sore, or lesion on hands, arms, or other exposed parts of the body, not excluded from contacting ingredients, products or product zone.	5	
B. Personnel in contact with unwrapped product or ingredients not using proper headwear including protection from facial hair.	4		I. Plant personnel not instructed in acceptable hygienic practices and proper sanitary rules of food handling.	5	
C. Gloves if worn not kept in a sanitary condition.	3		J. Personnel not prohibited from expectorating, eating, or smoking in product areas.	5	
D. Storage of employees personal effects in production rooms.	3				
E. Employees not washing hands after contamination.	Critical				
TOTAL					

PLANT OPERATIONS ARE CONSIDERED TO BE INSANITARY IF ONE OR MORE CRITICAL DEFECTS ARE FOUND OR IF THE PLANT RECEIVES A SIR SCORE OF 46 OR HIGHER.

REMARKS

SIGNATURE OF INSPECTOR	DATE

Exhibit 42.3
Sanitation Inspection Report (Overpacker)

SANITATION INSPECTION REPORT
(OVERPACKER)

SANITARY INSPECTION OF (Name and Address of Plant)	PHONE NO. OF PLANT
	INSPECTOR
	DATE INSPECTED
PLANT OWNED BY (Company or Individual)	INSPECTOR ACCOMPANIED BY (Name and Title)

GENERAL TYPES OF ITEMS INSPECTED

SANITARY DEFECTS			SAITNTARY DEFECTS		
I. PACKING MATERIALS			**IV. TOILET AND HAND - WASHING FACILITIES (Cont.)**		
A. Not free from adulteration.	Critical				
B. Shows evidence of insanitary conditions or deterioration.	5		D. Absence of hot or cold water, soap, or hand-drying facilities.	5	
C. Not stored under sanitary conditions.	5		E. Toilet rooms not properly vented to the outside.	5	
II. LIGHTING			**V. CONTROL OF INSECTS, BIRDS, AND ANIMALS**		
A. Insufficient lighting in areas where inspection and examination are performed (50-foot candles).	4		A. Rodent harborages or insect-breeding places present.	4	
III. DISPOSAL OF WASTES			B. Insects, birds, or animals present in plant.	5	
A. Floor drains not functional nor properly trapped.	3		C. Insecticides or rodenticides are handled so as to contaminate the product.	Critical	
B. Dry wastes not collected in suitable containers conveniently located throughout the plant.	3		**VI. STORAGE FACILITIES**		
C. All waste not collected and disposed of at frequent intervals nor in a sanitary condition.	4		A. Storing methods do not minimize deterioration nor contamination.	5	
IV. TOILET AND HAND-WASHING FACILITIES			B. Storage facilities not clean, sanitary, nor in good repair.	3	
A. Toilet room opens directly into packing or storage areas.	5		C. Shelves, cabinets, or dunnage not used where necessary to prevent contamination or deterioration.	5	
B. Doors not self-closing and tight fitting.	3				
C. Absence of sign directing employees to wash hands.	3				

SANITARY DEFECTS			SANITARY DEFECTS		
VII. PERSONNEL			**VII. PERSONNEL** (Cont.)		
A. Not free of communicable nor infectious disease.	Critical		B. Not free of infected cuts, open sores or other lesions on exposed parts of the body.	Critical	

TOTAL . . . .

PLANT OPERATIONS ARE CONSIDERED TO BE INSANITARY IF ONE OR MORE CRITICAL DEFECTS ARE FOUND OR IF THE PLANT RECEIVES A SIR SCORE OF 21 OR HIGHER.

REMARKS

SIGNATURE OF INSPECTOR	DATE

FORM FGIS-952-2 REVERSE

grounds or inside buildings should be neatly arranged on 12-inch-high or higher racks to permit routine cleanup of waste and debris. Covered or closed containers should be used for storage of ingredients.

Waste Disposal

Waste (trash, product spillage, garbage) must be adequately contained and disposed of frequently.

When an applicant disagrees with an inspector's findings, FDA may be requested to inspect the plant. If FDA is unable to perform an inspection, the decision of FGIS is final, subject to normal appeal procedures. FGIS cannot close an insanitary plant. However, the administrator, following the rules of practice governing denial or withdrawl of inspection services, can conditionally withdraw inspection services. Under the FGIS-FDA MOU, FGIS informs FDA of those plants subject to conditional withdrawal of inspection service. Inspection personnel complete the inspection of product lots being inspected at the time the processing plant receives a notice of withdrawl of inspection services. At that time, they must make a decision as to whether the lots being inspected or already inspected have been adversely affected by the insanitary condition(s).

Agricultural Marketing Service

The Agricultural Marketing Service (AMS) grades and inspects agricultural commodities under the Agricultural Marketing Act of 1946. The commodities include milk and dairy products, fresh and processed fruits, vegetables, and specialty crops, livestock and meat, and poultry and poultry products (a complete list may be obtained from AMS).

Milk and Dairy Products

With respect to milk and dairy products, the "General Specifications for Dairy Plants Approved for USDA Inspection and Grading Service," as well as the regulations governing the inspection and grading services of manufactured or processed dairy products, are contained in 7 CFR Part 58. Grading and inspection responsibilities can be divided into four major program areas: (a) Plant Surveys; (b) Inspection and Grading Service; (c) Laboratory Service; and (d) Resident Grading and Quality Control Service. These services are provided at the request of the applicant firm on a voluntary basis. Firms requesting these services must qualify and pay a fee commensurate with the cost of providing grading and inspection. Several of the services are offered cooperatively with state departments of agriculture or marketing.

Grading and inspection are designed to improve the quality, wholesomeness, manufacture, and distribution of dairy products and are widely used by dairy processors, wholesalers, buyers, restaurant owners, and others involved in the production, marketing, and use of dairy products. Only after an unannounced survey has indicated that a plant has met the requirements in the "General Specifications for Approved Dairy Plants" (1) can the plant qualify for the other services of grading, sampling, testing, and certification of its products.

Each plant survey contains detailed checks on more than 100 items. Approved plants are inspected at least twice yearly and plants manufacturing dried dairy products are surveyed quarterly. Dairy Division inspectors complete a plant survey report. Inspectors discuss the findings with plant management and submit their reports to the regional office where the reports are reviewed.

The regulations used by plant inspectors when performing surveys are found in the "General Specifications for Approved Dairy Plants and Standards for Grades of Dairy Products" (1). Plant approval may be denied or suspended if a determination is made that the plant is not performing satisfactorily in regard to control of insects, rodents, and other vermin. Topics covered in the "General Specifications" (1) common to all dairy plant inspections relative to pest control include the following items:

(a) *Surroundings.* The immediate surroundings must be free from refuse, rubbish, overgrown vegetation, and waste materials that might provide harborage for rodents, insects, and other vermin.

(b) *Buildings.* The building or buildings must be of sound construction and shall be kept in good repair to prevent the entrance or harboring of rodents, birds, insects, vermin, dogs, and cats. All service-pipe openings through outside walls must be effectively sealed or provided with tight metal collars.

(c) *Outside doors, windows, and similar openings.* All openings to the outer air, including doors, windows, skylights, and transoms, must be effectively protected or screened against the entrance of flies and other insects, rodents, birds, dust, and dirt. All outside doors opening into processing rooms must be in good condition and fit properly. All hinged, outside screen doors must open outward. All doors and windows should be kept clean and in good repair. Outside conveyor openings and other special-type outside openings must be effectively protected to prevent the entrance of flies and rodents by the use of doors, screens, flaps, fans, or tunnels. Outside openings for sanitary pipelines must be covered when not in use. On new construction, window sills should be slanted downward at approximately a 45-degree angle.

(d) *Coolers and freezers*. All refrigeration units must be free from rodents, insects, and other pests.

(e) *Supply room*. The supply room or areas used for storage of packaging materials, containers, and miscellaneous ingredients must be kept clean, dry, orderly, free from insects, rodents, and mold, and maintained in good repair. Such items stored therein must be adequately protected from dust, dirt, or other extraneous materials and so arranged on racks, shelves, or pallets to permit access to the supplies and to the cleaning and inspection of the room. Insecticides, rodenticides, cleaning compounds, and other nonfood products must be properly labeled, segregated, and stored in a separate room or cabinet away from milk, dairy products, ingredients, or packaging supplies.

(f) *Cleaning and sanitizing treatments*. Shelves and ledges should be wiped or vacuumed as often as necessary to keep them free of dust and debris. The material picked up by the vacuum clearner must be disposed of in sealed containers in order to prevent contamination by insects dispersing from the waste material.

(g) *Insect and rodent control program*. In addition to any commercial pest control service, if one is utilized, a specially designated employee should be made responsible for the performance of a regularly scheduled insect and rodent control program. Pesticides must be properly labeled, and shall be handled, stored, and used in such a manner as considered satisfactory by EPA.

(h) *Dry storage*. The products should be stored at least 18 inches from the wall in aisles, rows, or sections and lots, in such a manner as to be orderly and easily accessible for inspection. Rooms should be cleaned regularly. It is recommended that dunnage or pallets be used when practical. Care shall be taken in the storage of any other product foreign to dairy products or in the same room in order to prevent impairment or damage to the dairy product from mold, absorbed odors, vermin, or insect infestation. Storage rooms for the dry storage of products shall be adequate in size, kept clean, orderly, free from rodents, insects, and mold, and maintained in good repair. They shall be adequately lighted and ventilated. The ceiling, walls, beams, and floors shall be free from structural defects as well as from inaccessible false areas that may harbor insects.

(i) *Packaging room for bulk products*. Control panels shall be mounted a sufficient distance from the walls to facilitate cleaning or satisfactorily sealed to the wall, or shall be mounted in the wall and provided with tight-fitting, removable doors to facilitate cleaning. Unnecessary fixtures, equipment, or false areas that may collect dust and harbor insects should not be allowed in the packaging room.

Further information is available through the AMS instruction entitled *Dairy Plant Survey Guidelines* (5).

Fruits and Vegetables

With respect to fresh fruits, vegetables, and specialty crops, AMS inspects produce to facilitate marketing and to determine compliance with the Federal Export Fruit Act, federal import regulations, Federal or State Marketing Orders, and contract or purchase specifications.

Requirements for continuous inspection, associated with use of the "USDA-approved" shield, include conducting a survey to determine whether premises, buildings, and facilities are suitable and adequate. The premises shall be free from objectionable conditions including, but not limited to, litter, waste, and refuse. The packing plant buildings shall be properly constructed and maintained in a sanitary condition to ensure exclusion of dogs, cats, rodents, and other vermin from the rooms in which products are to be graded or stored.

The procedures followed to set up an in-plant inspection service for processed fruits, vegetables, and specialty crops are summarized here. Prior to the inauguration of in-plant inspection services, a survey is conducted to determine whether the plant and methods of operation are suitable and adequate. A plant survey shall be conducted at least yearly. The plant survey is based on the regulations issued under the Federal Food, Drug, and Cosmetic Act (FD&C Act) (6)—"Good Manufacturing Practice (Sanitation) in Manufacturing, Processing, Packing, or Holding Human Foods" (21 CFR Part 110). When inspection services are inaugurated, sanitation score sheets are used to record and report plant sanitation deficiencies. Sanitation is monitored during processing operations for each production shift. Publication 159-A-1, "Plant Sanitation, Sanitation Requirements," provides information on policy and procedures regarding plant sanitation requirements established by FDA.

FDA has entered into several MOUs with AMS. These agreements are directed toward the protection of public health by ensuring that foods are safe and wholesome and that products are honestly and informatively labeled. The MOUs currently in effect with FDA are:

(a) *Inspection and Grading of Food Products.* This agreement sets forth the general responsibilities and objectives of FDA and AMS in carrying out their respective regulatory and service activities;

(b) *Cooperative Efforts for Inspection, Sampling and Examination of Imported Raisins;*

(c) *Inspection, Sampling and Examination of Imported Dates and Date Material;*

(d) *Testing of Peanuts for Presence of Aflatoxin;*

(e) *Sampling and Aflatoxin Testing of Imported Pistachio Nuts;*

(f) *Aflatoxin Testing Program for In-shell Brazil Nuts.*

Inspection for Foreign Material

FDA has established food defect action levels (DALs) (4) for certain products. Under in-plant contracts or lot inspections, products are examined for the presence of foreign material to determine lot compliance with the DALs.

Poultry, Eggs, and Rabbits

AMS administers a mandatory inspection service for egg-product plants under authority of the Egg Products Inspection Act (EPIA) (7 CFR Part 59), and additionally, provides voluntary grading and acceptance services for poultry, rabbits, egg products, and shell eggs under authority of the Agricultural Marketing Act (AMA) of 1946, as amended. Pest control is one aspect of all these services.

With regard to egg-product plants, USDA is responsible for inspection to ensure that egg products distributed to consumers and used in products consumed by them are fit for human food and are not adulterated. Adulteration includes preparing, packaging, or holding shell eggs or egg products under insanitary conditions whereby they may become contaminated with filth. Instructions with regard to the duties and operating procedures for inspectors are contained in the *Egg Products Inspectors Handbook.* Pest control aspects of the EPIA and related regulations and instructions are as follows:

(a) Inspectors must approve for operation all plants where egg products are processed, including plant facilities, methods of operation, and sanitary procedures. Any conditions on the premises or area adjacent to the plant that create sanitary hazards, harborage for rodents, flies, and other insects, or undesirable off-odors are subject to correction by plant management. Inspectors check the premises daily for evidence of rodents and insects. If insects are present in a processing room, the inspector may not permit processing to start until the room is free of them.

(b) Products held or produced under insanitary conditions are subject to retention by the inspector until they are destroyed, denatured, or rendered inedible (so that they cannot be used for human food). Such operations are to be conducted in a manner that will maintain the sanitary levels required in plants and minimize objectionable odors, rodent and pest infestations, and so on.

(c) Inspectors record obervations of various sanitary procedures—fly, rodent, and odor control inside and outside the plant; use and storage of insecticides and rodenticides separately from other chemical compounds; and whether all chemicals are kept away from edible products.

(d) Shell-egg plants that do not receive voluntary grading service are subject to surveillance inspections under the mandatory EPIA to ensure that wasted or inedible eggs are disposed of properly. Pest control and sanitation, however, are monitored by FDA under the authority of the FD&C Act (6).

Under the Agricultural Marketing Act, AMS inspects, certifies, and identifies the class, quality, and condition of agricultural products. The pest control aspects of voluntary grading services for poultry, shell eggs, and rabbits, and voluntary inspection of egg products include the following:

(a) Poultry and rabbits may be graded only if they have first been inspected and certified by the poultry inspection service of USDA. The inspection regulations in 9 CFR Parts 311 and 382 provide detailed sanitation/pest control requirements. The grading regulations in 7 CFR Part 70 provide for the suspension or withdrawl of plant approval for failure to maintain grading facilities and equipment in a satisfactory state of repair, sanitation, or cleanliness. Duties and operating procedures for graders are contained in official publications (2, 9).

(b) The shell-egg grading regulation is in 7 CFR Part 56. Graders' duties and operating procedures are published by the U.S. Department of Agriculture (3, 8). Regulations state that buildings shall be of sound construction so as to prevent the entrance or harborage of vermin.

Meat and Meat Products

With respect to meat and meat products, AMS has been delegated retention authority by the Food Safety and Inspection Service (FSIS), USDA, which is the agency responsible for maintaining the wholesomeness and sanitation of these products. This means that when there is no FSIS inspection in an establishment where grading and certification duties are performed, AMS supervisors and graders are responsible for identifying apparent insanitary or other conditions that they believe could result in contaminated or unwholesome products. Upon finding such conditions, AMS must attach a tag indicating that the products or utensils are being rejected or retained by USDA. The local FSIS inspector must be notified of the action taken. If the insanitary conditions are minor and can be readily corrected, AMS personnel may remove the tag when the insanitary conditions have been corrected.

Virtually every significant law or regulation pertaining to the food industry in the USA has been noted in the compendium by Hui (7). Laws pertaining specifically to microbiological food safety have been briefly noted by Thompson et al. (11) for the USA and by Ryder (10) for the United Kingdom.

Cited References

1 Agricultural Marketing Service, U.S. Department of Agriculture. 1975. General specifications for approved dairy products. *Federal Register* 40(198)47910–47940.

2 Agricultural Marketing Service. 1989. *Poultry-Grading Manual*. Agriculture Handbook No. 31. U.S. Department of Agriculture, Washington, DC.

3 Agricultural Marketing Service. 1990. *Egg-Grading Manual*. Agriculture Handbook No. 75. U.S. Department of Agriculture, Washington, DC.

4 Center for Food Safety and Applied Nutrition. 1989. *The Food Defect Action Levels*. Food and Drug Administration, Washington, DC.

5 Dairy Grading Branch. 1979. *Dairy Plant Survey Guidelines*. Poultry and Dairy Quality Division, Food Safety and Quality Service, U.S. Department of Agriculture, Washington, DC.

6 Food and Drug Administration. 1986. *Federal Food, Drug, and Cosmetic Act, as Amended, and Related Laws*. HHS Publ. No. (FDA)86-1051. Public Health Service, Rockville, MD.

7 Hui, Y. H. 1986. *United States Food Laws, Regulations, and Standards*, 2nd ed., 2 vols. Wiley, New York, NY.

8 Poultry Division. 1989. *Shell Egg Grader's Handbook*. AMS PY Instruction No. 910 (Shell Eggs)-1. Agricultural Marketing Service, U.S. Department of Agriculture, Washington, DC.

9 Poultry Grading Branch. 1988. *Poultry Graders Handbook*. AMS PY Instruction No. 910 (Poultry Grading)-1. Agricultural Marketing Service, U.S. Department of Agriculture, Washington, DC.

10 Ryder, C. J. 1990. Foodborne illness. UK food legislation. *Lancet* 336(8730)1559–1562.

11 Thompson, P., et al. Foodborne illness. US food legislation. *Lancet* 336(8730)1557–1559.

43
Food Industry
Self-Inspection

Robert L. Hohman

For the most part, the food industry has recognized that an internal self-inspection program is an integral part of a successful sanitation management program. Self-inspection determines whether or not manufacturing, storage, or co-packer facilities are in compliance with the Good Manufacturing Practices (GMPs) and food laws and regulations. Equally important are concerns for consumer protection and industry reputation.

It must be emphasized, however, that a self-inspection program in itself is not a correctional tool but can only identify correctable conditions and practices. Self-inspections or audit systems act as management tools to ensure compliance with ever-changing food protection regulations.

The magnitude of the inspection program in industry is generally dictated by the size and resources of the corporation. Large corporations can usually afford a well-staffed corporate inspection branch with or without individual inspectors in each of their plant operations. Smaller corporations with more limited resources may need to rely on an individual who carries multiple responsibilities (i.e., safety, energy, the environment, quality control). Still others may procure the services of an outside professional inspection service or qualified consultant (see Chapter 48).

An effective industry inspection program usu-

ally contains the following essential features: management commitment, qualified personnel, inspectional tools and guidelines, effective reporting system, effective follow-up system, and motivational tools.

Management Commitment

The first and probably most important requirement for a successful inspection program is full commitment and involvement by all levels of management. Without such support, an inspection program will have little, if any, effectiveness. A manifestation of the management commitment is the reporting structure of the inspection group. This group must report to top management to avoid conflicts of responsibility and to permit inspection of the total physical facility and all departments within. For example, if inspectors are part of the production department, their responsibilities may be limited to that department and may not extend into other important parts of the facility. All levels of management must understand that top management ultimately bears the responsibility for compliance with the GMPs established by the Food and Drug Administration (FDA).

Qualified Personnel

As a source of qualified inspection personnel, industry can draw from individuals who are academically qualified in the fields of entomology, microbiology, and/or food sciences or those who have equivalent experience and specialized

Hohman: Borden, Inc., 990 Kingsmill Parkway, Columbus, OH 43229.

training in these fields (see Chapter 47). This training is usually obtained from special schools and workshops conducted by trade associations or regulatory agencies. It goes without saying that other attributes, such as being alert, observant, able to use good analytical judgment, as well as being honest and having the ability to communicate, are highly desirable.

Inspection Tools and Guidelines

For inspectors in industry to successfully perform their responsibilities, they must have a thorough knowledge of the standards from a regulatory as well as a corporate standpoint in order to determine compliance.

From the regulatory standpoint, inspectors must be guided by FDA's GMPs and, depending on the circumstances involved, they may require similar guidance from other regulatory agencies such as the U.S. Department of Agriculture, Department of Defense veterinarians, and state, county, or local health agencies.

From an industry viewpoint, most large and medium-sized companies have developed their own guidelines tailored not only to meet the requirements of the regulatory agencies, but also to fit their own special needs where required. Exhibit 43.1 is a taken from a typical industry inspection manual (Note: the numerous redundancies in this exhibit are intentional.)

During the course of an inspection, an inspector will need various devices for investigational purposes. A suitable flashlight is a must. A blacklight can be valuable for detecting rodent activity and for spotting milkstone on equipment in the dairy industry. The camera, if used properly, can be a valuable tool for pointing out insanitary conditions. A pyrethrum aerosol can be helpful to flush insects out of cracks and crevices. Spatulas, scrapers, and pliers are valuable aids for probing. A magnifying glass can be helpful for on-the-spot verification of the type of insect involved.

An Effective Reporting System

There are many varieties of inspection reporting systems that have been designed to fit the specific needs of a particular operation, the main objective being that the findings are reported in such a manner that effective corrective action can be accomplished. The effectiveness of a check-off form (Exhibits 43.2-43.4) versus a comprehensive reporting system may be debated, but the system ultimately selected should be the one that yields the best results. An industry reporting system would vary in most cases from a regulatory inspection format in that it frequently includes recommendations for correction of any violations.

An Effective Follow-Up System

There is a quote heard many times in industry: "You cannot inspect yourself out of trouble." What this implies is that an inspection is of little value if it is not supported by an effective follow-up system to ensure that corrective action has been implemented.

The nature and extent of the observations made and reported during an inspection dictates the urgency required for corrective action. For example, an active insect infestation would require immediate attention whereas an observed minor poor housekeeping practice that could eventually lead to such an infestation might not. The responsibility for follow-up action resides solely with plant management, whereas the inspector is responsible for continually alerting management to instances of noncompliance with the GMPs.

Motivational Tools

The need for an inspection program in the food industry is obvious. In nearly all instances, the motivation to operate a plant in compliance with the GMPs should surpass the fear of enforcement activities, even though one cannot discount the latter as another motivational factor. The two strongest motives for compliance with GMPs are the protection of the consumer and the maintenance of a respected trade name.

Internal mechanisms in industry for motivation often include the following:

(a) management incentive programs based on degree of compliance (this could be determined by a sanitation rating system);
(b) establishment of interdepartmental competition with visible awards;
(c) encouragement of key employee participation in sanitation workshops;
(d) conduct of scheduled routine sanitation training sessions with full employee participation; and
(e) demonstration of *sincere,* visible, continuing top-management interest in quality and sanitation.

Conclusion

In the inspection aspects of pest management for the food industry, the importance of a self-

Exhibit 43.1 Inspection Guides

SECTION A, INSECT CONTROL

The presence of any insects is considered as filth. Effective measures shall be taken to prevent the entrance, harborage, and breeding of insects in a plant or warehouse facility.

Insecticides must be approved by the Environmental Protection Agency (EPA) and applied according to label instructions. Private pest managers and plant personnel using such products shall maintain records of the use of such products, and such records shall be properly filed in the plant. No product may be used in a manner that might be harmful to personnel or that might contaminate products or product containers. All insecticides must be in isolated, locked storage.

Insects are capable of transmitting diseases to humans through contamination of food products and product contact surfaces. Accordingly, their presence in or on the premises of a food-processing establishment creates a potential public health hazard which can be guarded against only by effective control.

Inspection Guides

1. Are any live or dead insects noted?
2. Are excreta, trails in dust, or cast skins observed?
3. Is there evidence of insects under product belt conveyer system?
4. Is there evidence of insects in folds of ingredient bags?
5. Is there evidence of insects in product containers in storage?
6. Do rodent bait stations show evidence of insect infestation?
7. Are fly control units properly maintained and cleaned?
8. Is the plant insect control program assigned to a specific qualified and properly trained individual?
9. Where professional pest managers are used, are they licensed, are the insecticides they use approved, and do they leave a written report of materials used and locations of use? Are these reports on file in the plant?
10. Are pesticides used according to EPA directions, and are applicators certified in their use?
11. Are all insecticide materials stored in an isolated and locked area to prevent contamination of food products and food-packaging materials?
12. Are there cracks or damage in floors, walls, or doors?
13. Are loose moldings present on walls at floor or ceiling?
14. Are beams, ledges, or piping dusty?
15. Are switch boxes and switch panels clean and closed to prevent accumulations and possible harborage?
16. Are electrical conduit ends and terminals closed?
17. Can product dust get behind loose or creased pressure-sensitive tape?
18. Does product dust accumulate on lights and motors where warmth is an attractant for insects?
19. Are cabinets and lockers clean and in good order?
20. Are waste disposal containers emptied and cleaned regularly to eliminate possible breeding areas?
21. Are there openings in hollow walls?
22. Are false ceilings and areas over rooms built out into the plant clean?
23. Are uncleaned product belts held in storage?
24. Are elevator pits neglected or are they cleaned to prevent harborage?
25. Are there standing pools of water anywhere that may be an attractant?
26. Are there areas of dead storage on the floor that extend to the walls?
27. Have stored equipment and materials been cleaned prior to storage?
28. On the exterior premises, are deteriorating product, neglected waste, litter, or undrained water present?
29. Is the screening of doors, windows, and any outside wall openings effective?
30. Are doors self-closing?
31. Do outside doors open outward?
32. Are exhaust fans equipped with operative self-closing louvers?
33. Are ceiling vents and skylights screened?
34. Are worker ants traveling between the food sources and nesting sites?
35. Are there any termite tunnels in cracks of concrete floors or any flying adult termites during the mating season?
36. Examine each ventilator to see that louvers close when fan is off.
37. Inspect open pipes for insect harborage.
38. Inspect open cracks in the building structure for insect harborage.
39. Inspect behind wall signs, electrical installations, and in dark damp areas for cockroaches.
40. Are there cockroaches and infesting insects and/or larval cast skins on light fixtures and near electric motors?
41. Inspect hot areas such as around ovens or dry boxes for firebrats.
42. Inspect for silverfish in situations where starch or paper may provide food.
43. Inspect lockers, especially under paper, cardboard, or old shoes, for cockroaches.
44. Inspect for maggots in any liquid or semiliquid ingredient when drosophilid fruit flies are seen in the area. Look closely for maggots with a light for a minute or more.
45. Inspect for maggots and fly pupae in dusty corners near waste containers.

Exhibit 43.1 *Continued*

46. Inspect for insects around, on, and under idle equipment stored in the plant.

47. Inspect for insects in uncleaned cabinets or in cabinets that contain food.

48. Inspect for insects in and around waste containers.

49. Inspect for infesting insects in false ceilings or above suspended ceilings.

50. Inspect for insects in elevators and elevator pits.

51. Inspect for insects in standing pools of water.

52. Inspect for infesting insects in old rodent baits.

53. Inspect for carpet beetles in accumulations of dust.

54. Inspect for infesting insects on the roof around stacks and ventilators.

55. Inspect for infesting insects in uncleaned stored machine parts, electric motors, and conveyor belts.

56. Inspect for crawl marks made by infesting insects in dust on the floor, machinery, ducts, lights, or on ingredient bags.

57. Inspect for crawl marks made by infesting insects on top of ingredient drums.

58. Inspect for small holes in cereal-product ingredient bags, caused by the cigarette beetles that hatch inside the bag and bore out.

59. Inspect for cigarette beetles and their larvae around ventilators, on top of ducts, and lights. Remember, cigarette beetles have many tiny hairs on which dust accumulates so that they readily take on the looks of the surrounding media.

60. Inspect for infesting insects on floors where any object such as a machine or cabinet contacts the floor.

61. Inspect tailings daily (and record findings) for infesting insects. Flat grain beetles indicate possible contamination in silos or bins especially where excess moisture has caked the ingredient. Dome covers and air spaces in silos are susceptible areas.

62. Inspect breather bags and connecting socks (particularly at point where fastening band is located) for infesting insects.

63. Inspect vacuum cleaner collection containers. This may give you a clue to a possible problem.

64. Are there any infesting insects in lockers or on shoes in a plant where flour is an ingredient?

65. Are there any grain moths flying inside buildings?

66. Examine cloth bags or cloth material for grain moth pupae in attached cocoons.

67. Examine corn or popcorn grains for holes that indicate weevil or grain moth contamination.

68. Inspect return air ducts of air-conditioner units for infesting insects.

69. Look for insects in product debris that may have fallen onto pallets of ingredients or packing materials.

SECTION B, RODENT CONTROL

Effective control measures must be taken to prevent the entrance, harborage, and breeding of rodents (and other animals) in an establishment.

Approved rodenticides must be used in accordance with the label instructions. Private exterminators and plant personnel using rodenticides shall maintain records of their use, and such records shall be properly filed in the plant. No rodenticide may be used in a manner that might be harmful to personnel or that might contaminate products or product containers. All rodenticides must be isolated in special storage areas.

Rodents are capable of transmitting diseases to humans through contamination of food products and product contact surfaces.

Inspection Guides

1. Are any live rodents or carcasses noted?

2. Are excreta pellets present; are they old or fresh?

3. Is there evidence of runway smears, urine stains, or footmarks in dust?

4. Is there damage to food, materials, or property by gnawing?

5. Is there evidence of nests or efforts to obtain nesting materials?

6. In the examination of bait stations, are pellets present, stations damaged, bait spilled, mold or insect infestation present (all of which indicate lack of inspection and control)?

7. Are burrows present in embankments, along railroad sidings and docks, or at building foundations?

8. Is there evidence of other pests, such as cats, dogs, squirrels, chipmunks, and the like?

9. If rodenticides are used by plant personnel, are they in isolated, safe, locked storage, available only to designated personnel?

10. Where professional pest managers are used, are they licensed, are the rodenticides they use approved, and do they leave a written report of materials used and locations of use? Are these reports on file in the plant?

11. Is there a program of control (numbering stations and traps, a diagram giving locations, and a record of number of rodents caught)? Are service tags affixed to the underside of station lids to ensure that the stations were opened and inspected?

12. Are doors tight-fitting and rodentproof?

13. Are roof vents, skylights, and sluice drains screened and floor drains covered?

14. Are there any unprotected or unsealed openings in outside walls? This includes conveyor, pipe, and hose openings.

15. Is food or harborage available in uncleaned clothing and cabinet storage or in lunch or locker rooms?

16. Are all stocks arranged off the floor and away from walls?

Exhibit 43.1 *Continued*

17. Are clean conditions maintained in storage of little-used equipment and materials?

18. Is excess equipment cleaned prior to storage?

19. Are there dark and/or cluttered areas in storage rooms?

20. Are the outside plant premises free of clutter or are there piles of junk, pallets, lumber, or waste on the ground or along building walls?

21. Are grass or weeds uncut close to buildings?

22. In waste accumulation areas, is waste covered and stored off the ground? Is good housekeeping maintained?

23. Is incineration of wastes complete?

24. Is there a blacklight inspection of ingredients in storage for detection of urine stains?

25. Are there any conditions around the plant exterior that may attract and harbor rodents, such as junk and uncut grass or weeds?

26. Are there ill-fitting doors and openings in walls where rodents may enter? (An opening of ¼ inch in diameter will allow mice to enter.)

27. Are there any open floor drains or pipes leading to the outside?

28. Is stock poorly stored (directly on the floor or too close to walls)?

29. Is there rodent activity around slow-moving packaging material, advertising material, or ingredients?

30. Inspect for rodent activity inside a "well" or "chimney" of palletized product.

31. Are there signs of nests or nesting materials?

32. Are there any excreta pellets on and between pallets and product?

33. Is there rodent evidence around cluttered areas in storage?

34. Are there any rodent excreta pellets and gnawings on ingredient bags?

35. Is there rodent evidence in false ceilings and on mezzanines?

36. Is there rodent evidence in closets?

37. Is there rodent evidence in cabinets, switch boxes, and open lines?

38. Is there rodent evidence in bait boxes? Bait boxes should be inspected frequently.

39. Is there any rodent evidence around cloth materials?

40. Is there any rodent evidence in lockers and on old shoes?

41. Is there any rodent evidence on machine bases where warmth is provided by a motor?

SECTION C, BIRD CONTROL

Birds and food establishments are incompatible; therefore, effective measures must be taken to exclude the entrance of birds into plants and warehouses. They should also be excluded from the plant exterior by preventing roosting and nesting.

Birds are known to be carriers of fungal, protozoan, and bacterial diseases. Nests may be a source of mites that invade a building and bite the people inside.

Inspection Guides

1. Is there any evidence of live or dead birds?

2. Are droppings present?

3. Are loose feathers, twigs, or twine present?

4. Are birds roosting or are nests noted in eaves of buildings, under dock canopies, in covered can runs, outside sheds, or in little-used buildings?

5. Where signs on buildings are not mounted flush, do birds roost or nest on the signs?

6. Are all outside openings protected to prevent bird entrance?

7. Are there outside openings that may allow bird entry?

8. Are there any attractants for birds such as spilled food on the outside, particularly along docks?

9. Are there beams, lines, mounted building signs, and ledges where birds prefer to rest?

10. Control of pest birds must comply with the federal, state, or local regulations.

Exhibit 43.2 Sanitation Inspection Form (pest management evaluation)

Plant Location: Inspect Daily Fill this out Mon., Wed., Fri. Describe potential hazards on back	Date: INCOMING MATERIALS AREA YES / NO	PROCESS AREA YES / NO	PACKING AREA YES / NO	FINISHED PRODUCTS AREA YES / NO	SHIPPING AREA YES / NO
1. Covers on drains?					
2. Screens on windows?					
3. Doors have self-closing devices?					
4. Air-curtains (or fans) at openings?					
5. Is stock stored away from walls?					
6. Is waste disposal satisfactory?					
7. Cracks or holes in floors?					
8. Cracks or holes in walls?					
9. Cracks or holes in ceilings?					
10. Are there any flying insects?					
11. Are there any crawling insects?					
12. Are there any dead insects?					
13. Are there any rodent droppings?					
14. Rodent signs under black light?					
15. Are there any live or dead birds?					

16. Are there enough supplies and equipment for thorough insect control? YES _____ NO _____

17. Are there enough supplies and equipment for thorough rodent control? YES _____ NO _____

18. Is sweep-up equipment adequate and convenient? YES _____ NO _____

19. Outside: Are dock areas clean? YES _____ NO _____

 Are dock areas free from unused equipment? YES _____ NO _____

 Is there any harborage of insects? YES _____ NO _____

 Is there any harborage of rodents? YES _____ NO _____

 Are there bird or wasp nests? YES _____ NO _____

INSPECTOR _____

AUDITED BY _____

524 Robert L. Hohman

Exhibit 43.3 Bulk Car Cleaning, Loading, and Inspection Report

Shipping mill _____ Loading berth _____

Car No. _____ Order No. _____ Flour grade _____

Destination _____ Customer _____

CAR PREPARATION

Car returned from _____ Last loaded with _____

Cleaned by _____ Time spent cleaning _____ Cwt flour removed _____

Condition of linings: Good ____ Slight ____ Heavy ____ If chipping,
                              chipping   chipping   where in car? _____

Condition of gaskets: Good ____ Poor ____

Evidence of rust: No rust ____ Slight ____ Heavy ____ If rusting, where
                                    rusting   rusting   in car? _____

Evidence of mold: None ____ Comments _____

Infestation found? Yes ____ No ____ If infested, live ____ dead ____

Did car arrive fumigated? Yes ____ No ____ Was car fumigated? Yes ____ No ____

Airslide fabric checked ____ Swing valve checked and closed ____ Sanitary plate
                                                                  checked & secured ____

Car checker _____ Date inspected _____

CAR LOADING RECORD

Tags applied? Yes ____ No ____ If yes, name of flour on tag _____

Old tags removed? Yes ____ No ____ By _____

Started loading (hour) ____ Date _____ Approx. cwt specified on L/R _____

Finished loading (hour) ____ Date _____ Scale reading: Gross _____
                                                            Tare _____
                                                            Net _____

Kind of weather _____
Were all open hatches kept covered with insect-tight filters? _____

Signed _____ Seal numbers _____

CAR INSPECTION AFTER LOADING

All hatch covers equipped with gaskets? Yes ____ No ____
Were car interior and hatch covers free of condensation before sealing? Yes ____ No ____
Were hatch openings covered (paper or plastic)? Yes ____ No ____
Rust cleaned around hatches? Yes ____ No ____

Inspected by _____ Hatches tightly secured ____ Air cap closed ____

Exhibit 43.4 Quality Assurance Warehouse Audit (dry warehouse)

Division _____ Name _____

Address _____

City _____ State _____ Zip _____

Contact(s) _____ Title _____

_____ Title _____

Auditor(s) _____ Date(s) _____

HOUSEKEEPING & MAINTENANCE: TOTAL WEIGHT   (ACCEPTABLE) (UNACCEPTABLE)

Compliance Item	Weight	Compliance	Weight
Storage away from walls (24 inches prefer.; 4-inch min.)	3	No accumulated dirt on racks	2
		Wood pallets clean before re-use	2
Floor-wall juncture clean	2	Powered equipment clean	2
Sanitary strip along walls maintained	2	No excessive oil leaks from equipment	2
Walls, columns, ceiling, floor clean	2	Drinking fountains clean	1
No encrusted dirt on structure members	2	Angle-jet outlets on drinking fountains	1
Little or no cobwebs on ceiling	1	Drinking fountain cabinets in good repair	1
No water leaks in ceiling and walls	1	Lunchroom walls, floor, ceiling—clean	1
Walls in good repair	1	Lunchroom walls, floor, ceiling—good repair	2
Floors have no deep cracks	1	Lunchroom equipment clean	1
Expansion joints clean, filled	1	Lunchroom equipment in good repair	2
Doors, door tracks, and frames undamaged	1	Covered waste containers in lunchroom	2
Trackwell surfacing intact	1	Waste containers clean, no odors	2
Door guard posts capped or filled; tight	1	Lunchroom door self-closing	1
No loose products on floor	2	Lunches eaten only in designated areas	2
No dropped cases in traffic aisles	2	Restroom walls, floor, ceiling—clean	1
Storage bays swept when empty	2	Restroom walls, floor, ceiling—good repair	2
Cases on aprons & trackwell picked up daily	2	Restroom facilities adequate in number	2
No wide cracks or deep holes in aprons	1	Restroom equipment clean	1
No debris storage on loading docks	1	Restroom equipment in good repair	2
No debris accumulation on aprons/trackwells	1	Waste containers in restrooms	1
Bumper blocks on dock intact	1	Waste containers clean, no odors	2
Dock ladders, steps, handrails tight	2	Doors self-closing; handwash signs posted	1
Dock leveller pits clean	1	Vent fans in operating condition; clean	1
Exterior building strip in place; no growth	2	Hand towels, toilet tissue available	1
Weeds, grass kept cut	2	Toilets in separate stalls	2
Premises free of debris and paper	1	Stalls have door latches, clothes hanger	1
Exterior tracks free of warehouse debris	2	Outside waste in covered, metal containers	2
Traffic areas paved or treated	1	Inside waste containers clean, covered	2
Paved areas free of warehouse debris	1	Inside disposal containers metal, no leaks	2
No accumulated wastes stored over office	1	Disposal containers covered, free of odor	1
Bottom of shelves of storage racks off floor	1	Disposal containers stored in dock area	1
Shelves clean, orderly	1	Storage, boiler, telephone rooms clean/neat	1
Storage racks in good repair	1	Bay and aisle lines marked (white)	1
No spilled product on racks	2		

COMMENTS _____

_____

_____

inspection approach cannot be overemphasized. Its importance to industry in achieving the protection of the food supply is in the interest of both the industry and the consumer.

Selected References

Bryan, F. L. 1979. Prevention of foodborne diseases in food service establishments. *Journal of Environmental Health* 41(4)198-206.

Foley, V. 1963. How to inspect a food processing plant. *Journal of Milk and Food Technology* 26(3)94-96.

Foulk, J. D. 1990. Pest-related aspects of sanitation audits. *Pest Control Technology* 18(11)32, 33, 89; 18(12)69-72, 74.

Gentry, J. W. 1984. Inspection techniques. In *Insect Management for Food Storage and Processing*, F.

J. Baur, ed. American Association of Cereal Chemists, St. Paul, MN.

Gould, W. A., and R. W. Gould. 1988. *Total Quality Assurance for the Food Industries*. CTI Publications, Baltimore, MD.

Industry Programs Branch. 1984. *Do Your Own Establishment Inspection. A Guide to Self Inspection for the Smaller Food Processor and Warehouse.* HHS Publ. No. (FDA)82-2163. Food and Drug Administration, Washington, DC.

Marriott, N. G. 1989. *Principles of Food Sanitation.* AVI/Van Nostrand Reinhold, New York, NY.

Meekings, D. E. 1980. Regulatory inspections in perspective. *Association of Food and Drug Officials Quarterly Bulletin* 44(3)159-164.

Stauffer, J. E. 1988. *Quality Assurance of Food Ingredients, Processing and Distribution.* Food and Nutrition, Westport, CT.

44

Commentary on Regulatory Aspects

John V. Osmun

Laws at the federal and state levels that affect the food industry, as related to pest control, are of two different types—those pertaining to foods and drugs, and pesticide laws. It seems paradoxical that the control of pests that are classified as food contaminants frequently requires the use of pesticides, the concern for which, as contaminants in their own right, now looms greater than the pests themselves. In the first instance, it is a relatively straightforward procedure to interpret the laws and act accordingly. With pesticides, however, the matter is complicated and confused by definitions and interpretations of several laws and by apprehension on the part of the public.

From the earliest days of the Food & Drugs Act of 1906, it was clear that insects, for example, were sources of filth. This identity commenced with Harvey Wiley's concern for the house fly and its vectored bacteria, both as filth and as adulteration. There has been little challenge to the notion that insects, insect fragments, rodent feces and hairs, and bird feathers and feces are potential contaminants of foods and food products. Various sections of the Federal Food, Drug, and Cosmetic Act (FD&C Act) appear to define accurately what constitutes adulteration by pests and their potential hazard to health. Both sections 301 and 402 provide definitions. Paragraphs 402(a)(3) and (4) take the matter a step further by stating that the very presence of food pests in a food establishment constitutes a violation of the Act, and "if it [food] has been prepared, packed, or held under insanitary con-

ditions whereby it may have become contaminated . . . ," it may be deemed adulterated.

Various court decisions have interpreted the law rigorously, thus ensuring maximum consideration for the health and well-being of the consuming public. To make the law workable, guidelines were needed that balanced benefits against such risks that one might perceive when trace fragments were detected. This need has been addressed through the "Food Defect Action Levels" (DALs).

Federal and state food and drug officials have elected to undergird regulation with meaningful programs of education. This is viewed as a sensible and effective approach because there is a strong tendency among food processing managers to *want* to do things right. In addition, education is the most cost-effective and palatable way to achieve high levels of compliance.

We are faced with a different set of problems when we consider the potential for the presence of pesticides in foods. Consider the opportunities that exist for this to happen—in the field at time of initial production, in preprocessing storage, during processing or manufacture, in subsequent storage and, where relevant, when food-serving establishments are treated with pesticides that incidentally may reach food. Regulation of pesticides in these various situations is not governed by a single federal law; there are three basic ones involved. The Federal Insecticide, Fungicide, and Rodenticide Act (FIFRA) is heavy in agricultural orientation. It permits application of many registered pesticides to crops to ensure successful production and optimum quality. Tolerances are provided for under section 408 of the FD&C Act, but the responsibility

Osmun: Purdue University, West Lafayette, IN 47906.

529

for setting them was transferred years ago to the Environmental Protection Agency (EPA). Tolerances are intended to provide a large measure of safety while still permitting pesticide use to protect crops and animals from pests. The Food and Drug Administration (FDA) continues to enforce tolerances except in situations where the U.S. Department of Agriculture (USDA) does this through the Meat & Poultry Inspection Act. The latter also permits USDA to prescribe procedures for pesticide use in meat and poultry processing plants. The FD&C Act deals with food additives, intentional or otherwise, and thus it is concerned with minimizing the levels of pesticide contaminants that may be present in processed foods and feeds and thus possibly pose a threat to human health.

The problem lies in *definition* because pesticides that arrive in food through the route of an application to a raw agricultural commodity (and, therefore, have established tolerances) have tended to be treated more lightly than if they occurred as the result of an application made at the time food was being processed. In the latter case, the matter is a concern of section 409 (Food Additive Amendment, FD&C Act). In a similar manner, a field-applied pesticide concentrating to a higher level during processing is also a concern of FDA.

The infamous Delaney Clause, which prohibits the approval of a food additive that has been found to "induce cancer" in humans and animals, is also a provision of section 409 of the FD&C Act. Legally,

it has not been a factor when EPA has set tolerances, because tolerances are for pesticides applied to raw agricultural commodities. Yet, a pesticide that has been permitted tolerances on the basis of risk/benefit analysis becomes subject to the Delaney Clause if indeed it concentrates in processed foods. Scenarios of this type have the dual effect of influencing decisions within EPA and perpetuating concerns in the mind of the public. Some ask, if certain pesticides do concentrate during food processing, should they not be subject to the zero-risk standard of the Delaney Clause? Seemingly a "negligible-risk" standard, applicable to both raw and processed foods, would be a better solution. With such a procedure, some presently used pesticides, especially certain suspect fungicides, might eventually be withdrawn from use on those crops where their presence can be demonstrated in finished foods. At the same time, the use of many needed pesticides could be continued.

Laws and regulations governing the use of pesticides on and in foods and food products are adequate in some respects but less than perfect in others. FIFRA provides for registration and serves as a basis for restricting uses of pesticides, plus it requires training and certification of those applicators who use them. State laws in general provide parallel regulation. The FD&C Act, in terms of protecting the health of our nation, is excellent, but adjustments will be needed as our knowledge of the potential effects of trace chemicals becomes more sophisticated.

45

Commentary on Inspection Aspects

Austin M. Frishman

Integrated Pest Management programs attempt to minimize the use of pesticides by relying more heavily on other aspects of control. Inspections play an increasingly important role in such programs. By pointing out, for example, the need to modify construction so that floors, walls, ceilings, and machinery can be properly cleaned and by noting weaknesses in existing cleaning programs, quality assurance inspectors can minimize, if not totally eliminate, treating harborages with pesticides. This still results in eliminating the pests but often achieves that goal without the use of pesticides.

The mere fact of having to use a pesticide within a food facility generally signifies that the chain of events in the pest control program has a weak link that is permitting pests to enter and multiply. Inspectors are often able to find these deficiencies and point them out before a real problem develops.

No two inspections are alike. No two inspections even by the same inspector are identical. Try as one might to standarize the program, it is impossible, and rightfully so. However, the objective remains clear: To bring to the attention of responsible parties any actual or potential problems. These problems must then be acted upon promptly and in a positive manner.

Government inspectors, contract inspectors, and in-house inspectors all have the same objective. The bottom line is to help produce food fit for human consumption, wholesome and free from contaminants.

Frishman: AMF Pest Management Services, Inc., 30 Miller Road, Farmingdale, NY 11735.

Inspectors should think of themselves as part of a team. As a team player, they spark communication among all members of the team and bring a positive attitude to the job of inspecting.

A keen inspector looks not only for living pests but also for nonliving adulterants such as oil, glass, rust, paint chips, nuts, bolts, jewelry, and any other material that constitutes a contaminant.

The eye sees only what the brain tells it to. The material outlined in the chapters of this section will stimulate food inspectors to see more. Every time we do an inspection we learn something new, and that should help us to do that much better the next time out. With the advent of new technologies in the field—insect pheromones, radiology, and listening devices, for example—the findings of the inspector are becoming both more accurate and more valuable.

Think of the inspection report as a sales presentation. The inspection report attempts to bring out what needs to be done. How it is presented has a great deal to do with what gets done.

The information presented in this section gives us an excellent overview of an inspector's responsibilities and highlights many of the items to be considered during an inspection. The several checklists included and the regulatory aspects covered provide inspectors with a minimum set of requirements that can serve as a foundation for their investigations.

Great inspectors habitually aim higher than the minimum standards, but at the same time they are practical enough to balance their high standards against the realities of the business world.

531

VII
Management

46

Principles of Rearing Stored-Product Insects and of Insectary Management

R. W. Howe

There are numerous brief accounts of the methods used to rear storage insects for specific research projects and some excellent general directions for rearing the beetles and moths. The most useful accounts deal with storage beetles (13); beetles attacking whole cereals (30); pulse beetles (27); *Tenebroides mauritanicus,* cadelle (5); *Sitotroga cerealella,* Angoumois grain moth (30); *Plodia interpunctella,* Indianmeal moth (23); *Tineola bisselliella,* webbing clothes moth (21); *Galleria mellonella,* greater wax moth (14); *Hofmannophila pseudospretella,* brown house moth (9); and a range of moth species (2, 4, 28). There is also a vast literature on *S. cerealella* explaining how to harvest huge quantities of eggs for rearing parasitoids.

Two recent publications on insect rearing are particularly informative. One, a handbook (25), first deals with several general principles of culturing and then gives instructions on selected species that include a few storage beetles and moths. The other, the proceedings of a meeting in Atlanta, is mostly concerned with mass rearing (18). It is unlikely that sterile male techniques will ever be used in storage, but some species of moths and beetles could well be needed in abundance to emit sex pheromones and act as bait in traps. Also, very large numbers of insects might be needed occasionally in experiments with chemical control methods (16). Consequently, this series of papers that deals in detail with some of the universal problems of rearing will be cited frequently below.

Most of the papers noted above show that it is quite easy to rear many of the insects that commonly live indoors but that it is sometimes difficult to be sure of a consistent yield or regular supply of a particular stage of the life cycle. In this chapter, general principles that should be (but seldom are) considered for every species are discussed.

It is all too readily assumed that we can imitate the conditions that an insect meets in a building merely by confining it in a small jar. Inefficient methods of rearing can be tolerated when all that is needed are living or dead specimens for demonstration at an exhibition or in a classroom, but if the insects are needed for research, for experimental demonstrations, or for commercial use, such as animal food, some care is needed to ensure that suitable insects are obtained.

In general, rearing for experimental research illustrates all the features of management of an insectary producing numerous species. The first decisions to make are the numbers of insects needed, the stages of development required, their quality, and when and how often they will be needed. It is the demand for quality that raises the most problems.

Howe: Slough Laboratory, Ministry of Agriculture, Fisheries and Food, Slough SL3 7HJ, England. Current address: 28, Crossbush Road, Bognor Regis PO22 7LT, West Sussex, England.

Experimenters usually have two incompatible aims—to generalize from the results of their experiments and to minimize variation. The former requires the use of insects recently acquired from a range of representative sources; the latter, the use of inbred strains. Even recently collected insects rapidly adapt to culture conditions and some of their characteristics change. Thus, in the insectary, the ability to enter diapause is actively discouraged by breeding from individuals with rapid development; traits that are irrelevant there, such as cold-hardiness, are usually lost.

Genetic variability diminishes in culture (*1*) but there are techniques to maintain it (*17*). Storage species in general tend to retain their natural variability, possibly because the cultures are not so drastically different from the field conditions where they maintain multiple gene control of most characteristics. Cultures of *Sitophilus granarius* (granary weevil) have produced a steady yield over many generations (*15*).

Rearing policy must be governed by costs. The methods chosen depend upon the availability and relative costs of labor, culture media and other materials, space, and the intrinsic value of the insect stocks as well as on the demand for the species. In general, the system chosen is unlikely to be satisfactory from all points of view.

The basic necessities are a room or a chamber in which cultures can be held in a suitable environment, a supply of vessels to hold the cultures, and sufficient food to support the insect populations. The environment must be selected to ensure that the insects breed rapidly enough to meet the demand but not so fast as to be wasteful. The rearing system selected must prevent contamination by unwanted organisms and must be able to provide both the required numbers of insects at the right time and a way of disposing of unwanted stocks and waste. Finally, the safety of those working in the insectary must not be overlooked. Not only may they develop allergies to scales, frass, dust, and mites (*33*), but fungal spores may cause diseases such as farmer's lung (hypersensitivity pneumonitis). The risk of airborne hazards can be reduced by using filtration systems (*20, 34*).

Principles

General Points About Rearing Areas and Cultures

Researchers usually like to know the conditions under which their insects have been reared because these frequently influence experimental results. These influences are easier to assess if the stock has had a known constant history. Therefore, it is standard practice to rear insects at a fairly constant temperature, relative humidity (rh), and food moisture content, and to have a fixed schedule of light and dark periods.

However, even when the rearing room or environmental chamber is adequately controlled, temperature and humidity are seldom as constant in the cultures as might be supposed. This is because insect metabolism yields heat and water. Thus, according to the classic formula for respiratory metabolism, the breakdown of 2 grams of sugar yields just over 1 gram of water and well over 6 kilogram calories of heat. If this were retained in 100 grams of food, it would raise the moisture content by 1% and the temperature by well over 100°. Whether heat or water escapes more readily depends principally on the size of the culture. In a typical small culture in a jar, heat escapes and moisture does not, thus the medium tends to become moldy. Even so, the temperature in the center usually rises at least 2°.

In commercial bulks and in some larger-scale cultures, retention of heat raises the temperature sufficiently to kill insects that cannot migrate to safety. Mold develops in these circumstances because the temperature gradient causes a vapor pressure gradient, and moisture flows down it onto the cooler surfaces. Moldy cultures are a sign of overcrowding, probably the most common fault in rearing storage insects. A few species require molds in the diet and these should either be given damp food or have the food placed on top of a damp pad.

Insects from overcrowded cultures are usually smaller, lay fewer eggs, and grow more slowly than average and, in consequence, may not be fully satisfactory for research. In any event, having lived in a damp atmosphere, their behavioral patterns may be different from those of insects from a dry warehouse. Also, when there is a temperature gradient across a culture, the rearing conditions, especially for weevils, differ among individuals from a single culture.

Although heating is best avoided by rearing in sparse cultures, it can also be minimized by using shallow cultures and thus facilitating the escape of heat. Shallow cultures, of course, have larger surface areas and consequently take up more space (it may be feasible, however, to place the shelves closer together). Further, if the yield per culture is restricted, more cultures are needed to meet a stated demand. Sparse cultures, however, have a compensatory advantage where special age groups are required and double handling is necessary. Thus, small units are best, for instance, when pupae of a known age are needed and when there is a need to clear each culture completely of pupae twice and do it as quickly as possible.

Culture Environment

There is no need to discuss here the apparatus required to control light, temperature, and humidity (20, 24). Much is available, so there should always be something adaptable to local need at a suitable price. Light control needs to be cyclical, especially for species that enter diapause if they are not exposed to a long day. Many other species can be reared satisfactorily in darkness interrupted irregularly by staff switching on a light. Light control should be used for any species that appears to enter some kind of resting stage.

No universal constant temperature can be recommended since the suitable temperature ranges of stored-product species differ greatly. The most generally applicable temperature is 25°, though this is too low for some (e.g., *Trogoderma granarium*, khapra beetle) and too high for others (e.g., *Niptus hololeucus*, golden spider beetle). The most common fault is to choose too high a temperature. It is tempting to use the supposed constant temperature optimum for rapid multiplication when large numbers are wanted quickly, but when culture temperatures are some 2–3° above that of the room, a lower temperature is obviously preferable. If a cyclical regime is used, the highest temperature of the cycle should not exceed the optimum. When a continuous regular supply of stocks of a species is needed over a long period, a large number of cultures must be maintained simultaneously, each at a different stage of maturity. It is preferable to maintain these at a temperature of about 6° below the optimum for rapid multiplication even though this policy increases the number of cultures required. In general, the performance of insects so reared seems to be better than that of insects reared at the "optimum." This opinion contrasts with that expressed in some of the experimental literature where the use of suboptimal temperatures is described as subjecting the insects to "stress." When a stock is retained but not in current demand, a temperature some 10° below optimum should suffice, with concomitant savings on food media and labor.

Also, it is impossible to recommend a universal relative humidity for rearing areas, though for most species a range of 60–65% is satisfactory. At higher humidities, especially in crowded cultures, the development of fungi is inevitable and the risk of contamination by mites is high. However, 60% rh is too low for the satisfactory development of many species, especially if they are reared on a food of low equilibrium moisture content. This can be raised for a given relative humidity by adding glycerol to the food. It should be noted that this implies a lower relative humidity for the same food moisture content. It is probably advantageous for the wall of the room or environmental chamber to be moisture absorptive to buffer humidity changes and so decrease the sensitivity of the humidstat required. Condensation is likely on nonabsorbent surfaces, but such surfaces are easier to keep clean and unlikely to provide harborage for escaped or unwanted organisms.

Culture Vessels

Convenience and cost are the chief considerations in the choice of containers to hold cultures. Generally, clear (transparent) vessels are desirable because these permit light to reach the insects as well as allow for the culture inside to be seen. Glass jars are convenient and can be cheap if made in quantity for industry, but they are heavy, need to be washed thoroughly before and after use, and break easily. Various kinds of plastic vessels are lighter and less prone to break, but some become opaque quickly if insects score the inner surface; others cannot be put into an oven after use to kill off old cultures. Disposable containers overcome these problems because they do not need to be washed; if cheap enough, they are probably the best choice for a new venture where a routine for dealing with glass has not been designed. Burton and Perkins (7) discuss the merits of various kinds of containers.

The size of the vessel clearly determines many other aspects of a rearing program. Since it is desirable to limit both the density of culture populations and the depth of the culture medium to about 3 cm, it is best to use shallow vessels about 15 cm in diameter. If lack of space makes this impossible, it may be necessary to run the risks of heating and molds by using deeper cultures. However, small cultures are easier to handle and the insects in them can be collected and counted quickly. If organisms of a particular stage or age are needed, they can be cleared quickly from a small culture. If large numbers are needed, there might be quality or genetic differences between small cultures that would demand strict randomization of the division of insects into experimental groups.

The shape of the vessels is not important provided that a space can be left between them. Rectangular vessels fit most spaces more efficiently than round ones, but they should still have circular tops for ease of closure. Obviously, the vessels should be squat in shape (not tall) unless the cultured species requires a lot of air space above the food.

The surface of a culture on which adults walk can be increased by placing folded or crumpled paper or cotton wool in the culture. When used in this way, paper also makes it easier to remove a substantial proportion of the larvae or adults.

Cultures must be closed to confine the reared individuals and to exclude unwanted contaminants, but the closure must only minimally restrict air exchange. When the culture is handled frequently, the cover should be a porous cloth such as muslin or organdy held in place by a rubber band or a screwtop lid, or by a press-top lid in which a large hole has been cut. For a culture that will not be disturbed, the most useful closing material is filter paper, preferably black, or if there is a risk of microbiological contamination, a bacterial sporeproof paper. These can be sealed in position with paraffin. The pores of these papers allow free gas exchange but prevent the passage of mites. If mites are present on the outside of the vessels, they show up clearly against the black paper.

Almost all storage species are capable of boring through paper or cloth and usually make frequent attempts to do so when they are overcrowded. Should the species being reared be one of those prone to bore out even when not overcrowded, then these closures must be reinforced with fine metal gauze held in place by rubber bands or screwtop lids. The gauze can be soldered to metal lids.

Insects and mites that escape from cultures or invade the insectary from outside can be prevented from contaminating stock cultures. To achieve this, all cultures should be placed in trays of light mineral oil, preferably placing them on a stand, such as a can lid or saucer, so that they do not get oily. This will not prevent flying adults from reaching the cultures nor stop mites from falling onto them. The cultures in the rearing rooms should be grouped by species, each rack, shelf, chamber, or room devoted to a particular species, and each species arranged so that the technician can keep track of the age of each culture. It should be easy to identify the cultures that are to be replaced by new ones and those of the right age for use in experiments.

The insectary manager should have a wall planner that lists the dates when existing cultures were prepared and are due for use or destruction and when new cultures must be prepared.

Foodstuffs

Except for the parasitoids, stored-product insects tend to be scavengers and most eat almost any vegetable or animal product whether alive or dead. Almost the only materials to avoid as culture media are those that are very oily or deliquescent, and even these can be mixed with a material that will absorb the liquid. Oilseeds such as peanuts are good food for many species provided they are not minced. Insects in culture on dried fruits sometimes drown when the fructose deliquesces. Otherwise, most species tolerate a wide range of carbohydrate and protein content; the lack of B vitamins or sterols may restrict the success of some species that lack the gut symbionts to manufacture them. The addition of dried yeast (brewers' or bakers') at 5% by volume or weight usually provides the B vitamins, but where a species (e.g., *Dermestes maculatus*, hide beetle) requires a sterol allied to cholesterol, it is best to ensure that the diet includes an animal constituent such as meat meal.

Although the use of artificial diets is advocated for field insects (6, 24), for most storage species it is preferable to use a simple, natural food material or a mixture of several, possibly augmented with yeast. If a large-scale rearing program of a single species is envisaged, it may be worth creating a complex diet (23) or even a semisynthetic one. But where many species are bred, the simpler the food regimes and the fewer the commodities to be bought, the easier is the management. When new species have to be reared, or when a species proves difficult, it is instructive, especially when few individuals are available, to provide a mixture of all the available foodstuffs, plus any from which the species was collected, and to examine the spent culture to determine what was eaten.

The food materials are, however, not merely nutritive but may also constitute a part of the environment. The particle size and packing of the medium determine how easily the insects can eat the food and burrow into it. There is probably an ideal particle size for ease of feeding by individuals of each stage and each species. Some seeds may be too hard or too smooth for the insects to penetrate.

The packing of the food medium determines the extent to which insects remain near the surface or burrow into it, especially in deep cultures. If the particles are spherical and uniform in size, the food is easy to handle and sieve, but the food gradually compacts and the insects are more likely to remain on or near the surface. The incorporation of flaky particles (flaked maize, rolled oats) loosens the medium and enables insects to penetrate it, but this can make handling frustrating. If it is necessary to sieve off eggs, larvae, or pupae, it is better to use a shallow layer of fine powdery medium and increase the surface area for adults by placing paper concertinas on the surface.

It is desirable to purchase foodstuffs in bulk; all such materials must be free of insecticide contamination. The quality and moisture content must be checked upon receipt. A typical range of foods includes a cereal grain for *Sitophilus* species, pulses for bruchids, wholemeal flour or some other convenient milled product as a general-purpose food, and meat meal for insects needing cholesterol. Other useful ingredients are peanuts, rolled oats, dried yeast, honey, glycerol, and sultanas. Complex diets can be troublesome to prepare but can be worthwhile where vast numbers of only one or two species are reared and a routine of preparation can be organized.

When food materials are bought in large quantities, they must be disinfested immediately and stored so that they are protected from later infestation. Both ends may be achieved by keeping foods in a freezer at $-18°$ for at least 1 month before use. Food can also be kept in closed jars in cool rooms, but if there is then any risk of contamination by insects or mites, it should be heat treated at $60–70°$ for 3 or 4 hours. This, unfortunately, alters the moisture content/relative humidity relationship of the food; therefore, enough glycerol should be added to raise the moisture content by about 1% at the rearing humidity. In general, it is better to control moisture content by exposing food to a damp atmosphere than by adding water. The measures mentioned here are aimed at reducing and preventing contamination by pests. Autoclaving is the best way to eliminate fungi and bacteria in foods before they are used as culture media (22). The use of chemicals to inhibit growth of these microorganisms in insect cultures is undesirable because the chemicals may also affect the insects.

Organization
General Points
Rearing and research should be coordinated so that each helps the other. The objective of an insectary is to provide insects when they are needed. Thus, it is necessary to know what stages and how many of them will be available at various times in the future. In particular, the expected yield, its consistency, and how it will be spread over the life of the culture must be known. The research both depends upon and helps to supply the answers to these needs.

The results of research on life-cycle parameters can be used to plan the rearing of insects in numbers for projects such as insecticide testing (12, 16). This is eminently practical for storage insects for which there is a lot of the right kind of information readily available today. Given sufficient knowledge of this sort, the insectary manager can draw up a list of cultures to be prepared day by day and plot both the stage of development and the planned fate of each culture. Given rearing rooms at different temperatures, the manager should be able to meet emergency demands with relatively little notice if he keeps a small excess of cultures at a low temperature. These he can then speed up by moving them to a higher temperature. It is equally feasible to delay adult emergence by moving a culture to a lower temperature.

Identity of Cultures
In an insectary that handles many species, care should be taken not to mix up cultures of species that are very similar in appearance. Whenever a new culture is prepared, a few of the specimens should be taken out and positively identified. As a further precaution, where several very similar species occur within a genera, such as *Cryptolestes,* only one species should be handled in one room on any day and then all equipment should be thoroughly cleaned and sterilized to make sure that no insects remain to contaminate the next culture prepared.

Procedures
The easiest way to start a culture is to place a known number of young adults on a weighed amount of food. For many beetles, this procedure yields reasonable results. Some calculation of the number of offspring to be expected should be made. For moths and for beetles that live only a week or two and lay 50–100 eggs, the adults can be allowed to die in culture; usually it is reasonable to assume that about half of the adults are females. The number of adults and the quantity of food must be balanced to give the desired culture density and yield. If the beetles live for a long time, they must be removed

from the culture before they lay too many eggs; also, the likely yield of live offspring per female per week should be calculated. It is essential to avoid overcrowding. This leads to retarded development and small size. Tolerance limits should be set for the characteristics required in the insect populations (8).

The most uncertain facet of rearing is the fertility of eggs, especially for moths. A consistent culture yield can be obtained by putting freshly hatched larvae rather than adults on the food medium. Moths that live only a week or two generally lay eggs readily in the absence of food, but they also often require plenty of space for flying. These adults are, therefore, placed in a vessel such as a pint jar which is covered with a fine gauze and inverted over a petri dish in which a black filter paper is placed. Eggs laid through the gauze fall onto the filter paper. A wet pad for drinking may increase the number of eggs laid. With photoperiod control, egglaying can be concentrated into the early part of the period of darkness (3).

Cultures of other species that must be placed on the food can also be set up with eggs or freshly hatched larvae, though this is seldom necessary. With these, the eggs are removed from the food by using a sieve of suitable mesh. The eggs of some species (e.g., *Oryzaephilus surinamensis,* sawtoothed grain beetle) will not hatch after being subjected to harsh vibrations. To ease the task of finding the eggs among coarse food particles, the food must be ground and passed through a much finer sieve than the one used to remove the eggs. Many workers prefer to anesthetize adult moths and active beetles with carbon dioxide and handle them while they are quiescent. This may be essential if large numbers are involved. Robust beetles can be removed from foods with a suitable sieve and nearly all insects can be handled with a gentle suction tube. With a narrow-diameter tube, the insect can be held at the tip; with a wider tube, it can be gently trapped in a bypass reservoir. I prefer to handle all stages with a fine brush whenever practical.

Precautions

There are a few hazards that can destroy cultures or make them useless. Most storage species are scavengers and/or predators and will eat any food they find, including their own species. Vulnerable stages, especially eggs and pupae, must have refuges that provide protection. Corrugated cardboard or paper is satisfactory for moths, especially if narrow strips are formed into rolls. Sheets of cork or other soft material that wandering beetle larvae can bore into is another useful device (5).

Moth larvae spin a great deal of silk. This can hamper the movement of adults and the handling of cultures. To avoid accumulations of webbing, moth cultures should be limited to a single generation.

Although almost any storage species can invade cultures of another, the most common contaminants are mites and psocids. If the mite is a predator such as *Pyemotes* species, the cultures are best destroyed. The measures that must be taken to avoid the initial infestation from food, apparatus, or from other cultures in the rearing room have already been stressed. Contaminated cultures should be retained only if they are valuable. It is possible to wash robust beetles but this treatment is very drastic and not always successful. After being subjected to a stream of water in a sieve, the beetles are put on filter or blotting paper to dry and then placed on clean food for a couple of days. The whole process is repeated one or more times and each lot of food is kept. This food is discarded if mites appear or no larvae develop, but otherwise it can be used to support new cultures. The exterior surfaces of all culture jars should be washed frequently with alcohol.

Pesticides can be used against mites. An infestation of *Acaropsis* species in *Sitophilus* cultures was controlled by spraying shelves and walls with a 1:1 mixture of undiluted emulsifiable concentrate (EC) of dicofol and chlorbenzilate, followed by airing for 2 days (29). Cloth was soaked in 18.5% dicofol EC, dried, and used to cover the cultures and was spread out on the shelves. Liberal quantities of dicofol wettable powder were mixed with wheat, held for 24 hours, and then removed by sieving before the wheat was used.

The most serious hazards, however, are diseases caused by bacteria and sporozoans. If larvae turn black and die in cultures, disease may be suspected. Diagnosis can be confirmed by finding spores in a paste made by grinding larvae in a little water. More certain diagnosis can be achieved by starving the larvae for 2 days (to get rid of food in the gut), grinding them into a paste in a little water, and then digesting them with trypsin at 30° for 24 hours. The digested solution is strained through a 200-mesh gauze in a centrifuge at 14,000 rpm for 10 minutes, and then the solids are examined for spores.

If positive results are found, all glass apparatus should be washed in chromic acid, and subsequent cultures should be prepared with eggs that have been washed with a germicidal detergent such as benzalkonium chloride. Insects may be washed in a detergent solution to remove sporozoan spores (*31*). Eggs of *Plodia interpunctella* may be soaked in 2.5% formalin for 5 minutes to protect them from granulosis virus (*26*). Infected rooms can be sterilized with ultraviolet light. Goodwin (*11*) provides extensive keys to disease symptoms and the fungi causing them.

Diseases are better avoided than treated, so every insectary should have an effective quarantine system. Every new stock, whether from warehouses or another laboratory, should be put in an isolated quarantine insectary, thoroughly examined for diseases and contaminants, and reared there for at least two generations. It should not be passed to a central insectary until its health is ensured.

Insects Bred as Animal Food

The larvae of *Tenebrio* species and other large larvae are sometimes reared as food for birds and other animals. The marketed product is not the fertile stage, so part of the yield must be retained to lay eggs for the next generation. Rapid growth of *Tenebrio* larvae to a size suitable for food can be stimulated by rearing them at high temperature, but those then kept to pupate do not do so readily. For this kind of commercial breeding, two separate stocks must be kept, one at 30° and a much smaller one at 25°. All of the stock at 30° can be harvested; the stock at 25° is used to maintain egg laying for the next generation at both temperatures.

Contamination, especially with *Tribolium* species, can be a problem with large-scale rearing of *Tenebrio* species; good hygiene on this larger scale is not easy to achieve. The introduction of contaminant species can be largely avoided by keeping the food stocks at as low a temperature as possible and by protecting the rearing stock at 25° with the full list of insectary precautions. Because the manufacturing stream might become infested, this should be separated into units to provide a continuous stepped supply of mealworms, and each unit should be thoroughly treated as the mealworms are harvested to ensure that no unwanted insects remain.

Information on Species

The above account has not drawn attention to methods used for particular species, but the following notes are more specific. Rearing *Sitophilus* species, described in detail by Strong et al.

(*30*), is one of the most straightforward of tasks. However, the work of Strong and co-workers reveals the dependence of yield on the quantity of food used (yield from 7,000 *S. granarius* on wheat was only 10,000 each week for 3 weeks). Howe and Hole obtained 3,500 weevils from 160 adults on 320 grams of wheat in 3 weeks (*15*) and also about one egg per adult per day from 500–1,000 adults left on wheat for both 1- and 3-day periods (*16*). Large numbers of *Rhyzopertha dominica* (lesser grain borer) are apparently more difficult to rear, possibly because the females lay batches of eggs irregularly. Strong and colleagues (*30*) seeded cultures with 9,000 adults and did not remove them. Their eventual yield was 29,000. This species can be maintained on wholemeal flour, though it develops there less quickly than on wheat, but even on whole cereals, adult feeding generates a lot of floury dust.

Bruchid larvae, like weevils, develop within seeds, with cowpeas or kidney beans the most satisfactory. The adults are active but live only a week or two. Each female lays 50–100 eggs. Strong and co-workers (*27*) reared *Acanthoscelides obtectus* (bean weevil) on red kidney beans and *Callosobruchus maculatus* (cowpea weevil) on cowpeas, both at 27°.

There is surprisingly little critical literature about rearing the other important storage beetles. Bond and Monro (*5*) point out that *Tenebroides mauritanicus* needs yeast and moldy food to lay well and even then must be able to oviposit into crevices if the eggs are to escape being eaten. The pupae will also be eaten if the larvae are not provided cork into which to bore before pupating. Although *Araecerus fasciculatus* (coffee bean weevil) breeds as well in maize as in coffee or cocoa beans, Vitelli et al. (*32*) formulated a highly complex diet for this species.

Moths are more difficult to rear and have stimulated more interest. Again, Strong and colleagues (*28*) give a sound, basic method involving a complex food medium. For the phycitines, they seed cultures with eggs, but for *Sitotroga cerealella*, they release adults onto a large surface area of grain (*30*). The fertile eggs of this species turn red before hatching; if such eggs are placed on grain, we can be confident of their viability. Gonen and Donahaye (*10*) stipulate a culture depth of 3 cm for phycitines. Several authors advocate adding weighed quantities of eggs to the medium (*2, 23, 28*); volumetric egg scoops may also be useful. Some researchers (*2, 19*) recommend a simple diet of wheatfeed (fine wheat bran), but others depend on complex diets. Bell and Walker (*3*) give a good description of an apparatus for obtaining eggs. In general, there is a good agreement among authors on the principles of rearing moths; variations are confined mostly to minor details.

Perspectives from Slough Laboratory

The Slough Laboratory has maintained stocks of the beetle and moth pests of stored products for over 60 years and now has cultures of about 100 species. We rely on simple procedures and use only three rearing temperatures, 20°, 25°, and 30°, but put great emphasis on quarantine before admitting new stock into the rearing rooms and on preventing contamination by unwanted species in these rooms.

As far as possible, the basic foods used in quantity are restricted to wheat, wheatfeed, rolled oats, fishmeal, and yeast powder; a few other foods are needed for certain species. The most important element for successful rearing is getting the right moisture balance. Adults of some species must be able to drink water if they are to lay enough eggs for a successful culture. Water is usually provided as a corked tube with a paper wick or, occasionally, as damp blotting paper. This is essential for all spider beetles and *Dermestes* species, and is also given to *Caryedon serratus*, *Pharaxonotha kirschi*, *Tenebrio* species, cockroaches, and to all pyralid moths. Whenever the moisture content of the food is critical, glycerol is added to raise the moisture content at room humidity; this is standard practice for the wheatfeed on which the moths are reared.

For the many species that require or benefit from fungi in the diet, a damp pad of cotton wool is placed on the food or in the jar before the food is added. Species for which this is done are *Pyralis farinalis* (mealmoth), *Alphitobius* spp., *Alphitophagus bifasciatus* (twobanded fungus beetle), *Murmidius ovalis*, *Palorus subdepressus* (depressed flour beetle), and *Typhaea stercorea* (hairy fungus beetle).

The choice of diet depends upon the products commonly attacked in the field, wheatfeed being the mainstay for those found on vegetable products, fishmeal for those found on animal products, and a mixture of the two for those found on both. Wheat is used for those species that typically live within a cereal grain, but maize is needed for *Araecerus fasciculatus* and *Prostephanus truncatus* (larger grain borer). An appropriate pulse is needed for bruchid species

except for *C. serratus* which requires peanuts. Yeast powder is always added to diets based on fishmeal, and though it is less essential with wheatfeed, it is often useful in that medium for increasing the yield and rate of development.

Wholemeal flour plus yeast is used for the more common *Tribolium* species to facilitate handling; rolled oats are mixed with wheatfeed whenever the air spaces in the medium need to be increased. *Pyralis farinalis*, *Typhaea stercorea*, and the *Trogoderma* species are also given fishmeal. Other species that are bred on a wheatfeed/fishmeal/yeast mixture are *Gnatocerus cornutus* (broadhorned flour beetle), *Sitophagus hololeptoides*, *Anthrenus sarnicus*, *Attagenus* spp., *Tribolium* spp., spider beetles, and cockroaches. *Tenebroides mauritanicus* is given cork in which it can lay eggs and pupate. A mixture of wheatfeed, rolled oats, yeast, and peanuts is used for *Lophocateres pusillus*.

The fabric pests of the genera *Anthrenus*, *Tinea*, and *Tineola* get flannel with fishmeal and yeast. *Dermestes* and *Necrobia* species get fishmeal and yeast with pieces of bacon. For *Carpophilus hemipterus* (driedfruit beetle), the medium consists of sultanas mixed with rolled oats; *C. ligneus* gets the same mixture but the fruit is first allowed to ferment. Glucose and glycerol have been used for some of the strains of *Cadra cautella* (almond moth), *C. figulilella* (raisin moth), and *C. calidella*. Honey is added to the media for rearing species of the gallerine genera *Achroia* and *Galleria*.

It is a good idea to prepare a mixture of a variety of products for species whose tastes are unknown. I found that a mixture of wheat, maize, pulses, wheatfeed, rolled oats, peaflour, peanuts, fishmeal, yeast, and maize flakes supported many species of *Tinea* collected from a variety of habitats for long enough to decide that they were of minor importance. Some of the diets commonly used at Slough Laboratory are summarized in Table 46.1.

Cited References

1 Bartlett, A. C. 1984. Genetic changes during insect domestication. In *Advances and Challenges in Insect Rearing*, E. G. King and N. C. Leppla, eds. U.S. Department of Agriculture, New Orleans, LA.

2 Bell, C. H. 1976. Effect of cultural factors on the development of four stored-product moths. *Journal of Stored Products Research* 12(3)185–193.

3 Bell, C. H., and D. J. Walker. 1973. Diapause induction in *Ephestia elutella* (Hübner) and *Plo-*

dia interpunctella (Hübner) (Lepidoptera, Pyralidae) with a dawn-dusk lighting system. *Journal of Stored Products Research* 9(3)149–158.

4 Boles, H. P., and F. O. Marzke. 1966. Lepidoptera infesting stored products. In *Insect Colonization and Mass Production*, C. N. Smith, ed. Academic Press, New York, NY.

5 Bond, E. J., and H. A. U. Monro. 1954. Rearing the cadelle *Tenebroides mauritanicus* (L.) (Coleoptera: Ostomidae) as a test insect for insecticidal research. *Canadian Entomologist* 86(9)402–408.

6 Brewer, F. D., and O. Lindig. 1984. Ingredients for insect diets. Quality assurance, sources, and storage and handling. In *Advances and Challenges in Insect Rearing*, E. G. King and N. C. Leppla, eds. U.S. Department of Agriculture, New Orleans, LA.

7 Burton, R. L., and W. D. Perkins. 1984. Containerization for rearing insects. In *Advances and Challenges in Insect Rearing*, E. G. King and N. C. Leppla, eds. U.S. Department of Agriculture, New Orleans, LA.

8 Chambers, D. L., and T. R. Ashley. 1984. Putting the control in quality control in insect rearing. In *Advances and Challenges in Insect Rearing*, E. G. King and N. C. Leppla, eds. U.S. Department of Agriculture, New Orleans, LA.

9 Cole, J. H. 1962. *Hofmannophila pseudospretella* (Stnt.) (Lep., Oecophoridae), its status as a pest of woollen textiles, its laboratory culture and susceptibility to mothproofers. *Bulletin of Entomological Research* 53(1)83–89.

10 Gonen, M., and E. Donahaye. 1973. An improved technique for rearing the tropical warehouse moth *Ephestia cautella* (Walker) (Lepidoptera; Phycitidae). *Israel Journal of Entomology* 8:179–181.

11 Goodwin, R. H. 1984. Recognition and diagnosis of diseases in insectaries and the effects of disease agents on insect biology. In *Advances and Challenges in Insect Rearing*, E. G. King and N. C. Leppla, eds. U.S. Department of Agriculture, New Orleans, LA.

12 Haile, D. G., and D. E. Weidhaas. 1984. Systems analysis and modeling in mass rearing and control of insects. In *Advances and Challenges in Insect Rearing*, E. G. King and N. C. Leppla, eds. U.S. Department of Agriculture, New Orleans, LA.

13 Harein, P. K., and E. L. Soderstrom. 1966. Coleoptera infesting stored products. In *Insect Colonization and Mass Production*, C. N. Smith, ed. Academic Press, New York, NY.

14 Hartley, G. G., et al. 1977. *Equipment for Mass Rearing of the Greater Wax Moth and the Parasite* Lixophaga diatraeae. Report ARS-5-164. U.S. Department of Agriculture, Washington, DC.

15 Howe, R. W., and B. D. Hole. 1967. The yield of cultures of *Sitophilus granarius* at 25°C and

Table 46.1

Diets containing wheatfeed as the major constituent

Parts of

Wheatfeed	Additives	Species reared
10	1 yeast	*Pharaxonotha kirschi,* *Gnatocerus maxillosus* (slenderhorned flour beetle), *Lasioderma serricorne* (cigarette beetle), *Stegobium paniceum* (drugstore beetle), *Palorus ratzeburgii* (smalleyed flour beetle)
10	1 yeast (+ damp pad)	*Alphitobius diaperinus* (lesser mealworm), *Palorus subdepressus* (depressed flour beetle)
10	1 yeast 2 glycerol	phycitine moths
5	1 yeast 5 rolled oats	*Cryptolestes* spp., *Oryzaephilus* spp., *Tenebrio* spp., *Ahasverus advena* (foreign grain beetle), *Cathartus quadricollis* (squarenecked grain beetle)
5	1 yeast 5 rolled oats 1 peanuts	*Lophocateres pusillus,* *Tenebroides mauritanicus* (cadelle) (with cork)
10	5 rolled oats (+ damp pad)	*Murmidius ovalis*
8	4 fishmeal 1 yeast	ptinid beetles, *Gnatocerus cornutus* (broadhorned flour beetle), some *Tribolium* spp., some *Attagenus* spp.
5	5 rolled oats 2 fishmeal 1 yeast	*Sitophagus hololeptoides,* *Palembus* spp., *Attagenus fasciatus* (wardrobe beetle), some *Trogoderma* spp., cockroaches
8	8 fishmeal 1 yeast 1 cholesterol	*Anthrenus sarnicus*

NOTE: Wheatfeed = wheat shorts (fine bran plus small quantities of endosperm).

70 per cent relative humidity with some observations on rates of oviposition and development. *Journal of Stored Products Research* 2(4)257–272.

16 Howe, R. W., and B. D. Hole. 1967. Predicting the dosage of fumigant needed to eradicate insect pests from stored products. *Journal of Applied Ecology* 4(2)337–351.

17 Joslyn, D. J. 1984. Maintenance of genetic variability in reared insects. In *Advances and Challenges in Insect Rearing*, E. G. King and N. C. Leppla, eds. U.S. Department of Agriculture, New Orleans, LA.

18 King, E. G., and N. C. Leppla (eds.). 1984. *Advances and Challenges in Insect Rearing.* U.S. Department of Agriculture, New Orleans, LA.

19 Navarro, S., and M. Gonen. 1970. Some techniques for laboratory rearing and experimentation with *Ephestia cautella* (Wlk) (Lepidoptera, Phy-

citidae). *Journal of Stored Products Research* 6(2)187–189.

20 Owens, C. D. 1984. Controlled environments for insects and personnel in insect-rearing facilities. In *Advances and Challenges in Insect Rearing,* E. G. King and N. C. Leppla, eds. U.S. Department of Agriculture, New Orleans, LA.

21 Pence, R. J. 1958. A technique for rapid rearing of clothes moth eggs. *Journal of Economic Entomology* 51(6)919–921.

22 Shapiro, M. 1984. Micro-organisms as contaminants and pathogens in insect rearing. In *Advances and Challenges in Insect Rearing,* E. G. King and N. C. Leppla, eds. U.S. Department of Agriculture, New Orleans, LA.

23 Silhacek, D. L., and G. L. Miller. 1972. Growth and development of the Indian meal moth, *Plodia interpunctella* (Lepidoptera: Phycitidae), under laboratory mass-rearing conditions. *Annals of the Entomological Society of America* 65(5)1084–1087.

24 Singh, P. 1984. Insect diets: Historical developments, recent advances, and future prospects. In *Advances and Challenges in Insect Rearing,* E. G. King and N. C. Leppla, eds. U.S. Department of Agriculture, New Orleans, LA.

25 Singh, P., and R. F. Moore (eds.). 1985. *Handbook of Insect Rearing,* 2 vols. Elsevier, Amsterdam, The Netherlands.

26 Spitler, G. H. 1970. Protection of Indian-meal moth cultures from a granulosis virus. *Journal of Economic Entomology* 63(3)1024–1025.

27 Strong, R. G., G. J. Partida, and D. N. Warner. 1968. Rearing stored-product insects for laboratory studies: Bean and cowpea weevils. *Journal of Economic Entomology* 61(3)747–751.

28 Strong, R. G., G. J. Partida, and D. N. Warner. 1968. Rearing stored-product insects for laboratory studies: Six species of moths. *Journal of Economic Entomology* 61(5)1237–1249.

29 Strong, R. G., G. R. Pieper, and D. E. Sbur. 1959. Control and prevention of mites in granary and rice weevil cultures. *Journal of Economic Entomology* 52(3)443–446.

30 Strong, R. G., D. E. Sbur, and G. J. Partida. 1967. Rearing stored-product insects for laboratory studies: Lesser grain borer, granary weevil, rice weevil, *Sitophilus zeamais,* and Angoumois grain moth. *Journal of Economic Entomology* 60(4)1078–1082.

31 Tyler, P. S. 1962. On an infection of *Tribolium* spp. by the sporozoan *Triboliocystis garnhami* Dissanaike. *Journal of Insect Pathology* 4(1)270–272.

32 Vitelli, M. A., H. N. Nigg, and R. F. Brooks. 1976. Laboratory rearing of the coffee bean weevil. *Florida Entomologist* 59(3)301–303.

33 Wirtz, R. A. 1984. Health and safety in arthropod rearing. In *Advances and Challenges in Insect Rearing,* E. G. King and N. C. Leppla, eds. U.S. Department of Agriculture, New Orleans, LA.

34 Wolf, W. W. 1984. Controlling respiratory hazards in insectaries. In *Advances and Challenges in Insect Rearing,* E. G. King and N. C. Leppla, eds. U.S. Department of Agriculture, New Orleans, LA.

47
Education Strategies for Food-Pest Managers

Harold George Scott

Training of personnel in the principles and practices of stored-food pest control has been neglected in the United States as well as in other countries of the world. The resultant loss in human food supplies and in nutrition benefits is tragic, as is the needless exposure of humans and their domestic animals to unsafe and unsavory conditions (1).

Nutrition Loss

Half the nutritional and economic value of foods harvested by humans is lost between the farm and the table—during transport, storage, and preparation. These losses are caused by arthropods, rodents, birds, and molds, and by other factors including poor manufacturing practices, spillage, and waste. At a time of world food crisis—when Africa, Latin America, Asia, Europe, and the USSR all are dependent upon food imports—it is unthinkable that such largely preventable losses should be allowed to undermine efforts to increase world food supplies. Nevertheless, insufficient effort is being expended to effectively manage the problem, so the effect continues (2).

Direct Effects

The presence of stored-food pests and/or their by-products represents a highly undesirable kind of food pollution and causes significant dilution of food so that the contents are no longer representative of label claims.

Scott: *Program in Community Medicine, School of Medicine, Tulane University, New Orleans, LA 70112.*

Some pests, especially rats, can cause extensive damage to stored-food buildings, equipment, and packaging, whereas others, such as mites and rats, may bite humans. Rat-bite cases constitute a major health problem in many urban areas. In addition, certain fly, beetle, and moth larvae (as well as mites) may infest human tissues and/or body cavities.

Many stored-food pests cause extreme annoyance and some people are troubled by phobias—excessive fear of insects, rats, and other pests. At the very least, the presence of pests or their by-products in foods represents an unsavory and undesirable situation for the people destined to consume those foods (4).

Mechanical Transmission of Disease Organisms

Mechanical transmission is said to have occurred when stored-food pests carry "germs" from contaminated materials, active cases, or intermediate-reservoir animals to uninfected hosts, the "germs" neither changing life form nor multiplying significantly during the transfer.

Many stored-food pests are attracted to feces, urine, vomitus, sputum, feathers, lint, dust, and nesting materials as well as to garbage, spoiled or moldy foods, dead animals, and sewage. Their mouthparts, bodies, feet, hairs and feathers become contaminated from these materials. Later, they may walk over human food, utensils, or food-contact surfaces, depositing "germs." In addition, pests may transport "germs" in their digestive tracts and deposit contaminated feces as they move about. House flies, cockroaches, rodents, and birds are the stored-food pests most prominently involved in mechanical transmission of disease agents to humans, with enteric pathogens such as those causing dysenteries and diarrheas being the most commonly transmitted mechanically (4).

Biological Transmission of Disease Organisms

Biological transmission is said to have occurred when "germs" multiply, transform, or pass through a cycle in a vector, that vector serving as an essential host and part of the transmission cycle. Thus, the organisms causing plague and murine typhus multiply prodigiously in the alimentary tracts of vector fleas that carry these "germs" from reservoir rodents to humans. House mouse mites (*Lyponyssoides sanguineus*), which have fed upon and become infected from house mice, can transmit the rapidly reproducing organisms causing rickettsialpox from mice to humans. Cockroaches, as well as flea and beetle larvae, may ingest certain tapeworm eggs which can hatch and develop into infective stages within these insects. Humans and/or domestic pets can become infected when adult insects, present in infested foods, are accidentally ingested (4).

Training Requirements

General

Several areas of training emphasis are required to overcome inadequate preparation of personnel with regard to the principles and practices of stored-food pest control. Within each of these areas, massive amounts of technical information must be organized and presented in an understandable and usable form (6).

Awareness

Stored-food pests have a pronounced effect upon every human being. It is therefore essential that development of awareness by the general public be a major element in the training programs. Managerial, governmental, and technical personnel can only do so much to minimize the problem. It is up to members of the general public to do the rest.

Managerial

Individuals involved in food management—financing, production, transportation, storage, and utilization—should be fully aware of the severe problems created by stored-food pests and of how these problems can be mitigated. Since most managers have little or no knowledge of the subject and others exhibit incomplete and/or incorrect comprehension, a major element in food-pest control training must be education of all managerial personnel.

Governmental

Regulatory aspects of stored-food pest control are performed by government employees—federal, state, and local. In order to be effective and to stay on a sound legal footing, all government personnel involved in this activity should have in-depth knowledge of stored-food pest control. It is the responsibility of the agencies involved to be certain that their personnel are so trained.

Technological

Professional and subprofessional personnel who daily work in the technology of stored-food pest control must have the same in-depth knowledge required of government regulators. It is the responsibility of the employers of these individuals to ensure that such is actually the case.

Subject Matter

General

Depth of knowledge requirements differ for the various emphasis areas, but the following subjects should be covered for each (6):

- Extent, causes, and prevention of stored-food pollution and loss.
- Human health conditions associated with stored food and its loss.
- Systems of food management and their relationships to prevention of stored-food losses.

Nutrition Losses from Stored Food

Training in this aspect should explore losses occurring in stored food—those resulting from loss of food volume and most especially those resulting from nutritional degradation of the food. It should discuss methods for loss management and minimization. A sample schedule might include, but need not be limited to:

- The World Food Crisis
- Nutritional Values: The Pure and the Polluted
- From Farm to Table
- Losses Due to Insects and Other Arthropods
- Losses Due to Rodents and Other Mammals
- Losses Due to Birds
- Losses Due to Mold and Other Microorganisms
- Losses Due to Spillage and Waste
- Losses Due to Poor Employee and Manufacturing Practices
- Losses Due to Other Factors
- Measurement of Losses
- Minimization of Losses.

Stored-Food Management

Efficient and effective management is essential to stored-food protection, and the protection is essential in order to feed the human population of the Earth. Training in this area should explore management processes presently in use and compare them with more ideal systems that are being used in selected areas. A schedule might include but not necessarily be limited to the following:

- Food Production
- Farm Storage
- Elevators
- Warehouses
- Mills and Other Manufacturing Plants
- Storage of Rapidly Perishable Foods
- Food Transport
- Food Processing
- Food Packaging
- Food Standards and Grading
- Quality Control
- Food Regulation
- Epidemiology of Infestation and Pollution
- Control of Infestation and Pollution
- Organized Food Management
- Stored Food and World Hunger.

Stored-Food Pests

Essential to protection of stored foods is efficient, effective, and continuous detection, identification, evaluation, and control of stored-food pests. A schedule for training in these areas might include the following:

- Human Illnesses Associated with Stored-Food Pests
- Ecology of Food Storage and Stored-Food Pests
- Arthropod Pests: Morphology; Larval and adult taxonomy; Species habits, occurrence, distribution, and depredations; Beetles (pests, predators, scavengers); Moths; Flies (pests, predators, scavengers); Cockroaches; Other arthropods (e.g., mites, psocids, silverfish, wasps, ants, earwigs, springtails, bugs, millipedes)
- Vertebrate Pests: Morphology; Taxonomy; Species habits, occurrence, distribution, and depredations; Rodents; Birds; Other vertebrates (including bats)
- Molds and Other Microorganisms
- Regulatory and Forensic Aspects
- Stored-Food Pest Control: Agricultural barriers to infestation; Employee procedures and Good Manufacturing Practices; Sanitation, design, and construction; Rotation, transportation, and packaging; Physical, chemical, and biological control; Integrated pest management.

Basic Management Processes

Stored-food protection often comes down to effective handling of people, especially of those who work with food and are, therefore, in the best position to take action against losses. This also applies to the awareness training essential for the general public and for various employee echelons. A schedule for training in this aspect might include the following (5):

- Effective Listening
- Effective Speaking

- Time Management
- Coordination and Control
- Esprit de Corps
- Motivation
- Communication Up, Down, and Across
- Delegation
- Creativity
- Performance—Management and Measurement
- Applying Basic Management to Stored-Food Protection
- Why Managers Succeed.

Additional Areas of Training

To be fully effective, stored-food pest control must be grounded in scientific and managerial principles. Therefore, heavily involved personnel, especially technological and managerial, should be educated thoroughly in the fields of physics, chemistry (especially pesticide and physiological chemistry), biology (especially entomology, mammalogy, ornithology, parasitology, microbiology, nutrition, and ecology), as well as in economics and management.

Institutionalization of Training

Required training in stored-food pest control can, and should be, performed in a number of different institutional settings including:

- Universities and colleges that offer bachelor's, master's and doctoral degrees in the major involved fields such as entomology, environmental health, environmental engineering, and management;
- Technical schools that offer in-depth training in such areas as insect and rodent control, stored-food systems, environmental health technology, environmental engineering technology, and management;
- Government schools that offer intensive short-term training in specialized aspects of stored-food pest control as part of the overall regulatory effort;
- Industry schools that offer short- and long-term training in the various areas of stored-food pest control as part of the ongoing operational efforts;
- Trade associations and professional societies that offer various types of training and information exchange as part of periodic conferences, meetings, and publications.

Career Development

Basic and advanced training should be made available to all involved personnel in a logical and organized sequence so that the career potential of each employee will be enhanced. In addition, programs of continuing education should be provided to enable all personnel to keep refreshed in the field and apprised of new developments.

Training Technology and Management

Instuction in stored-food pest control can be approached through a number of training techniques including literature, films, tapes, demonstrations, courses, on-the-job training, apprenticeships, and externships. Training officers are advised to utilize several different techniques in each program in order to enhance learning and minimize monotony. In making technique selection, training officers should consider what the students are supposed to learn from the training; the competencies of the instructors; the capabilities of the students; what suitable mechanical, electronic, and audiovisual aids are available; and whether the approach coordinates with overall training activities.

Management

Provisions should be made in both long- and short-term planning for adequate *budgetary support* committed unequivocally to training. Training operations should take into account *physical requirements, human requirements,* and adequate *record keeping. Evaluation* of training should begin before actual training is under way and should be pursued continuously throughout the program.

Certification has become an extremely important element in stored-food pest control and a major motivation behind education programs. A careful investigation of legal requirements and professional recommendations should be made, and all control personnel should secure appropriate certifications and/or licensures (*3*).

Sustaining the Effort

Training needs in the stored-food field are immense, and a crash program alone will not suffice. Programs must be so organized and placed that they will provide continuous training services capable of sustaining the effort. The problem of stored-food loss will not be solved overnight, nor can the gains made be kept unless the effort is applied over long periods of time.

Cited References

1 Anonymous. 1974. The world food crisis. *Time,* Nov. 11, pp. 66–83.

2 Brody, J. E. 1974. Hungry pests gobble half of the world's food supply. *New York Times* News Service, Oct. 29.

3 Osmun, J. V., and G. T. Weekman. 1977. Pesticide applicator training and certification. In *Pesticide Management and Insecticide Resistance,* D. L. Watson and A. W. A. Brown, eds. Academic Press, New York, NY.

4 Pratt, H. D., K. S. Littig, and H. G. Scott. 1975. *Household and Stored-Food Insects of Public Health Importance and Their Control.* DHEW Publ. No. (CDC)75-8122. Center for Disease Control, Atlanta, GA.

5 Scott, H. G. 1968. *Basic Management Processes; Instructors Handbook.* Public Health Service, New Orleans, LA.

6 Scott, H. G. 1976. *Training Program in Maintaining Nutritional Levels in Stored Foods.* Tulane University, New Orleans, LA.

48

Professional and Consultant Services

Eugene J. Gerberg

Food-industry pest management is a complex, multifaceted task that reguires in-depth training, wide field experience, and the ability to communicate effectively. The food-industry sanitation consultant or pest management consultant must be able to provide adequate, accurate information to management and must be a problem-solver.

Why a Consultant Is Used

New and developing sanitation technologies often require food industries to rely upon outside sources to supply the necessary input of information in order to facilitate intelligent decisions. The consultant usually and preferably reports directly to top management, is not involved in "company politics," is unbiased in his or her sanitation audit, and provides a "second pair of eyes." He or she is a "trained stranger" who may be in the forest but can still see the trees.

For various reasons, people (including management) who tend to resist or ignore recommendations from in-house sanitarians or quality assurance personnel will often respond favorably to the recommendations of a consultant. It is clearly to the benefit of the industry involved to have such an outside consultant to demonstrate its sincere determination to maintain a safe, sanitary, pest-free operation.

The consultant brings in wide experience from a variety of food industries, thus enabling him or her

to offer several courses of action designed to reach desired goals. The consultant can set priorities, within physical and economic boundaries, outlining what should be done and why, and how and when it should be accomplished. A system of evaluating progress and a program of training may also be provided.

The consultant provides liaison with food-industry regulatory agencies, assists in interpretation of complex rules, and can keep management up to date on changes in regulations. The consultant provides an economic benefit because his or her annual contract fee is estimated on a per-diem basis; thus a salaried position with its payroll burden is avoided. In addition, the consultant tries to eliminate inefficient, wasteful, or costly practices that may not be obvious to personnel who work in the plant on a daily basis.

Qualifications of the Consultant

Ideally, the consultant should have biological training, preferably a degree in entomology. The consultant should be a Registered Professional Entomologist with at least a speciality in Urban and Industrial Entomology, Regulatory Entomology, or Medical/Veterinary Entomology. Additional training in microbiology, sanitary engineering and waste management, rodent control, food and drug regulation, pesticide regulation, plant maintenance, food processing, business administration, psychology, and public relations adds to the value of the consultant.

Aside from academic training, the consultant should have practical experience in pest control, food-industry inspection, and a working knowledge of the various federal regulations pertaining to food industries.

Gerberg: Insect Control & Research, Inc., 1330 Dillon Heights Avenue, Baltimore, MD 21228.

Also very important are an innate sense of curiosity, physical stamina, and a willingness to "get dirty" when making inspections.

Newcomers to consulting are advised to gain experience by working initially with an established professional, becoming familiar with field problems, limitations, and procedures.

Operations of the Consultant

In the performance of his or her duties, the consultant

(a) conducts a thorough sanitation inspection of raw ingredients, processing equipment, facilities, and grounds to detect existing and potential sources of infestations and contamination;

(b) prepares a written report or audit (not necessarily a scoring or numerical rating);

(c) meets with management personnel and provides a thorough critique of the situation (this should be a "no-holds-barred" discussion with emphasis on the pragmatic approach);

(d) brings management up to date on new advances in food-plant sanitation techniques, new regulations, and so on;

(e) reviews and advises on pest control programs, pesticide usage, and compliance with the Federal Insecticide, Fungicide, and Rodenticide Act and other applicable laws (a consultant may recommend a pesticide or a certain item of equipment, but it is ethically questionable for him or her to also sell such products);

(f) calls attention to actual or potential occupational safety and health violations as they pertain to sanitation;

(g) urges quick correction of deficiencies that can be readily corrected;

(h) discusses scheduling of other corrections that may be required;

(i) arranges for further checking of raw materials and finished products for extraneous matter and microbiological analysis;

(j) checks for label violations and brings these to the attention of the legal department;

(k) analyzes and advises on preventive sanitation of new equipment purchases and on new building constuction or renovation;

(l) assists the firm's legal staff or advisors to correct infractions and prepare a defense;

(m) serves as an expert witness in matters involving sanitation;

(n) provides in-house seminars and training on sanitation topics and participates in industrywide conferences;

(o) *solves problems* after a sanitation audit.

Food-plant sanitation must be an integrated pest management program, with emphasis on preventive methods rather than corrective measures. The consultant is in an ideal position to initiate, instigate, promulgate, harass (if necessary), and finally approve the required procedures and methods to implement a program that will result in a sanitary product, produced and stored in sanitary facilities, thus avoiding the negative publicity resulting from regulatory agency criticisms.

Finding a Competent Consultant

Perhaps the easiest method is to check with others in your industry. There are, however, other sources. The American Registry of Professional Entomologists can provide an unbiased list. Articles in trade journals and scientific journals may provide clues to competent consultants.

When a qualified person is found, the names of some of his or her clients should be requested. The fact that the consultant may perform services for competitors may be a "plus" and should not be an impediment to retaining his or her services. Perhaps one of the first things a consultant learns is that what he or she sees or learns is confidential. He or she would not last very long if this dictum was not observed religiously.

The consultant should normally be retained for a period of time, not a "one-shot" contract. Services that are tied to sales of materials should be avoided. Consultants provide an advisory service; salespeople provide materials, supplies, and other items necessary for correction or prevention.

A food-plant consulting service should be considered as an investment in improved production, as an insurance policy against adverse publicity, and as an improvement in public relations within and outside the plant.

Selected Reference

Anonymous. 1985. *Pest Control: A Guide to Specifying and Obtaining Services by Contract*. Property Management Association, Silver Spring, MD.

49

Integrated Pest Management for the Food Industry

Darrell F. Jones and Eugene G. Thompson

Recognizing, controlling, and eliminating pests associated with foods are activities that receive a great deal of attention from food processors, distributors, retail sales outlets, and government agencies. The pests (rodents, insects, mites, birds, and harmful microorganisms) are introduced to the food-processing or distribution chain as live, dead, or dormant forms within raw material shipments, packaging supplies, equipment, or finished products. They are also "out there" around the facilities, trying to gain access to the raw and processed foods. How to prevent these pests from gaining entry to facilities, and how to maintain control over the environment within which food is produced, distributed, or stored, requires a carefully planned and organized program. This approach has become known as Integrated Pest Management (IPM).

Situation Assessment

The IPM administrator is in very much the same situation as a combat commander charged with holding a position. A situation assessment for both would probably include such questions as

(a) Who is the enemy and what does he look like?
(b) What are his strengths and weaknesses?

Jones (Deceased): Quality Control Department, General Mills, Inc., Minneapolis, MN 55440.

Thompson: Nashoba Pest Management Consulting Service, Inc., 10509 Mourning Dove Drive, Austin, TX 78750. Formerly, Executive Director, Armed Forces Pest Management Board, Walter Reed Army Medical Center, Washington, DC 20307.

(c) Where, how, when, and in what form is he most likely to attack?
(d) What defense and intelligence-gathering systems are available to us?
(e) What countermeasures are available to meet the possibility that the enemy may overcome our defenses at some point or points and gain access to the position being defended?
(f) What is the most cost-effective method of achieving the objective?

The basic information needed to set up an IPM program is discussed in detail in the other chapters of this book. The answers to the first three questions are either generally known or can be found in this or other publications. IPM administrators must deal primarily with the last three questions. They must acquire and employ a basic knowledge of the pests when setting up effective defenses and countermeasures.

There are five primary steps to an IPM assessment:

(a) Know the basics of pest management.
(b) Build-out pests (exclusion).
(c) Perform good surveillance.
(d) Use nonchemical controls when feasible.
(e) Use chemical controls properly and only when needed.

Ideally, they should be considered in this order.

Basics of Pest Management

Assistance and Expertise

Most IPM administrators will need to bring in expertise and assistance from several sources. First, they must understand the pests they are confronting or are likely to confront. If they are not trained in pest management, then they must get help from an entomologist, sanitarian, or pest management consultant. These individuals are found within government agencies, universities, private consulting companies, and commercial pest control firms. Many of them are registered with the American Registry of Professional Entomologists. In larger organizations, this help will be available within the government agency or commercial activity. Manangers of most smaller organizations will have to seek such assistance from a source outside their own firms.

Administrators will also need help from engineers, especially in the development of plans to build-out pests. When pest problems have been identified through the efforts of an entomologist or pest management consultant, engineering expertise will be required to determine and design the most practical and cost-effective solutions. Maintenance, largely an engineering function, will also have to be brought into the picture to implement lasting solutions. IPM administrators will also require assistance from all plant or warehouse personnel and/or a contract service organization in applying the surveillance, nonchemical, and chemical portions of the IPM program.

Time and Transportation

Two basic factors are the key to the protection of food from pests: time and transportation. The longer raw commodities or finished products are stored and the more often such products are moved from one location to another, the more likely it is that the food will become exposed to infestation or contamination by one or more pests. By reducing storage time and eliminating multiple shipments, the likelihood of pest infestation is reduced.

A related factor is whether or not the product is stored in a comparatively open container (such as in a grain storage bin), protected in an insectproof container (such as a metal can or glass jar), or stored in something in between (such as a multilayer paper bag). Often, the degree of container protection provided is a result of evaluating the time and transportation factors required between production and utilization.

Characteristics of Pests

All pests require food, moisture, harborage, and time in order to reproduce. Warmth is an additional requirement for arthropods. Eliminating one or more of these requirements drastically reduces the capability of pests to survive and propagate. Food pests are all capable of reproducing in an environment that provides these basic needs and thus pose a continuous threat as potential infesters of food.

Food pests can be grouped in two categories: those that bore, gnaw, or peck a container to gain access to the food product; and those that can invade the food

only after an opening is provided. Thus, we have what are commonly referred to as penetrating, active, invading pests and passive pests. These facts are another key to the development of an IPM plan. You also know that pests enter a clean plant or warehouse in two ways: as attracted volunteers or carried in as passive captives. This is still another key factor to be considered in development of an IPM plan.

Full-Time IPM

IPM administrators must be aware constantly that pest management is not a "once done, then forgotten" program. It is an ongoing, day-to-day effort. They must continually reexamine the program for defects and be alert to the early signs of program failure. They must also keep up to date on the latest equipment and techniques that help to do the job efficiently and effectively.

This is the basis for requiring IPM plans to be formalized. The procedures and pesticides to be used, the people responsible for carrying out each procedure, and the departments (e.g., medical, fire, and security services) that must be coordinated prior to executing the plan should all be included. The IPM plan also serves as a basis for projecting associated financial and manpower requirements.

Building-Out Pests (Exclusion)
Outside

Good drainage reduces the availability of water, one of the requirements for most pests. Proper grading and paving aids drainage and reduces dust, harborages, and other attractants to pests. Plantings on the warehouse or plant grounds, especially in the immediate area of the building, must be carefully selected and maintained. Some food pests also feed on certain plants and trees often used in landscaping. These plants or trees should not be used on the grounds of a food-handling establishment in order to avoid the possibility that infestations of such plants or trees will spread to the food facility. Exterior trash disposal containers must receive special attention. Basic sanitation and timely removal of trash are essential in preventing infestations of such receptacles.

Structural Integrity

It is very important for buildings to be structurally sound. The exterior should be closely examined for cracks, crevices, and other defects that offer potential shelter or access to a building. Smooth-surface exterior walls are desirable. Building adjuncts, such as dock canopies, drainpipes, or equipment either on the roof or affixed to the building, must be installed and sometimes screened or modified in a manner that precludes pests from gaining entrance and eliminates harborages or bird roosting sites. The adjuncts must be constructed in such a manner as to make any potential harborages easy to clean and to examine for signs of pest activity. All doors and windows must fit tightly enough to exclude rodents, birds, and most insects. Screens or hardware cloth should be used over openings where ventilation is required.

Lighting Guidelines

Outside lighting should be located away from, but aimed at, work areas to minimize the potential of attracting flying insects to work or food-product zones. It is important to use only sodium vapor lamps (which are less attractive to flying insects) within the perimeter of the grounds surrounding a food plant. As a rule, mercury vapor lamps should not be used in any exterior location, but in the case of a facility with extensive grounds, mercury lamps might be appropriately used around the perimeter farthest from food-storage or food-processing situations.

New Construction

When new facilities or structural modifications are contemplated, the IPM administrator should play a key role in selecting the design of buildings and the types of building materials to be used. During remodeling, the person in charge of IPM should also anticipate periods when doors will be left open for extended periods of time or when outer wall sections must be temporarily removed. At such times, supplemental protection against potential pest entry must be provided.

Inside

Having discussed the general approach needed to minimize pest populations outside our facilities, we must realize that some pests will evade the best-planned exclusion programs. If the attractions, such as food odors, lights, and insect pheromones, are strong enough, some pests will gain access to the buildings. In addition, captive pests will be conveyed into facilities with incoming materials or personnel.

Elimination of Pest Access

The goal for this portion of the pest exclusion effort is to eliminate areas where pests can find harborage once they enter the buildings. Cracks, crevices, voids, and punctures in walls, floors, ceilings, and structural supports must be eliminated. False walls and ceilings or improperly installed insulation should be avoided where possible since they offer excellent rodent and insect harborage. Wet-processing areas must be well drained and adequately ventilated. Drains and sewer lines must be designed to prevent entry or movement of pests within the facility. Equipment must be sealed to walls and floors or positioned far enough from walls and off floors to permit easy access for inspection and cleaning.

Design for Exclusion

Equipment design is very important. Within each piece of equipment, the potential pest harborages must be known. Such problem areas must be identified so that, if possible, equipment can be modified or, at the very least, such trouble spots can receive regular attention. These problem areas must be kept in mind when ordering new equipment as well. The IPM administrator should have a major input during equipment selection. In some instances, harborage-free equipment is available or can be modified prior to purchase and installation.

Electrical installations, such as lighting fixtures, panels, and conduit chases or heating/air-conditioning ducts should not be overlooked, because they offer warmth and shelter for pests if poorly or improperly designed or installed. As with the exterior of the building, walls should have smooth surfaces that are easy to clean. In some situations, the use of a sealant on walls and floors will often provide such a surface.

Surveillance

Effective Inspections

It must be recognized that the soundness of our defenses and our ability to react to the unexpected or incipient pest infestation depend on thorough and regular inspections. Obviously, reliance cannot be placed solely on plant or warehouse personnel to uncover pest problems that indicate that defenses have been overcome or that a captive pest has invaded the facilities. The best inspections are those performed collectively by a team of individuals: the IPM administrator; personnel from a department other than that responsible for pest control; supplementary outside specialists, such as from a commercial pest control service or a professional inspection service; and responsible plant or warehouse personnel. This last group functions continuously in the "inspection mode."

All these services are necessary, and a reliable communications network must be established and maintained so that problems are anticipated and dealt with effectively. It is not enough to assign responsibilities for defenses against pests or to perform inspections without also assigning some real authority for accomplishing these objectives.

Comprehensive Inspections

Surveillance begins before the product is brought into the plant or warehouse. Pallets should always be checked because they are a common source of insect infestation. Inspectors should look for signs of pests, such as attached egg cases, and for infestible materials, such as grain kernels, on the pallet components. An infested ingredient or product should never be brought into a clean plant or warehouse. This includes products being held for return to a supplier. Products found to be infested at the time of receipt should be rejected. Surveillance must continue on a regularly scheduled basis until the ingredients are used or the finished products are shipped.

Attractants

Pheromones and food attractants may also be used as surveillance tools. Used as detection tools, pheromone traps can be especially helpful

(a) when visual inspections are difficult either because the species at hand normally elude visual inspection or because the insect population levels are very low;

(b) because they can be species specific;

(c) because they can lead to early implementation of control measures; and

(d) because they can be an important part of a pest-monitoring system.

Food attractants such as wheat germ oil can serve as pest detectors and also as a component of control measures. Both pheromones and food attractants are comparatively safe to use in sensitive situations. However, the proper placement of traps and the interpretation of the results may require a level of expertise beyond the capability of the average user.

Nonchemical Controls

Outside

Good housekeeping is always the first nonchemical control to be considered. Outside housekeeping programs must include an efficient refuse removal service, proper storage of clean spare equipment, adequate lawn and grounds care, and prompt removal of food spillage or other organic matter. A blend of good maintenance and common sense makes the subtle difference between effective and ineffective programs. For instance, keeping a truck ramp door tightly gasketed does little good if the door is left open unnecessarily. The value of rodent guard installations is lost when housekeeping is so poor that rodents can gain access to building interiors via junk or spare parts stacked next to the dock or building. A sound rodent trapping program outside the building, in combination with an effective baiting program, may also be required.

Inside

As with nonchemical controls around the exterior of premises, a sound housekeeping program is also essential inside the facility. Keeping a neat, clean, orderly building eliminates many potential pest harborages and helps focus on those potential harborages remaining that are due to structural or equipment conditions. This applies to every part of the building, including small rooms and closets in close proximity to the food-handling areas.

Good Housekeeping

Good sanitation is the first line of defense against pest infestation—it cannot be overemphasized. Keeping spare or standby equipment off the floor and away from the walls, and adequate cleaning of walls, ceilings, floors, and windows all contribute to good housekeeping. Spills must be cleaned up immediately and trash should be removed daily. This includes the removal of wastes from salvage operations. Leaking pipes must be repaired promptly. Infestible raw commodities should always be maintained in bins or silos that can be easily cleaned and fumigated. Finished food products subject to pest infestation should be stored off the floor and stacked away from walls so that all four sides of the product stack can be inspected. This also enables the product to be fumigated in place if fumigation becomes necessary. Good rotation of warehouse stock to include the implementation of a first-in/first-out system is a must.

Trapping and Repelling

Several techniques can be used for repelling and trapping pests as they attempt to enter structures. These include properly located snap and windup traps located just inside each entryway or other areas of concern; glue boards and light traps selectively used and well maintained; air curtains above doorways; and, for birds, the use of hardware cloth to preclude access to overhead beams or other roosting or nesting areas.

Heavy, overlapping, plastic vertical strips hung at entryways tend to prevent flying pests from entering doorways that must be left open. The units, hung from the top of the entryway, can be walked through or driven through with a loaded forklift, and then they automatically reclose after passage.

There are many electrical devices on the market today which, their manufacturers claim, repel or kill insects and rodents. Some of these devices have been found to be ineffective or even fraudulent. However, some can play a useful role in the overall IPM effort.

Light Traps

Electrocuting light traps are routinely used as a tool in IPM programs throughout the food-processing and food-distribution industries. The IPM administrator must insist that only quality, commercial-type light traps are purchased (backyard "zappers" cannot do the job). The key to whether or not light traps will be effective depends on their proper placement and maintenance (cleaning at least once every week and annual replacement of the ultraviolet lamp). If cleaned weekly, the trap collections may also be used as a part of the pest-monitoring program. Placement should be made with the advice of a professional consultant experienced in their use.

Several fraudulent versions of electronic, ultrasonic

rodent repellers have been marketed over the past several years; these must be avoided. Even the more reliable kinds have been shown to be ineffective inside most large warehouses. This is because the repellers require frequent relocation, are relatively expensive, and often lead to a false sense of security on the part of employees. Repellers of this sort can be effective in one situation: When beamed across entrances that must be left open, repellers can deter rodents from entering.

Rotating lights may repel birds when used in dark surroundings. However, some governmental entities prohibit the use of these lights except in very specific situations and only upon recommendation of a pest management consultant.

Personal Hygiene

Employees should be trained and encouraged to follow the highest standards of personal hygiene. In addition to how workers should dress (or not dress— wearing jewelry or not wearing a hair cover, for example), attention should be given to the acceptability of street clothes. It is possible that during off-duty time the street clothes of workers could become contaminated with microorganisms or other adulterants that during on-duty time might find their way into ingredients or products. FDA's Good Manufacturing Practices contain detailed directions on personal hygiene.

In-Transit

Transportation modes are probably the most common source of pest infestations, particularly by insects. Railcars, trailers, and trucks must be maintained in good repair and kept as clean as the plant and warehouse. All forms of transport must be inspected to ensure that they are clean, free of infestation, and free of holes, cracks, or gaps in the ceiling, walls, and floors. Small holes should be caulked or sealed before a food product is placed in the railcar or truck. If many holes are found, the transportation unit should be rejected or used to transport noninfestible items. The valuable time, money, and effort spent to protect a food product in a plant or warehouse can be wasted if a clean product is placed in an infested transport unit.

Packaging

Considerable protection is achieved through packaging. Food products packaged in glass or metal containers are noninfestible, but this results in additional weight, expense, and other problems such as breakage. The use of multiple layers of packaging materials, the selection of packaging materials with resistance to insect penetration, and proper sealing can all provide a great deal of protection for finished food prod-

ucts. Government and industry research continues to develop new materials that may someday provide insectproof containers other than cans and bottles.

Chemical Controls

Outside

Implementing the first three IPM steps of excluding pests, maintaining proper surveillance, and applying nonchemical controls provides considerable protection, but the presence of pests may still place too great a burden on the program. Thus, sometimes reliance must be placed on chemicals.

Anticoagulant rodenticides must be used in strategically located, tamperproof bait stations, both to help control the rodents and to indicate areas of rodent activity. Fly baits are effective to varying degrees depending on the species or number of breeding sites present.

Pesticides

Residual and contact insecticides must be judiciously applied for control of seasonal insects. Avicides and other bird control measures (e.g., exclusion by use of hardware cloth) must be selectively utilized to prevent birds from roosting and feeding around plant property. The services of commercial pest control firms are often needed for these or other specialized exterior problems such as the control of bats.

Inside

Baits of any kind are generally undesirable inside buildings. Therefore, only fumigants, contact and residual insecticides, liquid rodenticides, detergents, and sanitizers are generally recommended. The only exception to this policy is the use of anticoagulant rodenticide baits when all water sources cannot be eliminated, thus limiting the effective use of liquid rodenticides. Rodent bait stations must be monitored frequently (several times a week) to prevent the bait from serving as a source of insect infestations.

Identification of Pests and Application of Pesticides

It is extremely important to know the kinds of pests being dealt with and to carefully follow all pesticide label directions. Residual insecticides and dusts are directed into difficult-to-reach, potential harborages such as block walls, voids, and cracks and crevices that cannot be sealed off, around equipment bases, and along floor-wall junctures. Contact sprays, most commonly dispersed as ultralow-volume (ULV) space treatment, must be used on a schedule designed to control exposed flying and crawling insects. In the case of food pests, this is generally during the night.

Fumigants are used on ingredients, commodities, equipment, and structures, and on finished products to eliminate (or, in special situations, to prevent) infestations. Fumigants, if properly used, do not result in a residue or residual action. Therefore, the food ingredient or finished product can be reinfested almost immediately following fumigation.

Detergents and sanitizers must be used on equipment, walls, ceilings, hand tools, and hands.

In-Transit

Since railcars and trucks often have false floors, walls, and other areas not accessible for inspection, they often contain infested raw commodities or spilled food products. Fumigation or spraying may be necessary. For short trips or in situations where the shipping unit cannot be made relatively gas-tight, a residual spray applied to clean walls and floors, before the food product is loaded, provides a barrier against crawling insects. ULV space treatment is required if flying insects are involved.

Fumigants, properly applied, usually kill all pests. Trucks or tanker cars are usually the easiest to make gas-tight, and therefore successful fumigation can normally be achieved. The fumigation of trucks must be accomplished while the truck is in a static mode, and this results in lost time and demurrage charges.

Fumigants are available that allow railcar fumigation to be accomplished in transit, but the effects of air currents usually prevent gas levels from remaining sufficiently high to achieve successful fumigation of pests deep within the food product. The gas level, however, normally stays sufficiently high to kill exposed pests in the false walls, floors, and other hidden areas so that the primary objective of such a fumigation is achieved. The IPM administrator and the plant or warehouse personnel must be aware of this fact and must not reduce their surveillance efforts just because a product has arrived in a fumigated railcar.

Conclusions

Whether dealing with a commercial food processor, a commercial or governmental food distributor, or a food storage site, IPM is a basic requirement that must receive the same attention as any other required function in the operation of a plant, warehouse, or other food facility. Without a good IPM program, the basic product—food—may be lost.

The IPM concept is important because it results in a minimum use of chemical controls, can reduce long-term costs, and at the same time emphasizes the use of common sense. No single portion of an IPM program can stand on its own (e.g., the increased use of chemicals to substitute for good housekeeping practices or a sound building). To be successful, the program must be integrated and coordinated. The IPM objective is simple—to prevent pest entry and development in a facility. If this is done, manufactured or stored food products will be of high quality, and the potential contamination of those products with pests or pesticides will be minimized. All of this can be accomplished effectively and efficiently if done properly.

50

Commentary on Food-Pest Ecology and Management

Joseph D. Foulk

It would be very difficult to try to summarize the numerous and diverse contributions to this Bulletin. Some comments, however, are these: Our Editor's charge to the authors to present practical information that can be used by disparate food-industry personnel is commendable—as is his own conscientious and patient work at completing and publishing this significant treatise. This Technical Bulletin will become another one of those "core" references that all of us use throughout our careers. It will be beneficial not only to seasoned and experienced people, but also to younger workers with entry-level job functions and responsibilities in the academic, regulatory, and commercial areas.

We should always be aware that we are working with ever-changing biological systems (including humans) in pest management. Different levels of different pests in different locations are variables we must learn to evaluate by classic as well as novel approaches (2). It is unrealistic for us to brag that, as we enter the last decade of this century, we are now *primarily* involved with prevention of pest problems. Too many of us are still waging direct *control wars* against insect pests. But, thankfully, many pest managers are now working toward prevention.

Regrettably, some food firms continue to regard their sanitarians'—or well-paid consultants'—recommendations as pleas for "nonproduction expenditures." Corrective actions and the establishment of

structured and truly preventive programs too frequently occur only after expensive pest-control crises or after the presentation of uncontestable evidence of pest contamination of facilities, processing equipment, ingredients, and commodities in storage or at the retail levels—all resulting from "pure food" inspections (1).

Increasing numbers of firms are devoting efforts and monies toward effective sanitation programs and physical plant and equipment construction, renovation, and maintenance. There are three main reasons for this. First, they recognize that we are all consumers, and they take pride in supplying "A Clean Product from a Clean Plant." Next, they have become aware that the days of widespread use of pesticides are past and that negative publicity can cause both loss of reputation and reduced sales. Finally, because of the highly competitive nature of today's food industry, monies spent on prevention have been proven to be truly cost-effective.

Two intangibles relating to job performance in food-industry pest ecology and management should be mentioned: psychology and politics. In terms of psychology (relating more to individuals than corporations), successful characteristics for commercial food-industry sanitarians or "pest managers" include: patience, optimism, dedication, inquisitiveness, willingness to share information and "glories" attendant to a job well done, appreciation of the benefits of continuing education, reliability, and common sense.

From the political point of view, the need for and rewards of effective pest management absolutely must

Foulk: Food Safety Associates, P.O. Box 247, Edmonds, WA 98020.

557

be understood by top management. Today, short-range as well as long-range policies and procedures must be components of the driving force of food firms.

Finally, all the modern equipment, all the computerization, and all the various and sundry "meetings" we schedule and are required to attend won't rodentproof the door sills, regularly clean the elevator pits or insect-electrocutor catchtrays, or maintain *any* plant or warehouse so vertebrate or invertebrate pest problems are minimized. There are no "economic thresholds" for pest infestation in the food processing or warehousing industries. And sooner or later someone has to physically do the hard, often unpleasant work necessary to keep food plants free of vermin.

Cited References

1 Foulk, J. D. 1990. Pest-related aspects of sanitation audits. *Pest Control Technology* 18 (11)32, 33, 89; 18 (12)69–72, 74.

2 Foulk, J. D. 1990. Pest occurrence and prevention in the foodstuffs container manufacturing industry. *Dairy, Food and Environmental Sanitation* 10(12)725–730.

VIII
Glossary and Indexes

Glossary

Taxonomical Index

Subject Index

GLOSSARY

Abscission scar: Point where a kernel of grain was previously attached to the plant.

Adventive: An exotic species, introduced accidentally, that may be imperfectly naturalized.

Aerobe: An organism capable of living only in the presence of oxygen.

Anaerobe: An organism that does not require free oxygen to sustain its life.

Andropolymorphism: A phenomenon seen in certain mites in which occasional males have one pair of legs that is disproportionately large.

Anemotaxis: Orientation movement of a free-living organism in response to a current of air. Negative anemotaxis is flying with the wind; positive anemotaxis is flying across or against the wind.

Anthropobiocoenosis: A habitat or community created by human activity.

Atta: Unsorted wheat flour or meal.

Australasian Region: Biogeographic region including Australia, New Guinea, New Zealand, Tasmania, and the islands south and east of Wallace's line.

Autotrophs: Organisms capable of using carbon dioxide or carbonates as a sole source of carbon and simple inorganic nitrogen compounds for metabolic synthesis (e.g., green plants).

Biodeterioration: Process of disintegration and decay of materials by direct action of organisms, especially microorganisms, but also including such insects as seed weevils and carpenter ants.

Biological carrying capacity: Maximum population of a species capable of being sustained indefinitely on a substrate possessed of finite dimensions and resources.

Campodeiform: Beetle larva having the general shape of a bristletail of the genus *Campodea* (Machilidae, Thysanura).

Canola: Canadian cultivars of rapeseed (*Brassica campestris* L., *B. napus* L.) with very low erucic acid and glucosinolate levels.

Carnivory: Feeding on fresh animal protein.

Climax stage: Final stage of succession that permits the component organisms, the community, to live on indefinitely in a balanced environment.

Climatic plasticity index: Statistic describing the potential ability of a species to adapt in various climatic zones. The higher the value, the greater the climatic adaptability of the species.

Colony: Individuals of a single species living in close association, e.g., ant colony, coral colony, or bacterial colony on agar.

Community: Groups of coexisting, interacting populations of plants and animals in a given place; the groups may be of various sizes and degrees of integration.

Conjunctivitis: Inflammation of mucous membranes of the eyelids and eyeball.

Coprophagy: Feeding on human or animal excrement.

Cornicles: A pair of protruding horn-shaped dorsal tubes in aphids which may secrete a waxy fluid.

Coxa: Segment of the leg of an insect or other arthropod by which the leg articulates with the body.

Cuticle: An external membranous or hardened noncellular investment secreted by the outer surface of the body of many kinds of animals; exoskeleton of arthropods; outer skin of a vegetable or fruit.

Dejecta: All liquid or solid substances emanating or shed from the body of a living animal, such as hairs, cast skin, feces, webbing (as from spiders or moth larvae), urea, uric acid, feathers, and epidermal cells.

Diapause: A period of suspended growth or dormancy in certain insects and mites.

Diastema: Gap separating the chewing teeth (molars) from the incisors in rodents.

Dockage: Broken kernels, grain dust, chaff, and dirt.

Ecdysis: Molting or shedding of outer cuticle layer

by insects and other arthropods.

Eclosion: Emergence from an egg or pupa.

Ecology: Systematic study of interactions among plants, animals, and their environments.

Ecosystem: A biotic community in its abiotic environment, e.g., communities of microflora, insects, and mites as they live together and respond to changing temperature and relative humidity of the air; an assembly of interlocking life systems, the fundamental unit of ecology.

Ectoparasite: A parasite that lives on or feeds on the exterior of its host.

Elytra: Thickened sclerotized forewings of beetles.

Endophilous: Largely dependent upon anthropobiocenoses for both food and shelter.

Enteric acariasis: Accidental ingestion of mites.

Enteric arthropodiasis: Accidental ingestion of living arthropods.

Enteric canthariasis: Accidental ingestion of beetles.

Enteric myiasis: Accidental ingestion of fly larvae.

Ethiopian Region: Africa south of the Sahara, southern Arabia, and Madagascar and adjacent islands.

Facultative predation: Feeding on prey that does not constitute an essential part of the predator's diet but which may stimulate population growth.

Flax tow: Short broken fibers removed during scutching or hackling and used for yarn, twine, or stuffing.

Frass: Debris (e.g., insect fragments, excrement, bits of abraded substrate, and sawdust-like materials) produced by insects.

Fungus (plural, fungi): A microorganism without chlorophyll obtaining its nourishment from dead or living organic matter; molds and yeasts are fungi.

Gaster: Enlarged part of the abdomen of ants and other hymenopterous insects.

Genus (plural, genera): A group of related organisms (e.g., *Penicillium*) possessing common structural characteristics distinct from those of any other group (e.g., *Aspergillus*); the first term in a binomial name.

Glumes: Leaflike bracts surrounding a grass seed.

Good Manufacturing Practice Statements: Advisory protocols set forth to promote safe manufacture of food.

Halophiles: Microorganisms able to thrive in high concentrations of salt.

Hardpan: A densely compacted soil often containing a substantial proportion of clay.

Hastisetae: Specialized hairs of larval *Trogoderma* spp. (Dermestidae, Coleoptera) that are readily caducous and tend to penetrate surfaces (e.g., other insects) with which they come into contact.

Heliophilous: Attracted by or adapted to sunlight.

Hemisynanthropy: Existing independently of human habitats but benefiting from these habitats, especially in the production of large populations.

Heterotrophs: Requiring complex organic compounds of nitrogen and carbon for metabolic synthesis (i.e., most animals and plants that do not carry on photosynthesis).

Hibernaculum: A protected refuge where an insect hibernates, aestivates, or undergoes dormancy or diapause; the refuge may be a natural or artificial shelter, or it may be a cover constructed in part or entirely by the insect.

Holarctic Region: Northern parts of the Old and New Worlds, comprising the Palaearctic and Nearctic regions.

Holistic: A total or whole approach rather than in parts.

Hyperparasite: A parasite that is parasitic upon another parasite.

Hypopus (pl. hypopi or hypopodes)*:* The second stage (following the protonymphal stage) or deutonymph, presumably always nonfeeding and often skipped (in which case the protonymph molts directly into a tritonymph), of certain kinds of astigmatid mites; a stage often associated with transport (phoresy) on the body of an insect or other animal.

Imago: Insect in its final adult, sexually mature state.

Inquiline: An animal that lives as a "guest" in the nest or colony of another kind of animal.

Instar: Stage in the life of an insect or other arthropod between two successive molts.

Integrated pest management: A decisionmaking system in which all available interventions (treatments, actions) are brought to bear on a pest problem with the goal of providing the most effective, economical, and safest remedy possible. IPM is a process for determining *If* intervention is needed or justified; *When* intervention is needed; *Where* intervention is needed; and *What* intervention is needed.

Juvenile hormone: An insect hormone secreted by the corpora allata, which inhibits maturation to the imago; it has been used to control pest insects by disrupting their life cycle.

Kairomone: A chemical messenger released by one species that elicits a response in another species.

Keratin: Sulfur-containing fibrous protein that forms the chemical basis of epidermal tissues such as horn, hair, wool, and feathers.

Lamellation: Existing in layers.

Lenticel: Ventilating pore in plant tissue.

Life system: The component of an ecosystem that determines the existence, abundance, and evolution of a population; has a subject population and its effective environment which is composed of all external agencies affecting the population. Interlocking life systems make up the framework of an ecosystem.

Malocclusion: Excessive incisor growth in rodents occurring when an incisor is misaligned or broken and no longer opposes its counterpart in the other jaw. The condition may eventually force the mouth open, partially blocking the mouth cavity, and result in reduced food intake or starvation.

Mesophiles: Microorganisms thriving at moderate temperatures.

Midgut: Middle portion of the alimentary canal of an insect that lacks a cuticular lining.

Molds: Fungi that produce fine threads or hyphae through which nutrients are obtained and from which reproductive structures arise.

Morphogenesis: Formation and differentiation of tissues and organs.

Multivoltine: Having several broods in a season.

Mycotoxin: A fungal metabolite produced on food or feed that may cause illness or death when eaten by humans or animals.

Nearctic Region: Greenland, arctic America, and northern and mountainous parts of North America.

Negative phototaxis: Orientation away from light.

Necrophagy: Feeding on carrion.

Neotropic Region: South America, West Indies, and tropical North America.

Niche: Role of an organism in an ecological community involving its way of life and its effects on the environment as through its relations to other biotic or abiotic factors.

Oceanian Region: Islands and archipelagoes of the central and south Pacific, other than those included in the Australasian Region.

Ocellus (pl. ocelli): The simple eye of certain insects and other arthropods.

Oriental Region: Asia south and southeast of the Himalayas and the Malay Archipelago west of Wallace's line.

Osmophiles: Microorganisms able to thrive in high concentrations of sugar or salts, e.g., in brine or on brine pickles.

Overpacker: Factory specializing in combining small lots (packages) of food into large lots (e.g., pallet loads).

Oviposition: Laying of eggs.

Palaearctic Region: Region including Europe, Asia north of the Himalayas, northern Arabia, and Saharan Africa.

Palaeosubtropic Region: Those regions of the Eastern Hemisphere located immediately north and immediately south of the equatorial tropics.

Palaeotropic Region: Pertaining to the tropical regions of the Old World.

Palea: Upper enclosing bract of a grass flower.

Pedicel: The second segment of the antenna, usually elongate in the geniculate antennae of ants and wasps; the constricted region of the body between the thorax and the gaster of ants and wasps.

Pelshenke value: Grain-hardness measure, specifically of gluten strength.

Pericarp: The ripened and variously modified walls of a fruit (e.g., apple), derived from the ovary and consisting of three distinct layers.

Petiole: Slender abdominal segment joining the rest of the abdomen to the thorax of an insect.

pH: Hydrogen ion concentration; a pH from 0 to 7 is acidic and from 7 to 14 is alkaline; pH 7 is neutral; the numbers 1 through 14 represent the negative logarithm of the hydrogen ion concentration.

Pheromone: A chemical messenger released by one individual arthropod that elicits a response in another arthropod of the same species.

Photosynthesis: The conversion in plants of carbon dioxide and water into carbohydrates brought about by exposure to light (usually sunlight).

Pinkeye: An acute highly contagious conjunctivitis of humans and various domestic animals.

Plankton: Passively floating or weakly swimming microscopic animals and plants in a body of water; often utilized by certain fish as food.

Pleomorphic: Having more than one distinct form in a single life cycle.

Pollard: Coarse bran obtained from wheat, or finely ground bran together with the scourings obtained from wheat during milling and used for livestock feed.

Population: A group of organisms of the same species occupying a particular space at a particular time.

Postharvest fungi: Seedborne fungi that multiply and infect during transportation and storage up until the time that grain is processed for consumption, e.g., *Aspergillus* spp.

Preharvest fungi: Seedborne fungi that attack seeds on plants in the field before they are harvested, e.g., *Alternaria* spp.

Preimaginal development: All development in insects that precedes the final adult sexually mature, usually winged, state (imago).

Pronotum: Dorsal plate of an insect's prothorax.

Propodeum: Part of the thorax of a hymenopteran that lies immediately over and partly surrounding the insertion of the petiole of the abdomen and represents a basal abdominal segment which has become fused with the thorax.

Prothorax: First or anterior segment of the thorax of an insect bearing the first pair of legs.

Protonymphs: Any of various acarids in their first developmental stage.

Psychrophiles: Microorganisms thriving at relatively low temperatures.

Pulses: Edible seeds of various leguminous crops (e.g., soybeans).

Pygostyli: A pair of elongate appendages arising from the terminal abdominal segment of an insect.

Quinone: Yellow crystalline compound [CO(CHCH)$_2$CO] produced by certain tenebrionid beetles.

Robinin: Yellow crystalline glycoside derived from kaempferol and found in the flowers of a locust.

Rutin: Yellow crystalline flavonol glycoside found in rue leaves, tobacco leaves, buckwheat, flower buds of the Japanese pagoda tree, and other plants, which yields quercetin and rutinose on hydrolysis.

Saprophagy: Feeding on decaying organic matter.

Scarabeiform: Grub-like, resembling the body form of a larval scarab beetle.

Scotophase: The dark phase of a light:dark cycle.

Seres: Stages that follow one another in an ecological succession.

Setae: Fine, chitinous hairs or bristles of insects and other arthropods.

Speciation: Formation of biological species through gradual divergence from related groups or occurring abruptly by combination or transformation of genomes.

Species: A group of organisms that has certain genetically determined characteristics in common and is able to interbreed, e.g., *Penicillium notatum* Westling is a species.

Spermatheca: A sac in the female of an animal (e.g., flea) for receiving and storing sperm until fertilization.

Succession: Change in the composition of an ecosystem as the available competing organisms respond to and modify the environment; replacement of one community by another.

Supracoxal glands: Internal excretory organs opening near the leg bases of certain mites.

Synanthropy: As an ecological phenomenon, it refers to the tendency of animals such as flies to occupy human habitations over some period of time.

Tarsus: Part of the leg of an insect distal to the tibia, usually consisting of four or five segments and bearing two claws.

Tegula (pl. tegulae)*:* Small scale-like sclerite of the mesothorax of some insects that covers the base of the forewing.

Teneral: State of insect immediately after molting during which it is soft and immature in coloring.

Testa: Hard external coating or integument of a seed.

Thermophiles: Organisms that thrive at high temperatures.

Tritonymphs: Acarids in their third developmental stage.

Trochanter: Second segment, counting from the base of the leg of an insect, that is usually small and short and in some insects consists of two or rarely of several distinct parts.

Trophic levels: Steps by which organisms obtain their food from plants, with green plants, plant eaters, and carnivores occupying the first, second, and third trophic levels, respectively.

Univoltine: Producing one brood a year, especially a single brood of eggs capable of winter dormancy.

Water activity (A_w)*:* Amount of water available for microorganisms in a given substrate.

Wheatfeed: Fine bran plus a small quantity of endosperm.

Xeric habitat: Environment that is low or deficient in moisture.

Xerophiles: Organisms capable of thriving at lower water activities.

Xerotolerant: Ability of organisms to maintain life under dry conditions.

Yaws: Contagious tropical disease characterized by skin ulcers.

Yeasts: Simple fungi that generally produce cells by budding.

Zoophily: Insect preference for animals rather than humans as a source of food.

Taxonomic Index

Scientific Names

A

Absidia species, 22, 23, 40, 41
  lichtheimi, 41
Acanthoscelides obtectus, 25, 137–138, 363, 364, 375, 379–380, 541
Acaropsis species, 540
Acarus species, 68
  farris, 57–58, 73, 74, 402
  immobilis, 58, 74
  siro, 20, 23, 58–59, 70–71, 72, 73, 74, 75, 402, 464, 471, 472
Acetobacter species, 34
Achroia species, 542
  grisella, 181
Achromobacter species, 34
Adelina species, 170
Aeroglyphus robustus, 71, 73
Agathis species
  dammara, 142
  hawaiicola, 222
Ahasverus species, 20, 23
  advena, 140–141, 543
Aleurites moluccana, 203
Aleuroglyphus species, 20
  ovatus, 60, 67, 73, 74
Alphitobius species, 542
  diaperinus, 15, 158, 478, 543
Alphitophagus bifasciatus, 542
Alternaria species, 22, 34, 40, 60
  alternata, 24, 58
  citri, 49
  tenuis, 58, 66
Amira species, 218
Amomum subulatum, 135
Amyelois transitella, 363, 364
Anagasta kuehniella, 15, 25, 68, 159, 181, 223, 318, 363, 362, 400
Anastatus blattidarum, 110
Androlaelaps casalis, 23
Anisopteromalus calandrae, 168–169, 220, 366
Anoplolepis longipes, 208–209
Anthonomus grandis, 365
Anthrenus species, 542
  flavipes, 363, 366
  sarnicus, 542, 543
  verbasci, 14, 363, 366

Antrocephalus species
  aethiopicus, 218
  mahensis, 218
Apanteles species
  araeceri, 222
  carpatus, 222
  nephoptericis, 222
Aphelinoidea species, 221
Aplastomorpha vandinei, 220
Aprostocetus species, 219
  hagenowii, 219
Araecerus fasciculatus, 25, 134–135, 219, 220, 541, 542
Areca catechu, 131
Artocarpus species
  altilis, 209
  heterophyllus, 209
Aspergillus species, 20, 22–24, 34, 35, 40, 43, 46, 61, 74, 158, 316
  alliaceus, 41
  alternata, 72
  amstelodami, 41, 44, 60, 66, 141
  candidus, 22, 40, 41, 42, 44, 60, 66, 141
  chevalieri, 41, 62
  clavatus, 41
  conicus, 41
  echinulatus, 37, 41, 58
  flavipes, 41
  flavus, 21, 22, 34, 35, 40, 41, 42, 44, 45, 60
  fumigatus, 22, 41, 66, 479
  giganteus, 41
  glaucus, 22, 37, 40, 46
  nidulans, 41
  niger, 22, 41, 44, 50, 479
  ochraceus, 22, 40, 41, 44
  oryzae, 41
  repens, 41, 58, 60, 62
  restrictus, 41, 40, 60
  ruber, 41, 60
  sulphureus, 41
  sydowi, 22, 41
  tamarii, 41
  terreus, 22, 41, 44
  terricola, 41
  ustus, 41
  versicolor, 22, 24, 40, 41, 44, 66
  wentii, 41, 61
Astragalus cicer, 139

Attagenus species, 14, 542, 543
  brunneus, 363
  elongatulus, 363
  fasciatus, 543
  megatoma, 148, 362, 363
  pellio, 177
  piceus, 148
  unicolor, 148–149, 169, 362, 363, 364
Avetianella species, 218
Aximopsis species, 219
  fasciculatus, 220
  javensis, 219
  tephrosiae, 219

B

Bacillus species, 34
  cereus, 366
  thuringiensis, 189, 215, 365, 366, 388
  thuringiensis var. *tenebrionis*, 365
Bandicota bengalensis, 244, 249, 251–253, 255–256, 261–266, 268–276, 292, 295, 298
Bassus hawaiicola, 222
Beauveria species, 189
Blaberus craniifer, 87
Blastophaga psenes, 217
Blatta species, 121
  orientalis, 85, 87, 100–101, 221, 222
Blattella species
  asahinai, 89, 95, 121–129
  beybienkoi, 121
  germanica, 85, 87, 89, 98, 100–101, 121, 348, 388
  vaga, 89
Blatticola blattae, 109
Blattisocius species, 20
  dentriticus, 67–68
  keegani, 68
  tarsalis, 68, 189
Bombus terrestris, 199
Bordetella bronchiseptica, 478
Botrytis species, 34, 49
  cinerea, 49, 62, 66
Bracon species, 20
Brettanomyces species, 35
Bruchidius atrolineatus, 137
Bruchus species
  pisorum, 137
  rufimanus, 137

Common Names

Subject Index

Capsicum, 506
Caramels, 37
Carbamates, 108, 432, 437
Carbaryl, 348, 353, 433
Carbon dioxide
    atmospheric modification for pest control, 402
    effect on mites, 60
    fumigation with, 94, 95, 388, 444, 448, 451
    and microbial growth, 49
    pest resistance to, 388
Carbon monoxide, 425
Carbon tetrachloride, 454
Cardamom, 135
Caribbean, rodent control, 245
Carobs, 181, 182, 184, 268
Carrots, 49, 132
Casein, mite development on, 60
Cassava, 7, 134, 142, 160
Cassia, 506
Cats, domestic, disease transmission by, 478
Catsup, 33, 35, 491
Caves, 455
Cayenne pepper, 132, 133
Celery seed, 506
Centipedes, 209, 431
Central America
    bats, 238
    beetles, 135, 137–138, 140, 377
    flies, 202
    see also specific countries
Cereal grains and products, 3
    beetle pests, 135, 140–145, 148, 151, 152, 154, 157–160, 162, 177, 328, 374, 377
    carbon/nitrogen ratio, 20
    fungi on, 40–41
    hardness of, 375
    inspection of, 507
    losses to pests, 26, 253
    microbial decomposers, 36–39, 40
    mites in, 23, 58, 64, 72, 471
    moisture content, 36
    moth pests, 182, 186, 328, 377
    packaging, 403
    production in Canada and USA, 43
    psocid pests, 82, 328
    resistance to insects, 377–380
    rodent consumption, 268, 269
    spoilage organisms, 33–34, 37
    see also Barley; Corn; Maize; Rice; Sorghum; Triticale; Wheat
Certification of pest control operators, 435, 460
Chagas disease, 478
Champ, Bruce R., 385–397
Cheese rooms, 455
Cheeses
    beetle pests of, 140, 149, 150

enteric diseases from, 470, 480
insect infestation in transit, 354
fly pests, 198, 202, 204
mite development on, 57–60, 64, 66, 71, 471
spoilage organisms, 37, 51
Chemical control of pests
    alternatives to, 429
    application procedures, 417–418, 434–436
    background information, 429, 431
    bin wall, ceiling, and floor treatments, 415–416
    choice of chemicals, 431–433
    dust formulations, 417, 436
    effectiveness, 416
    empty metal grain bins, 416
    insects in bulk-stored grains, 415–418, 460
    insects in processed foods, 427–440, 460
    in integrated pest management, 245, 357, 427–429, 460, 555–556
    laws applicable to, 502
    liquid formulations, 417, 436
    maintenance of, 426
    mite control, 440
    moth control, 418
    need for, 427
    personnel, 435
    planning, 419
    prestorage treatment, 415–416
    preventive aspect of, 428–429
    reasons for, 427–428
    rodents in bulk-stored grains, 419–426, 460
    safety precautions, 417–418, 425–426, 439
    space treatment, 435–436
    surface treatment of grain, 418
    target species, 433
    tracking powders, 257, 422, 425
    with traps, 97–98, 440
    treatment during storage, 416–418
    see also Baits; Fumigants; Insecticides; Pesticides; Rodenticides
Chemoreception, 95–98, 105
Chemosterilants, 232, 293–294
Cherries, 48, 49
Chicken, see Poultry/poultry houses
Chick-peas, 131, 135, 138, 145
Chili powder, 132, 133
Chilies, 142
China
    crop yields, 7
    grain shipments to Japan, 43
    mold infection of grain shipments, 44
    population growth, 6
    rodents, 292
    rotaviral enteritis, 479

Chitin inhibitors, 367
Chlordane, 102, 103, 108
Chlorobenzilate, 440, 540
Chlorophacinone, 421, 422
Chloropicrin, 425, 432, 441
Chlorpyrifos, 93, 94, 98, 101, 106, 107, 108, 430, 431, 433, 437
Chlorpyrifos-methyl, 415–416
Chocolate, 37, 162
Choice tests
    cockroaches, 86, 94–95, 101–103, 107
    of host-plant resistance, 376, 459
    mite food preferences, 60
    rodent baits, 423
Cholcalciferol, 420, 421
Cholesterol, 149
Cholesteryl chloride, 61, 66
Cider, 198
Cinnamon, 133
Citrus fruits, 48, 49, 50, 68, 182, 268, 318
Climate
    classifications, 20–21
    and microflora and arthropods in stored grains, 40
    plasticity index for insects, 24–25
    regulation of storage ecosystems, 20–21
Cloves, 133
Cockroaches, 14, 85–112, 121–129
    aggregation pheromones, 97, 100, 105–106, 124
    attitudes toward, 85, 111
    attraction to food, 92, 95–97, 98, 100
    baits, 93, 95–98, 100, 103, 126–128
    behavior modification in, 88, 90–91, 101, 102, 104, 107
    boric acid effects, 90, 93, 94, 97, 101–106
    bread attractancy, 96–97
    cannibalism, 109
    chemoreception, 95–98, 105
    choice tests, 86, 94–95, 101–103, 107
    complexity of habitat, 92, 95, 104, 111–112
    crack-and-crevice treatment, 92–93
    critical thermal maximum, 87
    in cultures, 88, 542, 543
    cuticle permeability, 87–88
    as disease agents and vectors, 86, 125, 472, 473, 477–480, 484, 485, 546
    distribution and dispersal, 15, 85, 88, 89, 90, 92, 101, 104, 121, 319
    droppings, 252
    dry-ice fumigation, 94
    dusts, 93
    exclusion of, 94, 341
    exploratory activity, 94–95, 96–97, 98, 103

mold problems, 45–46
moth pests, 182, 184, 185–187, 317, 318, 433
oil repellency and toxicity, 367
production in USA, 43
rodent consumption, 268
safe moisture levels, 45
Pears, 47, 49
Peas
    inspection of, 507, 508, 511–512
    pests of, 132, 133, 152, 162, 354, 375
Pecans, 403
Pepper, black, 133, 160
Peppers, 47, 439
Perfetti, Randolph B., 501–504
Permethrin, 347, 348, 353, 434
Pest birds
    alimentation, 230
    auditory responses, 231
    biological pressures, 340
    brain structure and behavior, 230–232
    in bulk-stored grain ecosystems, 20, 26–27
    as carriers of pests, 27, 229–230
    cerebrum, 231
    chemical controls, 230, 231, 232, 233, 234, 555
    chemosterilants, 232
    contaminants from, 229–230, 352
    crop losses to, 9, 26–27
    disease associations, 230, 340, 352, 477
    exclusion from facilities and equipment, 232, 234, 331, 339–341, 357, 555
    feathers, 229–230, 352
    feeding behavior, 26–27, 231
    fire hazards, 232
    inspection for, 523
    leg structure, 230
    management strategies, 229–234, 323
    movement patterns, 231–232
    nest location, 27, 232, 340, 341
    odor perception, 231
    physiology and behavior, 229–232
    population dynamics, 26–27
    protected status, 234
    repellents, 230, 231, 233
    roosting and nesting, 232, 340, 341
    seasonal stress, 232–233
    stored-grain losses to, 26, 27
    toxic wick perches, 230, 232
    transportation concerns, 352, 355
    uric acid production, 230
Pest control, 3–4, 17
    and animal rights/animal welfare, 324
    CIA concept, 128
    commentary on, 459–466
    community programs, 283–284, 292–293

computer models in, 316, 318
costs of, 3–4
diapause and, 188, 366, 444
and ecology, 315
failures, 319, 323
Federal Grain Inspection Service practices, 508
habitat considerations in, 89–90
harmful effects on humans, 484–485
holistic, 112, 243–244, 280, 323
light and, 20, 126, 366, 407–408, 553
manuals, 260
regional strategies, 129
research needs for, 315, 316, 317
see also Chemical control of pests; Fumigation/fumigants; Integrated pest management; Physical methods of pest management
Pesticides, 4, 46
    action levels, 285–286, 439–440
    analytical methods for residue analysis, 503–504
    application of, 280, 386, 555–556
    and biological control, 217
    bird control with, 230, 231, 232, 233, 234, 555
    for conveyances, 353–355
    costs of, 399, 440
    dosages, 386
    general-use, 429, 435
    health risks, 125, 279, 358, 361, 399, 420, 421, 425, 447–448, 484–485, 501–502, 529
    in insectary management, 541
    in integrated pest management, 245, 555–556
    labels/labeling, 355, 451, 460, 502
    laws applicable to, 502
    microencapsulated, 90, 93–94, 107, 108, 422
    misuse of, 285–286, 484
    mite control in cultures, 540
    nonresiduals, 429
    nontarget species killed with, 245, 420, 421, 422, 423, 428
    public attitudes on, 245, 361
    registration, 285–286, 429, 438–439, 502–504
    regulation, 415, 438–439, 501–504, 529
    residuals, 214, 355, 385, 386, 388, 502
    resistance of pests to, 385–389
    restricted-use, 323, 373, 399, 425, 429, 435
    storage and disposal, 439, 449
    temperature considerations, 87
    tolerances, 285–286, 438–439, 445, 502–504, 529–530
    toxicity data requirements, 503

volume of use, 5, 279
    see also Insecticides; Pest resistance to pesticides; Rodenticides
Pest resistance to pesticides, 385–389, 399
    anticoagulant resistance in rodents, 280, 297, 389, 422
    biological pesticides, 365, 388
    cockroaches, 85, 107–108, 385, 388–389, 432
    cross-resistance, 108
    delay of, 108
    development of, 108
    distinguishing from other phenomena, 386–387
    fumigants, 385, 386, 388–389, 432
    and IPM, 428
    mechanisms and modeling, 386
    monitoring methods, 387–388, 460
    nature of, 385–386
    occurrence of, 361, 388–389, 459–460
    organochlorine compounds, 107–108
    and population growth rate among pests, 386
    residual pesticides, 385, 386, 388, 418
    survey of, 388
    tests of, 387, 388, 389
Pest STRIP, 437
Pests, see Storage pests; and specific pests
Pets
    and rodents, 252, 289
    see also Animal/pet feed
pH, and microbial growth, 38
d-Phenothrin, 353
Pheromones
    aggregation, 74, 97, 100, 105–106, 124, 362, 364, 367
    alarm, 67
    in baits, 366
    beetle, 362–364, 440
    cockroach, 92, 97, 100, 105–107, 124
    combined with pathogens, 364–366
    macrolide molecules, 364
    mating disruption with, 364–365
    mite, 67, 74
    monitoring pests with, 361–364, 367, 459
    moth, 189, 362–367
    rearing insects for, 535
    reduction of insecticide repellency with, 97, 105–107
    sex, 100, 105, 106, 189, 362–364, 366, 535
    in traps, 189, 318, 362–364, 440
    see also specific components
Philippines
    crop losses to pests, 9
    hymenopterous parasites, 220

physical pest control methods, 400,
402, 403, 404, 406, 407
prestorage insecticide treatment, 415
production in Canada and USA, 43
resistance to insects, 378, 379
rodent consumption, 253, 254, 268
shorts, 151
spoilage of, 45
taste changes due to pests, 472
triglycerides, 367
whole grain, 185
Wheatfeed
beetle development on, 132, 133, 143,
144, 155, 163–164, 178, 541, 543
moth development on, 184, 185, 187
Wheatings, 185
Wine, 35, 68, 198
Wirtz, Robert A., 469–475

Wisconsin, monitoring beetle pests in,
362
World Food Council, 7
World Health Organization, 284, 293,
387, 389

Y

Yams, 134
Yaws, 198
Yeasts, 33, 96
attractants in, 367
bakers', 64
beetle development on, 132, 141, 144,
149, 151, 152, 159, 160, 163, 541,
543

brewers', 144, 149, 151, 152, 367, 538
mite development on, 58, 60, 61–62,
63, 64, 67, 69, 70, 72, 73
moth development on, 184, 185, 187
pH range for growth, 38
spoilage of foods, 33–35
standards, 498
succession in grains, 22
water activity, 36
Yogurt, 35

Z

ZETA, 362
ZETOH, 362
Zinc phosphide, 420, 421
Zoophily, 195, 196
ZTA, 362